LINEAR ALGEBRA, GEODESY, AND GPS

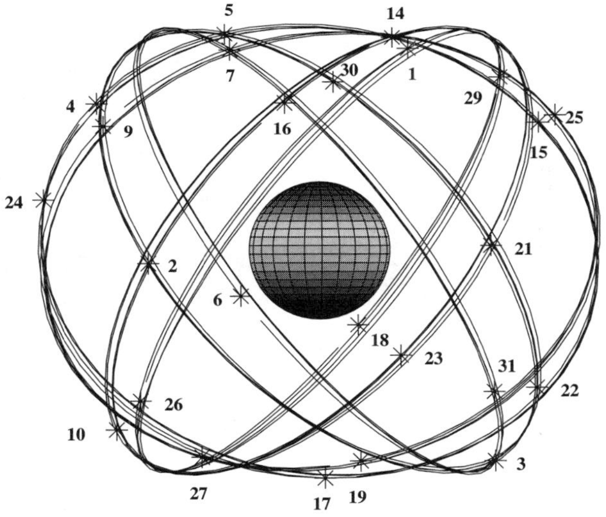

LINEAR ALGEBRA, GEODESY, AND GPS

GILBERT STRANG

Massachusetts Institute of Technology

and

KAI BORRE

Aalborg University

WELLESLEY-CAMBRIDGE PRESS

Library of Congress Cataloging-in-Publication Data

Strang, Gilbert.
Linear algebra, geodesy, and GPS / Gilbert Strang and Kai Borre.
Includes bibliographical references and index.
ISBN 0-9614088-6-3 (hardcover)
1. Algebras, Linear. 2. Geodesy–Mathematics. 3. Global Positioning System.
I. Borre, K. (Kai) II. Title.
TA347.L5 S87 1997
526′.1′015125–dc20 96-44288

Designed by Frank Jensen
Cover photograph by Michael Bevis at Makapu'u on Oahu, Hawaii
Cover design by Tracy Baldwin

Printed in the United States of America 8 7 6 5 4 3 2 1

Other texts from Wellesley-Cambridge Press

Wavelets and Filter Banks, Gilbert Strang and Truong Nguyen, ISBN 0-9614088-7-1.

Introduction to Applied Mathematics, Gilbert Strang, ISBN 0-9614088-0-4.

An Analysis of the Finite Element Method, Gilbert Strang and George Fix, ISBN 0-9614088-8-X.

Calculus, Gilbert Strang, ISBN 0-9614088-2-0.

Introduction to Linear Algebra, Gilbert Strang, ISBN 0-9614088-5-5.

Wellesley-Cambridge Press
Box 812060
Wellesley MA 02181 USA
(617) 431-8488 FAX (617) 253-4358
http://www-math.mit.edu/~gs
email: gs@math.mit.edu

All books may be ordered by email.

TABLE OF CONTENTS

Preface **ix**

The Mathematics of GPS **xiii**

Part I Linear Algebra

1	**Vectors and Matrices**	**3**
1.1	Vectors	3
1.2	Lengths and Dot Products	11
1.3	Planes	20
1.4	Matrices and Linear Equations	28
2	**Solving Linear Equations**	**37**
2.1	The Idea of Elimination	37
2.2	Elimination Using Matrices	46
2.3	Rules for Matrix Operations	54
2.4	Inverse Matrices	65
2.5	Elimination = Factorization: $A = LU$	75
2.6	Transposes and Permutations	87
3	**Vector Spaces and Subspaces**	**101**
3.1	Spaces of Vectors	101
3.2	The Nullspace of A: Solving $A\boldsymbol{x} = \boldsymbol{0}$	109
3.3	The Rank of A: Solving $A\boldsymbol{x} = \boldsymbol{b}$	122
3.4	Independence, Basis, and Dimension	134
3.5	Dimensions of the Four Subspaces	146
4	**Orthogonality**	**157**
4.1	Orthogonality of the Four Subspaces	157
4.2	Projections	165
4.3	Least-Squares Approximations	174
4.4	Orthogonal Bases and Gram-Schmidt	184
5	**Determinants**	**197**
5.1	The Properties of Determinants	197
5.2	Cramer's Rule, Inverses, and Volumes	206

6 Eigenvalues and Eigenvectors **211**
6.1 Introduction to Eigenvalues 211
6.2 Diagonalizing a Matrix 221
6.3 Symmetric Matrices 233
6.4 Positive Definite Matrices 237
6.5 Stability and Preconditioning 248

7 Linear Transformations **251**
7.1 The Idea of a Linear Transformation 251
7.2 Choice of Basis: Similarity and SVD 258

Part II Geodesy

8 Leveling Networks **275**
8.1 Heights by Least Squares 275
8.2 Weighted Least Squares 280
8.3 Leveling Networks and Graphs 282
8.4 Graphs and Incidence Matrices 288
8.5 One-Dimensional Distance Networks 305

9 Random Variables and Covariance Matrices **309**
9.1 The Normal Distribution and χ^2 309
9.2 Mean, Variance, and Standard Deviation 319
9.3 Covariance 320
9.4 Inverse Covariances as Weights 322
9.5 Estimation of Mean and Variance 326
9.6 Propagation of Means and Covariances 328
9.7 Estimating the Variance of Unit Weight 333
9.8 Confidence Ellipses 337

10 Nonlinear Problems **343**
10.1 Getting Around Nonlinearity 343
10.2 Geodetic Observation Equations 349
10.3 Three-Dimensional Model 362

11 Linear Algebra for Weighted Least Squares **369**
11.1 Gram-Schmidt on A and Cholesky on $A^{\mathrm{T}}A$ 369
11.2 Cholesky's Method in the Least-Squares Setting 372
11.3 SVD: The Canonical Form for Geodesy 375
11.4 The Condition Number 377
11.5 Regularly Spaced Networks 379
11.6 Dependency on the Weights 391
11.7 Elimination of Unknowns 394
11.8 Decorrelation and Weight Normalization 400

12 Constraints for Singular Normal Equations **405**
12.1 Rank Deficient Normal Equations 405
12.2 Representations of the Nullspace 406
12.3 Constraining a Rank Deficient Problem 408
12.4 Linear Transformation of Random Variables 413
12.5 Similarity Transformations 414
12.6 Covariance Transformations 421
12.7 Variances at Control Points 423

13 Problems With Explicit Solutions **431**
13.1 Free Stationing as a Similarity Transformation 431
13.2 Optimum Choice of Observation Site 434
13.3 Station Adjustment 438
13.4 Fitting a Straight Line 441

Part III Global Positioning System (GPS)

14 Global Positioning System **447**
14.1 Positioning by GPS 447
14.2 Errors in the GPS Observables 453
14.3 Description of the System 458
14.4 Receiver Position From Code Observations 460
14.5 Combined Code and Phase Observations 463
14.6 Weight Matrix for Differenced Observations 465
14.7 Geometry of the Ellipsoid 467
14.8 The Direct and Reverse Problems 470
14.9 Geodetic Reference System 1980 471
14.10 Geoid, Ellipsoid, and Datum 472
14.11 World Geodetic System 1984 476
14.12 Coordinate Changes From Datum Changes 477

15 Processing of GPS Data **481**
15.1 Baseline Computation and M-Files 481
15.2 Coordinate Changes and Satellite Position 482
15.3 Receiver Position from Pseudoranges 487
15.4 Separate Ambiguity and Baseline Estimation 488
15.5 Joint Ambiguity and Baseline Estimation 494
15.6 The LAMBDA Method for Ambiguities 495
15.7 Sequential Filter for Absolute Position 499
15.8 Additional Useful Filters 505

16 Random Processes **515**
16.1 Random Processes in Continuous Time 515
16.2 Random Processes in Discrete Time 523
16.3 Modeling 527

17 Kalman Filters **543**
17.1 Updating Least Squares 543
17.2 Static and Dynamic Updates 548
17.3 The Steady Model 552
17.4 Derivation of the Kalman Filter 558
17.5 Bayes Filter for Batch Processing 566
17.6 Smoothing 569
17.7 An Example from Practice 574

The Receiver Independent Exchange Format **585**

Glossary **601**

References **609**

Index of M-files **615**

Index **617**

PREFACE

Geodesy begins with measurements from control points. Geometric geodesy measures heights and angles and distances on the Earth. For the Global Positioning System (GPS), the control points are satellites and the accuracy is phenomenal. But even when the measurements are reliable, we make more than absolutely necessary. Mathematically, the positioning problem is overdetermined.

There are errors in the measurements. The data are nearly consistent, but not exactly. An algorithm must be chosen—very often it is *least squares*—to select the output that best satisfies all the inconsistent and overdetermined and redundant (but still accurate!) measurements. This book is about algorithms for geodesy and global positioning.

The starting point is least squares. The equations $A\boldsymbol{x} = \boldsymbol{b}$ are overdetermined. No vector $\boldsymbol{x}$ gives agreement with all measurements $\boldsymbol{b}$, so a "best solution" $\hat{\boldsymbol{x}}$ must be found. This fundamental linear problem can be understood in different ways, and all of these ways are important:

1 (Calculus) Choose $\hat{\boldsymbol{x}}$ to minimize $\|\boldsymbol{b} - A\boldsymbol{x}\|^2$.

2 (Geometry) Project $\boldsymbol{b}$ onto the "column space" containing all vectors $A\boldsymbol{x}$.

3 (Linear algebra) Solve the normal equations $A^{\mathrm{T}}A\hat{\boldsymbol{x}} = A^{\mathrm{T}}\boldsymbol{b}$.

Chapter 4 develops these ideas. We emphasize especially how least squares is a projection: The residual error $\boldsymbol{r} = \boldsymbol{b} - A\hat{\boldsymbol{x}}$ is orthogonal to the columns of A. That means $A^{\mathrm{T}}\boldsymbol{r} = \boldsymbol{0}$, which is the same as $A^{\mathrm{T}}A\hat{\boldsymbol{x}} = A^{\mathrm{T}}\boldsymbol{b}$. This is basic linear algebra, and we follow the exposition in the book by Gilbert Strang (1993). We hope that each reader will find new insights into this fundamental problem.

Another source of information affects the best answer. The measurement errors have probability distributions. When data are more reliable (with smaller variance), they should be weighted more heavily. By using *statistical information on means and variances*, the output is improved. Furthermore the statistics may change with time—we get new information as measurements come in.

The classical unweighted problem $A^{\mathrm{T}}A\hat{\boldsymbol{x}} = A^{\mathrm{T}}\boldsymbol{b}$ becomes more dynamic and realistic (and more subtle) in several steps:

- ***Weighted*** least squares (using the covariance matrix Σ to assign weights)
- ***Recursive*** least squares (for fast updating without recomputing)
- ***Dynamic*** least squares (using sequential filters as the state of the system changes).

The Kalman filter updates not only $\hat{\boldsymbol{x}}$ itself (the estimated state vector) but also its variance.

Chapter 17 develops the theory of filtering in detail, with examples of positioning problems for a GPS receiver. We describe the Kalman filter and also its variant the Bayes filter—which computes the updates in a different (and sometimes faster) order. The formulas for filtering are here based directly on matrix algebra, not on the theory of conditional probability—because more people understand matrices!

Throughout geodesy and global positioning are two other complications that cannot be ignored. This subject requires

- ***Nonlinear*** least squares (distance $\sqrt{x^2 + y^2}$ and angle $\arctan \frac{y}{x}$ are not linear)
- ***Integer*** least squares (to count wavelengths from satellite to receiver).

Nonlinearity is handled incrementally by small linearized steps. Chapter 10 shows how to compute and use the gradient vector, containing the derivatives of measurements with respect to coordinates. This gives the (small) change in position estimates due to a (small) adjustment in the measurements.

Integer least squares resolves the "ambiguity" in counting wavelengths—because the receiver sees only the fractional part. This could be quite a difficult problem. A straightforward approach usually succeeds, and we describe (with MATLAB software) the LAMBDA method that preconditions and decorrelates harder problems.

Inevitably we must deal with numerical error, in the solution procedures as well as the data. The condition number of the least squares problem may be large—the normal equations may be nearly singular. Many applications are actually rank deficient, and we require extra constraints to produce a unique solution. The key tool from matrix analysis is the ***Singular Value Decomposition*** (SVD), which is described in Chapter 7. It is a choice of orthogonal bases in which the matrix becomes diagonal. It applies to all rectangular matrices A, by using the (orthogonal) eigenvectors of $A^{\mathrm{T}}A$ and AA^{T}.

The authors hope very much that this book will be useful to its readers. We all have a natural desire to know where we are! Positioning is absolutely important (and relatively simple). GPS receivers are not expensive. You could control a fleet of trucks, or set out new lots, or preserve your own safety in the wild, by quick and accurate knowledge of position. From The Times of 11 July 1996, GPS enables aircraft to shave up to an hour off the time from Chicago to Hong Kong. This is one of the world's longest non-stop scheduled flights—now a little shorter.

The GPS technology is moving the old science of geodesy into new and completely unexpected applications. This is a fantastic time for everyone who deals with measurements of the Earth. We think Gauss would be pleased.

We hope that the friends who helped us will be pleased too. Our debt is gladly acknowledged, and it is a special pleasure to thank Clyde C. Goad. Discussions with him have opened up new aspects of geodesy and important techniques for GPS. He sees ideas and good algorithms, not just formulas. We must emphasize that algorithms and software are an integral part of this book.

Our algorithms are generally expressed in MATLAB. The reader can obtain all the M-files from http://www.i4.auc.dk/borre/matlab. Those M-files execute the techniques (and examples) that the book describes. The first list of available M-files is printed at the end of the book, and is continuously updated in our web homepages. Computation is now an essential part of this subject.

This book separates naturally into three parts. The first is basic linear algebra. The second is the application to the (linear and also nonlinear) science of measurement. The third is the excitement of GPS. You will see how the theory is immediately needed and used. *Measurements are all around us, today and tomorrow. The goal is to extract the maximum information from those measurements.*

Gilbert Strang
MIT
gs@math.mit.edu
http://www-math.mit.edu/~gs

Kai Borre
Aalborg University
borre@i4.auc.dk
http://www.i4.auc.dk/borre

Acknowledgements

One of the greatest pleasures in writing about GPS is to make friends all over the world. This is such a community of cooperation; the whole system depends on it. We mentioned first the help from Clyde Goad in Columbus. For the design and layout of the book, Frank Jensen in Aalborg gave crucial support. And we thank ten others for their tremendous generosity: Rick Bennett at Harvard, Mike Bevis in Hawaii (for the cover photograph also!), Yehuda Bock at Scripps, Tom Herring and Bob King at MIT, Stephen Hilla at the National Geodetic Survey, Richard Langley in Fredericton, Steven Lee at Oak Ridge, Peter Teunissen and Christian Tiberius in Delft. We remember a letter long ago from Helmut Moritz, who connected geodesy to the most basic framework of applied mathematics. And we are specially grateful to Søren Lundbye-Christensen for his advice on our presentation of random variables and covariance matrices and filtering.

The second author gladly acknowledges the support of the Danish Technical Research Council, the COWI Foundation, and the Faculty of Sciences at Aalborg University. We appreciate the substantial help of the Danish Ministry of Education to publish this book.

We thank you all.

This popular article is reprinted from SIAM News *of June 1997.*
Then Chapter 14 begins an in-depth description of the whole GPS system and its applications.

THE MATHEMATICS OF GPS

It is now possible to find out where you are. You can even discover which way you are moving and how fast. These measurements of position and velocity come from GPS (the Global Positioning System). A handheld GPS receiver gives your position on the Earth within 80 meters, and usually better, by measuring the range to four or more satellites. For centimeter accuracy we need two receivers and more careful mathematics; it is a problem in weighted least squares. Tracking the movement of tectonic plates requires much greater precision than locating trucks or ships or hikers, but GPS can be used for them all.

This article describes some of the mathematics behind GPS. The overall least squares problem has interesting complications. A GPS receiver has several satellites in view, maybe six for good accuracy. The satellites broadcast their positions. By using the time delays of the signals to calculate its distance to each satellite, the receiver knows where it is. In principle three distance measurements should be enough. They specify spheres around three satellites, and the receiver lies at their point of intersection. (Three spheres intersect at another point, but it's not on the Earth.)

In reality we need a minimum of four satellites. The four measurements determine not only position but clock time. *In GPS, time really is the fourth dimension!* The problem is that the receiver clock is not perfectly in sync with the satellite clock. This would cause a major error—the uncorrected reading is called a pseudorange. The receiver clock is not top quality, and the military intentionally dithers the satellite clock and coordinates. (There was a sensational rescue of a downed pilot in Serbia, because he carried a GPS receiver and could say exactly where he was. The dithering is to prevent others from using or sabotaging the system.) The President has proposed that this selective availability should end within ten years, and meanwhile GPS scientists have a way around it.

Clock Errors

The key is to work with differences. If we have two receivers, one in a known and fixed position, the errors in the satellite clock (and the extra delay as signals come through the ionosphere, which is serious) will virtually cancel. This is ***Differential*** GPS. Similarly an extra satellite compensates for error in the receiver clock. Double differencing can give centimeter and even millimeter accuracy, when the other sources of error are removed and measurements are properly averaged (often by a Kalman filter). Southern California will soon have 250 continuous GPS receivers, watching for movements of the Earth's crust.

The carrier signal from the satellite is modulated by a known code (either a coarse C/A code of zeros and ones, or a precise P code). The receiver computes the travel time by synchronizing with that sequence of bits. When the receiver code and satellite code are correlated, the travel time is known—except for the discrepancy between the clocks.

That clock error Δt, multiplied by the speed of light, produces an unknown error $c\Delta t$ in the measurement distance—the same error for all satellites. Suppose a handheld receiver locks onto four satellites. (You could buy one that locks onto only three, but don't do it.) The receiver solves a nonlinear problem in geometry. What it knows is the difference d_{ij} between its distance to satellite i and to satellite j. In a plane, when we know the difference d_{12} between distances to two points, the receiver is located on a hyperbola. In space this becomes a hyperboloid (a hyperbola of revolution). Then the receiver lies at the intersection of three hyperboloids, determined by d_{12} and d_{13} and d_{14}.

Two hyperboloids are likely to intersect in a simple closed curve. The third probably cuts that curve at two points. But again, one point is near the Earth and the other is far away. The error in position might be 50 meters, which is good for a hiker or sailor but not close enough for a pilot who needs to land a plane. For \$ 250 you could recognize the earthquake but not the tiny movements that foreshadow it. The serious mathematics of GPS is in using the phase of the signal and postprocessing the data, to reduce the errors from meters to millimeters.

Clocks Versus Lunar Angles

It is fascinating to meet this world of extremely accurate measurers. In a past century, they would have been astronomers or perhaps clockmakers. A little book called *Longitude*, by Dava Sobel, describes one of the first great scientific competitions—to provide ship captains with their position at sea. This was after the loss of two thousand men in 1707, when British warships ran aground entering the English Channel. The competition was between the Astronomers Royal, using distance to the moon and its angle with the stars, and a man named John Harrison, who made four amazing clocks. It was GPS all over again! Or more chronologically, GPS solved the same positioning problem later—by transmitting its own codes from its own satellites and using cesium and rubidium clocks.

The accuracy demanded in the 18th century was a modest $\frac{1}{2}^\circ$ in longitude. The Earth rotates that much in two minutes. For a six-week voyage this allows a clock error of three seconds per day. Newton recommended the moon, and a German named Mayer won 3000 English pounds for his lunar tables. Even Euler got 300 for providing the right equations. (There is just more than I can say here.) But lunar angles had to be measured, on a rolling ship at sea, within 1.5 minutes of arc.

The big prize was practically in sight, when Harrison came from nowhere and built clocks that could do better. You can see them at Greenwich, outside London. H-1 and H-2 and H-3 are big sea clocks and still running. Harrison's greatest Watch, the H-4 that weighs three pounds and could run if the curators allowed it, is resting. It lost only five seconds on a long trip to Jamaica, and eventually it won the prize.

The modern version of this same competition was between VLBI and GPS. Very Long Baseline Interferometry uses "God's satellites", the distant quasars. The clock at the receiver has to be very accurate and expensive. The equipment can be moved on a flatbed but it is certainly not handheld. There are valuable applications of VLBI, but it is GPS that will appear everywhere. You can find it in a rental car, and now it is optional in your Mercedes or BMW. The display tells you where you are, and the way to the hotel. The reader will understand, without any attempt at reviewing all the possible applications of GPS, that this system is going to affect our lives. GPS is perhaps the second most important military contribution to civilian science, after the Internet. But the key is the atomic clock in the satellite, designed by university physicists to confirm Einstein's prediction of the gravitational red shift.

Integer Least Squares

I would like to return to the basic measurement of distance, because an interesting and difficult mathematical problem is involved. The key is to count radio wavelengths between satellites and receiver. This number (the phase) is an integer plus a fraction. The integer part (called the *ambiguity*) is the tricky problem. It has to be right, because one missing wavelength means an error of 19 cm or 24 cm (the satellite transmits on two frequencies).

Once the integer is known we try to hold on to it. A loss-of-lock from buildings or trees may create cycle slips. Then the phase must be initialized again. Determining this integer is like swimming laps in a pool—after an hour, the fractional part is obvious, but it is easy to forget the number of laps completed. You could estimate it by dividing total swim time by approximate lap time. For a short swim, the integer is probably reliable. But the longer you swim, the greater the variance in the ratio. In GPS, the longer the baseline between receivers, the harder it is to find this whole number.

In reality, we have a network of receivers, each counting wavelengths to multiple satellites (on two frequencies). There might be 100 integer ambiguities to determine simultaneously. It is a problem in *integer least squares*. This is identical to the nearest lattice vector problem in computational combinatorics:

$$\text{Minimize} \quad (\boldsymbol{x}-\boldsymbol{x}_0)^{\mathrm{T}} A (\boldsymbol{x}-\boldsymbol{x}_0) \quad \text{for } \boldsymbol{x} \text{ in } \mathbf{Z}^n.$$

The minimum over $\mathbf{R}^n$ is clearly zero, at $\boldsymbol{x} = \boldsymbol{x}_0$. We are looking for the lattice point $\boldsymbol{x}$ (the ambiguity vector) that is closest to $\boldsymbol{x}_0$ in the metric of A. This minimization over $\mathbf{Z}^n$ can be such a difficult problem (for large random matrices A) that its solution has been used by cryptographers to encode messages. In GPS the weighting matrix A involves distances between receivers, and the problem is hardest for a global network.

The minimization is easy when A is diagonal, because the variables are uncoupled. Each component of $\boldsymbol{x}$ will be the nearest integer to the corresponding component of $\boldsymbol{x}_0$. But an ill-conditioned A severely stretches the lattice. A direct search for the best $\boldsymbol{x}$ becomes horrible. The natural idea is to precondition A by diagonalizing as nearly as possible—always keeping the change of basis matrices Z and Z^{-1} integral. Then $\boldsymbol{y}^{\mathrm{T}}(Z^{\mathrm{T}}AZ)\boldsymbol{y}$ is more nearly uncoupled than $\boldsymbol{x}^{\mathrm{T}}A\boldsymbol{x}$, and $\boldsymbol{y} = Z^{-1}\boldsymbol{x}$ is integral exactly when $\boldsymbol{x}$ is.

If you think of the usual row operations, it is no longer possible to get exact zeros below the pivots. But subtracting integer multiples will make each off-diagonal entry less than half the pivot. Working on the GPS problem in Delft, Peter Teunissen and his group found that one further twist is important. The pivots of $Z^{\mathrm{T}}AZ$ should come out ordered in size. Then this preconditioned problem will have a relatively small number of candidates $\boldsymbol{y}$, and a direct search will find the best one. (A good reference is *GPS for Geodesy*, edited by Kleusberg and Teunissen (1996). Our book and web page provide MATLAB code for this "LAMBDA method" and a fresh treatment of the Kalman filter for processing the data.) A potentially difficult problem has been made tractable for most GPS applications.

Shall we try to predict the future? The number of users will continue to increase. The 24 dedicated satellites will gradually be upgraded. Probably a new civilian frequency will be added. (The military might remove their frequency L_2. They will certainly not reveal the top secret that transforms P code into a secure Y code.) In ordinary life, applications to navigation are certain to be enormous. Ships and planes and cars will have GPS built in. Maybe the most important prediction is the simplest to state: *You will eventually get one*.

It is nice to think that applied mathematics has a part in this useful step forward.

Part I

Linear Algebra

1

VECTORS AND MATRICES

The heart of linear algebra is in two operations—both with vectors. Chapter 1 explains these central ideas, on which everything builds. We start with two-dimensional vectors and three-dimensional vectors, which are reasonable to draw. Then we move into higher dimensions. The really impressive feature of linear algebra is how smoothly it takes that step into n-dimensional space. Your mental picture stays completely correct, even if drawing a ten-dimensional vector is impossible.

You can add vectors $(\boldsymbol{v} + \boldsymbol{w})$. You can multiply a vector $\boldsymbol{v}$ by a number $(c\boldsymbol{v})$. This is where the book is going, and the first steps are the two operations in Sections 1.1 and 1.2:

1.1 *Linear combinations* $c\boldsymbol{v} + d\boldsymbol{w}$.

1.2 The *dot product* $\boldsymbol{v} \cdot \boldsymbol{w}$.

Then a matrix appears, with vectors in its columns and vectors in its rows. We take linear combinations of the columns. We form dot products with the rows. And one step further—there will be *linear equations*. The equations involve an unknown vector called $\boldsymbol{x}$, and they come into Sections 1.3 and 1.4:

1.3 The dot product is in the equation for a plane.

1.4 Linear combinations are in the all-important system $A\boldsymbol{x} = \boldsymbol{b}$.

Those equations are algebra (linear algebra). Behind them lies geometry (linear geometry!). A system of equations is the same thing as an intersection of planes (in n dimensions). At the same time, the equation $A\boldsymbol{x} = \boldsymbol{b}$ looks for ***the combination of column vectors that produces*** $\boldsymbol{b}$.

These are the two ways to look at linear algebra. They are the "row picture" and the "column picture." Chapter 1 makes a direct start on them both.

1.1 Vectors

"You can't add apples and oranges." That sentence might not be news, but it still contains some truth. In a strange way, it is the reason for vectors! If we keep the number of apples separate from the number of oranges, we have a pair of numbers. That pair is a ***two-dimensional vector*** $\boldsymbol{v}$:

$$\boldsymbol{v} = \begin{bmatrix} v_1 \\ v_2 \end{bmatrix} \qquad \begin{array}{l} v_1 = \text{number of apples} \\ v_2 = \text{number of oranges.} \end{array}$$

We wrote $\boldsymbol{v}$ as a column vector. The numbers v_1 and v_2 are its "components." The main point so far is to have a single letter $\boldsymbol{v}$ (in boldface) for this pair of numbers (in lightface).

Even if we don't add v_1 to v_2, we do *add vectors*. The first components of $\boldsymbol{v}$ and $\boldsymbol{w}$ stay separate from the second components:

$$\boldsymbol{v} = \begin{bmatrix} v_1 \\ v_2 \end{bmatrix} \quad \text{and} \quad \boldsymbol{w} = \begin{bmatrix} w_1 \\ w_2 \end{bmatrix} \quad \text{add to} \quad \boldsymbol{v} + \boldsymbol{w} = \begin{bmatrix} v_1 + w_1 \\ v_2 + w_2 \end{bmatrix}.$$

You see the reason. The total number of apples is $v_1 + w_1$. The number of oranges is $v_2 + w_2$. Vector addition is basic and important. Subtraction of vectors follows the same idea: *The components of* $\boldsymbol{v} - \boldsymbol{w}$ *are* _____ *and* _____.

Vectors can be multiplied by 2 or by -1 or by any number c. There are two ways to double a vector. One way is to add $\boldsymbol{v} + \boldsymbol{v}$. The other way (the usual way) is to multiply each component by 2:

$$2\boldsymbol{v} = \begin{bmatrix} 2v_1 \\ 2v_2 \end{bmatrix} \quad \text{and} \quad -\boldsymbol{v} = \begin{bmatrix} -v_1 \\ -v_2 \end{bmatrix}.$$

The components of $c\boldsymbol{v}$ are cv_1 and cv_2. The number c is called a "scalar."

Notice that the sum of $-\boldsymbol{v}$ and $\boldsymbol{v}$ is the zero vector. This is $\mathbf{0}$, which is not the same as the number zero! The vector $\mathbf{0}$ has components 0 and 0. Forgive us for hammering away at the difference between a vector and its components. Linear algebra is built on these operations $\boldsymbol{v} + \boldsymbol{w}$ and $c\boldsymbol{v}$—***adding vectors and multiplying by scalars***.

There is another way to see a vector, that shows all its components at once. The vector $\boldsymbol{v}$ can be represented by an arrow. When $\boldsymbol{v}$ has two components, the arrow is in two-dimensional space (a plane). If the components are v_1 and v_2, the arrow goes v_1 units to the right and v_2 units up. This vector is drawn twice in Figure 1.1. First, it starts at the origin (where the axes meet). This is the usual picture. Unless there is a special reason,

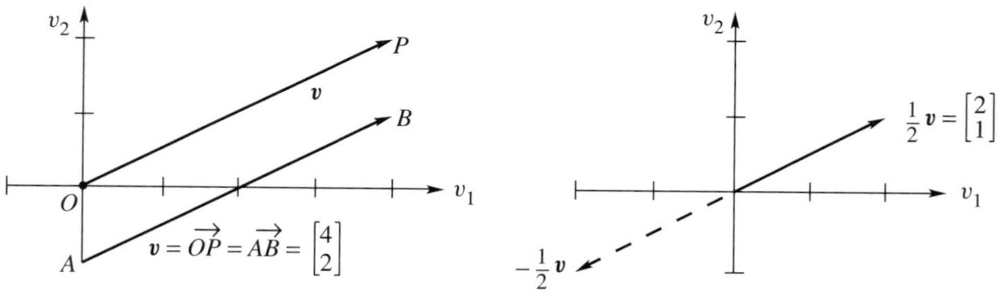

Figure 1.1 The arrow usually starts at the origin; $c\boldsymbol{v}$ is always parallel to $\boldsymbol{v}$.

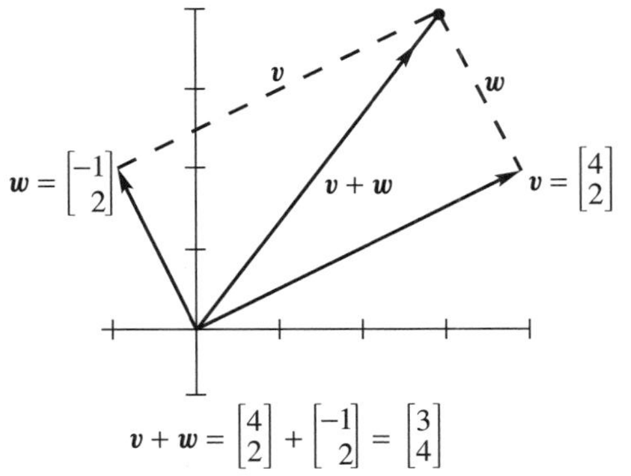

Figure 1.2 Vector addition $\boldsymbol{v} + \boldsymbol{w}$ produces a parallelogram. Add the components separately.

our vectors will begin at (0, 0). But the second arrow shows the starting point shifted over to A. The arrows $\overrightarrow{OP}$ and $\overrightarrow{AB}$ represent the same vector. One reason for allowing all starting points is to visualize the sum $\boldsymbol{v} + \boldsymbol{w}$:

> ***Vector addition*** (head to tail) At the end of $\boldsymbol{v}$, place the start of $\boldsymbol{w}$.

In Figure 1.2, $\boldsymbol{v} + \boldsymbol{w}$ goes from the beginning of $\boldsymbol{v}$ to the end of $\boldsymbol{w}$.

We travel along $\boldsymbol{v}$ and then along $\boldsymbol{w}$. Or we take the shortcut along $\boldsymbol{v} + \boldsymbol{w}$. We could also go along $\boldsymbol{w}$ and then $\boldsymbol{v}$. In other words, $\boldsymbol{w} + \boldsymbol{v}$ gives the same answer as $\boldsymbol{v} + \boldsymbol{w}$. These are different ways along the parallelogram (in this example it is a rectangle). The endpoint is still the diagonal $\boldsymbol{v} + \boldsymbol{w}$.

Check that by algebra: The first component is $v_1 + w_1$ which equals $w_1 + v_1$. The order makes no difference, and $\boldsymbol{v} + \boldsymbol{w} = \boldsymbol{w} + \boldsymbol{v}$:

$$\begin{bmatrix} 1 \\ 5 \end{bmatrix} + \begin{bmatrix} 3 \\ 3 \end{bmatrix} = \begin{bmatrix} 4 \\ 8 \end{bmatrix} \quad \text{and} \quad \begin{bmatrix} 3 \\ 3 \end{bmatrix} + \begin{bmatrix} 1 \\ 5 \end{bmatrix} = \begin{bmatrix} 4 \\ 8 \end{bmatrix}.$$

The zero vector has $v_1 = 0$ and $v_2 = 0$. It is too short to draw a decent arrow, but you know that $\boldsymbol{v} + \boldsymbol{0} = \boldsymbol{v}$. For $2\boldsymbol{v}$ we double the length of the arrow. We reverse its direction for $-\boldsymbol{v}$. This reversing gives a geometric way to subtract vectors.

> ***Vector subtraction*** To draw $\boldsymbol{v} - \boldsymbol{w}$, go forward along $\boldsymbol{v}$ and then backward along $\boldsymbol{w}$ (Figure 1.3). The components are $v_1 - w_1$ and $v_2 - w_2$.

We will soon meet a "product" of vectors. It is not the vector whose components are $v_1 w_1$ and $v_2 w_2$.

Linear Combinations

We have added vectors, subtracted vectors, and multiplied by scalars. The answers $\boldsymbol{v} + \boldsymbol{w}$, $\boldsymbol{v} - \boldsymbol{w}$, and $c\boldsymbol{v}$ are computed a component at a time. By combining these operations, we

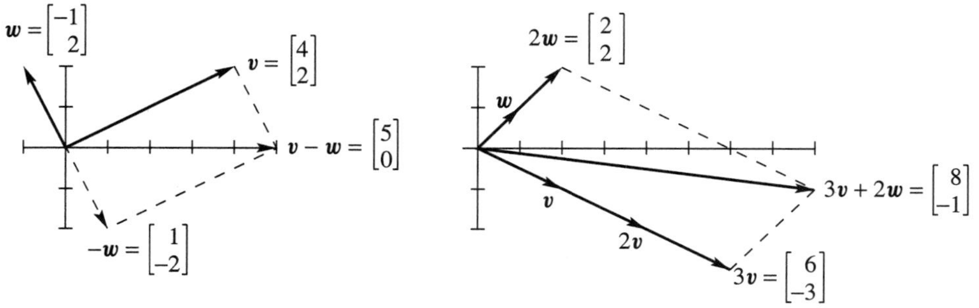

Figure 1.3 Vector subtraction $\boldsymbol{v} - \boldsymbol{w}$. The linear combination $3\boldsymbol{v} + 2\boldsymbol{w}$.

now form *"linear combinations"* of $\boldsymbol{v}$ and $\boldsymbol{w}$. Apples still stay separate from oranges—the combination is a new vector $c\boldsymbol{v} + d\boldsymbol{w}$.

Definition The sum of $c\boldsymbol{v}$ and $d\boldsymbol{w}$ is a ***linear combination*** of $\boldsymbol{v}$ and $\boldsymbol{w}$. Figure 1.3b shows the combination $3\boldsymbol{v} + 2\boldsymbol{w}$:

$$3\begin{bmatrix} 2 \\ -1 \end{bmatrix} + 2\begin{bmatrix} 1 \\ 1 \end{bmatrix} = \begin{bmatrix} 8 \\ -1 \end{bmatrix}.$$

This is the fundamental construction of linear algebra: *multiply and add.* The sum $\boldsymbol{v} + \boldsymbol{w}$ is a special combination, when $c = d = 1$. The multiple $2\boldsymbol{v}$ is the particular case with $c = 2$ and $d = 0$. Soon you will be looking at all linear combinations of $\boldsymbol{v}$ and $\boldsymbol{w}$—a whole family of vectors at once. It is this big view, going from two vectors to a "plane of vectors," that makes the subject work.

In the forward direction, a combination of $\boldsymbol{v}$ and $\boldsymbol{w}$ is supremely easy. We are given the multipliers $c = 3$ and $d = 2$, so we multiply. Then add $3\boldsymbol{v} + 2\boldsymbol{w}$. The serious problem is the opposite question, when c and d are "unknowns." In that case we are only given the answer $c\boldsymbol{v} + d\boldsymbol{w}$ (with components 8 and -1). We look for the right multipliers c and d. The two components give two equations in these two unknowns.

When 100 unknowns multiply 100 vectors each with 100 components, the best way to find those unknowns is explained in Chapter 2.

Vectors in Three Dimensions

A vector with two components corresponds to a point in the xy plane. The components of the vector are the coordinates of the point: $x = v_1$ and $y = v_2$. The arrow ends at this point (v_1, v_2), when it starts from $(0, 0)$. Now we allow vectors to have three components. The xy plane is replaced by three-dimensional space.

Here are typical vectors (still column vectors but with three components):

$$\boldsymbol{v} = \begin{bmatrix} 1 \\ 2 \\ 2 \end{bmatrix} \quad \text{and} \quad \boldsymbol{w} = \begin{bmatrix} 2 \\ 3 \\ -1 \end{bmatrix} \quad \text{and} \quad \boldsymbol{v} + \boldsymbol{w} = \begin{bmatrix} 3 \\ 5 \\ 1 \end{bmatrix}.$$

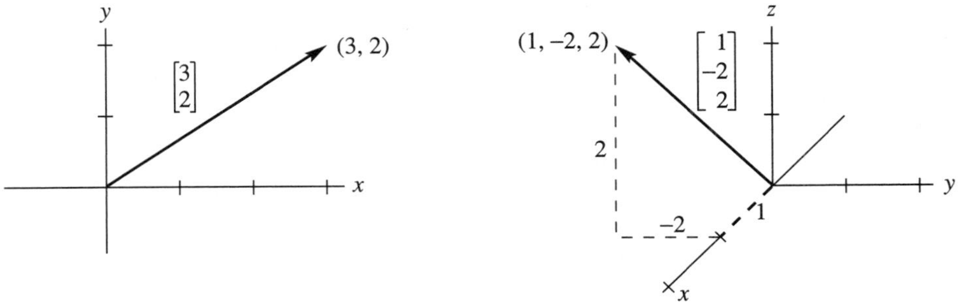

Figure 1.4 Vectors $\begin{bmatrix} x \\ y \end{bmatrix}$ and $\begin{bmatrix} x \\ y \\ z \end{bmatrix}$ correspond to points (x, y) and (x, y, z).

The vector $\boldsymbol{v}$ corresponds to an arrow in 3-space. Usually the arrow starts at the origin, where the xyz axes meet and the coordinates are $(0, 0, 0)$. The arrow ends at the point with coordinates $x = 1$, $y = 2$, $z = 2$. There is a perfect match between the column vector and the arrow from the origin and the point where the arrow ends. Those are three ways to describe the same vector:

$$\textbf{\textit{From now on}} \quad \boldsymbol{v} = \begin{bmatrix} 1 \\ 2 \\ 2 \end{bmatrix} \quad \textbf{\textit{is also written as}} \quad \boldsymbol{v} = (1, 2, 2).$$

The reason for the first form (in a column) is to fit next to a matrix. The reason for the second form is to save space. This becomes essential for vectors with many components. To print $(1, 2, 2, 4, 4, 6)$ in a column would waste the environment.

Important note $(1, 2, 2)$ is not a row vector. The row vector $[\,1 \quad 2 \quad 2\,]$ is absolutely different, even though it has the same three components. A column vector can be printed vertically (with brackets). It can also be printed horizontally (with commas and parentheses). Thus $(1, 2, 2)$ is in actuality a column vector. It is just temporarily lying down.

In three dimensions, vector addition is still done a component at a time. The result $\boldsymbol{v} + \boldsymbol{w}$ has components $v_1 + w_1$ and $v_2 + w_2$ and $v_3 + w_3$—maybe apples, oranges, and pears. You see already how to add vectors in 4 or 5 or n dimensions—this is the end of linear algebra for groceries.

The addition $\boldsymbol{v} + \boldsymbol{w}$ is represented by arrows in space. When $\boldsymbol{w}$ starts at the end of $\boldsymbol{v}$, the third side is $\boldsymbol{v} + \boldsymbol{w}$. When $\boldsymbol{w}$ follows $\boldsymbol{v}$, we get the other sides of a parallelogram. Question: Do the four sides all lie in the same plane? *Yes*. And the sum $\boldsymbol{v} + \boldsymbol{w} - \boldsymbol{v} - \boldsymbol{w}$ goes around the parallelogram to produce ______.

Summary This first section of the book explained vector addition $\boldsymbol{v} + \boldsymbol{w}$ and scalar multiplication $c\boldsymbol{v}$. Then we moved to other linear combinations $c\boldsymbol{v} + d\boldsymbol{w}$. A typical combination of three vectors in three dimensions is $\boldsymbol{u} + 4\boldsymbol{v} - 2\boldsymbol{w}$:

$$\begin{bmatrix} 1 \\ 0 \\ 3 \end{bmatrix} + 4 \begin{bmatrix} 1 \\ 2 \\ 1 \end{bmatrix} - 2 \begin{bmatrix} 2 \\ 3 \\ -1 \end{bmatrix} = \begin{bmatrix} 1 \\ 2 \\ 9 \end{bmatrix}.$$

We end with this question: What surface do you get from all the linear combinations of $\boldsymbol{u}$ and $\boldsymbol{v}$? The surface includes the line through $\boldsymbol{u}$ and the line through $\boldsymbol{v}$. It includes the zero vector (which is the combination $0\boldsymbol{u} + 0\boldsymbol{v}$). The surface also includes the diagonal line through $\boldsymbol{v} + \boldsymbol{w}$—and every other combination $c\boldsymbol{v} + d\boldsymbol{w}$. *This whole surface is a plane.*

Note on Computing Suppose the components of $\boldsymbol{v}$ are $v(1), \ldots, v(N)$, and similarly for $\boldsymbol{w}$. In a language like FORTRAN, the sum $\boldsymbol{v} + \boldsymbol{w}$ requires a loop to add components separately:

```
   DO 10 I = 1,N
10 VPLUSW(I) = v(I) + w(I)
```

MATLAB works directly with vectors and matrices. When $\boldsymbol{v}$ and $\boldsymbol{w}$ have been defined, $\boldsymbol{v} + \boldsymbol{w}$ is immediately understood. It is *printed* unless the line ends in a semicolon. Input two specific vectors as rows—the prime ′ at the end changes them to columns. Then print $\boldsymbol{v} + \boldsymbol{w}$ and another linear combination:

```
v = [2 3 4]';
w = [1 1 1]';
u = v + w
2 * v - 3 * w
```

The sum will print with "u = ". The unnamed combination prints with "ans = ":

```
u =          ans =
     3              1
     4              3
     5              5
```

Problem Set 1.1

Problems 1–9 are about addition of vectors and linear combinations.

1 Draw the vectors $\boldsymbol{v} = \begin{bmatrix} 4 \\ 1 \end{bmatrix}$ and $\boldsymbol{w} = \begin{bmatrix} -2 \\ 2 \end{bmatrix}$ and $\boldsymbol{v} + \boldsymbol{w}$ and $\boldsymbol{v} - \boldsymbol{w}$ in a single xy plane.

2 If $\boldsymbol{v} + \boldsymbol{w} = \begin{bmatrix} 3 \\ 1 \end{bmatrix}$ and $\boldsymbol{v} - \boldsymbol{w} = \begin{bmatrix} 1 \\ 3 \end{bmatrix}$, compute and draw $\boldsymbol{v}$ and $\boldsymbol{w}$.

3 From $\boldsymbol{v} = \begin{bmatrix} 2 \\ 1 \end{bmatrix}$ and $\boldsymbol{w} = \begin{bmatrix} 1 \\ 2 \end{bmatrix}$, find the components of $3\boldsymbol{v}+\boldsymbol{w}$ and $\boldsymbol{v}-3\boldsymbol{w}$ and $c\boldsymbol{v}+d\boldsymbol{w}$.

4 Compute $\boldsymbol{u} + \boldsymbol{v}$ and $\boldsymbol{u} + \boldsymbol{v} + \boldsymbol{w}$ and $2\boldsymbol{u} + 2\boldsymbol{v} + \boldsymbol{w}$ when

$$\boldsymbol{u} = \begin{bmatrix} 1 \\ 2 \\ 3 \end{bmatrix}, \quad \boldsymbol{v} = \begin{bmatrix} -3 \\ 1 \\ -2 \end{bmatrix}, \quad \boldsymbol{w} = \begin{bmatrix} 2 \\ -3 \\ -1 \end{bmatrix}.$$

5 (a) Every multiple of $\boldsymbol{v} = (1, -2, 1)$ has components that add up to ______.

(b) Every linear combination of $\boldsymbol{v}$ and $\boldsymbol{w} = (0, 1, -1)$ has components that add to ______.

(c) Find c and d so that $c\boldsymbol{v} + d\boldsymbol{w} = (4, 2, -6)$.

6 In the xy plane mark these nine linear combinations:

$$c\begin{bmatrix}3\\1\end{bmatrix}+d\begin{bmatrix}0\\1\end{bmatrix} \quad \text{with} \quad c=0,1,2 \quad \text{and} \quad d=0,1,2.$$

7 (a) The subtraction $\boldsymbol{v}-\boldsymbol{w}$ goes forward along $\boldsymbol{v}$ and backward on $\boldsymbol{w}$. Figure 1.3 also shows a second route to $\boldsymbol{v}-\boldsymbol{w}$. What is it?

(b) If you look at all combinations of $\boldsymbol{v}$ and $\boldsymbol{w}$, what "surface of vectors" do you see?

8 The parallelogram in Figure 1.2 has diagonal $\boldsymbol{v}+\boldsymbol{w}$. What is its other diagonal? What is the sum of the two diagonals? Draw that vector sum.

9 If three corners of a parallelogram are $(1,1)$, $(4,2)$, and $(1,3)$, what are all the possible fourth corners? Draw two of them.

Problems 10–13 involve the length of vectors. Compute (length of $\boldsymbol{v}$)^2 as $v_1^2+v_2^2$.

10 The parallelogram with sides $\boldsymbol{v}=(4,2)$ and $\boldsymbol{w}=(-1,2)$ looks like a rectangle (Figure 1.2). Check the Pythagoras formula $a^2+b^2=c^2$ which is for right triangles only:

$$(\text{length of } \boldsymbol{v})^2+(\text{length of } \boldsymbol{w})^2=(\text{length of } \boldsymbol{v}+\boldsymbol{w})^2.$$

11 In this $90°$ case, $a^2+b^2=c^2$ also works for $\boldsymbol{v}-\boldsymbol{w}$. Check that

$$(\text{length of } \boldsymbol{v})^2+(\text{length of } \boldsymbol{w})^2=(\text{length of } \boldsymbol{v}-\boldsymbol{w})^2.$$

Give an example of $\boldsymbol{v}$ and $\boldsymbol{w}$ for which this formula fails.

12 To emphasize that right triangles are special, construct $\boldsymbol{v}$ and $\boldsymbol{w}$ and $\boldsymbol{v}+\boldsymbol{w}$ without a $90°$ angle. Compare (length of $\boldsymbol{v}$)^2 + (length of $\boldsymbol{w}$)^2 with (length of $\boldsymbol{v}+\boldsymbol{w}$)^2.

13 In Figure 1.2 check that (length of $\boldsymbol{v}$)+(length of $\boldsymbol{w}$) is larger than (length of $\boldsymbol{v}+\boldsymbol{w}$). This is true for every triangle, except the very thin triangle when $\boldsymbol{v}$ and $\boldsymbol{w}$ are ______ . Notice that these lengths are not squared.

Problems 14–18 are about special vectors on cubes and clocks.

14 Copy the cube and draw the vector sum of $\boldsymbol{i}=(1,0,0)$ and $\boldsymbol{j}=(0,1,0)$ and $\boldsymbol{k}=(0,0,1)$. The addition $\boldsymbol{i}+\boldsymbol{j}$ yields the diagonal of ______ .

15 Three edges of the unit cube are $\boldsymbol{i},\boldsymbol{j},\boldsymbol{k}$. Its eight corners are $(0,0,0)$, $(1,0,0)$, $(0,1,0)$, ______ . What are the coordinates of the center point? The center points of the six faces are ______ .

16 How many corners does a cube have in 4 dimensions? How many faces? How many edges? A typical corner is $(0,0,1,0)$.

17 (a) What is the sum $\boldsymbol{V}$ of the twelve vectors that go from the center of a clock to the hours 1:00, 2:00, ..., 12:00?

(b) If the vector to 4:00 is removed, find the sum of the eleven remaining vectors.

(c) Suppose the 1:00 vector is cut in half. Add it to the other eleven vectors.

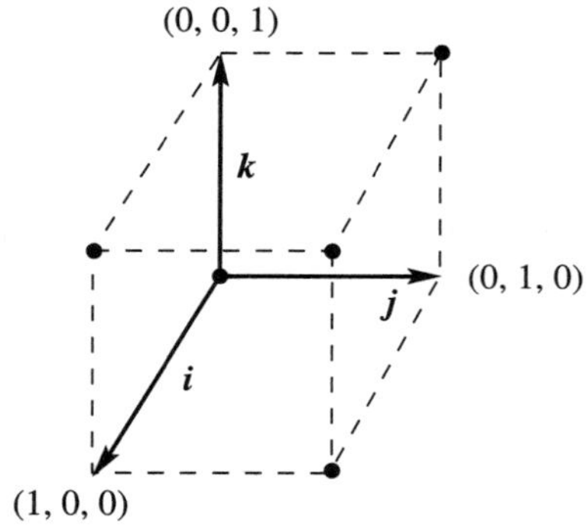

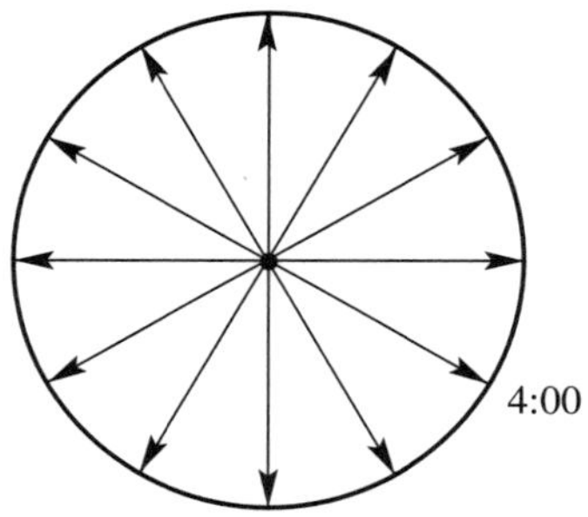

18 Suppose the twelve vectors start from the top of the clock (at 12:00) instead of the center. The vector to 6:00 is doubled and the vector to 12:00 is now zero. Add the new twelve vectors.

Problems 19–27 go further with linear combinations of $\boldsymbol{v}$ and $\boldsymbol{w}$ (and $\boldsymbol{u}$).

19 The first figure shows $\boldsymbol{u} = \frac{1}{2}\boldsymbol{v} + \frac{1}{2}\boldsymbol{w}$. This is halfway between $\boldsymbol{v}$ and $\boldsymbol{w}$ because $\boldsymbol{v} - \boldsymbol{u} = \frac{1}{2}(\_\_\_\_)$. Mark the points $\frac{3}{4}\boldsymbol{v} + \frac{1}{4}\boldsymbol{w}$ and $\frac{1}{4}\boldsymbol{v} + \frac{1}{4}\boldsymbol{w}$.

20 Mark the point $-\boldsymbol{v} + 2\boldsymbol{w}$ and one other combination $c\boldsymbol{v} + d\boldsymbol{w}$ with $c + d = 1$. Draw the line of all these "affine" combinations that have $c + d = 1$.

21 Locate $\frac{1}{3}\boldsymbol{v}+\frac{1}{3}\boldsymbol{w}$ and $\frac{2}{3}\boldsymbol{v}+\frac{2}{3}\boldsymbol{w}$. The combinations $c\boldsymbol{v}+c\boldsymbol{w}$ fill out what line? Restricted by $c \geq 0$ those combinations with $c = d$ fill what ray?

22 (a) Mark $\frac{1}{2}\boldsymbol{v} + \boldsymbol{w}$ and $\boldsymbol{v} + \frac{1}{2}\boldsymbol{w}$. Restricted by $0 \leq c \leq 1$ and $0 \leq d \leq 1$, shade in all combinations $c\boldsymbol{v} + d\boldsymbol{w}$.

(b) Restricted only by $0 \leq c$ and $0 \leq d$ draw the "cone" of all combinations.

23 The second figure shows vectors $\boldsymbol{u}$, $\boldsymbol{v}$, $\boldsymbol{w}$ in three-dimensional space.

(a) Locate $\frac{1}{3}\boldsymbol{u} + \frac{1}{3}\boldsymbol{v} + \frac{1}{3}\boldsymbol{w}$ and $\frac{1}{2}\boldsymbol{u} + \frac{1}{2}\boldsymbol{w}$.

(b) Challenge problem: Under what restrictions on c, d, e, and $c + d + e$ will the points $c\boldsymbol{u} + d\boldsymbol{v} + e\boldsymbol{w}$ fill in the dotted triangle with corners $\boldsymbol{u}$, $\boldsymbol{v}$, $\boldsymbol{w}$?

24 The three sides of the dotted triangle are $\boldsymbol{v} - \boldsymbol{u}$ and $\boldsymbol{w} - \boldsymbol{v}$ and $\boldsymbol{u} - \boldsymbol{w}$. Their sum is ____. Draw the head-to-tail addition around a plane triangle of (3, 1) plus (−1, 1) plus (−2, −2).

25 Shade in the pyramid of combinations $c\boldsymbol{u} + d\boldsymbol{v} + e\boldsymbol{w}$ with $c \geq 0$, $d \geq 0$, $e \geq 0$ and $c + d + e \leq 1$. Mark the vector $\frac{1}{2}(\boldsymbol{u} + \boldsymbol{v} + \boldsymbol{w})$ as inside or outside this pyramid.

26 In 3-dimensional space, which vectors are combinations of $\boldsymbol{u}$, $\boldsymbol{v}$, and $\boldsymbol{w}$? Which vectors are in the plane of $\boldsymbol{u}$ and $\boldsymbol{v}$, and *also* in the plane of $\boldsymbol{v}$ and $\boldsymbol{w}$?

27 (a) Choose $\boldsymbol{u}$, $\boldsymbol{v}$, $\boldsymbol{w}$ so that their combinations fill only a line.

(b) Choose $\boldsymbol{u}$, $\boldsymbol{v}$, $\boldsymbol{w}$ so that their combinations fill only a plane.

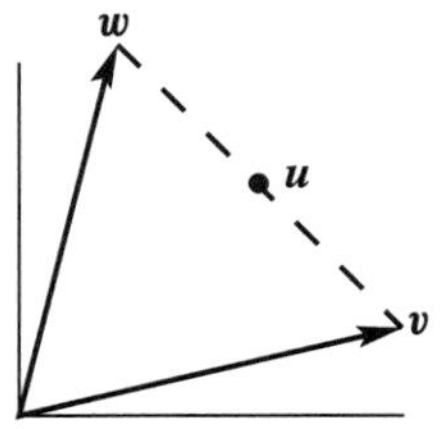

Problems 19–22

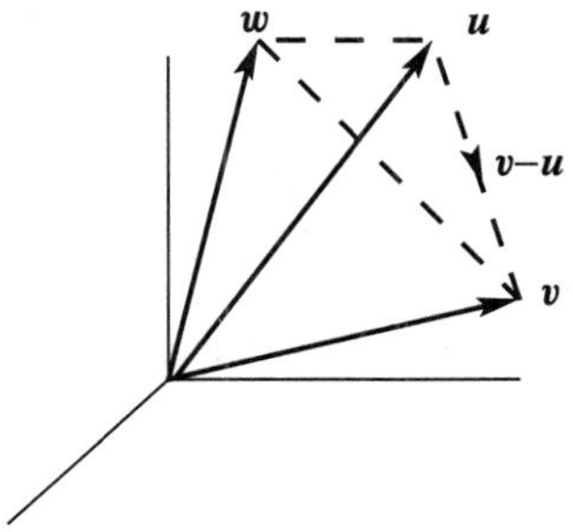

Problems 23–26

1.2 Lengths and Dot Products

The first section mentioned multiplication of vectors, but it backed off. Now we go forward to define the *"dot product"* of $\boldsymbol{v}$ and $\boldsymbol{w}$. This multiplication involves the separate products $v_1 w_1$ and $v_2 w_2$, but it doesn't stop there. Those two numbers are added to produce the single number $\boldsymbol{v} \cdot \boldsymbol{w}$.

Definition The ***dot product*** or ***inner product*** of $\boldsymbol{v} = (v_1, v_2)$ and $\boldsymbol{w} = (w_1, w_2)$ is the number

$$\boldsymbol{v} \cdot \boldsymbol{w} = v_1 w_1 + v_2 w_2. \tag{1.1}$$

Example 1.1 The vectors $\boldsymbol{v} = (4, 2)$ and $\boldsymbol{w} = (-1, 2)$ happen to have a zero dot product:

$$\begin{bmatrix} 4 \\ 2 \end{bmatrix} \cdot \begin{bmatrix} -1 \\ 2 \end{bmatrix} = -4 + 4 = 0.$$

In mathematics, zero is always a special number. For dot products, it means that these vectors are perpendicular. The angle between them is 90°. When we drew them in Figure 1.2, we saw a rectangle (not just a parallelogram). The clearest example of perpendicular vectors is $\boldsymbol{i} = (1, 0)$ along the x axis and $\boldsymbol{j} = (0, 1)$ up the y axis. Again the dot product is $\boldsymbol{i} \cdot \boldsymbol{j} = 0 + 0 = 0$. Those vectors form a right angle.

The vectors $\boldsymbol{v} = (1, 2)$ and $\boldsymbol{w} = (2, 1)$ are not perpendicular. Their dot product is 4. Soon that will reveal the angle between them (not 90°).

Example 1.2 Put a weight of 4 at the point $x = -1$ and a weight of 2 at the point $x = 2$. If the x axis is a see-saw, it will balance on the center point $x = 0$. The weight at $x = -1$ balances the child at $x = 2$. They balance because the dot product is $(4)(-1) + (2)(2) = 0$.

This example is typical of engineering and science. The vector of weights is $(w_1, w_2) = (4, 2)$. The vector of distances from the center is $(v_1, v_2) = (-1, 2)$. The "moment" of the first weight is w_1 times v_1, force times distance. The moment of the second weight is $w_2 v_2$. The total moment is the dot product $w_1 v_1 + w_2 v_2$.

First comment ***The dot product $\boldsymbol{w} \cdot \boldsymbol{v}$ equals $\boldsymbol{v} \cdot \boldsymbol{w}$.*** The order of $\boldsymbol{v}$ and $\boldsymbol{w}$ makes no difference.

Second comment The "vector $\boldsymbol{w}$" is just a list of weights. The "vector $\boldsymbol{v}$" is just a list of distances. With a giant child sitting at the center point, the weight vector would be $(4, 2, 100)$. The distance vector would be $(-1, 2, 0)$. The dot product would still be zero and the see-saw would still balance:

$$\begin{bmatrix} 4 \\ 2 \\ 100 \end{bmatrix} \cdot \begin{bmatrix} -1 \\ 2 \\ 0 \end{bmatrix} = (4)(-1) + (2)(2) + (100)(0) = 0.$$

With six weights, the vectors will be six-dimensional. This is how linear algebra goes quickly into high dimensions. The x axis or the see-saw is only one-dimensional, but you have to make that leap to a vector of six weights.

Example 1.3 Dot products enter in economics and business. We have five products to buy and sell. The unit prices are $(p_1, p_2, p_3, p_4, p_5)$—the "price vector." The quantities we buy or sell are $(q_1, q_2, q_3, q_4, q_5)$—positive when we sell, negative when we buy. Selling the quantity q_1 at the price p_1 brings in $q_1 p_1$. The total income is the dot product $\boldsymbol{q} \cdot \boldsymbol{p}$:

$$\text{Income} = (q_1, q_2, \ldots, q_5) \cdot (p_1, p_2, \ldots, p_5) = q_1 p_1 + q_2 p_2 + \cdots + q_5 p_5.$$

A zero dot product means that "the books balance." Total sales equal total purchases if $\boldsymbol{q} \cdot \boldsymbol{p} = 0$. Then $\boldsymbol{p}$ is perpendicular to $\boldsymbol{q}$ (in five-dimensional space).

Small note: Spreadsheets have become essential computer software in management. What does a spreadsheet actually do? It computes linear combinations and dot products. What you see on the screen is a matrix.

Question: How are linear combinations related to dot products? *Answer*: Look at the first component of $3\boldsymbol{v} - 2\boldsymbol{w}$.

$$\text{A typical linear combination is} \quad 3\begin{bmatrix} 4 \\ 0 \\ 4 \end{bmatrix} - 2\begin{bmatrix} 1 \\ 2 \\ 3 \end{bmatrix} = \begin{bmatrix} 10 \\ -4 \\ 6 \end{bmatrix}.$$

The number 10 is 3 times 4 minus 2 times 1. Think of that as $(3, -2) \cdot (4, 1) = 10$. This is a dot product. The numbers -4 and 6 are also dot products. *When vectors have m components, a linear combination uses m dot products.*

Main point To compute the dot product, multiply each v_i times w_i. Then add.

Lengths and Unit Vectors

An important case is the dot product of a vector with itself. In this case $\boldsymbol{v} = \boldsymbol{w}$. When the vector is $\boldsymbol{v} = (1, 2, 3)$, the dot product is $\boldsymbol{v} \cdot \boldsymbol{v} = 14$:

$$\begin{bmatrix} 1 \\ 2 \\ 3 \end{bmatrix} \cdot \begin{bmatrix} 1 \\ 2 \\ 3 \end{bmatrix} = 1 + 4 + 9 = 14.$$

The answer is not zero because $\boldsymbol{v}$ is not perpendicular to itself. Instead of a 90° angle we have 0°. In this special case, the dot product $\boldsymbol{v} \cdot \boldsymbol{v}$ gives the *length squared.*

Definition The ***length*** (or ***norm***) of a vector $\boldsymbol{v}$ is the square root of $\boldsymbol{v} \cdot \boldsymbol{v}$:

$$\text{length} = \|\boldsymbol{v}\| = \sqrt{\boldsymbol{v} \cdot \boldsymbol{v}}.$$

In two dimensions the length is $\|\boldsymbol{v}\| = \sqrt{v_1^2 + v_2^2}$. In three dimensions it is $\|\boldsymbol{v}\| = \sqrt{v_1^2 + v_2^2 + v_3^2}$. By the calculation above, the length of $\boldsymbol{v} = (1, 2, 3)$ is $\|\boldsymbol{v}\| = \sqrt{14}$.

We can explain this definition. $\|\boldsymbol{v}\|$ is just the ordinary length of the arrow that represents the vector. In two dimensions, the arrow is in a plane. If the components are 1 and 2, the arrow is the third side of a right triangle (Figure 1.5). The formula $a^2 + b^2 = c^2$, which connects the three sides, is $1^2 + 2^2 = \|\boldsymbol{v}\|^2$. That formula gives $\|\boldsymbol{v}\| = \sqrt{5}$. Lengths are always positive, except for the zero vector with $\|\mathbf{0}\| = 0$.

For the length of $\boldsymbol{v} = (1, 2, 3)$, we used the right triangle formula twice. First, the vector in the base has components 1, 2, 0 and length $\sqrt{5}$. This base vector is perpendicular to the vector (0, 0, 3) that goes straight up. So the diagonal of the box has length $\|\boldsymbol{v}\| = \sqrt{5+9} = \sqrt{14}$.

The length of a four-dimensional vector would be $\sqrt{v_1^2 + v_2^2 + v_3^2 + v_4^2}$. Thus (1, 1, 1, 1) has length $\sqrt{1^2 + 1^2 + 1^2 + 1^2} = 2$. This is the diagonal through a unit cube in four-dimensional space. The diagonal in three dimensions has length $\sqrt{3}$.

The word "unit" is always indicating that some measurement equals "one." The unit price is the price for one item. A unit cube has sides of length one. A unit circle is a circle with radius one. Now we define the idea of a "unit vector."

Definition A ***unit vector u*** is a vector whose length equals one. Then $\boldsymbol{u} \cdot \boldsymbol{u} = 1$.

An example in four dimensions is $\boldsymbol{u} = (\frac{1}{2}, \frac{1}{2}, \frac{1}{2}, \frac{1}{2})$. Then $\boldsymbol{u} \cdot \boldsymbol{u}$ is $\frac{1}{4} + \frac{1}{4} + \frac{1}{4} + \frac{1}{4} = 1$. We divided the vector $\boldsymbol{v} = (1, 1, 1, 1)$ by its length $\|\boldsymbol{v}\| = 2$ to get the unit vector.

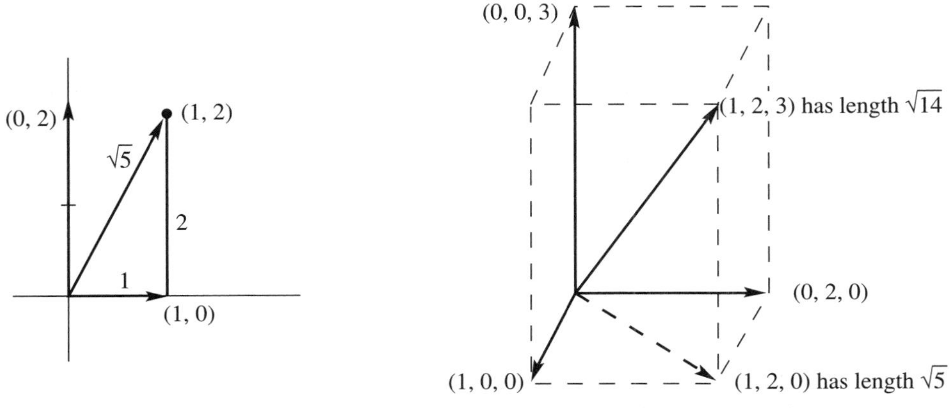

Figure 1.5 The length of two-dimensional and three-dimensional vectors.

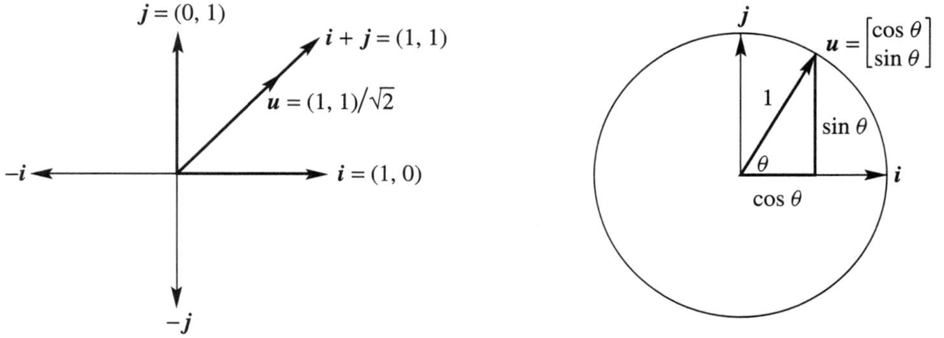

Figure 1.6 The coordinate vectors $\boldsymbol{i}$ and $\boldsymbol{j}$. The unit vector $\boldsymbol{u}$ at angle 45° and the unit vector at angle θ.

Example 1.4 The standard unit vectors along the x and y axes are written $\boldsymbol{i}$ and $\boldsymbol{j}$. In the xy plane, the unit vector that makes an angle "theta" with the x axis is $\boldsymbol{u}$:

$$\boldsymbol{i} = \begin{bmatrix} 1 \\ 0 \end{bmatrix} \quad \text{and} \quad \boldsymbol{j} = \begin{bmatrix} 0 \\ 1 \end{bmatrix} \quad \text{and} \quad \boldsymbol{u} = \begin{bmatrix} \cos\theta \\ \sin\theta \end{bmatrix}.$$

When $\theta = 0$, the vector $\boldsymbol{u}$ is the same as $\boldsymbol{i}$. When $\theta = 90°$ (or $\frac{\pi}{2}$ radians), $\boldsymbol{u}$ is the same as $\boldsymbol{j}$. At any angle, the components $\cos\theta$ and $\sin\theta$ produce $\boldsymbol{u} \cdot \boldsymbol{u} = 1$ because $\cos^2\theta + \sin^2\theta = 1$. These vectors reach out to the unit circle in Figure 1.6. Thus $\cos\theta$ and $\sin\theta$ are simply the coordinates of that point at angle θ on the unit circle.

In three dimensions, the unit vectors along the axes are $\boldsymbol{i}$, $\boldsymbol{j}$, and $\boldsymbol{k}$. Their components are $(1, 0, 0)$ and $(0, 1, 0)$ and $(0, 0, 1)$. Notice how every three-dimensional vector is a linear combination of $\boldsymbol{i}$, $\boldsymbol{j}$, and $\boldsymbol{k}$. The vector $\boldsymbol{v} = (2, 2, 1)$ is equal to $2\boldsymbol{i} + 2\boldsymbol{j} + \boldsymbol{k}$. Its length is $\sqrt{2^2 + 2^2 + 1^2}$. This is the square root of 9, so $\|\boldsymbol{v}\| = 3$.

Since $(2, 2, 1)$ has length 3, the vector $(\frac{2}{3}, \frac{2}{3}, \frac{1}{3})$ has length 1. To create a unit vector, just divide $\boldsymbol{v}$ by its length $\|\boldsymbol{v}\|$.

1A Unit vectors Divide any nonzero vector $\boldsymbol{v}$ by its length. The result is $\boldsymbol{u} = \boldsymbol{v}/\|\boldsymbol{v}\|$. This is a unit vector in the same direction as $\boldsymbol{v}$.

Figure 1.6 shows the components of $\boldsymbol{i} + \boldsymbol{j} = (1, 1)$ divided by the length $\sqrt{2}$. Then $u_1^2 = \frac{1}{2}$ and $u_2^2 = \frac{1}{2}$ and $\|\boldsymbol{u}\|^2 = \frac{1}{2} + \frac{1}{2} = 1$. When we divide the vector by $\|\boldsymbol{v}\|$, we divide its length by $\|\boldsymbol{v}\|$. This leaves a unit length $\|\boldsymbol{u}\| = 1$.

In three dimensions we found $\boldsymbol{u} = (\frac{2}{3}, \frac{2}{3}, \frac{1}{3})$. Check that $\boldsymbol{u} \cdot \boldsymbol{u} = \frac{4}{9} + \frac{4}{9} + \frac{1}{9} = 1$. Then $\boldsymbol{u}$ reaches out to the "unit sphere" centered at the origin. Unit vectors correspond to points on the sphere of radius one.

The Angle Between Two Vectors

We stated that perpendicular vectors have $\boldsymbol{v} \cdot \boldsymbol{w} = 0$. The dot product is zero when the angle is 90°. To give a reason, we have to connect right angles to dot products. Then we show how use $\boldsymbol{v} \cdot \boldsymbol{w}$ to find the angle between two nonzero vectors, perpendicular or not.

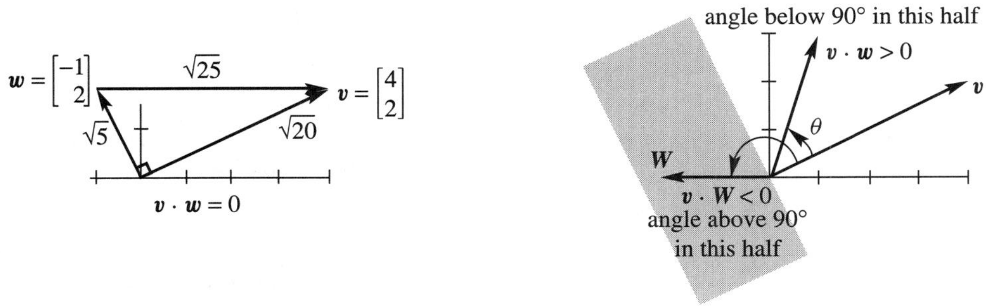

Figure 1.7 Perpendicular vectors have $v \cdot w = 0$. The angle θ is below 90° when $v \cdot w > 0$.

1B Right angles The dot product $v \cdot w$ is zero when v is perpendicular to w.

Proof When v and w are perpendicular, they form two sides of a right triangle. The third side (the hypotenuse going across in Figure 1.7) is $v - w$. So the law $a^2 + b^2 = c^2$ for their lengths becomes

$$\|v\|^2 + \|w\|^2 = \|v - w\|^2 \quad \text{(for perpendicular vectors only).} \tag{1.2}$$

Writing out the formulas for those lengths (in two dimensions), this equation is

$$v_1^2 + v_2^2 + w_1^2 + w_2^2 = (v_1 - w_1)^2 + (v_2 - w_2)^2. \tag{1.3}$$

The right side begins with $v_1^2 - 2v_1w_1 + w_1^2$. Then v_1^2 and w_1^2 are on both sides of the equation and they cancel. Similarly $(v_2 - w_2)^2$ contains v_2^2 and w_2^2 and the cross term $-2v_2w_2$. All terms cancel except $-2v_1w_1$ and $-2v_2w_2$. (In three dimensions there would also be $-2v_3w_3$.) The last step is to divide by -2:

$$0 = -2v_1w_1 - 2v_2w_2 \quad \text{which leads to} \quad v_1w_1 + v_2w_2 = 0. \tag{1.4}$$

Conclusion Right angles produce $v \cdot w = 0$. We have proved **Theorem 1B**. The dot product is zero when the angle is $\theta = 90°$. Then $\cos\theta = 0$.

The zero vector $v = \mathbf{0}$ is perpendicular to every vector w because $\mathbf{0} \cdot w$ is always zero.

Now suppose $v \cdot w$ is not zero. It may be positive, it may be negative. The sign of $v \cdot w$ immediately tells whether we are below or above a right angle. The angle is less than 90° when $v \cdot w$ is positive. The angle is above 90° when $v \cdot w$ is negative. Figure 1.7 shows a typical vector $w = (1, 3)$ in the white half-plane, with $v \cdot w > 0$. The vector $W = (-2, 0)$ in the screened half-plane has $v \cdot W < 0$.

The vectors we drew have $v \cdot w = 10$ and $v \cdot W = -8$. The borderline is where vectors are perpendicular to v. On that dividing line between plus and minus, the dot product is zero.

The next page takes one more step with the geometry of dot products. We find the exact angle θ. This is not necessary for linear algebra—you could stop here! Once we have matrices and linear equations, we won't come back to θ. But while we are on the subject

Figure 1.8 The dot product of unit vectors is the cosine of the angle θ.

of angles, this is the place for the formula. It will show that the angle θ in Figure 1.7 is exactly 45°.

Start with unit vectors $\boldsymbol{u}$ and $\boldsymbol{U}$. The sign of $\boldsymbol{u} \cdot \boldsymbol{U}$ tells whether $\theta < 90°$ or $\theta > 90°$. Because the vectors have length 1, we learn more than that. *The dot product* $\boldsymbol{u} \cdot \boldsymbol{U}$ *is the* **cosine** *of* θ.

1C (a) If $\boldsymbol{u}$ and $\boldsymbol{U}$ are unit vectors then $\boldsymbol{u} \cdot \boldsymbol{U} = \cos\theta$.

(b) If $\boldsymbol{u}$ and $\boldsymbol{U}$ are unit vectors then $|\boldsymbol{u} \cdot \boldsymbol{U}| \le 1$.

Statement (b) follows directly from (a). Remember that $\cos\theta$ is never greater than 1. It is never less than -1. So we discover an extra fact about unit vectors. Their dot product is between -1 and 1.

Figure 1.8 shows this clearly when the vectors are $(\cos\theta, \sin\theta)$ and $(1, 0)$. The dot product of those vectors is $\cos\theta$. That is the cosine of the angle between them.

After rotation through any angle α, these are still unit vectors. The angle between them is still θ. The vectors are $\boldsymbol{u} = (\cos\alpha, \sin\alpha)$ and $\boldsymbol{U} = (\cos\beta, \sin\beta)$. Their dot product is $\cos\alpha\cos\beta + \sin\alpha\sin\beta$. From trigonometry this is the same as $\cos(\beta - \alpha)$. Since $\beta - \alpha$ equals θ we have reached the formula $\boldsymbol{u} \cdot \boldsymbol{U} = \cos\theta$. Then the fact that $|\cos\theta| \le 1$ tells us immediately that $|\boldsymbol{u} \cdot \boldsymbol{U}| \le 1$.

Problem 24 proves the inequality $|\boldsymbol{u} \cdot \boldsymbol{U}| \le 1$ directly, without mentioning angles. Problem 22 applies another formula from trigonometry, the "Law of Cosines." The inequality and the cosine formula are true in any number of dimensions. The dot product does not change when vectors are rotated—the angle between them stays the same.

What if $\boldsymbol{v}$ and $\boldsymbol{w}$ are not unit vectors? Their dot product is not generally $\cos\theta$. We have to divide by their lengths to get unit vectors $\boldsymbol{u} = \boldsymbol{v}/\|\boldsymbol{v}\|$ and $\boldsymbol{U} = \boldsymbol{w}/\|\boldsymbol{w}\|$. Then the dot product of those rescaled vectors $\boldsymbol{u}$ and $\boldsymbol{U}$ gives $\cos\theta$.

Whatever the angle, this dot product of $\boldsymbol{v}/\|\boldsymbol{v}\|$ with $\boldsymbol{w}/\|\boldsymbol{w}\|$ never exceeds one. That is the "Schwarz inequality" for dot products—or more correctly the Cauchy-Schwarz-Buniakowsky inequality. It was found in France and Germany and Russia (and maybe elsewhere—it is the most important inequality in mathematics). With the extra factor $\|\boldsymbol{v}\|\,\|\boldsymbol{w}\|$ from rescaling to unit vectors, we have $\cos\theta$:

1D (a) **COSINE FORMULA** If $\boldsymbol{v}$ and $\boldsymbol{w}$ are nonzero vectors then $\dfrac{\boldsymbol{v} \cdot \boldsymbol{w}}{\|\boldsymbol{v}\|\,\|\boldsymbol{w}\|} = \cos\theta$.

(b) **SCHWARZ INEQUALITY** If $\boldsymbol{v}$ and $\boldsymbol{w}$ are any vectors then $|\boldsymbol{v} \cdot \boldsymbol{w}| \le \|\boldsymbol{v}\|\,\|\boldsymbol{w}\|$.

Example 1.5 Find the angle between $\boldsymbol{v} = \left[\begin{smallmatrix}4\\2\end{smallmatrix}\right]$ and $\boldsymbol{w} = \left[\begin{smallmatrix}1\\3\end{smallmatrix}\right]$ in Figure 1.7b.

Solution The dot product is $\boldsymbol{v} \cdot \boldsymbol{w} = 10$. The length of $\boldsymbol{v}$ is $\|\boldsymbol{v}\| = \sqrt{20}$. The length of $\boldsymbol{w}$ is $\|\boldsymbol{w}\| = \sqrt{10}$. Therefore the cosine of the angle is

$$\cos\theta = \frac{\boldsymbol{v} \cdot \boldsymbol{w}}{\|\boldsymbol{v}\|\,\|\boldsymbol{w}\|} = \frac{10}{\sqrt{20}\sqrt{10}} = \frac{1}{\sqrt{2}}.$$

The angle that has this cosine is 45°. It is below 90° because $\boldsymbol{v} \cdot \boldsymbol{w} = 10$ is positive. Of course $\|\boldsymbol{v}\|\,\|\boldsymbol{w}\| = \sqrt{200}$ is larger than 10.

Example 1.6 For $\boldsymbol{v} = (a, b)$ and $\boldsymbol{w} = (b, a)$ the Schwarz inequality says that $2ab \leq a^2 + b^2$. For example $2(3)(4) = 24$ is less than $3^2 + 4^2 = 25$.

The dot product of (a, b) and (b, a) is $2ab$. The lengths are $\|\boldsymbol{v}\| = \|\boldsymbol{w}\| = \sqrt{a^2 + b^2}$. Then $\boldsymbol{v} \cdot \boldsymbol{w} = 2ab$ never exceeds $\|\boldsymbol{v}\|\,\|\boldsymbol{w}\| = a^2 + b^2$. The difference between them can never be negative:

$$a^2 + b^2 - 2ab = (a - b)^2 \geq 0.$$

This is more famous if we write $x = a^2$ and $y = b^2$. Then the "geometric mean" $\sqrt{xy}$ is not larger than the "arithmetic mean," which is the average of x and y:

$$ab \leq \frac{a^2 + b^2}{2} \quad \text{becomes} \quad \sqrt{xy} \leq \frac{x + y}{2}.$$

Computing dot products and lengths and angles It is time for a moment of truth. The dot product $\boldsymbol{v} \cdot \boldsymbol{w}$ is usually seen as ***a row times a column***:

$$\text{Instead of} \quad \begin{bmatrix}1\\2\end{bmatrix} \cdot \begin{bmatrix}3\\4\end{bmatrix} \quad \text{we more often see} \quad \begin{bmatrix}1 & 2\end{bmatrix}\begin{bmatrix}3\\4\end{bmatrix}.$$

In FORTRAN we multiply components and add (using a loop):

```
   DO 10 I = 1,N
10 VDOTW = VDOTW + V(I) * W(I)
```

MATLAB works with whole vectors, not their components. If v and w are column vectors then v′ is a row as above:

```
dot = v' * w
```

The length of $\boldsymbol{v}$ is already known to MATLAB as norm(v). We could define it ourselves as sqrt(v′ ∗ v), using the square root function—also known. The cosine and the angle (in radians) we do have to define ourselves:

```
cos = v' * w/(norm(v) * norm(w));
angle = acos(cos)
```

We used the *arc cosine* (acos) function to find the angle from its cosine. We have ***not*** created a new function cos(v,w) for future use. That would become an M-file, and Chapter 2 will show its format. (Quite a few M-files have been created especially for this book. They are listed at the end.) The instructions above will cause the numbers dot and angle to be printed. The cosine will not be printed because of the semicolon.

Problem Set 1.2

1 Calculate the dot products $\boldsymbol{u}\cdot\boldsymbol{v}$ and $\boldsymbol{u}\cdot\boldsymbol{w}$ and $\boldsymbol{v}\cdot\boldsymbol{w}$ and $\boldsymbol{w}\cdot\boldsymbol{v}$:

$$\boldsymbol{u}=\begin{bmatrix}-.6\\ .8\end{bmatrix} \qquad \boldsymbol{v}=\begin{bmatrix}3\\4\end{bmatrix} \qquad \boldsymbol{w}=\begin{bmatrix}4\\3\end{bmatrix}.$$

2 Compute the lengths $\|\boldsymbol{u}\|$ and $\|\boldsymbol{v}\|$ and $\|\boldsymbol{w}\|$ of those vectors. Check the Schwarz inequalities $|\boldsymbol{u}\cdot\boldsymbol{v}|\le\|\boldsymbol{u}\|\,\|\boldsymbol{v}\|$ and $|\boldsymbol{v}\cdot\boldsymbol{w}|\le\|\boldsymbol{v}\|\,\|\boldsymbol{w}\|$.

3 Write down unit vectors in the directions of $\boldsymbol{v}$ and $\boldsymbol{w}$ in Problem 1. Find the cosine of the angle θ between them.

4 Find unit vectors $\boldsymbol{u}_1$ and $\boldsymbol{u}_2$ that are parallel to $\boldsymbol{v}=(3,1)$ and $\boldsymbol{w}=(2,1,2)$. Also find unit vectors $\boldsymbol{U}_1$ and $\boldsymbol{U}_2$ that are perpendicular to $\boldsymbol{v}$ and $\boldsymbol{w}$.

5 For any unit vectors $\boldsymbol{v}$ and $\boldsymbol{w}$, find the angle θ between

(a) $\boldsymbol{v}$ and $\boldsymbol{v}$ (b) $\boldsymbol{w}$ and $-\boldsymbol{w}$ (c) $\boldsymbol{v}+\boldsymbol{w}$ and $\boldsymbol{v}-\boldsymbol{w}$.

6 Find the angle θ (from its cosine) between

(a) $\boldsymbol{v}=\begin{bmatrix}1\\\sqrt{3}\end{bmatrix}$ and $\boldsymbol{w}=\begin{bmatrix}1\\0\end{bmatrix}$ (b) $\boldsymbol{v}=\begin{bmatrix}2\\2\\-1\end{bmatrix}$ and $\boldsymbol{w}=\begin{bmatrix}2\\-1\\2\end{bmatrix}$

(c) $\boldsymbol{v}=\begin{bmatrix}1\\\sqrt{3}\end{bmatrix}$ and $\boldsymbol{w}=\begin{bmatrix}-1\\\sqrt{3}\end{bmatrix}$ (d) $\boldsymbol{v}=\begin{bmatrix}3\\1\end{bmatrix}$ and $\boldsymbol{w}=\begin{bmatrix}-1\\-2\end{bmatrix}$.

7 Describe all vectors that are perpendicular to $\boldsymbol{v}=(2,-1)$. Repeat for $\boldsymbol{V}=(1,1,1)$.

8 True or false (give a reason if true or a counterexample if false):

(a) If $\boldsymbol{u}$ is perpendicular (in three dimensions) to $\boldsymbol{v}$ and $\boldsymbol{w}$, then $\boldsymbol{v}$ and $\boldsymbol{w}$ are parallel.

(b) If $\boldsymbol{u}$ is perpendicular to $\boldsymbol{v}$ and $\boldsymbol{w}$, then $\boldsymbol{u}$ is perpendicular to every combination $c\boldsymbol{v}+d\boldsymbol{w}$.

(c) There is always a nonzero combination $c\boldsymbol{v}+d\boldsymbol{w}$ that is perpendicular to $\boldsymbol{u}$.

9 The slopes of the arrows from $(0,0)$ to (v_1,v_2) and (w_1,w_2) are v_2/v_1 and w_2/w_1. If the product of those slopes is $v_2w_2/v_1w_1=-1$, show that $\boldsymbol{v}\cdot\boldsymbol{w}=0$ and the vectors are perpendicular.

10 Draw arrows from $(0,0)$ to the points $\boldsymbol{v}=(1,2)$ and $\boldsymbol{w}=(-2,1)$. Write down the two slopes and multiply them. That answer is a signal that $\boldsymbol{v}\cdot\boldsymbol{w}=0$ and the arrows are ______ .

11 If $\boldsymbol{v}\cdot\boldsymbol{w}$ is negative, what does this say about the angle between $\boldsymbol{v}$ and $\boldsymbol{w}$? Draw a 3-dimensional vector $\boldsymbol{v}$ (an arrow), and show where to find all $\boldsymbol{w}$'s with $\boldsymbol{v}\cdot\boldsymbol{w}<0$.

12 With $\boldsymbol{v}=(1,1)$ and $\boldsymbol{w}=(1,5)$ choose a number c so that $\boldsymbol{w}-c\boldsymbol{v}$ is perpendicular to $\boldsymbol{v}$. Then find the formula that gives this number c for any $\boldsymbol{v}$ and $\boldsymbol{w}$.

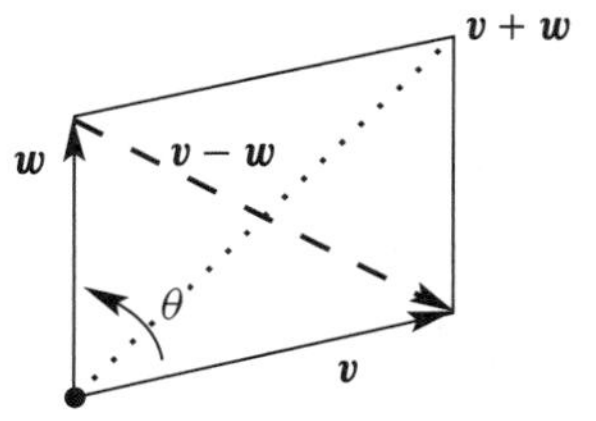

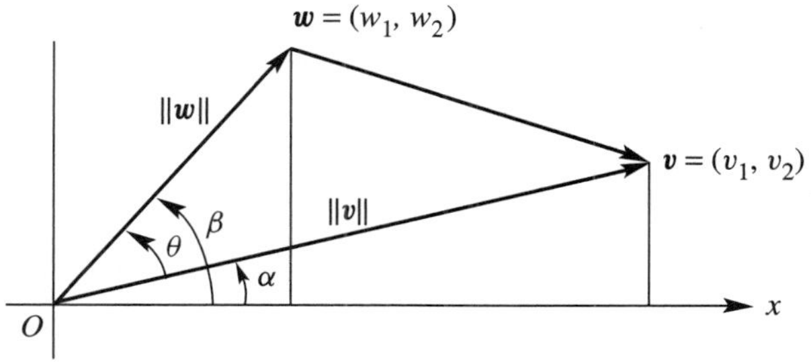

13 Find two vectors $\boldsymbol{v}$ and $\boldsymbol{w}$ that are perpendicular to $(1, 1, 1)$ and to each other.

14 Find three vectors $\boldsymbol{u}, \boldsymbol{v}, \boldsymbol{w}$ that are perpendicular to $(1, 1, 1, 1)$ and to each other.

15 The geometric mean of $x = 2$ and $y = 8$ is $\sqrt{xy} = 4$. The arithmetic mean is larger: $\frac{1}{2}(x + y) =$ _____. This came in Example 1.6 from the Schwarz inequality for $\boldsymbol{v} = (\sqrt{2}, \sqrt{8})$ and $\boldsymbol{w} = (\sqrt{8}, \sqrt{2})$. Find $\cos\theta$ for this $\boldsymbol{v}$ and $\boldsymbol{w}$.

16 How long is the vector $\boldsymbol{v} = (1, 1, \ldots, 1)$ in 9 dimensions? Find a unit vector $\boldsymbol{u}$ in the same direction as $\boldsymbol{v}$ and a vector $\boldsymbol{w}$ that is perpendicular to $\boldsymbol{v}$.

17 What are the cosines of the angles α, β, θ between the vector $(1, 0, -1)$ and the unit vectors $\boldsymbol{i}, \boldsymbol{j}, \boldsymbol{k}$ along the axes? Check the formula $\cos^2\alpha + \cos^2\beta + \cos^2\theta = 1$.

Problems 18–24 lead to the main facts about lengths and angles. Never again will we have several proofs in a row.

18 (Rules for dot products) These equations are simple but useful: (1) $\boldsymbol{v} \cdot \boldsymbol{w} = \boldsymbol{w} \cdot \boldsymbol{v}$
(2) $\boldsymbol{u} \cdot (\boldsymbol{v} + \boldsymbol{w}) = \boldsymbol{u} \cdot \boldsymbol{v} + \boldsymbol{u} \cdot \boldsymbol{w}$ (3) $(c\boldsymbol{v}) \cdot \boldsymbol{w} = c(\boldsymbol{v} \cdot \boldsymbol{w})$
Use Rules (1–2) with $\boldsymbol{u} = \boldsymbol{v} + \boldsymbol{w}$ to prove that $\|\boldsymbol{v} + \boldsymbol{w}\|^2 = \boldsymbol{v} \cdot \boldsymbol{v} + 2\boldsymbol{v} \cdot \boldsymbol{w} + \boldsymbol{w} \cdot \boldsymbol{w}$.

19 (The triangle inequality: length of $\boldsymbol{v} + \boldsymbol{w} \le$ length of $\boldsymbol{v}$+length of $\boldsymbol{w}$) Problem 18 found $\|\boldsymbol{v} + \boldsymbol{w}\|^2 = \|\boldsymbol{v}\|^2 + 2\boldsymbol{v} \cdot \boldsymbol{w} + \|\boldsymbol{w}\|^2$. Show how the Schwarz inequality leads to

$$\|\boldsymbol{v} + \boldsymbol{w}\|^2 \le (\|\boldsymbol{v}\| + \|\boldsymbol{w}\|)^2 \quad \text{or} \quad \|\boldsymbol{v} + \boldsymbol{w}\| \le \|\boldsymbol{v}\| + \|\boldsymbol{w}\|.$$

20 (The perpendicularity test $\boldsymbol{v} \cdot \boldsymbol{w} = 0$ in three dimensions)

(a) The right triangle still obeys $\|\boldsymbol{v}\|^2 + \|\boldsymbol{w}\|^2 = \|\boldsymbol{v} - \boldsymbol{w}\|^2$. Show how this leads as in equations (1–3) to $v_1w_1 + v_2w_2 + v_3w_3 = 0$.

(b) Is it also true that $\|\boldsymbol{v}\|^2 + \|\boldsymbol{w}\|^2 = \|\boldsymbol{v} + \boldsymbol{w}\|^2$? Draw this right triangle too.

21 (Cosine of θ from the formula $\cos(\beta - \alpha) = \cos\beta\cos\alpha + \sin\beta\sin\alpha$)
The figure shows that $\cos\alpha = v_1/\|\boldsymbol{v}\|$ and $\sin\alpha = v_2/\|\boldsymbol{v}\|$. Similarly $\cos\beta$ is _____ and $\sin\beta$ is _____. The angle θ is $\beta - \alpha$. Substitute into the formula for $\cos(\beta - \alpha)$ to find $\cos\theta = \boldsymbol{v} \cdot \boldsymbol{w}/\|\boldsymbol{v}\| \|\boldsymbol{w}\|$.

22 (The formula for $\cos\theta$ from the Law of Cosines)
With $\boldsymbol{v}$ and $\boldsymbol{w}$ at angle θ, the Law of Cosines gives the length of the third side:

$$\|\boldsymbol{v}-\boldsymbol{w}\|^2 = \|\boldsymbol{v}\|^2 - 2\|\boldsymbol{v}\|\,\|\boldsymbol{w}\|\cos\theta + \|\boldsymbol{w}\|^2.$$

Compare with $(\boldsymbol{v}-\boldsymbol{w})\cdot(\boldsymbol{v}-\boldsymbol{w}) =$ ______ to find the formula for $\cos\theta$.

23 (The Schwarz inequality by algebra instead of trigonometry)

(a) Multiply out both sides of $(v_1w_1+v_2w_2)^2 \le (v_1^2+v_2^2)(w_1^2+w_2^2)$.

(b) Show that the difference between those sides equals $(v_1w_2 - v_2w_1)^2$. This cannot be negative since it is a square—so the inequality is true.

24 (One-line proof of the Schwarz inequality $|\boldsymbol{u}\cdot\boldsymbol{U}| \le 1$)
If (u_1, u_2) and (U_1, U_2) are unit vectors, pick out the step that uses Example 1.6:

$$|\boldsymbol{u}\cdot\boldsymbol{U}| \le |u_1|\,|U_1| + |u_2|\,|U_2| \le \frac{u_1^2+U_1^2}{2} + \frac{u_2^2+U_2^2}{2} = \frac{1+1}{2} = 1.$$

Put $(u_1, u_2) = (.6, .8)$ and $(U_1, U_2) = (.8, .6)$ in that whole line and find $\cos\theta$.

25 Why is $|\cos\theta|$ never greater than 1 in the first place?

26 Pick any numbers that add to $x+y+z=0$. Find the angle between your vector $\boldsymbol{v} = (x, y, z)$ and the vector $\boldsymbol{w} = (z, x, y)$. Challenge question: Explain why $\boldsymbol{v}\cdot\boldsymbol{w}/\|\boldsymbol{v}\|\,\|\boldsymbol{w}\|$ is always $-\frac{1}{2}$.

1.3 Planes

The first sections mostly discussed two-dimensional vectors. But everything in those sections—dot products, lengths, angles, and linear combinations $c\boldsymbol{v}+d\boldsymbol{w}$—follows the same pattern when vectors have three components. The ideas also apply in 4 and 5 and n dimensions. We start with ordinary "3-space," and our goal is to find the *equation of a plane*.

One form of the equation involves dot products. The other form involves linear combinations. The dot products and the combinations now appear in ***equations***. The equations are the new feature—and planes are the perfect examples.

Imagine a plane (like a wall of your room). Many vectors are parallel to that plane. But only one direction is perpendicular to the plane. This perpendicular direction is the best way to describe the plane. The vectors parallel to the plane are identified by a vector $\boldsymbol{n}$ to which they are all perpendicular.

"Perpendicular" means the same as "orthogonal." Another word is "normal," and its first letter is the reason we use $\boldsymbol{n}$. You can visualize $\boldsymbol{n}$ pointing straight out from the plane, with any nonzero length. If $\boldsymbol{n}$ is a normal vector, so is $2\boldsymbol{n}$. So is $-\boldsymbol{n}$.

Start with a plane through the origin $(0, 0, 0)$. If the point (x, y, z) lies on the plane, then the vector $\boldsymbol{v}$ to that point is perpendicular to $\boldsymbol{n}$. The vector $\boldsymbol{v}$ has components (x, y, z). The normal vector $\boldsymbol{n}$ has components (a, b, c). The geometry of "perpendicular" converts to the algebra of "dot product equals zero." That is exactly the equation we want: The plane through (0, 0, 0) perpendicular to $\boldsymbol{n} = (a, b, c)$ has the linear equation

$$\boldsymbol{n} \cdot \boldsymbol{v} = 0 \quad \text{or} \quad ax + by + cz = 0. \tag{1.5}$$

This is the first plane in Figure 1.9—it goes through the origin. Now look at other planes, all perpendicular to the same vector $\boldsymbol{n}$. They are parallel to the first plane, but they don't go through (0, 0, 0).

Suppose the particular point (x_0, y_0, z_0) lies in the plane. Other points in the plane have other coordinates (x, y, z). Not all points are allowed—that would fill the whole space. The normal vector is $\boldsymbol{n} = (a, b, c)$, the vector to (x_0, y_0, z_0) is $\boldsymbol{v}_0$, the vector to (x, y, z) is $\boldsymbol{v}$. Look at the vector $\boldsymbol{v} - \boldsymbol{v}_0$ in Figure 1.9. It is parallel to the plane, which makes it perpendicular to $\boldsymbol{n}$. The dot product of $\boldsymbol{n}$ with $\boldsymbol{v} - \boldsymbol{v}_0$ produces our equation.

1E The plane that passes through $\boldsymbol{v}_0 = (x_0, y_0, z_0)$ perpendicular to $\boldsymbol{n}$ has the linear equation

$$\boldsymbol{n} \cdot (\boldsymbol{v} - \boldsymbol{v}_0) = 0 \quad \text{or} \quad a(x - x_0) + b(y - y_0) + c(z - z_0) = 0. \tag{1.6}$$

The equation expresses two facts of geometry. The plane goes through the given point (x_0, y_0, z_0) and it is perpendicular to $\boldsymbol{n}$. Remember that x, y, z are variables—they describe all points in the plane. The numbers x_0, y_0, z_0 are constants—they describe one point. One possible choice of x, y, z is x_0, y_0, z_0. Then $\boldsymbol{v} = \boldsymbol{v}_0$ satisfies the equation and is a point on the plane.

To emphasize the difference between $\boldsymbol{v}$ and $\boldsymbol{v}_0$, move the constants in $\boldsymbol{v}_0$ to the other side. The normal vector $\boldsymbol{n}$ is still (a, b, c):

1F Every plane perpendicular to $\boldsymbol{n}$ has a linear equation with coefficients a, b, c:

$$ax + by + cz = ax_0 + by_0 + cz_0 \quad \text{or} \quad ax + by + cz = d. \tag{1.7}$$

Different values of d give parallel planes. The value $d = 0$ gives a plane through the origin.

Example 1.7 The plane $x + 2y + 3z = 0$ goes through the origin. Check: $x = 0$, $y = 0$, $z = 0$ satisfies the equation. The normal vector is $\boldsymbol{n} = (1, 2, 3)$. Another normal vector is

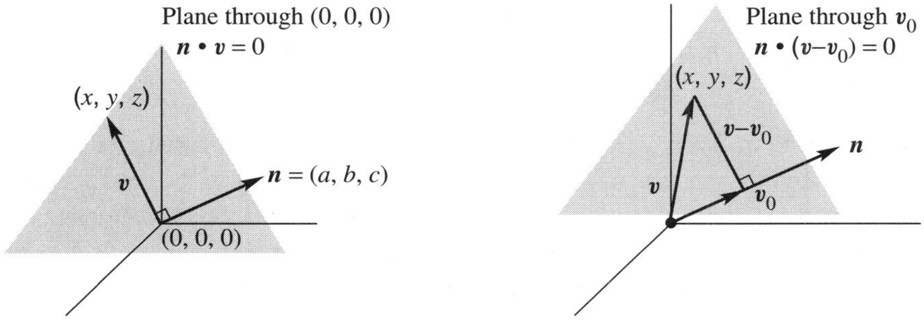

Figure 1.9 The planes perpendicular to $\boldsymbol{n}$ are $\boldsymbol{n} \cdot \boldsymbol{v} = 0$ and $\boldsymbol{n} \cdot (\boldsymbol{v} - \boldsymbol{v}_0) = 0$ (which is $\boldsymbol{n} \cdot \boldsymbol{v} = d$ and also $ax + by + cz = d$).

$\boldsymbol{n} = (2, 4, 6)$. The equation $2x + 4y + 6z = 0$ produces the same plane. Two points in this plane are $(x, y, z) = (2, -1, 0)$ and $(x, y, z) =$ _____.

Example 1.8 Keep the same normal vector $(1, 2, 3)$. Find the parallel plane through $(x_0, y_0, z_0) = (1, 1, 1)$.

Solution The equation is still $x + 2y + 3z = d$. Choose $d = 6$, so the point $x = 1$, $y = 1$, $z = 1$ satisfies the equation. This number 6 is the same as $ax_0 + by_0 + cz_0 = 1 + 2 + 3$. The number d is always $\boldsymbol{n} \cdot \boldsymbol{v}_0$.

Example 1.9 The "xy plane" has the equation $z = 0$. Its normal vector is $(0, 0, 1)$—this is the standard unit vector $\boldsymbol{k}$, pointing straight up. A typical point on the xy plane has coordinates $(x, y, 0)$. The dot product of the perpendicular vector $(0, 0, 1)$ and the in-plane vector $(x, y, 0)$ is zero.

Distance to a Plane

How far from the origin is the plane $x + 2y + 3z = 6$? We know that the particular point $(1, 1, 1)$ is on this plane. Its distance from $(0, 0, 0)$ is the square root of $1^2 + 1^2 + 1^2$, which is near 1.73. But $(1, 1, 1)$ might not be the point on the plane that is closest to the origin. It isn't.

Geometry comes to the rescue, if you look at Figure 1.9. The vector $\boldsymbol{v}_0$ happened to be chosen in the same direction as $\boldsymbol{n}$. Other choices were available, because $\boldsymbol{v}_0$ can be any point on the plane. But this $\boldsymbol{v}_0$ is the closest point, and it is a multiple of $\boldsymbol{n}$. The length of this $\boldsymbol{v}_0$ is the distance to the plane.

1G The shortest distance from $(0, 0, 0)$ to the plane $\boldsymbol{n} \cdot \boldsymbol{v} = d$ is $|d|/\|\boldsymbol{n}\|$. The vector $\boldsymbol{v}$ that gives this shortest distance is $d/\|\boldsymbol{n}\|^2$ times $\boldsymbol{n}$. It is perpendicular to the plane:

$$\boldsymbol{n} \cdot \boldsymbol{v} = \frac{d\boldsymbol{n} \cdot \boldsymbol{n}}{\|\boldsymbol{n}\|^2} = d \qquad \text{and the distance is} \qquad \|\boldsymbol{v}\| = \frac{\|d\boldsymbol{n}\|}{\|\boldsymbol{n}\|^2} = \frac{|d|}{\|\boldsymbol{n}\|}. \tag{1.8}$$

Our example has $d = 6$ and $\|\boldsymbol{n}\| = \sqrt{14}$. The distance $6/\sqrt{14}$ is about 1.6, smaller than 1.73. The closest point is $\boldsymbol{v} = \frac{6}{14}\boldsymbol{n}$. Notice that $\boldsymbol{n} \cdot \boldsymbol{v} = \frac{6}{14}\boldsymbol{n} \cdot \boldsymbol{n}$ which is equal to 6. So $\boldsymbol{v}$ reaches the plane.

Note 1 If the normal vector is a unit vector ($\|\boldsymbol{n}\| = 1$), then the distance is $|d|/1$ which is $|d|$. To make this happen for the plane $x + 2y + 3z = 6$, divide the equation by $\sqrt{14}$. Then the normal vector has length 1, and the new right side is exactly the distance $6/\sqrt{14}$.

If we multiply an equation $\boldsymbol{n} \cdot \boldsymbol{v} = d$ by 2, the plane doesn't change. Its equation is now $2\boldsymbol{n} \cdot \boldsymbol{v} = 2d$. The distance from $(0, 0, 0)$ is $|2d|/\|2\boldsymbol{n}\|$. We cancel the 2's to see that the distance didn't change.

Note 2 Why is the shortest $\boldsymbol{v}_0$ always in the direction of $\boldsymbol{n}$? If $\boldsymbol{v} = (x, y, z)$ is any vector to the plane, its dot product with $\boldsymbol{n}$ is $\boldsymbol{n} \cdot \boldsymbol{v} = d$. That is the equation of the plane! We are trying to make the length $\|\boldsymbol{v}\|$ as small as possible. But the Cauchy-Schwarz inequality says that

$$|\boldsymbol{n} \cdot \boldsymbol{v}| \le \|\boldsymbol{n}\|\,\|\boldsymbol{v}\| \quad \text{or} \quad |d| \le \|\boldsymbol{n}\|\,\|\boldsymbol{v}\|. \tag{1.9}$$

Divide by $\|\boldsymbol{n}\|$. The length $\|\boldsymbol{v}\|$ is at least $|d|/\|\boldsymbol{n}\|$. That distance is the shortest.

Example 1.10 If $\boldsymbol{v} = (2, 1, 2)$ is the shortest vector from $(0, 0, 0)$ to a plane, find the plane.

Solution The shortest vector is always perpendicular to the plane. Therefore $(2, 1, 2)$ is a normal vector $\boldsymbol{n}$. The equation of the plane must be $2x + y + 2z = d$. The point $(2, 1, 2)$ also lies on the plane—it must satisfy the equation. This requires $d = 9$. So the plane is $2x + y + 2z = 9$. The distance to it is ______.

A Plane in Four-Dimensional Space

Sooner or later, this book has to move to n dimensions. Vectors will have n components. Planes will have "dimension" $n-1$. This step might as well come now, starting with $n = 4$:

$$\boldsymbol{v} = \begin{bmatrix} 3 \\ 0 \\ 0 \\ 1 \end{bmatrix} \quad \text{and} \quad \boldsymbol{n} = \begin{bmatrix} 1 \\ -1 \\ -1 \\ -3 \end{bmatrix} \quad \text{are vectors in 4-dimensional space.}$$

The length of $\boldsymbol{v}$ is $\sqrt{3^2 + 0^2 + 0^2 + 1^2}$ which is $\sqrt{10}$. Similarly $\|\boldsymbol{n}\|$ equals $\sqrt{12}$. Most important, $\boldsymbol{v} \cdot \boldsymbol{n}$ is computed as before—it equals $v_1n_1 + v_2n_2 + v_3n_3 + v_4n_4$. For these particular vectors the dot product is $3 + 0 + 0 - 3 = 0$. Since $\boldsymbol{v} \cdot \boldsymbol{n} = 0$, the vectors are perpendicular (in four-dimensional space). Then $\boldsymbol{v}$ lies on the plane through $(0, 0, 0, 0)$ that is perpendicular to $\boldsymbol{n}$.

You see how linear algebra extends into four dimensions—the rules stay the same. That applies to vector addition ($\boldsymbol{v} + \boldsymbol{n}$ has components $4, -1, -1, -2$). It applies to $2\boldsymbol{v}$ and $-\boldsymbol{v}$ and the linear combination $2\boldsymbol{v} + 3\boldsymbol{n}$. The only difference in four dimensions is that there are four components.

There is a big difference in the geometry: we can't draw pictures. It is hard enough to represent 3-dimensional space on a flat page. Four or five dimensions are virtually impossible. But no problem to do the algebra—the equation is $\boldsymbol{n} \cdot \boldsymbol{v} = 0$ or $\boldsymbol{n} \cdot \boldsymbol{v} = d$ or $x - y - z - 3t = d$.

Definition A plane contains all points $\boldsymbol{v}$ that satisfy one linear equation $\boldsymbol{n} \cdot \boldsymbol{v} = d$.

The plane goes through the origin if $d = 0$, because $\boldsymbol{v} = \boldsymbol{0}$ satisfies the equation. To visualize a plane in high dimensions, think first of $\boldsymbol{n}$. This vector sticks out perpendicular to a plane that is "one dimension lower."

By this definition, a plane in two-dimensional space is an ordinary straight line. (We suppose a plane in one-dimensional space is an ordinary point.) The equation $x + 2y = 3$

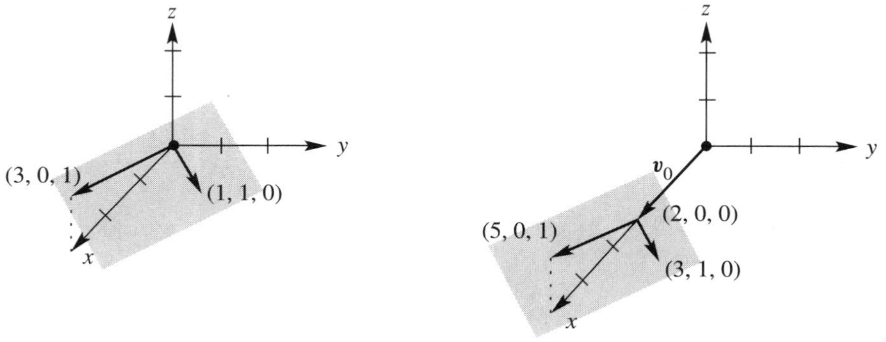

Figure 1.10 Linear combinations produce a plane.

gives a line perpendicular to $\boldsymbol{n} = (1, 2)$. A plane in five dimensions is four-dimensional and in some way this plane is flat. We can't really see that (maybe Einstein could). When our vision fails we fall back on $\boldsymbol{n} \cdot (\boldsymbol{v} - \boldsymbol{v}_0) = 0$ or $\boldsymbol{n} \cdot \boldsymbol{v} = d$.

Linear Combinations Produce Planes

So far planes have come from dot products. They will now come from linear combinations. As before, we start with a plane through $(0, 0, 0)$. Instead of working with the normal vector $\boldsymbol{n}$, we work with the vectors $\boldsymbol{v}$ that are *in the plane*.

Example 1.11 Describe all points in the plane $x - y - 3z = 0$.

Solution Two points in this plane are $(1, 1, 0)$ and $(3, 0, 1)$. A third point is $(0, 0, 0)$. To describe all the points, the short answer is by dot products. We require $\boldsymbol{n} \cdot \boldsymbol{v} = 0$ or $(1, -1, -3) \cdot (x, y, z) = 0$. The new and different answer is to describe every vector $\boldsymbol{v}$ that satisfies this equation.

The vectors in the plane $x - y - 3z = 0$ are linear combinations of two "basic vectors":

$$\boldsymbol{v} = y \begin{bmatrix} 1 \\ 1 \\ 0 \end{bmatrix} + z \begin{bmatrix} 3 \\ 0 \\ 1 \end{bmatrix} \quad \text{or} \quad \boldsymbol{v} = \begin{bmatrix} y + 3z \\ y \\ z \end{bmatrix}. \qquad (1.10)$$

The first component of $\boldsymbol{v}$ is $x = y + 3z$. This is the original equation! The linear combination (1.10) displays all points in the plane. Any two numbers y and z are allowed in the combination, so the plane in Figure 1.10 is two-dimensional. The free choice does not extend to x, which is determined by $y + 3z$.

Now move the plane away from $(0, 0, 0)$. Suppose its equation is $x - y - 3z = 2$. That is $\boldsymbol{n} \cdot \boldsymbol{v} = d$. It describes the plane by a dot product. The other way to describe the plane is to display every $\boldsymbol{v}$ that satisfies this equation. Here is a systematic way to display *all points $\boldsymbol{v}$ on the plane*.

Advice on $\boldsymbol{v}_0$ Choose one particular point $\boldsymbol{v}_0$ that has $y = z = 0$. The equation $x - y - 3z = 2$ requires that $x = 2$. So the point is $\boldsymbol{v}_0 = (2, 0, 0)$. In Chapter 2 this is called a *"particular solution."*

Advice on $\boldsymbol{v}$ Starting from $\boldsymbol{v}_0$, add to it all vectors that satisfy $x - y - 3z = 0$. The right side is now zero! This is the equation for the difference $\boldsymbol{v} - \boldsymbol{v}_0$. When we find the solutions, we add them to $\boldsymbol{v}_0 = (2, 0, 0)$. The original plane through $(0, 0, 0)$ is just being shifted out parallel to itself, to go through the new point $(2, 0, 0)$.

We already solved $x - y - 3z = 0$. A first solution is $(1, 1, 0)$, when $y = 1$ and $z = 0$. For the second solution switch to $y = 0$ and $z = 1$. The equation gives $x = 3$. A second solution is $(3, 0, 1)$. Now comes the linear algebra: Take all linear combinations of these two solutions.

The points $\boldsymbol{v} - \boldsymbol{v}_0$ on the plane $x - y - 3z = 0$ (notice the zero) ***are the combinations***

$$\boldsymbol{v} - \boldsymbol{v}_0 = y \begin{bmatrix} 1 \\ 1 \\ 0 \end{bmatrix} + z \begin{bmatrix} 3 \\ 0 \\ 1 \end{bmatrix}. \tag{1.11}$$

The points $\boldsymbol{v}$ on the plane $x - y - 3z = 2$ (notice the 2) ***move out by $\boldsymbol{v}_0 = (2, 0, 0)$:***

$$\boldsymbol{v} = \begin{bmatrix} 2 \\ 0 \\ 0 \end{bmatrix} + y \begin{bmatrix} 1 \\ 1 \\ 0 \end{bmatrix} + z \begin{bmatrix} 3 \\ 0 \\ 1 \end{bmatrix}. \tag{1.12}$$

This is the description by linear combinations. ***Go out to $\boldsymbol{v}_0$ and then travel along the plane***. The particular vector $\boldsymbol{v}_0$ is multiplied by the fixed number one. The vector $3\boldsymbol{v}_0$ would not be on the plane. The two in-plane vectors are multiplied by any numbers y and z. All those linear combinations fill out the plane.

Problem Set 1.3

1 Find two points on the plane $3x + y - z = 6$ and also find the normal vector $\boldsymbol{n}$. Verify that the vector between the two points is perpendicular to $\boldsymbol{n}$.

2 The plane $4x - y - 2z = 1$ is parallel to the plane _____ and perpendicular to the vector _____ .

3 True or false (give an example in either case):

(a) Two planes perpendicular to the same vector $\boldsymbol{n}$ are parallel.

(b) Two planes containing the same vector $\boldsymbol{v}$ are parallel.

(c) Two planes that do not intersect are parallel.

4 Find equations $ax + by + cz = d$ for these three planes:

(a) Through the point $(2, 2, 1)$ and perpendicular to $\boldsymbol{n} = (1, 5, 2)$.

(b) Through the point $(1, 5, 2)$ and perpendicular to $\boldsymbol{n} = (1, 5, 2)$.

(c) The "xz plane" containing all points $(x, 0, z)$. See Example 1.3.

5 If you reverse the sign of d, what does that do to the plane? Describe the position of $x + y + z = -2$ compared to $x + y + z = 2$.

6 Find the equation of the plane through $(0, 0, 0)$ that contains all linear combinations of $\boldsymbol{v} = (1, 1, 0)$ and $\boldsymbol{w} = (1, 0, 1)$.

7 Find the equation $ax + by = c$ of the line through $(0, 0)$ perpendicular to $\boldsymbol{n} = (1, 4)$. What is the equation of the parallel line through $(2, 3)$?

8 Choose three convenient points $\boldsymbol{u}, \boldsymbol{v}, \boldsymbol{w}$ on the plane $x + y + z = 2$. Their combinations $c\boldsymbol{u} + d\boldsymbol{v} + e\boldsymbol{w}$ lie on the same plane if $c + d + e =$ ______. (Those are *affine* combinations.)

9 Find the equation $ax + by + cz + dt = e$ of the plane through $(1, 1, 1, 1)$ in four-dimensional space. Choose the plane to be perpendicular to $\boldsymbol{n} = (1, 4, 1, 2)$.

10 Find the equation of the plane through $(0, 0, 0)$ that contains the vectors $(4, 1, 0)$ and $(-2, 0, 1)$.

Questions 11–16 ask for all points $\boldsymbol{v}$ that satisfy a linear equation. Follow the advice on $\boldsymbol{v}$ and $\boldsymbol{v}_0$ in the last paragraphs of this section.

11 Choose a particular point $\boldsymbol{v}_0$ on the plane $x - 3y - z = 6$. Then choose two in-plane vectors that satisfy $x - 3y - z = 0$. Combine them with $\boldsymbol{v}_0$ as in equation (1.12) to display all points on the plane.

12 What is the particular solution $\boldsymbol{v}_0$ to $2x + 4y - z = 0$, when you follow the advice in the text? Write all vectors in this plane as combinations of two vectors.

13 Find a very particular solution $\boldsymbol{v}_0$ to $x + y + z = 0$. Then choose two nonzero solutions. Combine them to display all solutions.

14 Choose a particular point on the line $x - 3y = 9$, and also a solution to $x - 3y = 0$. Combine them as in (1.12) to display all points on the line.

15 Choose a particular solution $\boldsymbol{v}_0$ (four components) to $x + 2y + 3z + 4t = 24$. Then find solutions to $x + 2y + 3z + 4t = 0$ of the form $(x, 1, 0, 0)$ and $(x, 0, 1, 0)$ and $(x, 0, 0, 1)$. Combine $\boldsymbol{v}_0$ with these three solutions to display all points $\boldsymbol{v}$ on the plane in four dimensions.

16 The advice on choosing $\boldsymbol{v}_0$ will fail if the equation of the plane is $0x + 2y + 3z = 12$. When we choose $y = z = 0$ we can't solve for x. Give advice on this case: Choose $\boldsymbol{v}_0$ and then choose two solutions to $0x + 2y + 3z = 0$ and take their combinations.

Questions 17–22 are about the distance to a plane.

17 The text found $6/\sqrt{14}$ as the distance from $(0, 0, 0)$ to the plane $x + 2y + 3z = 6$. Find the distance to $2x + 4y + 6z = 12$ from the formula $|d|/\|\boldsymbol{n}\|$. Explain why this still equals $6/\sqrt{14}$.

18 How far is the origin from the following planes? Find the nearest point $\boldsymbol{v} = d\boldsymbol{n}/\|\boldsymbol{n}\|^2$:

(a) $2x + 2y + z = 18$ (b) $x - 3y - 7z = 0$ (c) $x - z = 6$

19 This question asks for the distance from $(0, 0)$ to the line $x - 2y = 25$.

(a) Why is the shortest vector in the direction of $(1, -2)$?

(b) What point $(t, -2t)$ in that direction lies on the line?

(c) What is the distance from $(0, 0)$ to the line?

20 This follows Problem 19 for a general line $ax + by = c$.

(a) The shortest vector from $(0, 0)$ is in the direction of $\boldsymbol{n} =$ _____.

(b) The point $t\boldsymbol{n}$ lies on the line if $t =$ _____.

(c) The shortest distance is $|c|/\sqrt{a^2 + b^2}$ because _____.

21 The shortest distance from $(1, 0, 5)$ to the plane $x + 2y - 2z = 27$ is in the direction $\boldsymbol{n} = (1, 2, -2)$. The point $(1, 0, 5) + t(1, 2, -2)$ lies on the plane when $t =$ _____. Therefore the shortest distance is $\|t\boldsymbol{n}\| =$ _____.

22 The shortest distance from a point $\boldsymbol{w}$ to the plane $\boldsymbol{n} \cdot \boldsymbol{v} = d$ is in the direction of _____. The point $\boldsymbol{v} = \boldsymbol{w} + t\boldsymbol{n}$ lies on the plane when $t =$ _____. Therefore the shortest distance is $\|t\boldsymbol{n}\| =$ _____. When $\boldsymbol{w} = \boldsymbol{0}$ this distance should be $|d|/\|\boldsymbol{n}\|$.

23 The two planes $x + 2y + 3z = 12$ and $x - y - z = 2$ intersect in a line (in three dimensions). The two vectors _____ are perpendicular to the line. The two points _____ are on the line.

24 The equation $x + y + z - t = 1$ represents a plane in 4-dimensional space. Find

(a) its normal vector $\boldsymbol{n}$

(b) its distance to $(0, 0, 0, 0)$ from the formula $|d|/\|\boldsymbol{n}\|$

(c) the nearest point to $(0, 0, 0, 0)$

(d) a point $\boldsymbol{v}_0 = (x, 0, 0, 0)$ on the plane

(e) three vectors in the parallel plane $x + y + z - t = 0$

(f) all points that satisfy $x + y + z - t = 1$, following equation (1.12).

25 For which $\boldsymbol{n}$ will the plane $\boldsymbol{n} \cdot \boldsymbol{v} = 0$ fail to intersect the plane $x + y + z = 1$?

26 At what angle does the plane $y + z = 0$ meet the plane $x + z = 0$?

1.4 Matrices and Linear Equations

The central problem of linear algebra is to solve linear equations. There are two ways to describe that problem—first by *rows* and then by *columns*. In this chapter we explain the problem, in the next chapter we solve it.

Start with a system of *three equations in three unknowns*. Let the unknowns be x, y, z, and let the linear equations be

$$\begin{aligned} x + 2y + 3z &= 6 \\ 2x + 5y + 2z &= 4 \\ 6x - 3y + z &= 2. \end{aligned} \tag{1.13}$$

We look for numbers x, y, z that solve all three equations at once. Those numbers might or might not exist. For this system, they do exist. When the number of equations matches the number of unknowns, there is *usually* one solution. The immediate problem is how to visualize the three equations. There is a row picture and a column picture.

R ***The row picture shows three planes meeting at a single point.***

The first plane comes from the first equation $x+2y+3z = 6$. That plane crosses the x, y, z axes at $(6, 0, 0)$ and $(0, 3, 0)$ and $(0, 0, 2)$. The plane does not go through the origin—the right side of its equation is 6 and not zero.

The second plane is given by the equation $2x + 5y + 2z = 4$ from the second row. The numbers 0, 0, 2 satisfy both equations, so the point $(0, 0, 2)$ lies on both planes. *Those two planes intersect in a line* **L**, which goes through $(0, 0, 2)$.

The third equation gives a third plane. It cuts the line **L** at a single point. ***That point lies on all three planes***. This is the row picture of equation (1.13)—three planes meeting at a single point (x, y, z). The numbers x, y, z solve all three equations.

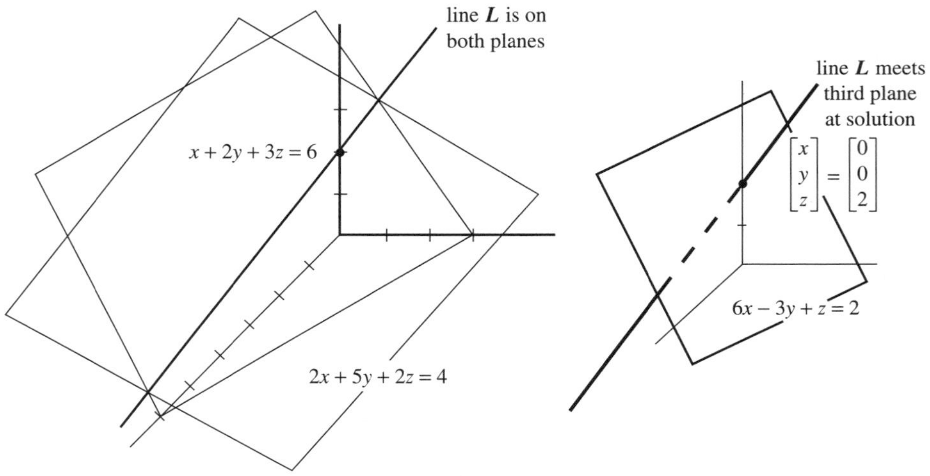

Figure 1.11 Row picture of three equations: Three planes meet at a point.

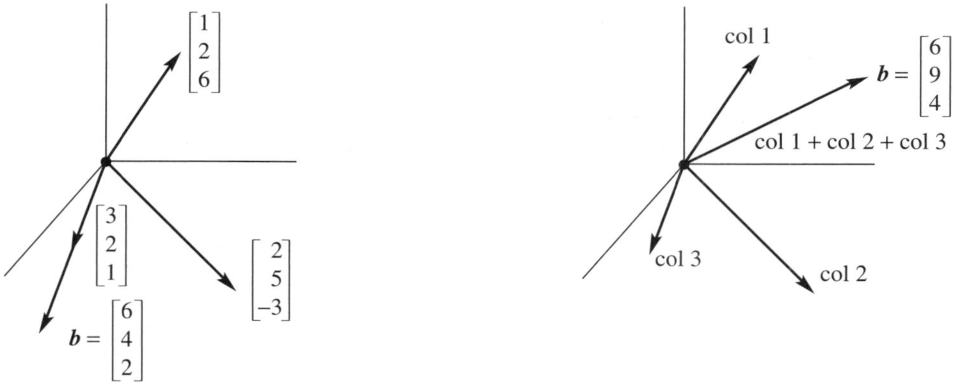

Figure 1.12 Column picture of three equations: Combinations of columns are $\boldsymbol{b} = 2 \times$ column 3 so $(x, y, z) = (0, 0, 2)$; $\boldsymbol{b} =$ sum of columns so $(x, y, z) = (1, 1, 1)$.

C ***The column picture combines the columns on the left side to produce the right side.***

Write the three equations as one vector equation based on columns:

$$x \begin{bmatrix} 1 \\ 2 \\ 6 \end{bmatrix} + y \begin{bmatrix} 2 \\ 5 \\ -3 \end{bmatrix} + z \begin{bmatrix} 3 \\ 2 \\ 1 \end{bmatrix} = \begin{bmatrix} 6 \\ 4 \\ 2 \end{bmatrix}. \tag{1.14}$$

The row picture has planes given by *dot products*. The column picture has this *linear combination*. The unknown numbers x, y, z are the coefficients in the combination. We multiply the columns by the correct numbers x, y, z to give the column $(6, 4, 2)$.

For this particular equation we know the right combination (we made up the problem). If x and y are zero, and z equals 2, then 2 times the third column agrees with the column on the right. The solution is $x = 0$, $y = 0$, $z = 2$. That point $(0, 0, 2)$ lies on all three planes in the row picture. It solves all three equations. The row and column pictures show the same solution in different ways.

For one moment, change to a new right side $(6, 9, 4)$. This vector equals column $1 +$ column $2 +$ column 3. The solution with this new right side is $(x, y, z) =$ ______. ***The numbers*** x, y, z ***multiply the columns to give*** $\boldsymbol{b}$.

The Matrix Form of the Equations

We have three rows in the row picture and three columns in the column picture (plus the right side). The three rows and columns contain nine numbers. *These nine numbers fill a 3 by 3 matrix.* We are coming to the matrix picture. The "coefficient matrix" has the rows and columns that have so far been kept separate:

$$\text{The coefficient matrix is} \quad A = \begin{bmatrix} 1 & 2 & 3 \\ 2 & 5 & 2 \\ 6 & -3 & 1 \end{bmatrix}.$$

The capital letter A stands for all nine coefficients (in this square array). The letter $\boldsymbol{b}$ will denote the column vector of right hand sides. The components of $\boldsymbol{b}$ are 6, 4, 2. The unknown is also a column vector, with components x, y, z. We call it $\boldsymbol{x}$ (boldface because it is a vector, $\boldsymbol{x}$ because it is unknown). By rows the equations were (1.13), by columns they were (1.14), and now by matrices they are (1.15). The shorthand is $A\boldsymbol{x} = \boldsymbol{b}$:

$$\textbf{\textit{Matrix notation}}: \quad \begin{bmatrix} 1 & 2 & 3 \\ 2 & 5 & 2 \\ 6 & -3 & 1 \end{bmatrix} \begin{bmatrix} x \\ y \\ z \end{bmatrix} = \begin{bmatrix} 6 \\ 4 \\ 2 \end{bmatrix} \quad \text{is} \quad A\boldsymbol{x} = \boldsymbol{b}. \tag{1.15}$$

We multiply the matrix A times the unknown vector $\boldsymbol{x}$ to get the right side $\boldsymbol{b}$.

Basic question: What does it mean to "multiply A times $\boldsymbol{x}$"? We can do it by rows or we can do it by columns. Either way, $A\boldsymbol{x} = \boldsymbol{b}$ must be a correct representation of the three equations. So there are two ways to multiply A times $\boldsymbol{x}$:

Multiplication by rows $A\boldsymbol{x}$ comes from *dot products*, a row at a time:

$$A\boldsymbol{x} = \begin{bmatrix} (\textbf{row 1}) \cdot \boldsymbol{x} \\ (\textbf{row 2}) \cdot \boldsymbol{x} \\ (\textbf{row 3}) \cdot \boldsymbol{x} \end{bmatrix}. \tag{1.16}$$

Multiplication by columns $A\boldsymbol{x}$ is a *linear combination* of the columns:

$$A\boldsymbol{x} = x\ (\text{column } 1) + y\ (\text{column } 2) + z\ (\text{column } 3). \tag{1.17}$$

However you do it, all five of our equations (1.13)–(1.17) show A times $\boldsymbol{x}$.

Examples

$$A\boldsymbol{x} = \begin{bmatrix} 1 & 0 & 0 \\ 1 & 0 & 0 \\ 1 & 0 & 0 \end{bmatrix} \begin{bmatrix} 4 \\ 5 \\ 6 \end{bmatrix} = \begin{bmatrix} 4 \\ 4 \\ 4 \end{bmatrix} \qquad A\boldsymbol{x} = \begin{bmatrix} 1 & 0 & 0 \\ 0 & 1 & 0 \\ 0 & 0 & 1 \end{bmatrix} \begin{bmatrix} 4 \\ 5 \\ 6 \end{bmatrix} = \begin{bmatrix} 4 \\ 5 \\ 6 \end{bmatrix}.$$

In the first example $A\boldsymbol{x}$ is $(4, 4, 4)$. If you are a row person, the dot product of every row with $(4, 5, 6)$ is 4. If you are a column person, the linear combination is 4 times the first column $(1, 1, 1)$. In that matrix the second and third columns are zero vectors.

The second example deserves a careful look, because the matrix is special. It has 1's on the "main diagonal." Off that diagonal are 0's. *Whatever vector this matrix multiplies, that vector is not changed.* This is like multiplication by 1, but for matrices and vectors. The exceptional matrix in this example is the 3 by 3 ***identity matrix***, and it is denoted by I:

$$I = \begin{bmatrix} 1 & 0 & 0 \\ 0 & 1 & 0 \\ 0 & 0 & 1 \end{bmatrix} \quad \text{always yields the multiplication} \quad I\boldsymbol{v} = \boldsymbol{v}.$$

Remark The two ways of looking at a matrix (*rows or columns*) are the keys to linear algebra. We can choose between them (when we compute), or we can hold onto both (when we think). The computing decision depends on how the matrix is stored. A vector machine

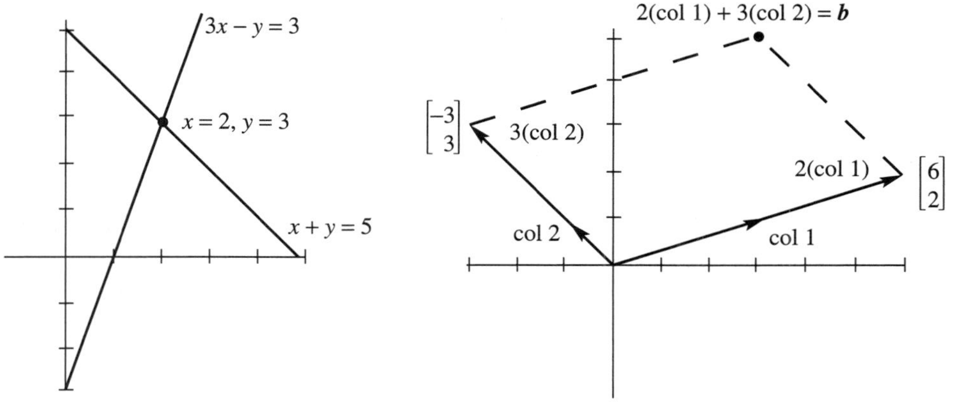

Figure 1.13 Row picture for 2 by 2 (intersecting lines). Column picture (linear combination produces $\boldsymbol{b}$).

like the CRAY uses columns. Also FORTRAN is slightly more efficient with columns. But equations are usually solved by row operations—this is the heart of the next chapter.

We now review the row and column and matrix pictures when there are *two* equations:

$$\begin{aligned} 3x - y &= 3 \\ x + y &= 5 \end{aligned} \quad \text{or} \quad \begin{bmatrix} 3 & -1 \\ 1 & 1 \end{bmatrix} \begin{bmatrix} x \\ y \end{bmatrix} = \begin{bmatrix} 3 \\ 5 \end{bmatrix}.$$

Those give two lines in the xy plane. They also give two columns, to be combined into $\boldsymbol{b} = (3, 5)$. Figure 1.13 shows both pictures.

R (rows) The first row gives the line $3x - y = 3$. The second row gives the second line $x + y = 5$. Those lines cross at the solution $x = 2$, $y = 3$.

C (columns) A combination of the columns gives the vector on the right side:

$$2\begin{bmatrix} 3 \\ 1 \end{bmatrix} + 3\begin{bmatrix} -1 \\ 1 \end{bmatrix} = \begin{bmatrix} 3 \\ 5 \end{bmatrix}.$$

M (matrix) The two by two coefficient matrix times the solution vector (2, 3) equals $\boldsymbol{b}$:

$$A\boldsymbol{x} = \boldsymbol{b} \quad \text{is} \quad \begin{bmatrix} 3 & -1 \\ 1 & 1 \end{bmatrix} \begin{bmatrix} 2 \\ 3 \end{bmatrix} = \begin{bmatrix} 3 \\ 5 \end{bmatrix}.$$

Multiply by dot products if you prefer rows. Combine columns if you like vectors. This example is a small review of the whole chapter. Dot products and linear combinations are the two routes to $A\boldsymbol{x}$.

This book sees Ax as a combination of columns of A. Even if you compute by dot products, please hold on to multiplication by columns.

Possible Breakdown We hate to end this chapter on a negative note, but the equation $A\boldsymbol{x} = \boldsymbol{b}$ might not have a solution. The two lines in the xy plane might be parallel. The three planes in xyz space can fail to intersect in a single point. There are two types of failure, either no solution or too many solutions:

(a) The two lines or three planes intersect at *no* points.

(b) The lines or planes intersect at *infinitely many* points.

These cases are "exceptional" but very possible. If you write down equations at random, breakdown won't happen (almost certainly). The probability of failure is zero. But the equations of applied mathematics are not random—exceptions that have zero probability happen every day. When we solve $A\boldsymbol{x} = \boldsymbol{b}$ in Chapter 2, we watch closely for what can go wrong.

Matrix Notation The four entries in a 2 by 2 matrix are a_{11}, a_{12} in the first row and a_{21}, a_{22} in the second row. The first index gives the row number. Thus a_{ij} or $A(i, j)$ is an entry in row i. The second index j gives the column number (we see this in detail in Section 2.2). The entry a_{12} is in row 1, column 2:

$$A = \begin{bmatrix} a_{11} & a_{12} \\ a_{21} & a_{22} \end{bmatrix} \quad \text{or} \quad A = \begin{bmatrix} A(1,1) & A(1,2) \\ A(2,1) & A(2,2) \end{bmatrix}.$$

For an m by n matrix, the row number i goes from 1 to m and j stops at n.

MATLAB enters a matrix by rows, with a semicolon between: A = [3 −1; 1 1]. If it knows the columns a = [3 1]′ and c = [−1 1]′, then also A = [a c]. The capital letter A is different from the small letter a.

Matrix times Vector In MATLAB this is A ∗ v. The software might multiply by rows or by columns, who knows? (We think it is by columns.) In a language that works with entries A(I, J) and components V(J), the decision is ours. This short program takes the dot product with row I in the inner loop, where it sums over J:

```
      DO 10 I = 1,2
      DO 10 J = 1,2
   10 B(I) = B(I) + A(I,J) * V(J)
```

Problem 23 has the loops in the opposite order. Then the inner loop (with J fixed) goes down the column. That code computes $A\boldsymbol{v}$ as a combination of the columns.

This one line is to reveal that MATLAB solves $A\boldsymbol{v} = \boldsymbol{b}$ by v = A\b.

Problem Set 1.4

Problems 1–8 are about the row and column pictures of $Ax = b$.

1 With $A = I$ (the identity matrix) draw the planes in the row picture. Three sides of a box meet at the solution:

$$\begin{aligned} 1x + 0y + 0z &= 2 \\ 0x + 1y + 0z &= 3 \\ 0x + 0y + 1z &= 4 \end{aligned} \quad \text{or} \quad \begin{bmatrix} 1 & 0 & 0 \\ 0 & 1 & 0 \\ 0 & 0 & 1 \end{bmatrix} \begin{bmatrix} x \\ y \\ z \end{bmatrix} = \begin{bmatrix} 2 \\ 3 \\ 4 \end{bmatrix}.$$

2 Draw the vectors in the column picture of Problem 1. Two times column 1 plus three times column 2 plus four times column 3 equals the right side $\boldsymbol{b}$.

3 If the equations in Problem 1 are multiplied by 1, 2, 3 they become

$$\begin{aligned} 1x + 0y + 0z &= 2 \\ 0x + 2y + 0z &= 6 \\ 0x + 0y + 3z &= 12 \end{aligned} \qquad \text{or} \qquad \begin{bmatrix} 1 & 0 & 0 \\ 0 & 2 & 0 \\ 0 & 0 & 3 \end{bmatrix} \begin{bmatrix} x \\ y \\ z \end{bmatrix} = \begin{bmatrix} 2 \\ 6 \\ 12 \end{bmatrix}$$

Why is the row picture the same? Is the solution the same? The column picture is not the same—draw it.

4 If equation 1 is added to equation 2, which of these are changed: the row picture, the column picture, the coefficient matrix, the solution? The new equations in Problem 1 would be $x = 2$, $x + y = 5$, $z = 4$.

5 Find any point on the line of intersection of the planes $x+y+3z=6$ and $x-y+z=4$. By trial and error find another point on the line.

6 The first of these equations plus the second equals the third:

$$\begin{aligned} x + y + z &= 2 \\ x + 2y + z &= 3 \\ 2x + 3y + 2z &= 5. \end{aligned}$$

The first two planes meet along a line. The third plane contains that line, because if x, y, z satisfy the first two equations then they also ______. The equations have infinitely many solutions (the whole line). Find three solutions.

7 Move the third plane in Problem 6 to a parallel plane $2x + 3y + 2z = 9$. Now the three equations have no solution—why not? The first two planes meet along a line, but the third plane doesn't cross that line (artists please draw).

8 In Problem 6 the columns are $(1, 1, 2)$ and $(1, 2, 3)$ and $(1, 1, 2)$. This is a "singular case" because the third column is ______. Find two combinations of the three columns that give the right side $(2, 3, 5)$.

Problems 9–14 are about multiplying matrices and vectors.

9 Compute each Ax by dot products of the rows with the column vector:

$$\text{(a)} \quad \begin{bmatrix} 1 & 2 & 4 \\ -2 & 3 & 1 \\ -4 & 1 & 2 \end{bmatrix} \begin{bmatrix} 2 \\ 2 \\ 3 \end{bmatrix} \qquad \text{(b)} \quad \begin{bmatrix} 2 & 1 & 0 & 0 \\ 1 & 2 & 1 & 0 \\ 0 & 1 & 2 & 1 \\ 0 & 0 & 1 & 2 \end{bmatrix} \begin{bmatrix} 1 \\ 1 \\ 1 \\ 2 \end{bmatrix}$$

10 Compute each Ax in Problem 9 as a combination of the columns:

$$9\text{(a) becomes}\quad A\boldsymbol{x} = 2\begin{bmatrix}1\\-2\\-4\end{bmatrix} + 2\begin{bmatrix}2\\3\\1\end{bmatrix} + 3\begin{bmatrix}4\\1\\2\end{bmatrix} = \begin{bmatrix}\ \\ \ \\ \ \end{bmatrix}.$$

How many separate multiplications for $A\boldsymbol{x}$, when the matrix is "3 by 3"?

11 Find the two components of $A\boldsymbol{x}$ by rows or by columns:

$$\begin{bmatrix}2 & 3\\5 & 1\end{bmatrix}\begin{bmatrix}4\\2\end{bmatrix} \quad\text{and}\quad \begin{bmatrix}3 & 6\\6 & 12\end{bmatrix}\begin{bmatrix}2\\-1\end{bmatrix} \quad\text{and}\quad \begin{bmatrix}1 & 2 & 4\\2 & 0 & 1\end{bmatrix}\begin{bmatrix}3\\1\\1\end{bmatrix}.$$

12 Multiply A times $\boldsymbol{x}$ to find three components of $A\boldsymbol{x}$:

$$\begin{bmatrix}0 & 0 & 1\\0 & 1 & 0\\1 & 0 & 0\end{bmatrix}\begin{bmatrix}x\\y\\z\end{bmatrix} \quad\text{and}\quad \begin{bmatrix}2 & 1 & 3\\1 & 2 & 3\\3 & 3 & 6\end{bmatrix}\begin{bmatrix}1\\1\\-1\end{bmatrix} \quad\text{and}\quad \begin{bmatrix}2 & 1\\1 & 2\\3 & 3\end{bmatrix}\begin{bmatrix}1\\1\end{bmatrix}.$$

13 (a) A matrix with m rows and n columns multiplies a vector with _____ components to produce a vector with _____ components.

(b) The planes from the m equations $A\boldsymbol{x} = \boldsymbol{b}$ are in _____ -dimensional space. The combination of the columns of A is in _____ -dimensional space.

14 (a) How would you define a *linear* equation in three unknowns x, y, z?

(b) If $\boldsymbol{v}_0 = (x_0, y_0, z_0)$ and $\boldsymbol{v}_1 = (x_1, y_1, z_1)$ both satisfy the linear equation, then so does $c\boldsymbol{v}_0 + d\boldsymbol{v}_1$, provided $c + d =$ _____ .

(c) All combinations of $\boldsymbol{v}_0$ and $\boldsymbol{v}_1$ satisfy the equation when the right side is _____ .

Problems 15–22 ask for matrices that act in special ways on vectors.

15 (a) What is the 2 by 2 identity matrix? I times $\begin{bmatrix}x\\y\end{bmatrix}$ equals $\begin{bmatrix}x\\y\end{bmatrix}$.

(b) What is the 2 by 2 exchange matrix? P times $\begin{bmatrix}x\\y\end{bmatrix}$ equals $\begin{bmatrix}y\\x\end{bmatrix}$.

16 (a) What 2 by 2 matrix R rotates every vector by 90°? R times $\begin{bmatrix}x\\y\end{bmatrix}$ is $\begin{bmatrix}y\\-x\end{bmatrix}$.

(b) What 2 by 2 matrix rotates every vector by 180°?

17 What 3 by 3 matrix P permutes the vector (x, y, z) to (y, z, x)? What matrix P^{-1} permutes (y, z, x) back to (x, y, z)?

18 What 2 by 2 matrix E subtracts the first component from the second component? What 3 by 3 matrix does the same?

$$E\begin{bmatrix}3\\5\end{bmatrix} = \begin{bmatrix}3\\2\end{bmatrix} \quad\text{and}\quad E\begin{bmatrix}3\\5\\7\end{bmatrix} = \begin{bmatrix}3\\2\\7\end{bmatrix}.$$

19 What 3 by 3 matrix E multiplies (x, y, z) to give $(x, y, z+x)$? What matrix E^{-1} multiplies (x, y, z) to give $(x, y, z-x)$? If you multiply $(3, 4, 5)$ by E and then multiply by E^{-1}, the two results are (_____) and (_____).

20 What 2 by 2 matrix P_1 projects the vector (x, y) onto the x axis to produce $(x, 0)$? What matrix P_2 projects onto the y axis to produce $(0, y)$? If you multiply $(5, 7)$ by P_1 and then multiply by P_2, the two results are (_____) and (_____).

21 What 2 by 2 matrix R rotates every vector through 45°? The vector $(1, 0)$ goes to $(\sqrt{2}/2, \sqrt{2}/2)$. The vector $(0, 1)$ goes to $(-\sqrt{2}/2, \sqrt{2}/2)$. Those determine the matrix. Draw these particular vectors in the xy plane and find R.

22 Write the dot product of $(1, 4, 5)$ and (x, y, z) as a matrix multiplication $A\boldsymbol{x}$. The matrix A has one row. The solutions to $A\boldsymbol{x} = \mathbf{0}$ lie on a _____. The columns of A are only in _____ -dimensional space.

23 Which code finds the dot product of V with row 1 and then row 2? Which code finds column 1 times V(1) plus column 2 times V(2)? Read the paragraph just before this problem set. If A has 4 rows and 3 columns, what changes are needed in the codes?

```
   DO 10 I = 1,2                      DO 10 J = 1,2
   DO 10 J = 1,2                      DO 10 I = 1,2
10 B(I) = B(I) + A(I,J) * V(J)     10 B(I) = B(I) + A(I,J) * V(J)
```

24 In both codes the first step is B(1) = A(1, 1) ∗ V(1). Write the other three steps in the order executed by each code.

25 In three lines of MATLAB, enter a matrix and a column vector and multiply them.

Questions 26–28 are a review of the row and column pictures.

26 Draw each picture in a plane for the equations $x - 2y = 0$, $x + y = 6$.

27 For two linear equations in three unknowns x, y, z, the row picture will show (2 or 3) (lines or planes). Those lie in (2 or 3)-dimensional space. The column picture is in (2 or 3)-dimensional space.

28 For four linear equations in two unknowns x and y, the row picture shows four _____. The column picture is in _____ -dimensional space. The equations have no solution unless the vector on the right side is a combination of _____ .

29 (Markov matrix) Start with the vector $\boldsymbol{u}_0 = (1, 0)$. Multiply again and again by the same matrix A. The next three vectors are $\boldsymbol{u}_1, \boldsymbol{u}_2, \boldsymbol{u}_3$:

$$\boldsymbol{u}_1 = \begin{bmatrix} .8 & .3 \\ .2 & .7 \end{bmatrix} \begin{bmatrix} 1 \\ 0 \end{bmatrix} = \begin{bmatrix} .8 \\ .2 \end{bmatrix} \qquad \boldsymbol{u}_2 = A\boldsymbol{u}_1 = \_\_\_\_\_ \qquad \boldsymbol{u}_3 = A\boldsymbol{u}_2 = \_\_\_\_\_ .$$

What property do you notice for all four vectors $\boldsymbol{u}_0, \boldsymbol{u}_1, \boldsymbol{u}_2, \boldsymbol{u}_3$?

30 With a computer continue Problem 29 as far as the vector $\boldsymbol{u}_7$. Then from $\boldsymbol{v}_0 = (0, 1)$ compute $\boldsymbol{v}_1 = A\boldsymbol{v}_0$ and $\boldsymbol{v}_2 = A\boldsymbol{v}_1$ on to $\boldsymbol{v}_7$. Do the same from $\boldsymbol{w}_0 = (.5, .5)$ as far as $\boldsymbol{w}_7$. What do you notice about $\boldsymbol{u}_7$, $\boldsymbol{v}_7$ and $\boldsymbol{w}_7$? Extra: Plot the results by hand or computer. Here is a MATLAB code—you can use other languages:

```
u = [1; 0]; A = [.8 .3; .2 .7];
x = u; k = [0:1:7];
while length(x) <= 7
  u = A * u; x = [x u];
end
plot(k, x)
```

31 The $\boldsymbol{u}$'s and $\boldsymbol{v}$'s and $\boldsymbol{w}$'s are approaching a steady state vector $\boldsymbol{s}$. Guess that vector and check that $A\boldsymbol{s} = \boldsymbol{s}$. This vector is "steady" because if you start with $\boldsymbol{s}$, you stay with $\boldsymbol{s}$.

32 This MATLAB code allows you to input the starting column $\boldsymbol{u}_0 = [a; b; 1 - a - b]$ with a mouse click. These three components should be positive. (Click the left button on your point (a, b) or on several points. Add the instruction disp(u′) after the first end to see the steady state in the text window. Right button aborts.) The matrix A is entered by rows—its columns add to 1.

```
A = [.8 .2 .1; .1 .7 .3; .1 .1 .6]
axis([0 1 0 1]); axis('square')
plot(0, 0); hold on
title('Markov−your name'); xlabel('a'); ylabel('b'); grid
but = 1;
while but == 1
  [a,b,but] = ginput(1)
  u = [a; b; 1−a−b];
  x = u; k = [0:1:7];
  while length(x) <= 7
    u = A * u; x = [x u];
  end
  plot(x(1,:), x(2,:), x(1,:), x (2,:), 'o');
end
hold off
```

33 Invent a 3 by 3 **magic matrix** M_3 with entries $1, 2, \ldots, 9$. All rows and columns and diagonals add to 15. The first row could be 8, 3, 4. What is M_3 times $(1, 1, 1)$? What is M_4 times $(1, 1, 1, 1)$ if this magic matrix has entries $1, \ldots, 16$?

2

SOLVING LINEAR EQUATIONS

2.1 The Idea of Elimination

This chapter explains a systematic way to solve linear equations. The method is called ***"elimination"***, and you can see it immediately in a 2 by 2 example. (2 by 2 means two equations and two unknowns.) Before elimination, x and y appear in both equations. After elimination, the first unknown x has disappeared from the second equation:

$$\text{Before:} \quad \begin{aligned} x - 2y &= 1 \\ 3x + 2y &= 11 \end{aligned} \qquad \text{After:} \quad \begin{aligned} x - 2y &= 1 \\ 8y &= 8. \end{aligned}$$

Solving the second pair is much easier. The last equation $8y = 8$ instantly gives $y = 1$. Substituting for y in the first equation leaves $x - 2 = 1$. Therefore $x = 3$ and the solution is complete.

Elimination produces an ***upper triangular system***—this is the goal. The nonzero coefficients $1, -2, 8$ form a triangle. To solve this system, x and y are computed in reverse order (bottom to top). The last equation $8y = 8$ yields the last unknown, and we go upward to x. This quick process is called ***back substitution***. It is used for upper triangular systems of any size, after forward elimination is complete.

Important point: The original equations have the same solution $x = 3$ and $y = 1$. The combination $3x + 2y$ equals 11 as it should. Figure 2.1 shows this original system as a pair of lines, intersecting at the solution point $(3, 1)$. After elimination, the lines still meet at the same point. One line is horizontal because its equation $8y = 8$ does not contain x.

Also important: ***How did we get from the first pair of lines to the second pair?*** We subtracted 3 times the first equation from the second equation. The step that eliminates x is the fundamental operation in this chapter. We use it so often that we look at it closely:

To eliminate x: Subtract a multiple of equation 1 from equation 2.

Three times $x - 2y = 1$ gives $3x - 6y = 3$. When this is subtracted from $3x + 2y = 11$, the right side becomes 8. The main point is that $3x$ cancels $3x$. What remains on the left side is $2y - (-6y)$ or $8y$. Therefore $8y = 8$, and x is eliminated.

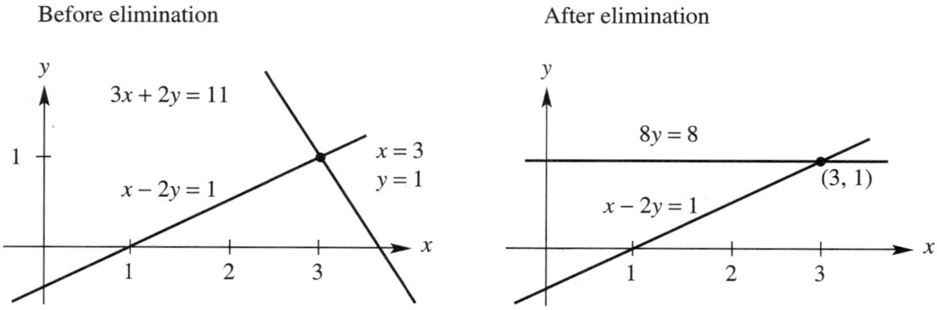

Figure 2.1 Two lines meet at the solution. So does the new line $8y = 8$.

Ask yourself how that multiplier $l = 3$ was found. The first equation contained x. ***The first pivot was*** 1 (the coefficient of x). The second equation contained $3x$, so the first equation was multiplied by 3. Then subtraction produced the zero.

You will see the general rule if we change the first equation to $5x - 10y = 5$. (Same straight line.) The first pivot is now 5. The multiplier is now $l = \frac{3}{5}$. *To find the multiplier, divide by the pivot*:

$$\begin{array}{llcr} \text{Multiply equation 1 by } \frac{3}{5} & 5x - 10y = 5 & \text{becomes} & 5x - 10y = 5 \\ \text{Subtract from equation 2} & 3x + 2y = 11 & & 8y = 8. \end{array}$$

The system is triangular and the last equation still gives $y = 1$. Back substitution produces $5x - 10 = 5$ and $5x = 15$ and $x = 3$. Multiplying the first equation by 5 changed the numbers but not the lines or the solution. Here is the rule to eliminate a coefficient below a pivot:

$$\textbf{\textit{Multiplier}}\ l = \frac{\text{coefficient to eliminate}}{\text{pivot}} = \frac{3}{5}.$$

The new second equation contains the second pivot, which is 8. It is the coefficient of y. We would use it to eliminate y from the third equation if there were one. *To solve n equations we want n pivots.*

You could have solved those equations for x and y without reading this book. It is an extremely humble problem, but we stay with it a little longer. Elimination might break down and we have to see how.

Breakdown of Elimination

Normally, elimination produces the pivots that take us to the solution. But failure is possible. The method might go forward up to a certain point, and then ask us to *divide by zero.* We can't do it. The process has to stop. There might be a way to adjust and continue—or failure may be unavoidable. Example 2.1 fails with no solution, Example 2.2 fails with too many solutions, and Example 2.3 succeeds after a temporary breakdown.

If the equations have no solution, then elimination must certainly have trouble.

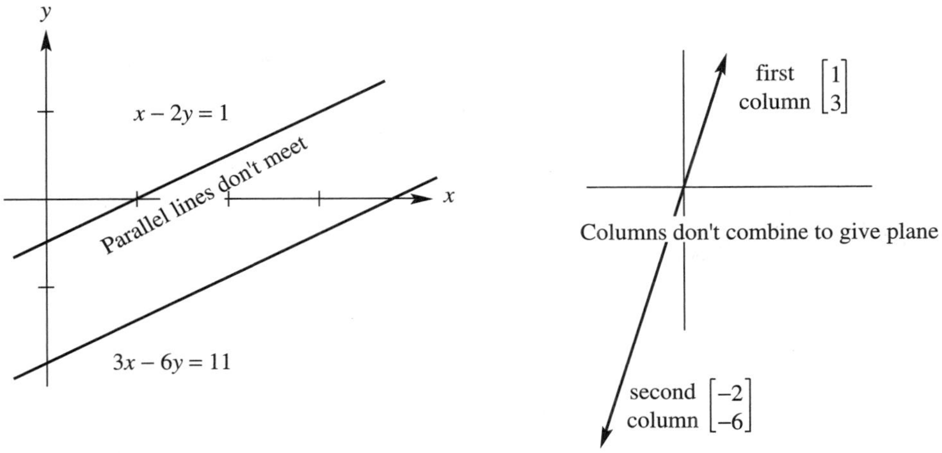

Figure 2.2 Row picture and column picture when elimination fails.

Example 2.1 Permanent failure with no solution. Elimination makes this clear:

$$\begin{array}{rcr} x - 2y = 1 & \text{subtract 3 times} & x - 2y = 1 \\ 3x - 6y = 11 & \text{eqn. 1 from eqn. 2} & 0y = 8. \end{array}$$

The last equation is $0y = 8$. There is *no* solution. Normally we divide the right side 8 by the second pivot, but *this system has no second pivot.* ***(Zero is never allowed as a pivot!)*** The row and column pictures of this 2 by 2 system show that failure was unavoidable.

The row picture in Figure 2.2 shows parallel lines—which never meet. A solution must lie on both lines. Since there is no meeting point, the equations have no solution.

The column picture shows the two columns in the same direction. *All combinations of the columns lie in that direction.* But the column from the right side is in a different direction $(1, 11)$. No combination of the columns can produce this right side—therefore no solution.

With a different right side, failure shows as a whole line of solutions. Instead of no solution there are infinitely many:

Example 2.2 Permanent failure with infinitely many solutions:

$$\begin{array}{rcr} x - 2y = 1 & \text{subtract 3 times} & x - 2y = 1 \\ 3x - 6y = 3 & \text{eqn. 1 from eqn. 2} & 0y = 0. \end{array}$$

Now the last equation is $0y = 0$. *Every* y satisfies this equation. There is really only one equation, namely $x - 2y = 1$. The unknown y is ***"free"***. After y is freely chosen, then x is determined as $x = 1 + 2y$.

In the row picture, the parallel lines have become the same line. Every point on that line satisfies both equations.

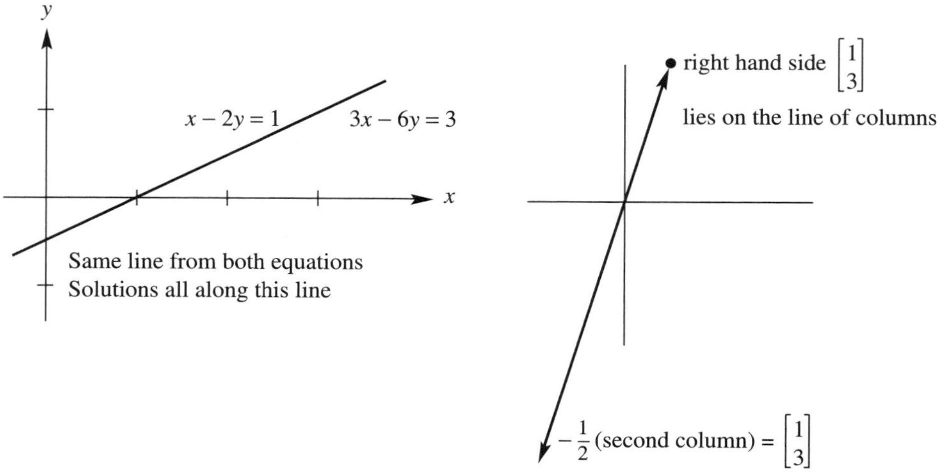

Figure 2.3 Row picture and column picture with infinitely many solutions.

In the column picture, the right side now lines up with both columns from the left side. We can choose $x = 1$ and $y = 0$—the first column equals the right side. We can also choose $x = 0$ and $y = -\frac{1}{2}$; the second column times $-\frac{1}{2}$ equals the right side. There are infinitely many other solutions. Every (x, y) that solves the row problem also solves the column problem.

There is another way elimination can go wrong—but this time it can be fixed. *Suppose the first pivot position contains zero.* We refuse to allow zero as a pivot. When the first equation has no term involving x, we can exchange it with an equation below. With an acceptable pivot the process goes forward:

Example 2.3 Temporary failure, but a row exchange produces two pivots:

$$\begin{array}{rcr} 0x + 2y = 4 & \text{exchange} & 3x - 2y = 5 \\ 3x - 2y = 5 & \text{equations} & 2y = 4. \end{array}$$

The new system is already triangular. This small example is ready for back substitution. The last equation gives $y = 2$, and then the first equation gives $x = 3$. The row picture is normal (two intersecting lines). The column picture is also normal (column vectors not in the same direction). The pivots 3 and 2 are normal—but a row exchange was required.

Examples 2.1–2.2 are ***singular***—there is no second pivot. Example 2.3 is ***nonsingular***—there is a full set of pivots. Singular equations have no solution or infinitely many solutions. Nonsingular equations have one solution. Pivots must be nonzero because we have to divide by them.

Three Equations in Three Unknowns

To understand Gaussian elimination, you have to go beyond 2 by 2 systems. Three by three is enough to see the pattern. For now the matrices are square—an equal number of rows and columns. Here is a 3 by 3 system, specially constructed so that all steps lead to whole numbers and not fractions:

$$\begin{aligned} 2x + 4y - 2z &= 2 \\ 4x + 9y - 3z &= 8 \\ -2x - 3y + 7z &= 10. \end{aligned} \tag{2.1}$$

What are the steps? The first pivot is 2 in the upper left corner. Below that pivot we want to create zeros. The first multiplier is the ratio $l = 4/2 = 2$. Subtraction removes the $4x$ from the second equation:

1 Subtract 2 times equation 1 from equation 2. Then equation 2_{new} is $0x + y + z = 4$.

We also eliminate x from equation 3—still using the first pivot. The quick way is to add equation 1 to equation 3. Then $2x$ cancels $-2x$. We do exactly that, but the rule in this book is to *subtract rather than add.* The systematic pattern has multiplier $-2/2 = -1$. Subtracting -1 times an equation is the same as adding that equation:

2 Subtract -1 times equation 1 from equation 3. Equation 3_{new} is $0x + y + 5z = 12$.

Now look at the situation. The two new equations involve only y and z. *We have reached a 2 by 2 system,* with $y + z = 4$ and $y + 5z = 12$. The final step eliminates y to reach a 1 by 1 problem:

3 Subtract equation 2_{new} from equation 3_{new}. Then $0x + 0y + 4z = 8$.

The original 3 by 3 system has been converted into a triangular 3 by 3 system:

$$\begin{aligned} 2x + 4y - 2z &= 2 \\ 4x + 9y - 3z &= 8 \\ -2x - 3y + 7z &= 10 \end{aligned} \quad \text{has become} \quad \begin{aligned} 2x + 4y - 2z &= 2 \\ y + z &= 4 \\ 4z &= 8. \end{aligned} \tag{2.2}$$

The goal is achieved—forward elimination is complete. ***Notice the pivots 2, 1, 4 along the diagonal.*** Those pivots 1 and 4 were hidden in the original system! Elimination brought them out.

The triangular system is zero below the pivots—three elimination steps produced three zeros. This triangle is ready for back substitution, which is quick:

$$4z = 8 \text{ gives } z = 2, \quad y + z = 4 \text{ gives } y = 2, \quad \text{equation 1 gives } x = -1.$$

The row picture shows planes, starting with the first equation $2x + 4y - 2z = 2$. The planes from all our equations go through the solution $(-1, 2, 2)$. The original three planes are sloping, but the very last plane $4z = 8$ is horizontal.

The column picture shows a linear combination of column vectors producing the right side. The coefficients in that combination are $-1, 2, 2$ (the solution):

$$(-1)\begin{bmatrix}2\\4\\-2\end{bmatrix}+2\begin{bmatrix}4\\9\\-3\end{bmatrix}+2\begin{bmatrix}-2\\-3\\7\end{bmatrix} \text{ equals } \begin{bmatrix}2\\8\\10\end{bmatrix}. \qquad (2.3)$$

The numbers x, y, z multiply columns 1, 2, 3 in the original system and also in the triangular system.

For a 4 by 4 problem, or an n by n problem, elimination proceeds the same way:

1 Use the first equation to create zeros below the first pivot.

2 Use the new second equation to create zeros below the second pivot.

3 Keep going to find the nth pivot.

The result of forward elimination is an upper triangular system. It is nonsingular if there is a full set of n pivots (never zero!). Here is a final example to show the original system, the triangular system, and the solution from back substitution:

$$\begin{aligned} x+\ y+\ z&=6 \\ x+2y+2z&=9 \\ x+2y+3z&=10 \end{aligned} \qquad \begin{aligned} x+y+z&=6 \\ y+z&=3 \\ z&=1 \end{aligned} \qquad \begin{aligned} x&=3 \\ y&=2 \\ z&=1. \end{aligned}$$

All multipliers are 1. All pivots are 1. All planes meet at the solution $(3, 2, 1)$. The columns combine with coefficients 3, 2, 1 to give the right side.

Problem Set 2.1

Problems 1–10 are about elimination on 2 by 2 systems.

1 What multiple l of equation 1 should be subtracted from equation 2?

$$\begin{aligned} 2x+3y&=1 \\ 10x+9y&=11. \end{aligned}$$

After this elimination step, write down the upper triangular system and circle the two pivots. The numbers 1 and 11 have no influence on those pivots.

2 Solve the triangular system of Problem 1 by back substitution, y before x. Verify that x times $(2, 10)$ plus y times $(3, 9)$ equals $(1, 11)$. If the right side changes to $(4, 44)$, what is the new solution?

3 What multiple of equation 1 should be *subtracted* from equation 2?

$$\begin{aligned} 2x-4y&=6 \\ -x+5y&=0. \end{aligned}$$

After this elimination step, solve the triangular system. If the right side changes to $(-6, 0)$, what is the new solution?

4 What multiple l of equation 1 should be subtracted from equation 2?

$$\begin{aligned} ax + by &= f \\ cx + dy &= g. \end{aligned}$$

The first pivot is a (assumed nonzero). Elimination produces what formula for the second pivot? The second pivot is missing when $ad = bc$.

5 Choose a right side which gives no solution and another right side which gives infinitely many solutions. What are two of those solutions?

$$\begin{aligned} 3x + 2y &= 7 \\ 6x + 4y &= \end{aligned}$$

6 Choose a coefficient b that makes this system singular. Then choose a right side g that makes it solvable.

$$\begin{aligned} 2x + by &= 13 \\ 4x + 8y &= g. \end{aligned}$$

7 For which numbers a does elimination break down (1) permanently (2) temporarily?

$$\begin{aligned} ax + 3y &= -3 \\ 4x + 6y &= 6. \end{aligned}$$

Solve for x and y after fixing the second breakdown by a row exchange.

8 For which three numbers k does elimination break down? In each case, is the number of solutions 0 or 1 or ∞?

$$\begin{aligned} kx + 3y &= 6 \\ 3x + ky &= -6. \end{aligned}$$

9 What is the test on b_1 and b_2 to decide whether these two equations have a solution? How many solutions will they have?

$$\begin{aligned} 3x - 2y &= b_1 \\ 6x - 4y &= b_2. \end{aligned}$$

10 In the xy plane, draw the lines $x+y = 5$ and $x+2y = 6$ and the equation $y =$ _____ that comes from elimination. The line $5x - 4y = c$ will go through the solution of these equations if $c =$ _____.

Problems 11–20 study elimination on 3 by 3 systems (and possible failure).

11 Reduce this system to upper triangular form:

$$\begin{aligned} 2x + 3y + z &= 1 \\ 4x + 7y + 5z &= 7 \\ -2y + 2z &= 6. \end{aligned}$$

Circle the pivots. Solve by back substitution for z, y, x. Two row operations are enough if a zero coefficient appears in which positions?

12 Apply elimination and back substitution to solve

$$\begin{aligned} 2x - 3y \phantom{{}+z} &= 3 \\ 4x - 5y + z &= 7 \\ 2x - y - 3z &= 5. \end{aligned}$$

Circle the pivots. List the three row operations which subtract a multiple of the pivot row from a lower row.

13 Which number d will force a row exchange, and what is the triangular system for that d?

$$\begin{aligned} 2x + 5y + z &= 0 \\ 4x + dy + z &= 2 \\ y - z &= 3. \end{aligned}$$

Which number d makes this system singular (no third pivot)?

14 Which number b leads later to a row exchange? Which b leads to a missing pivot? In that singular case find a nonzero solution x, y, z.

$$\begin{aligned} x + by \phantom{{}-z} &= 0 \\ x - 2y - z &= 0 \\ y + z &= 0. \end{aligned}$$

15 (a) Construct a 3 by 3 system that needs two row exchanges to reach a triangular form and a solution.

(b) Construct a 3 by 3 system that needs a row exchange to keep going, but breaks down later.

16 If rows 1 and 2 are the same, how far can you get with elimination? If columns 1 and 2 are the same, which pivot is missing?

$$\begin{aligned} 2x - y + z &= 0 \qquad & 2x + 2y + z &= 0 \\ 2x - y + z &= 0 \qquad & 4x + 4y + z &= 0 \\ 4x + y + z &= 2 \qquad & 6x + 6y + z &= 2. \end{aligned}$$

17 Construct a 3 by 3 example that has 9 different coefficients on the left side, but rows 2 and 3 become zero in elimination.

18 Which number q makes this system singular and which right side t gives it infinitely many solutions? Find the solution that has $z = 1$.

$$\begin{aligned} x + 4y - 2z &= 1 \\ x + 7y - 6z &= 6 \\ 3y + qz &= t. \end{aligned}$$

19 (Recommended) It is impossible for a system of linear equations to have exactly two solutions. *Explain why.*

(a) If (x, y, z) and (X, Y, Z) are two solutions, what is another one?

(b) If three planes meet at two points, where else do they meet?

20 How can three planes fail to have an intersection point, when no two planes are parallel? Draw your best picture. Find a third equation that can't be solved if $x + y + z = 0$ and $x - y - z = 0$.

Problems 21–23 move up to 4 by 4 and *n* by *n*.

21 Find the pivots and the solution for these four equations:

$$\begin{aligned} 2x + y \qquad\qquad &= 0 \\ x + 2y + z \qquad &= 0 \\ y + 2z + t &= 0 \\ z + 2t &= 5. \end{aligned}$$

22 This system has the same pivots and right side as Problem 21. How is the solution different (if it is)?

$$\begin{aligned} 2x - y \qquad\qquad &= 0 \\ -x + 2y - z \qquad &= 0 \\ - y + 2z - t &= 0 \\ - z + 2t &= 5. \end{aligned}$$

23 If you extend Problems 21–22 following the $1, 2, 1$ pattern or the $-1, 2, -1$ pattern, what is the fifth pivot? What is the nth pivot?

24 If elimination leads to these equations, what are all possible original matrices A? Row exchanges not allowed:

$$\begin{aligned} x + y + z &= 0 \\ y + z &= 0 \\ 3z &= 0. \end{aligned}$$

25 For which three numbers a will elimination fail to give three pivots?

$$A = \begin{bmatrix} a & 2 & 3 \\ a & a & 4 \\ a & a & a \end{bmatrix}.$$

26 Look for a matrix with row sums 4, 8 and column sums 2, s:

$$\begin{bmatrix} a & b \\ c & d \end{bmatrix} \quad \begin{matrix} a+b=4 & a+c=2 \\ c+d=8 & b+d=s \end{matrix}$$

The four equations are solvable only if $s =$ _____. Then two solutions are in the matrices _____.

2.2 Elimination Using Matrices

We now combine two ideas—elimination and matrices. The goal is to express all the steps of elimination (and the final result) in the clearest possible way. In a 3 by 3 example, elimination could be described in words. For larger systems, a long list of steps would be hopeless. You will see how to subtract a multiple of one row from another row—*using matrices.*

The matrix form of a linear system is $A\boldsymbol{x} = \boldsymbol{b}$. Here are $\boldsymbol{b}$, $\boldsymbol{x}$, and A:

1 The vector of right sides is $\boldsymbol{b}$.

2 The vector of unknowns is $\boldsymbol{x}$. (The unknowns change from $x, y, z, \ldots$ to $x_1, x_2, x_3, \ldots$ because we run out of letters before we run out of numbers.)

3 The coefficient matrix is A.

The example in the previous section has the beautifully short form $A\boldsymbol{x} = \boldsymbol{b}$:

$$\begin{aligned} 2x_1 + 4x_2 - 2x_3 &= 2 \\ 4x_1 + 9x_2 - 3x_3 &= 8 \\ -2x_1 - 3x_2 + 7x_3 &= 10 \end{aligned} \quad \text{is the same as} \quad \begin{bmatrix} 2 & 4 & -2 \\ 4 & 9 & -3 \\ -2 & -3 & 7 \end{bmatrix} \begin{bmatrix} x_1 \\ x_2 \\ x_3 \end{bmatrix} = \begin{bmatrix} 2 \\ 8 \\ 10 \end{bmatrix}. \tag{2.4}$$

The nine numbers on the left go into the matrix A. That matrix not only sits beside $\boldsymbol{x}$, it *multiplies* $\boldsymbol{x}$. The rule for "A times $\boldsymbol{x}$" is exactly chosen to yield the three equations.

Review of A times $\boldsymbol{x}$ A matrix times a vector gives a vector. The matrix is square when the number of equations (three) matches the number of unknowns (three). Our matrix is 3 by 3. A general square matrix is n by n. Then the vector $\boldsymbol{x}$ is in n-dimensional space. This example is in 3-dimensional space:

$$\textbf{\textit{The unknown is}} \quad \boldsymbol{x} = \begin{bmatrix} x_1 \\ x_2 \\ x_3 \end{bmatrix} \quad \textbf{\textit{and the solution is}} \quad \boldsymbol{x} = \begin{bmatrix} -1 \\ 2 \\ 2 \end{bmatrix}.$$

Key point: $Ax = b$ represents the row form and also the column form of the equations. We can multiply Ax a column at a time:

$$Ax = (-1)\begin{bmatrix} 2 \\ 4 \\ -2 \end{bmatrix} + 2\begin{bmatrix} 4 \\ 9 \\ -3 \end{bmatrix} + 2\begin{bmatrix} -2 \\ -3 \\ 7 \end{bmatrix} = \begin{bmatrix} 2 \\ 8 \\ 10 \end{bmatrix}. \tag{2.5}$$

This rule is used so often that we repeat it for emphasis.

2A The product Ax is a ***combination of the columns*** of A. Columns are multiplied by components of x. Then Ax is x_1 times (column 1) $+ \cdots + x_n$ times (column n).

One more point about matrix notation: The entry in row 1, column 1 (the top left corner) is called a_{11}. The entry in row 1, column 3 is a_{13}. The entry in row 3, column 1 is a_{31}. (Row number comes before column number.) The word "entry" for a matrix corresponds to the word "component" for a vector. General rule: ***The entry in row i, column j of the matrix A is a_{ij}.***

Example 2.4 This matrix has $a_{ij} = 2i + j$. Then $a_{11} = 3$. Also $a_{12} = 4$ and $a_{21} = 5$. Here is Ax with numbers and letters:

$$\begin{bmatrix} 3 & 4 \\ 5 & 6 \end{bmatrix}\begin{bmatrix} 2 \\ 1 \end{bmatrix} = \begin{bmatrix} 3 \cdot 2 + 4 \cdot 1 \\ 5 \cdot 2 + 6 \cdot 1 \end{bmatrix} \qquad \begin{bmatrix} a_{11} & a_{12} \\ a_{21} & a_{22} \end{bmatrix}\begin{bmatrix} x_1 \\ x_2 \end{bmatrix} = \begin{bmatrix} a_{11}x_1 + a_{12}x_2 \\ a_{21}x_1 + a_{22}x_2 \end{bmatrix}.$$

For variety we multiplied a row at a time. The first component of Ax is $6+4 = 10$. That is the dot product of the row (3, 4) with the column (2, 1). Using letters, it is the dot product of (a_{11}, a_{12}) with (x_1, x_2). The first component of Ax uses the first row of A. So the row number in a_{11} and a_{12} stays at 1.

The ith component of Ax involves a_{i1} and a_{i2} and ... from row i. The short formula uses "sigma notation":

2B The ith component of Ax is $a_{i1}x_1 + a_{i2}x_2 + \cdots + a_{in}x_n$. This is $\sum_{j=1}^{n} a_{ij}x_j$.

The symbol $\sum$ is an instruction to add. Start with $j = 1$ and stop with $j = n$. Start with $a_{i1}x_1$ and stop with $a_{in}x_n$.*

The Matrix Form of One Elimination Step

$Ax = b$ is a convenient form for the original equation (2.4). What about the elimination steps? The first step in this example subtracts 2 times the first equation from the second equation. On the right side, 2 times the first component is subtracted from the second:

$$b = \begin{bmatrix} 2 \\ 8 \\ 10 \end{bmatrix} \quad \text{changes to} \quad b' = \begin{bmatrix} 2 \\ 4 \\ 10 \end{bmatrix}.$$

*Einstein shortened this even more by omitting the $\sum$. The repeated j in $a_{ij}x_j$ automatically meant addition. He also wrote the sum as $a_i^j x_j$. Not being Einstein, we include the $\sum$. For summation Gauss used the notation [ ].

We want to do that subtraction with a matrix! The same result is achieved when we multiply E times $\boldsymbol{b}$:

$$\textbf{\textit{The elimination matrix is}} \quad E = \begin{bmatrix} 1 & 0 & 0 \\ -2 & 1 & 0 \\ 0 & 0 & 1 \end{bmatrix}.$$

With numbers and letters, multiplication by E subtracts 2 times row 1 from row 2:

$$\begin{bmatrix} 1 & 0 & 0 \\ -2 & 1 & 0 \\ 0 & 0 & 1 \end{bmatrix}\begin{bmatrix} 2 \\ 8 \\ 10 \end{bmatrix} = \begin{bmatrix} 2 \\ 4 \\ 10 \end{bmatrix} \qquad \begin{bmatrix} 1 & 0 & 0 \\ -2 & 1 & 0 \\ 0 & 0 & 1 \end{bmatrix}\begin{bmatrix} b_1 \\ b_2 \\ b_3 \end{bmatrix} = \begin{bmatrix} b_1 \\ b_2 - 2\,b_1 \\ b_3 \end{bmatrix}$$

Notice how 2 and 10 stay the same. Those are b_1 and b_3. The first and third rows of E are the first and third rows of the identity matrix I. That matrix leaves all vectors unchanged. The new second component 4 is the number that appeared after the elimination step.

It is easy to describe all the "elementary matrices" or "elimination matrices" like E. Start with the identity matrix I and change one of its zeros to $-l$:

2C The ***identity matrix*** has 1's on the diagonal and 0's everywhere else. Then $I\boldsymbol{b} = \boldsymbol{b}$. The ***elimination matrix*** that subtracts a multiple l of row j from row i has the extra nonzero entry $-l$ in the i, j position.

Example 2.5

$$I = \begin{bmatrix} 1 & 0 & 0 \\ 0 & 1 & 0 \\ 0 & 0 & 1 \end{bmatrix} \quad \text{and} \quad E_{31} = \begin{bmatrix} 1 & 0 & 0 \\ 0 & 1 & 0 \\ -l & 0 & 1 \end{bmatrix}.$$

If you multiply I times $\boldsymbol{b}$, you get $\boldsymbol{b}$ again. If you multiply E_{31} times $\boldsymbol{b}$, then l times the first component is subtracted from the third component. Here we get $9 - 4 = 5$:

$$\begin{bmatrix} 1 & 0 & 0 \\ 0 & 1 & 0 \\ 0 & 0 & 1 \end{bmatrix}\begin{bmatrix} 1 \\ 3 \\ 9 \end{bmatrix} = \begin{bmatrix} 1 \\ 3 \\ 9 \end{bmatrix} \quad \text{and} \quad \begin{bmatrix} 1 & 0 & 0 \\ 0 & 1 & 0 \\ -4 & 0 & 1 \end{bmatrix}\begin{bmatrix} 1 \\ 3 \\ 9 \end{bmatrix} = \begin{bmatrix} 1 \\ 3 \\ 5 \end{bmatrix}.$$

This is on the right side of $A\boldsymbol{x} = \boldsymbol{b}$. What about the left side? The multiplier $l = 4$ was chosen to produce a zero, by subtracting 4 times the pivot. ***The purpose of E_{31} is to create a zero in the* (3, 1) *position.***

The notation fits the purpose. Start with A. Apply E's to produce zeros below the pivots (the first E is E_{21}). End with a triangular system. We now look in detail at the left side—elimination applied to $A\boldsymbol{x}$.

First a small point. The vector $\boldsymbol{x}$ stays the same. The solution is not changed by elimination. (That may be more than a small point.) It is the coefficient matrix that is changed! When we start with $A\boldsymbol{x} = \boldsymbol{b}$ and multiply by E, the result is $EA\boldsymbol{x} = E\boldsymbol{b}$. We want the new matrix EA—the result of multiplying E times A.

Matrix Multiplication

The question is: ***How do we multiply two matrices?*** When the first matrix is E (an elimination matrix E_{21}), there is already an important clue. We know A, and we know what it becomes after the elimination step. To keep everything right, we hope and expect that

$$E = \begin{bmatrix} 1 & 0 & 0 \\ -2 & 1 & 0 \\ 0 & 0 & 1 \end{bmatrix} \quad \text{times} \quad A = \begin{bmatrix} 2 & 4 & -2 \\ 4 & 9 & -3 \\ -2 & -3 & 7 \end{bmatrix}$$

$$\text{gives} \quad EA = \begin{bmatrix} 2 & 4 & -2 \\ 0 & 1 & 1 \\ -2 & -3 & 7 \end{bmatrix}.$$

This step does not change rows 1 and 3. Those rows are repeated in EA—only row 2 is different. *Twice the first row has been subtracted from the second row.* Matrix multiplication agrees with elimination—and the new system of equations is $EA\boldsymbol{x} = E\boldsymbol{b}$.

That is simple but it hides a subtle idea. Multiplying both sides of the original equation gives $E(A\boldsymbol{x}) = E\boldsymbol{b}$. With our proposed multiplication of matrices, this is also $(EA)\boldsymbol{x} = E\boldsymbol{b}$. The first was E times $A\boldsymbol{x}$, the second is EA times $\boldsymbol{x}$. They are the same! The parentheses are not needed. We just write $EA\boldsymbol{x} = E\boldsymbol{b}$.

When multiplying ABC, you can do BC first or you can do AB first. This is the point of an "associative law" like $3 \times (4 \times 5) = (3 \times 4) \times 5$. We multiply 3 times 20, or we multiply 12 times 5. Both answers are 60. That law seems so obvious that it is hard to imagine it could be false. But the "commutative law" $3 \times 4 = 4 \times 3$ looks even more obvious. For matrices, EA is different from AE.

2D Associative Law but not Commutative Law
It is true that $A(BC) = (AB)C$. It is not usually true that AB equals BA.

There is another requirement on matrix multiplication. We know how to multiply a matrix times a vector (A times $\boldsymbol{x}$ or E times $\boldsymbol{b}$). The new matrix-matrix law should be consistent with the old matrix-vector law. When B has only one column (this column is $\boldsymbol{b}$), the matrix-matrix product EB should agree with $E\boldsymbol{b}$. Even more, we should be able to multiply matrices *a column at a time*:

If the matrix B contains several columns $\boldsymbol{b}_1, \boldsymbol{b}_2, \boldsymbol{b}_3$, then the columns of EB should be $E\boldsymbol{b}_1, E\boldsymbol{b}_2, E\boldsymbol{b}_3$.

This holds true for the matrix multiplication above (where we have A instead of B). If you multiply column 1 by E, you get column 1 of the answer:

$$\begin{bmatrix} 1 & 0 & 0 \\ -2 & 1 & 0 \\ 0 & 0 & 1 \end{bmatrix} \begin{bmatrix} 2 \\ 4 \\ -2 \end{bmatrix} = \begin{bmatrix} 2 \\ 0 \\ -2 \end{bmatrix} \quad \text{and similarly for columns 2 and 3.}$$

This requirement deals with columns, while elimination deals with rows. A third approach (in the next section) describes each individual entry of the product. The beauty of matrix multiplication is that all three approaches (rows, columns, whole matrices) come out right.

The Matrix P_{ij} for a Row Exchange

To subtract row j from row i we use E_{ij}. To exchange or "permute" those rows we use another matrix P_{ij}. Remember the situation when row exchanges are needed: Zero is in the pivot position. Lower down that column is a nonzero. By exchanging the two rows, we have a pivot (nonzero!) and elimination goes forward.

What matrix P_{23} exchanges row 2 with row 3? We can find it by exchanging rows of I:

$$P_{23} = \begin{bmatrix} 1 & 0 & 0 \\ 0 & 0 & 1 \\ 0 & 1 & 0 \end{bmatrix}.$$

This is a ***row exchange matrix***. Multiplying by P_{23} exchanges components 2 and 3 of any column vector. Therefore it exchanges rows 2 and 3 of any matrix:

$$\begin{bmatrix} 1 & 0 & 0 \\ 0 & 0 & 1 \\ 0 & 1 & 0 \end{bmatrix} \begin{bmatrix} 1 \\ 3 \\ 5 \end{bmatrix} = \begin{bmatrix} 1 \\ 5 \\ 3 \end{bmatrix} \quad \text{and} \quad \begin{bmatrix} 1 & 0 & 0 \\ 0 & 0 & 1 \\ 0 & 1 & 0 \end{bmatrix} \begin{bmatrix} 2 & 4 & 1 \\ 0 & 0 & 3 \\ 0 & 6 & 5 \end{bmatrix} = \begin{bmatrix} 2 & 4 & 1 \\ 0 & 6 & 5 \\ 0 & 0 & 3 \end{bmatrix}.$$

On the right, P_{23} is doing what it was created for. With zero in the second pivot position and "6" below it, the exchange puts 6 into the pivot.

Notice how matrices *act*. They don't just sit there. We will soon meet other permutation matrices, which can put any number of rows into a different order. Rows 1, 2, 3 can be moved to 3, 1, 2. Our P_{23} is one particular permutation matrix—it works on two rows.

2E Permutation Matrix P_{ij} exchanges rows i and j when it multiplies a matrix. P_{ij} is the identity matrix with rows i and j reversed.

$$\textbf{\textit{To exchange equations 1 and 3 multiply by}} \quad P_{13} = \begin{bmatrix} 0 & 0 & 1 \\ 0 & 1 & 0 \\ 1 & 0 & 0 \end{bmatrix}.$$

Usually no row exchanges are needed. The odds are good that elimination uses only the E_{ij}. But the P_{ij} are ready if needed, to put a new pivot into position.

The Augmented Matrix A'

This book eventually goes far beyond elimination. Matrices have all kinds of practical applications—in which they are multiplied. Our best starting point was a square E times a square A, because we met this in elimination—and we know what answer to expect for EA. The next step is to allow a *rectangular matrix* A'. It still comes from our original equations, so we still have a check on the product EA'.

Key idea: The equal signs and the letters x, y, z are not really involved in elimination. They don't change and they don't move. Elimination acts on A and $\boldsymbol{b}$, and it does the same thing to both. ***We can include b as an extra column and follow it through elimination.*** The matrix A is enlarged or "augmented" by $\boldsymbol{b}$:

$$\textbf{\textit{Augmented matrix}} \quad A' = \begin{bmatrix} A & \boldsymbol{b} \end{bmatrix} = \begin{bmatrix} 2 & 4 & -2 & 2 \\ 4 & 9 & -3 & 8 \\ -2 & -3 & 7 & 10 \end{bmatrix}.$$

A' places the left side and right side next to each other. *Elimination acts on whole rows of* A'. The left side and right side are both multiplied by E, to subtract 2 times equation 1 from equation 2. With A' those steps happen together:

$$\begin{bmatrix} 1 & 0 & 0 \\ -2 & 1 & 0 \\ 0 & 0 & 1 \end{bmatrix} \begin{bmatrix} 2 & 4 & -2 & 2 \\ 4 & 9 & -3 & 8 \\ -2 & -3 & 7 & 10 \end{bmatrix} = \begin{bmatrix} 2 & 4 & -2 & 2 \\ 0 & 1 & 1 & 4 \\ -2 & -3 & 7 & 10 \end{bmatrix} = EA'.$$

The new second row contains 0, 1, 1, 4. The new second equation is $x_2 + x_3 = 4$. Both requirements on matrix multiplication are obeyed:

R (by rows): Each row of E acts on A' to give a row of EA'.

C (by columns): E acts on each column of A' to give a column of EA'.

Notice again that word "acts." This is essential. Matrices do something! The matrix A acts on $\boldsymbol{x}$ to give $\boldsymbol{b}$. The matrix E operates on A to give EA. The whole process of elimination is a sequence of row operations, alias matrix multiplications. A goes to $E_{21}A$ which goes to $E_{31}E_{21}A$ which goes to $E_{32}E_{31}E_{21}A$ which is a triangular matrix.

The right side is included when we work with A'. Then A' goes to $E_{21}A'$ which goes to $E_{31}E_{21}A'$ which goes to $E_{32}E_{31}E_{21}A'$ which is a triangular system of equations.

We stop for exercises on multiplication by E, before writing down the rules for all matrix multiplications.

Problem Set 2.2

Problems 1–14 are about elimination matrices.

1 Write down the 3 by 3 matrices that produce these elimination steps:

(a) E_{21} subtracts 5 times row 1 from row 2.

(b) E_{32} subtracts -7 times row 2 from row 3.

(c) P_{12} exchanges rows 1 and 2.

2 In Problem 1, applying E_{21} and then E_{32} to the column $\boldsymbol{b} = (1, 0, 0)$ gives $E_{32}E_{21}\boldsymbol{b} =$ _____. Applying E_{32} before E_{21} gives $E_{21}E_{32}\boldsymbol{b} =$ _____. $E_{21}E_{32}$ is different from $E_{32}E_{21}$ because when E_{32} comes first, row _____ feels no effect from row _____.

3 Which three matrices E_{21}, E_{31}, E_{32} put A into triangular form U?

$$A = \begin{bmatrix} 1 & 1 & 0 \\ 4 & 6 & 1 \\ -2 & 2 & 0 \end{bmatrix} \quad \text{and} \quad E_{32}E_{31}E_{21}A = U.$$

Multiply those E's to get one matrix M that does elimination: $MA = U$.

4 Include $\boldsymbol{b} = (1, 0, 0)$ as a fourth column in Problem 3 to produce A'. Carry out the elimination steps on this augmented matrix A' to solve $A\boldsymbol{x} = \boldsymbol{b}$.

5 Suppose a 3 by 3 matrix has $a_{33} = 7$ and its third pivot is 2. If you change a_{33} to 11, the third pivot is ______. If you change a_{33} to ______, there is no third pivot.

6 If every column of A is a multiple of $(1, 1, 1)$, then $A\boldsymbol{x}$ is always a multiple of $(1, 1, 1)$. Do a 3 by 3 example. How many pivots are produced by elimination?

7 Suppose E_{31} subtracts 7 times row 1 from row 3. To reverse that step you should ______ 7 times row ______ to row ______. The matrix to do this reverse step (the inverse matrix) is $R_{31} =$ ______.

8 Suppose E_{31} subtracts 7 times row 1 from row 3. What matrix R_{31} is changed into I? Then $E_{31}R_{31} = I$ where Problem 7 has $R_{31}E_{31} = I$.

9 (a) E_{21} subtracts row 1 from row 2 and then P_{23} exchanges rows 2 and 3. What matrix $M = P_{23}E_{21}$ does both steps at once?

(b) P_{23} exchanges rows 2 and 3 and then E_{31} subtracts row 1 from row 3. What matrix $M = E_{31}P_{23}$ does both steps at once? Explain why the M's are the same but the E's are different.

10 Create a matrix that has $a_{11} = a_{22} = a_{33} = 1$ but elimination produces two negative pivots. (The first pivot is 1.)

11 Multiply these matrices:

$$\begin{bmatrix} 1 & 0 \\ 5 & 1 \end{bmatrix} \begin{bmatrix} 2 & 2 \\ 0 & 0 \end{bmatrix} \qquad \begin{bmatrix} 1 & 0 & 0 \\ -1 & 1 & 0 \\ -1 & 0 & 1 \end{bmatrix} \begin{bmatrix} 1 & 2 & 3 \\ 1 & 3 & 1 \\ 1 & 4 & 0 \end{bmatrix}.$$

12 Explain these facts. If the third column of B is all zero, the third column of EB is all zero (for any E). If the third row of B is all zero, the third row of EB might not be zero.

13 This 4 by 4 matrix will need elimination matrices E_{21} and E_{32} and E_{43}. What are those matrices?

$$A = \begin{bmatrix} 2 & -1 & 0 & 0 \\ -1 & 2 & -1 & 0 \\ 0 & -1 & 2 & -1 \\ 0 & 0 & -1 & 2 \end{bmatrix}.$$

14 Write down the 3 by 3 matrix that has $a_{ij} = 2i - 3j$. This matrix has $a_{32} = 0$, but elimination still needs E_{32} to produce a zero. Which previous step destroys the original zero and what is E_{32}?

Problems 15–22 are about creating and multiplying matrices.

15 Write these ancient problems in a 2 by 2 matrix form $A\boldsymbol{x} = \boldsymbol{b}$:

(a) X is twice as old as Y and their ages add to 33.

(b) The line $y = mx + c$ goes through $(x, y) = (2, 5)$ and $(3, 7)$. Find m and c.

16 The parabola $y = a + bx + cx^2$ goes through the points $(x, y) = (1, 4)$ and $(2, 8)$ and $(3, 14)$. Find a matrix equation for the unknowns (a, b, c). Solve by elimination.

17 Multiply these matrices in the orders EF and FE:

$$E = \begin{bmatrix} 1 & 0 & 0 \\ a & 1 & 0 \\ b & 0 & 1 \end{bmatrix} \qquad F = \begin{bmatrix} 1 & 0 & 0 \\ 0 & 1 & 0 \\ 0 & c & 1 \end{bmatrix}.$$

Also compute $E^2 = EE$ and $F^2 = FF$.

18 Multiply these row exchange matrices in the orders PQ and QP:

$$P = \begin{bmatrix} 0 & 1 & 0 \\ 1 & 0 & 0 \\ 0 & 0 & 1 \end{bmatrix} \quad \text{and} \quad Q = \begin{bmatrix} 0 & 0 & 1 \\ 0 & 1 & 0 \\ 1 & 0 & 0 \end{bmatrix}.$$

Also compute $P^2 = PP$ and $(PQ)^2 = PQPQ$.

19 (a) Suppose all columns of B are the same. Then all columns of EB are the same, because each one is E times ______ .

(b) Suppose all rows of B are the same. Show by example that all rows of EB are *not* the same.

20 If E adds row 1 to row 2 and F adds row 2 to row 1, does EF equal FE?

21 The entries of A and $\boldsymbol{x}$ are a_{ij} and x_j. The matrix E_{21} subtracts row 1 from row 2. Write a formula for

(a) the third component of $A\boldsymbol{x}$

(b) the $(2, 1)$ entry of EA

(c) the $(2, 1)$ component of $E\boldsymbol{x}$

(d) the first component of $EA\boldsymbol{x}$.

22 The elimination matrix $E = \begin{bmatrix} 1 & 0 \\ -2 & 1 \end{bmatrix}$ subtracts 2 times row 1 of A from row 2 of A. The result is EA. In the opposite order AE, we are subtracting 2 times ______ of A from ______ . (Do an example.)

Problems 23–26 include the column b in the augmented matrix A'.

23 Apply elimination to the 2 by 3 augmented matrix A'. What is the triangular system $Ux = c$? What is the solution x?

$$Ax = \begin{bmatrix} 2 & 3 \\ 4 & 1 \end{bmatrix} \begin{bmatrix} x_1 \\ x_2 \end{bmatrix} = \begin{bmatrix} 1 \\ 17 \end{bmatrix}.$$

24 Apply elimination to the 3 by 4 augmented matrix A'. How do you know this system has no solution?

$$Ax = \begin{bmatrix} 1 & 2 & 3 \\ 2 & 3 & 4 \\ 3 & 5 & 7 \end{bmatrix} \begin{bmatrix} x \\ y \\ z \end{bmatrix} = \begin{bmatrix} 1 \\ 2 \\ 6 \end{bmatrix}.$$

Change the last number 6 so that there *is* a solution.

25 The equations $Ax = b$ and $Ax^* = b^*$ have the same matrix A.

(a) What double augmented matrix A'' should you use in elimination to solve both equations at once?

(b) Solve both of these equations by working on a 2 by 4 matrix A'':

$$\begin{bmatrix} 1 & 4 \\ 2 & 7 \end{bmatrix} \begin{bmatrix} x \\ y \end{bmatrix} = \begin{bmatrix} 1 \\ 0 \end{bmatrix} \quad \text{and} \quad \begin{bmatrix} 1 & 4 \\ 2 & 7 \end{bmatrix} \begin{bmatrix} x \\ y \end{bmatrix} = \begin{bmatrix} 0 \\ 1 \end{bmatrix}.$$

26 Choose the numbers a, b, c, d in this augmented matrix so that there is (a) no solution (b) infinitely many solutions.

$$A' = \begin{bmatrix} 1 & 2 & 3 & a \\ 0 & 4 & 5 & b \\ 0 & 0 & d & c \end{bmatrix}$$

Which of the numbers a, b, c, or d have no effect on the solvability?

27 Challenge question: E_{ij} is the 4 by 4 identity matrix with an extra 1 in the (i, j) position, $i > j$. Describe the matrix $E_{ij}E_{kl}$. When does it equal $E_{kl}E_{ij}$? *Try examples first.*

2.3 Rules for Matrix Operations

A matrix is a rectangular array of numbers or "entries." When A has m rows and n columns, it is an "m by n" matrix. The upper left entry is a_{11}, the upper right entry is a_{1n}. (The lower left is a_{m1} and the lower right is a_{mn}.) Matrices can be added if their shapes are the same. They can be multiplied by any constant c. Here are examples of $A + B$ and $2A$, for 3 by 2 matrices:

$$\begin{bmatrix} 1 & 2 \\ 3 & 4 \\ 0 & 0 \end{bmatrix} + \begin{bmatrix} 2 & 2 \\ 4 & 4 \\ 9 & 9 \end{bmatrix} = \begin{bmatrix} 3 & 4 \\ 7 & 8 \\ 9 & 9 \end{bmatrix} \quad \text{and} \quad 2\begin{bmatrix} 1 & 2 \\ 3 & 4 \\ 0 & 0 \end{bmatrix} = \begin{bmatrix} 2 & 4 \\ 6 & 8 \\ 0 & 0 \end{bmatrix}.$$

$$\begin{bmatrix} * & & & \\ a_{i1} & a_{i2} & \cdots & a_{i5} \\ * & & & \\ * & & & \end{bmatrix} \begin{bmatrix} * & * & b_{1j} & * & * & * \\ & & b_{2j} & & & \\ & & \vdots & & & \\ & & b_{5j} & & & \end{bmatrix} = \begin{bmatrix} & & * & & & \\ * & * & (AB)_{ij} & * & * & * \\ & & * & & & \\ & & * & & & \end{bmatrix}$$

$$\text{4 by 5} \qquad\qquad \text{5 by 6} \qquad\qquad \text{4 by 6}$$

Figure 2.4 With $i = 2$ and $j = 3$, $(AB)_{23}$ is (row 2) · (column 3).

Matrices are added exactly as vectors are—one entry at a time. We could even regard vectors as special matrices with only one column (so $n = 1$). The matrix $-A$ comes from multiplication by $c = -1$ (reversing all the signs). Adding A to $-A$ leaves the *zero matrix*, with all entries zero.

The 3 by 2 zero matrix is different from the 2 by 3 zero matrix. Even 0 has a shape (several shapes). All this is only common sense.

The serious question is ***matrix multiplication.*** When can we multiply A times B, and what is the product AB? We cannot multiply A and B above (both 3 by 2). They don't pass the following test, which makes the multiplication possible. ***The number of columns of A must equal the numbers of rows of B.*** If A has two columns, B must have two rows. When A is 3 by 2, the matrix B can be 2 by 1 (a vector) or 2 by 2 (square) or 2 by 20. When there are twenty columns in B, each column is ready to be multiplied by A. Then AB is 3 by 20.

Suppose A is m by n and B is n by p. We can multiply. The product AB is m by p.

$$\begin{bmatrix} \textbf{m rows} \\ n \text{ columns} \end{bmatrix} \begin{bmatrix} n \text{ rows} \\ \textbf{p columns} \end{bmatrix} = \begin{bmatrix} \textbf{m rows} \\ \textbf{p columns} \end{bmatrix}.$$

The dot product is an extreme case. Then 1 by n multiplies n by 1. The result is 1 by 1. That single number is the dot product.

In every case AB is filled with dot products. For the top corner, (row 1 of A) · (column 1 of B) gives the (1, 1) entry of AB. To multiply matrices, take all these dot products: ***(each row of** A**)** · **(each column of** B**)**.

2F The entry in row i and column j of AB is (row i of A) · (column j of B).

Figure 2.4 picks out the second row ($i = 2$) of a 4 by 5 matrix A. It picks out the third column ($j = 3$) of a 5 by 6 matrix B. Their dot product goes into row 2 and column 3 of AB. The matrix AB has *as many rows as* A (4 rows), and *as many columns as* B.

Example 2.6 Square matrices can be multiplied if and only if they have the same size:

$$\begin{bmatrix} 1 & 1 \\ 2 & -1 \end{bmatrix} \begin{bmatrix} 2 & 2 \\ 3 & 4 \end{bmatrix} = \begin{bmatrix} 5 & 6 \\ 1 & 0 \end{bmatrix}.$$

The first dot product is $1 \cdot 2 + 1 \cdot 3 = 5$. Three more dot products give 6, 1, and 0. Each dot product requires two multiplications—thus eight in all.

If A and B are n by n, so is AB. It contains n^2 dot products, each needing n multiplications. The computation of AB uses n^3 separate multiplications. For $n = 100$ we multiply a million times. For $n = 2$ we have $n^3 = 8$.

Mathematicians thought until recently that AB absolutely needed $2^3 = 8$ multiplications. Then somebody found a way to do it with 7. By breaking n by n matrices into 2 by 2 blocks, this idea also reduced the count for large matrices. Instead of n^3 it went below $n^{2.8}$, and the exponent keeps falling.* The best at this moment is $n^{2.376}$. But the algorithm is so awkward that scientific computing is done the regular way—with n^2 dot products and n multiplications each.

Example 2.7 Suppose A is a row vector (1 by 3) and B is a column vector (3 by 1). Then AB is 1 by 1 (only one entry, the dot product). On the other hand B times A is a full matrix:

$$\begin{bmatrix} 1 & 2 & 3 \end{bmatrix} \begin{bmatrix} 0 \\ 1 \\ 2 \end{bmatrix} = \begin{bmatrix} 8 \end{bmatrix} \qquad \begin{bmatrix} 0 \\ 1 \\ 2 \end{bmatrix} \begin{bmatrix} 1 & 2 & 3 \end{bmatrix} = \begin{bmatrix} 0 & 0 & 0 \\ 1 & 2 & 3 \\ 2 & 4 & 6 \end{bmatrix}.$$

A row times a column is an *"inner"* product—another name for dot product. A column times a row is an *"outer"* product. These are extreme cases of matrix multiplication, with very thin matrices. They follow the rule for shapes in multiplication: (n by 1) times (1 by n) is (n by n).

Rows and Columns of AB

In the big picture, A multiplies each column of B. When A multiplies a vector, we are combining the columns of A. ***Each column of AB is a combination of the columns of A.*** That is the column picture of matrix multiplication:

Column of AB is* (matrix A) *times* (column of B) = *combination of columns of A.

Look next at the row picture—which is reversed. Each row of A multiplies the whole matrix B. The result is a row of AB. It is a combination of the rows of B:

$$\begin{bmatrix} \text{row } i \text{ of } A \end{bmatrix} \begin{bmatrix} 1 & 2 & 3 \\ 4 & 5 & 6 \\ 7 & 8 & 9 \end{bmatrix} = \begin{bmatrix} \text{row } i \text{ of } AB \end{bmatrix}.$$

We see row operations in elimination (E times A). We see columns in A times $\boldsymbol{x}$. The "row-column picture" is the usual one—dot products of rows with columns. Believe it or not, ***there is also a "column-row picture."*** Not everybody knows that columns multiply rows to give the same answer AB. This is in Example 2.8 below.

*Maybe the exponent won't stop falling before 2. No number in between looks special.

The Laws for Matrix Operations

May we put on record six laws that matrices do obey, while emphasizing an equation they don't obey? The matrices can be square or rectangular, and the laws involve three operations:

1 Multiply A by a scalar to get cA.

2 Add matrices (same size) to get $A + B$.

3 Multiply matrices (right sizes) to get AB or BA.

You know the right sizes for AB: (m by n) multiplies (n by p) to produce (m by p).

The laws involving $A + B$ are all simple and all obeyed. Here are three addition laws:

$$\begin{aligned} A + B &= B + A && \text{(commutative law)} \\ c(A + B) &= cA + cB && \text{(distributive law)} \\ A + (B + C) &= (A + B) + C && \text{(associative law).} \end{aligned}$$

Three more laws hold for multiplication, but $AB = BA$ is not one of them:

$$\begin{aligned} AB &\neq BA && \text{(the commutative “law” is \textit{usually broken})} \\ C(A + B) &= CA + CB && \text{(distributive law from the left)} \\ (A + B)C &= AC + BC && \text{(distributive law from the right)} \\ A(BC) &= (AB)C && \text{(associative law)(parentheses not needed).} \end{aligned}$$

When A and B are not square, AB is a different size from BA. These matrices can't be equal—even if both multiplications are allowed. For square matrices, almost any example shows that AB is different from BA:

$$AB = \begin{bmatrix} 0 & 0 \\ 1 & 0 \end{bmatrix} \begin{bmatrix} 0 & 1 \\ 0 & 0 \end{bmatrix} = \begin{bmatrix} 0 & 0 \\ 0 & 1 \end{bmatrix} \quad \text{but} \quad BA = \begin{bmatrix} 0 & 1 \\ 0 & 0 \end{bmatrix} \begin{bmatrix} 0 & 0 \\ 1 & 0 \end{bmatrix} = \begin{bmatrix} 1 & 0 \\ 0 & 0 \end{bmatrix}.$$

It is true that $AI = IA$. All square matrices commute with I and also with $B = cI$. Only these matrices cI commute with all other matrices.

The law $A(B + C) = AB + AC$ is proved a column at a time. Start with $A(\boldsymbol{b} + \boldsymbol{c}) = A\boldsymbol{b} + A\boldsymbol{c}$ for the first column. That is the key to everything—***linearity***. Say no more.

The law $A(BC) = (AB)C$ means that you can multiply BC first or AB first (Problems 4 and 14). Look at the special case when $A = B = C =$ square matrix. Then (A *times* A^2) $=$ (A^2 *times* A). The product in either order is A^3. All the matrix powers A^p commute with each other, and they follow the same rules as numbers:

$$A^p = AAA \cdots A \ (p \text{ factors}) \qquad (A^p)(A^q) = A^{p+q} \qquad (A^p)^q = A^{pq}.$$

Those are the ordinary laws for exponents. A^3 times A^4 is A^7 (seven factors). A^3 to the fourth power is A^{12} (twelve A's). When p and q are zero or negative these rules still hold, provided A has a "-1 power"—which is the *inverse matrix* A^{-1}. Then $A^0 = I$ is the identity matrix (no factors).

For a number, a^{-1} is $1/a$. For a matrix, the inverse is written A^{-1}. (It is *never* I/A; MATLAB would accept $A\backslash I$.) Every number has an inverse except $a = 0$. To decide when A has an inverse is a central problem in linear algebra. Section 2.4 will start on the answer. This section is a Bill of Rights for matrices, to say when A and B can be multiplied and how.

Block Matrices and Block Multiplication

We have to say one more thing about matrices. They can be broken into ***blocks*** (which are smaller matrices). This often happens naturally. Here is a 4 by 6 matrix broken into blocks of size 2 by 2—and each block is just I:

$$A = \left[\begin{array}{cc|cc|cc} 1 & 0 & 1 & 0 & 1 & 0 \\ 0 & 1 & 0 & 1 & 0 & 1 \\ \hline 1 & 0 & 1 & 0 & 1 & 0 \\ 0 & 1 & 0 & 1 & 0 & 1 \end{array}\right] = \begin{bmatrix} I & I & I \\ I & I & I \end{bmatrix}.$$

If B is also 4 by 6 and its block sizes match the block sizes in A, you can add $A + B$ *a block at a time*.

We have seen block matrices before. The right side vector $\boldsymbol{b}$ was placed next to A in the "augmented matrix." Then $A' = [\,A \quad \boldsymbol{b}\,]$ has two column blocks. Multiplying A' by an elimination matrix gave $[\,EA \quad E\boldsymbol{b}\,]$. No problem to multiply blocks times blocks, when their shapes permit:

2G Block multiplication If the cuts between columns of A match the cuts between rows of B, then block multiplication of AB is allowed:

$$\begin{bmatrix} A_{11} & A_{12} \\ A_{21} & A_{22} \end{bmatrix} \begin{bmatrix} B_{11} & \cdots \\ B_{21} & \cdots \end{bmatrix} = \begin{bmatrix} A_{11}B_{11} + A_{12}B_{21} & \cdots \\ A_{21}B_{11} + A_{22}B_{21} & \cdots \end{bmatrix}. \tag{2.6}$$

This equation is the same as if the blocks were numbers (which are 1 by 1 blocks). We are careful to keep A's in front of B's, because BA can be different. The cuts between rows of A give cuts between rows of AB. Any column cuts in B are also column cuts in AB. Here are the block sizes for equation (2.6) to go through:

$$\begin{bmatrix} m_1 \text{ by } n_1 & m_1 \text{ by } n_2 \\ m_2 \text{ by } n_1 & m_2 \text{ by } n_2 \end{bmatrix} \begin{bmatrix} n_1 \text{ by } p_1 \cdots \\ n_2 \text{ by } p_1 \cdots \end{bmatrix} = \begin{bmatrix} m_1 \text{ by } p_1 \cdots \\ m_2 \text{ by } p_1 \cdots \end{bmatrix}.$$

The column cuts in A must match the row cuts in B. Then the blocks will multiply.

Main point When matrices split into blocks, it is often simpler to see how they act. The block matrix of I's above is much clearer than the original 4 by 6 matrix.

Example 2.8 (Important) Let the blocks of A be its columns. Let the blocks of B be its rows. Then block multiplication AB is ***columns times rows***:

$$AB = \begin{bmatrix} | & & | \\ \boldsymbol{a}_1 & \cdots & \boldsymbol{a}_n \\ | & & | \end{bmatrix} \begin{bmatrix} — & \boldsymbol{b}_1 & — \\ & \vdots & \\ — & \boldsymbol{b}_n & — \end{bmatrix} = \begin{bmatrix} \\ \boldsymbol{a}_1\boldsymbol{b}_1 + \cdots + \boldsymbol{a}_n\boldsymbol{b}_n \\ \\ \end{bmatrix}. \tag{2.7}$$

This is another way to multiply matrices! Compare it with the usual way, rows times columns. Row 1 of A times column 1 of B gave the $(1, 1)$ entry in AB. Now *column* 1 of A times *row* 1 of B gives a full matrix—not just a single number. Look at this example:

$$\begin{bmatrix} 1 & 4 \\ 1 & 5 \end{bmatrix} \begin{bmatrix} 3 & 2 \\ 1 & 0 \end{bmatrix} = \begin{bmatrix} 1 \\ 1 \end{bmatrix} \begin{bmatrix} 3 & 2 \end{bmatrix} + \begin{bmatrix} 4 \\ 5 \end{bmatrix} \begin{bmatrix} 1 & 0 \end{bmatrix}$$

$$= \begin{bmatrix} 3 & 2 \\ 3 & 2 \end{bmatrix} + \begin{bmatrix} 4 & 0 \\ 5 & 0 \end{bmatrix}. \tag{2.8}$$

We stop there so you can see columns multiplying rows. If an m by 1 matrix (a column) multiplies a 1 by p matrix (a row), the result is m by p (a matrix). That is what we found. Dot products are "inner products," these are "outer products."

When you add the two matrices at the end of equation (2.8), you get the correct answer AB. In the top left corner the answer is $3 + 4 = 7$. This agrees with the row-column dot product of $(1, 4)$ with $(3, 1)$.

Summary The usual way, rows times columns, gives four dot products (8 multiplications). The new way, columns times rows, gives two full matrices (8 multiplications). The eight multiplications are the same, just the order is different.

Example 2.9 (Elimination by blocks) Suppose the first column of A contains $2, 6, 8$. To change 6 and 8 to 0 and 0, multiply the pivot row by 3 and 4 and subtract. Each elimination step is really a multiplication by an elimination matrix E_{ij}:

$$E_{21} = \begin{bmatrix} 1 & 0 & 0 \\ -3 & 1 & 0 \\ 0 & 0 & 1 \end{bmatrix} \quad \text{and} \quad E_{31} = \begin{bmatrix} 1 & 0 & 0 \\ 0 & 1 & 0 \\ -4 & 0 & 1 \end{bmatrix}.$$

The "block idea" is to do both eliminations with one matrix E. Multiply E_{21} and E_{31} to find that matrix E, which clears out the whole first column below the pivot:

$$E = \begin{bmatrix} 1 & 0 & 0 \\ -3 & 1 & 0 \\ -4 & 0 & 1 \end{bmatrix} \quad \text{multiplies } A \text{ to give} \quad EA = \begin{bmatrix} 2 & x & x \\ 0 & x & x \\ 0 & x & x \end{bmatrix}.$$

Block multiplication gives a formula for those x's in EA. The matrix A has four blocks a, $\boldsymbol{b}$, $\boldsymbol{c}$, D: a number, the rest of a row, the rest of a column, and the rest of the matrix. Watch how they multiply:

$$EA = \left[\begin{array}{c|c} 1 & 0 \\ \hline -\boldsymbol{c}/a & I \end{array}\right] \left[\begin{array}{c|c} a & \boldsymbol{b} \\ \hline \boldsymbol{c} & D \end{array}\right] = \left[\begin{array}{c|c} a & \boldsymbol{b} \\ \hline 0 & D - \boldsymbol{cb}/a \end{array}\right]. \tag{2.9}$$

The pivot is $a = 2$. The column under it is $\boldsymbol{c} = \begin{bmatrix} 6 \\ 8 \end{bmatrix}$. Elimination multiplies a by $\boldsymbol{c}/a$ and subtracts from $\boldsymbol{c}$ to get zeros in the first column. It multiplies the column vector $\boldsymbol{c}/a$ times the row vector $\boldsymbol{b}$, and subtracts to get $D - \boldsymbol{cb}/a$. This is ordinary elimination, written with vectors—a column at a time.

Problem Set 2.3

Problems 1–17 are about the laws of matrix multiplication.

1 Suppose A is 3 by 5, B is 5 by 3, C is 5 by 1, and D is 3 by 1. Which of these matrix operations are allowed, and what are the shapes of the results?

(a) BA

(b) $A(B + C)$

(c) ABD

(d) $AC + BD$

(e) $ABABD$.

2 What rows or columns and what matrices do you multiply to find

(a) the third column of AB?

(b) the first row of AB?

(c) the entry in row 3, column 4 of AB?

(d) the entry in row 1, column 1 of CDE?

3 Compute $AB + AC$ and separately $A(B + C)$ and compare:

$$A = \begin{bmatrix} 1 & 5 \\ 2 & 3 \end{bmatrix} \quad \text{and} \quad B = \begin{bmatrix} 0 & 2 \\ 0 & 1 \end{bmatrix} \quad \text{and} \quad C = \begin{bmatrix} 3 & 1 \\ 0 & 0 \end{bmatrix}.$$

4 In Problem 3, multiply A times BC. Then multiply AB times C.

5 Compute A^2 and A^3. Make a prediction for A^5 and A^n:

$$A = \begin{bmatrix} 1 & b \\ 0 & 1 \end{bmatrix} \quad \text{and} \quad A = \begin{bmatrix} 2 & 2 \\ 0 & 0 \end{bmatrix}.$$

6 Show that $(A + B)^2$ is different from $A^2 + 2AB + B^2$, when

$$A = \begin{bmatrix} 1 & 2 \\ 0 & 0 \end{bmatrix} \quad \text{and} \quad B = \begin{bmatrix} 1 & 0 \\ 3 & 0 \end{bmatrix}.$$

Write down the correct rule for $(A + B)(A + B) = A^2 + AB + \_\_\_\_\_$.

7 True or false. Give a specific example when false:

(a) If columns 1 and 3 of B are the same, so are columns 1 and 3 of AB.

(b) If rows 1 and 3 of B are the same, so are rows 1 and 3 of AB.

(c) If rows 1 and 3 of A are the same, so are rows 1 and 3 of AB.

(d) $(AB)^2 = A^2B^2$.

8 How are the rows of DA and EA related to the rows of A, when

$$D = \begin{bmatrix} 3 & 0 \\ 0 & 5 \end{bmatrix} \quad \text{and} \quad E = \begin{bmatrix} 0 & 1 \\ 0 & 0 \end{bmatrix}?$$

How are the columns of AD and AE related to the columns of A?

9 Row 1 of A is added to row 2. This gives EA below. Then column 1 of EA is added to column 2 to produce $(EA)F$:

$$EA = \begin{bmatrix} 1 & 0 \\ 1 & 1 \end{bmatrix} \begin{bmatrix} a & b \\ c & d \end{bmatrix} = \begin{bmatrix} a & b \\ a+c & b+d \end{bmatrix}$$

$$\text{and} \quad (EA)F = (EA)\begin{bmatrix} 1 & 1 \\ 0 & 1 \end{bmatrix} = \begin{bmatrix} a & a+b \\ a+c & a+c+b+d \end{bmatrix}.$$

(a) Do those steps in the opposite order. First add column 1 of A to column 2 by AF, then add row 1 of AF to row 2 by $E(AF)$.

(b) Compare with $(EA)F$. What law is or is not obeyed by matrix multiplication?

10 Row 1 of A is added to row 2 to produce EA. Then adding row 2 of EA to row 1 gives $F(EA)$:

$$F(EA) = \begin{bmatrix} 1 & 1 \\ 0 & 1 \end{bmatrix} \begin{bmatrix} a & b \\ a+c & b+d \end{bmatrix} = \begin{bmatrix} 2a+c & 2b+d \\ a+c & b+d \end{bmatrix}.$$

(a) Do those steps in the opposite order: first row 2 to row 1 by FA, then row 1 of FA to row 2.

(b) What law is or is not obeyed by matrix multiplication?

11 (3 by 3 matrices) Choose B so that for every matrix A (if possible)

(a) $BA = 4A$

(b) $BA = 4B$

(c) BA has rows 1 and 3 of A reversed

(d) All rows of BA are the same.

12 Suppose $AB = BA$ and $AC = CA$ for two particular matrices B and C:

$$A = \begin{bmatrix} a & b \\ c & d \end{bmatrix} \quad \text{commutes with} \quad B = \begin{bmatrix} 1 & 0 \\ 0 & 0 \end{bmatrix} \quad \text{and} \quad C = \begin{bmatrix} 0 & 1 \\ 0 & 0 \end{bmatrix}.$$

Prove that $a = d$ and $b = c = 0$. Then A is a multiple of I. Only the matrices $A = aI$ commute with all other 2 by 2 matrices (like B and C).

13 Which of the following matrices are guaranteed to equal $(A - B)^2$: $A^2 - B^2$, $(B - A)^2$, $A^2 - 2AB + B^2$, $A(A - B) - B(A - B)$, $A^2 - AB - BA + B^2$?

14 True or false:

(a) If A^2 is defined then A is square.

(b) If AB and BA are defined then A and B are square.

(c) If AB and BA are defined then AB and BA are square.

(d) If $AB = B$ then $A = I$.

15 If A is m by n, how many separate multiplications are involved when

(a) A multiplies a vector $\boldsymbol{x}$ with n components?

(b) A multiplies an n by p matrix B?

(c) A multiplies itself to produce A^2? Here $m = n$.

16 To prove that $(AB)C = A(BC)$, use the column vectors $\boldsymbol{b}_1, \ldots, \boldsymbol{b}_n$ of B. First suppose that C has only one column $\boldsymbol{c}$ with entries $c_1, \ldots, c_n$:

AB has columns $A\boldsymbol{b}_1, \ldots, A\boldsymbol{b}_n$ and $B\boldsymbol{c}$ has one column $c_1\boldsymbol{b}_1 + \cdots + c_n\boldsymbol{b}_n$.

Then $(AB)\boldsymbol{c} = c_1A\boldsymbol{b}_1 + \cdots + c_nA\boldsymbol{b}_n$ while $A(B\boldsymbol{c}) = A(c_1\boldsymbol{b}_1 + \cdots + c_n\boldsymbol{b}_n)$.

Linearity makes those last two equal: $(AB)\boldsymbol{c} =$ ______. The same is true for all other ______ of C. Therefore $(AB)C = A(BC)$.

17 For $A = \begin{bmatrix} 2 & -1 \\ 3 & -2 \end{bmatrix}$ and $B = \begin{bmatrix} 1 & 0 & 4 \\ 1 & 0 & 6 \end{bmatrix}$, compute these answers *and nothing more*:

(a) column 2 of AB

(b) row 2 of AB

(c) row 2 of AA

(d) row 2 of AAA

Problems 18–20 use a_{ij} for the entry in row i, column j of A.

18 Write down the 3 by 3 matrix A whose entries are

(a) $a_{ij} = i + j$

(b) $a_{ij} = (-1)^{i+j}$

(c) $a_{ij} = i/j$.

19 What words would you use to describe each of these classes of matrices? Give an example in each class. Which matrices belong to all four classes?

(a) $a_{ij} = 0$ if $i \neq j$

(b) $a_{ij} = 0$ if $i < j$

(c) $a_{ij} = a_{ji}$

(d) $a_{ij} = a_{1j}$.

20 The entries of A are a_{ij}. Assuming that zeros don't appear, what is

(a) the first pivot?

(b) the multiplier of row 1 to be subtracted from row i?

(c) the new entry that replaces a_{i2} after that subtraction?

(d) the second pivot?

Problems 21–25 involve powers of A.

21 Compute A^2, A^3, A^4 and also $A\boldsymbol{v}$, $A^2\boldsymbol{v}$, $A^3\boldsymbol{v}$, $A^4\boldsymbol{v}$ for

$$A = \begin{bmatrix} 0 & 1 & 0 & 0 \\ 0 & 0 & 1 & 0 \\ 0 & 0 & 0 & 1 \\ 0 & 0 & 0 & 0 \end{bmatrix} \quad \text{and} \quad \boldsymbol{v} = \begin{bmatrix} x \\ y \\ z \\ t \end{bmatrix}.$$

22 Find all the powers A^2, A^3, ... and AB, $(AB)^2$, ... for

$$A = \begin{bmatrix} .5 & .5 \\ .5 & .5 \end{bmatrix} \quad \text{and} \quad B = \begin{bmatrix} 1 & 0 \\ 0 & -1 \end{bmatrix}.$$

23 By trial and error find 2 by 2 matrices (of real numbers) such that

(a) $A^2 = -I$

(b) $BC = -CB$ (not allowing $BC = 0$).

24 (a) Find a nonzero matrix A for which $A^2 = 0$.

(b) Find a matrix that has $A^2 \neq 0$ but $A^3 = 0$.

25 By experiment with $n = 2$ and $n = 3$ predict A^n for

$$A_1 = \begin{bmatrix} 2 & 1 \\ 0 & 1 \end{bmatrix} \quad \text{and} \quad A_2 = \begin{bmatrix} 1 & 1 \\ 1 & 1 \end{bmatrix} \quad \text{and} \quad A_3 = \begin{bmatrix} a & b \\ 0 & 0 \end{bmatrix}.$$

Problems 26–34 use column-row multiplication and block multiplication.

26 Multiply AB using columns times rows:

$$AB = \begin{bmatrix} 1 & 0 \\ 2 & 4 \\ 2 & 1 \end{bmatrix} \begin{bmatrix} 3 & 3 & 0 \\ 1 & 2 & 1 \end{bmatrix} = \begin{bmatrix} 1 \\ 2 \\ 2 \end{bmatrix} \begin{bmatrix} 3 & 3 & 0 \end{bmatrix} + \underline{\qquad} = \underline{\qquad}.$$

27 The product of upper triangular matrices is upper triangular:

$$AB = \begin{bmatrix} x & x & x \\ 0 & x & x \\ 0 & 0 & x \end{bmatrix} \begin{bmatrix} x & x & x \\ 0 & x & x \\ 0 & 0 & x \end{bmatrix} = \begin{bmatrix} & & \\ 0 & & \\ 0 & 0 & \end{bmatrix}.$$

Row times column is dot product (Row 2 of A) $\cdot$ (column 1 of B) $= 0$. Which other dot products give zeros?

Column times row is full matrix Draw x's and 0's in (column 2 of A) (row 2 of B) and in (column 3 of A) (row 3 of B).

28 Draw the cuts in A (2 by 3) and B (3 by 4) and AB to show how each of the four multiplication rules is really a block multiplication:

(1) Matrix A times columns of B.

(2) Rows of A times matrix B.

(3) Rows of A times columns of B.

(4) Columns of A times rows of B.

29 Draw the cuts in A and $\boldsymbol{x}$ to multiply $A\boldsymbol{x}$ a column at a time: x_1 times column $1 + \cdots$.

30 Which matrices E_{21} and E_{31} produce zeros in the (2, 1) and (3, 1) positions of $E_{21}A$ and $E_{31}A$?

$$A = \begin{bmatrix} 2 & 1 & 0 \\ -2 & 0 & 1 \\ 8 & 5 & 3 \end{bmatrix}.$$

Find the single matrix $E = E_{31}E_{21}$ that produces both zeros at once. Multiply EA.

31 Block multiplication says in Example 2.9 that

$$EA = \begin{bmatrix} 1 & \boldsymbol{0} \\ -\boldsymbol{c}/a & I \end{bmatrix} \begin{bmatrix} a & \boldsymbol{b} \\ \boldsymbol{c} & D \end{bmatrix} = \begin{bmatrix} a & \boldsymbol{b} \\ \boldsymbol{0} & D - \boldsymbol{cb}/a \end{bmatrix}.$$

In Problem 30, what are $\boldsymbol{c}$ and D and what is $D - \boldsymbol{cb}/a$?

32 With $i^2 = -1$, the product of $(A + iB)$ and $(\boldsymbol{x} + i\boldsymbol{y})$ is $A\boldsymbol{x} + iB\boldsymbol{x} + iA\boldsymbol{y} - B\boldsymbol{y}$. Use blocks to separate the real part without i from the imaginary part that multiplies i:

$$\begin{bmatrix} A & -B \\ \cdots & \cdots \end{bmatrix} \begin{bmatrix} \boldsymbol{x} \\ \boldsymbol{y} \end{bmatrix} = \begin{bmatrix} A\boldsymbol{x} - B\boldsymbol{y} \\ \cdots \end{bmatrix} \begin{matrix} \text{real part} \\ \text{imaginary part} \end{matrix}$$

33 Each complex multiplication $(a + ib)(c + id)$ seems to need ac, bd, ad, bc. But notice that $ad+bc = (a+b)(c+d)-ac-bd$; those 3 multiplications are enough. For matrices A, B, C, D this **"3M method"** becomes $3n^3$ versus $4n^3$. Check additions:

(a) How many additions for a dot product? For n by n matrix multiplication?

(b) (Old method) How many additions for $R = AC - BD$ and $S = AD + BC$?

(c) (**3M method**) How many additions for R and $S = (A+B)(C+D)-AC-BD$? For $n = 2$ the multiplications are 24 versus 32. Additions are 32 versus 24. For $n > 2$ the 3M method clearly wins.

34 Which cut is *necessary* in B? Which two cuts are optional in B?

$$AB = \left[\begin{array}{cc|c} x & x & x \\ x & x & x \\ x & x & x \end{array}\right]\begin{bmatrix} x & x & x \\ x & x & x \\ x & x & x \end{bmatrix}. \quad \text{(Keep } A \text{ as is.)}$$

35 Suppose you solve $A\boldsymbol{x} = \boldsymbol{b}$ for three special right sides $\boldsymbol{b}$:

$$\boldsymbol{b} = \begin{bmatrix} 1 \\ 0 \\ 0 \end{bmatrix} \quad \text{and} \quad \begin{bmatrix} 0 \\ 1 \\ 0 \end{bmatrix} \quad \text{and} \quad \begin{bmatrix} 0 \\ 0 \\ 1 \end{bmatrix}.$$

If the three solutions are the columns of a matrix X, what is A times X?

36 If the three solutions in Question 35 are $\boldsymbol{x}_1 = (1, 1, 1)$ and $\boldsymbol{x}_2 = (0, 1, 1)$ and $\boldsymbol{x}_3 = (0, 0, 1)$, what is the solution when $\boldsymbol{b} = (3, 5, 8)$? Challenge problem: What is the matrix A?

2.4 Inverse Matrices

Suppose A is a square matrix. We look for a matrix A^{-1} of the same size, such that A^{-1} ***times*** A ***equals*** I. Whatever A does, A^{-1} undoes. Their product is the identity matrix—which does nothing. But this ***"inverse matrix"*** might not exist.

What a matrix mostly does is to multiply a vector $\boldsymbol{x}$. That gives $A\boldsymbol{x}$. Multiplying $A\boldsymbol{x}$ by A^{-1} gives $A^{-1}(A\boldsymbol{x})$, which is also $(A^{-1}A)\boldsymbol{x}$. We are back to $I\boldsymbol{x} = \boldsymbol{x}$. The product $A^{-1}A$ is like multiplying by a number and then dividing by that number. An ordinary number has an inverse if it is not zero—matrices are more complicated and more interesting. The matrix A^{-1} is called "A inverse."

Not all matrices have inverses. This is the first question we ask about a square matrix—is it invertible? We don't mean that our first calculation is immediately to find A^{-1}. In most problems we never compute it! The inverse exists if and only if elimination produces n pivots. (This is proved later. We must allow row exchanges to help find those pivots.)

Elimination solves $Ax = b$ without knowing or using A^{-1}. But the idea of invertibility is fundamental.

Definition The matrix A is ***invertible*** if there exists a matrix A^{-1} such that

$$A^{-1}A = I \quad \text{and} \quad AA^{-1} = I. \tag{2.10}$$

Note 1 The matrix A cannot have two different inverses. Suppose $BA = I$ and also $AC = I$. Then

$$B(AC) = (BA)C \quad \text{gives} \quad BI = IC \quad \text{or} \quad B = C.$$

This shows that a left-inverse B (multiplying from the left) and a right-inverse C (multiplying from the right) must be the same matrix.

Note 2 If A is invertible then the one and only solution to $Ax = b$ is $x = A^{-1}b$:

$$\textit{Multiply} \quad \boldsymbol{Ax = b} \quad \textit{by} \quad \boldsymbol{A^{-1}} \quad \textit{to find} \quad \boldsymbol{x = A^{-1}Ax = A^{-1}b}.$$

Note 3 (Important) Suppose there is a nonzero vector x such that $Ax = \mathbf{0}$. ***Then A cannot have an inverse.*** No matrix can bring $\mathbf{0}$ back to x. When A is invertible, multiply $Ax = \mathbf{0}$ by A^{-1} to find the one and only solution $x = \mathbf{0}$.

Note 4 A 2 by 2 matrix is invertible if and only if $ad - bc$ is not zero:

$$\begin{bmatrix} a & b \\ c & d \end{bmatrix}^{-1} = \frac{1}{ad - bc}\begin{bmatrix} d & -b \\ -c & a \end{bmatrix}. \tag{2.11}$$

This number $ad - bc$ is the *determinant* of A. A matrix is invertible if its determinant is not zero (Chapter 5). The test for n pivots is usually decided before the determinant appears.

Note 5 A diagonal matrix is invertible when none of its diagonal entries are zero:

$$\text{If} \quad A = \begin{bmatrix} d_1 & & \\ & \ddots & \\ & & d_n \end{bmatrix} \quad \text{then} \quad A^{-1} = \begin{bmatrix} 1/d_1 & & \\ & \ddots & \\ & & 1/d_n \end{bmatrix}.$$

Note 6 The 2 by 2 matrix $A = \left[\begin{smallmatrix} 1 & 2 \\ 1 & 2 \end{smallmatrix}\right]$ is not invertible. It fails the test in Note 4, because $ad - bc$ equals $2 - 2 = 0$. It fails the test in Note 3, because $Ax = \mathbf{0}$ when $x = (2, -1)$. It also fails to have two pivots. Elimination turns the second row of A into a zero row.

The Inverse of a Product

For two nonzero numbers a and b, the sum $a+b$ might or might not be invertible. The numbers $a=3$ and $b=-3$ have inverses $\frac{1}{3}$ and $-\frac{1}{3}$. Their sum $a+b=0$ has no inverse. But the product $ab=-9$ does have an inverse, which is $\frac{1}{3}$ times $-\frac{1}{3}$.

For two matrices A and B, the situation is similar. It is hard to say much about the invertibility of $A+B$. But the *product* AB has an inverse, whenever the factors A and B are separately invertible. The important point is that A^{-1} and B^{-1} come in ***reverse order***:

2H If A and B are invertible then so is AB. The inverse of AB is

$$(AB)^{-1}=B^{-1}A^{-1}. \tag{2.12}$$

To see why the order is reversed, start with AB. Multiplying on the left by A^{-1} leaves B. Multiplying by B^{-1} leaves the identity matrix I:

$$B^{-1}A^{-1}AB \quad \text{equals} \quad B^{-1}IB=B^{-1}B=I.$$

Similarly AB times $B^{-1}A^{-1}$ equals $AIA^{-1}=AA^{-1}=I$. This illustrates a basic rule of mathematics: Inverses come in reverse order. It is also common sense: If you put on socks and then shoes, the first to be taken off are the ______. The same idea applies to three or more matrices:

$$(ABC)^{-1}=C^{-1}B^{-1}A^{-1}. \tag{2.13}$$

Example 2.10 If E is the elementary matrix that subtracts 5 times row 1 from row 2, then E^{-1} *adds* 5 times row 1 to row 2:

$$E=\begin{bmatrix}1&0&0\\-5&1&0\\0&0&1\end{bmatrix} \quad \text{and} \quad E^{-1}=\begin{bmatrix}1&0&0\\5&1&0\\0&0&1\end{bmatrix}.$$

Multiply EE^{-1} to get the identity matrix I. Also multiply $E^{-1}E$ to get I. We are adding and subtracting the same 5 times row 1. Whether we add and then subtract (this is EE^{-1}) or subtract first and then add (this is $E^{-1}E$), we are back at the start.

Note For square matrices, an inverse on one side is automatically an inverse on the other side. If $AB=I$ then automatically $BA=I$. In that case B is A^{-1}. This is very useful to know but we are not ready to prove it.

Example 2.11 Suppose F subtracts 2 times row 2 from row 3, and F^{-1} adds it back:

$$F=\begin{bmatrix}1&0&0\\0&1&0\\0&-2&1\end{bmatrix} \quad \text{and} \quad F^{-1}=\begin{bmatrix}1&0&0\\0&1&0\\0&2&1\end{bmatrix}.$$

Again $FF^{-1} = I$ and also $F^{-1}F = I$. Now multiply F by the matrix E in Example 2.10 to find FE. Also multiply E^{-1} times F^{-1} to find $(FE)^{-1}$. Notice the order of those inverses!

$$FE = \begin{bmatrix} 1 & 0 & 0 \\ -5 & 1 & 0 \\ 10 & -2 & 1 \end{bmatrix} \quad \text{is inverted by} \quad E^{-1}F^{-1} = \begin{bmatrix} 1 & 0 & 0 \\ 5 & 1 & 0 \\ 0 & 2 & 1 \end{bmatrix}. \tag{2.14}$$

This is strange but correct. The product FE contains "10" but its inverse doesn't. You can check that FE times $E^{-1}F^{-1}$ gives I.

There must be a reason why 10 appears in FE but not in its inverse. E subtracts 5 times row 1 from row 2. Then F subtracts 2 times the *new* row 2 (which contains -5 times row 1) from row 3. ***In this order FE, row 3 feels an effect from row 1.*** The effect is to subtract -10 times row 1 from row 3—this accounts for the 10.

In the order $E^{-1}F^{-1}$, that effect does not happen. First F^{-1} adds 2 times row 2 to row 3. After that, E^{-1} adds 5 times row 1 to row 2. There is no 10. The example makes two points:

1 Usually we cannot find A^{-1} from a quick look at A.

2 When elementary matrices come in their normal order GFE, we *can* find the product of inverses $E^{-1}F^{-1}G^{-1}$ quickly. The multipliers fall into place below the diagonal.

This special property of $E^{-1}F^{-1}G^{-1}$ will be useful in the next section. We will explain it again, more completely. In this section our job is A^{-1}, and we expect some serious work to compute it. Here is a way to organize that computation.

The Calculation of A^{-1} by Gauss-Jordan Elimination

It was hinted earlier that A^{-1} might not be explicitly needed. The equation $A\boldsymbol{x} = \boldsymbol{b}$ is solved by $\boldsymbol{x} = A^{-1}\boldsymbol{b}$. But it is not necessary or efficient to compute A^{-1} and multiply it times $\boldsymbol{b}$. *Elimination goes directly to* $\boldsymbol{x}$. In fact elimination is also the way to calculate A^{-1}, as we now show.

The idea is to solve $AA^{-1} = I$ *a column at a time.* A multiplies the first column of A^{-1} (call that $\boldsymbol{x}_1$) to give the first column of I (call that $\boldsymbol{e}_1$). This is our equation $A\boldsymbol{x}_1 = \boldsymbol{e}_1$. Each column of A^{-1} is multiplied by A to produce a column of I:

$$AA^{-1} = A\begin{bmatrix} \boldsymbol{x}_1 & \boldsymbol{x}_2 & \boldsymbol{x}_3 \end{bmatrix} = \begin{bmatrix} \boldsymbol{e}_1 & \boldsymbol{e}_2 & \boldsymbol{e}_3 \end{bmatrix} = I. \tag{2.15}$$

To invert a 3 by 3 matrix A, we have to solve three systems of equations: $A\boldsymbol{x}_1 = \boldsymbol{e}_1$ and $A\boldsymbol{x}_2 = \boldsymbol{e}_2$ and $A\boldsymbol{x}_3 = \boldsymbol{e}_3$. Then $\boldsymbol{x}_1, \boldsymbol{x}_2, \boldsymbol{x}_3$ are the columns of A^{-1}.

Note This already shows why computing A^{-1} is expensive. We must solve n equations for its n columns. To solve $A\boldsymbol{x} = \boldsymbol{b}$ directly, we deal only with *one* column.

In defense of A^{-1}, we want to say that its cost is not n times the cost of one system $A\boldsymbol{x} = \boldsymbol{b}$. Surprisingly, the cost for n columns is only multiplied by 3. This saving is because the n equations $A\boldsymbol{x}_i = \boldsymbol{e}_i$ all involve the same matrix A. When we have many different right sides, elimination only has to be done once on A. Working with the right sides is relatively cheap. The complete A^{-1} needs n^3 elimination steps, whereas solving for a single $\boldsymbol{x}$ needs $n^3/3$. The next section calculates these costs.

The ***Gauss-Jordan method*** computes A^{-1} by solving all n equations together. Usually the "augmented matrix" has one extra column $\boldsymbol{b}$, from the right side of the equations. Now we have three right sides $\boldsymbol{e}_1, \boldsymbol{e}_2, \boldsymbol{e}_3$ (when A is 3 by 3). They are the columns of I, so the augmented matrix is really just $[\,A \quad I\,]$. Here is a worked-out example when A has 2's on the main diagonal and -1's next to the 2's:

$$\begin{bmatrix} A & \boldsymbol{e}_1 & \boldsymbol{e}_2 & \boldsymbol{e}_3 \end{bmatrix} = \begin{bmatrix} 2 & -1 & 0 & 1 & 0 & 0 \\ -1 & 2 & -1 & 0 & 1 & 0 \\ 0 & -1 & 2 & 0 & 0 & 1 \end{bmatrix}$$

$$\rightarrow \begin{bmatrix} 2 & -1 & 0 & 1 & 0 & 0 \\ 0 & \frac{3}{2} & -1 & \frac{1}{2} & 1 & 0 \\ 0 & -1 & 2 & 0 & 0 & 1 \end{bmatrix} \qquad (\tfrac{1}{2}\text{row } 1 + \text{row } 2)$$

$$\rightarrow \begin{bmatrix} 2 & -1 & 0 & 1 & 0 & 0 \\ 0 & \frac{3}{2} & -1 & \frac{1}{2} & 1 & 0 \\ 0 & 0 & \frac{4}{3} & \frac{1}{3} & \frac{2}{3} & 1 \end{bmatrix} \qquad (\tfrac{2}{3}\text{ row } 2 + \text{row } 3)$$

We are halfway. The matrix in the first three columns is now U (upper triangular). The pivots $2, \frac{3}{2}, \frac{4}{3}$ are on its diagonal. Gauss would finish by back substitution on each of the three equations. The contribution of Jordan is to continue with elimination, all the way to the ***"reduced echelon form"*** R. Rows are added to rows above them, to produce ***zeros above the pivots***:

$$\rightarrow \begin{bmatrix} 2 & -1 & 0 & 1 & 0 & 0 \\ 0 & \frac{3}{2} & 0 & \frac{3}{4} & \frac{3}{2} & \frac{3}{4} \\ 0 & 0 & \frac{4}{3} & \frac{1}{3} & \frac{2}{3} & 1 \end{bmatrix} \qquad (\tfrac{3}{4}\text{ row } 3 + \text{row } 2)$$

$$\rightarrow \begin{bmatrix} 2 & 0 & 0 & \frac{3}{2} & 1 & \frac{1}{2} \\ 0 & \frac{3}{2} & 0 & \frac{3}{4} & \frac{3}{2} & \frac{3}{4} \\ 0 & 0 & \frac{4}{3} & \frac{1}{3} & \frac{2}{3} & 1 \end{bmatrix} \qquad (\tfrac{2}{3}\text{ row } 2 + \text{row } 1)$$

Now divide each row by its pivot. (That can be done earlier.) The new pivots are 1. The result is to produce I in the first half of the matrix. (This reduced matrix is $R = I$ because A is invertible. More later.) ***The columns of A^{-1} are in the second half.***

$$\begin{matrix} (\text{divide by } 2) \\ (\text{divide by } \frac{3}{2}) \\ (\text{divide by } \frac{4}{3}) \end{matrix} \qquad \begin{bmatrix} 1 & 0 & 0 & \frac{3}{4} & \frac{1}{2} & \frac{1}{4} \\ 0 & 1 & 0 & \frac{1}{2} & 1 & \frac{1}{2} \\ 0 & 0 & 1 & \frac{1}{4} & \frac{1}{2} & \frac{3}{4} \end{bmatrix} = \begin{bmatrix} I & \boldsymbol{x}_1 & \boldsymbol{x}_2 & \boldsymbol{x}_3 \end{bmatrix}.$$

Starting from the 3 by 6 matrix $[\,A \quad I\,]$, we have reached $[\,I \quad A^{-1}\,]$. Here is the whole Gauss-Jordan process on one line:

$$\textit{Multiply} \quad \begin{bmatrix} A & I \end{bmatrix} \quad \textit{by} \quad A^{-1} \quad \textit{to get} \quad \begin{bmatrix} I & A^{-1} \end{bmatrix}.$$

The elimination steps gradually multiply by A^{-1}. They end up with the actual inverse matrix. Again, we probably don't need A^{-1} at all. But for small matrices, it can be worthwhile to know the inverse. We add three observations about this particular A^{-1} because it is an important example:

1 A is *symmetric* across its main diagonal. So is A^{-1}.

2 The product of the pivots is $2(\frac{3}{2})(\frac{4}{3}) = 4$. This number 4 is the *determinant* of A. Chapter 5 will show why 4 is also the denominator in A^{-1}:

$$A^{-1} = \frac{1}{4}\begin{bmatrix} 3 & 2 & 1 \\ 2 & 4 & 2 \\ 1 & 2 & 3 \end{bmatrix}. \tag{2.16}$$

3 The matrix A is *tridiagonal* (three nonzero diagonals). But A^{-1} is a full matrix with no zeros. That is another reason we don't often compute A^{-1}.

Example 2.12 Find A^{-1} by Gauss-Jordan elimination with $A = \left[\begin{smallmatrix} 2 & 3 \\ 4 & 7 \end{smallmatrix}\right]$.

$$\begin{bmatrix} A & I \end{bmatrix} = \begin{bmatrix} 2 & 3 & 1 & 0 \\ 4 & 7 & 0 & 1 \end{bmatrix} \rightarrow \begin{bmatrix} 2 & 3 & 1 & 0 \\ 0 & 1 & -2 & 1 \end{bmatrix}$$

$$\rightarrow \begin{bmatrix} 2 & 0 & 7 & -3 \\ 0 & 1 & -2 & 1 \end{bmatrix} \rightarrow \begin{bmatrix} 1 & 0 & \frac{7}{2} & -\frac{3}{2} \\ 0 & 1 & -2 & 1 \end{bmatrix}.$$

The last matrix shows A^{-1}. The reduced echelon form of $[\,A \quad I\,]$ is $[\,I \quad A^{-1}\,]$. The code for X = inverse(A) has only three important lines! The teaching code uses ref where MATLAB itself would use rref (or go directly to inv):

```
I = eye(n,n);
R = ref([A I]);
X = R(:,n + 1:n + n)
```

The last line discards columns 1 to n, in the left half of R. It picks out $X = A^{-1}$ from the right half. Of course A must be invertible, or the left half will not be I.

Singular versus Invertible

We come back to the central question. Which matrices have inverses? At the start of this section we proposed the pivot test: ***The inverse exists exactly when A has a full set of n pivots.*** (Row exchanges are allowed!) Now we can prove that fact, by looking carefully at elimination:

1 With n pivots, elimination solves all the equations $A\boldsymbol{x}_i = \boldsymbol{e}_i$. The solutions $\boldsymbol{x}_i$ go into A^{-1}. Then $AA^{-1} = I$ and A^{-1} is at least a right-inverse.

2 Gauss-Jordan elimination is really a long sequence of matrix multiplications:

$$(D^{-1}\cdots E\cdots P\cdots E)A = I.$$

The factor D^{-1} divides by the pivots. The matrices E produce zeros, the matrices P exchange rows (details in Section 2.6). The product matrix in parentheses is evidently a left-inverse that gives $A^{-1}A = I$. It equals the right-inverse by Note 1. *With a full set of pivots A is invertible.*

The converse is also true. *If A^{-1} exists, then A must have a full set of n pivots.* Again the argument has two steps:

1 If A has a whole row of zeros, the inverse cannot exist. Every matrix product AB will have that zero row, and cannot be equal to I.

2 Suppose a column has no pivot. At that point the matrix can look like

$$\begin{bmatrix} d_1 & x & x & x \\ 0 & d_2 & x & x \\ 0 & 0 & 0 & d_3 \\ 0 & 0 & 0 & x \end{bmatrix}.$$

In case $d_3 = 0$, we have a row of zeros. Otherwise subtract a multiple of row 3 so that row 4 is all zero. We have reached a matrix with no inverse. Since each elimination step is invertible, the original A was not invertible. ***Without pivots A^{-1} fails.***

2I Elimination gives a complete test for A^{-1} to exist. There must be n pivots.

Example 2.13 If L is lower triangular with 1's on the diagonal, so is L^{-1}.

Use the Gauss-Jordan method to construct L^{-1}. Start by subtracting multiples of pivot rows from rows *below*. Normally this gets us halfway to the inverse, but for L it gets us all the way:

$$\begin{bmatrix} L & I \end{bmatrix} = \begin{bmatrix} 1 & 0 & 0 & 1 & 0 & 0 \\ a & 1 & 0 & 0 & 1 & 0 \\ b & c & 1 & 0 & 0 & 1 \end{bmatrix}$$

$$\begin{matrix} \\ \to \\ \to \end{matrix} \begin{bmatrix} 1 & 0 & 0 & 1 & 0 & 0 \\ 0 & 1 & 0 & -a & 1 & 0 \\ 0 & c & 1 & -b & 0 & 1 \end{bmatrix} \quad \begin{matrix} \\ (a \text{ times row 1 from row 2}) \\ (b \text{ times row 1 from row 3}) \end{matrix}$$

$$\begin{matrix} \\ \\ \to \end{matrix} \begin{bmatrix} 1 & 0 & 0 & 1 & 0 & 0 \\ 0 & 1 & 0 & -a & 1 & 0 \\ 0 & 0 & 1 & ac-b & -c & 1 \end{bmatrix} = \begin{bmatrix} I & L^{-1} \end{bmatrix}.$$

When L goes to I by elimination, I goes to L^{-1}. In other words, the product of elimination matrices $E_{32}E_{31}E_{21}$ is L^{-1}. All the pivots are 1's (a full set) and nothing appears above the diagonal.

Problem Set 2.4

1 Find the inverses (directly or from the 2 by 2 formula) of A, B, C:

$$A = \begin{bmatrix} 0 & 3 \\ 4 & 0 \end{bmatrix} \quad \text{and} \quad B = \begin{bmatrix} 2 & 0 \\ 4 & 2 \end{bmatrix} \quad \text{and} \quad C = \begin{bmatrix} 3 & 4 \\ 5 & 7 \end{bmatrix}.$$

2 For these "permutation matrices" find P^{-1} by trial and error (with 1's and 0's):

$$P = \begin{bmatrix} 0 & 0 & 1 \\ 0 & 1 & 0 \\ 1 & 0 & 0 \end{bmatrix} \quad \text{and} \quad P = \begin{bmatrix} 0 & 1 & 0 \\ 0 & 0 & 1 \\ 1 & 0 & 0 \end{bmatrix}.$$

3 Solve for the columns of $A^{-1} = \left[\begin{smallmatrix} a & b \\ c & d \end{smallmatrix}\right]$:

$$\begin{bmatrix} 10 & 20 \\ 20 & 50 \end{bmatrix} \begin{bmatrix} a \\ c \end{bmatrix} = \begin{bmatrix} 1 \\ 0 \end{bmatrix} \quad \text{and} \quad \begin{bmatrix} 10 & 20 \\ 20 & 50 \end{bmatrix} \begin{bmatrix} b \\ d \end{bmatrix} = \begin{bmatrix} 0 \\ 1 \end{bmatrix}.$$

4 Show that $\left[\begin{smallmatrix} 1 & 2 \\ 3 & 6 \end{smallmatrix}\right]$ has no inverse by trying to solve for the columns (a, c) and (b, d):

$$\begin{bmatrix} 1 & 2 \\ 3 & 6 \end{bmatrix} \begin{bmatrix} a & b \\ c & d \end{bmatrix} = \begin{bmatrix} 1 & 0 \\ 0 & 1 \end{bmatrix}.$$

5 Find three 2 by 2 matrices (not $A = I$) that are their own inverses: $A^2 = I$.

6 (a) If A is invertible and $AB = AC$, prove quickly that $B = C$.

(b) If $A = \left[\begin{smallmatrix} 1 & 1 \\ 1 & 1 \end{smallmatrix}\right]$, find two matrices $B \neq C$ such that $AB = AC$.

7 (Important) If the 3 by 3 matrix A has row 1 + row 2 = row 3, show that it is not invertible:

(a) Explain why $A\boldsymbol{x} = (1, 0, 0)$ cannot have a solution.

(b) Which right sides (b_1, b_2, b_3) might allow a solution to $A\boldsymbol{x} = \boldsymbol{b}$?

(c) What happens to row 3 in elimination?

8 Suppose A is invertible and you exchange its first two rows. Is the new matrix B invertible and how would you find B^{-1} from A^{-1}?

9 Find the inverses (in any legal way) of

$$A = \begin{bmatrix} 0 & 0 & 0 & 2 \\ 0 & 0 & 3 & 0 \\ 0 & 4 & 0 & 0 \\ 5 & 0 & 0 & 0 \end{bmatrix} \quad \text{and} \quad B = \begin{bmatrix} 3 & 2 & 0 & 0 \\ 4 & 3 & 0 & 0 \\ 0 & 0 & 6 & 5 \\ 0 & 0 & 7 & 6 \end{bmatrix}.$$

10 (a) Find invertible matrices A and B such that $A + B$ is not invertible.

(b) Find singular matrices A and B such that $A + B$ is invertible.

11 If the product $C = AB$ is invertible (A and B are square), then A itself is invertible. Find a formula for A^{-1} that involves C^{-1} and B.

12 If the product $M = ABC$ of three square matrices is invertible, then B is invertible. (So are A and C.) Find a formula for B^{-1} that involves M^{-1} and A and C.

13 If A is invertible and you add row 1 to row 2 to get B, how do you find B^{-1} from A^{-1}?

$$\text{The inverse of} \quad B = \begin{bmatrix} 1 & 0 \\ 1 & 1 \end{bmatrix} \begin{bmatrix} & A & \end{bmatrix} \quad \text{is} \quad \underline{\qquad}.$$

14 Prove that a matrix with a column of zeros cannot have an inverse.

15 Multiply $\left[\begin{smallmatrix} a & b \\ c & d \end{smallmatrix}\right]$ times $\left[\begin{smallmatrix} d & -b \\ -c & a \end{smallmatrix}\right]$. What is the inverse of each matrix if $ad \neq bc$?

16 (a) What single matrix E has the same effect as these three steps? Subtract row 1 from row 2, subtract row 1 from row 3, then subtract row 2 from row 3.

(b) What single matrix L has the same effect as these three steps? Add row 2 to row 3, add row 1 to row 3, then add row 1 to row 2.

17 If the 3 by 3 matrix A has column 1 + column 2 = column 3, show that it is not invertible:

(a) Find a nonzero solution to $A\boldsymbol{x} = \mathbf{0}$.

(b) Does elimination keep column 1 + column 2 = column 3? Explain why there is no third pivot.

18 If B is the inverse of A^2, show that AB is the inverse of A.

19 Find the numbers a and b that give the correct inverse:

$$\begin{bmatrix} 4 & -1 & -1 & -1 \\ -1 & 4 & -1 & -1 \\ -1 & -1 & 4 & -1 \\ -1 & -1 & -1 & 4 \end{bmatrix}^{-1} = \begin{bmatrix} a & b & b & b \\ b & a & b & b \\ b & b & a & b \\ b & b & b & a \end{bmatrix}.$$

20 There are sixteen 2 by 2 matrices whose entries are 1's and 0's. How many of them are invertible?

Questions 21–27 are about the Gauss-Jordan method for calculating A^{-1}.

21 Change I into A^{-1} as you reduce A to I (by row operations):

$$\begin{bmatrix} A & I \end{bmatrix} = \begin{bmatrix} 1 & 3 & 1 & 0 \\ 2 & 7 & 0 & 1 \end{bmatrix} \quad \text{and} \quad \begin{bmatrix} A & I \end{bmatrix} = \begin{bmatrix} 1 & 3 & 1 & 0 \\ 3 & 8 & 0 & 1 \end{bmatrix}$$

22 Follow the 3 by 3 text example but with plus signs in A. Eliminate above and below the pivots to reduce $[\,A \quad I\,]$ to $R = [\,I \quad A^{-1}\,]$:

$$\begin{bmatrix} A & I \end{bmatrix} = \begin{bmatrix} 2 & 1 & 0 & 1 & 0 & 0 \\ 1 & 2 & 1 & 0 & 1 & 0 \\ 0 & 1 & 2 & 0 & 0 & 1 \end{bmatrix}.$$

23 Use Gauss-Jordan elimination with I next to A to solve $AA^{-1} = I$:

$$\begin{bmatrix} 1 & a & b \\ 0 & 1 & c \\ 0 & 0 & 1 \end{bmatrix} \begin{bmatrix} \\ \boldsymbol{x}_1 & \boldsymbol{x}_2 & \boldsymbol{x}_3 \\ \\ \end{bmatrix} = \begin{bmatrix} 1 & 0 & 0 \\ 0 & 1 & 0 \\ 0 & 0 & 1 \end{bmatrix}.$$

24 Find A^{-1} (*if it exists*) by Gauss-Jordan elimination on $[\,A \quad I\,]$:

$$A = \begin{bmatrix} 2 & 1 & 1 \\ 1 & 2 & 1 \\ 1 & 1 & 2 \end{bmatrix} \quad \text{and} \quad A = \begin{bmatrix} 2 & -1 & -1 \\ -1 & 2 & -1 \\ -1 & -1 & 2 \end{bmatrix}.$$

25 What three matrices E_{21} and E_{12} and D^{-1} reduce $A = \left[\begin{smallmatrix} 1 & 2 \\ 2 & 6 \end{smallmatrix}\right]$ to the identity matrix? Multiply $D^{-1}E_{12}E_{21}$ to find A^{-1}.

26 Invert these matrices by the Gauss-Jordan method starting with $[\,A \quad I\,]$:

$$A = \begin{bmatrix} 1 & 0 & 0 \\ 2 & 1 & 3 \\ 0 & 0 & 1 \end{bmatrix} \quad \text{and} \quad A = \begin{bmatrix} 1 & 1 & 1 \\ 1 & 2 & 2 \\ 1 & 2 & 3 \end{bmatrix}.$$

27 Exchange rows and continue with Gauss-Jordan to find A^{-1}:

$$\begin{bmatrix} A & I \end{bmatrix} = \begin{bmatrix} 0 & 2 & 1 & 0 \\ 2 & 2 & 0 & 1 \end{bmatrix}.$$

28 True or false (with a counterexample if false):

(a) A 4 by 4 matrix with _____ zeros is not invertible.

(b) Every matrix with 1's down the main diagonal is invertible.

(c) If A is invertible then A^{-1} is invertible.

(d) If A is invertible then A^2 is invertible.

29 For which numbers c is this matrix not invertible, and why not?

$$A = \begin{bmatrix} 2 & c & c \\ c & c & c \\ 8 & 7 & c \end{bmatrix}.$$

30 Prove that this matrix is invertible if $a \neq 0$ and $a \neq b$:

$$A = \begin{bmatrix} a & b & b \\ a & a & b \\ a & a & a \end{bmatrix}.$$

31 This matrix has a remarkable inverse. Find A^{-1} and guess the 5 by 5 inverse and multiply AA^{-1} to confirm:

$$A = \begin{bmatrix} 1 & -1 & 1 & -1 \\ 0 & 1 & -1 & 1 \\ 0 & 0 & 1 & -1 \\ 0 & 0 & 0 & 1 \end{bmatrix}.$$

32 Use the inverses in Question 31 to solve $A\boldsymbol{x} = (1, 1, 1, 1)$ and $A\boldsymbol{x} = (1, 1, 1, 1, 1)$.

33 The Hilbert matrices have $a_{ij} = 1/(i + j - 1)$. Ask MATLAB for the exact inverse invhilb(6). Then ask for inv(hilb(6)). How can these be different, when the computer never makes mistakes?

34 Find the inverses (assuming they exist) of these block matrices:

$$\begin{bmatrix} I & 0 \\ C & I \end{bmatrix} \quad \begin{bmatrix} A & 0 \\ C & D \end{bmatrix} \quad \begin{bmatrix} 0 & I \\ I & D \end{bmatrix}.$$

2.5 Elimination = Factorization: $A = LU$

Students often say that mathematics courses are too theoretical. Well, not this section. It is almost purely practical. The goal is to describe Gaussian elimination in the most useful way. Many key ideas of linear algebra, when you look at them closely, are really *factorizations* of a matrix. The original matrix A becomes the product of two or three special matrices. The first factorization—also the most important in practice—comes now from elimination. ***The factors are triangular matrices.***

Start with a 2 by 2 example. There is only one nonzero to eliminate from A (it is the number 6). Subtract 3 times row 1 from row 2. That step is E_{21} in the forward direction (a subtraction). The return step from U to A is E_{21}^{-1} (an addition):

$$E_{21}A = \begin{bmatrix} 1 & 0 \\ -3 & 1 \end{bmatrix} \begin{bmatrix} 2 & 1 \\ 6 & 8 \end{bmatrix} = \begin{bmatrix} 2 & 1 \\ 0 & 5 \end{bmatrix} = U$$

$$E_{21}^{-1}U = \begin{bmatrix} 1 & 0 \\ 3 & 1 \end{bmatrix} \begin{bmatrix} 2 & 1 \\ 0 & 5 \end{bmatrix} = \begin{bmatrix} 2 & 1 \\ 6 & 8 \end{bmatrix} = A.$$

The second line is our factorization. Instead of $A = E_{21}^{-1}U$ we write $A = LU$. Move now to larger matrices with many E's. ***Then L will include all their inverses.***

Each step from A to U is a multiplication by a simple matrix. The elimination matrices E_{ij} produce zeros below the pivots. (E_{ij} subtracts a multiple of row j from row i.) The row exchange matrices P_{ij} move nonzeros up to the pivot position. (P_{ij} exchanges rows i and j. It comes in the next section.) A long sequence of E's and P's multiplies A to produce U. When we put each E^{-1} and P^{-1} on the other side of the equation, they multiply U and bring back A:

$$A = (E^{-1} \cdots P^{-1} \cdots E^{-1})U. \tag{2.17}$$

This is a factorization, but with too many factors. The matrices E^{-1} and P^{-1} are too simple. The good factorization has a single matrix L to account for all the separate factors E_{ij}^{-1}. Another matrix P accounts for all the row exchanges. Then A is built out of L, P, and U.

To keep this clear, we want to start with the most frequent case—when A is invertible and $P = I$. ***No row exchanges are involved.*** If A is 3 by 3, we multiply it by E_{21} and then by E_{31} and E_{32}. That produces zero in the (2, 1) and (3, 1) and (3, 2) positions—all below the diagonal. We end with U. Now move those E's onto the other side, *where their inverses multiply* U:

$$(E_{32}E_{31}E_{21})A = U \quad \text{becomes} \quad A = (E_{21}^{-1}E_{31}^{-1}E_{32}^{-1})U \quad \text{which is} \quad A = LU. \tag{2.18}$$

The inverses go in opposite order, as they must. That product of three inverses is L. ***We have reached*** $\boldsymbol{A = LU}$. Now we stop to understand it.

First point Every matrix E_{ij}^{-1} is *lower triangular* with 1's on the diagonal. Its off-diagonal entry is $+l$, to undo the subtraction in E_{ij} with $-l$. These numbers multiply the pivot row j to change row i. Subtraction goes from A toward U, and addition brings back A. The main diagonals of E and E^{-1} contain all 1's.

Second point The product of several E's is still lower triangular. Equation (2.18) shows a lower triangular matrix (in parentheses) multiplying A to get U. It also shows a lower triangular matrix (the product of E_{ij}^{-1}) multiplying U to bring back A. ***This product of inverses is L.***

There are two good reasons for working with the inverses. One reason is that we want to factor A, not U. By choosing the "inverse form" of equation (2.18), we get $A = LU$. The second reason is that we get something extra, almost more than we deserve. This is the third point, which shows that L is exactly right.

Third point Each multiplier l_{ij} goes directly into its i, j position—*unchanged*—in the product L. Usually a matrix multiplication will mix up and combine all the numbers. Here that doesn't happen. The order is right for the inverse matrices, to keep the l's separate. We checked that directly for the two matrices E^{-1} and F^{-1} in the last section. The explanation below is more general.

Since each E^{-1} has 1's down its diagonal, the final good point is that L does too.

2J ($\boldsymbol{A = LU}$) Elimination without row exchanges factors A into LU. The upper triangular U has the pivots on its diagonal. The lower triangular L has 1's on its diagonal, with the multipliers l_{ij} in position below the diagonal.

Example 2.14 For the matrix A with $-1, 2, -1$ on its diagonals, elimination subtracts $-\frac{1}{2}$ times row 1 from row 2. Then $l_{21} = -\frac{1}{2}$. The last step subtracts $-\frac{2}{3}$ times row 2 from row 3, to reach U:

$$A = \begin{bmatrix} 2 & -1 & 0 \\ -1 & 2 & -1 \\ 0 & -1 & 2 \end{bmatrix} = \begin{bmatrix} 1 & 0 & 0 \\ -\frac{1}{2} & 1 & 0 \\ 0 & -\frac{2}{3} & 1 \end{bmatrix} \begin{bmatrix} 2 & -1 & 0 \\ 0 & \frac{3}{2} & -1 \\ 0 & 0 & \frac{4}{3} \end{bmatrix} = LU.$$

The (3, 1) multiplier is zero because the (3, 1) entry in A is zero. No operation needed.

Example 2.15 Change the top left entry from 2 to 1. The multipliers all become -1. The pivots are all $+1$. That pattern continues when A is 4 by 4 or n by n:

$$A = \begin{bmatrix} 1 & -1 & 0 & 0 \\ -1 & 2 & -1 & 0 \\ 0 & -1 & 2 & -1 \\ 0 & 0 & -1 & 2 \end{bmatrix}$$

$$= \begin{bmatrix} 1 & & & \\ -1 & 1 & & \\ 0 & -1 & 1 & \\ 0 & 0 & -1 & 1 \end{bmatrix} \begin{bmatrix} 1 & -1 & 0 & 0 \\ & 1 & -1 & 0 \\ & & 1 & -1 \\ & & & 1 \end{bmatrix} = LU.$$

These examples are showing something extra about L and U, which is very important in practice. Assume no row exchanges. When can we predict *zeros* in L and U?

When rows of A start with zeros, so do the corresponding rows of L.

When columns of A start with zeros, so do the corresponding columns of U.

If a row below the pivot starts with zero, we don't need an elimination step. The three zeros below the diagonal in A gave the same three zeros in L. The multipliers l_{ij} were zero. That saves computer time. Similarly, zeros at the *start* of a column are not touched by elimination. They survive into the columns of U. But please realize: Zeros in the *middle* of a matrix are likely to be filled in, while elimination sweeps forward.

We now explain why L has the multipliers l_{ij} in position, with no mix-up.

The key reason why A equals LU: Ask yourself about the rows that are subtracted from lower rows. Are they the original rows of A? *No*, elimination probably changed them. Are they rows of U? *Yes*, the pivot rows never get changed again. For the third row of U, we are subtracting multiples of rows 1 and 2 of U (*not rows of* A!):

$$\text{Row 3 of } U = \text{row 3 of } A - l_{31}(\text{row 1 of } U) - l_{32}(\text{row 2 of } U). \tag{2.19}$$

Rewrite that equation:

$$\text{Row 3 of } A = l_{31}(\text{row 1 of } U) + l_{32}(\text{row 2 of } U) + 1(\text{row 3 of } U). \tag{2.20}$$

The right side shows the row $[\,l_{31} \quad l_{32} \quad 1\,]$ multiplying the matrix U. *This is row 3 of $A = LU$.* All rows look like this, whatever the size of A. As long as there are no row exchanges, we have $A = LU$.

Remark The LU factorization is "unsymmetric" in one respect. U has the pivots on its diagonal where L has 1's. This is easy to change. ***Divide U by a diagonal matrix D that contains the pivots.*** Then U equals D times a matrix with 1's on the diagonal:

$$U = \begin{bmatrix} d_1 & & & \\ & d_2 & & \\ & & \ddots & \\ & & & d_n \end{bmatrix} \begin{bmatrix} 1 & u_{12}/d_1 & u_{13}/d_1 & \vdots \\ & 1 & u_{23}/d_2 & \\ & & \ddots & \vdots \\ & & & 1 \end{bmatrix}.$$

It is convenient (but confusing) to keep the same letter U for this new upper triangular matrix. It has 1's on the diagonal (like L). Instead of the normal LU, the new form has LDU. ***Lower triangular times diagonal times upper triangular***:

The triangular factorization can be written $\boldsymbol{A = LU}$ ***or*** $\boldsymbol{A = LDU}$.

Whenever you see LDU, it is understood that U has 1's on the diagonal. *Each row is divided by its first nonzero entry—the pivot.* Then L and U are treated evenly. Here is LU and then also LDU:

$$\begin{bmatrix} 1 & 0 \\ 3 & 1 \end{bmatrix} \begin{bmatrix} 2 & 8 \\ 0 & 5 \end{bmatrix} = \begin{bmatrix} 1 & 0 \\ 3 & 1 \end{bmatrix} \begin{bmatrix} 2 & \\ & 5 \end{bmatrix} \begin{bmatrix} 1 & 4 \\ 0 & 1 \end{bmatrix}. \tag{2.21}$$

The pivots 2 and 5 went into D. Dividing the rows by 2 and 5 left the rows $[\,1 \quad 4\,]$ and $[\,0 \quad 1\,]$ in the new U. The multiplier 3 is still in L.

One Square System = Two Triangular Systems

We emphasized that L contains our memory of Gaussian elimination. It holds all the numbers that multiplied the pivot rows, before subtracting them from lower rows. When do we need this record and how do we use it?

We need to remember L as soon as there is a *right side* $\boldsymbol{b}$. The factors L and U were completely decided by the left side (the matrix A). Now work on the right side. Most computer programs for linear equations separate the problem $A\boldsymbol{x} = \boldsymbol{b}$ this way, into two different subroutines:

1 ***Factor*** (from A find L and U)

2 ***Solve*** (from L and U and $\boldsymbol{b}$ find $\boldsymbol{x}$).

Up to now, we worked on $\boldsymbol{b}$ while we were working on A. No problem with that—just augment the matrix by an extra column. The Gauss-Jordan method for A^{-1} worked on n right sides at once. But most computer codes keep the two sides separate. The memory of forward elimination is held in L and U, at no extra cost in storage. Then we process $\boldsymbol{b}$ whenever we want to. The User's Guide to LINPACK remarks that "This situation is so common and the savings are so important that no provision has been made for solving a single system with just one subroutine."

How do we process $\boldsymbol{b}$? First, apply the forward elimination steps (which are stored in L). Second, apply back substitution (using U). The first part changes $\boldsymbol{b}$ to a new right side $\boldsymbol{c}$—*we are really solving* $L\boldsymbol{c} = \boldsymbol{b}$. Then back substitution solves $U\boldsymbol{x} = \boldsymbol{c}$. The original system $A\boldsymbol{x} = \boldsymbol{b}$ is factored into ***two triangular systems***:

$$\textit{Solve} \quad L\boldsymbol{c} = \boldsymbol{b} \quad \textit{and then solve} \quad U\boldsymbol{x} = \boldsymbol{c}. \tag{2.22}$$

To see that $\boldsymbol{x}$ is the correct solution, multiply $U\boldsymbol{x} = \boldsymbol{c}$ by L. Then LU equals A and $L\boldsymbol{c}$ equals $\boldsymbol{b}$. We have $A\boldsymbol{x} = \boldsymbol{b}$.

To emphasize: There is *nothing new* about equation (2.22). This is exactly what we have done all along. Forward elimination changed $\boldsymbol{b}$ into $\boldsymbol{c}$ as it changed A into U, by working on both sides of the equation. We were really solving the triangular system $L\boldsymbol{c} = \boldsymbol{b}$ by "forward substitution." Whether done at the same time or later, this prepares the right side for back substitution. An example shows it all:

Example 2.16 Forward elimination ends at $U\boldsymbol{x} = \boldsymbol{c}$. Here $\boldsymbol{c}$ is $\begin{bmatrix} 8 \\ 5 \end{bmatrix}$:

$$\begin{aligned} 2u + 2v &= 8 \\ 4u + 9v &= 21 \end{aligned} \qquad \text{becomes} \qquad \begin{aligned} 2u + 2v &= 8 \\ 5v &= 5. \end{aligned}$$

The multiplier is 2. It goes into L. We have just solved $L\boldsymbol{c} = \boldsymbol{b}$:

$$L\boldsymbol{c} = \boldsymbol{b} \ \text{The lower triangular system} \quad \begin{bmatrix} 1 & 0 \\ 2 & 1 \end{bmatrix} \begin{bmatrix} \boldsymbol{c} \end{bmatrix} = \begin{bmatrix} 8 \\ 21 \end{bmatrix} \quad \text{gives} \quad \boldsymbol{c} = \begin{bmatrix} 8 \\ 5 \end{bmatrix}.$$

$$U\boldsymbol{x} = \boldsymbol{c} \ \text{The upper triangular system} \quad \begin{bmatrix} 2 & 2 \\ 0 & 5 \end{bmatrix} \begin{bmatrix} \boldsymbol{x} \end{bmatrix} = \begin{bmatrix} 8 \\ 5 \end{bmatrix} \quad \text{gives} \quad \boldsymbol{x} = \begin{bmatrix} 3 \\ 1 \end{bmatrix}.$$

It is satisfying that L and U replace A with no extra storage space. The triangular matrices can take the n^2 storage locations that originally held A. The l's fit in where U has zeros (below the diagonal). The whole discussion is only looking to see what elimination actually did.

The Cost of Elimination

A very practical question is cost—or computing time. Can we solve 1000 equations on a PC? What if $n = 10,000$? Large systems come up all the time in scientific computing, where a three-dimensional problem can easily lead to a million unknowns. We can let the calculation run overnight, but we can't leave it for 100 years.

It is quite easy to estimate the number of individual multiplications that go into Gaussian elimination. Concentrate on the left side (the expensive side, where A changes to U).

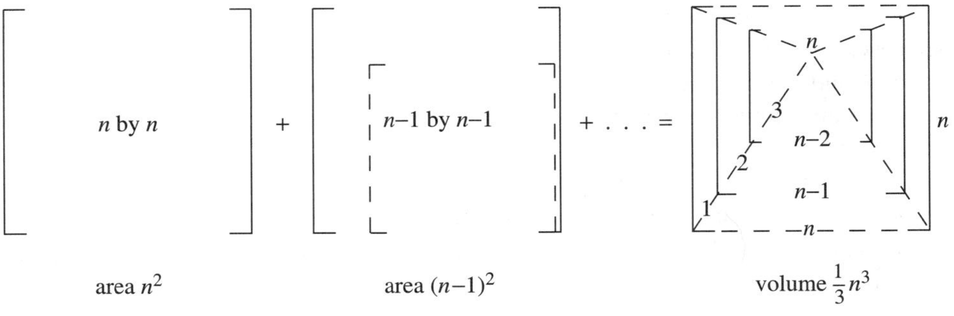

Figure 2.5 Elimination needs about $\frac{1}{3}n^3$ multiplications and subtractions. Exact count $(n^2 - n) + \cdots + (1^2 - 1) = \frac{1}{3}n^3 - \frac{1}{3}n$.

The first stage produces zeros below the first pivot—it clears out the first column. To change an entry in A requires one multiplication and one subtraction. We will count the whole first stage as n^2 multiplications and n^2 subtractions. It is actually less, $n^2 - n$, because row 1 does not change. Now column 1 is set.

The next stage clears out the second column below the second pivot. The matrix is now of size $n - 1$. Estimate it by $(n-1)^2$ multiplications and subtractions. The matrices are getting smaller as elimination goes forward. Since the matrix has n columns, the rough count to reach U is $n^2 + (n-1)^2 + \cdots + 2^2 + 1^2$.

There is an exact formula for this sum. It happens to be $\frac{1}{3}n(n + \frac{1}{2})(n + 1)$. For large n, the $\frac{1}{2}$ and the 1 are not important. The number that matters is $\frac{1}{3}n^3$, and it comes immediately from the square pyramid in Figure 2.5. The base shows the n^2 steps for stage 1. The height shows the n stages as the working matrix gets smaller. The volume is a good estimate for the total number of steps. *The volume of an n by n by n pyramid is* $\frac{1}{3}n^3$:

Elimination on A requires about $\frac{1}{3}n^3$ multiplications and $\frac{1}{3}n^3$ subtractions.

How long does it take to factor A into LU with matrices of order $n = 100$? We used the MATLAB command t = clock; lu(A); etime(clock, t). The elapsed time on a SUN Sparcstation 1 was *one second.* For $n = 200$ the elapsed time was eight seconds. (This follows the n^3 rule! The time t is multiplied by 8 when n is multiplied by 2.) The matrix A was random. Starting with the command for n = 10:10:100, A = rand(n, n); you could measure times for $n = 10, 20, \ldots, 100$.

According to the n^3 rule, matrices of order $n = 1000$ will take 10^3 seconds. Matrices of order 10,000 will take 10^6 seconds. This is very expensive but remember that these matrices are full. Most matrices in practice are sparse (many zero entries). In that case $A = LU$ is very much faster. For tridiagonal matrices of order 10,000, solving $A\boldsymbol{x} = \boldsymbol{b}$ is a breeze.

We also ran lu(A) with the Student Version of MATLAB on a 12 megahertz 386 PC. For the maximum size $n = 32$, the elapsed time was $\frac{1}{3}$ second. That includes the time to format the answer. If you experiment with for n = 5:30, A = rand(n, n); you will see the fixed cost for small matrices.

These are 1993 times, and speeds are going up. The day we started writing this was the day that IBM announced it would open a new laboratory, to design massively parallel computers including their RISC technology. After "fierce internal debate," they decided to tie together thousands of processors—as the Connection Machine and the Intel Hypercube are doing now. Every numerical algorithm, even Gaussian elimination, has to be reorganized to run well on a parallel machine.

What about the right side $\boldsymbol{b}$? The first stage subtracts multiples of b_1 from the lower components $b_2, \ldots, b_n$. This is $n-1$ steps. The second stage takes only $n-2$ steps, because b_1 is not involved. The last stage of *forward* elimination (down to a 2 by 2 system) takes one step on the right side and produces $\boldsymbol{c}$.

Now start back substitution. Computing x_n uses one step (divide by the last pivot). The next unknown uses two steps. The first unknown x_1 requires n steps ($n-1$ substitutions of the other unknowns, then division by the first pivot). The total count on the right side—*forward to the bottom and back to the top*—is exactly n^2:

$$(n-1)+(n-2)+\cdots+1+1+2+\cdots+(n-1)+n=n^2. \tag{2.23}$$

To see that sum, pair off $(n-1)$ with 1 and $(n-2)$ with 2. The pairings finally leave n terms each equal to n. That makes n^2. The right side costs a lot less than the left side!

Each right side (from $\boldsymbol{b}$ to $\boldsymbol{c}$ to $\boldsymbol{x}$) ***needs n^2 multiplications and n^2 subtractions.***

Here are the MATLAB codes slu(A) to factor $A = LU$ and slv(A,b) to solve $A\boldsymbol{x} = \boldsymbol{b}$. The program slu stops if zero appears in a pivot position (that fact is printed). These M-files are on the diskette that is available at no cost from MathWorks, to run with the Student Version of MATLAB.

M-file: slu

```
function [L, U] = slu(A)
% SLU  Square LU factorization with no row exchanges.

[n,n] = size(A);
tol = 1.e-6;
for k = 1:n
  if abs(A(k,k)) < tol
    disp(['Small pivot in column ' int2str(k)])
  end
  L(k,k) = 1;
  for i = k+1:n
    L(i,k) = A(i,k)/A(k,k);
    for j = k+1:n
      A(i,j) = A(i,j) - L(i,k) * A(k,j);
    end
  end
```

```
  for j = k:n
    U(k,j) = A(k,j);
  end
end
```

```
function x = slv(A,b)
% Solve Ax = b using L and U from slu(A). No row exchanges!

[L,U] = slu(A);

% Forward elimination to solve Lc = b.
% L is lower triangular with 1's on the diagonal.

[n,n] = size(A);
for k = 1:n
  s = 0;
  for j = 1:k-1
    s = s + L(k,j) * c(j);
  end
  c(k) = b(k) - s;
end

% Back substitution to solve Ux = c.
% U is upper triangular with nonzeros on the diagonal.

for k = n:-1:1
  t = 0;
  for j = k+1:n
    t = t + U(k,j) * x(j);
  end
  x(k) = (c(k) - t)/U(k,k);
end
x = x';
```

M-file: slv

Problem Set 2.5

Problems 1–8 compute the factorization $A = LU$ (and also $A = LDU$).

1 What matrix E puts A into triangular form $EA = U$? Multiply by $E^{-1} = L$ to factor A into LU:

$$A = \begin{bmatrix} 2 & 1 & 0 \\ 0 & 4 & 2 \\ 6 & 3 & 5 \end{bmatrix}.$$

2 What two elimination matrices E_{21} and E_{32} put A into upper triangular form $E_{32}E_{21}A = U$? Multiply by E_{32}^{-1} and E_{21}^{-1} to factor A into $E_{21}^{-1}E_{32}^{-1}U$ which is LU. Find L and U for

$$A = \begin{bmatrix} 1 & 1 & 1 \\ 2 & 4 & 5 \\ 0 & 4 & 0 \end{bmatrix}.$$

3 What three elimination matrices E_{21}, E_{31}, E_{32} put A into upper triangular form $E_{32}E_{31}E_{21}A = U$? Multiply by E_{32}^{-1}, E_{31}^{-1} and E_{21}^{-1} to factor A into $E_{21}^{-1}E_{31}^{-1}E_{32}^{-1}U$ which is LU. Find L and U:

$$A = \begin{bmatrix} 1 & 0 & 1 \\ 2 & 2 & 2 \\ 3 & 4 & 5 \end{bmatrix}.$$

4 Suppose A is already lower triangular with 1's on the diagonal. Then $U = I$!

$$A = \begin{bmatrix} 1 & 0 & 0 \\ a & 1 & 0 \\ b & c & 1 \end{bmatrix}.$$

The elimination matrices E_{21}, E_{31}, E_{32} contain $-a$ then $-b$ then $-c$.

(a) Multiply $E_{32}E_{31}E_{21}$ to find the single matrix E that produces $EA = I$.

(b) Multiply $E_{21}^{-1}E_{31}^{-1}E_{32}^{-1}$ to find the single matrix L that gives $A = LU$ (or LI).

5 When zero appears in a pivot position, the factorization $A = LU$ *is not possible*! (We are requiring nonzero pivots in U.) Show directly why these are both impossible:

$$\begin{bmatrix} 0 & 1 \\ 2 & 3 \end{bmatrix} = \begin{bmatrix} 1 & 0 \\ l & 1 \end{bmatrix}\begin{bmatrix} d & e \\ 0 & f \end{bmatrix} \qquad \begin{bmatrix} 1 & 1 & 0 \\ 1 & 1 & 2 \\ 1 & 2 & 1 \end{bmatrix} = \begin{bmatrix} 1 & & \\ l & 1 & \\ m & n & 1 \end{bmatrix}\begin{bmatrix} d & e & g \\ & f & h \\ & & i \end{bmatrix}.$$

This difficulty is fixed by a row exchange.

6 Which number c leads to zero in the second pivot position? A row exchange is needed and $A = LU$ is not possible. Which c produces zero in the third pivot position? Then a row exchange can't help and elimination fails:

$$A = \begin{bmatrix} 1 & c & 0 \\ 2 & 4 & 1 \\ 3 & 5 & 1 \end{bmatrix}.$$

7 What are L and D for this matrix A? What is U in $A = LU$ and what is the new U in $A = LDU$?

$$A = \begin{bmatrix} 2 & 4 & 8 \\ 0 & 3 & 9 \\ 0 & 0 & 7 \end{bmatrix}.$$

8 A and B are symmetric across the diagonal (because $4 = 4$). Find their factorizations LDU and say how U is related to L:

$$A = \begin{bmatrix} 2 & 4 \\ 4 & 11 \end{bmatrix} \quad \text{and} \quad B = \begin{bmatrix} 1 & 4 & 0 \\ 4 & 12 & 4 \\ 0 & 4 & 0 \end{bmatrix}.$$

9 (Recommended) Compute L and U for the symmetric matrix

$$A = \begin{bmatrix} a & a & a & a \\ a & b & b & b \\ a & b & c & c \\ a & b & c & d \end{bmatrix}.$$

Find four conditions on a, b, c, d to get $A = LU$ with four pivots.

10 Find L and U for the nonsymmetric matrix

$$A = \begin{bmatrix} a & r & r & r \\ a & b & s & s \\ a & b & c & t \\ a & b & c & d \end{bmatrix}.$$

Find the four conditions on a, b, c, d, r, s, t to get $A = LU$ with four pivots.

Problems 11-12 use L and U (without needing A) to solve $A\boldsymbol{x} = \boldsymbol{b}$.

11 Solve the triangular system $L\boldsymbol{c} = \boldsymbol{b}$ to find $\boldsymbol{c}$. Then solve $U\boldsymbol{x} = \boldsymbol{c}$ to find $\boldsymbol{x}$:

$$L = \begin{bmatrix} 1 & 0 \\ 4 & 1 \end{bmatrix} \quad \text{and} \quad U = \begin{bmatrix} 2 & 4 \\ 0 & 1 \end{bmatrix} \quad \text{and} \quad \boldsymbol{b} = \begin{bmatrix} 2 \\ 11 \end{bmatrix}.$$

For safety find $A = LU$ and solve $A\boldsymbol{x} = \boldsymbol{b}$ as usual. Circle $\boldsymbol{c}$ when you see it.

12 Solve $L\boldsymbol{c} = \boldsymbol{b}$ to find $\boldsymbol{c}$. Then solve $U\boldsymbol{x} = \boldsymbol{c}$ to find $\boldsymbol{x}$. What was A?

$$L = \begin{bmatrix} 1 & 0 & 0 \\ 1 & 1 & 0 \\ 1 & 1 & 1 \end{bmatrix} \quad \text{and} \quad U = \begin{bmatrix} 1 & 1 & 1 \\ 0 & 1 & 1 \\ 0 & 0 & 1 \end{bmatrix} \quad \text{and} \quad \boldsymbol{b} = \begin{bmatrix} 4 \\ 5 \\ 6 \end{bmatrix}.$$

13 (a) When you apply the usual elimination steps to L, what matrix do you reach?

$$L = \begin{bmatrix} 1 & 0 & 0 \\ l_{21} & 1 & 0 \\ l_{31} & l_{32} & 1 \end{bmatrix}.$$

(b) When you apply the same steps to I, what matrix do you get?

(c) When you apply the same steps to LU, what matrix do you get?

14 If $A = LDU$ and also $A = L_1D_1U_1$ with all factors invertible, then $L = L_1$ and $D = D_1$ and $U = U_1$. *"The factors are unique."*

(a) Derive the equation $L_1^{-1}LD = D_1U_1U^{-1}$. Are the two sides lower or upper triangular or diagonal?

(b) Show that the main diagonals in that equation give $D = D_1$. Why does $L = L_1$?

15 *Tridiagonal matrices* have zero entries except on the main diagonal and the two adjacent diagonals. Factor these into $A = LU$ and $A = LDL^{\mathrm{T}}$:

$$A = \begin{bmatrix} 1 & 1 & 0 \\ 1 & 2 & 1 \\ 0 & 1 & 2 \end{bmatrix} \quad \text{and} \quad A = \begin{bmatrix} a & a & 0 \\ a & a+b & b \\ 0 & b & b+c \end{bmatrix}.$$

16 When T is tridiagonal, its L and U factors have only two nonzero diagonals. How would you take advantage of the zeros in T in a computer code for Gaussian elimination? Find L and U.

$$T = \begin{bmatrix} 1 & 2 & 0 & 0 \\ 2 & 3 & 1 & 0 \\ 0 & 1 & 2 & 3 \\ 0 & 0 & 3 & 4 \end{bmatrix}.$$

17 If A and B have nonzeros in the positions marked by x, which zeros (marked by 0) are still zero in their factors L and U?

$$A = \begin{bmatrix} x & x & x & x \\ x & x & x & 0 \\ 0 & x & x & x \\ 0 & 0 & x & x \end{bmatrix} \quad \text{and} \quad B = \begin{bmatrix} x & x & x & 0 \\ x & x & 0 & x \\ x & 0 & x & x \\ 0 & x & x & x \end{bmatrix}.$$

18 After elimination has produced zeros below the first pivot, put x's to show which blank entries are known in L and U:

$$\begin{bmatrix} x & x & x \\ x & x & x \\ x & x & x \end{bmatrix} = \begin{bmatrix} 1 & 0 & 0 \\ & 1 & 0 \\ & & 1 \end{bmatrix} \begin{bmatrix} & & \\ 0 & & \\ 0 & & \end{bmatrix}.$$

19 Suppose you eliminate upwards (almost unheard of). Use the last row to produce zeros in the last column (the pivot is 1). Then use the second row to produce zero above the second pivot. Find the factors in $A = UL(!)$:

$$A = \begin{bmatrix} 5 & 3 & 1 \\ 3 & 3 & 1 \\ 1 & 1 & 1 \end{bmatrix}.$$

20 Collins uses elimination in both directions, forward (down) and backward (up), meeting at the center. Substitution goes out from the center. After eliminating both 2's in A, one from above and one from below, what 4 by 4 matrix is left? Solve $A\boldsymbol{x} = \boldsymbol{b}$ his way.

$$A = \begin{bmatrix} 1 & 1 & 0 & 0 \\ 2 & 1 & 1 & 0 \\ 0 & 1 & 3 & 2 \\ 0 & 0 & 1 & 1 \end{bmatrix} \quad \text{and} \quad \boldsymbol{b} = \begin{bmatrix} 5 \\ 8 \\ 8 \\ 2 \end{bmatrix}.$$

21 (Important) If A has pivots 2, 7, 6 with no row exchanges, what are the pivots for the upper left 2 by 2 submatrix B (without row 3 and column 3)? Explain why.

22 Starting from a 3 by 3 matrix A with pivots 2, 7, 6, add a fourth row and column to produce M. What are the first three pivots for M, and why? What fourth row and column are sure to produce 9 as the fourth pivot?

23 MATLAB knows the n by n matrix pascal(n). Find its LU factors for $n = 4$ and 5 and describe their pattern. Use chol(pascal(n)) or slu(A) above or work by hand. The row exchanges in MATLAB's lu code spoil the pattern, but Cholesky (= chol) doesn't:

$$A = \text{pascal(4)} = \begin{bmatrix} 1 & 1 & 1 & 1 \\ 1 & 2 & 3 & 4 \\ 1 & 3 & 6 & 10 \\ 1 & 4 & 10 & 20 \end{bmatrix}.$$

24 (Careful review) For which c is $A = LU$ impossible—with three pivots?

$$A = \begin{bmatrix} 1 & 2 & 0 \\ 3 & c & 1 \\ 0 & 1 & 1 \end{bmatrix}$$

25 Change the program slu(A) into sldu(A), so that it produces the three matrices L, D, and U.

26 Rewrite slu(A) so that the factors L and U appear in the same n^2 storage locations that held the original A. The extra storage used for L is not required.

27 Explain in words why $x(k)$ is $(c(k) - t)/U(k,k)$ at the end of slv(A, b).

28 Write a program that multiplies triangular matrices, L times U. Don't loop from 1 to n when you know there are zeros in the matrices. Somehow L times U should undo the operations in slu.

2.6 Transposes and Permutations

We need one more matrix, and fortunately it is much simpler than the inverse. It is the ***"transpose"*** of A, which is denoted by A^{T}. *Its columns are the rows of* A. When A is an m by n matrix, the transpose is n by m:

$$\text{If} \quad A = \begin{bmatrix} 1 & 2 & 3 \\ 0 & 0 & 4 \end{bmatrix} \quad \text{then} \quad A^{\mathrm{T}} = \begin{bmatrix} 1 & 0 \\ 2 & 0 \\ 3 & 4 \end{bmatrix}.$$

You can write the rows of A into the columns of A^{T}. Or you can write the columns of A into the rows of A^{T}. The matrix "flips over" its main diagonal. The entry in row i, column j of A^{T} comes from row j, column i of the original A:

$$(A^{\mathrm{T}})_{ij} = A_{ji}.$$

The transpose of a lower triangular matrix is upper triangular. The transpose of A^{T} is ______.

Note MATLAB's symbol for the transpose is A′. To enter a column vector, type v = [1 2 3]′. To enter a matrix with second column w = [4 5 6]′ you could define M = [v w]. Quicker to enter by rows and then transpose the whole matrix: M = [1 2 3; 4 5 6]′.

The rules for transposes are very direct. We can transpose $A + B$ to get $(A + B)^{\mathrm{T}}$. Or we can transpose A and B separately, and then add $A^{\mathrm{T}} + B^{\mathrm{T}}$—same result. The serious questions are about the transpose of a product AB and an inverse A^{-1}:

$$\text{The transpose of} \quad A + B \quad \text{is} \quad A^{\mathrm{T}} + B^{\mathrm{T}}. \tag{2.24}$$

$$\text{The transpose of} \quad AB \quad \text{is} \quad (AB)^{\mathrm{T}} = B^{\mathrm{T}}A^{\mathrm{T}}. \tag{2.25}$$

$$\text{The transpose of} \quad A^{-1} \quad \text{is} \quad (A^{-1})^{\mathrm{T}} = (A^{\mathrm{T}})^{-1}. \tag{2.26}$$

Notice especially how $B^{\mathrm{T}}A^{\mathrm{T}}$ comes in reverse order like $B^{-1}A^{-1}$. The proof for the inverse is quick: $B^{-1}A^{-1}$ times AB produces I. To see this reverse order for $(AB)^{\mathrm{T}}$, start with $(A\boldsymbol{x})^{\mathrm{T}}$:

$A\boldsymbol{x}$ combines the columns of A while $\boldsymbol{x}^{\mathrm{T}}A^{\mathrm{T}}$ combines the rows of A^{T}.

It is the same combination of the same vectors! In A they are columns, in A^{T} they are rows. So the transpose of the column $A\boldsymbol{x}$ is the row $\boldsymbol{x}^{\mathrm{T}}A^{\mathrm{T}}$. That fits our formula $(A\boldsymbol{x})^{\mathrm{T}} = \boldsymbol{x}^{\mathrm{T}}A^{\mathrm{T}}$. Now we can prove the formulas for $(AB)^{\mathrm{T}}$ and $(A^{-1})^{\mathrm{T}}$.

When $B = [\boldsymbol{x}_1 \ \boldsymbol{x}_2]$ has two columns, apply the same idea to each column. The columns of AB are $A\boldsymbol{x}_1$ and $A\boldsymbol{x}_2$. Their transposes are $\boldsymbol{x}_1^{\mathrm{T}}A^{\mathrm{T}}$ and $\boldsymbol{x}_2^{\mathrm{T}}A^{\mathrm{T}}$. Those are the rows of $B^{\mathrm{T}}A^{\mathrm{T}}$:

$$\text{Transposing} \quad AB = \begin{bmatrix} & & & \\ A\boldsymbol{x}_1 & A\boldsymbol{x}_2 & \cdots \\ & & & \end{bmatrix} \quad \text{gives} \quad \begin{bmatrix} \boldsymbol{x}_1^{\mathrm{T}}A^{\mathrm{T}} \\ \boldsymbol{x}_2^{\mathrm{T}}A^{\mathrm{T}} \\ \vdots \end{bmatrix} \quad \text{which is } B^{\mathrm{T}}A^{\mathrm{T}}. \tag{2.27}$$

The right answer $B^{\mathrm{T}}A^{\mathrm{T}}$ comes out a row at a time. Maybe numbers are the easiest:

$$AB = \begin{bmatrix} 1 & 0 \\ 1 & 1 \end{bmatrix}\begin{bmatrix} 5 & 0 \\ 4 & 1 \end{bmatrix} = \begin{bmatrix} 5 & 0 \\ 9 & 1 \end{bmatrix} \quad \text{and} \quad B^{\mathrm{T}}A^{\mathrm{T}} = \begin{bmatrix} 5 & 4 \\ 0 & 1 \end{bmatrix}\begin{bmatrix} 1 & 1 \\ 0 & 1 \end{bmatrix} = \begin{bmatrix} 5 & 9 \\ 0 & 1 \end{bmatrix}.$$

The rule extends to three or more factors: $(ABC)^{\mathrm{T}}$ equals $C^{\mathrm{T}}B^{\mathrm{T}}A^{\mathrm{T}}$.

Now apply this product rule to both sides of $A^{-1}A = I$. On one side, I^{T} is just I. On the other side, we discover that $(A^{-1})^{\mathrm{T}}$ is the inverse of A^{T}:

$$A^{-1}A = I \quad \text{leads to} \quad A^{\mathrm{T}}(A^{-1})^{\mathrm{T}} = I. \tag{2.28}$$

Similarly $AA^{-1} = I$ leads to $(A^{-1})^{\mathrm{T}}A^{\mathrm{T}} = I$. Notice especially: A^{T} *is invertible exactly when A is invertible.* We can invert first or transpose first, it doesn't matter.

A and A^{T} have the same pivots, when there are no row exchanges:

If $A = LDU$ then $A^{\mathrm{T}} = U^{\mathrm{T}}D^{\mathrm{T}}L^{\mathrm{T}}$. The pivot matrix $D = D^{\mathrm{T}}$ is the same.

Example 2.17 The inverse of $A = \left[\begin{smallmatrix} 1 & 0 \\ 6 & 1 \end{smallmatrix}\right]$ is $A^{-1} = \left[\begin{smallmatrix} 1 & 0 \\ -6 & 1 \end{smallmatrix}\right]$. The transpose is $A^{\mathrm{T}} = \left[\begin{smallmatrix} 1 & 6 \\ 0 & 1 \end{smallmatrix}\right]$.

$$(A^{-1})^{\mathrm{T}} \quad \text{equals} \quad \begin{bmatrix} 1 & -6 \\ 0 & 1 \end{bmatrix} \quad \text{which is} \quad (A^{\mathrm{T}})^{-1}.$$

Before leaving these rules, we call attention to dot products. The following statement looks extremely simple, but it actually contains the deep purpose for the transpose. For any vectors $\boldsymbol{x}$ and $\boldsymbol{y}$,

$$(A\boldsymbol{x})^{\mathrm{T}}\boldsymbol{y} \quad \text{equals} \quad \boldsymbol{x}^{\mathrm{T}}A^{\mathrm{T}}\boldsymbol{y} \quad \text{equals} \quad \boldsymbol{x}^{\mathrm{T}}(A^{\mathrm{T}}\boldsymbol{y}). \tag{2.29}$$

We can put in parentheses or leave them out. In electrical networks, $\boldsymbol{x}$ is the vector of potentials and $\boldsymbol{y}$ is the vector of currents. $A\boldsymbol{x}$ gives potential differences and $A^{\mathrm{T}}\boldsymbol{y}$ is the current into nodes. Every vector has a meaning—the dot product in equation (2.29) is the heat loss.

In engineering and physics, $\boldsymbol{x}$ can be the displacement of a structure under a load. Then $A\boldsymbol{x}$ is the stretching (the strain). When $\boldsymbol{y}$ is the internal stress, $A^{\mathrm{T}}\boldsymbol{y}$ is the external force. Equation (2.29) is a statement about dot products and also a balance equation:

Internal work (**strain · stress**) = *external work* (**displacement · force**).

In economics, $\boldsymbol{x}$ gives the amounts of n outputs. The m inputs to produce them are $A\boldsymbol{x}$. The costs per input go into $\boldsymbol{y}$. Then the values per output are the components of $A^{\mathrm{T}}\boldsymbol{y}$:

Total input cost $A\boldsymbol{x} \cdot \boldsymbol{y}$ equals total output value $\boldsymbol{x} \cdot A^{\mathrm{T}}\boldsymbol{y}$.

Problems 31–33 bring out applications. Here we only emphasize the transpose. When A moves from one side of a dot product to the other side, it becomes A^{T}.

Symmetric Matrices

For some matrices—these are the most important matrices—transposing A to A^{T} produces no change. Then $A^{\mathrm{T}} = A$. The matrix is *symmetric* across the main diagonal. A symmetric matrix is necessarily square. Its (j, i) and (i, j) entries are equal.

Definition A ***symmetric matrix*** has $A^{\mathrm{T}} = A$. This means that $a_{ji} = a_{ij}$.

Example 2.18 $A = \begin{bmatrix} 1 & 2 \\ 2 & 5 \end{bmatrix} = A^{\mathrm{T}}$ and $D = \begin{bmatrix} 1 & 0 \\ 0 & 10 \end{bmatrix} = D^{\mathrm{T}}$.

A is symmetric because of the 2's on opposite sides of the diagonal. In D those 2's are zeros. A diagonal matrix is automatically symmetric.

The inverse of a symmetric matrix is also symmetric. (We have to add: If $A = A^{\mathrm{T}}$ has an inverse.) When A^{-1} is transposed, equation (2.26) gives $(A^{\mathrm{T}})^{-1}$ which is A^{-1} again. The inverses of A and D are certainly symmetric:

$$A^{-1} = \begin{bmatrix} 5 & -2 \\ -2 & 1 \end{bmatrix} = (A^{-1})^{\mathrm{T}} \quad \text{and} \quad D^{-1} = \begin{bmatrix} 1 & 0 \\ 0 & 0.1 \end{bmatrix} = (D^{-1})^{\mathrm{T}}.$$

Now we show how symmetric matrices come from nonsymmetric matrices and their transposes.

Symmetric Products $R^{\mathrm{T}}R$ and RR^{T} and LDL^{T}

Choose any matrix R, probably rectangular. Multiply R^{T} times R. Then the product $R^{\mathrm{T}}R$ is automatically a symmetric matrix:

$$\textit{The transpose of} \quad \boldsymbol{R^{\mathrm{T}}R} \quad \textit{is} \quad \boldsymbol{R^{\mathrm{T}}(R^{\mathrm{T}})^{\mathrm{T}}} \quad \textit{which is} \quad \boldsymbol{R^{\mathrm{T}}R}. \tag{2.30}$$

In words, the (i, j) entry of $R^{\mathrm{T}}R$ is the dot product of column i of R with column j of R. The (j, i) entry is the same dot product. So $R^{\mathrm{T}}R$ is symmetric.

The matrix RR^{T} is also symmetric. (The shapes of R and R^{T} allow multiplication.) But RR^{T} is a different matrix from $R^{\mathrm{T}}R$. In our experience, most scientific problems that start with a rectangular matrix R end up with $R^{\mathrm{T}}R$ or RR^{T} or both.

Example 2.19 $R = \begin{bmatrix} 1 & 2 & 2 \end{bmatrix}$ and $R^{\mathrm{T}}R = \begin{bmatrix} 1 & 2 & 2 \\ 2 & 4 & 4 \\ 2 & 4 & 4 \end{bmatrix}$ and $RR^{\mathrm{T}} = [\,9\,]$.

The product $R^{\mathrm{T}}R$ is n by n. In the opposite order, RR^{T} is m by m. Even if $m = n$, it is not likely that $R^{\mathrm{T}}R = RR^{\mathrm{T}}$. Equality can happen, but it is abnormal.

When elimination is applied to a symmetric matrix, we hope that $A^{\mathrm{T}} = A$ is an advantage. That depends on the smaller matrices staying symmetric as elimination proceeds—which they do. It is true that the upper triangular U cannot be symmetric. The symmetry is not in LU, it is in LDU. Remember how the diagonal matrix of pivots can be divided out of U, to leave 1's on the diagonal.

$$\begin{bmatrix}1&2\\2&7\end{bmatrix}=\begin{bmatrix}1&0\\2&1\end{bmatrix}\begin{bmatrix}1&2\\0&3\end{bmatrix} \qquad (LU \text{ misses the symmetry})$$

$$=\begin{bmatrix}1&0\\2&1\end{bmatrix}\begin{bmatrix}1&0\\0&3\end{bmatrix}\begin{bmatrix}1&2\\0&1\end{bmatrix} \qquad (LDU \text{ captures the symmetry because } U=L^{\mathrm{T}}).$$

When A is symmetric, the usual form $A = LDU$ becomes $A = LDL^{\mathrm{T}}$. The final U (with 1's on the diagonal) is the transpose of L (also with 1's on the diagonal). The diagonal D—the matrix of pivots—is symmetric by itself.

2K If $A = A^{\mathrm{T}}$ can be factored into LDU with no row exchanges, then $U = L^{\mathrm{T}}$. ***The symmetric factorization is*** $\boldsymbol{A = LDL^{\mathrm{T}}}$.

Notice that the transpose of LDL^{T} is automatically $(L^{\mathrm{T}})^{\mathrm{T}}D^{\mathrm{T}}L^{\mathrm{T}}$ which is LDL^{T} again. We have a symmetric factorization for a symmetric matrix. The work of elimination is cut essentially in half, from $n^3/3$ multiplications to $n^3/6$. The storage is also cut essentially in half. We don't have to store entries above the diagonal, or work on them, because they are known from entries below the diagonal.

Permutation Matrices

The transpose plays a special role for *permutation matrices*. These are matrices P with a single "1" in every row and every column. Then P^{T} is also a permutation matrix—maybe the same or maybe different. Any product P_1P_2 is again a permutation matrix. We now create every P from the identity matrix, by reordering the rows.

The simplest permutation matrix is $P = I$ (*no exchanges*). The next simplest are the row exchanges P_{ij}. Those are constructed by exchanging two rows i and j of I. Other permutations exchange three rows, ijk to jki or jik (those are different). By doing all possible row exchanges to I, we get all possible permutation matrices:

Definition An n by n ***permutation matrix*** P has the rows of I (n rows) in any order.

Example 2.20 There are six 3 by 3 permutation matrices. Here they are without the zeros:

$$I=\begin{bmatrix}1&&\\&1&\\&&1\end{bmatrix} \quad P_{21}=\begin{bmatrix}&1&\\1&&\\&&1\end{bmatrix} \quad P_{32}P_{21}=\begin{bmatrix}&1&\\&&1\\1&&\end{bmatrix}$$

$$P_{31}=\begin{bmatrix}&&1\\&1&\\1&&\end{bmatrix} \quad P_{32}=\begin{bmatrix}1&&\\&&1\\&1&\end{bmatrix} \quad P_{21}P_{32}=\begin{bmatrix}&&1\\1&&\\&1&\end{bmatrix}.$$

There are $n!$ permutation matrices of order n. The symbol $n!$ stands for "n factorial," the product of the numbers $(1)(2)\cdots(n)$. Thus $3! = (1)(2)(3)$ which is 6.

There are only two permutation matrices of order 2, namely $\left[\begin{smallmatrix}1&0\\0&1\end{smallmatrix}\right]$ and $\left[\begin{smallmatrix}0&1\\1&0\end{smallmatrix}\right]$.

Important: The *inverse* of a permutation matrix is also a permutation matrix. In the example above, the four matrices on the left are their own inverses. The two matrices on the right are inverses of each other. In all cases, a single row exchange P_{21} or P_{ij} is its own inverse. It exchanges the rows back again. But for a product like $P_{32}P_{21}$, the inverses go in opposite order (of course). The inverse is $P_{21}P_{32}$.

More important: $\boldsymbol{P^{-1}}$ ***is the same as*** $\boldsymbol{P^{\mathrm{T}}}$. The four matrices on the left above are their own transposes. The two matrices on the right are transposes—and inverses—of each other. You can check directly that P times P^{T} equals I. The "1" in the first row of P hits the "1" in the first column of P^{T}. It misses the ones in all the other columns. So $PP^{\mathrm{T}} = I$.

Another proof of $P^{\mathrm{T}} = P^{-1}$ is in the following reasoning:

P is a product of row exchanges. A row exchange is its own transpose and its own inverse. P^{T} is the product in the opposite order. So is P^{-1}. Therefore $P^{\mathrm{T}} = P^{-1}$.

Symmetric matrices led to $\boldsymbol{A = LDL^{\mathrm{T}}}$***. Now permutations lead to*** $\boldsymbol{PA = LU}$***.***

The LU Factorization with Row Exchanges

We hope you remember $A = LU$. It started with $A = (E_{21}^{-1} \cdots E_{ij}^{-1} \cdots)U$. Every elimination step was carried out by an E_{ij} and inverted by an E_{ij}^{-1}. Those inverses were compressed into one matrix L. The lower triangular L has 1's on the diagonal, and the result is $A = LU$.

This is a great factorization, but it doesn't always work—because of row exchanges. If exchanges are needed then $A = (E^{-1} \cdots P^{-1} \cdots E^{-1} \cdots P^{-1} \cdots)U$. Every row exchange is carried out by a P_{ij} and inverted by that P_{ij}. We now compress those row exchanges into a *single permutation matrix*. This gives a factorization for every invertible matrix A—which we naturally want.

The main question is where to collect the P_{ij}'s. There are two good possibilities—do them before elimination, or do them after the E_{ij}'s. One way gives $PA = LU$, the other way has the permutation in the middle.

1 The row exchanges can be moved to the left side, onto A. We think of them as done *in advance*. Their product P puts the rows of A in the right order, so that no exchanges are needed for PA. Then PA factors into LU.

2 We can hold all row exchanges until *after forward elimination*. This leaves the pivot rows in a strange order. P_1 puts them in the right order which produces U_1 (upper triangular as usual). Then $A = L_1P_1U_1$.

$PA = LU$ is used in almost all computing (and always in MATLAB). The form $A = L_1P_1U_1$ is the right one for theoretical algebra—it is definitely more elegant. If we give space to both, it is because the difference is not well known. Probably you will not spend a long time on either one. Please don't. The most important case by far has $P = I$, when A equals LU with no exchanges.

The algebraist's form $A = L_1P_1U_1$ Suppose $a_{11} = 0$ but a_{21} is nonzero. Then row 2 is the pivot row—the row exchange waits! Produce zeros below the pivot a_{21} in the usual way. Now choose the first available pivot in column 2. Row 1 is available (use it if a_{12} is nonzero). Row 2 is not available (it was already a pivot row). By accepting the first available pivot, every step subtracts a pivot row from a *lower* row.

$$A = \begin{bmatrix} 0 & 1 & 1 \\ 1 & 2 & 1 \\ 2 & 7 & 9 \end{bmatrix} \xrightarrow[l_{32}=2]{} \begin{bmatrix} 0 & 1 & 1 \\ 1 & 2 & 1 \\ 0 & 3 & 7 \end{bmatrix} \xrightarrow[l_{31}=3]{} \begin{bmatrix} 0 & 1 & 1 \\ 1 & 2 & 1 \\ 0 & 0 & 4 \end{bmatrix}. \tag{2.31}$$

Forward elimination ends there. The last matrix is not U, but by exchanging two rows it becomes U_1. So that last matrix is P_1U_1. The two elimination steps are undone by L_1. The whole factorization is

$$A = \begin{bmatrix} 1 & & \\ 0 & 1 & \\ 3 & 2 & 1 \end{bmatrix} \begin{bmatrix} & 1 & \\ 1 & & \\ & & 1 \end{bmatrix} \begin{bmatrix} 1 & 2 & 1 \\ 0 & 1 & 1 \\ 0 & 0 & 4 \end{bmatrix} = L_1P_1U_1. \tag{2.32}$$

The computer's form $PA = LU$ To put the pivots on the diagonal, do the row exchanges first. For this matrix, exchange rows 1 and 2 to put the first pivot in its usual place. Then go through elimination:

$$PA = \begin{bmatrix} 1 & 2 & 1 \\ 0 & 1 & 1 \\ 2 & 7 & 9 \end{bmatrix} \xrightarrow[l_{31}=2]{} \begin{bmatrix} 1 & 2 & 1 \\ 0 & 1 & 1 \\ 0 & 3 & 7 \end{bmatrix} \xrightarrow[l_{32}=3]{} \begin{bmatrix} 1 & 2 & 1 \\ 0 & 1 & 1 \\ 0 & 0 & 4 \end{bmatrix}.$$

In this case P equals P_1 and U equals U_1 (not always). The matrix L is different:

$$PA = \begin{bmatrix} 1 & 0 & 0 \\ 0 & 1 & 0 \\ 2 & 3 & 1 \end{bmatrix} \begin{bmatrix} 1 & 2 & 1 \\ 0 & 1 & 1 \\ 0 & 0 & 4 \end{bmatrix} = LU. \tag{2.33}$$

When P comes in advance, the l's are moved around. The numbers 2 and 3 are reversed. We could establish a general rule to connect P, L, U to P_1, L_1, U_1 but we won't. We feel safer when we do elimination on PA to find LU.

2L ($PA = LU$) If A is invertible, a permutation P will put its rows in the right order to factor PA into LU. There is a full set of pivots and $A\boldsymbol{x} = \boldsymbol{b}$ can be solved.

In the MATLAB code, watch for these lines that exchange row k with row r below it (where the kth pivot has been found). The notation A(k,1:n) picks out row k and all columns 1 to n—to produce the whole row. The permutation P starts as the identity matrix, which is eye(n,n):

```
A([r k],1:n) = A([k r], 1:n);
L([r k],1:k−1) = L([k r],1:k−1);
P([r k],1:n) = P([k r],1:n);
sign = −sign
```

The **"sign"** of P tells whether the number of row exchanges is even (sign $= +1$) or odd (sign $= -1$). At the start, P is I and sign $= +1$. When there is a row exchange, the sign is reversed. It gives the **"determinant of P"** and it does not depend on the order of the row exchanges.

Summary $A = L_1 P_1 U_1$ is more elegant, because L_1 is constructed as you go. The first available nonzero in each column becomes the pivot. It may not be on the main diagonal, but pivot rows are still subtracted from lower rows. At the end, P_1 reorders the rows to put the pivots on the diagonal of U_1.

For PA we get back to the familiar LU. This is the usual factorization. We should tell you: The computer does *not* always use the first available pivot. An algebraist accepts a small pivot—anything but zero. A computer looks down the column for the largest pivot. P may contain more row exchanges than are algebraically necessary, but we still have $PA = LU$.

Our advice is to understand permutations but let MATLAB do the computing. Calculations of $A = LU$ are enough to do by hand, without P. Here are the codes splu(A) to factor $PA = LU$z and splv(A,b) to solve $A\boldsymbol{x} = \boldsymbol{b}$ for any square invertible A. The program splu stops if no pivot can be found in column k. That fact is printed.

These M-files are on the diskette of Teaching Codes available from MathWorks.

The program plu in Chapter 3 will drop the s and allow the matrix to be rectangular. We still get $PA = LU$. But there may not be n pivots on the diagonal of U, and the matrix A is generally not invertible.

M-file: splu

```
function [P, L, U, sign] = splu(A)
% SPLU Square PA = LU factorization with row exchanges.

[m,n] = size(A);
if m ~= n
  error('Matrix must be square.')
end
P = eye(n,n);
L = eye(n,n);
U = zeros(n,n);
tol = 1.e-6;
sign = 1;

for k = 1:n
  if abs(A(k,k)) < tol
    for r = k:n
      if abs(A(r,k)) >= tol
        break
      end
      if r == n
        disp('A is singular')
```

```
          error(['No pivot in column ' int2str(k)])
        end
      end
      A([r k],1:n) = A([k r],1:n);
      if k > 1, L([r k],1:k-1) = L([k r],1:k-1); end
      P([r k],1:n) = P([k r],1:n);
      sign = -sign;
    end

    for i = k + 1:n
      L(i,k) = A(i,k)/A(k,k);
      for j = k + 1:n
        A(i,j) = A(i,j) - L(i,k) * A(k,j);
      end
    end
    for j = k:n
      U(k,j) = A(k,j) * (abs(A(k,j)) >= tol);
    end
  end

  if nargout < 4
    roworder = P * (1:m)';
    disp('Pivots in rows:'), disp(roworder'); end
  end
```

M-file: splv

```
function x = splv(A,b)
% SPLV   Solve Ax = b using P,L,U from splu(A).
%        Actually solve PAx = Pb.  A is invertible!
%        The MATLAB backslash operator A\b also finds x.

[P,L,U] = splu(A);
[n,n] = size(A);
b = P * b;

% Forward elimination to solve Lc = b. Really Lc = Pb.

c = zeros(n,1);
for k = 1:n
  s = 0;
  for j = 1:k-1
     s = s + L(k,j) * c(j);
  end
  c(k) = b(k) - s;
end
```

```
% Back substitution to solve Ux = c. Then PAx = LUx = Pb.

x = zeros(n,1);
for k = n:-1:1
  t = 0;
  for j = k+1:n
    t = t + U(k,j) * x(j);
  end
  x(k) = (c(k) - t)/U(k,k);
end
```

Problem Set 2.6

Questions 1–8 are about the rules for transpose matrices.

1 Find A^{T} and A^{-1} and $(A^{-1})^{\mathrm{T}}$ and $(A^{\mathrm{T}})^{-1}$ for

$$A = \begin{bmatrix} 1 & 0 \\ 8 & 2 \end{bmatrix} \quad \text{and also} \quad A = \begin{bmatrix} 1 & 1 \\ 1 & 0 \end{bmatrix}.$$

2 Verify that $(AB)^{\mathrm{T}}$ equals $B^{\mathrm{T}}A^{\mathrm{T}}$ but does not equal $A^{\mathrm{T}}B^{\mathrm{T}}$:

$$A = \begin{bmatrix} 1 & 0 \\ 2 & 1 \end{bmatrix} \quad B = \begin{bmatrix} 1 & 3 \\ 0 & 1 \end{bmatrix} \quad AB = \begin{bmatrix} 1 & 3 \\ 2 & 7 \end{bmatrix}.$$

In case $AB = BA$ (not generally true!) prove that $B^{\mathrm{T}}A^{\mathrm{T}} = A^{\mathrm{T}}B^{\mathrm{T}}$.

3 The matrix $((AB)^{-1})^{\mathrm{T}}$ comes from $(A^{-1})^{\mathrm{T}}$ and $(B^{-1})^{\mathrm{T}}$. *In what order*?

4 Show that $A^2 = 0$ is possible but $A^{\mathrm{T}}A = 0$ is not possible (unless $A = 0$.)

* **Transparent proof that $(AB)^{\mathrm{T}} = B^{\mathrm{T}}A^{\mathrm{T}}$.** Matrices can be transposed by looking through the page from the other side. Hold up to the light and practice with B below. Its column becomes a row in B^{T}. To see better, draw the figure with heavy lines on thin paper and turn it over so the symbol B^{T} is upright.

The three matrices are in position for matrix multiplication: the row of A times the column of B gives the entry in AB. Looking from the reverse side, the row of B^{T} times the column of A^{T} gives the correct entry in $B^{\mathrm{T}}A^{\mathrm{T}} = (AB)^{\mathrm{T}}$.

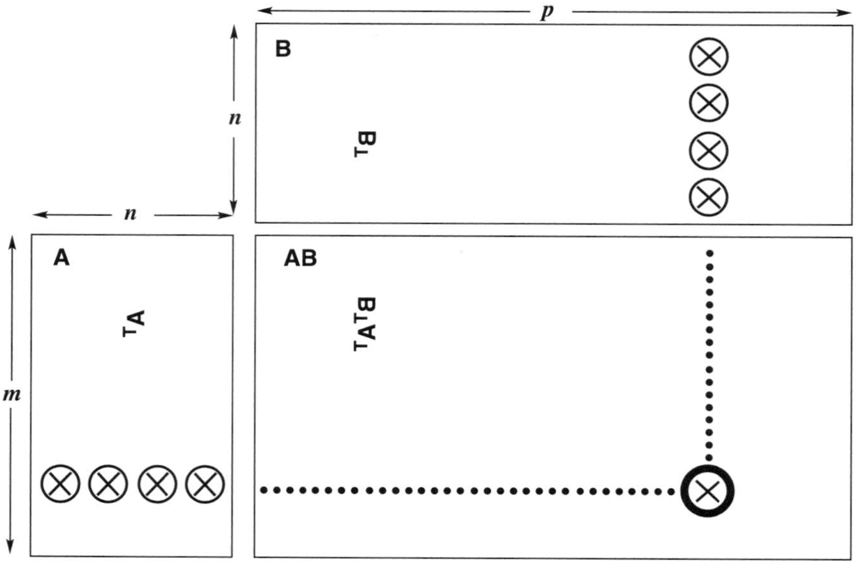

5 (a) The row vector $\boldsymbol{x}^{\mathrm{T}}$ times A times the column $\boldsymbol{y}$ produces what number?

$$\boldsymbol{x}^{\mathrm{T}}A\boldsymbol{y} = \begin{bmatrix} 0 & 1 \end{bmatrix} \begin{bmatrix} 1 & 2 & 3 \\ 4 & 5 & 6 \end{bmatrix} \begin{bmatrix} 0 \\ 1 \\ 0 \end{bmatrix} = \underline{\qquad}.$$

(b) This is the row $\boldsymbol{x}^{\mathrm{T}}A =$ ____ times the column $\boldsymbol{y} = (0, 1, 0)$.

(c) This is the row $\boldsymbol{x}^{\mathrm{T}} = [\,0 \quad 1\,]$ times the column $A\boldsymbol{y} =$ ____.

6 The components of $A\boldsymbol{x}$ are $\sum_{j=1}^{n} a_{ij}x_j$. Its dot product with $\boldsymbol{y}$ is

$$\sum_{i=1}^{m}\left(\sum_{j=1}^{n} a_{ij}x_j\right)y_i.$$

The components of $A^{\mathrm{T}}\boldsymbol{y}$ are $\sum_{i=1}^{m} a_{ij}y_i$. Its dot product with $\boldsymbol{x}$ is

$$\sum_{j=1}^{n}\left(\sum_{i=1}^{m} a_{ij}y_i\right)x_j.$$

The double sum can have j first or i first. Either way a_{ij} multiplies x_j and ____. The key is $(A\boldsymbol{x})^{\mathrm{T}}(\quad) = \boldsymbol{x}^{\mathrm{T}}(\quad)$.

7 When you transpose a block matrix $M = \left[\begin{smallmatrix} A & B \\ C & D \end{smallmatrix}\right]$ the result is $M^{\mathrm{T}} =$ ____. Test it.

8 True or false:

(a) The block matrix $\left[\begin{smallmatrix} 0 & A \\ A & 0 \end{smallmatrix}\right]$ is automatically symmetric.

(b) If A and B are symmetric then their product AB is symmetric.

(c) If A is not symmetric then A^{-1} is not symmetric.

(d) When A, B, C are symmetric, the transpose of $(ABC)^{\mathrm{T}}$ is ABC.

Questions 9–14 are about permutation matrices.

9 If P_1 and P_2 are permutation matrices, so is $P_1 P_2$. After two permutations we still have the rows of I in some order. Give examples with $P_1 P_2 \neq P_2 P_1$ and $P_3 P_4 = P_4 P_3$.

10 There are 12 *"even"* permutations of $(1, 2, 3, 4)$, including $(1, 2, 3, 4)$ with no exchanges and $(4, 3, 2, 1)$ with two exchanges. What are the other ten with an even number of exchanges? Instead of writing a 4 by 4 matrix, the numbers 4, 3, 2, 1 give the position of the 1 in each row.

11 Which permutation matrix makes PA upper triangular? Which permutations make $P_1 A P_2$ lower triangular?

$$A = \begin{bmatrix} 0 & 0 & 6 \\ 1 & 2 & 3 \\ 0 & 4 & 5 \end{bmatrix}.$$

12 (a) Explain why the dot product of $\boldsymbol{x}$ and $\boldsymbol{y}$ equals the dot product of $P\boldsymbol{x}$ and $P\boldsymbol{y}$. P is any permutation matrix.

(b) With $\boldsymbol{x} = (1, 2, 3)$ and $\boldsymbol{y} = (1, 1, 2)$ show that $P\boldsymbol{x} \cdot \boldsymbol{y}$ is not always equal to $\boldsymbol{x} \cdot P\boldsymbol{y}$.

13 (a) If you take powers of a permutation matrix, why is some power P^k equal to I?

(b) Find a 5 by 5 permutation matrix so that the smallest power to equal I is P^6. (This is a challenge question. Combine a 2 by 2 with a 3 by 3.)

14 Some permutation matrices are symmetric: $P^{\mathrm{T}} = P$. Then $P^{\mathrm{T}} P = I$ becomes $P^2 = I$.

(a) Find a 4 by 4 example with $P^{\mathrm{T}} = P$ that is not just an exchange of two rows.

(b) If P sends row 1 to row 4, then P^{T} sends row ______ to row ______. When $P^{\mathrm{T}} = P$ the row exchanges come in pairs with no overlap.

15 Find 2 by 2 symmetric matrices with these properties:

(a) A is not invertible.

(b) A is invertible but cannot be factored into LU.

(c) A can be factored into LU but not into LL^{T}.

16 If $A = A^{\mathrm{T}}$ and $B = B^{\mathrm{T}}$, which of these matrices are certainly symmetric?

(a) $A^2 - B^2$

(b) $(A + B)(A - B)$

(c) ABA

(d) $ABAB$.

17 (a) How many entries can be chosen independently, if $A = A^{\mathrm{T}}$ is 5 by 5?

(b) How do L and D (still 5 by 5) give the same number of choices?

(c) How many entries can be chosen if A is *skew-symmetric*? This means that $A^{\mathrm{T}} = -A$.

18 Suppose R is rectangular (m by n) and A is symmetric (m by m).

(a) Prove that $R^{\mathrm{T}}AR$ is symmetric. What shape is this matrix?

(b) Prove that $R^{\mathrm{T}}R$ has no negative numbers on its diagonal.

19 Factor these symmetric matrices into $A = LDL^{\mathrm{T}}$. The pivot matrix D is diagonal:

$$A = \begin{bmatrix} 1 & 3 \\ 3 & 2 \end{bmatrix} \quad \text{and} \quad A = \begin{bmatrix} 1 & b \\ b & c \end{bmatrix} \quad \text{and} \quad A = \begin{bmatrix} 2 & -1 & 0 \\ -1 & 2 & -1 \\ 0 & -1 & 2 \end{bmatrix}.$$

20 After elimination clears out column 1 below the pivot, find the symmetric 2 by 2 matrix that remains:

$$A = \begin{bmatrix} 2 & 4 & 8 \\ 4 & 3 & 9 \\ 8 & 9 & 0 \end{bmatrix} \quad \text{and} \quad A = \begin{bmatrix} 1 & b & c \\ b & d & e \\ c & e & f \end{bmatrix}.$$

Questions 21–29 are about the factorizations $PA = LU$ and $A = L_1P_1U_1$.

21 Find the $PA = LU$ factorizations (and check them) for

$$A = \begin{bmatrix} 0 & 1 & 1 \\ 1 & 0 & 1 \\ 2 & 3 & 4 \end{bmatrix} \quad \text{and} \quad A = \begin{bmatrix} 1 & 2 & 0 \\ 2 & 4 & 1 \\ 1 & 1 & 1 \end{bmatrix}.$$

22 Find a 3 by 3 permutation matrix (call it A) that needs two row exchanges to reach the end of elimination. What are its factors P, L, and U?

23 Factor this matrix into $PA = LU$. Factor it also into $A = L_1P_1U_1$ (hold the row exchange until forward elimination is complete):

$$A = \begin{bmatrix} 0 & 1 & 2 \\ 0 & 3 & 8 \\ 2 & 1 & 1 \end{bmatrix}.$$

24 Write out P after each step of the MATLAB code splu, when

$$A = \begin{bmatrix} 0 & 1 \\ 2 & 3 \end{bmatrix} \quad \text{and} \quad A = \begin{bmatrix} 0 & 0 & 1 \\ 2 & 3 & 4 \\ 0 & 5 & 6 \end{bmatrix}.$$

25 Write out P and L after each step of the code splu when

$$A = \begin{bmatrix} 0 & 1 & 2 \\ 1 & 1 & 0 \\ 2 & 5 & 4 \end{bmatrix}.$$

26 Extend the MATLAB code splu to a code spldu which factors PA into LDU.

27 What is the matrix L_1 in $A = L_1P_1U_1$?

$$A = \begin{bmatrix} 1 & 1 & 1 \\ 1 & 1 & 3 \\ 2 & 5 & 8 \end{bmatrix} \rightarrow \begin{bmatrix} 1 & 1 & 1 \\ 0 & 0 & 2 \\ 0 & 3 & 6 \end{bmatrix} = P_1U_1 = \begin{bmatrix} 1 & 0 & 0 \\ 0 & 0 & 1 \\ 0 & 1 & 0 \end{bmatrix} \begin{bmatrix} 1 & 1 & 1 \\ 0 & 3 & 6 \\ 0 & 0 & 2 \end{bmatrix}.$$

28 Suppose A is a permutation matrix. Then L, U, L_1, U_1 all equal I. Explain why P is A^{T} (in $PA = LU$) but P_1 is A.

29 Show that the second pivots are different in U and U_1 when

$$A = \begin{bmatrix} 0 & 1 & 0 \\ 0 & 2 & 1 \\ 1 & 0 & 0 \end{bmatrix}.$$

30 (a) Choose E_{21} to remove the 3 below the first pivot. Then multiply $E_{21}AE_{21}^{\mathrm{T}}$ to remove both 3's:

$$A = \begin{bmatrix} 1 & 3 & 0 \\ 3 & 11 & 4 \\ 0 & 4 & 9 \end{bmatrix} \quad \text{is going toward} \quad D = \begin{bmatrix} 1 & 0 & 0 \\ 0 & 2 & 0 \\ 0 & 0 & 1 \end{bmatrix}.$$

(b) Choose E_{32} to remove the 4 below the second pivot. Then A is reduced to D by $E_{32}E_{21}AE_{21}^{\mathrm{T}}E_{32}^{\mathrm{T}} = D$. Invert the E's to find L in $A = LDL^{\mathrm{T}}$.

The final questions are about applications of the identity $(A\boldsymbol{x})^{\mathrm{T}}\boldsymbol{y} = \boldsymbol{x}^{\mathrm{T}}(A^{\mathrm{T}}\boldsymbol{y})$.

31 Wires go between Boston, Chicago, and Seattle. Those cities are at voltages x_B, x_C, and x_S. With unit resistances between cities, the currents are in $\boldsymbol{y}$:

$$\boldsymbol{y} = A\boldsymbol{x} \quad \text{is} \quad \begin{bmatrix} y_{BC} \\ y_{CS} \\ y_{BS} \end{bmatrix} = \begin{bmatrix} 1 & -1 & 0 \\ 0 & 1 & -1 \\ 1 & 0 & -1 \end{bmatrix} \begin{bmatrix} x_B \\ x_C \\ x_S \end{bmatrix}.$$

(a) Find the total currents $A^{\mathrm{T}}\boldsymbol{y}$ out of the three cities.

(b) Verify that $(A\boldsymbol{x})^{\mathrm{T}}\boldsymbol{y}$ agrees with $\boldsymbol{x}^{\mathrm{T}}(A^{\mathrm{T}}\boldsymbol{y})$—six terms in both.

32 Producing x_1 trucks and x_2 planes needs $x_1 + 50x_2$ tons of steel, $40x_1 + 1000x_2$ pounds of rubber, and $2x_1 + 50x_2$ months of work. If the unit costs y_1, y_2, y_3 are \$700 per ton, \$3 per pound, and \$3000 per month, what are the values of a truck and a plane? Those are the components of $A^{\mathrm{T}}\boldsymbol{y}$.

33 $A\boldsymbol{x}$ gives the amounts of steel, rubber, and work to produce $\boldsymbol{x}$ in Problem 32. Find A. Then $A\boldsymbol{x} \cdot \boldsymbol{y}$ is the ______ of inputs while $\boldsymbol{x} \cdot A^{\mathrm{T}}\boldsymbol{y}$ is the value of ______ .

34 The matrix P that multiplies (x, y, z) to give (z, x, y) is also a rotation matrix. Find P and P^3. The rotation axis $\boldsymbol{a} = (1, 1, 1)$ equals $P\boldsymbol{a}$. What is the angle of rotation from $\boldsymbol{v} = (2, 3, -5)$ to $P\boldsymbol{v} = (-5, 2, 3)$?

35 Write $A = \begin{bmatrix} 1 & 2 \\ 4 & 9 \end{bmatrix}$ as the product EH of an elementary row operation matrix E and a symmetric matrix H.

36 Suppose D is a diagonal matrix and U is upper triangular. Choose two matrices C that make $H = CDU$ into a symmetric matrix.

37 This chapter ends with a great new factorization $A = EH$. Start from $A = LDU$, with 1's on the diagonals of L and U. Insert C^{-1} and C to find E and H:

$$A = (LC^{-1})(CDU) = (\text{lower triangular } E)\ (\text{symmetric matrix } H)$$

with 1's on the diagonal of E. What is C?

3

VECTOR SPACES AND SUBSPACES

3.1 Spaces of Vectors

To a newcomer, matrix calculations involve a lot of numbers. To you, they involve vectors. The columns of $A\boldsymbol{x}$ and AB are linear combinations of n vectors—the columns of A. This chapter moves from numbers and vectors to a third level of understanding (the highest level). Instead of individual columns, we look at ***spaces*** of vectors. Without seeing vector spaces and especially their ***subspaces***, you haven't understood everything about $A\boldsymbol{x} = \boldsymbol{b}$.

Since this chapter goes a little deeper, it may seem a little harder. That is natural. We are looking inside the calculations, to find the mathematics. The authors' job is to make it clear. These pages go to the heart of linear algebra.

We begin with the most important vector spaces. They are denoted by $\mathbf{R}^1$, $\mathbf{R}^2$, $\mathbf{R}^3$, $\mathbf{R}^4, \ldots$. Each space $\mathbf{R}^n$ consists of a whole collection of vectors. $\mathbf{R}^5$ contains all column vectors with five components. This is called "5-dimensional space."

Definition ***The space $\mathbf{R}^n$ consists of all column vectors with n components.***
The components are real numbers, which is the reason for the letter $\mathbf{R}$. A vector whose n components are complex numbers lies in the space $\mathbf{C}^n$.

The space $\mathbf{R}^2$ is represented by the usual xy plane. The two components of the vector give the x and y coordinates of a point, and the vector goes out from $(0, 0)$. Similarly the vectors in $\mathbf{R}^3$ correspond to points in three-dimensional space. The one-dimensional space $\mathbf{R}^1$ is a line (like the x axis). As before, we print vectors as a column between brackets, or along a line using commas and parentheses:

$$\begin{bmatrix} 4 \\ 0 \\ 1 \end{bmatrix} \text{ is in } \mathbf{R}^3, \quad (1, 1, 0, 1, 1) \text{ is in } \mathbf{R}^5, \quad \begin{bmatrix} 1+i \\ 1-i \end{bmatrix} \text{ is in } \mathbf{C}^2.$$

The great thing about linear algebra is that it deals easily with five-dimensional space. We don't draw the vectors, we just need the five numbers (or n numbers). To multiply $\boldsymbol{v}$

by 7, multiply every component by 7. Here 7 is a "scalar." To add vectors in $\mathbf{R}^5$, add them a component at a time. The two essential vector operations go on *inside the vector space*:

We can add two vectors in $\mathbf{R}^n$, and we can multiply any vector by any scalar.

"Inside the vector space" means that ***the result stays in the same space.*** If $\boldsymbol{v}$ is the vector in $\mathbf{R}^4$ with components $1, 0, 0, 1$, then $2\boldsymbol{v}$ is the vector in $\mathbf{R}^4$ with components $2, 0, 0, 2$. (In this case 2 is the "scalar.") A whole series of properties can be verified in every $\mathbf{R}^n$. The commutative law is $\boldsymbol{v} + \boldsymbol{w} = \boldsymbol{w} + \boldsymbol{v}$, the distributive law is $c(\boldsymbol{v} + \boldsymbol{w}) = c\boldsymbol{v} + c\boldsymbol{w}$, and there is a unique "zero vector" satisfying $\mathbf{0} + \boldsymbol{v} = \boldsymbol{v}$. Those are three of the eight conditions listed at the start of the problem set. These conditions are required of every vector space. The point of the eight conditions is that there are vector spaces other than $\mathbf{R}^n$, and they have to obey reasonable rules.

A real vector space is a set of "vectors" *together with rules for vector addition and for multiplication by real numbers.* The addition and the multiplication must produce vectors that are in the space. And the eight conditions must be satisfied (which is usually no problem). Here are three vector spaces other than $\mathbf{R}^n$:

M The vector space of *all real* 2 *by* 2 *matrices.*

F The vector space of *all real functions* $f(x)$.

Z The vector space that consists only of a *zero vector.*

In **M** the "vectors" are really matrices; in **F** the vectors are functions; in **Z** the only addition is $\mathbf{0} + \mathbf{0} = \mathbf{0}$. In each case we can add: matrices to matrices, functions to functions, zero vector to zero vector. We can multiply a matrix by 4 or a function by 4 or the zero vector by 4. The result is still in **M** or **F** or **Z**. The eight conditions (all easily checked) are discussed in the exercises.

The space **Z** is zero-dimensional (by any reasonable definition of dimension). It is the smallest possible vector space. We hesitate to call it $\mathbf{R}^0$, which means no components—you might think there was no vector. The vector space **Z** contains exactly one vector (zero). No space can do without that zero vector. In fact each space has its own zero vector—the zero matrix, the zero function, the vector $(0, 0, 0)$ in $\mathbf{R}^3$.

Subspaces

At different times, we will ask you to think of matrices and functions as vectors. But at all times, the vectors that we need most are ordinary column vectors. They are vectors with n components—but *maybe not all* of the vectors with n components. There are important vector spaces ***inside*** $\mathbf{R}^n$.

Start with the usual three-dimensional space $\mathbf{R}^3$. Choose a plane through the origin $(0, 0, 0)$. ***That plane is a vector space in its own right.*** If we add two vectors in the plane, their sum is in the plane. If we multiply an in-plane vector by 2 or -5, it is still in the

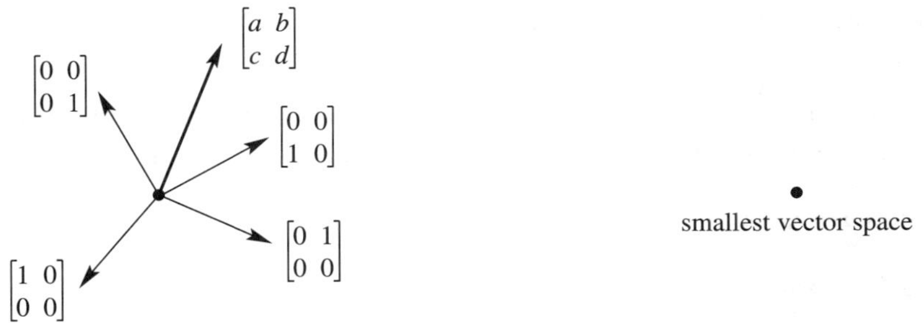

Figure 3.1 The "four-dimensional" matrix space $\mathbf{M}$. The "zero-dimensional" space $\mathbf{Z}$.

plane. The plane is not $\mathbf{R}^2$ (even if it is like $\mathbf{R}^2$). The vectors have three components and they belong to $\mathbf{R}^3$. The plane is a vector space inside $\mathbf{R}^3$.

This illustrates one of the most fundamental ideas in linear algebra. The plane is a ***subspace*** of the full vector space $\mathbf{R}^3$.

Definition A ***subspace*** of a vector space is a set of vectors (including $\mathbf{0}$) that satisfies two requirements: ***If v and w are vectors in the subspace and c is any scalar, then***

(i) $\boldsymbol{v} + \boldsymbol{w}$ is in the subspace and **(ii)** $c\boldsymbol{v}$ is in the subspace.

In other words, the set of vectors is "closed" under addition and multiplication. Those operations leave us in the subspace. They follow the rules of the host space, so the eight required conditions are automatic—they are already satisfied in the larger space. We just have to check the requirements (i) and (ii) for a subspace.

First fact: Every subspace contains the zero vector. The plane in $\mathbf{R}^3$ has to go through $(0, 0, 0)$. We mention this separately, for extra emphasis, but it follows directly from rule (ii). Choose $c = 0$, and the rule requires $0\boldsymbol{v}$ to be in the subspace. Also $-\boldsymbol{v}$ is in the subspace.

Planes that don't contain the origin fail those tests. When $\boldsymbol{v}$ is on the plane, $-\boldsymbol{v}$ and $0\boldsymbol{v}$ are *not* on the plane. A plane that misses the origin is not a subspace.

Lines through the origin are also subspaces. When we multiply by 5, or add two vectors on the line, we stay on the line. The whole space is a subspace (*of itself*). Here is a list of all the possible subspaces of $\mathbf{R}^3$:

(**L**) Any line through $(0, 0, 0)$ ($\mathbf{R}^3$) the whole space

(**P**) Any plane through $(0, 0, 0)$ (**Z**) the single vector $(0, 0, 0)$

If we try to keep only *part* of a plane or line, the requirements for a subspace don't hold. Look at these examples.

Example 3.1 Keep only the vectors (x, y) whose components are positive or zero (a quarter-plane). The vector $(2, 3)$ is included but $(-2, -3)$ is not. So rule (ii) is violated when we try to multiply by $c = -1$. The quarter-plane is not a subspace.

Example 3.2 Include also the vectors whose components are both negative. Now we have two quarter-planes. Requirement (ii) is satisfied; we can multiply by any c. But rule (i)

now fails. The sum of $\boldsymbol{v} = (2, 3)$ and $\boldsymbol{w} = (-3, -2)$ is $(-1, 1)$, and this vector is outside the quarter-planes.

Rules (i) and (ii) involve vector addition $\boldsymbol{v} + \boldsymbol{w}$ and multiplication by scalars like c and d. The rules can be combined into a single requirement—*the rule for subspaces*:

A subspace containing v and w must contain all linear combinations $c\boldsymbol{v} + d\boldsymbol{w}$.

Example 3.3 Inside the vector space **M** of all 2 by 2 matrices, here are two subspaces:

$$(\mathbf{U}) \text{ All upper triangular matrices } \begin{bmatrix} a & b \\ 0 & d \end{bmatrix} \quad (\mathbf{D}) \text{ All diagonal matrices } \begin{bmatrix} a & 0 \\ 0 & d \end{bmatrix}.$$

Add any two matrices in **U**, and the sum is inD **U**. The same is true for diagonal matrices in **D**. In this case **D** is also a subspace of **U**! Of course the zero matrix is in these subspaces, when a, b, and d all equal zero.

To find a smaller subspace of diagonal matrices, we could require $a = d$. The matrices are multiples of the identity matrix I. The sum $2I + 3I$ is in this subspace, and so is 3 times $4I$. It is a "line of matrices" inside **M** and **U** and **D**.

Is the matrix I a subspace by itself? Certainly not. Only the zero matrix is. Your mind will invent more subspaces of 2 by 2 matrices—write them down for Problem 5.

The Column Space of A

We come to the most important subspaces, which are tied directly to a matrix A. We are trying to solve $A\boldsymbol{x} = \boldsymbol{b}$. If A is not invertible, the system is solvable for some $\boldsymbol{b}$ and not solvable for other $\boldsymbol{b}$. How can we describe the good right sides—the vectors that *can* be written as A times some vector $\boldsymbol{x}$?

Remember the key to $A\boldsymbol{x}$. It is a combination of the columns of A. To get every possible $\boldsymbol{b}$, we use every possible $\boldsymbol{x}$. So start with the columns of A, and ***take all their linear combinations. This produces the column space of A.*** It is a vector space made up of column vectors—not just the columns of A, but all combinations $A\boldsymbol{x}$ of those columns.

By taking all combinations, we fill out a vector space.

Definition The ***column space*** of A consists of all linear combinations of the columns. The combinations are the vectors $A\boldsymbol{x}$.

For the solvability of $A\boldsymbol{x} = \boldsymbol{b}$, the question is whether $\boldsymbol{b}$ is a combination of the columns. The vector $\boldsymbol{b}$ on the right side has to be in the column space produced by A on the left side.

3A The system $A\boldsymbol{x} = \boldsymbol{b}$ is solvable if and only if $\boldsymbol{b}$ is in the column space of A.

When $\boldsymbol{b}$ is in the column space, it is a combination of the columns. The coefficients in that combination give us a solution $\boldsymbol{x}$ to the system $A\boldsymbol{x} = \boldsymbol{b}$.

Suppose A is an m by n matrix. Its columns have m components (not n). So the columns belong to $\mathbf{R}^m$. ***The column space of A is a subspace of*** $\mathbf{R}^m$. The set of all

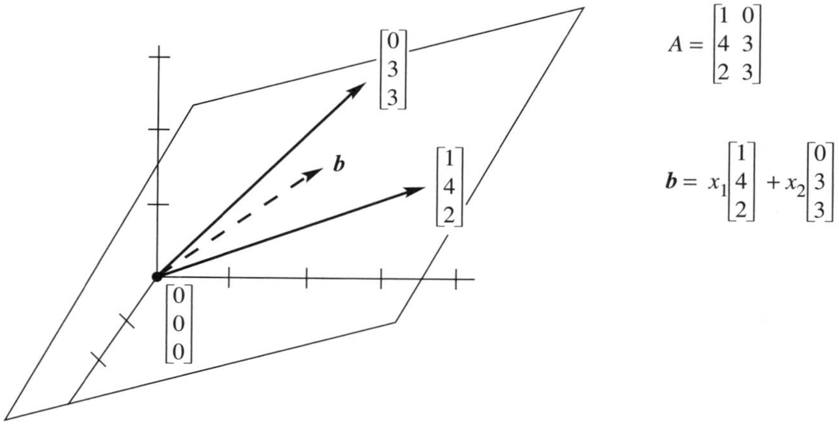

Figure 3.2 The column space $\boldsymbol{R}(A)$ is a plane. $A\boldsymbol{x} = \boldsymbol{b}$ is solvable when $\boldsymbol{b}$ is on that plane.

column combinations $A\boldsymbol{x}$ satisfies rules (i) and (ii): When we add linear combinations or multiply by scalars, we still produce combinations of the columns. The word "subspace" is justified. Here is a 3 by 2 matrix, whose column space is a subspace of $\mathbf{R}^3$.

Example 3.4

$$A\boldsymbol{x} \quad \text{is} \quad \begin{bmatrix} 1 & 0 \\ 4 & 3 \\ 2 & 3 \end{bmatrix} \begin{bmatrix} x_1 \\ x_2 \end{bmatrix} \quad \text{which is} \quad x_1 \begin{bmatrix} 1 \\ 4 \\ 2 \end{bmatrix} + x_2 \begin{bmatrix} 0 \\ 3 \\ 3 \end{bmatrix}.$$

The column space consists of all combinations of the two columns—any x_1 times the first column plus any x_2 times the second column. *Those combinations fill up a plane in* $\mathbf{R}^3$ (Figure 3.2). If the right side $\boldsymbol{b}$ lies on that plane, then it is one of the combinations and (x_1, x_2) is a solution to $A\boldsymbol{x} = \boldsymbol{b}$. The plane has zero thickness, so it is more likely that $\boldsymbol{b}$ is not in the column space. Then there is no solution to our 3 equations in 2 unknowns.

Of course (0, 0, 0) is in the column space. The plane passes through the origin. There is certainly a solution to $A\boldsymbol{x} = \mathbf{0}$. That solution, always available, is $\boldsymbol{x} =$ ______.

To repeat, the attainable right sides $\boldsymbol{b}$ are exactly the vectors in the column space. One possibility is the first column itself—take $x_1 = 1$ and $x_2 = 0$. Another combination is the second column—take $x_1 = 0$ and $x_2 = 1$. The new level of understanding is to see *all* combinations—the whole subspace is generated by those two columns.

Notation The column space of A is denoted by $\boldsymbol{R}(A)$. This time $\boldsymbol{R}$ stands for ***"range"***. We write $\boldsymbol{R}(A)$ and $\boldsymbol{R}(B)$ and $\boldsymbol{R}(C)$ for the column spaces of A and B and C. Start with the columns and take all their linear combinations.

Example 3.5 Describe the column spaces for

$$A = \begin{bmatrix} 1 & 0 \\ 0 & 1 \end{bmatrix} \quad \text{and} \quad B = \begin{bmatrix} 1 & 2 \\ 2 & 4 \end{bmatrix} \quad \text{and} \quad C = \begin{bmatrix} 1 & 2 & 3 \\ 0 & 0 & 4 \end{bmatrix}.$$

Solution The column space of $A = I$ is the *whole space* $\mathbf{R}^2$. Every vector is a combination of the columns of I. In vector space language, $\boldsymbol{R}(I)$ is $\mathbf{R}^2$.

The column space of B is only a line. The second column $(2, 4)$ is a multiple of the first column $(1, 2)$. Those vectors are different, but our eye is on vector ***spaces***. The column space contains $(1, 2)$ and $(2, 4)$ and all other vectors $(c, 2c)$ along that line. The equation $A\boldsymbol{x} = \boldsymbol{b}$ is only solvable when $\boldsymbol{b}$ is on the line.

The third matrix (with three columns) places no restriction on $\boldsymbol{b}$. The column space $\boldsymbol{R}(C)$ is all of $\mathbf{R}^2$. Every $\boldsymbol{b}$ is attainable. The vector $\boldsymbol{b} = (5, 4)$ is column 2 plus column 3, so $\boldsymbol{x}$ can be $(0, 1, 1)$. The same vector $(5, 4)$ is also 2(column 1) + column 3, so another possible $\boldsymbol{x}$ is $(2, 0, 1)$. This matrix has the same column space as $A = I$—any $\boldsymbol{b}$ is allowed. But now $\boldsymbol{x}$ has more components and there are more solutions.

The next section creates another vector space, to describe all the possible solutions of $A\boldsymbol{x} = \boldsymbol{0}$. This section created the column space, to describe all the attainable right sides $\boldsymbol{b}$.

Problem Set 3.1

The first problems are about vector spaces in general. The vectors in those spaces are not necessarily column vectors. In the definition of a *vector space*, vector addition $\boldsymbol{x} + \boldsymbol{y}$ and scalar multiplication $c\boldsymbol{x}$ are required to obey the following eight rules:

(1) $\boldsymbol{x} + \boldsymbol{y} = \boldsymbol{y} + \boldsymbol{x}$

(2) $\boldsymbol{x} + (\boldsymbol{y} + \boldsymbol{z}) = (\boldsymbol{x} + \boldsymbol{y}) + \boldsymbol{z}$

(3) There is a unique "zero vector" such that $\boldsymbol{x} + \boldsymbol{0} = \boldsymbol{x}$ for all $\boldsymbol{x}$

(4) For each $\boldsymbol{x}$ there is a unique vector $-\boldsymbol{x}$ such that $\boldsymbol{x} + (-\boldsymbol{x}) = \boldsymbol{0}$

(5) 1 times $\boldsymbol{x}$ equals $\boldsymbol{x}$

(6) $(c_1 c_2)\boldsymbol{x} = c_1(c_2\boldsymbol{x})$

(7) $c(\boldsymbol{x} + \boldsymbol{y}) = c\boldsymbol{x} + c\boldsymbol{y}$

(8) $(c_1 + c_2)\boldsymbol{x} = c_1\boldsymbol{x} + c_2\boldsymbol{x}$.

1 Suppose the sum $(x_1, x_2) + (y_1, y_2)$ is defined to be $(x_1 + y_2, x_2 + y_1)$. With the usual multiplication $c\boldsymbol{x} = (cx_1, cx_2)$, which of the eight conditions are not satisfied?

2 Suppose the multiplication $c\boldsymbol{x}$ is defined to produce $(cx_1, 0)$ instead of (cx_1, cx_2). With the usual addition in $\mathbf{R}^2$, are the eight conditions satisfied?

3 (a) Which rules are broken if we keep only the positive numbers $x > 0$ in $\mathbf{R}^1$? Every c must be allowed.

(b) The positive numbers with $\boldsymbol{x} + \boldsymbol{y}$ and $c\boldsymbol{x}$ redefined to equal the usual xy and x^c *do* satisfy the eight rules. Test rule 7 when $c = 3, x = 2, y = 1$. (Then $\boldsymbol{x} + \boldsymbol{y} = 2$ and $c\boldsymbol{x} = 8$.) Which number is the "zero vector"? The vector "**–2**" is the number ______ .

4 The matrix $A = \begin{bmatrix} 2 & -2 \\ 2 & -2 \end{bmatrix}$ is a "vector" in the space $\mathbf{M}$ of all 2 by 2 matrices. Write down the zero vector in this space, the vector $\frac{1}{2}A$, and the vector $-A$. What matrices are in the smallest subspace containing A?

5 (a) Describe a subspace of $\mathbf{M}$ that contains $A = \begin{bmatrix} 1 & 0 \\ 0 & 0 \end{bmatrix}$ but not $B = \begin{bmatrix} 0 & 0 \\ 0 & -1 \end{bmatrix}$.

(b) If a subspace of $\mathbf{M}$ contains A and B, must it contain I?

(c) Describe a subspace of $\mathbf{M}$ that contains no nonzero diagonal matrices.

6 The functions $\boldsymbol{f}(x) = x^2$ and $\boldsymbol{g}(x) = 5x$ are "vectors" in $\mathbf{F}$. This is the vector space of all real functions. (The functions are defined for $-\infty < x < \infty$.) The combination $3\boldsymbol{f}(x) - 4\boldsymbol{g}(x)$ is the function $\boldsymbol{h}(x) =$ ______.

7 Which rule is broken if multiplying $\boldsymbol{f}(x)$ by c gives the function $f(cx)$? Keep the usual addition $\boldsymbol{f}(x) + \boldsymbol{g}(x)$.

8 If the sum of the "vectors" $\boldsymbol{f}(x)$ and $\boldsymbol{g}(x)$ is defined to be the function $f(g(x))$, then the "zero vector" is $\boldsymbol{g}(x) = x$. Keep the usual scalar multiplication $c\boldsymbol{f}(x)$ and find two rules that are broken.

Questions 9–18 are about the "subspace requirements": $\boldsymbol{x} + \boldsymbol{y}$ and $c\boldsymbol{x}$ must stay in the subspace.

9 One requirement can be met while the other fails. Show this with

(a) A set of vectors in $\mathbf{R}^2$ for which $\boldsymbol{x} + \boldsymbol{y}$ stays in the set but $\frac{1}{2}\boldsymbol{x}$ may be outside.

(b) A set of vectors in $\mathbf{R}^2$ (other than two quarter-planes) for which every $c\boldsymbol{x}$ stays in the set but $\boldsymbol{x} + \boldsymbol{y}$ may be outside.

10 Which of the following subsets of $\mathbf{R}^3$ are actually subspaces?

(a) The plane of vectors (b_1, b_2, b_3) with $b_1 = 0$.

(b) The plane of vectors with $b_1 = 1$.

(c) The vectors with $b_1 b_2 b_3 = 0$.

(d) All linear combinations of $\boldsymbol{v} = (1, 4, 0)$ and $\boldsymbol{w} = (2, 2, 2)$.

(e) All vectors that satisfy $b_1 + b_2 + b_3 = 0$.

(f) All vectors with $b_1 \le b_2 \le b_3$.

11 Describe the smallest subspace of the matrix space $\mathbf{M}$ that contains

(a) $\begin{bmatrix} 1 & 0 \\ 0 & 0 \end{bmatrix}$ and $\begin{bmatrix} 0 & 1 \\ 0 & 0 \end{bmatrix}$

(b) $\begin{bmatrix} 1 & 1 \\ 0 & 0 \end{bmatrix}$ (c) $\begin{bmatrix} 1 & 0 \\ 0 & 0 \end{bmatrix}$ and $\begin{bmatrix} 1 & 0 \\ 0 & 1 \end{bmatrix}$.

12 Let P be the plane in $\mathbf{R}^3$ with equation $x + y - 2z = 4$. Find two vectors in P and check that their sum is not in P.

13 Let $\mathbf{P}_0$ be the plane through $(0, 0, 0)$ parallel to the previous plane P. What is the equation for $\mathbf{P}_0$? Find two vectors in $\mathbf{P}_0$ and check that their sum is in $\mathbf{P}_0$.

14 The subspaces of $\mathbf{R}^3$ are planes, lines, $\mathbf{R}^3$ itself, or $\mathbf{Z}$ containing $(0, 0, 0)$.

(a) Describe the three types of subspaces of $\mathbf{R}^2$.

(b) Describe the five types of subspaces of $\mathbf{R}^4$.

15 (a) The intersection of two planes through $(0, 0, 0)$ is probably a ______.

(b) The intersection of a plane through $(0, 0, 0)$ with a line through $(0, 0, 0)$ is probably a ______.

(c) If $\mathbf{S}$ and $\mathbf{T}$ are subspaces of $\mathbf{R}^5$, prove that $\mathbf{S} \cap \mathbf{T}$ (the set of vectors in both subspaces) is a subspace of $\mathbf{R}^5$. *Check the requirements on* $\boldsymbol{x} + \boldsymbol{y}$ *and* $c\boldsymbol{x}$.

16 Suppose $\mathbf{P}$ is a plane through $(0, 0, 0)$ and $\mathbf{L}$ is a line through $(0, 0, 0)$. The smallest vector space containing both $\mathbf{P}$ and $\mathbf{L}$ is either ______ or ______.

17 (a) Show that the set of *invertible* matrices in $\mathbf{M}$ is not a subspace.

(b) Show that the set of *singular* matrices in $\mathbf{M}$ is not a subspace.

18 True or false (check addition in each case by an example):

(a) The symmetric matrices in $\mathbf{M}$ (with $A^{\mathrm{T}} = A$) form a subspace.

(b) The skew-symmetric matrices in $\mathbf{M}$ (with $A^{\mathrm{T}} = -A$) form a subspace.

(c) The unsymmetric matrices in $\mathbf{M}$ (with $A^{\mathrm{T}} \neq A$) form a subspace.

Questions 19–27 are about column spaces $R(A)$ and the equation $A\boldsymbol{x} = \boldsymbol{b}$.

19 Describe the column spaces (lines or planes) of these particular matrices:

$$A = \begin{bmatrix} 1 & 2 \\ 0 & 0 \\ 0 & 0 \end{bmatrix} \quad \text{and} \quad B = \begin{bmatrix} 1 & 0 \\ 0 & 2 \\ 0 & 0 \end{bmatrix} \quad \text{and} \quad C = \begin{bmatrix} 1 & 0 \\ 2 & 0 \\ 0 & 0 \end{bmatrix}.$$

20 For which right sides $\boldsymbol{b}$ do these systems have solutions?

$$\text{(a)} \quad \begin{bmatrix} 1 & 4 & 2 \\ 2 & 8 & 4 \\ -1 & -4 & -2 \end{bmatrix} \begin{bmatrix} x_1 \\ x_2 \\ x_3 \end{bmatrix} = \begin{bmatrix} b_1 \\ b_2 \\ b_3 \end{bmatrix} \qquad \text{(b)} \quad \begin{bmatrix} 1 & 4 \\ 2 & 9 \\ -1 & -4 \end{bmatrix} \begin{bmatrix} x_1 \\ x_2 \end{bmatrix} = \begin{bmatrix} b_1 \\ b_2 \\ b_3 \end{bmatrix}.$$

21 Adding row 1 of A to row 2 produces B. Adding column 1 to column 2 produces C. A combination of the columns of ______ is also a combination of the columns of A. Those two matrices have the same column ______:

$$A = \begin{bmatrix} 1 & 2 \\ 2 & 4 \end{bmatrix} \quad \text{and} \quad B = \begin{bmatrix} 1 & 2 \\ 3 & 6 \end{bmatrix} \quad \text{and} \quad C = \begin{bmatrix} 1 & 3 \\ 2 & 6 \end{bmatrix}.$$

22 For which vectors (b_1, b_2, b_3) do these systems have a solution?

$$\begin{bmatrix} 1 & 1 & 1 \\ 0 & 1 & 1 \\ 0 & 0 & 1 \end{bmatrix} \begin{bmatrix} x_1 \\ x_2 \\ x_3 \end{bmatrix} = \begin{bmatrix} b_1 \\ b_2 \\ b_3 \end{bmatrix} \quad \text{and} \quad \begin{bmatrix} 1 & 1 & 1 \\ 0 & 1 & 1 \\ 0 & 0 & 0 \end{bmatrix} \begin{bmatrix} x_1 \\ x_2 \\ x_3 \end{bmatrix} = \begin{bmatrix} b_1 \\ b_2 \\ b_3 \end{bmatrix}$$

$$\text{and} \quad \begin{bmatrix} 1 & 1 & 1 \\ 0 & 0 & 1 \\ 0 & 0 & 1 \end{bmatrix} \begin{bmatrix} x_1 \\ x_2 \\ x_3 \end{bmatrix} = \begin{bmatrix} b_1 \\ b_2 \\ b_3 \end{bmatrix}.$$

23 If we add an extra column $\boldsymbol{b}$ to a matrix A, then the column space gets larger unless ______. Give an example where the column space gets larger and an example where it doesn't.

24 The columns of AB are combinations of the columns of A. The column *space* of AB is contained in (possibly equal to) the column space of ______. Give an example where those two column spaces are not equal.

25 Suppose $A\boldsymbol{x} = \boldsymbol{b}$ and $A\boldsymbol{y} = \boldsymbol{b}^*$ are both solvable. Then $A\boldsymbol{z} = \boldsymbol{b} + \boldsymbol{b}^*$ is solvable. What is $\boldsymbol{z}$? This translates into: If $\boldsymbol{b}$ and $\boldsymbol{b}^*$ are in the column space $\boldsymbol{R}(A)$, then ______.

26 If A is any 5 by 5 invertible matrix, then its column space is ______. Why?

27 True or false (with a counterexample if false):

(a) The vectors $\boldsymbol{b}$ that are not in the column space $\boldsymbol{R}(A)$ form a subspace.

(b) If $\boldsymbol{R}(A)$ contains only the zero vector, then A is the zero matrix.

(c) The column space of $2A$ equals the column space of A.

(d) The column space of $A - I$ equals the column space of A.

28 Construct a 3 by 3 matrix whose column space contains $(1, 1, 0)$ and $(1, 0, 1)$ but not $(1, 1, 1)$.

3.2 The Nullspace of A: Solving $A\boldsymbol{x} = \boldsymbol{0}$

This section is about $A\boldsymbol{x} = \boldsymbol{0}$ for rectangular matrices. You will see the crucial role of the pivots—and especially the importance of ***missing pivots***. Note first: $A\boldsymbol{x} = \boldsymbol{0}$ can always be solved. *One immediate solution is* $\boldsymbol{x} = \boldsymbol{0}$*—the zero combination.* Elimination will decide if there are other solutions to $A\boldsymbol{x} = \boldsymbol{0}$, and find out what they are.

Start with an important subspace. It contains every solution $\boldsymbol{x}$. The columns of A have m components, but now $\boldsymbol{x}$ has n components. Please notice that difference!

Definition The ***nullspace*** of A consists of ***all solutions to*** $A\boldsymbol{x} = \boldsymbol{0}$. These vectors $\boldsymbol{x}$ are in $\mathbf{R}^n$. The nullspace containing the solutions $\boldsymbol{x}$ is denoted by $\boldsymbol{N}(A)$.

Check that the solution vectors form a subspace. Suppose $\boldsymbol{x}$ and $\boldsymbol{y}$ are in the nullspace (this means $A\boldsymbol{x} = \mathbf{0}$ and $A\boldsymbol{y} = \mathbf{0}$). The rules of matrix multiplication give $A(\boldsymbol{x} + \boldsymbol{y}) = \mathbf{0} + \mathbf{0}$. The rules also give $A(c\boldsymbol{x}) = c\mathbf{0}$. The right sides are still zero. Therefore $\boldsymbol{x} + \boldsymbol{y}$ and $c\boldsymbol{x}$ are in the nullspace $\boldsymbol{N}(A)$. Since we can add and multiply without leaving the nullspace, it is a subspace.

To repeat: The solution vectors $\boldsymbol{x}$ are in $\mathbf{R}^n$. The nullspace is a subspace of $\mathbf{R}^n$, while the column space $\boldsymbol{R}(A)$ is a subspace of $\mathbf{R}^m$. If the right side $\boldsymbol{b}$ is not zero, the solutions of $A\boldsymbol{x} = \boldsymbol{b}$ do *not* form a subspace. The vector $\boldsymbol{x} = \mathbf{0}$ is not a solution if $\boldsymbol{b} \neq \mathbf{0}$.

Example 3.6 The solutions to $x + 2y + 3z = 0$ form a plane through the origin. It is a subspace of $\mathbf{R}^3$. The plane is the nullspace of the 1 by 3 matrix $A = [1 \;\; 2 \;\; 3]$. The solutions to $x + 2y + 3z = 6$ also form a plane, but this plane is not a subspace.

For many matrices, the only solution to $A\boldsymbol{x} = \mathbf{0}$ is $\boldsymbol{x} = \mathbf{0}$. Their nullspaces contain only that single vector $\boldsymbol{x} = \mathbf{0}$. We call this space $\mathbf{Z}$, for zero. The only combination of the columns that produces $\boldsymbol{b} = \mathbf{0}$ is then the "zero combination" or "trivial combination." The solution is trivial (just $\boldsymbol{x} = \mathbf{0}$) but the idea is not trivial.

This case of a zero nullspace and a unique solution is of the greatest importance.

Example 3.7 Describe the nullspaces of $A = \begin{bmatrix} 1 & 2 \\ 2 & 4 \end{bmatrix}$ and $B = \begin{bmatrix} 1 & 2 \\ 3 & 4 \end{bmatrix}$.

Solution Apply elimination to the linear equations $A\boldsymbol{x} = \mathbf{0}$:

$$\begin{aligned} x_1 + 2x_2 &= 0 \\ 2x_1 + 4x_2 &= 0 \end{aligned} \quad \rightarrow \quad \begin{aligned} x_1 + 2x_2 &= 0 \\ 0 &= 0 \end{aligned}$$

There is really only one equation. The second equation is twice the first equation. In the row picture, the line $x_1 + 2x_2 = 0$ is the same as the line $2x_1 + 4x_2 = 0$. That line is the nullspace $\boldsymbol{N}(A)$.

To describe a typical point in the nullspace, give any value to x_2. *This unknown is "free".* Then x_1 equals $-2x_2$. If $x_2 = 4$ then $x_1 = -8$. If $x_2 = 5$ then $x_1 = -10$. Those vectors $(-8, 4)$ and $(-10, 5)$ and $(-2x_2, x_2)$ lie in the nullspace $\boldsymbol{N}(A)$.

The nullspace of this A is a line in the direction of $(-2, 1)$. The line goes through $(0, 0)$, which belongs to every nullspace.

For the matrix B, the nullspace contains only the zero vector:

$$\begin{aligned} x_1 + 2x_2 &= 0 \\ 3x_1 + 4x_2 &= 0 \end{aligned} \quad \rightarrow \quad \begin{aligned} x_1 + 2x_2 &= 0 \\ -2x_2 &= 0 \end{aligned} \quad \rightarrow \quad \begin{aligned} x_1 &= 0 \\ x_2 &= 0 \end{aligned}$$

The only solution to $B\boldsymbol{x} = \mathbf{0}$ is $\boldsymbol{x} = \mathbf{0}$. B is invertible so automatically $\boldsymbol{x} = B^{-1}\mathbf{0} = \mathbf{0}$.

Elimination easily found the nullspaces for these 2 by 2 systems. The same method extends to m by n systems, square or rectangular. But there are new possibilities for columns without pivots. You have to see the elimination process one more time.

Solving $Ax = \mathbf{0}$ by Elimination

This is important. We solve m equations in n unknowns—and the right sides are all zero. The left sides are simplified by row operations, after which we read off the solution (or solutions). Remember the two stages in solving $Ax = \mathbf{0}$:

1 Forward elimination from A to a triangular U.

2 Back substitution in $U\boldsymbol{x} = \mathbf{0}$ to find $\boldsymbol{x}$.

You will notice a difference in back substitution, when A and U have fewer than n pivots. We are allowing all matrices in this section, not just the nice ones (with inverses).

Pivots are still nonzero. The columns below them are still zero. But it might happen that a column has no pivot. In that case, don't stop the calculation. Go on to the next column. The first example is a 3 by 4 matrix:

$$A = \begin{bmatrix} 1 & 2 & 3 & 4 \\ 2 & 4 & 8 & 10 \\ 3 & 6 & 11 & 14 \end{bmatrix}.$$

Certainly $a_{11} = 1$ is the first pivot. Clear out the 2 and 3 below it:

$$A \to \begin{bmatrix} 1 & 2 & 3 & 4 \\ 0 & 0 & 2 & 2 \\ 0 & 0 & 2 & 2 \end{bmatrix} \qquad \begin{matrix} \\ \text{(subtract } 2 \times \text{ row 1)} \\ \text{(subtract } 3 \times \text{ row 1)} \end{matrix}$$

The second column has a zero in the pivot position. We look below the zero for a nonzero entry, ready to do a row exchange. *The entry below that position is also zero.* Elimination can do nothing with the second column. This signals trouble, which we expect anyway for a rectangular matrix. There is no reason to quit, and we go on to the third column.

The second pivot is 2 (but it is in the third column). Subtracting row 2 from row 3 clears out that column. We arrive at

$$U = \begin{bmatrix} 1 & 2 & 3 & 4 \\ 0 & 0 & 2 & 2 \\ 0 & 0 & 0 & 0 \end{bmatrix} \qquad \begin{matrix} (\textit{only two pivots}) \\ (\textit{the last equation} \\ \textit{became } 0 = 0). \end{matrix}$$

The fourth column also has a zero in the pivot position—but nothing can be done. There is no row below it to exchange, and forward elimination is complete. The matrix has three rows, four columns, and *only two pivots*. The original $A\boldsymbol{x} = \mathbf{0}$ seemed to involve three different equations, but the third equation is the sum of the first two. It is automatically satisfied ($0 = 0$) when the first two equations are satisfied. Elimination reveals the inner truth about a system of equations.

Now comes back substitution, to find all solutions to $U\boldsymbol{x} = \mathbf{0}$. With four unknowns and only two pivots, there are many solutions. The question is how to write them all down.

A good method is to separate the basic variables or ***pivot variables*** from the ***free variables***.

P The ***pivot*** variables are x_1 and x_3, because columns 1 and 3 contain pivots.

F The ***free*** variables are x_2 and x_4, because columns 2 and 4 have no pivots.

The free variables x_2 and x_4 can be given any values whatsoever. Then back substitution finds the pivot variables x_1 and x_3. (In Chapter 2 no variables were free. When A is invertible, all variables are pivot variables.) You could think of the pivot variables on the left side and the free variables moved to the right side:

$$U\boldsymbol{x} = \begin{bmatrix} 1 & 2 & 3 & 4 \\ 0 & 0 & 2 & 2 \\ 0 & 0 & 0 & 0 \end{bmatrix} \begin{bmatrix} x_1 \\ x_2 \\ x_3 \\ x_4 \end{bmatrix} = \begin{bmatrix} 0 \\ 0 \\ 0 \end{bmatrix} \quad \text{gives} \quad \begin{aligned} x_1 + 3x_3 &= -2x_2 - 4x_4 \\ 2x_3 &= -2x_4. \end{aligned}$$

To describe all the solutions, the best way is to find two *special solutions*. Choose the free variables to be 1 or 0. Then the pivot variables are determined by the equations $U\boldsymbol{x} = \boldsymbol{0}$.

Special Solutions

1 Set $x_2 = 1$ and $x_4 = 0$. By back substitution $x_3 = 0$ and $x_1 = -2$.

2 Set $x_2 = 0$ and $x_4 = 1$. By back substitution $x_3 = -1$ and $x_1 = -1$.

These special solutions solve $U\boldsymbol{x} = \boldsymbol{0}$ and therefore $A\boldsymbol{x} = \boldsymbol{0}$. They are in the nullspace. The good thing is that *every solution is a combination of the special solutions.*

Complete Solution

$$\boldsymbol{x} = x_2 \underset{\text{special}}{\begin{bmatrix} -2 \\ 1 \\ 0 \\ 0 \end{bmatrix}} + x_4 \underset{\text{special}}{\begin{bmatrix} -1 \\ 0 \\ -1 \\ 1 \end{bmatrix}} = \underset{\text{complete}}{\begin{bmatrix} -2x_2 - x_4 \\ x_2 \\ -x_4 \\ x_4 \end{bmatrix}}. \tag{3.1}$$

Please look again at that answer. It is the main goal of this section. The vector $(-2, 1, 0, 0)$ is the special solution when $x_2 = 1$ and $x_4 = 0$. The second special solution has $x_2 = 0$ and $x_4 = 1$. ***All solutions are linear combinations of those two.*** The special solutions are in the nullspace $\boldsymbol{N}(A)$, and their combinations fill out the whole nullspace.

The MATLAB code null computes these special solutions. They go into the columns of a ***nullspace matrix*** N. The complete solution to $A\boldsymbol{x} = \boldsymbol{0}$ is a combination of those columns. Once we have the special solutions, we have everything.

There is a special solution for each free variable. If no variables are free—this means there are n pivots—then the only solution to $U\boldsymbol{x} = \boldsymbol{0}$ and $A\boldsymbol{x} = \boldsymbol{0}$ is the trivial solution $\boldsymbol{x} = \boldsymbol{0}$. This is the uniqueness case, when all variables are pivot variables. In that case the nullspaces of A and U contain only the zero vector. With no free variables, the output from null is an empty matrix N.

Example 3.8 Find the nullspaces of $U = \begin{bmatrix} 1 & 5 & 7 \\ 0 & 0 & 9 \end{bmatrix}$ and $A = \begin{bmatrix} 3 & 8 \\ 0 & 4 \end{bmatrix}$.

The second column of U has no pivot. So x_2 is free. The special solution has $x_2 = 1$. Back substitution into $9x_3 = 0$ gives $x_3 = 0$. Then $x_1 + 5x_2 = 0$ or $x_1 = -5$. The solutions to $U\boldsymbol{x} = \mathbf{0}$ are

$$\boldsymbol{x} = x_2 \begin{bmatrix} -5 \\ 1 \\ 0 \end{bmatrix} = \begin{bmatrix} -5x_2 \\ x_2 \\ 0 \end{bmatrix}$$

The nullspace of U is a line through the special solution. One variable is free. N has one column.

For the matrix A, *no* variables are free. Both columns have pivots. The equation $4x_2 = 0$ gives $x_2 = 0$, and then $3x_1 = 0$ forces $x_1 = 0$. The nullspace consists of the point $\boldsymbol{x} = \mathbf{0}$ (the zero vector). The nullspace matrix N is empty (but the nullspace is never empty!).

Echelon Matrices

Forward elimination goes from A to U. The process starts with an m by n matrix A. It acts by row operations, including row exchanges. It goes on to the next column when no pivot is available in the current column. The process ends with an m by n matrix U, which has a special form. U is an ***echelon matrix***, or *"staircase matrix"*.

Here is a 4 by 7 echelon matrix with the three pivots boxed:

$$U = \begin{bmatrix} \boxed{x} & x & x & x & x & x & x \\ 0 & \boxed{x} & x & x & x & x & x \\ 0 & 0 & 0 & 0 & 0 & \boxed{x} & x \\ 0 & 0 & 0 & 0 & 0 & 0 & 0 \end{bmatrix}$$

three pivot variables x_1, x_2, x_6
four free variables x_3, x_4, x_5, x_7
four special solutions

Question What are the column space $\boldsymbol{R}(U)$ and the nullspace $\boldsymbol{N}(U)$ for this matrix?

Answer The columns have four components so they lie in $\mathbf{R}^4$. The fourth component of every column is zero. Every combination of the columns—every vector in $\boldsymbol{R}(U)$—has fourth component zero. *The column space of U consists of all vectors of the form* $(b_1, b_2, b_3, 0)$. For those vectors we can solve $U\boldsymbol{x} = \boldsymbol{b}$ by back substitution. These vectors $\boldsymbol{b}$ are all possible combinations of the seven columns.*

The nullspace $\boldsymbol{N}(U)$ is a subspace of $\mathbf{R}^7$. The solutions to $U\boldsymbol{x} = \mathbf{0}$ are combinations of the four special solutions—*one for each free variable*:

> The free variables are x_3, x_4, x_5, x_7 because columns 3, 4, 5, 7 have no pivots. Set one free variable to 1 and set the other free variables to zero. Solve $U\boldsymbol{x} = \mathbf{0}$ for the pivot variables. Then $(x_1, \ldots, x_7)$ is one of the four special solutions in the nullspace matrix N.

*We really only need the three columns with pivots. The free variables can all be set to zero, to find one particular solution of $U\boldsymbol{x} = \boldsymbol{b}$.

Since echelon matrices U are so useful, we spell out their rules. The nonzero rows come first. The pivots are the first nonzero entries in those rows, and they go down in a staircase pattern. The usual row operations (in the code plu) produce a column of zeros below every pivot.

Counting the pivots leads to an extremely important theorem. Suppose A has more columns than rows. ***With $n > m$ there is at least one free variable.*** The system $Ax = \mathbf{0}$ has a special solution. This solution is *not zero*!

3B If $Ax = \mathbf{0}$ has more unknowns than equations ($n > m$: more columns than rows), then it has nonzero solutions.

A short wide matrix has nonzero vectors in its nullspace. There must be at least $n - m$ free variables, since the number of pivots cannot exceed m. (The matrix only has m rows, and a row never has two pivots.) Of course a row might have *no* pivot—which means an extra free variable. But here is the point: When a free variable can be set to 1, the equation $Ax = \mathbf{0}$ has a nonzero solution.

To repeat: There are at most m pivots. With $n > m$, the system $Ax = \mathbf{0}$ has a free variable and a nonzero solution. Actually there are infinitely many solutions, since any multiple cx is also a solution. The nullspace contains at least a line of solutions. With two free variables, there are two special solutions and the nullspace is even larger.

The nullspace is a subspace whose "dimension" is the number of free variables. This central idea—the ***dimension*** of a subspace—is defined and explained in this chapter.

The Reduced Echelon Matrix R

From the echelon matrix U we can go one more step. Continue onward from

$$U = \begin{bmatrix} 1 & 2 & 3 & 4 \\ 0 & 0 & 2 & 2 \\ 0 & 0 & 0 & 0 \end{bmatrix}.$$

This matrix can be simplified by further row operations (which must also be done to the right side). *We can divide the second row by* 2. Then both pivots equal 1. *We can subtract* 3 *times this new row* $[0\ \ 0\ \ 1\ \ 1]$ *from the row above.* That produces a zero above the second pivot as well as below. The ***"reduced"*** echelon matrix is

$$R = \begin{bmatrix} 1 & 2 & 0 & 1 \\ 0 & 0 & 1 & 1 \\ 0 & 0 & 0 & 0 \end{bmatrix}.$$

A reduced echelon matrix has 1's as pivots. It has 0's everywhere else in the pivot columns. Zeros above pivots come from ***upward elimination***.

If A is invertible, its echelon form is a triangular U. Its *reduced* echelon form is the identity matrix $R = I$. This is the ultimate in row reduction.

Some classes will take the extra step from U to R, others won't. It is optional. Whether you do it or not, you know how. The equations look simpler but the solutions to $R\boldsymbol{x} = \mathbf{0}$ are the same as the solutions to $U\boldsymbol{x} = \mathbf{0}$. Dividing an equation by 2, or subtracting one equation from another, has no effect on the nullspace. This is ***Gauss-Jordan elimination***, which produces the extra zeros in R (above the pivots). Gaussian elimination stops at U.

The zeros in R make it easy to find the special solutions (the same as before):

1 Set $x_2 = 1$ and $x_4 = 0$. Solve $R\boldsymbol{x} = \mathbf{0}$. Then $x_1 = -2$ and $x_3 = 0$.

2 Set $x_2 = 0$ and $x_4 = 1$. Solve $R\boldsymbol{x} = \mathbf{0}$. Then $x_1 = -1$ and $x_3 = -1$.

The numbers -2 and 0 are sitting in column 2 of R (with plus signs). The numbers -1 and -1 are sitting in column 4 (with plus signs). By reversing signs we can read off the special solutions from the reduced matrix R. The general solution to $A\boldsymbol{x} = \mathbf{0}$ or $U\boldsymbol{x} = \mathbf{0}$ or $R\boldsymbol{x} = \mathbf{0}$ is a combination of those two special solutions: ***The nullspace $N(A) = N(U) = N(R)$ contains***

$$\boldsymbol{x} = x_2 \begin{bmatrix} -2 \\ 1 \\ 0 \\ 0 \end{bmatrix} + x_4 \begin{bmatrix} -1 \\ 0 \\ -1 \\ 1 \end{bmatrix} = \textit{complete solution of } A\boldsymbol{x} = \mathbf{0}.$$

To summarize: The pivot columns of R hold the identity matrix (with zero rows below). The free columns show the special solutions. Back substitution is very quick but it costs more to reach R. Most computer programs stop at U, but ref goes on:

$$A = \begin{bmatrix} 1 & 3 & 3 \\ 2 & 6 & 9 \\ -1 & -3 & 3 \end{bmatrix} \rightarrow U = \begin{bmatrix} 1 & 3 & 3 \\ 0 & 0 & 3 \\ 0 & 0 & 0 \end{bmatrix} \rightarrow R = \begin{bmatrix} 1 & 3 & 0 \\ 0 & 0 & 1 \\ 0 & 0 & 0 \end{bmatrix}.$$

A note about the theory The echelon matrix U depends on the order of the elimination steps. A row exchange gives new pivots and a different U. But the pivots in R are all 1's. This reduced matrix stays exactly the same, even if you get to it with extra row exchanges. The theory of linear algebra notices four results that do *not* depend on the order of the elimination steps:

1 The vectors $\boldsymbol{x}$ in the nullspace of A.

2 The selection of free variables and pivot variables.

3 The special solutions to $A\boldsymbol{x} = \mathbf{0}$ (each with a free variable equal to 1).

4 The reduced echelon form R.

The key is to realize what it means when a column has no pivot. ***That** free **column is a combination of** previous* pivot ***columns***. This is exactly the special solution, with free variable 1. The free column and pivot columns (times the right numbers from the special solution) solve $A\boldsymbol{x} = \mathbf{0}$. Those numbers are also in R. So the original A controls 1–4 above for these reasons:

1 The nullspace is determined by $A\boldsymbol{x} = \mathbf{0}$, before elimination starts.

2 x_j is free exactly when column j of A is a combination of columns $1, \ldots, j-1$.

3 The special solutions express the free columns as combinations of the pivot columns.

4 R has those special combinations in its free columns. Its pivot columns contain I.

You can do any row operations in any order, and you never reach a different R.

The code plu computes the echelon matrix U. Notice the differences from splu for square matrices. If a column has no pivot, the algorithm goes on to the next column. The program calls findpiv to locate the first entry larger than *tol* in the available columns and rows. This subroutine findpiv needs only four strange lines:

```
[m,n] = size(A); r = find(abs(A(:)) > tol);
if isempty(r), return, end
r = r(1); j = fix((r-1)/m) + 1;
p = p+j-1; k = k+r-(j-1)*m-1;
```

The pivot location is called $[r, p]$ and p is added to the list of pivot columns. This list can be printed. The program ref computes the reduced echelon form R, and null puts the special solutions (one for each free variable) into the nullspace matrix N.

M-file: plu

```
function [P, L, U, pivcol] = plu(A)
[m,n] = size(A);
P = eye(m,m);
L = eye(m,m);
U = zeros(m,n);
pivcol = [];
tol = 1.e-6;
p = 1;
for k = 1:min(m,n)
  [r, p] = findpiv(A(k:m,p:n),k,p,tol);
  if r ~= k
    A([r k],1:n) = A([k r],1:n);
    if k > 1, L([r k],1:k-1) = L([k r],1:k-1); end
    P([r k],1:m) = P([k r],1:m);
  end
  if abs(A(k,p)) >= tol
    pivcol = [pivcol p];
    for i = k+1:m
```

```
        L(i,k) = A(i,p)/A(k,p);
        for j = k+1:n
          A(i,j) = A(i,j) - L(i,k) * A(k,j);
        end
      end
    end
    for j = k:n
      U(k,j) = A(k,j) * (abs(A(k,j)) >= tol);
    end
    if p < n, p = p+1; end
  end
```

M-file: ref

```
function [R, pivcol] = ref(A)
% Scale rows of U so that all pivots equal 1.
% Eliminate nonzeros above the pivots to reach R.

[P,L,U,pivcol] = plu(A);
R = U;
[m,n] = size(R);
for k = 1:length(pivcol);
  p = pivcol(k);
  for j = p+1:n
    R(k,j) = R(k,j)/R(k,p);
  end
  R(k,p) = 1;
  for i = 1:k-1
    for j = p+1:n
      R(i,j) = R(i,j) - R(i,p) * R(k,j);
    end
    R(i,p) = 0;
  end
end
```

M-file: null

```
function N = null(A)
% The n-r columns of N are special solutions to Ax = 0.
% N combines I with the nonpivot columns of R.

[R,pivcol] = ref(A);
[m,n] = size(A);
r = length(pivcol);
nopiv = 1:n;
nopiv(pivcol) = [];
N = zeros(n,n-r);
if n > r
  N(nopiv,:) = eye(n-r,n-r);
```

```
  if r > 0
    N(pivcol,:) = -R(1:r,nopiv);
  end
end
```

Summary of the Situation

From $A\boldsymbol{x} = \mathbf{0}$ — *We can't see* how many special solutions there are. *We can't see* the actual numbers in the special solutions.

If $m < n$, there is at least one free variable and one special solution. The matrix has more columns than rows. There are nonzero vectors in its nullspace. If $m \geq n$ we cannot determine from looking at $A\boldsymbol{x} = \mathbf{0}$ whether $\boldsymbol{x} = \mathbf{0}$ is the only solution.

From $U\boldsymbol{x} = \mathbf{0}$ — *We* ***can*** *see* how many special solutions there are. *We can't see* the actual numbers in the special solutions.

The number of special solutions is the number of free variables. This is the total number of variables (n) minus the number of pivots (r). Then $n - r$ free variables have columns without pivots. To find the $n - r$ special solutions, do back substitutions in $U\boldsymbol{x} = \mathbf{0}$—after assigning the value 1 to a free variable.

From $R\boldsymbol{x} = \mathbf{0}$ — *We* ***can*** *see* how many special solutions there are. *We* ***can*** *see* the actual numbers in the special solutions.

In the reduced echelon matrix R, every pivot equals 1—with zeros above and below. If it happens that these r pivot columns come first, the reduced form looks like

$$R = \begin{bmatrix} I & F \\ 0 & 0 \end{bmatrix} \begin{matrix} r \\ m-r \end{matrix} \qquad \begin{matrix} r & n-r \end{matrix}$$

The pivot part is the identity matrix. The free part F can contain any numbers. They come from elimination on A—downward to U and upward to R. When A is invertible, F is empty and $R = I$.

The special solutions to $R\boldsymbol{x} = \mathbf{0}$ (also to $U\boldsymbol{x} = \mathbf{0}$ and $A\boldsymbol{x} = \mathbf{0}$) can be found directly from R. These $n - r$ solutions go into the columns of a **nullspace matrix** N. Notice how these block matrices give $RN = 0$:

$$N = \begin{bmatrix} -F \\ I \end{bmatrix} \begin{matrix} r \\ n-r \end{matrix} \qquad n-r$$

Example 3.9

$$A = \begin{bmatrix} 1 & 1 & 0 \\ 1 & 3 & 2 \end{bmatrix} \quad \rightarrow \quad U = \begin{bmatrix} 1 & 1 & 0 \\ 0 & 2 & 2 \end{bmatrix} \quad \rightarrow \quad R = \begin{bmatrix} 1 & 0 & -1 \\ 0 & 1 & 1 \end{bmatrix} = \begin{bmatrix} I & F \end{bmatrix}.$$

A has $r = 2$ pivots and $n - r = 3 - 2$ free variables. The special solution has free variable $x_3 = 1$. Then $x_2 = -1$ and $x_1 = 1$:

$$N = \begin{bmatrix} -F \\ I \end{bmatrix} = \begin{bmatrix} 1 \\ -1 \\ 1 \end{bmatrix}.$$

The code null constructs N from R. The two parts of N are $-F$ and I as above, but the pivot variables may be mixed in with the free variables—so I may not come last.

Problem Set 3.2

Questions 1–8 are about the matrices in Problems 1 and 5.

1 Reduce these matrices to their ordinary echelon forms U:

$$\text{(a)} \quad A = \begin{bmatrix} 1 & 2 & 2 & 4 & 6 \\ 1 & 2 & 3 & 6 & 9 \\ 0 & 0 & 1 & 2 & 3 \end{bmatrix} \qquad \text{(b)} \quad A = \begin{bmatrix} 2 & 4 & 2 \\ 0 & 4 & 4 \\ 0 & 8 & 8 \end{bmatrix}.$$

Which are the free variables and which are the pivot variables?

2 For the matrices in Problem 1, find a special solution for each free variable. (Set the free variable to 1. Set the other free variables to zero.)

3 By combining the special solutions in Problem 2, describe every solution to $Ax = \mathbf{0}$. The nullspace of A contains only the vector $x = \mathbf{0}$ when ______.

4 By further row operations on each U in Problem 1, find the reduced echelon form R. The nullspace of R is ______ the nullspace of U.

5 By row operations reduce A to its echelon form U. Write down a 2 by 2 lower triangular L such that $A = LU$.

$$\text{(a)} \quad A = \begin{bmatrix} -1 & 3 & 5 \\ -2 & 6 & 10 \end{bmatrix} \qquad \text{(b)} \quad A = \begin{bmatrix} -1 & 3 & 5 \\ -2 & 6 & 7 \end{bmatrix}.$$

6 For the matrices in Problem 5, find the special solutions to $Ax = \mathbf{0}$. For an m by n matrix, the number of pivot variables plus the number of free variables is ______.

7 In Problem 5, describe the nullspace of each A in two ways. Give the equations for the plane or line $N(A)$, and give all vectors x that satisfy those equations (combinations of the special solutions).

8 Reduce the echelon forms U in Problem 5 to R. For each R draw a box around the identity matrix that is in the pivot rows and pivot columns.

Questions 9–17 are about free variables and pivot variables.

9 True or false (with reason if true and example if false):

(a) A square matrix has no free variables.

(b) An invertible matrix has no free variables.

(c) An m by n matrix has no more than n pivot variables.

(d) An m by n matrix has no more than m pivot variables.

10 Construct 3 by 3 matrices A to satisfy these requirements (if possible):

(a) A has no zero entries but $U = I$.

(b) A has no zero entries but $R = I$.

(c) A has no zero entries but $R = U$.

(d) $A = U = 2R$.

11 Put 0's and x's (for zeros and nonzeros) in a 4 by 7 echelon matrix U so that the pivot variables are

(a) 2, 4, 5 (b) 1, 3, 6, 7 (c) 4 and 6.

12 Put 0's and 1's and x's (zeros, ones, and nonzeros) in a 4 by 8 reduced echelon matrix R so that the free variables are

(a) 2, 4, 5, 6 (b) 1, 3, 6, 7, 8.

13 Suppose column 4 of a 3 by 5 matrix is all zero. Then x_4 is certainly a _____ variable. The special solution for this variable is the vector $\boldsymbol{x} =$ _____.

14 Suppose the first and last columns of a 3 by 5 matrix are the same (not zero). Then _____ is a free variable. The special solution for this variable is $\boldsymbol{x} =$ _____.

15 Suppose an m by n matrix has r pivots. The number of special solutions is _____. The nullspace contains only $\boldsymbol{x} = \mathbf{0}$ when $r =$ _____. The column space is all of $\mathbf{R}^m$ when $r =$ _____.

16 The nullspace of a 5 by 5 matrix contains only $\boldsymbol{x} = \mathbf{0}$ when the matrix has _____ pivots. The column space is $\mathbf{R}^5$ when there are _____ pivots. Explain why.

17 The equation $x - 3y - z = 0$ determines a plane in $\mathbf{R}^3$. What is the matrix A in this equation? Which are the free variables? The special solutions are $(3, 1, 0)$ and _____.

18 The plane $x - 3y - z = 12$ is parallel to the plane $x - 3y - z = 0$ in Problem 17. One particular point on this plane is $(12, 0, 0)$. All points on the plane have the form (fill in the first components)

$$\begin{bmatrix} x \\ y \\ z \end{bmatrix} = \begin{bmatrix} \ \\ 0 \\ 0 \end{bmatrix} + y \begin{bmatrix} \ \\ 1 \\ 0 \end{bmatrix} + z \begin{bmatrix} \ \\ 0 \\ 1 \end{bmatrix}.$$

19 If $\boldsymbol{x}$ is in the nullspace of B, prove that $\boldsymbol{x}$ is in the nullspace of AB. This means: If $B\boldsymbol{x} = \mathbf{0}$ then ______. Give an example in which these nullspaces are different.

20 If A is invertible then $\boldsymbol{N}(AB)$ *equals* $\boldsymbol{N}(B)$. Following Problem 19, prove this second part: If $AB\boldsymbol{x} = \mathbf{0}$ then $B\boldsymbol{x} = \mathbf{0}$.

This means that $U\boldsymbol{x} = \mathbf{0}$ whenever $LU\boldsymbol{x} = \mathbf{0}$ (same nullspace). The key is not that L is triangular but that L is ______.

Questions 21–28 ask for matrices (if possible) with specific properties.

21 Construct a matrix whose nullspace consists of all combinations of $(2, 2, 1, 0)$ and $(3, 1, 0, 1)$.

22 Construct a matrix whose nullspace consists of all multiples of $(4, 3, 2, 1)$.

23 Construct a matrix whose column space contains $(1, 1, 5)$ and $(0, 3, 1)$ and whose nullspace contains $(1, 1, 2)$.

24 Construct a matrix whose column space contains $(1, 1, 0)$ and $(0, 1, 1)$ and whose nullspace contains $(1, 0, 1)$ and $(0, 0, 1)$.

25 Construct a matrix whose column space contains $(1, 1, 1)$ and whose nullspace is the line of multiples of $(1, 1, 1, 1)$.

26 Construct a 2 by 2 matrix whose nullspace equals its column space. This is possible.

27 Why does no 3 by 3 matrix have a nullspace that equals its column space?

28 If $AB = 0$ then the column space of B is contained in the ______ of A. Give an example.

29 The reduced form R of a 3 by 3 matrix with randomly chosen entries is almost sure to be ______. What R is most likely if the random A is 4 by 3?

30 Show by example that these three statements are generally *false*:

(a) A and A^{T} have the same nullspace.

(b) A and A^{T} have the same free variables.

(c) A and A^{T} have the same pivots. (The matrix may need a row exchange.) A and A^{T} do have the same *number* of pivots. This will be important.

31 What is the nullspace matrix N (containing the special solutions) for A, B, C?

$$A = \begin{bmatrix} I & I \end{bmatrix} \quad \text{and} \quad B = \begin{bmatrix} I & I \\ 0 & 0 \end{bmatrix} \quad \text{and} \quad C = I.$$

32 If the nullspace of A consists of all multiples of $\boldsymbol{x} = (2, 1, 0, 1)$, how many pivots appear in U?

33 If the columns of N are the special solutions to $R\boldsymbol{x} = \mathbf{0}$, what are the nonzero rows of R?

$$N = \begin{bmatrix} 2 & 3 \\ 1 & 0 \\ 0 & 1 \end{bmatrix} \quad \text{and} \quad N = \begin{bmatrix} 0 \\ 0 \\ 1 \end{bmatrix} \quad \text{and} \quad N = \begin{bmatrix} \quad \end{bmatrix}.$$

34 (a) What are the five 2 by 2 reduced echelon matrices R whose entries are all 0's and 1's?

(b) What are the eight 1 by 3 matrices containing only 0's and 1's? Are all eight of them reduced echelon matrices?

3.3 The Rank of A: Solving $A\boldsymbol{x} = \boldsymbol{b}$

This section moves forward with calculation and also with theory. While solving $A\boldsymbol{x} = \boldsymbol{b}$, we answer one question about the column space and a different question about the nullspace. Here are the questions:

1 Does $\boldsymbol{b}$ belong to the column space? *Yes*, if $A\boldsymbol{x} = \boldsymbol{b}$ has a solution.

2 Is $\boldsymbol{x} = \mathbf{0}$ alone in the nullspace? *Yes*, if the solution is unique.

If there are solutions to $A\boldsymbol{x} = \boldsymbol{b}$, we find them all. The complete solution is $\boldsymbol{x} = \boldsymbol{x}_p + \boldsymbol{x}_n$. To a ***particular solution*** $\boldsymbol{x}_p$ we add all solutions of $A\boldsymbol{x}_n = \mathbf{0}$.

Elimination will get one last workout. The essential facts about A can be seen from the pivots. A solution exists when there is a pivot in every row. The solution is unique when there is a pivot in every column. The ***number of pivots***, which is the controlling number for $A\boldsymbol{x} = \boldsymbol{b}$, is given a name. This number is the ***rank*** of A.

The Complete Solution to $A\boldsymbol{x} = \boldsymbol{b}$

The last section totally solved $A\boldsymbol{x} = \mathbf{0}$. Elimination converted the problem to $U\boldsymbol{x} = \mathbf{0}$. The free variables were given special values (one and zero). Then the pivot variables were found by back substitution. We paid no attention to the right side $\boldsymbol{b}$ because it started and ended as zero. The solution $\boldsymbol{x}$ was in the nullspace of A.

Now $\boldsymbol{b}$ is not zero. Row operations on the left side must act also on the right side. One way to organize that is to add $\boldsymbol{b}$ as an extra column of the matrix. We keep the same example as before. But we ***"augment"*** A with the right side $(b_1, b_2, b_3) = (1, 6, 7)$:

$$\begin{bmatrix} 1 & 2 & 3 & 4 \\ 2 & 4 & 8 & 10 \\ 3 & 6 & 11 & 14 \end{bmatrix} \begin{bmatrix} x_1 \\ x_2 \\ x_3 \\ x_4 \end{bmatrix} = \begin{bmatrix} 1 \\ 6 \\ 7 \end{bmatrix} \quad \text{has the augmented matrix} \quad \begin{bmatrix} 1 & 2 & 3 & 4 & 1 \\ 2 & 4 & 8 & 10 & 6 \\ 3 & 6 & 11 & 14 & 7 \end{bmatrix}.$$

Take the usual steps. Subtract 2 times row 1 from row 2, and 3 times row 1 from row 3:

$$\begin{bmatrix} 1 & 2 & 3 & 4 \\ 0 & 0 & 2 & 2 \\ 0 & 0 & 2 & 2 \end{bmatrix} \begin{bmatrix} x_1 \\ x_2 \\ x_3 \\ x_4 \end{bmatrix} = \begin{bmatrix} 1 \\ 4 \\ 4 \end{bmatrix} \quad \text{has the augmented matrix} \quad \begin{bmatrix} 1 & 2 & 3 & 4 & 1 \\ 0 & 0 & 2 & 2 & 4 \\ 0 & 0 & 2 & 2 & 4 \end{bmatrix}.$$

Now subtract the new row 2 from the new row 3 to reach $U\boldsymbol{x} = \boldsymbol{c} =$ **last column**:

$$\begin{bmatrix} 1 & 2 & 3 & 4 \\ 0 & 0 & 2 & 2 \\ 0 & 0 & 0 & 0 \end{bmatrix} \begin{bmatrix} x_1 \\ x_2 \\ x_3 \\ x_4 \end{bmatrix} = \begin{bmatrix} 1 \\ 4 \\ 0 \end{bmatrix} \quad \text{has the augmented matrix} \quad \begin{bmatrix} 1 & 2 & 3 & 4 & 1 \\ 0 & 0 & 2 & 2 & 4 \\ 0 & 0 & 0 & 0 & 0 \end{bmatrix}.$$

Are we making the point? It is not necessary to keep writing the letters x_1, x_2, x_3, x_4 and the equality sign. The augmented matrix contains all information for *both sides* of the equation. While A goes to U, $\boldsymbol{b}$ goes to $\boldsymbol{c}$. The operations on A and $\boldsymbol{b}$ are the same, so they are done together in the augmented matrix.

We worked with the specific vector $\boldsymbol{b} = (1, 6, 7)$, and we reached $\boldsymbol{c} = (1, 4, 0)$. Here are the same augmented matrices for a general vector (b_1, b_2, b_3):

$$\begin{bmatrix} 1 & 2 & 3 & 4 & b_1 \\ 2 & 4 & 8 & 10 & b_2 \\ 3 & 6 & 11 & 14 & b_3 \end{bmatrix} \rightarrow \begin{bmatrix} 1 & 2 & 3 & 4 & b_1 \\ 0 & 0 & 2 & 2 & b_2 - 2b_1 \\ 0 & 0 & 2 & 2 & b_3 - 3b_1 \end{bmatrix}$$

$$\rightarrow \begin{bmatrix} 1 & 2 & 3 & 4 & b_1 \\ 0 & 0 & 2 & 2 & b_2 - 2b_1 \\ 0 & 0 & 0 & 0 & b_3 - b_2 - b_1 \end{bmatrix}.$$

For the specific vector $\boldsymbol{b} = (1, 6, 7)$, the final entry is $b_3 - b_2 - b_1 = 7 - 6 - 1$. The third equation becomes $0 = 0$. Therefore the system has a solution. This particular $\boldsymbol{b}$ is in the column space of A.

For the general vector (b_1, b_2, b_3), ***the equations might or might not be solvable***. The third equation has all zeros on the left side. So we must have $b_3 - b_2 - b_1 = 0$ on the right side. Elimination has identified this exact requirement for solvability. *It is the test for* $\boldsymbol{b}$ *to be in the column space*. The requirement was satisfied by $(1, 6, 7)$ because $7 - 6 - 1 = 0$.

Particular Solution and Homogeneous Solution

Now find the complete solution for $\boldsymbol{b} = (1, 6, 7)$. The last equation is $0 = 0$. That leaves two equations in four unknowns. The two free variables are as free as ever. There is a good way to write down all the solutions to $A\boldsymbol{x} = \boldsymbol{b}$:

Particular solution Set both free variables x_2 and x_4 to zero. Solve for x_1 and x_3:

$$\begin{aligned} x_1 + 0 + 3x_3 + 0 &= 1 \\ 2x_3 + 0 &= 4 \end{aligned} \quad \text{gives} \quad \begin{aligned} x_1 &= -5 \\ x_3 &= \ \ 2. \end{aligned}$$

This particular solution is $\boldsymbol{x}_p = (-5, 0, 2, 0)$. For safety, check that $A\boldsymbol{x} = \boldsymbol{b}$.

Homogeneous solution The word "homogeneous" means that the right side is zero. But $A\boldsymbol{x} = \boldsymbol{0}$ is already solved (last section). There are two special solutions, first with $x_2 = 1$ and second with $x_4 = 1$. The solutions to $A\boldsymbol{x} = \boldsymbol{0}$ come from the nullspace so we call them $\boldsymbol{x}_n$:

$$\boldsymbol{x}_n = \boldsymbol{x}_{\text{homogeneous}} = x_2 \begin{bmatrix} -2 \\ 1 \\ 0 \\ 0 \end{bmatrix} + x_4 \begin{bmatrix} -1 \\ 0 \\ -1 \\ 1 \end{bmatrix}. \tag{3.2}$$

3C ***The complete solution is one particular solution plus all homogeneous solutions***:

$$\boldsymbol{x} = \boldsymbol{x}_p + \boldsymbol{x}_n \qquad \textit{is} \qquad \boldsymbol{x}_{\text{complete}} = \boldsymbol{x}_{\text{particular}} + \boldsymbol{x}_{\text{homogeneous}}.$$

To find every solution to $A\boldsymbol{x} = \boldsymbol{b}$, start with one particular solution. Add to it every solution to $A\boldsymbol{x}_n = \boldsymbol{0}$. This homogeneous part $\boldsymbol{x}_n$ comes from the nullspace. Our particular solution $\boldsymbol{x}_p$ comes from solving with all free variables set to zero. Another book could choose another particular solution.

We write the complete solution $\boldsymbol{x}_p + \boldsymbol{x}_n$ to this example as

$$\boldsymbol{x} = \begin{bmatrix} -5 \\ 0 \\ 2 \\ 0 \end{bmatrix} + x_2 \begin{bmatrix} -2 \\ 1 \\ 0 \\ 0 \end{bmatrix} + x_4 \begin{bmatrix} -1 \\ 0 \\ -1 \\ 1 \end{bmatrix}. \tag{3.3}$$

There is a "double infinity" of solutions, because there are two free variables.

In the special case $\boldsymbol{b} = \boldsymbol{0}$, the particular solution is the zero vector! The complete solution to $A\boldsymbol{x} = \boldsymbol{0}$ is just like equation (3.3), but we didn't print $\boldsymbol{x}_p = (0, 0, 0, 0)$. What we printed was equation (3.2). The solution plane in Figure 3.3 goes through $\boldsymbol{x}_p = \boldsymbol{0}$ when $\boldsymbol{b} = \boldsymbol{0}$.

All linear equations—matrix equations or differential equations—fit this pattern $\boldsymbol{x} = \boldsymbol{x}_p + \boldsymbol{x}_n$.

Here are the five steps to the complete solution $\boldsymbol{x}_p + \boldsymbol{x}_n$ of $A\boldsymbol{x} = \boldsymbol{b}$:

1 Add $\boldsymbol{b}$ as an extra column next to A. Reduce A to U by row operations.

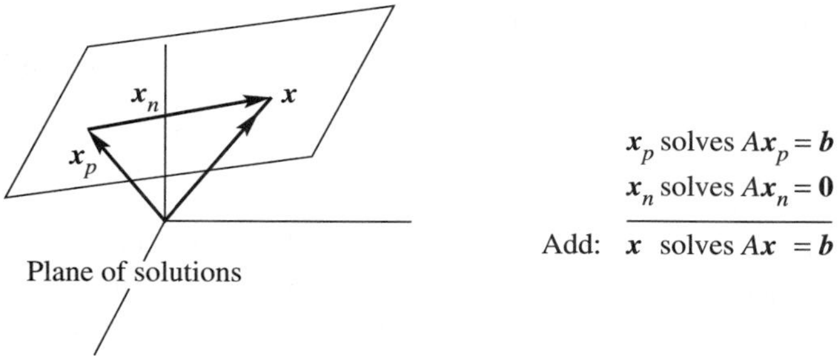

Figure 3.3 The complete $x = x_p + x_n$ is a particular solution plus any solution to $Ax = 0$.

2 Zero rows in U must have zeros also in the extra column. Those equations are $0 = 0$.

3 Set the free variables to zero to find a particular solution x_p.

4 Set each free variable, in turn, equal to 1. Find the special solutions to $Ax = 0$. Their combinations give x_n.

5 The complete solution to $Ax = b$ is x_p plus the complete solution to $Ax = 0$.

Reduction from $Ax = b$ to $Ux = c$ to $Rx = d$

By staying longer with forward elimination, we can make back substitution quicker. Instead of stopping at U, go on to R. This is the ***reduced*** echelon form (still optional). All its pivots equal 1. *Below and above each pivot are zeros.* We met this ***Gauss-Jordan elimination*** for $Ax = 0$, and now we apply it to $Ax = b$.

Go back to our 3 by 4 example, with the right side b in column 5. For this augmented matrix, continue elimination from U to R. First divide row 2 by its pivot, so the new pivot is 1:

$$\underbrace{\begin{bmatrix} 1 & 2 & 3 & 4 \\ 0 & 0 & 2 & 2 \\ 0 & 0 & 0 & 0 \end{bmatrix}}_{U}\!\!\begin{matrix} 1 \\ 4 \\ \underset{c}{0} \end{matrix} \rightarrow \begin{bmatrix} 1 & 2 & 3 & 4 & 1 \\ 0 & 0 & 1 & 1 & 2 \\ 0 & 0 & 0 & 0 & 0 \end{bmatrix} \rightarrow \underbrace{\begin{bmatrix} 1 & 2 & 0 & 1 \\ 0 & 0 & 1 & 1 \\ 0 & 0 & 0 & 0 \end{bmatrix}}_{R}\!\!\begin{matrix} -5 \\ 2 \\ \underset{d}{0} \end{matrix}.$$

The last step subtracted 3 times row 2 from row 1—the row above. The identity matrix is now in columns 1 and 3, with zeros beneath it. The particular solution x_p only uses those pivot columns—the free variables are set to zero. We can read off $x_p = (-5, 0, 2, 0)$ directly from the new right side. *When the matrix is I, the particular solution is right there.*

The nullspace matrix N is also easy from R. Change column 5 to zero. The special solution with $x_2 = 1$ has $x_1 = -2$ and $x_3 = 0$ (exactly as before). The special solution with $x_4 = 1$ has $x_1 = -1$ and $x_3 = -1$ (also as before). These numbers come directly from R—with signs reversed. Solutions are not changed—this is the point of elimination! The solutions become clearer as we go from $A\boldsymbol{x} = \boldsymbol{b}$ to $U\boldsymbol{x} = \boldsymbol{c}$ to $R\boldsymbol{x} = \boldsymbol{d}$.

Consistent = Solvable

With (b_1, b_2, b_3) on the right side, the equations might be solvable or not. Elimination produced a third equation $0 = b_3 - b_2 - b_1$. To be solvable, this equation must be $0 = 0$.

Translate that into: ***The column space of A contains all vectors with*** $\boldsymbol{b_3 - b_2 - b_1 = 0}$. Here is the common sense reason. On the left side, row 3 minus row 2 minus row 1 leaves a row of zeros. (Elimination discovered that.) The same combination of right sides must give zero, if $A\boldsymbol{x} = \boldsymbol{b}$ is solvable.

In different words: Each column of A has component 3 minus component 2 minus component 1 equal to zero. So if $\boldsymbol{b}$ is a combination of the columns, it also has $b_3 - b_2 - b_1 = 0$. Then the equations are ***"consistent"***. They have a solution.

Important for the future The three rows were multiplied by $-1, -1, +1$ to give the zero row. This vector $\boldsymbol{y} = (-1, -1, +1)$ is in ***the nullspace of*** A^{T}:

$$A^{\mathrm{T}}\boldsymbol{y} = \begin{bmatrix} 1 & 2 & 3 \\ 2 & 4 & 6 \\ 3 & 8 & 11 \\ 4 & 10 & 14 \end{bmatrix} \begin{bmatrix} -1 \\ -1 \\ +1 \end{bmatrix} = \begin{bmatrix} 0 \\ 0 \\ 0 \\ 0 \end{bmatrix} \qquad \begin{array}{c} -\text{ row 1 of } A \\ -\text{ row 2 of } A \\ +\text{ row 3 of } A \\ \hline \text{total: zero} \end{array}$$

When a combination of rows gives the zero row, the same combination must give zero on the right side. The dot product of $\boldsymbol{y} = (-1, -1, +1)$ with $\boldsymbol{b} = (b_1, b_2, b_3)$ is $-b_1 - b_2 + b_3$. This dot product must be zero, when $A\boldsymbol{x} = \boldsymbol{b}$ is solvable. That is the hidden relation between $\boldsymbol{y}$ in the nullspace of A^{T} and $\boldsymbol{b}$ in the column space of A.

Example 3.10 (Extra practice in reducing $A\boldsymbol{x} = \boldsymbol{b}$ to $U\boldsymbol{x} = \boldsymbol{c}$ and solving for $\boldsymbol{x}$)

$$\begin{bmatrix} 1 & 3 & 5 \\ 2 & 6 & 10 \end{bmatrix} \begin{bmatrix} x_1 \\ x_2 \\ x_3 \end{bmatrix} = \begin{bmatrix} b_1 \\ b_2 \end{bmatrix} \quad \text{reduces to} \quad \begin{bmatrix} 1 & 3 & 5 \\ 0 & 0 & 0 \end{bmatrix} \begin{bmatrix} x_1 \\ x_2 \\ x_3 \end{bmatrix} = \begin{bmatrix} b_1 \\ b_2 - 2b_1 \end{bmatrix}.$$

This is solvable only if $0 = b_2 - 2b_1$. So the column space of A is the line of vectors with $b_2 = 2b_1$. Row 2 of the matrix is $2 \times$ row 1. The free variables are x_2 and x_3. There is only one pivot (here $R\boldsymbol{x} = \boldsymbol{d}$ is the same as $U\boldsymbol{x} = \boldsymbol{c}$):

The particular solution has $x_2 = 0$ and $x_3 = 0$. Then $x_1 = b_1$.

The first special solution with $\boldsymbol{b} = 0$ has $x_2 = 1$ and $x_3 = 0$. Then $x_1 = -3$.

The second special solution with $\boldsymbol{b} = 0$ has $x_2 = 0$ and $x_3 = 1$. Then $x_1 = -5$.

The complete solution (assuming $b_2 = 2b_1$, otherwise no solution) is

$$\boldsymbol{x} = \boldsymbol{x}_p + \boldsymbol{x}_n = \begin{bmatrix} b_1 \\ 0 \\ 0 \end{bmatrix} + x_2 \begin{bmatrix} -3 \\ 1 \\ 0 \end{bmatrix} + x_3 \begin{bmatrix} -5 \\ 0 \\ 1 \end{bmatrix}.$$

The Rank of a Matrix

The numbers m and n give the size of a matrix—but not necessarily the *true* size of a linear system. An equation like $0 = 0$ should not count. In our 3 by 4 example, the third row of A was the sum of the first two. After elimination, the third row of U was zero. There were only two pivots—therefore two pivot variables. This number $r = 2$ is the ***rank of the matrix*** A. It is also the rank of U.

The rank r is emerging as the key to the true size of the system.

Definition 1 The ***rank*** of A is the ***number of pivots*** (nonzero of course).

Definition 2 The ***rank*** of A is the ***number of independent rows***.

The first definition is computational. The pivots spring from the elimination steps—we watch for nonzeros in certain positions. Counting those pivots is at the lowest level of linear algebra. A computer can do it for us (which is good).

Actually the computer has a hard time to decide whether a small number is really zero. When it subtracts 3 times $.33\cdots3$ from 1, does it obtain zero? Our teaching codes use the tolerance 10^{-6}, but that is not entirely safe.

The second definition of rank is at a higher level. It deals with entire rows—vectors and not just numbers. We have to say exactly what it means for rows to be "independent." That crucial idea comes in the next section.

A third definition, at the top level of linear algebra, will deal with *spaces* of vectors. The rank r gives the size or "dimension" of the row space. The great thing is that r also reveals the dimension of all other important subspaces—including the nullspace.

Example 3.11 $\begin{bmatrix} 1 & 3 & 4 \\ 2 & 6 & 8 \end{bmatrix}$ and $\begin{bmatrix} 1 & 3 \\ 2 & 6 \end{bmatrix}$ and $\begin{bmatrix} 1 \\ 2 \end{bmatrix}$ and $[\,1\,]$ all have rank 1.

This much we can already say about the rank: *There are r pivot variables* because there are r pivots. This leaves $n - r$ free variables. The pivot variables correspond to the r columns with pivots. The free variables correspond to the $n - r$ columns without pivots.

Certainly the rank satisfies $r \le m$, since m rows can't contain more than m pivots. Also $r \le n$, since n columns can't contain more than n pivots. *The number of pivots* (which is r) *cannot exceed m or n*. The extreme cases, when r is as large as possible, are in many ways the best for $Ax = b$:

3D If $r = n$ there are no free variables. $Ax = b$ cannot have two different solutions (***uniqueness***).

If $r = m$ there are no zero rows in U. $Ax = b$ has at least one solution (***existence***).

Start with $r = n$ pivots. There are no free variables ($n - r = 0$). Back substitution gives all n components of x. The nullspace contains only $x = \mathbf{0}$. The shape of A is *tall and thin*—it has at least n rows and maybe more. The solution is unique if it exists, but *it might not exist*. The extra rows might not lead to $0 = 0$.

The other extreme has $r = m$ pivots. There are no zero rows in U. The column space contains every vector b. The shape of A is *short and wide*—it has at least m columns and maybe more. A solution to $Ax = b$ always exists, but *it might not be unique*. Any extra columns lead to free variables which give more solutions.

With $r = n$ there are 0 or 1 solutions. The columns are "independent." The matrix has *"full column rank."* With $r = m$ there is 1 solution, or infinitely many. The rows are independent. The matrix has *"full row rank."* With $m = n = r$ there is exactly one solution. The matrix is square and invertible.

The most important case is $m = n = r$. The whole of Chapter 2 was devoted to this square invertible case—when $Ax = b$ has exactly one solution $x = A^{-1}b$. Here are four examples, all with two pivots. The rank of each matrix is $r = 2$:

$$A = \begin{bmatrix} 1 & 3 \\ 2 & 8 \\ 1 & 7 \end{bmatrix} \quad \begin{matrix} m = 3 \\ n = 2 \\ r = 2 \end{matrix} \quad \begin{matrix} \text{one zero row in } U \\ \text{no free variables} \\ \text{unique solution (if it exists)} \end{matrix} \quad U = \begin{bmatrix} 1 & 3 \\ 0 & 2 \\ 0 & 0 \end{bmatrix}$$

$$A = \begin{bmatrix} 1 & 2 & 1 \\ 3 & 8 & 7 \end{bmatrix} \quad \begin{matrix} m = 2 \\ n = 3 \\ r = 2 \end{matrix} \quad \begin{matrix} \text{no zero rows in } U \\ \text{one free variable} \\ \text{solution exists (not unique)} \end{matrix} \quad U = \begin{bmatrix} 1 & 2 & 1 \\ 0 & 2 & 4 \end{bmatrix}$$

$$A = \begin{bmatrix} 1 & 2 \\ 3 & 8 \end{bmatrix} \quad \begin{matrix} m = 2 \\ n = 2 \\ r = 2 \end{matrix} \quad \begin{matrix} \text{solution exists} \\ \text{solution is unique} \\ A \text{ is invertible} \end{matrix} \quad U = \begin{bmatrix} 1 & 2 \\ 0 & 2 \end{bmatrix}$$

$$A = \begin{bmatrix} 1 & 2 & 1 \\ 3 & 8 & 7 \\ 3 & 8 & 7 \end{bmatrix} \quad \begin{matrix} m = 3 \\ n = 3 \\ r = 2 \end{matrix} \quad \begin{matrix} \text{one zero row in } U \\ \text{one free variable} \\ \text{solution may not exist} \\ \text{solution is not unique} \end{matrix} \quad U = \begin{bmatrix} 1 & 2 & 1 \\ 0 & 2 & 4 \\ 0 & 0 & 0 \end{bmatrix}$$

We have a first definition of rank, by counting pivots. Now we need the concept of independence. Then we are ready for the cornerstone of this subject—the Fundamental Theorem of Linear Algebra.

Summary of the Situation

From $Ax = b$ — *We can't see* if solutions exist; *We can't see* what the solutions are

The vector b might not be in the column space of A. Forward elimination takes A to U. The right side goes from b to c:

From $Ux = c$ — *We **can** see* if solutions exist; *We can't see* what the solutions are

Every zero row in U must be matched by a zero in c. These are the last $m - r$ rows of U and the last $m - r$ components of c. Then ***solve*** finds x by back substitution, with all free variables set to zero. Now eliminate upward to reach R, as the right side changes to d:

From $Rx = d$ — *We **can** see* if solutions exist; *We **can** see* what the solutions are

The reduced matrix R has the identity matrix I in its pivot columns. If those columns come first, the equations $R\boldsymbol{x} = \boldsymbol{d}$ look like

$$R\boldsymbol{x} = \begin{bmatrix} I & F \\ 0 & 0 \end{bmatrix} \begin{bmatrix} \boldsymbol{x}_p \\ \boldsymbol{x}_f \end{bmatrix} = \begin{bmatrix} \boldsymbol{d}_p \\ \boldsymbol{d}_z \end{bmatrix} \begin{matrix} r & \text{pivot rows} \\ m-r & \text{zero rows} \end{matrix}$$

$$r \quad n-r$$

The zero rows in R must be matched by $\boldsymbol{d}_z = \boldsymbol{0}$. Then solve for $\boldsymbol{x}$. The particular solution has all free variables zero: $\boldsymbol{x}_f = \boldsymbol{0}$ and $\boldsymbol{x}_p = \boldsymbol{d}_p$ (because R contains I).

$$\boldsymbol{x}_{\text{complete}} = \boldsymbol{x}_{\text{particular}} + \boldsymbol{x}_{\text{homogeneous}} = \begin{bmatrix} \boldsymbol{d}_p \\ \boldsymbol{0} \end{bmatrix} + \begin{bmatrix} -F \\ I \end{bmatrix} \boldsymbol{x}_f.$$

That is the nullspace matrix N multiplying $\boldsymbol{x}_f$. It contains the special solutions, with 1's and 0's for free variables in I. All solutions to $R\boldsymbol{x} = \boldsymbol{0}$ and $A\boldsymbol{x} = \boldsymbol{0}$ are combinations $N\boldsymbol{x}_f$ of the special solutions. For example:

$$\begin{bmatrix} A & \boldsymbol{b} \end{bmatrix} = \begin{bmatrix} 1 & 1 & 0 & 5 \\ 1 & 3 & 2 & 9 \end{bmatrix} \rightarrow \begin{bmatrix} U & \boldsymbol{c} \end{bmatrix} = \begin{bmatrix} 1 & 1 & 0 & 5 \\ 0 & 2 & 2 & 4 \end{bmatrix}$$

$$\rightarrow \begin{bmatrix} R & \boldsymbol{d} \end{bmatrix} = \begin{bmatrix} 1 & 0 & -1 & 3 \\ 0 & 1 & 1 & 2 \end{bmatrix}.$$

There are $r = 2$ pivots and $n - r = 3 - 2$ free variables. There are no zero rows in U and R, so the $\boldsymbol{d}_z$ part of the right side is empty. Solutions exist. The particular solution contains 3 and 2 from the last right side $\boldsymbol{d}$. The homogeneous solution contains -1 and 1 from the free column of R, *with the signs reversed*:

$$\boldsymbol{x}_{\text{complete}} = \boldsymbol{x}_{\text{particular}} + \boldsymbol{x}_{\text{nullspace}} = \begin{bmatrix} 3 \\ 2 \\ 0 \end{bmatrix} + \begin{bmatrix} 1 \\ -1 \\ 1 \end{bmatrix} x_3.$$

Problem Set 3.3

Questions 1–12 are about the solution of $A\boldsymbol{x} = \boldsymbol{b}$. Follow the five steps in the text to $\boldsymbol{x}_p$ and $\boldsymbol{x}_n$.

1 Write the complete solution in the form of equation (3.3):

$$\begin{aligned} x + 3y + 3z &= 1 \\ 2x + 6y + 9z &= 5 \\ -x - 3y + 3z &= 5. \end{aligned}$$

2 Find the complete solution (also called the *general solution*) to

$$\begin{bmatrix} 1 & 3 & 1 & 2 \\ 2 & 6 & 4 & 8 \\ 0 & 0 & 2 & 4 \end{bmatrix} \begin{bmatrix} x \\ y \\ z \\ t \end{bmatrix} = \begin{bmatrix} 1 \\ 3 \\ 1 \end{bmatrix}.$$

3 Under what condition on b_1, b_2, b_3 is this system solvable? Include $\boldsymbol{b}$ as a fourth column in elimination. Find all solutions:

$$\begin{aligned} x + 2y - 2z &= b_1 \\ 2x + 5y - 4z &= b_2 \\ 4x + 9y - 8z &= b_3. \end{aligned}$$

4 Under what conditions on b_1, b_2, b_3, b_4 is each system solvable? Find $\boldsymbol{x}$ in that case.

$$\begin{bmatrix} 1 & 2 \\ 2 & 4 \\ 2 & 5 \\ 3 & 9 \end{bmatrix} \begin{bmatrix} x_1 \\ x_2 \end{bmatrix} = \begin{bmatrix} b_1 \\ b_2 \\ b_3 \\ b_4 \end{bmatrix} \quad \text{and} \quad \begin{bmatrix} 1 & 2 & 3 \\ 2 & 4 & 6 \\ 2 & 5 & 7 \\ 3 & 9 & 12 \end{bmatrix} \begin{bmatrix} x_1 \\ x_2 \\ x_3 \end{bmatrix} = \begin{bmatrix} b_1 \\ b_2 \\ b_3 \\ b_4 \end{bmatrix}.$$

5 Show by elimination that (b_1, b_2, b_3) is in the column space of A if $b_3 - 2b_2 + 4b_1 = 0$.

$$A = \begin{bmatrix} 1 & 3 & 1 \\ 3 & 8 & 2 \\ 2 & 4 & 0 \end{bmatrix}.$$

What combination of the rows of A gives the zero row?

6 Which vectors (b_1, b_2, b_3) are in the column space of A? Which combinations of the rows of A give zero?

$$\text{(a)} \quad A = \begin{bmatrix} 1 & 2 & 1 \\ 2 & 6 & 3 \\ 0 & 2 & 5 \end{bmatrix} \qquad \text{(b)} \quad A = \begin{bmatrix} 1 & 1 & 1 \\ 1 & 2 & 4 \\ 2 & 4 & 8 \end{bmatrix}.$$

7 Construct a 2 by 3 system $A\boldsymbol{x} = \boldsymbol{b}$ with particular solution $\boldsymbol{x}_p = (2, 4, 0)$ and homogeneous solution $\boldsymbol{x}_n =$ any multiple of $(1, 1, 1)$.

8 Why can't a 1 by 3 system have $\boldsymbol{x}_p = (2, 4, 0)$ and $\boldsymbol{x}_n =$ any multiple of $(1, 1, 1)$?

9 (a) If $A\boldsymbol{x} = \boldsymbol{b}$ has two solutions $\boldsymbol{x}_1$ and $\boldsymbol{x}_2$, find two solutions to $A\boldsymbol{x} = \boldsymbol{0}$.

(b) Then find another solution to $A\boldsymbol{x} = \boldsymbol{0}$ and another solution to $A\boldsymbol{x} = \boldsymbol{b}$.

10 Explain why these are all false:

(a) The complete solution is any linear combination of $\boldsymbol{x}_p$ and $\boldsymbol{x}_n$.

(b) A system $A\boldsymbol{x} = \boldsymbol{b}$ has at most one particular solution.

(c) The solution $\boldsymbol{x}_p$ with all free variables zero is the shortest solution (minimum length $\|\boldsymbol{x}\|$). Find a 2 by 2 counterexample.

(d) If A is invertible there is no homogeneous solution $\boldsymbol{x}_n$.

11 Suppose column 5 of U has no pivot. Then x_5 is a ______ variable. The zero vector is ______ the only solution to $A\boldsymbol{x} = \boldsymbol{0}$. If $A\boldsymbol{x} = \boldsymbol{b}$ has a solution, then it has ______ solutions.

12 Suppose row 3 of U has no pivot. Then that row is ______. The equation $U\boldsymbol{x} = \boldsymbol{c}$ is only solvable provided ______. The equation $A\boldsymbol{x} = \boldsymbol{b}$ (*is*) (*is not*) (*might not be*) solvable.

Questions 13–18 are about matrices of "full rank" $r = m$ or $r = n$.

13 The largest possible rank of a 3 by 5 matrix is ______. Then there is a pivot in every ______ of U. The solution to $A\boldsymbol{x} = \boldsymbol{b}$ (*always exists*) (*is unique*). The column space of A is ______. An example is $A =$ ______.

14 The largest possible rank of a 6 by 4 matrix is ______. Then there is a pivot in every ______ of U. The solution to $A\boldsymbol{x} = \boldsymbol{b}$ (*always exists*) (*is unique*). The nullspace of A is ______. An example is $A =$ ______.

15 Find by elimination the rank of A and also the rank of A^{T}:

$$A = \begin{bmatrix} 1 & 4 & 0 \\ 2 & 11 & 5 \\ -1 & 2 & 10 \end{bmatrix} \quad \text{and} \quad A = \begin{bmatrix} 1 & 0 & 1 \\ 1 & 1 & 2 \\ 1 & 1 & q \end{bmatrix} \quad \text{(rank depends on } q\text{)}.$$

16 Find the rank of A and also of $A^{\mathrm{T}}A$ and also of AA^{T}:

$$A = \begin{bmatrix} 1 & 1 & 5 \\ 1 & 0 & 1 \end{bmatrix} \quad \text{and} \quad A = \begin{bmatrix} 2 & 0 \\ 1 & 1 \\ 1 & 2 \end{bmatrix}.$$

17 Reduce A to its echelon form U. Then find a triangular L so that $A = LU$.

$$A = \begin{bmatrix} 3 & 4 & 1 & 0 \\ 6 & 5 & 2 & 1 \end{bmatrix} \quad \text{and} \quad A = \begin{bmatrix} 1 & 0 & 1 & 0 \\ 2 & 2 & 0 & 3 \\ 0 & 6 & 5 & 4 \end{bmatrix}.$$

18 Find the complete solution in the form (3.3) to these full rank systems:

(a) $x + y + z = 4$ (b) $\begin{aligned} x + y + z &= 4 \\ x - y + z &= 4. \end{aligned}$

19 If $A\boldsymbol{x} = \boldsymbol{b}$ has infinitely many solutions, why is it impossible for $A\boldsymbol{x} = \boldsymbol{B}$ (new right side) to have only one solution? Could $A\boldsymbol{x} = \boldsymbol{B}$ have no solution?

20 Choose the number q so that (if possible) the rank is (a) 1, (b) 2, (c) 3:

$$A = \begin{bmatrix} 6 & 4 & 2 \\ -3 & -2 & -1 \\ 9 & 6 & q \end{bmatrix} \quad \text{and} \quad B = \begin{bmatrix} 3 & 1 & 3 \\ q & 2 & q \end{bmatrix}.$$

Questions 21–26 are about matrices of rank $r = 1$.

21 Fill out these matrices so that they have rank 1:

$$A = \begin{bmatrix} 1 & 2 & 4 \\ 2 & & \\ 4 & & \end{bmatrix} \quad \text{and} \quad B = \begin{bmatrix} 2 & & \\ 1 & & \\ 2 & 6 & -3 \end{bmatrix} \quad \text{and} \quad M = \begin{bmatrix} a & b \\ c & \end{bmatrix}.$$

22 If A is an m by n matrix with $r = 1$, its columns are multiples of one column and its rows are multiples of one row. The column space is a ______ in $\mathbf{R}^m$. The nullspace is a ______ in $\mathbf{R}^n$. Also the column space of A^{T} is a ______ in $\mathbf{R}^n$.

23 Choose vectors $\boldsymbol{u}$ and $\boldsymbol{v}$ so that $A = \boldsymbol{u}\boldsymbol{v}^{\mathrm{T}}$ = column times row:

$$A = \begin{bmatrix} 3 & 6 & 6 \\ 1 & 2 & 2 \\ 4 & 8 & 8 \end{bmatrix} \quad \text{and} \quad A = \begin{bmatrix} 2 & 2 & 6 & 4 \\ -1 & -1 & -3 & -2 \end{bmatrix}.$$

$A = \boldsymbol{u}\boldsymbol{v}^{\mathrm{T}}$ is the natural form for every matrix that has rank $r = 1$.

24 If A is a rank one matrix, the second row of U is ______. Do an example.

25 Multiply a rank one matrix times a rank one matrix, to find the rank of AB and AM:

$$A = \begin{bmatrix} 1 & 2 \\ 2 & 4 \end{bmatrix} \quad \text{and} \quad B = \begin{bmatrix} 2 & 1 & 4 \\ 3 & 1.5 & 6 \end{bmatrix} \quad \text{and} \quad M = \begin{bmatrix} 1 & b \\ c & bc \end{bmatrix}.$$

26 The rank one matrix $\boldsymbol{u}\boldsymbol{v}^{\mathrm{T}}$ times the rank one matrix $\boldsymbol{w}\boldsymbol{z}^{\mathrm{T}}$ is $\boldsymbol{u}\boldsymbol{z}^{\mathrm{T}}$ times the number ______. This has rank one unless ______ $= 0$.

27 Give examples of matrices A for which the number of solutions to $A\boldsymbol{x} = \boldsymbol{b}$ is

(a) 0 or 1, depending on $\boldsymbol{b}$

(b) ∞, regardless of $\boldsymbol{b}$

(c) 0 or ∞, depending on $\boldsymbol{b}$

(d) 1, regardless of $\boldsymbol{b}$.

28 Write down all known relations between r and m and n if $A\boldsymbol{x} = \boldsymbol{b}$ has

(a) no solution for some $\boldsymbol{b}$

(b) infinitely many solutions for every $\boldsymbol{b}$

(c) exactly one solution for some $\boldsymbol{b}$, no solution for other $\boldsymbol{b}$

(d) exactly one solution for every $\boldsymbol{b}$.

Questions 29–33 are about Gauss-Jordan elimination and the reduced echelon matrix R.

29 Continue elimination from U to R. Divide rows by pivots so the new pivots are all 1. Then produce zeros *above* those pivots to reach R:

$$U = \begin{bmatrix} 2 & 4 & 4 \\ 0 & 3 & 6 \\ 0 & 0 & 0 \end{bmatrix} \quad \text{and} \quad U = \begin{bmatrix} 2 & 4 & 4 \\ 0 & 3 & 6 \\ 0 & 0 & 5 \end{bmatrix}.$$

30 Suppose U is square with n pivots (an invertible matrix). *Explain why $R = I$.*

31 Apply Gauss-Jordan elimination to $Ux = \mathbf{0}$ and $Ux = c$. Reach $Rx = \mathbf{0}$ and $Rx = d$:

$$\begin{bmatrix} U & \mathbf{0} \end{bmatrix} = \begin{bmatrix} 1 & 2 & 3 & 0 \\ 0 & 0 & 4 & 0 \end{bmatrix} \quad \text{and} \quad \begin{bmatrix} U & c \end{bmatrix} = \begin{bmatrix} 1 & 2 & 3 & 5 \\ 0 & 0 & 4 & 8 \end{bmatrix}.$$

Solve $Rx = \mathbf{0}$ to find x_n (its free variable is $x_2 = 1$). Solve $Rx = d$ to find x_p (its free variable is $x_2 = 0$).

32 Gauss-Jordan elimination yields the reduced matrix R. Find $Rx = \mathbf{0}$ and $Rx = d$:

$$\begin{bmatrix} U & \mathbf{0} \end{bmatrix} = \begin{bmatrix} 3 & 0 & 6 & 0 \\ 0 & 0 & 2 & 0 \\ 0 & 0 & 0 & 0 \end{bmatrix} \quad \text{and} \quad \begin{bmatrix} U & c \end{bmatrix} = \begin{bmatrix} 3 & 0 & 6 & 9 \\ 0 & 0 & 2 & 4 \\ 0 & 0 & 0 & 5 \end{bmatrix}.$$

Solve $Ux = \mathbf{0}$ or $Rx = \mathbf{0}$ to find x_n (free variable $= 1$). What are the solutions to $Rx = d$?

33 Reduce $Ax = b$ to $Ux = c$ (Gaussian elimination) and then to $Rx = d$ (Gauss-Jordan):

$$Ax = \begin{bmatrix} 1 & 0 & 2 & 3 \\ 1 & 3 & 2 & 0 \\ 2 & 0 & 4 & 9 \end{bmatrix} \begin{bmatrix} x_1 \\ x_2 \\ x_3 \\ x_4 \end{bmatrix} = \begin{bmatrix} 2 \\ 5 \\ 10 \end{bmatrix} = b.$$

Find a particular solution x_p and all homogeneous solutions x_n.

34 Find matrices A and B with the given property or explain why you can't: The only solution of $Ax = \begin{bmatrix} 1 \\ 2 \\ 3 \end{bmatrix}$ is $x = \begin{bmatrix} 0 \\ 1 \end{bmatrix}$. The only solution of $Bx = \begin{bmatrix} 0 \\ 1 \end{bmatrix}$ is $x = \begin{bmatrix} 1 \\ 2 \\ 3 \end{bmatrix}$.

35 Find the LU factorization of A and the complete solution to $Ax = b$:

$$A = \begin{bmatrix} 1 & 3 & 1 \\ 1 & 2 & 3 \\ 2 & 4 & 6 \\ 1 & 1 & 5 \end{bmatrix} \quad \text{and} \quad b = \begin{bmatrix} 1 \\ 3 \\ 6 \\ 5 \end{bmatrix} \quad \text{and then} \quad b = \begin{bmatrix} 1 \\ 0 \\ 0 \\ 0 \end{bmatrix}.$$

36 The complete solution to $Ax = \begin{bmatrix} 1 \\ 3 \end{bmatrix}$ is $x = \begin{bmatrix} 1 \\ 0 \end{bmatrix} + c\begin{bmatrix} 0 \\ 1 \end{bmatrix}$. Find A.

3.4 Independence, Basis, and Dimension

This important section is about the true size of a subspace. The columns of A have m components. But the true "dimension" of the column space is not necessarily m (unless the subspace is the whole space $\mathbf{R}^m$). The dimension is measured by counting ***independent columns***—and we have to say what that means.

The idea of independence applies to any vectors $\boldsymbol{v}_1, \ldots, \boldsymbol{v}_n$ in any vector space. Most of this section concentrates on the subspaces that we know and use—especially the column space and nullspace. In the last part we also study "vectors" that are not column vectors. Matrices and functions can be linearly independent (or not). First come the key examples using column vectors, and then extra examples with matrices and functions.

The final goal is to understand a *basis* for a vector space—like $\boldsymbol{i} = (1, 0)$ and $\boldsymbol{j} = (0, 1)$ in the xy plane. We are at the heart of our subject, and we cannot go on without a basis. If you include matrix examples and function examples, allow extra time. The four essential ideas in this section are

1 Independent vectors

2 Basis for a space

3 Spanning a space

4 Dimension of a space.

Linear Independence

The two vectors $(3, 1)$ and $(6, 2)$ are not independent. One vector is a multiple of the other. They lie on a line—which is a one-dimensional subspace. The vectors $(3, 1)$ and $(4, 2)$ *are* independent—they go in different directions. Also the vectors $(3, 1)$ and $(7, 3)$ are independent. But the three vectors $(3, 1)$ and $(4, 2)$ and $(7, 3)$ are *not* independent—because the first vector plus the second vector equals the third.

The key question is: Which combinations of the vectors give zero?

Definition The sequence of vectors $\boldsymbol{v}_1, \ldots, \boldsymbol{v}_n$ is ***linearly independent*** if the only combination that gives the zero vector is $0\boldsymbol{v}_1 + 0\boldsymbol{v}_2 + \cdots + 0\boldsymbol{v}_n$. Thus linear independence means that

$$c_1\boldsymbol{v}_1 + c_2\boldsymbol{v}_2 + \cdots + c_n\boldsymbol{v}_n = \mathbf{0} \quad \textit{\textbf{only happens when all }} c\textit{'s} \textit{\textbf{ are zero.}}$$

If a combination $\sum c_j\boldsymbol{v}_j$ gives $\mathbf{0}$, when the c's are not all zero, the vectors are ***dependent***.

Correct language: "The sequence of vectors is linearly independent." Acceptable shortcut: "The vectors are independent."

A collection of vectors is either dependent or independent. They can be combined to give the zero vector (with nonzero c's) or they can't.

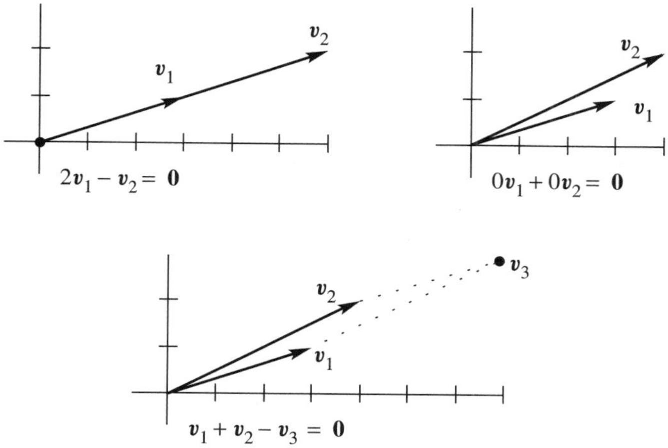

Figure 3.4 (a) Dependent (on a line) (b) Independent (c) A combination of the vectors equals zero.

The examples with (3, 1) and (6, 2) and (4, 2) are drawn in Figure 3.4:

Dependent:	$2(3,1) - (6,2) = (0,0)$	$c_1 = 2$ and $c_2 = -1$
Independent:	Only $0(3,1) + 0(4,2) = (0,0)$	$c_1 = 0$ and $c_2 = 0$
Dependent:	$(3,1) + (4,2) - (7,3) = (0,0)$	$c_1 = 1, c_2 = 1, c_3 = -1$

The c's in the first part of Figure 3.4 could also be 4 and -2. For independent vectors the only choice is 0 and 0. The c's in the third case could be 7, 7, and -7.

The test for independent vectors is: Which combinations equal zero? For n columns in a matrix A, the combinations are exactly $A\boldsymbol{x}$. We are asking about $A\boldsymbol{x} = \mathbf{0}$. ***Are there any nonzero vectors in the nullspace?*** If so, those columns are dependent.

3E ***The columns of A are independent if the only solution to $A\boldsymbol{x} = \mathbf{0}$ is $\boldsymbol{x} = \mathbf{0}$.*** Elimination produces no free variables. The matrix has rank $r = n$. The nullspace contains only $\boldsymbol{x} = \mathbf{0}$.

That is the definition of linear independence, written specifically for column vectors. The vectors go into the columns of A, and the c's go into $\boldsymbol{x}$. Then solve $A\boldsymbol{x} = \mathbf{0}$ by elimination. If there are any free variables, there will be nonzero solutions—the vectors are *dependent.* ***The components of $\boldsymbol{x}$ are the numbers $c_1, \ldots, c_n$ that we are looking for.***

Example 3.12 The columns of this matrix are dependent:

$$A\boldsymbol{x} = \begin{bmatrix} 2 & 1 & 7 \\ 1 & 2 & 8 \\ 2 & 1 & 7 \end{bmatrix} \begin{bmatrix} 2 \\ 3 \\ -1 \end{bmatrix} = \begin{bmatrix} 0 \\ 0 \\ 0 \end{bmatrix} \quad \text{so} \quad 2\begin{bmatrix} \\ \\ \\ \end{bmatrix} + 3\begin{bmatrix} \\ \\ \\ \end{bmatrix} - 1\begin{bmatrix} \\ \\ \\ \end{bmatrix} = \begin{bmatrix} 0 \\ 0 \\ 0 \end{bmatrix}.$$

The rank of A is only $r = 2$. *Independent columns would give full rank $r = n = 3$.*

In that matrix the rows are also dependent. You can see a combination of those rows that gives the zero row (it is row 1 minus row 3). For a *square matrix*, we will show that dependent columns imply dependent rows and vice versa.

Another way to describe linear dependence is this: *"One of the vectors is a combination of the other vectors."* That sounds clear. Why don't we say this from the start? Instead our definition was longer: *"Some combination gives the zero vector, other than the trivial combination with every $c = 0$."* We must rule out the easy way to get the zero vector. That trivial combination of zeros gives every author a headache. In the first statement, the vector that is a combination of the others has coefficient $c = 1$.

The point is, our definition doesn't pick out one particular vector as guilty. All columns of A are treated the same. We look at $A\boldsymbol{x} = \mathbf{0}$, and it has a nonzero solution or it hasn't. In the end that is better than asking if the last column (or the first, or a column in the middle) is a combination of the others.

One case is of special importance. Suppose seven columns have five components each ($m = 5$ is less than $n = 7$). Then the columns must be dependent! Any seven vectors from $\mathbf{R}^5$ are dependent. The rank of A cannot be larger than 5. There cannot be more than five pivots in five rows. The system $A\boldsymbol{x} = \mathbf{0}$ has at least $7 - 5 = 2$ free variables, so it has nonzero solutions—which means that the columns are dependent.

3F Any set of n vectors in $\mathbf{R}^m$ must be linearly dependent if $n > m$.

This is exactly the statement in Section 3.2, when $A\boldsymbol{x} = \mathbf{0}$ has more unknowns than equations. The matrix has more columns than rows—it is short and wide. The columns are dependent if $n > m$, because $A\boldsymbol{x} = \mathbf{0}$ has a nonzero solution.

Vectors That Span a Subspace

The first subspace in this book was the column space. Starting with n columns $\boldsymbol{v}_1, \ldots, \boldsymbol{v}_n$, the subspace was filled out by including all combinations $x_1\boldsymbol{v}_1 + \cdots + x_n\boldsymbol{v}_n$. *The column space consists of all linear combinations of the columns.* We now introduce the single word "span" to describe this: The column space is ***spanned*** by the columns.

Definition A set of vectors ***spans*** a space if their linear combinations fill the space.

To repeat: ***The columns of A span the column space.*** Don't say "the matrix spans the column space." The idea of taking linear combinations is familiar, and the word *span* says it more quickly. If a space $\mathbf{V}$ consists of all linear combinations of the particular vectors $\boldsymbol{v}_1, \ldots, \boldsymbol{v}_n$, then these vectors span $\mathbf{V}$.

The smallest space containing those vectors is the space $\mathbf{V}$ that they span. We have to be able to add vectors and multiply by scalars. We must include all linear combinations to produce a vector space.

Example 3.13 The vectors $\boldsymbol{v}_1 = \left[\begin{smallmatrix}1\\0\end{smallmatrix}\right]$ and $\boldsymbol{v}_2 = \left[\begin{smallmatrix}0\\1\end{smallmatrix}\right]$ span the two-dimensional space $\mathbf{R}^2$.

Example 3.14 The three vectors $\boldsymbol{v}_1 = \left[\begin{smallmatrix}1\\0\end{smallmatrix}\right]$, $\boldsymbol{v}_2 = \left[\begin{smallmatrix}0\\1\end{smallmatrix}\right]$, $\boldsymbol{v}_3 = \left[\begin{smallmatrix}4\\7\end{smallmatrix}\right]$ also span the same space $\mathbf{R}^2$.

Example 3.15 The vectors $w_1 = \begin{bmatrix} 1 \\ 1 \end{bmatrix}$ and $w_2 = \begin{bmatrix} -1 \\ -1 \end{bmatrix}$ only span a line in $\mathbf{R}^2$. So does w_1 by itself. So does w_2 by itself.

Think of two vectors coming out from (0, 0, 0) in 3-dimensional space. Generally they span a plane. Your mind fills in that plane by taking linear combinations. Mathematically you know other possibilities: two vectors spanning a line, three vectors spanning all of $\mathbf{R}^3$, three vectors spanning only a plane. It is even possible that three vectors span only a line, or ten vectors span only a plane. They are certainly not independent!

The columns span the column space. Here is a new subspace—*which begins with the rows. **The combinations of the rows produce the "row space."***

Definition The ***row space*** of a matrix is the subspace of $\mathbf{R}^n$ spanned by the rows.

The rows of an m by n matrix have n components. They are vectors in $\mathbf{R}^n$—or they would be if they were written as column vectors. There is a quick way to do that: *Transpose the matrix.* Instead of the rows of A, look at the columns of A^{T}. Same numbers, but now in columns.

The row space of A is $R(A^{\mathrm{T}})$. It is the column space of A^{T}. It is a subspace of $\mathbf{R}^n$. The vectors that span it are the columns of A^{T}, which are the rows of A.

Example 3.16 $A = \begin{bmatrix} 1 & 4 \\ 2 & 7 \\ 3 & 5 \end{bmatrix}$ and $A^{\mathrm{T}} = \begin{bmatrix} 1 & 2 & 3 \\ 4 & 7 & 5 \end{bmatrix}$. Here $m = 3$ and $n = 2$.

The column space of A is spanned by the two columns of A. It is a plane in $\mathbf{R}^3$. The row space of A is spanned by the three rows of A (columns of A^{T}). It is all of $\mathbf{R}^2$. Remember: The rows are in $\mathbf{R}^n$. The columns are in $\mathbf{R}^m$. Same numbers, different vectors, different spaces.

A Basis for a Vector Space

In the xy plane, a set of independent vectors could be small—just one single vector. A set that spans the xy plane could be large—three vectors, or four, or infinitely many. One vector won't span the plane. Three vectors won't be independent. A ***"basis"*** for the plane is just right.

Definition A ***basis*** for a vector space is a sequence of vectors that has two properties at once:

1 The vectors are ***linearly independent***.

2 The vectors ***span the space***.

This combination of properties is fundamental to linear algebra. Every vector v in the space is a combination of the basis vectors, because they span the space. More than that, the combination that produces v is *unique*, because the basis vectors $v_1, \ldots, v_n$ are independent:

There is one and only one way to write v as a combination of the basis vectors.

Reason: Suppose $\boldsymbol{v} = a_1\boldsymbol{v}_1 + \cdots + a_n\boldsymbol{v}_n$ and also $\boldsymbol{v} = b_1\boldsymbol{v}_1 + \cdots + b_n\boldsymbol{v}_n$. By subtraction $(a_1 - b_1)\boldsymbol{v}_1 + \cdots + (a_n - b_n)\boldsymbol{v}_n$ is the zero vector. From the independence of the $\boldsymbol{v}$'s, each $a_i - b_i = 0$. Hence $a_i = b_i$.

Example 3.17 The columns of $I = \left[\begin{smallmatrix} 1 & 0 \\ 0 & 1 \end{smallmatrix}\right]$ are a basis for $\mathbf{R}^2$. This is the "standard basis."

$$\text{The basis vectors are } \boldsymbol{i} = \begin{bmatrix} 1 \\ 0 \end{bmatrix} \text{ and } \boldsymbol{j} = \begin{bmatrix} 0 \\ 1 \end{bmatrix}. \text{They are independent. They span } \mathbf{R}^2.$$

Everybody thinks of this basis first. The vector $\boldsymbol{i}$ goes across and $\boldsymbol{j}$ goes straight up. Similarly the columns of the 3 by 3 identity matrix are the standard basis $\boldsymbol{i}, \boldsymbol{j}, \boldsymbol{k}$. The columns of the n by n identity matrix give the "standard basis" for $\mathbf{R}^n$. Now we find other bases.

Example 3.18 (Important) The columns of *any invertible n by n matrix* give a basis for $\mathbf{R}^n$:

$$A = \begin{bmatrix} 1 & 2 \\ 2 & 5 \end{bmatrix} \quad \text{and} \quad A = \begin{bmatrix} 1 & 0 & 0 \\ 1 & 1 & 0 \\ 1 & 1 & 1 \end{bmatrix} \quad \text{but not} \quad A = \begin{bmatrix} 1 & 2 \\ 2 & 4 \end{bmatrix}.$$

When A is invertible, its columns are independent. The only solution to $A\boldsymbol{x} = \boldsymbol{0}$ is $\boldsymbol{x} = \boldsymbol{0}$. The columns span the whole space $\mathbf{R}^n$—because every vector $\boldsymbol{b}$ is a combination of the columns. $A\boldsymbol{x} = \boldsymbol{b}$ can always be solved by $\boldsymbol{x} = A^{-1}\boldsymbol{b}$. Do you see how everything comes together? Here it is in one sentence:

3G The vectors $\boldsymbol{v}_1, \ldots, \boldsymbol{v}_n$ are a ***basis for*** $\mathbf{R}^n$ exactly when they are ***the columns of an n by n invertible matrix***. Thus $\mathbf{R}^n$ has infinitely many different bases.

When the columns are independent, they are a basis for the column space. When the columns are dependent, we keep only the *pivot columns*—the r columns with pivots. The picture is clearest for an echelon matrix U.

The pivot columns of A are one basis for the column space of A—which is different from the column space of U. Section 3.5 will study these bases carefully.

3H ***The pivot columns are a basis for the column space.*** The pivot rows are a basis for the row space.

Example 3.19 For an echelon matrix those bases are clear:

$$U = \begin{bmatrix} 1 & 2 & 1 & 4 & 7 \\ 0 & 0 & 5 & 3 & 8 \\ 0 & 0 & 0 & 6 & 9 \\ 0 & 0 & 0 & 0 & 0 \end{bmatrix} \quad \text{has pivot columns} \quad \begin{bmatrix} 1 \\ 0 \\ 0 \\ 0 \end{bmatrix} \begin{bmatrix} 1 \\ 5 \\ 0 \\ 0 \end{bmatrix} \begin{bmatrix} 4 \\ 3 \\ 6 \\ 0 \end{bmatrix}.$$

Columns 1, 3, 4 are a basis for the column space of U. So are columns 2, 3, 4 and 1, 3, 5 and 2, 4, 5. (But not columns 1 and 4 by themselves, and not 1, 2, 4.) There are infinitely many bases (not just these columns!). But the pivot columns are the natural choice.

How to show that the pivot columns are independent? Start with a combination that gives zero. *Then show that all the c's must be zero*:

$$c_1\begin{bmatrix}1\\0\\0\\0\end{bmatrix}+c_2\begin{bmatrix}1\\5\\0\\0\end{bmatrix}+c_3\begin{bmatrix}4\\3\\6\\0\end{bmatrix}=\begin{bmatrix}0\\0\\0\\0\end{bmatrix}. \tag{3.4}$$

The third component gives $6c_3 = 0$. Therefore $c_3 = 0$. Then the second component gives $5c_2 = 0$ so $c_2 = 0$. Now the first component gives $c_1 = 0$. The pivots "stick out." We use them in order (back substitution) to prove that *the c's are zero.* Please note how a proof of independence starts, with a combination (3.4) that gives zero.

Important: Every vector $\boldsymbol{b}$ in the column space can be produced out of the three pivot columns 1, 3, 4—by solving $A\boldsymbol{x} = \boldsymbol{b}$. The pivot columns span the column space. Those columns are independent. So they are a basis.

The three nonzero *rows* of U are a basis for the row space. They span the row space (the zero row adds nothing). The three rows are independent (again the pivots stick out). Are they the only basis? *Never.*

Question 1 If $\boldsymbol{v}_1, \ldots, \boldsymbol{v}_n$ is a basis for $\mathbf{R}^n$, can we use some of the $\boldsymbol{v}$'s in a basis for the row space?

Answer Not always. The $\boldsymbol{v}$'s might not be in the row space. To find a basis for a subspace $\mathbf{S}$, we have to look at $\mathbf{S}$. One thing is sure: The subspace won't have more than n basis vectors.

Question 2 ***Given five vectors, how do you find a basis for the space they span?***

Answer Make them the rows of A, and eliminate to find the nonzero rows of U.

Second way Put the five vectors into the columns of A. Eliminate to find the pivot columns (of A not of U!). The program basis uses the column numbers from pivcol.

M-file: basis

```
function B = basis(A)
[P,L,U,pivcol] = plu(A)
B = A(:,pivcol);
```

The column space of U had $r = 3$ basis vectors. Could another basis have more than r vectors, or fewer? This is a crucial question with a good answer. ***All bases for a vector space contain the same*** *number* ***of vectors. This number is the "dimension."***

Dimension of a Vector Space

We have to prove what was just stated. There are many choices for the basis vectors, but the *number* of basis vectors doesn't change.

3I If $\boldsymbol{v}_1, \ldots, \boldsymbol{v}_m$ and $\boldsymbol{w}_1, \ldots, \boldsymbol{w}_n$ are both bases for the same vector space, then $m = n$.

Proof Suppose there are more $\boldsymbol{w}$'s than $\boldsymbol{v}$'s. From $n > m$ we want to reach a contradiction. The $\boldsymbol{v}$'s are a basis, so $\boldsymbol{w}_1$ must be a combination of the $\boldsymbol{v}$'s. If $\boldsymbol{w}_1$ equals $a_{11}\boldsymbol{v}_1 + \cdots + a_{m1}\boldsymbol{v}_m$, this is the first column of a matrix multiplication VA:

$$W = \begin{bmatrix} \boldsymbol{w}_1 & \boldsymbol{w}_2 & \ldots & \boldsymbol{w}_n \end{bmatrix} = \begin{bmatrix} \boldsymbol{v}_1 & \ldots & \boldsymbol{v}_m \end{bmatrix} \begin{bmatrix} a_{11} & \\ \vdots & \\ a_{m1} & \end{bmatrix} = VA. \qquad (3.5)$$

We don't know the a's, but we know the shape of A (it is m by n). The second vector $\boldsymbol{w}_2$ is also a combination of the $\boldsymbol{v}$'s. The coefficients in that combination fill the second column of A. The key is that A has a row for every $\boldsymbol{v}$ and a column for every $\boldsymbol{w}$. It is a short wide matrix, since $n > m$. ***There is a nonzero solution to*** $A\boldsymbol{x} = \boldsymbol{0}$. But then $VA\boldsymbol{x} = \boldsymbol{0}$ and $W\boldsymbol{x} = \boldsymbol{0}$ and a combination of the $\boldsymbol{w}$'s gives zero! The $\boldsymbol{w}$'s could not be a basis—which is the contradiction we wanted.

If $m > n$ we exchange the $\boldsymbol{v}$'s and $\boldsymbol{w}$'s and repeat the same steps. The only way to avoid a contradiction is to have $m = n$. This completes the proof.

The number of basis vectors depends on the space—not on a particular basis. The number is the same for every basis, and it tells how many "degrees of freedom" the vector space allows. For the space $\mathbf{R}^n$, the number is n. This is the "dimension" of $\mathbf{R}^n$. We now introduce the important word ***dimension*** for other spaces too.

Definition The ***dimension of a vector space*** is the number of vectors in every basis.

This matches our intuition. The line through $\boldsymbol{v} = (1, 5, 2)$ has dimension one. It is a subspace with that one vector in its basis. Perpendicular to that line is the plane $x + 5y + 2z = 0$. This plane has dimension 2. To prove it, we find a basis $(-5, 1, 0)$ and $(-2, 0, 1)$. The dimension is 2 because the basis contains two vectors.

The plane is the nullspace of the matrix $A = [\,1 \;\; 5 \;\; 2\,]$, which has two free variables. Our basis vectors $(-5, 1, 0)$ and $(-2, 0, 1)$ are the "special solutions" to $A\boldsymbol{x} = \boldsymbol{0}$. The next section studies other nullspaces, and here we emphasize only this: The basis is not unique (unless the rank is n and the basis is empty). But all bases contain the same number of vectors.

Summary The key words of this section are "independence" and "span" and "basis" and "dimension." The connections are clearest for ***independent columns***:

3J A matrix with ***full column rank*** has all these properties:

1 The n columns are independent.

2 The only solution to $A\boldsymbol{x} = \boldsymbol{0}$ is $\boldsymbol{x} = \boldsymbol{0}$.

3 Rank of the matrix = dimension of the column space = n.

4 The columns are a basis for the column space.

The next chapter adds Property **5** to this list: ***The square matrix $A^{\mathrm{T}}A$ is invertible.*** These are the only matrices which Chapter 4 allows.

Note about the language of linear algebra We never say "rank of a space" or "dimension of a basis" or "basis of a matrix." Those terms have no meaning. It is the ***dimension of the column space*** that equals the ***rank of the matrix.***

Bases for Matrix Spaces and Function Spaces

The words "independence" and "basis" and "dimension" are not at all restricted to column vectors. We can ask whether matrices A_1, A_2, A_3 are independent. We can find a basis for the solutions to $d^2y/dx^2 = y$. That basis contains functions, maybe $y = e^x$ and $y = e^{-x}$. Counting the basis functions gives the dimension 2.

We think matrix spaces and function spaces are optional. Your class can go past this part—no problem. But in some way, you haven't got the ideas straight until you can apply them to "vectors" other than column vectors.

Matrix spaces The vector space **M** contains all 2 by 2 matrices. Its dimension is 4 and here is a basis:

$$A_1, A_2, A_3, A_4 = \begin{bmatrix} 1 & 0 \\ 0 & 0 \end{bmatrix}, \begin{bmatrix} 0 & 1 \\ 0 & 0 \end{bmatrix}, \begin{bmatrix} 0 & 0 \\ 1 & 0 \end{bmatrix}, \begin{bmatrix} 0 & 0 \\ 0 & 1 \end{bmatrix}.$$

Those matrices are linearly independent. We are not looking at their columns, but at the whole matrix. Combinations of those four matrices can produce any matrix in **M**, so they span the space:

$$c_1A_1 + c_2A_2 + c_3A_3 + c_4A_4 = \begin{bmatrix} c_1 & c_2 \\ c_3 & c_4 \end{bmatrix}.$$

This is zero only if the c's are all zero—which proves independence.

The matrices A_1, A_2, A_4 are a basis for a subspace—the upper triangular matrices. Its dimension is 3. A_1 and A_4 are a basis for a two-dimensional subspace—the diagonal matrices. What is a basis for the symmetric matrices? Keep A_1 and A_4, and throw in $A_2 + A_3$.

To push this further, think about the space of all n by n matrices. For a basis, choose matrices that have only a single nonzero entry (that entry is 1). There are n^2 positions for that 1, so there are n^2 basis matrices:

The dimension of the whole matrix space is n^2.

The dimension of the subspace of *upper triangular* matrices is $\frac{1}{2}n^2 + \frac{1}{2}n$.

The dimension of the subspace of *diagonal* matrices is n.

The dimension of the subspace of *symmetric* matrices is $\frac{1}{2}n^2 + \frac{1}{2}n$.

Function spaces $d^2y/dx^2 = 0$ and $d^2y/dx^2 + y = 0$ and $d^2y/dx^2 - y = 0$ involve the second derivative. In calculus we solve these equations to find the functions $y(x)$:

$$\begin{array}{ll} y'' = 0 & \text{is solved by any linear function } y = cx + d \\ y'' = -y & \text{is solved by any combination } y = c\sin x + d\cos x \\ y'' = y & \text{is solved by any combination } y = ce^x + de^{-x}. \end{array}$$

The second solution space has two basis functions: $\sin x$ and $\cos x$. The third solution space has basis functions e^x and e^{-x}. The first space has x and 1. It is the "nullspace" of the second derivative! The dimension is 2 in each case (these are second-order equations).

What about $y'' = 1$? Its solutions do *not* form a subspace—there is a nonzero right side $\boldsymbol{b} = 1$. A particular solution is $y(x) = \frac{1}{2}x^2$. The complete solution is $y(x) = \frac{1}{2}x^2 + cx + d$. All those functions satisfy $y'' = 1$. Notice the particular solution plus any function $cx + d$ in the nullspace.

A linear differential equation is like a linear matrix equation $A\boldsymbol{x} = \boldsymbol{b}$. But we solve it by calculus instead of linear algebra. The particular solution is a function, the special solutions are functions (most often they are $y = e^{cx}$).

We end here with the space $\mathbf{Z}$ that contains only the zero vector. The dimension of this space is zero. The empty set (containing no vectors) is a basis. We can never allow the zero vector into a basis, because then a combination of the vectors gives zero—and linear independence is lost.

Problem Set 3.4

Questions 1–10 are about linear independence and linear dependence.

1 Show that $\boldsymbol{v}_1, \boldsymbol{v}_2, \boldsymbol{v}_3$ are independent but $\boldsymbol{v}_1, \boldsymbol{v}_2, \boldsymbol{v}_3, \boldsymbol{v}_4$ are dependent:

$$\boldsymbol{v}_1 = \begin{bmatrix} 1 \\ 0 \\ 0 \end{bmatrix} \quad \boldsymbol{v}_2 = \begin{bmatrix} 1 \\ 1 \\ 0 \end{bmatrix} \quad \boldsymbol{v}_3 = \begin{bmatrix} 1 \\ 1 \\ 1 \end{bmatrix} \quad \boldsymbol{v}_4 = \begin{bmatrix} 2 \\ 3 \\ 4 \end{bmatrix}.$$

Solve either $c_1\boldsymbol{v}_1 + c_2\boldsymbol{v}_2 + c_3\boldsymbol{v}_3 = \mathbf{0}$ or $A\boldsymbol{x} = \mathbf{0}$. The $\boldsymbol{v}$'s go in the columns of A.

2 (Recommended) Find the largest possible number of independent vectors among

$$\boldsymbol{v}_1 = \begin{bmatrix} 1 \\ -1 \\ 0 \\ 0 \end{bmatrix} \quad \boldsymbol{v}_2 = \begin{bmatrix} 1 \\ 0 \\ -1 \\ 0 \end{bmatrix} \quad \boldsymbol{v}_3 = \begin{bmatrix} 1 \\ 0 \\ 0 \\ -1 \end{bmatrix} \quad \boldsymbol{v}_4 = \begin{bmatrix} 0 \\ 1 \\ -1 \\ 0 \end{bmatrix} \quad \boldsymbol{v}_5 = \begin{bmatrix} 0 \\ 1 \\ 0 \\ -1 \end{bmatrix} \quad \boldsymbol{v}_6 = \begin{bmatrix} 0 \\ 0 \\ 1 \\ -1 \end{bmatrix}.$$

3 Prove that if $a = 0$ or $d = 0$ or $f = 0$ (3 cases), the columns of U are dependent:

$$U = \begin{bmatrix} a & b & c \\ 0 & d & e \\ 0 & 0 & f \end{bmatrix}.$$

4 If a, d, f in Question 3 are all nonzero, show that the only solution to $U\boldsymbol{x} = \mathbf{0}$ is $\boldsymbol{x} = \mathbf{0}$. Then U has independent columns.

5 Decide the dependence or independence of

(a) the vectors $(1, 3, 2)$ and $(2, 1, 3)$ and $(3, 2, 1)$

(b) the vectors $(1, -3, 2)$ and $(2, 1, -3)$ and $(-3, 2, 1)$.

6 Choose three independent columns of U. Then make two other choices. Do the same for A.

$$U = \begin{bmatrix} 2 & 3 & 4 & 1 \\ 0 & 6 & 7 & 0 \\ 0 & 0 & 0 & 9 \\ 0 & 0 & 0 & 0 \end{bmatrix} \quad \text{and} \quad A = \begin{bmatrix} 2 & 3 & 4 & 1 \\ 0 & 6 & 7 & 0 \\ 0 & 0 & 0 & 9 \\ 4 & 6 & 8 & 2 \end{bmatrix}.$$

7 If $\boldsymbol{w}_1, \boldsymbol{w}_2, \boldsymbol{w}_3$ are independent vectors, show that the differences $\boldsymbol{v}_1 = \boldsymbol{w}_2 - \boldsymbol{w}_3$ and $\boldsymbol{v}_2 = \boldsymbol{w}_1 - \boldsymbol{w}_3$ and $\boldsymbol{v}_3 = \boldsymbol{w}_1 - \boldsymbol{w}_2$ are *dependent*. Find a combination of the $\boldsymbol{v}$'s that gives zero.

8 If $\boldsymbol{w}_1, \boldsymbol{w}_2, \boldsymbol{w}_3$ are independent vectors, show that the sums $\boldsymbol{v}_1 = \boldsymbol{w}_2 + \boldsymbol{w}_3$, $\boldsymbol{v}_2 = \boldsymbol{w}_1 + \boldsymbol{w}_3$, and $\boldsymbol{v}_3 = \boldsymbol{w}_1 + \boldsymbol{w}_2$ are *independent*. (Write $c_1\boldsymbol{v}_1 + c_2\boldsymbol{v}_2 + c_3\boldsymbol{v}_3 = \mathbf{0}$ in terms of the $\boldsymbol{w}$'s. Find and solve equations for the c's.)

9 Suppose $\boldsymbol{v}_1, \boldsymbol{v}_2, \boldsymbol{v}_3, \boldsymbol{v}_4$ are vectors in $\mathbf{R}^3$.

(a) These four vectors are dependent because ______.

(b) The two vectors $\boldsymbol{v}_1$ and $\boldsymbol{v}_2$ will be dependent if ______.

(c) The vectors $\boldsymbol{v}_1$ and $(0, 0, 0)$ are dependent because ______.

10 Find two independent vectors on the plane $x+2y-3z-t = 0$ in $\mathbf{R}^4$. Then find three independent vectors. Why not four? This plane is the nullspace of what matrix?

Questions 11–15 are about the space *spanned* by a set of vectors. Take all linear combinations of the vectors.

11 Describe the subspace of $\mathbf{R}^3$ (is it a line or plane or $\mathbf{R}^3$?) spanned by

(a) the two vectors $(1, 1, -1)$ and $(-1, -1, 1)$

(b) the three vectors $(0, 1, 1)$ and $(1, 1, 0)$ and $(0, 0, 0)$

(c) the columns of a 3 by 5 echelon matrix with 2 pivots

(d) all vectors with positive components.

12 The vector $\boldsymbol{b}$ is in the subspace spanned by the columns of A when there is a solution to ______. The vector $\boldsymbol{c}$ is in the row space of A when there is a solution to ______. True or false: If the zero vector is in the row space, the rows are dependent.

13 Find the dimensions of these 4 spaces. Which two of the spaces are the same? (a) column space of A, (b) column space of U, (c) row space of A, (d) row space of U:

$$A = \begin{bmatrix} 1 & 1 & 0 \\ 1 & 3 & 1 \\ 3 & 1 & -1 \end{bmatrix} \quad \text{and} \quad U = \begin{bmatrix} 1 & 1 & 0 \\ 0 & 2 & 1 \\ 0 & 0 & 0 \end{bmatrix}.$$

14 Choose $\boldsymbol{x} = (x_1, x_2, x_3, x_4)$ in $\mathbf{R}^4$. It has 24 rearrangements like (x_2, x_1, x_3, x_4) and (x_4, x_3, x_1, x_2). Those 24 vectors, including $\boldsymbol{x}$ itself, span a subspace $\mathbf{S}$. Find specific vectors $\boldsymbol{x}$ so that the dimension of $\mathbf{S}$ is: (a) zero, (b) one, (c) three, (d) four.

15 $\boldsymbol{v} + \boldsymbol{w}$ and $\boldsymbol{v} - \boldsymbol{w}$ are combinations of $\boldsymbol{v}$ and $\boldsymbol{w}$. Write $\boldsymbol{v}$ and $\boldsymbol{w}$ as combinations of $\boldsymbol{v} + \boldsymbol{w}$ and $\boldsymbol{v} - \boldsymbol{w}$. The two pairs of vectors _____ the same space. When are they a basis for the same space?

Questions 16–26 are about the requirements for a basis.

16 If $\boldsymbol{v}_1, \ldots, \boldsymbol{v}_n$ are linearly independent, the space they span has dimension _____. These vectors are a _____ for that space. If the vectors are the columns of an m by n matrix, then m is _____ than n.

17 Find a basis for each of these subspaces of $\mathbf{R}^4$:

(a) All vectors whose components are equal.

(b) All vectors whose components add to zero.

(c) All vectors that are perpendicular to $(1, 1, 0, 0)$ and $(1, 0, 1, 1)$.

(d) The column space and the nullspace of $U = \left[\begin{smallmatrix} 1 & 0 & 1 & 0 & 1 \\ 0 & 1 & 0 & 1 & 0 \end{smallmatrix}\right]$.

18 Find three different bases for the column space of U above. Then find two different bases for the row space of U.

19 Suppose $\boldsymbol{v}_1, \boldsymbol{v}_2, \ldots, \boldsymbol{v}_6$ are six vectors in $\mathbf{R}^4$.

(a) Those vectors (do)(do not)(might not) span $\mathbf{R}^4$.

(b) Those vectors (are)(are not)(might be) linearly independent.

(c) Any four of those vectors (are)(are not)(might be) a basis for $\mathbf{R}^4$.

20 The columns of A are n vectors from $\mathbf{R}^m$. If they are linearly independent, what is the rank of A? If they span $\mathbf{R}^m$, what is the rank? If they are a basis for $\mathbf{R}^m$, what then?

21 Find a basis for the plane $x - 2y + 3z = 0$ in $\mathbf{R}^3$. Then find a basis for the intersection of that plane with the xy plane. Then find a basis for all vectors perpendicular to the plane.

22 Suppose the columns of a 5 by 5 matrix A are a basis for $\mathbf{R}^5$.

(a) The equation $A\boldsymbol{x} = \mathbf{0}$ has only the solution $\boldsymbol{x} = \mathbf{0}$ because ______.

(b) If $\boldsymbol{b}$ is in $\mathbf{R}^5$ then $A\boldsymbol{x} = \boldsymbol{b}$ is solvable because ______.

Conclusion: A is invertible. Its rank is 5.

23 Suppose $\mathbf{S}$ is a 5-dimensional subspace of $\mathbf{R}^6$. True or false:

(a) Every basis for $\mathbf{S}$ can be extended to a basis for $\mathbf{R}^6$ by adding one more vector.

(b) Every basis for $\mathbf{R}^6$ can be reduced to a basis for $\mathbf{S}$ by removing one vector.

24 U comes from A by subtracting row 1 from row 3:

$$A = \begin{bmatrix} 1 & 3 & 2 \\ 0 & 1 & 1 \\ 1 & 3 & 2 \end{bmatrix} \quad \text{and} \quad U = \begin{bmatrix} 1 & 3 & 2 \\ 0 & 1 & 1 \\ 0 & 0 & 0 \end{bmatrix}.$$

Find bases for the two column spaces. Find bases for the two row spaces. Find bases for the two nullspaces.

25 True or false (give a good reason):

(a) If the columns of a matrix are dependent, so are the rows.

(b) The column space of a 2 by 2 matrix is the same as its row space.

(c) The column space of a 2 by 2 matrix has the same dimension as its row space.

(d) The columns of a matrix are a basis for the column space.

26 For which numbers c and d do these matrices have rank 2?

$$A = \begin{bmatrix} 1 & 2 & 5 & 0 & 5 \\ 0 & 0 & c & 2 & 2 \\ 0 & 0 & 0 & d & 2 \end{bmatrix} \quad \text{and} \quad B = \begin{bmatrix} c & d \\ d & c \end{bmatrix}.$$

Questions 27–32 are about spaces where the "vectors" are matrices.

27 Find a basis for each of these subspaces of 3 by 3 matrices:

(a) All diagonal matrices.

(b) All symmetric matrices ($A^{\mathrm{T}} = A$).

(c) All skew-symmetric matrices ($A^{\mathrm{T}} = -A$).

28 Construct six linearly independent 3 by 3 echelon matrices $U_1, \ldots, U_6$.

29 Find a basis for the space of all 2 by 3 matrices whose columns add to zero. Find a basis for the subspace whose rows also add to zero.

30 Show that the six 3 by 3 permutation matrices (Section 2.6) are linearly dependent.

31 What subspace of 3 by 3 matrices is spanned by

(a) all invertible matrices?

(b) all echelon matrices?

(c) the identity matrix?

32 Find a basis for the space of 2 by 3 matrices whose nullspace contains (2, 1, 1).

Questions 33–37 are about spaces where the "vectors" are functions.

33 (a) Find all functions that satisfy $\frac{dy}{dx} = 0$.

(b) Choose a particular function that satisfies $\frac{dy}{dx} = 3$.

(c) Find all functions that satisfy $\frac{dy}{dx} = 3$.

34 The cosine space $\mathbf{F}_3$ contains all combinations $y(x) = A\cos x + B\cos 2x + C\cos 3x$. Find a basis for the subspace with $y(0) = 0$.

35 Find a basis for the space of functions that satisfy

(a) $\frac{dy}{dx} - 2y = 0$ (b) $\frac{dy}{dx} - \frac{y}{x} = 0$.

36 Suppose $y_1(x)$, $y_2(x)$, $y_3(x)$ are three different functions of x. The vector space they span could have dimension 1, 2, or 3. Give an example y_1, y_2, y_3 to show each possibility.

37 Find a basis for the space of polynomials $p(x)$ of degree ≤ 3. Find a basis for the subspace with $p(1) = 0$.

38 Find a basis for the space $\mathbf{S}$ of vectors (a, b, c, d) with $a + c + d = 0$ and also for the space $\mathbf{T}$ with $a + b = 0$ and $c = 2d$. What is the dimension of the intersection $\mathbf{S} \cap \mathbf{T}$?

3.5 Dimensions of the Four Subspaces

The main theorem in this chapter connects ***rank*** and ***dimension***. The ***rank*** of a matrix is the number of pivots. The ***dimension*** of a subspace is the number of vectors in a basis. We count pivots or we count basis vectors. *The rank of A reveals the dimensions of all four fundamental subspaces.* Here are the subspaces:

1 The ***row space*** is $\boldsymbol{R}(A^{\mathrm{T}})$, a subspace of $\mathbf{R}^n$.

2 The ***column space*** is $\boldsymbol{R}(A)$, a subspace of $\mathbf{R}^m$.

3 The ***nullspace*** is $\boldsymbol{N}(A)$, a subspace of $\mathbf{R}^n$.

4 The ***left nullspace*** is $\boldsymbol{N}(A^{\mathrm{T}})$, a subspace of $\mathbf{R}^m$. This is our new space.

In this book the column space and nullspace came first. We know $\boldsymbol{R}(A)$ and $\boldsymbol{N}(A)$ pretty well. The other two subspaces were barely mentioned—now they come forward. For the row space we take all combinations of the rows. *This is also the column space of* A^{T}. For the left nullspace we solve $A^{\mathrm{T}}\boldsymbol{y} = \boldsymbol{0}$—that system is n by m. The vectors $\boldsymbol{y}$ go on the *left* side of A when the equation is written as $\boldsymbol{y}^{\mathrm{T}}A = \boldsymbol{0}^{\mathrm{T}}$.

The matrices A and A^{T} are usually very different. So are their column spaces and so are their nullspaces. But those spaces are connected in an absolutely beautiful way.

Part 1 of the Fundamental Theorem finds the dimensions of the four subspaces. One fact will stand out: ***The row space and column space have the same dimension.*** That dimension is r (the rank of the matrix). The other important fact involves the two nullspaces: *Their dimensions are $n - r$ and $m - r$, to make up the full dimensions n and m.*

Part 2 of the Theorem will describe how the four subspaces fit together (two in $\mathbf{R}^n$ and two in $\mathbf{R}^m$). That completes the "right way" to understand $A\boldsymbol{x} = \boldsymbol{b}$. Stay with it—you are doing real mathematics.

The Four Subspaces for U

Suppose we have an *echelon matrix*. It is upper triangular, with pivots in a staircase pattern. We call it U, not A, to emphasize this special form. Because of that form, the four subspaces are easy to identify. We will find a basis for each subspace and check its dimension. Then for any other matrix A, we watch how the subspaces change (or don't change) as elimination takes us to U.

As a specific 3 by 5 example, look at the four subspaces for the echelon matrix U:

$$\begin{matrix} m=3 \\ n=5 \\ r=2 \end{matrix} \qquad \begin{bmatrix} 1 & 3 & 5 & 7 & 9 \\ 0 & 0 & 0 & 4 & 8 \\ 0 & 0 & 0 & 0 & 0 \end{bmatrix} \qquad \begin{matrix} \text{pivot rows 1 and 2} \\ \\ \text{pivot columns 1 and 4} \end{matrix}$$

The pivots are 1 and 4. The rank is $r = 2$ (*two pivots*). Take the subspaces in order:

1 The ***row space*** of U has dimension 2, matching the rank. Reason: The first two rows are a basis. The row space contains combinations of all three rows, but the third row (the zero row) adds nothing new. So rows 1 and 2 span the row space.

Rows 1 and 2 are also independent. Certainly $(1, 3, 5, 7, 9)$ and $(0, 0, 0, 4, 8)$ are not parallel. But we give a proof of independence which applies to the nonzero rows of any echelon matrix. Start the proof with a combination of rows 1 and 2:

$$c_1 \text{ (row 1)} + c_2 \text{ (row 2)} = (c_1, 3c_1, 5c_1, 7c_1 + 4c_2, 9c_1 + 8c_2).$$

The key is this: *Suppose this combination equals* $(0, 0, 0, 0, 0)$. *You have to show that both c's must be zero.* Look at the first component; it gives $c_1 = 0$. With c_1 gone, look at the fourth component; it forces $c_2 = 0$.

If there were r nonzero rows, we would start with any combination. We would look at the first pivot position. We would discover that $c_1 = 0$. Then $c_2 = 0$ and $c_3 = 0$. All the c's are forced to be zero, which means that the rows are independent. Conclusion: The r pivot rows are a basis.

The dimension of the row space is r. The nonzero rows of U form a basis.

2 The ***column space*** of U also has dimension $r = 2$. Reason: The columns with pivots form a basis. In the example, the pivot columns 1 and 4 are independent. For a proof that applies to any echelon matrix, start the same way as for the rows. Suppose c_1 times the first column $(1, 0, 0)$ plus c_2 times the fourth column $(7, 4, 0)$ gives the zero vector: $(c_1 + 7c_2, 4c_2, 0) = (0, 0, 0)$. This time we work backwards: The second component $4c_2 = 0$ forces c_2 to be zero. Then the first component forces c_1 to be zero. This is nothing but back substitution, and all the c's come out zero.

So the pivot columns are independent:

The dimension of the column space is r. The pivot columns form a basis.

Those pivot columns produce any $\boldsymbol{b}$ in the column space. Set all the free variables to zero. Then back substitute in $U\boldsymbol{x} = \boldsymbol{b}$, to find the pivot variables x_1 and x_4. This makes $\boldsymbol{b}$ a combination of columns 1 and 4. The pivot columns span the column space.

You can see directly that column 2 is 3(column 1). Also column 3 is 5(column 1). Note for future reference: Column 5 equals 2(column 4)−5(column 1).

3 The ***nullspace*** of this U has dimension $n - r = 5 - 2$. There are $n - r = 3$ free variables. Here x_2, x_3, x_5 are free (no pivots in those columns). Those 3 free variables yield 3 special solutions to $U\boldsymbol{x} = \boldsymbol{0}$. Set a free variable to 1, and solve for x_1 and x_4:

$$s_2 = \begin{bmatrix} -3 \\ 1 \\ 0 \\ 0 \\ 0 \end{bmatrix} \quad s_3 = \begin{bmatrix} -5 \\ 0 \\ 1 \\ 0 \\ 0 \end{bmatrix} \quad s_5 = \begin{bmatrix} 5 \\ 0 \\ 0 \\ -2 \\ 1 \end{bmatrix} \qquad \begin{array}{l} U\boldsymbol{x} = \boldsymbol{0} \text{ has the} \\ \text{complete solution} \\ \boldsymbol{x} = x_2 s_2 + x_3 s_3 + x_5 s_5. \end{array}$$

There is a special solution for each free variable. With n variables and r pivot variables, the count is easy:

The nullspace has dimension n − r. The special solutions form a basis.

We have to remark: Those solutions tell again how to write columns 2, 3, 5 as combinations of 1 and 4. The special solution $(-3, 1, 0, 0, 0)$ says that column 2 equals 3(column 1).

The serious step is to prove that we have a basis. First, the special solutions are *independent*: Any combination $x_2 s_2 + x_3 s_3 + x_5 s_5$ has the numbers x_2, x_3, x_5 in components 2, 3, 5. So this combination gives the zero vector only if $x_2 = x_3 = x_5 = 0$. The special solutions *span the nullspace*: If $U\boldsymbol{x}$ equals zero then $\boldsymbol{x} = x_2 s_2 + x_3 s_3 + x_5 s_5$. In this equation, the components x_2, x_3, x_5 are the same on both sides. The pivot variables x_1 and x_4 must also agree—they are totally determined by $U\boldsymbol{x} = \boldsymbol{0}$. This proves what we strongly believed—by combining the special solutions we get all solutions.

4 The ***left nullspace*** of this U has dimension $m - r = 3 - 2$. Reason: $U^{\mathrm{T}}\boldsymbol{y} = \mathbf{0}$ has one free variable. There is a *line* of solutions $\boldsymbol{y} = (0, 0, y_3)$. This left nullspace is 1-dimensional.

$U^{\mathrm{T}}\boldsymbol{y}$ is a combination of the rows of U, just as $U\boldsymbol{x}$ is a combination of the columns. So $U^{\mathrm{T}}\boldsymbol{y} = \mathbf{0}$ means:

$$y_1 \text{ times row 1 plus } y_2 \text{ times row 2 plus } y_3 \text{ times row 3 equals the zero row.}$$

Conclusion: y_3 can be any number, because row 3 of U is all zeros. But $y_1 = y_2 = 0$ (because rows 1 and 2 are independent). The special solution is $\boldsymbol{y} = (0, 0, 1)$.

An echelon matrix U has $m - r$ zero rows. We are solving $U^{\mathrm{T}}\boldsymbol{y} = \mathbf{0}$. The last $m - r$ components of $\boldsymbol{y}$ are free:

$$U^{\mathrm{T}}\boldsymbol{y} = \mathbf{0}: \quad \begin{array}{l} \textbf{\textit{The left nullspace has dimension m} - \textit{r}.} \\ \textbf{\textit{The first r components of y are zero.}} \end{array}$$

The first r rows of U are independent. The other rows are zero. To produce a zero combination, $\boldsymbol{y}$ must start with r zeros. This leaves dimension $m - r$.

Why is this a "*left* nullspace"? The reason is that $U^{\mathrm{T}}\boldsymbol{y} = \mathbf{0}$ can be transposed to $\boldsymbol{y}^{\mathrm{T}}U = \mathbf{0}^{\mathrm{T}}$. Now $\boldsymbol{y}^{\mathrm{T}}$ is a row vector to the *left* of U. This subspace came fourth, and some linear algebra books omit it—but that misses the beauty of the whole subject.

In $\mathbf{R}^n$ ***the row space and nullspace have dimensions*** r and $n - r$ (adding to n).

In $\mathbf{R}^m$ ***the column space and left nullspace have dimensions*** r and $m - r$ (total m).

So far this is proved for echelon matrices. Figure 3.5 shows the same for A.

The Four Subspaces for A

We have a small job still to do. ***The dimensions for A are the same as for U.*** The job is to explain why. Suppose for example that all multipliers in elimination are 1. Then A is L times U:

$$LU = \begin{bmatrix} 1 & 0 & 0 \\ 1 & 1 & 0 \\ 1 & 1 & 1 \end{bmatrix} \begin{bmatrix} 1 & 3 & 5 & 7 & 9 \\ 0 & 0 & 0 & 4 & 8 \\ 0 & 0 & 0 & 0 & 0 \end{bmatrix} = \begin{bmatrix} 1 & 3 & 5 & 7 & 9 \\ 1 & 3 & 5 & 11 & 17 \\ 1 & 3 & 5 & 11 & 17 \end{bmatrix} = A.$$

This matrix A still has $m = 3$ and $n = 5$ and especially $r = 2$. We can go quickly through its four subspaces, finding a basis and checking the dimension. The reasoning doesn't depend on this particular L or U.

1 ***A has the same row space as U.*** Same dimension r and same basis.

Reason: Every row of A is a combination of the rows of U. Also every row of U is a combination of the rows of A. In one direction the combinations are given by L, in the other direction by L^{-1}. Elimination changes the rows of A to the rows of U, but the row *spaces* are identical.

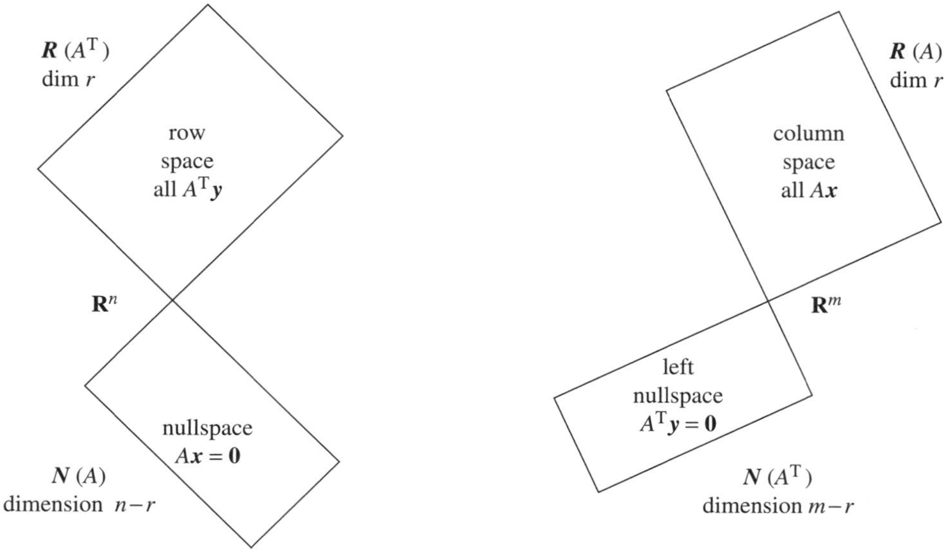

Figure 3.5 The dimensions of the four fundamental subspaces (for U and for A).

Since A has the same row space as U, we can choose the first r rows of U as a basis. Or we could choose r suitable rows of the original A. They might not always be the *first* r rows of A, because those could be dependent. (Then there will be row exchanges. The exchanges are made by P.) The first r rows of PA do form a basis for the row space.

2 *The column space of A has dimension r.* For every matrix this fact is essential: ***The number of independent columns equals the number of independent rows.***

Wrong reason: "A and U have the same column space." This is false—look at A and U in the example above. The columns of U end in zeros. The columns of A don't end in zeros. The column spaces are different, but their *dimensions* are the same—equal to r.

Right reason: A combination of the columns of A is zero exactly when the same combination of the columns of U is zero. Say that another way: $A\boldsymbol{x}=\mathbf{0}$ *exactly when* $U\boldsymbol{x}=\mathbf{0}$. Columns 1 and 4 were independent in U, so columns 1 and 4 are independent in A. Look at the matrix. Columns 1 and 2 were dependent in U so they are dependent in A.

Conclusion The r pivot columns in U are a basis for its column space. So the r corresponding columns of A are a basis for *its* column space.

3 ***A has the same nullspace as U.*** Same dimension $n-r$ and same basis as U.

Reason: $A\boldsymbol{x}=\mathbf{0}$ exactly when $U\boldsymbol{x}=\mathbf{0}$. The elimination steps don't change the solutions. The special solutions are a basis for this nullspace. There are $n-r$ free variables, so the dimension is $n-r$. Notice that $r+(n-r)$ equals n:

$$(\textbf{dimension of column space}) + (\textbf{dimension of nullspace}) = \textbf{dimension of } \mathbf{R}^n.$$

4 ***The left nullspace of A*** (the nullspace of A^{T}) ***has dimension m − r.***

Reason: A^{T} is just as good a matrix as A. When we know the dimensions for every A, we also know them for A^{T}. Its column space was proved to have dimension r. Since A^{T} is n by m, the "whole space" is now $\mathbf{R}^m$. The counting rule for A was $r + (n - r) = n$. The counting rule for A^{T} is $r + (m - r) = m$. So the nullspace of A^{T} has dimension $m - r$. We now have all details of the main theorem:

***Fundamental Theorem of Linear Algebra*, Part 1**

The column space and row space both have dimension r.
The nullspaces have dimensions n − r and m − r.

By concentrating on *spaces* of vectors, not on individual numbers or vectors, we get these clean rules. You will soon take them for granted—eventually they begin to look obvious. But if you write down an 11 by 17 matrix with 187 nonzero entries, we don't think most people would see why these facts are true:

$$\text{dimension of } \boldsymbol{R}(A) = \text{dimension of } \boldsymbol{R}(A^{\mathrm{T}})$$
$$\text{dimension of } \boldsymbol{R}(A) + \text{dimension of } \boldsymbol{N}(A) = 17.$$

Example 3.20 $A = [\,1 \quad 2 \quad 3\,]$ has $m = 1$ and $n = 3$ and rank $r = 1$.

The row space is a line in $\mathbf{R}^3$. The nullspace is the plane $Ax = x_1 + 2x_2 + 3x_3 = 0$. This plane has dimension 2 (which is $3 - 1$). The dimensions add to $1 + 2 = 3$.

The columns of this 1 by 3 matrix are in $\mathbf{R}^1$! The column space is all of $\mathbf{R}^1$. The left nullspace contains only the zero vector. The only solution to $A^{\mathrm{T}}\boldsymbol{y} = \mathbf{0}$ is $\boldsymbol{y} = \mathbf{0}$. That is the only combination of the row that gives the zero row. Thus $\boldsymbol{N}(A^{\mathrm{T}})$ is $\mathbf{Z}$, the zero space with dimension 0 (which is $m - r$). In $\mathbf{R}^m$ the dimensions add to $1 + 0 = 1$.

Example 3.21 $A = \begin{bmatrix} 1 & 2 & 3 \\ 1 & 2 & 3 \end{bmatrix}$ has $m = 2$ with $n = 3$ and $r = 1$.

The row space is the same line through $(1, 2, 3)$. The nullspace is the same plane $x_1 + 2x_2 + 3x_3 = 0$. The dimensions still add to $1 + 2 = 3$.

The columns are multiples of the first column $(1, 1)$. But there is more than the zero vector in the left nullspace. The first row minus the second row is the zero row. Therefore $A^{\mathrm{T}}\boldsymbol{y} = \mathbf{0}$ has the solution $\boldsymbol{y} = (1, -1)$. The column space and left nullspace are perpendicular lines in $\mathbf{R}^2$. Their dimensions are 1 and 1, adding to 2:

$$\text{column space} = \text{line through } \begin{bmatrix} 1 \\ 1 \end{bmatrix} \qquad \text{left nullspace} = \text{line through } \begin{bmatrix} 1 \\ -1 \end{bmatrix}.$$

If A has three equal rows, its rank is ______. What are two of the $\boldsymbol{y}$'s in its left nullspace? ***The y's combine the rows to give the zero row.***

Matrices of Rank One

That last example had rank $r = 1$—and rank one matrices are special. We can describe them all. You will see again that dimension of row space = dimension of column space. When $r = 1$, every row is a multiple of the same row:

$$A = \begin{bmatrix} 1 & 2 & 3 \\ 2 & 4 & 6 \\ -3 & -6 & -9 \\ 0 & 0 & 0 \end{bmatrix} \quad \text{equals} \quad \begin{bmatrix} 1 \\ 2 \\ -3 \\ 0 \end{bmatrix} \quad \text{times} \quad \begin{bmatrix} 1 & 2 & 3 \end{bmatrix}.$$

A column times a row (4 by 1 times 1 by 3) produces a matrix (4 by 3). All rows are multiples of the row $(1, 2, 3)$. All columns are multiples of the column $(1, 2, -3, 0)$. The row space is a line in $\mathbf{R}^n$, and the column space is a line in $\mathbf{R}^m$.

Every rank one matrix has the special form $\boldsymbol{A = uv^{\mathrm{T}}}$ ***= column times row.***

The columns are multiples of $\boldsymbol{u}$. The rows are multiples of $\boldsymbol{v}^{\mathrm{T}}$. *The nullspace is the plane perpendicular to* $\boldsymbol{v}$. ($A\boldsymbol{x} = \mathbf{0}$ means that $\boldsymbol{u}(\boldsymbol{v}^{\mathrm{T}}\boldsymbol{x}) = \mathbf{0}$ and then $\boldsymbol{v}^{\mathrm{T}}\boldsymbol{x} = 0$.) It is this perpendicularity of the subspaces that will be Part 2 of the Fundamental Theorem.

We end this section with a 4 by 3 example. Its rank is 2. Its echelon form is U:

$$A = \begin{bmatrix} 1 & 1 & 3 \\ 2 & 5 & 6 \\ 3 & 6 & 9 \\ 0 & 3 & 0 \end{bmatrix} \quad \rightarrow \quad U = \begin{bmatrix} 1 & 1 & 3 \\ 0 & 3 & 0 \\ 0 & 0 & 0 \\ 0 & 0 & 0 \end{bmatrix}.$$

Rows 1 and 2 of A or U give a basis for the (same) row space. Columns 1 and 2 give bases for the (different) column spaces. The special solution $\boldsymbol{s} = (-3, 0, 1)$ is a basis for the nullspace of A and U. The vectors $(0, 0, 1, 0)$ and $(0, 0, 0, 1)$ satisfy $U^{\mathrm{T}}\boldsymbol{y} = \mathbf{0}$. The vectors $\boldsymbol{y} = (1, 1, -1, 0)$ and $(-2, 1, 0, 1)$ satisfy $A^{\mathrm{T}}\boldsymbol{y} = \mathbf{0}$.

The nullspace has dimension $3 - 2 = 1$. Both left nullspaces have dimension $4 - 2 = 2$.

Problem Set 3.5

1 (a) If a 7 by 9 matrix has rank 5, what are the dimensions of the four subspaces?

(b) If a 3 by 4 matrix has rank 3, what are its column space and left nullspace?

2 Find bases for the four subspaces associated with A and B:

$$A = \begin{bmatrix} 1 & 2 & 4 \\ 2 & 4 & 8 \end{bmatrix} \quad \text{and} \quad B = \begin{bmatrix} 1 & 2 & 4 \\ 2 & 5 & 8 \end{bmatrix}.$$

3 Find a basis for each of the four subspaces associated with

$$A = \begin{bmatrix} 0 & 1 & 2 & 3 & 4 \\ 0 & 1 & 2 & 4 & 6 \\ 0 & 0 & 0 & 1 & 2 \end{bmatrix} = \begin{bmatrix} 1 & 0 & 0 \\ 1 & 1 & 0 \\ 0 & 1 & 1 \end{bmatrix} \begin{bmatrix} 0 & 1 & 2 & 3 & 4 \\ 0 & 0 & 0 & 1 & 2 \\ 0 & 0 & 0 & 0 & 0 \end{bmatrix}.$$

4 Construct a matrix with the required property or explain why no such matrix exists:

(a) Column space contains $\begin{bmatrix} 1 \\ 1 \\ 0 \end{bmatrix}, \begin{bmatrix} 0 \\ 0 \\ 1 \end{bmatrix}$, row space contains $\begin{bmatrix} 1 \\ 2 \end{bmatrix}, \begin{bmatrix} 2 \\ 5 \end{bmatrix}$.

(b) Column space has basis $\begin{bmatrix} 1 \\ 1 \\ 3 \end{bmatrix}$, nullspace has basis $\begin{bmatrix} 3 \\ 1 \\ 1 \end{bmatrix}$.

(c) Dimension of nullspace $= 1 +$ dimension of left nullspace.

(d) Left nullspace contains $\begin{bmatrix} 1 \\ 3 \end{bmatrix}$, row space contains $\begin{bmatrix} 3 \\ 1 \end{bmatrix}$.

(e) Row space $=$ column space, nullspace $\neq$ left nullspace.

5 If $\mathbf{V}$ is the subspace spanned by $(1, 1, 1)$ and $(2, 1, 0)$, find a matrix A that has $\mathbf{V}$ as its row space and a matrix B that has $\mathbf{V}$ as its nullspace.

6 Without elimination, find dimensions and bases for the four subspaces for

$$A = \begin{bmatrix} 0 & 3 & 3 & 3 \\ 0 & 0 & 0 & 0 \\ 0 & 1 & 0 & 1 \end{bmatrix} \quad \text{and} \quad B = \begin{bmatrix} 1 \\ 4 \\ 5 \end{bmatrix}.$$

7 Suppose the 3 by 3 matrix A is invertible. Write down bases for the four subspaces for A, and also for the 3 by 6 matrix $B = [\, A \;\; A \,]$.

8 What are the dimensions of the four subspaces for A, B, and C, if I is the 3 by 3 identity matrix and 0 is the 3 by 2 zero matrix?

$$A = \begin{bmatrix} I & 0 \end{bmatrix} \quad \text{and} \quad B = \begin{bmatrix} I & I \\ 0^{\mathrm{T}} & 0^{\mathrm{T}} \end{bmatrix} \quad \text{and} \quad C = \begin{bmatrix} 0 \end{bmatrix}.$$

9 Which subspaces are the same for these matrices of different sizes?

(a) $[\, A \,]$ and $\begin{bmatrix} A \\ A \end{bmatrix}$ (b) $\begin{bmatrix} A \\ A \end{bmatrix}$ and $\begin{bmatrix} A & A \\ A & A \end{bmatrix}$.

Prove that all three matrices have the same rank r.

10 If the entries of a 3 by 3 matrix are chosen randomly between 0 and 1, what are the most likely dimensions of the four subspaces? What if the matrix is 3 by 5?

11 (Important) A is an m by n matrix of rank r. Suppose there are right sides $\boldsymbol{b}$ for which $A\boldsymbol{x} = \boldsymbol{b}$ has *no solution*.

(a) What are all inequalities ($<$ or $\leq$) that must be true between m, n, and r?

(b) How do you know that $A^{\mathrm{T}}\boldsymbol{y} = \mathbf{0}$ has solutions other than $\boldsymbol{y} = \mathbf{0}$?

12 Construct a matrix with $(1, 0, 1)$ and $(1, 2, 0)$ as a basis for its row space and its column space. Why can't this be a basis for the row space and nullspace?

13 True or false:

(a) If $m = n$ then the row space of A equals the column space.

(b) The matrices A and $-A$ share the same four subspaces.

(c) If A and B share the same four subspaces then A is a multiple of B.

14 Without computing A, find bases for the four fundamental subspaces:

$$A = \begin{bmatrix} 1 & 0 & 0 \\ 6 & 1 & 0 \\ 9 & 8 & 1 \end{bmatrix} \begin{bmatrix} 1 & 2 & 3 & 4 \\ 0 & 1 & 2 & 3 \\ 0 & 0 & 1 & 2 \end{bmatrix}.$$

15 If you exchange the first two rows of A, which of the four subspaces stay the same? If $\boldsymbol{v} = (1, 2, 3, 4)$ is in the column space of A, write down a vector in the column space of the new matrix.

16 Explain why $\boldsymbol{v} = (1, 2, 3)$ cannot be a row of A and also be in the nullspace of A.

17 Describe the four subspaces of $\mathbf{R}^3$ associated with

$$A = \begin{bmatrix} 0 & 1 & 0 \\ 0 & 0 & 1 \\ 0 & 0 & 0 \end{bmatrix} \quad \text{and} \quad I + A = \begin{bmatrix} 1 & 1 & 0 \\ 0 & 1 & 1 \\ 0 & 0 & 1 \end{bmatrix}.$$

18 (Left nullspace) Add the extra column $\boldsymbol{b}$ and reduce A to echelon form:

$$\begin{bmatrix} A & \boldsymbol{b} \end{bmatrix} = \begin{bmatrix} 1 & 2 & 3 & b_1 \\ 4 & 5 & 6 & b_2 \\ 7 & 8 & 9 & b_3 \end{bmatrix} \quad \to \quad \begin{bmatrix} 1 & 2 & 3 & b_1 \\ 0 & -3 & -6 & b_2 - 4b_1 \\ 0 & 0 & 0 & b_3 - 2b_2 + b_1 \end{bmatrix}.$$

A combination of the rows of A has produced the zero row. What combination is it? (Look at $b_3 - 2b_2 + b_1$ on the right side.) Which vectors are in the nullspace of A^{T} and which are in the nullspace of A?

19 Following the method of Problem 18, reduce A to echelon form and look at zero rows. The $\boldsymbol{b}$ column tells which combinations you have taken of the rows:

$$\text{(a)} \quad \begin{bmatrix} 1 & 2 & b_1 \\ 3 & 4 & b_2 \\ 4 & 6 & b_3 \end{bmatrix} \qquad \text{(b)} \quad \begin{bmatrix} 1 & 2 & b_1 \\ 2 & 3 & b_2 \\ 2 & 4 & b_3 \\ 2 & 5 & b_4 \end{bmatrix}$$

From the $\boldsymbol{b}$ column after elimination, read off vectors in the left nullspace of A (combinations of rows that give zero rows). Check that you have $m - r$ basis vectors for $\boldsymbol{N}(A^{\mathrm{T}})$.

20 (a) Describe all solutions to $A\boldsymbol{x} = \mathbf{0}$ if

$$A = \begin{bmatrix} 1 & 0 & 0 \\ 2 & 1 & 0 \\ 3 & 4 & 1 \end{bmatrix} \begin{bmatrix} 4 & 2 & 0 & 1 \\ 0 & 0 & 1 & 3 \\ 0 & 0 & 0 & 0 \end{bmatrix}.$$

(b) How many independent solutions are there to $A^{\mathrm{T}}\boldsymbol{y} = \mathbf{0}$?

(c) Give a basis for the column space of A.

21 Suppose A is the sum of two matrices of rank one: $A = \boldsymbol{u}\boldsymbol{v}^{\mathrm{T}} + \boldsymbol{w}\boldsymbol{z}^{\mathrm{T}}$.

(a) Which vectors span the column space of A?

(b) Which vectors span the row space of A?

(c) The rank is less than 2 if ______ or if ______ .

(d) Compute A and its rank if $\boldsymbol{u} = \boldsymbol{z} = (1, 0, 0)$ and $\boldsymbol{v} = \boldsymbol{w} = (0, 0, 1)$.

22 Construct a matrix whose column space has basis $(1, 2, 4)$, $(2, 2, 1)$ and whose row space has basis $(1, 0, 0)$, $(0, 1, 1)$.

23 Without multiplying matrices, find bases for the row and column spaces of A:

$$A = \begin{bmatrix} 1 & 2 \\ 4 & 5 \\ 2 & 7 \end{bmatrix} \begin{bmatrix} 3 & 0 & 3 \\ 1 & 1 & 2 \end{bmatrix}.$$

How do you know from these shapes that A is not invertible?

24 $A^{\mathrm{T}}\boldsymbol{y} = \boldsymbol{d}$ is solvable when the right side $\boldsymbol{d}$ is in which subspace? The solution is unique when the ______ contains only the zero vector.

25 True or false (with a reason or a counterexample):

(a) A and A^{T} have the same number of pivots.

(b) A and A^{T} have the same left nullspace.

(c) If the row space equals the column space then $A^{\mathrm{T}} = A$.

(d) If $A^{\mathrm{T}} = -A$ then the row space of A equals the column space.

26 If $AB = C$, the rows of C are combinations of the rows of ______ . So the rank of C is not greater than the rank of ______ . Since $B^{\mathrm{T}}A^{\mathrm{T}} = C^{\mathrm{T}}$, the rank of C is also not greater than the rank of ______ .

27 If a, b, c are given with $a \neq 0$, how would you choose d so that $A = \left[\begin{smallmatrix} a & b \\ c & d \end{smallmatrix}\right]$ has rank one? Find a basis for the row space and nullspace.

28 Find the ranks of the 8 by 8 checkerboard matrix B and chess matrix C:

$$B = \begin{bmatrix} 1 & 0 & 1 & 0 & 1 & 0 & 1 & 0 \\ 0 & 1 & 0 & 1 & 0 & 1 & 0 & 1 \\ 1 & 0 & 1 & 0 & 1 & 0 & 1 & 0 \\ \bullet & \bullet & \bullet & \bullet & \bullet & \bullet & \bullet & \bullet \\ 0 & 1 & 0 & 1 & 0 & 1 & 0 & 1 \end{bmatrix} \quad \text{and} \quad C = \begin{bmatrix} r & n & b & q & k & b & n & r \\ p & p & p & p & p & p & p & p \\ & & & \text{four zero rows} & & & & \\ p & p & p & p & p & p & p & p \\ r & n & b & q & k & b & n & r \end{bmatrix}$$

The numbers r, n, b, q, k, p are all different. Find bases for the row space and left nullspace of B and C. Challenge problem: Find a basis for the nullspace of C.

4

ORTHOGONALITY

4.1 Orthogonality of the Four Subspaces

Vectors are orthogonal when their dot product is zero: $\boldsymbol{v} \cdot \boldsymbol{w} = 0$ or $\boldsymbol{v}^{\mathrm{T}}\boldsymbol{w} = 0$. This chapter moves up a level, from orthogonal vectors to ***orthogonal subspaces***. Orthogonal means the same as perpendicular.

Subspaces entered Chapter 3 with a specific purpose—to throw light on $A\boldsymbol{x} = \boldsymbol{b}$. Right away we needed the column space (for $\boldsymbol{b}$) and the nullspace (for $\boldsymbol{x}$). Then the light turned onto A^{T}, uncovering two more subspaces. Those four fundamental subspaces reveal what a matrix really does.

A matrix multiplies a vector: *A times $\boldsymbol{x}$*. At the first level this is only numbers. At the second level $A\boldsymbol{x}$ is a combination of column vectors. The third level shows subspaces. But we don't think you have seen the whole picture until you study Figure 4.1. It fits the subspaces together, to show the hidden reality of A times $\boldsymbol{x}$. The right angles between subspaces are something new—and we have to say what they mean.

The row space is perpendicular to the nullspace. Every row of A is perpendicular to every solution of $A\boldsymbol{x} = \mathbf{0}$. Similarly every column is perpendicular to every solution of $A^{\mathrm{T}}\boldsymbol{y} = \mathbf{0}$. That gives the 90° angle on the right side of the figure. This perpendicularity of subspaces is Part 2 of the Fundamental Theorem of Linear Algebra.

May we add a word about the left nullspace? It is never reached by $A\boldsymbol{x}$, so it might seem useless. But when $\boldsymbol{b}$ is outside the column space—when we want to solve $A\boldsymbol{x} = \boldsymbol{b}$ and can't do it—then this nullspace of A^{T} comes into its own. It contains the error in the "least-squares" solution. That is the key application of linear algebra in this chapter.

Part 1 of the Fundamental Theorem gave the dimensions of the subspaces. The row and column spaces have the same dimension r (they are drawn the same size). The nullspaces have the remaining dimensions $n - r$ and $m - r$. Now we will show that the row space and nullspace are actually perpendicular.

Definition Two subspaces $\boldsymbol{V}$ and $\boldsymbol{W}$ of a vector space are ***orthogonal*** if every vector $\boldsymbol{v}$ in $\boldsymbol{V}$ is perpendicular to every vector $\boldsymbol{w}$ in $\boldsymbol{W}$:

$$\boldsymbol{v} \cdot \boldsymbol{w} = 0 \;\; \textit{or} \;\; \boldsymbol{v}^{\mathrm{T}}\boldsymbol{w} = 0 \;\; \textit{for all } \boldsymbol{v} \text{ in } \boldsymbol{V} \text{ and all } \boldsymbol{w} \text{ in } \boldsymbol{W}.$$

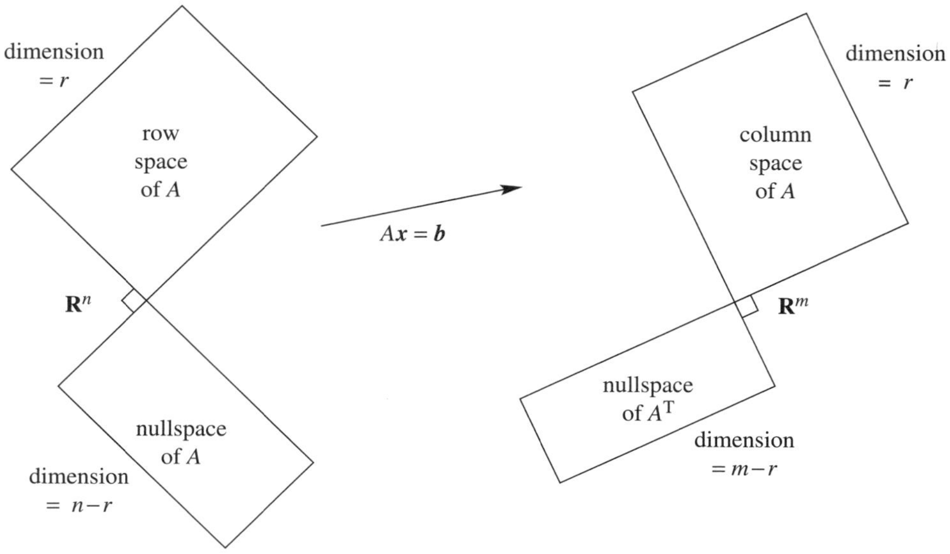

Figure 4.1 Two pairs of orthogonal subspaces. Dimensions add to n and m.

Example 4.1 The floor of your room (extended to infinity) is a subspace $\boldsymbol{V}$. The line where two walls meet is a subspace $\boldsymbol{W}$ (one-dimensional). Those subspaces are orthogonal. Every vector up the meeting line is perpendicular to every vector in the floor. The origin $(0, 0, 0)$ is in the corner. We assume you don't live in a tent.

Example 4.2 Suppose $\boldsymbol{V}$ is still the floor but $\boldsymbol{W}$ is one of the walls (a two-dimensional space). The wall and floor look orthogonal but they are not! You can find vectors in $\boldsymbol{V}$ and $\boldsymbol{W}$ that are not perpendicular. In fact a vector running along the bottom of the wall is also in the floor. This vector is in both $\boldsymbol{V}$ and $\boldsymbol{W}$—and it is not perpendicular to itself.

Example 4.3 If a vector is in two orthogonal subspaces, it must be perpendicular to itself. It is $\boldsymbol{v}$ and it is $\boldsymbol{w}$, so $\boldsymbol{v}^{\mathrm{T}}\boldsymbol{v} = 0$. This has to be the zero vector. Zero is the only point where the nullspace meets the row space.

The crucial examples for linear algebra come from the fundamental subspaces.

4A Every vector $\boldsymbol{x}$ in the nullspace of A is perpendicular to every row of A. ***The nullspace and row space are orthogonal subspaces.***

To see why $\boldsymbol{x}$ is perpendicular to the rows, look at $A\boldsymbol{x} = \mathbf{0}$. Each row multiplies $\boldsymbol{x}$:

$$A\boldsymbol{x} = \begin{bmatrix} \text{row } 1 \\ \vdots \\ \text{row } m \end{bmatrix} \begin{bmatrix} \\ \boldsymbol{x} \\ \\ \end{bmatrix} = \begin{bmatrix} 0 \\ \vdots \\ 0 \end{bmatrix}. \tag{4.1}$$

The first equation says that row 1 is perpendicular to $\boldsymbol{x}$. The last equation says that row m is perpendicular to $\boldsymbol{x}$. *Every row has a zero dot product with* $\boldsymbol{x}$. Then $\boldsymbol{x}$ is perpendicular to every combination of the rows. The whole row space $\boldsymbol{R}(A^{\mathrm{T}})$ is orthogonal to the whole nullspace $\boldsymbol{N}(A)$.

Here is a second proof of that orthogonality for readers who like matrix shorthand. The vectors in the row space are combinations $A^{\mathrm{T}}\boldsymbol{y}$ of the rows. Take the dot product of $A^{\mathrm{T}}\boldsymbol{y}$ with any $\boldsymbol{x}$ in the nullspace. *These vectors are perpendicular*:

$$\boldsymbol{x}^{\mathrm{T}}(A^{\mathrm{T}}\boldsymbol{y}) = (A\boldsymbol{x})^{\mathrm{T}}\boldsymbol{y} = \boldsymbol{0}^{\mathrm{T}}\boldsymbol{y} = 0. \tag{4.2}$$

We like the first proof. You can see those rows of A multiplying $\boldsymbol{x}$ to produce zeros in equation (4.1). The second proof shows why A and A^{T} are both in the Fundamental Theorem. A^{T} goes with $\boldsymbol{y}$ and A goes with $\boldsymbol{x}$. At the end we used $A\boldsymbol{x} = \boldsymbol{0}$.

In this next example, the rows are perpendicular to $(1, 1, -1)$ in the nullspace:

$$A\boldsymbol{x} = \begin{bmatrix} 1 & 3 & 4 \\ 5 & 2 & 7 \end{bmatrix} \begin{bmatrix} 1 \\ 1 \\ -1 \end{bmatrix} = \begin{bmatrix} 0 \\ 0 \end{bmatrix} \quad \text{gives the dot products} \quad \begin{matrix} 1+3-4=0 \\ 5+2-7=0 \end{matrix}$$

Now we turn to the other two subspaces. They are also orthogonal, but in $\mathbf{R}^m$.

4B Every vector $\boldsymbol{y}$ in the nullspace of A^{T} is perpendicular to every column of A. ***The left nullspace and column space are orthogonal.***

Apply the original proof to A^{T}. Its nullspace is orthogonal to its row space—which is the column space of A. Q.E.D. For a visual proof, look at $\boldsymbol{y}^{\mathrm{T}}A = \boldsymbol{0}$. The row vector $\boldsymbol{y}^{\mathrm{T}}$ multiplies each column of A:

$$\boldsymbol{y}^{\mathrm{T}}A = \begin{bmatrix} \quad \boldsymbol{y}^{\mathrm{T}} \quad \end{bmatrix} \begin{bmatrix} \mathrm{c} & & \mathrm{c} \\ \mathrm{o} & & \mathrm{o} \\ 1 & & 1 \\ & \cdots & \\ 1 & & n \end{bmatrix} = \begin{bmatrix} 0 & \ldots & 0 \end{bmatrix}. \tag{4.3}$$

The dot product with every column is zero. Then $\boldsymbol{y}$ is perpendicular to each column—and to the whole column space.

Very important The fundamental subspaces are more than just orthogonal (in pairs). Their dimensions are also right. Two lines could be perpendicular in 3-dimensional space, but they *could not be* the row space and nullspace of a matrix. The lines have dimensions 1 and 1, adding to 2. The correct dimensions r and $n-r$ must add to $n = 3$. Our subspaces have dimensions 2 and 1, or 3 and 0. The fundamental subspaces are not only orthogonal, they are *orthogonal complements*.

Definition The ***orthogonal complement*** of a subspace $\boldsymbol{V}$ contains *every* vector that is perpendicular to $\boldsymbol{V}$. This orthogonal subspace is denoted by $\boldsymbol{V}^{\perp}$ (pronounced "$\boldsymbol{V}$ perp").

By this definition, the nullspace is the orthogonal complement of the row space. *Every* $\boldsymbol{x}$ that is perpendicular to the rows satisfies $A\boldsymbol{x} = \boldsymbol{0}$, and is included in the nullspace.

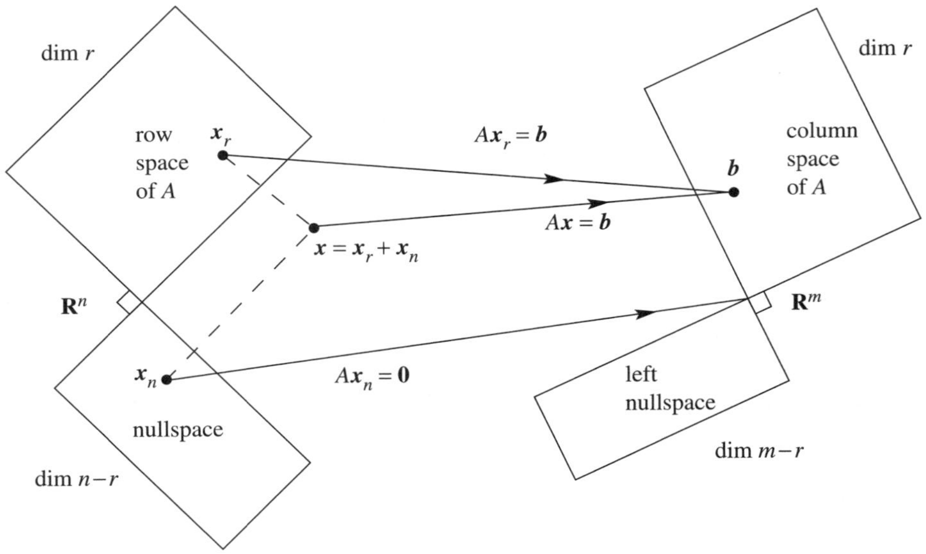

Figure 4.2 The true action of A times $\boldsymbol{x}$: row space to column space, nullspace to zero.

The reverse is also true (automatically). If $\boldsymbol{v}$ is orthogonal to the nullspace, it must be in the row space. Otherwise we could add this $\boldsymbol{v}$ as an extra row of the matrix, without changing the nullspace. The row space and its dimension would grow, which breaks the law $r + (n - r) = n$. We conclude that $\boldsymbol{N}(A)^{\perp}$ is exactly the row space $\boldsymbol{R}(A^{\mathrm{T}})$.

The left nullspace and column space are not only orthogonal in $\mathbf{R}^m$, they are orthogonal complements. The 90° angles are marked in Figure 4.2. Their dimensions add to the full dimension m.

***Fundamental Theorem of Linear Algebra*, Part 2**

The nullspace is the orthogonal complement of the row space (in $\boldsymbol{R}^n$).
The left nullspace is the orthogonal complement of the column space (in $\boldsymbol{R}^m$).

Part 1 gave the dimensions of the subspaces, Part 2 gives their orientation. They are perpendicular (in pairs). The point of "complements" is that every $\boldsymbol{x}$ can be split into a *row space component* $\boldsymbol{x}_r$ and a *nullspace component* $\boldsymbol{x}_n$. When A multiplies $\boldsymbol{x} = \boldsymbol{x}_r + \boldsymbol{x}_n$, Figure 4.2 shows what happens:

The nullspace component goes to zero: $A\boldsymbol{x}_n = \boldsymbol{0}$.

The row space component goes to the column space: $A\boldsymbol{x}_r = A\boldsymbol{x}$.

Every vector goes to the column space! Multiplying by A cannot do anything else. But more than that: *Every vector in the column space comes from one and only one vector* $\boldsymbol{x}_r$ *in the row space*. Proof: If $A\boldsymbol{x}_r = A\boldsymbol{x}_r'$, the difference $\boldsymbol{x}_r - \boldsymbol{x}_r'$ is in the nullspace. It is also in the row space, where $\boldsymbol{x}_r$ and $\boldsymbol{x}_r'$ came from. This difference must be the zero vector, because the spaces are perpendicular. Therefore $\boldsymbol{x}_r = \boldsymbol{x}_r'$.

There is an invertible matrix hiding inside A, if we throw away the two nullspaces. From the row space to the column space, A is invertible. The "pseudoinverse" will invert it in Section 7.2.

Example 4.4 Every diagonal matrix has an r by r invertible submatrix:

$$A = \begin{bmatrix} 3 & 0 & 0 & 0 & 0 \\ 0 & 5 & 0 & 0 & 0 \\ 0 & 0 & 0 & 0 & 0 \end{bmatrix} \quad \text{contains} \quad \begin{bmatrix} 3 & 0 \\ 0 & 5 \end{bmatrix}.$$

The rank is $r = 2$. The 2 by 2 submatrix in the upper corner is certainly invertible. The other eleven zeros are responsible for the nullspaces.

Section 7.2 will show how every A becomes a diagonal matrix, when we choose the right bases for $\mathbf{R}^n$ and $\mathbf{R}^m$. This ***Singular Value Decomposition*** is a part of the theory that has become extremely important in applications.

Combining Bases from Subspaces

What follows are some valuable facts about bases. They could have come earlier, when bases were defined, but they were saved until now—when we are ready to use them. After a week you have a clearer sense of what a basis is (independent vectors that span the space). When the count is right, one of those two properties implies the other:

4C Any n linearly independent vectors in $\mathbf{R}^n$ must span $\mathbf{R}^n$. Any n vectors that span $\mathbf{R}^n$ must be independent.

Normally we have to check both properties: First, that the vectors are linearly independent, and second, that they span the space. For n vectors in $\mathbf{R}^n$, *either* independence *or* spanning is enough by itself. Starting with the correct number of vectors, one property of a basis implies the other.

This is true in any vector space, but we care most about $\mathbf{R}^n$. When the vectors go into the columns of an n by n matrix A, here are the same two facts. Remember that A is *square*:

4D If the n columns are independent, they must span all of $\mathbf{R}^n$. If the n columns span $\mathbf{R}^n$, they must be independent.

A square system $A\boldsymbol{x} = \boldsymbol{b}$ always has one solution if it never has two solutions, and vice versa. Uniqueness implies existence and existence implies uniqueness. The square matrix A is invertible. Its columns are a basis for $\mathbf{R}^n$.

Our standard method of proof is elimination. If there are no free variables (uniqueness), there must be n pivots. Then back substitution solves $A\boldsymbol{x} = \boldsymbol{b}$ (existence). In the opposite direction, suppose $A\boldsymbol{x} = \boldsymbol{b}$ can always be solved (existence of solutions). Then elimination produced no zero rows. There are n pivots and no free variables. The nullspace contains only $\boldsymbol{x} = \mathbf{0}$ (uniqueness of solutions).

The count is always right for the row space of A and its nullspace. They have dimensions r and $n-r$. With a basis for the row space and a basis for the nullspace, we have $r+(n-r)=n$ vectors—the right number. Those n vectors are independent.† *Therefore they span* $\mathbf{R}^n$. They are a basis:

Every $\boldsymbol{x}$ in $\mathbf{R}^n$ is the sum $\boldsymbol{x}_r+\boldsymbol{x}_n$ of a row space vector $\boldsymbol{x}_r$ and a nullspace vector $\boldsymbol{x}_n$.

This confirms the splitting in Figure 4.2. It is the key point of orthogonal complements—the dimensions add to n and no vectors are missing.

Example 4.5 For $A=\begin{bmatrix} I & I \end{bmatrix}=\left[\begin{smallmatrix}1&0&1&0\\0&1&0&1\end{smallmatrix}\right]$, write any vector $\boldsymbol{x}$ as $\boldsymbol{x}_r+\boldsymbol{x}_n$.

$(1,0,1,0)$ and $(0,1,0,1)$ are a basis for the row space. $(1,0,-1,0)$ and $(0,1,0,-1)$ are a basis for the nullspace. Those four vectors are a basis for $\mathbf{R}^4$. Any $\boldsymbol{x}=(a,b,c,d)$ can be split into $\boldsymbol{x}_r+\boldsymbol{x}_n$:

$$\begin{bmatrix} a\\b\\c\\d \end{bmatrix}=\frac{a+c}{2}\begin{bmatrix}1\\0\\1\\0\end{bmatrix}+\frac{b+d}{2}\begin{bmatrix}0\\1\\0\\1\end{bmatrix}+\frac{a-c}{2}\begin{bmatrix}1\\0\\-1\\0\end{bmatrix}+\frac{b-d}{2}\begin{bmatrix}0\\1\\0\\-1\end{bmatrix}.$$

Problem Set 4.1

Questions 1–10 grow out of Figures 4.1 and 4.2.

1 Suppose A is a 2 by 3 matrix of rank one. Draw Figure 4.1 to match the sizes of the subspaces.

2 Redraw Figure 4.2 for a 3 by 2 matrix of rank $r=2$. What are the two parts $\boldsymbol{x}_r$ and $\boldsymbol{x}_n$?

3 Construct a matrix with the required property or say why that is impossible:

(a) Column space contains $\begin{bmatrix}1\\2\\-3\end{bmatrix}$ and $\begin{bmatrix}2\\-3\\5\end{bmatrix}$, nullspace contains $\begin{bmatrix}1\\1\\1\end{bmatrix}$

(b) Row space contains $\begin{bmatrix}1\\2\\-3\end{bmatrix}$ and $\begin{bmatrix}2\\-3\\5\end{bmatrix}$, nullspace contains $\begin{bmatrix}1\\1\\1\end{bmatrix}$

(c) Column space is perpendicular to nullspace

(d) Row 1 + row 2 + row 3 = $\mathbf{0}$, column space contains $(1,2,3)$

(e) Columns add up to zero column, rows add to a row of 1's.

4 It is possible for the row space to contain the nullspace. Find an example.

†If a combination of the vectors gives $\boldsymbol{x}_r+\boldsymbol{x}_n=\mathbf{0}$, then $\boldsymbol{x}_r=-\boldsymbol{x}_n$ is in both subspaces. It is orthogonal to itself and must be zero. All coefficients of the row space basis and nullspace basis must be zero—which proves independence of the n vectors together.

5 (a) If $Ax = b$ has a solution and $A^T y = 0$, then y is perpendicular to ______.

(b) If $Ax = b$ has no solution and $A^T y = 0$, explain why y is not perpendicular to ______.

6 In Figure 4.2, how do we know that Ax_r is equal to Ax? How do we know that this vector is in the column space?

7 If Ax is in the nullspace of A^T then Ax must be ______. Why? Which other subspace is Ax in? This is important: *A^TA has the same nullspace as A.*

8 Suppose A is a symmetric matrix ($A^T = A$).

(a) Why is its column space perpendicular to its nullspace?

(b) If $Ax = 0$ and $Az = 5z$, why is x perpendicular to z? These are "eigenvectors."

9 (Recommended) Draw Figure 4.2 to show each subspace correctly for

$$A = \begin{bmatrix} 1 & 2 \\ 3 & 6 \end{bmatrix} \quad \text{and} \quad B = \begin{bmatrix} 1 & 0 \\ 3 & 0 \end{bmatrix}.$$

10 Find the pieces x_r and x_n and draw Figure 4.2 properly if

$$A = \begin{bmatrix} 1 & -1 \\ 0 & 0 \\ 0 & 0 \end{bmatrix} \quad \text{and} \quad x = \begin{bmatrix} 2 \\ 0 \end{bmatrix}.$$

Questions 11–19 are about orthogonal subspaces.

11 Prove that every y in $N(A^T)$ is perpendicular to every Ax in the column space, using the matrix shorthand of equation (4.2). Start from $A^T y = 0$.

12 The Fundamental Theorem is also stated in the form of Fredholm's alternative: For any A and b, exactly one of these two problems has a solution:

(a) $Ax = b$

(b) $A^T y = 0$ with $b^T y \neq 0$.

Either b is in the column space of A or else b is not orthogonal to the nullspace of A^T. Choose A and b so that (a) has no solution. Find a solution to (b).

13 If S is the subspace of $\mathbf{R}^3$ containing only the zero vector, what is $S^\perp$? If S is spanned by $(1, 1, 1)$, what is $S^\perp$? If S is spanned by $(2, 0, 0)$ and $(0, 0, 3)$, what is $S^\perp$?

14 Suppose S only contains two vectors $(1, 5, 1)$ and $(2, 2, 2)$ (not a subspace). Then $S^\perp$ is the nullspace of the matrix $A =$ ______. Therefore $S^\perp$ is a ______ even if S is not.

15 Suppose $\boldsymbol{L}$ is a one-dimensional subspace (a line) in $\mathbf{R}^3$. Its orthogonal complement $\boldsymbol{L}^\perp$ is the ______ perpendicular to $\boldsymbol{L}$. Then $(\boldsymbol{L}^\perp)^\perp$ is a ______ perpendicular to $\boldsymbol{L}^\perp$. In fact $(\boldsymbol{L}^\perp)^\perp$ is the same as ______.

16 Suppose $\boldsymbol{V}$ is the whole space $\mathbf{R}^4$. Then $\boldsymbol{V}^\perp$ contains only the vector ______. Then $(\boldsymbol{V}^\perp)^\perp$ contains ______. So $(\boldsymbol{V}^\perp)^\perp$ is the same as ______.

17 Suppose $\boldsymbol{S}$ is spanned by the vectors $(1, 2, 2, 3)$ and $(1, 3, 3, 2)$. Find two vectors that span $\boldsymbol{S}^\perp$.

18 If $\boldsymbol{P}$ is the plane of vectors in $\mathbf{R}^4$ satisfying $x_1 + x_2 + x_3 + x_4 = 0$, write a basis for $\boldsymbol{P}^\perp$. Construct a matrix that has $\boldsymbol{P}$ as its nullspace.

19 If a subspace $\boldsymbol{S}$ is contained in a subspace $\boldsymbol{V}$, prove that the subspace $\boldsymbol{S}^\perp$ contains $\boldsymbol{V}^\perp$.

Questions 20–25 are about perpendicular columns and rows.

20 Suppose an n by n matrix is invertible: $AA^{-1} = I$. Then the first column of A^{-1} is orthogonal to the space spanned by ______.

21 Suppose the columns of A are unit vectors, all mutually perpendicular. What is $A^{\mathrm{T}}A$?

22 Construct a 3 by 3 matrix A with no zero entries whose columns are mutually perpendicular. Compute $A^{\mathrm{T}}A$. Why is it a diagonal matrix?

23 The lines $3x + y = b_1$ and $6x + 2y = b_2$ are ______. They are the same line if ______. In that case (b_1, b_2) is perpendicular to the vector ______. The nullspace of the matrix is the line $3x + y =$ ______. One particular vector in that nullspace is ______.

24 Why is each of these statements false?

(a) $(1, 1, 1)$ is perpendicular to $(1, 1, -2)$ so the planes $x + y + z = 0$ and $x + y - 2z = 0$ are orthogonal subspaces.

(b) The lines from $(0, 0, 0)$ through $(2,4,5)$ and $(1, -3, 2)$ are orthogonal complements.

(c) If two subspaces meet only in the zero vector, the subspaces are orthogonal.

25 Find a matrix with $\boldsymbol{v} = (1, 2, 3)$ in the row space and column space. Find another matrix with $\boldsymbol{v}$ in the nullspace and column space. Which pairs of subspaces can $\boldsymbol{v}$ *not* be in?

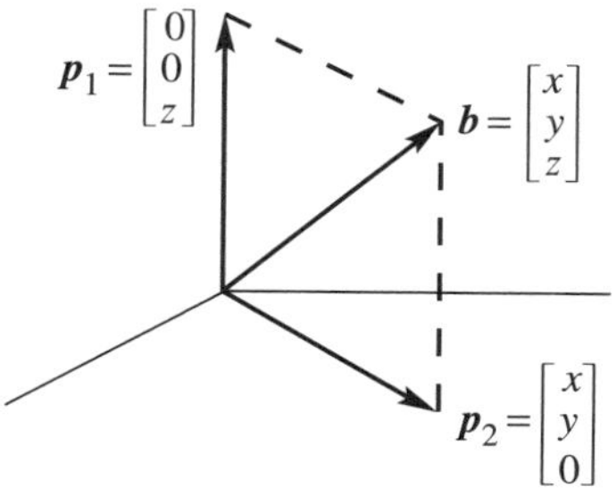

Figure 4.3 The projection of $\boldsymbol{b}$ onto a line and a plane.

4.2 Projections

May we start this section with two questions? (In addition to that one.) The first question aims to show that projections are easy to visualize. The second question is about matrices:

1 What are the projections of $\boldsymbol{b} = (2, 3, 4)$ onto the z axis and the xy plane?

2 What matrices produce those projections?

If $\boldsymbol{b}$ is projected onto a line, its projection $\boldsymbol{p}$ is ***the part of b along that line***. When $\boldsymbol{b}$ is projected onto a plane, $\boldsymbol{p}$ is the part in that plane.

There is a matrix P that multiplies $\boldsymbol{b}$ to give $\boldsymbol{p}$. ***The projection is*** $\boldsymbol{p} = P\boldsymbol{b}$.

The picture in your mind should be Figure 4.3. For the first projection we go across to the z axis. For the second projection we drop straight down to the xy plane. One way gives $\boldsymbol{p}_1 = (0, 0, 4)$ and the other way gives $\boldsymbol{p}_2 = (2, 3, 0)$. Those are the parts of $\boldsymbol{b} = (2, 3, 4)$ along the line and in the plane.

The matrices P_1 and P_2 are 3 by 3. They multiply $\boldsymbol{b}$ with 3 components to produce $\boldsymbol{p}$ with 3 components. Projection onto a line comes from a rank one matrix. Projection onto a plane comes from a rank two matrix:

$$\text{Onto the } z \text{ axis:} \quad P_1 = \begin{bmatrix} 0 & 0 & 0 \\ 0 & 0 & 0 \\ 0 & 0 & 1 \end{bmatrix} \qquad \text{Onto the } xy \text{ plane:} \quad P_2 = \begin{bmatrix} 1 & 0 & 0 \\ 0 & 1 & 0 \\ 0 & 0 & 0 \end{bmatrix}.$$

P_1 picks out the z component of every vector. P_2 picks out the "in plane" component. To find $\boldsymbol{p}_1$ and $\boldsymbol{p}_2$, multiply by P_1 and P_2 (small $\boldsymbol{p}$ for the vector, capital P for the matrix that produces it):

$$\boldsymbol{p}_1 = P_1\boldsymbol{b} = \begin{bmatrix} 0 & 0 & 0 \\ 0 & 0 & 0 \\ 0 & 0 & 1 \end{bmatrix}\begin{bmatrix} x \\ y \\ z \end{bmatrix} = \begin{bmatrix} 0 \\ 0 \\ z \end{bmatrix} \qquad \boldsymbol{p}_2 = P_2\boldsymbol{b} = \begin{bmatrix} 1 & 0 & 0 \\ 0 & 1 & 0 \\ 0 & 0 & 0 \end{bmatrix}\begin{bmatrix} x \\ y \\ z \end{bmatrix} = \begin{bmatrix} x \\ y \\ 0 \end{bmatrix}.$$

In this case the two projections are perpendicular. The xy plane and z axis are *orthogonal subspaces*, like the floor of a room and the line between two walls. More than that, the subspaces are orthogonal ***complements***. Their dimensions add to $1+2=3$—every vector in the whole space is the sum of its parts in the two subspaces. The projections $\boldsymbol{p}_1$ and $\boldsymbol{p}_2$ are exactly those parts:

$$\text{The vectors give } \boldsymbol{p}_1 + \boldsymbol{p}_2 = \boldsymbol{b}. \quad \text{The matrices give } P_1 + P_2 = I. \tag{4.4}$$

This is perfect. Our goal is reached—for this example. We have the same goal for any line and any plane and any n-dimensional subspace. The object is to find the part $\boldsymbol{p}$ in each subspace, and the projection matrix P that produces $\boldsymbol{p} = P\boldsymbol{b}$. Every subspace of $\mathbf{R}^m$ has an m by m (square) projection matrix. To compute P, we absolutely need a good description of the subspace.

The best description is to have a basis. The basis vectors go into the columns of A. ***We are projecting onto the column space of A.*** Certainly the z axis is the column space of the following matrix A_1. The xy plane is the column space of A_2. That plane is also the column space of A_3 (a subspace has many bases):

$$A_1 = \begin{bmatrix} 0 \\ 0 \\ 1 \end{bmatrix} \quad \text{and} \quad A_2 = \begin{bmatrix} 1 & 0 \\ 0 & 1 \\ 0 & 0 \end{bmatrix} \quad \text{and} \quad A_3 = \begin{bmatrix} 1 & 2 \\ 2 & 3 \\ 0 & 0 \end{bmatrix}.$$

The problem is to project onto the column space of any m by n matrix A. Start with a line. Then $n = 1$.

Projection Onto a Line

We are given a point $\boldsymbol{b} = (b_1, \ldots, b_m)$ in m-dimensional space. We are also given a line through the origin, in the direction of $\boldsymbol{a} = (a_1, \ldots, a_m)$. We are looking along that line, to find the point $\boldsymbol{p}$ closest to $\boldsymbol{b}$. The key is orthogonality: ***The line connecting b to p is perpendicular to the vector a.*** This is the dotted line in Figure 4.4—which we now compute by algebra.

The projection $\boldsymbol{p}$ is some multiple of $\boldsymbol{a}$ (call it $\boldsymbol{p} = \hat{x}\boldsymbol{a}$). Our first step is to compute this unknown number $\hat{x}$. That will give the vector $\boldsymbol{p}$. Then from the formula for $\boldsymbol{p}$, we read off the projection matrix P. These three steps will lead to all projection matrices: find $\hat{x}$, then find $\boldsymbol{p}$, then find P.

The dotted line $\boldsymbol{b} - \boldsymbol{p}$ is $\boldsymbol{b} - \hat{x}\boldsymbol{a}$. It is perpendicular to $\boldsymbol{a}$—this will determine $\hat{x}$. Use the fact that two vectors are perpendicular when their dot product is zero:

$$\boldsymbol{a} \cdot (\boldsymbol{b} - \hat{x}\boldsymbol{a}) = 0 \quad \text{or} \quad \boldsymbol{a} \cdot \boldsymbol{b} - \hat{x}\boldsymbol{a} \cdot \boldsymbol{a} = 0 \quad \text{or} \quad \hat{x} = \frac{\boldsymbol{a} \cdot \boldsymbol{b}}{\boldsymbol{a} \cdot \boldsymbol{a}} = \frac{\boldsymbol{a}^{\mathrm{T}}\boldsymbol{b}}{\boldsymbol{a}^{\mathrm{T}}\boldsymbol{a}}. \tag{4.5}$$

For vectors the multiplication $\boldsymbol{a}^{\mathrm{T}}\boldsymbol{b}$ is the same as $\boldsymbol{a} \cdot \boldsymbol{b}$. Using the transpose is better, because it applies also to matrices. (We will soon meet $A^{\mathrm{T}}\boldsymbol{b}$.) Our formula for $\hat{x}$ immediately gives the formula for $\boldsymbol{p}$:

4E The projection of $\boldsymbol{b}$ onto the line through $\boldsymbol{a}$ is the vector $\boldsymbol{p} = \hat{x}\boldsymbol{a} = (\boldsymbol{a}^{\mathrm{T}}\boldsymbol{b}/\boldsymbol{a}^{\mathrm{T}}\boldsymbol{a})\boldsymbol{a}$.

Special case 1: If $\boldsymbol{b} = \boldsymbol{a}$ then $\hat{x} = 1$. The projection of $\boldsymbol{a}$ onto $\boldsymbol{a}$ is $\boldsymbol{a}$ itself.
Special case 2: If $\boldsymbol{b}$ is perpendicular to $\boldsymbol{a}$ then $\boldsymbol{a}^{\mathrm{T}}\boldsymbol{b} = 0$ and $\boldsymbol{p} = \mathbf{0}$.

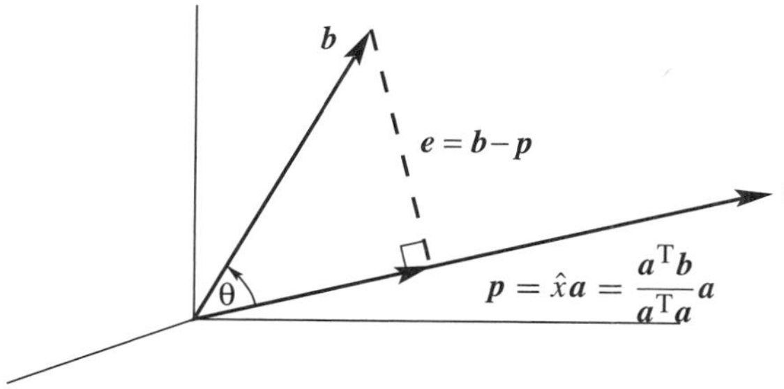

Figure 4.4 The projection of $\boldsymbol{b}$ onto a line has length $\|\boldsymbol{p}\| = \|\boldsymbol{b}\| \cos\theta$.

Example 4.6 Project $\boldsymbol{b} = \begin{bmatrix} 1 \\ 1 \\ 1 \end{bmatrix}$ onto $\boldsymbol{a} = \begin{bmatrix} 1 \\ 2 \\ 2 \end{bmatrix}$ to find $\boldsymbol{p} = \hat{x}\boldsymbol{a}$ in Figure 4.4.

Solution The formula for $\hat{x}$ is $\frac{\boldsymbol{a}^{\mathrm{T}}\boldsymbol{b}}{\boldsymbol{a}^{\mathrm{T}}\boldsymbol{a}} = \frac{5}{9}$. The projection is $\boldsymbol{p} = \frac{5}{9}\boldsymbol{a} = \left(\frac{5}{9}, \frac{10}{9}, \frac{10}{9}\right)$. The error vector between $\boldsymbol{b}$ and $\boldsymbol{p}$ is $\boldsymbol{e} = \boldsymbol{b} - \boldsymbol{p}$. The error $\boldsymbol{e} = \left(\frac{4}{9}, -\frac{1}{9}, -\frac{1}{9}\right)$ should be perpendicular to $\boldsymbol{a}$ and it is: $\boldsymbol{e} \cdot \boldsymbol{a} = \boldsymbol{e}^{\mathrm{T}}\boldsymbol{a} = \frac{4}{9} - \frac{2}{9} - \frac{2}{9} = 0$.

Look at the right triangle of $\boldsymbol{b}$, $\boldsymbol{p}$, and $\boldsymbol{e}$. The vector $\boldsymbol{b}$ is split into two parts—its component along the line is $\boldsymbol{p}$, its perpendicular component is $\boldsymbol{e}$. Those two sides of a right triangle have length $\|\boldsymbol{b}\| \cos\theta$ and $\|\boldsymbol{b}\| \sin\theta$. The trigonometry matches the dot product:

$$\boldsymbol{p} = \frac{\boldsymbol{a}^{\mathrm{T}}\boldsymbol{b}}{\boldsymbol{a}^{\mathrm{T}}\boldsymbol{a}}\boldsymbol{a} \quad \text{so its length is} \quad \|\boldsymbol{p}\| = \frac{\|\boldsymbol{a}\|\,\|\boldsymbol{b}\| \cos\theta}{\|\boldsymbol{a}\|^2}\|\boldsymbol{a}\| = \|\boldsymbol{b}\| \cos\theta. \tag{4.6}$$

The dot product is simpler than getting involved with $\cos\theta$ and the length of $\boldsymbol{b}$. The example has square roots in $\cos\theta = 5/3\sqrt{3}$ and $\|\boldsymbol{b}\| = \sqrt{3}$. There are no square roots in the projection $\boldsymbol{p} = \frac{5}{9}\boldsymbol{a}$.

Now comes the ***projection matrix***. In the formula for $\boldsymbol{p}$, what matrix is multiplying $\boldsymbol{b}$? You can see it better if the number $\hat{x}$ is on the right side of $\boldsymbol{a}$:

$$\boldsymbol{p} = \boldsymbol{a}\hat{x} = \boldsymbol{a}\frac{\boldsymbol{a}^{\mathrm{T}}\boldsymbol{b}}{\boldsymbol{a}^{\mathrm{T}}\boldsymbol{a}} = P\boldsymbol{b} \quad \text{when } P \text{ is the matrix} \quad \frac{\boldsymbol{a}\boldsymbol{a}^{\mathrm{T}}}{\boldsymbol{a}^{\mathrm{T}}\boldsymbol{a}}.$$

P is a column times a row! The column is $\boldsymbol{a}$, the row is $\boldsymbol{a}^{\mathrm{T}}$. Then divide by the number $\boldsymbol{a}^{\mathrm{T}}\boldsymbol{a}$. The matrix P is m by m, but ***its rank is one***. We are projecting onto a one-dimensional subspace, the line through $\boldsymbol{a}$.

Example 4.7 Find the projection matrix $P = \dfrac{\boldsymbol{a}\boldsymbol{a}^{\mathrm{T}}}{\boldsymbol{a}^{\mathrm{T}}\boldsymbol{a}}$ onto the line through $\boldsymbol{a} = \begin{bmatrix} 1 \\ 2 \\ 2 \end{bmatrix}$.

Solution Multiply column times row and divide by $\boldsymbol{a}^{\mathrm{T}}\boldsymbol{a} = 9$:

$$P = \frac{\boldsymbol{a}\boldsymbol{a}^{\mathrm{T}}}{\boldsymbol{a}^{\mathrm{T}}\boldsymbol{a}} = \frac{1}{9}\begin{bmatrix} 1 \\ 2 \\ 2 \end{bmatrix}\begin{bmatrix} 1 & 2 & 2 \end{bmatrix} = \begin{bmatrix} \frac{1}{9} & \frac{2}{9} & \frac{2}{9} \\ \frac{2}{9} & \frac{4}{9} & \frac{4}{9} \\ \frac{2}{9} & \frac{4}{9} & \frac{4}{9} \end{bmatrix}.$$

This matrix projects *any* vector $\boldsymbol{b}$ onto $\boldsymbol{a}$. Check $\boldsymbol{p} = P\boldsymbol{b}$ for the particular $\boldsymbol{b} = (1, 1, 1)$ in Example 4.6:

$$\boldsymbol{p} = P\boldsymbol{b} = \begin{bmatrix} \frac{1}{9} & \frac{2}{9} & \frac{2}{9} \\ \frac{2}{9} & \frac{4}{9} & \frac{4}{9} \\ \frac{2}{9} & \frac{4}{9} & \frac{4}{9} \end{bmatrix} \begin{bmatrix} 1 \\ 1 \\ 1 \end{bmatrix} = \begin{bmatrix} \frac{5}{9} \\ \frac{10}{9} \\ \frac{10}{9} \end{bmatrix} \quad \text{which is correct.}$$

Here are various remarks about P. If the vector $\boldsymbol{a}$ is doubled, the matrix P stays the same (it still projects onto the same line). If the matrix is squared, P^2 equals P (because projecting a second time doesn't change anything). The diagonal entries of P add up to $\frac{1}{9} + \frac{4}{9} + \frac{4}{9} = 1$.

The matrix $I - P$ should be a projection too. It produces the other side $\boldsymbol{e}$ of the triangle—the perpendicular part of $\boldsymbol{b}$. Note that $(I - P)\boldsymbol{b}$ equals $\boldsymbol{b} - \boldsymbol{p}$ which is $\boldsymbol{e}$. When P projects onto one subspace, $I - P$ projects onto the perpendicular subspace. Here $I - P$ projects onto the plane perpendicular to $\boldsymbol{a}$.

Now we move from a line in $\mathbf{R}^3$ to an n-dimensional subspace of $\mathbf{R}^m$. Projecting onto a subspace takes more effort. The crucial formulas are collected in equations (4.8)–(4.10).

Projection Onto a Subspace

Start with n vectors $\boldsymbol{a}_1, \ldots, \boldsymbol{a}_n$. Assume they are linearly independent. ***Find the combination*** $\hat{x}_1\boldsymbol{a}_1 + \cdots + \hat{x}_n\boldsymbol{a}_n$ ***that is closest to a given vector b.*** This is our problem: To project every vector $\boldsymbol{b}$ onto the n-dimensional subspace spanned by the $\boldsymbol{a}$'s.

With $n = 1$ (only one vector) this is projection onto a line. The line is the column space of a matrix A, which has just one column. In general the matrix A has n columns $\boldsymbol{a}_1, \ldots, \boldsymbol{a}_n$. Their combinations in $\mathbf{R}^m$ are the vectors $A\boldsymbol{x}$ in the column space. We are looking for the particular combination $\boldsymbol{p} = A\hat{\boldsymbol{x}}$ (***the projection***) that is closest to $\boldsymbol{b}$. The hat over $\boldsymbol{x}$ indicates the *best* choice, to give the closest vector in the column space. That choice is $\hat{x} = \boldsymbol{a}^{\mathrm{T}}\boldsymbol{b}/\boldsymbol{a}^{\mathrm{T}}\boldsymbol{a}$ when $n = 1$.

We solve this problem for an n-dimensional subspace in three steps: *Find the vector* $\hat{\boldsymbol{x}}$, *find the projection* $\boldsymbol{p} = A\hat{\boldsymbol{x}}$, *find the matrix* P.

The key is in the geometry. The dotted line in Figure 4.5 goes from $\boldsymbol{b}$ to the nearest point $A\hat{\boldsymbol{x}}$ in the subspace. ***This error vector*** $\boldsymbol{b} - A\hat{\boldsymbol{x}}$ ***is perpendicular to the subspace.*** The error $\boldsymbol{b} - A\hat{\boldsymbol{x}}$ makes a right angle with all the vectors $\boldsymbol{a}_1, \ldots, \boldsymbol{a}_n$. That gives the n equations we need to find $\hat{\boldsymbol{x}}$:

$$\begin{matrix} \boldsymbol{a}_1^{\mathrm{T}}(\boldsymbol{b} - A\hat{\boldsymbol{x}}) = 0 \\ \vdots \\ \boldsymbol{a}_n^{\mathrm{T}}(\boldsymbol{b} - A\hat{\boldsymbol{x}}) = 0 \end{matrix} \quad \text{or} \quad \begin{bmatrix} - \boldsymbol{a}_1^{\mathrm{T}} - \\ \vdots \\ - \boldsymbol{a}_n^{\mathrm{T}} - \end{bmatrix} \begin{bmatrix} \\ \boldsymbol{b} - A\hat{\boldsymbol{x}} \\ \\ \end{bmatrix} = \begin{bmatrix} \\ \boldsymbol{0} \\ \\ \end{bmatrix}. \tag{4.7}$$

The matrix in those equations is A^{T}. The n equations are exactly $A^{\mathrm{T}}(\boldsymbol{b} - A\hat{\boldsymbol{x}}) = \boldsymbol{0}$.

Rewrite $A^{\mathrm{T}}(\boldsymbol{b} - A\hat{\boldsymbol{x}}) = \boldsymbol{0}$ in its famous form $A^{\mathrm{T}}A\hat{\boldsymbol{x}} = A^{\mathrm{T}}\boldsymbol{b}$. This is the equation for $\hat{\boldsymbol{x}}$, and the coefficient matrix is $A^{\mathrm{T}}A$. Now we can find $\hat{\boldsymbol{x}}$ and $\boldsymbol{p}$ and P:

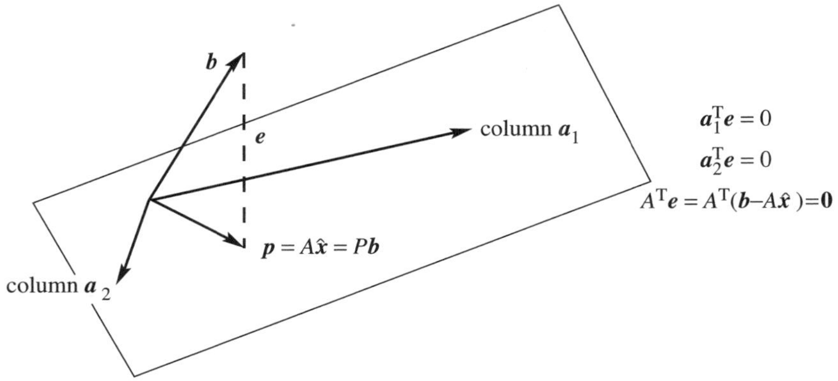

Figure 4.5 The projection $\boldsymbol{p}$ is the nearest point to $\boldsymbol{b}$ in the column space of A. The error $\boldsymbol{e}$ is in the nullspace of A^{T}.

4F The combination $\hat{x}_1\boldsymbol{a}_1 + \cdots + \hat{x}_n\boldsymbol{a}_n = A\hat{\boldsymbol{x}}$ that is closest to $\boldsymbol{b}$ comes from

$$A^{\mathrm{T}}(\boldsymbol{b} - A\hat{\boldsymbol{x}}) = \mathbf{0} \quad \text{or} \quad A^{\mathrm{T}}A\hat{\boldsymbol{x}} = A^{\mathrm{T}}\boldsymbol{b}. \tag{4.8}$$

The matrix $A^{\mathrm{T}}A$ is n by n and invertible. The solution is $\hat{\boldsymbol{x}} = (A^{\mathrm{T}}A)^{-1}A^{\mathrm{T}}\boldsymbol{b}$. The ***projection*** of $\boldsymbol{b}$ onto the subspace is

$$\boldsymbol{p} = A\hat{\boldsymbol{x}} = A(A^{\mathrm{T}}A)^{-1}A^{\mathrm{T}}\boldsymbol{b}. \tag{4.9}$$

This formula shows the ***projection matrix*** that produces $\boldsymbol{p} = P\boldsymbol{b}$:

$$P = A(A^{\mathrm{T}}A)^{-1}A^{\mathrm{T}}. \tag{4.10}$$

Compare with projection onto a line, when A has only one column $\boldsymbol{a}$:

$$\hat{x} = \frac{\boldsymbol{a}^{\mathrm{T}}\boldsymbol{b}}{\boldsymbol{a}^{\mathrm{T}}\boldsymbol{a}} \quad \text{and} \quad \boldsymbol{p} = \boldsymbol{a}\frac{\boldsymbol{a}^{\mathrm{T}}\boldsymbol{b}}{\boldsymbol{a}^{\mathrm{T}}\boldsymbol{a}} \quad \text{and} \quad P = \frac{\boldsymbol{a}\boldsymbol{a}^{\mathrm{T}}}{\boldsymbol{a}^{\mathrm{T}}\boldsymbol{a}}.$$

The formulas are identical! The number $\boldsymbol{a}^{\mathrm{T}}\boldsymbol{a}$ becomes the matrix $A^{\mathrm{T}}A$. When it is a number, we divide by it. When it is a matrix, we invert it. The new formulas contain $(A^{\mathrm{T}}A)^{-1}$ instead of $1/\boldsymbol{a}^{\mathrm{T}}\boldsymbol{a}$. The linear independence of the columns $\boldsymbol{a}_1, \ldots, \boldsymbol{a}_n$ will guarantee that this inverse matrix exists.

The key step was the equation $A^{\mathrm{T}}(\boldsymbol{b} - A\hat{\boldsymbol{x}}) = \mathbf{0}$. We used geometry (perpendicular vectors). Linear algebra gives this "normal equation" too, in a very quick way:

1 Our subspace is the column space of A.

2 The error vector $\boldsymbol{b} - A\hat{\boldsymbol{x}}$ is perpendicular to that subspace.

3 Therefore $\boldsymbol{b} - A\hat{\boldsymbol{x}}$ is in the left nullspace. This means $A^{\mathrm{T}}(\boldsymbol{b} - A\hat{\boldsymbol{x}}) = \mathbf{0}$.

The left nullspace is important in projections. This nullspace of A^{T} contains the error vector $\boldsymbol{e} = \boldsymbol{b} - A\hat{\boldsymbol{x}}$. The vector $\boldsymbol{b}$ is being split into the projection $\boldsymbol{p}$ and the error $\boldsymbol{e} = \boldsymbol{b} - \boldsymbol{p}$. Figure 4.5 shows these two parts in the two subspaces.

Example 4.8 If $A = \begin{bmatrix} 1 & 0 \\ 1 & 1 \\ 1 & 2 \end{bmatrix}$ and $\boldsymbol{b} = \begin{bmatrix} 6 \\ 0 \\ 0 \end{bmatrix}$ find $\hat{\boldsymbol{x}}$ and $\boldsymbol{p}$ and P.

Solution Compute the square matrix $A^{\mathrm{T}}A$ and also the vector $A^{\mathrm{T}}\boldsymbol{b}$:

$$\begin{bmatrix} 1 & 1 & 1 \\ 0 & 1 & 2 \end{bmatrix} \begin{bmatrix} 1 & 0 \\ 1 & 1 \\ 1 & 2 \end{bmatrix} = \begin{bmatrix} 3 & 3 \\ 3 & 5 \end{bmatrix} \quad \text{and} \quad \begin{bmatrix} 1 & 1 & 1 \\ 0 & 1 & 2 \end{bmatrix} \begin{bmatrix} 6 \\ 0 \\ 0 \end{bmatrix} = \begin{bmatrix} 6 \\ 0 \end{bmatrix}.$$

Now solve the normal equation $A^{\mathrm{T}}A\hat{\boldsymbol{x}} = A^{\mathrm{T}}\boldsymbol{b}$ to find $\hat{\boldsymbol{x}}$:

$$\begin{bmatrix} 3 & 3 \\ 3 & 5 \end{bmatrix} \begin{bmatrix} \hat{x}_1 \\ \hat{x}_2 \end{bmatrix} = \begin{bmatrix} 6 \\ 0 \end{bmatrix} \quad \text{gives} \quad \hat{\boldsymbol{x}} = \begin{bmatrix} \hat{x}_1 \\ \hat{x}_2 \end{bmatrix} = \begin{bmatrix} 5 \\ -3 \end{bmatrix}. \tag{4.11}$$

The combination $\boldsymbol{p} = A\hat{\boldsymbol{x}}$ is the projection of $\boldsymbol{b}$ onto the column space of A:

$$\boldsymbol{p} = 5 \begin{bmatrix} 1 \\ 1 \\ 1 \end{bmatrix} - 3 \begin{bmatrix} 0 \\ 1 \\ 2 \end{bmatrix} = \begin{bmatrix} 5 \\ 2 \\ -1 \end{bmatrix}. \quad \text{The error is} \quad \boldsymbol{b} - \boldsymbol{p} = \begin{bmatrix} 1 \\ -2 \\ 1 \end{bmatrix}. \tag{4.12}$$

That solves the problem for one particular $\boldsymbol{b}$. To solve it for every $\boldsymbol{b}$, compute the matrix $P = A(A^{\mathrm{T}}A)^{-1}A^{\mathrm{T}}$. Find the determinant of $A^{\mathrm{T}}A$ (which is $15 - 9 = 6$) and find $(A^{\mathrm{T}}A)^{-1}$. Then multiply A times $(A^{\mathrm{T}}A)^{-1}$ times A^{T} to reach P:

$$(A^{\mathrm{T}}A)^{-1} = \frac{1}{6} \begin{bmatrix} 5 & z-3 \\ -3 & 3 \end{bmatrix} \quad \text{and} \quad P = \frac{1}{6} \begin{bmatrix} 5 & 2 & -1 \\ 2 & 2 & 2 \\ -1 & 2 & 5 \end{bmatrix}. \tag{4.13}$$

Two checks on the calculation. First, the error $\boldsymbol{e} = (1, -2, 1)$ is perpendicular to both columns $(1, 1, 1)$ and $(0, 1, 2)$. Second, the final P times $\boldsymbol{b} = (6, 0, 0)$ correctly gives $\boldsymbol{p} = (5, 2, -1)$. We must also have $P^2 = P$, because a second projection doesn't change the first projection.

Warning The matrix $P = A(A^{\mathrm{T}}A)^{-1}A^{\mathrm{T}}$ is deceptive. You might try to split $(A^{\mathrm{T}}A)^{-1}$ into A^{-1} times $(A^{\mathrm{T}})^{-1}$. If you make that mistake, and substitute it into P, you will find $P = AA^{-1}(A^{\mathrm{T}})^{-1}A^{\mathrm{T}}$. Apparently everything cancels. This looks like $P = I$, the identity matrix. We want to say why this is wrong.

The matrix A is rectangular. It has no inverse matrix. We cannot replace AA^{-1} by I, because there is no A^{-1} in the first place.

In our experience, a problem that involves a rectangular matrix almost always leads to $A^{\mathrm{T}}A$. We cannot split its inverse into A^{-1} and $(A^{\mathrm{T}})^{-1}$, which don't exist. What does exist is the inverse of the square matrix $A^{\mathrm{T}}A$. This fact is so crucial that we state it clearly and give a proof.

4G $A^{\mathrm{T}}A$ is invertible if and only if the columns of A are linearly independent.

Proof $A^{\mathrm{T}}A$ is a square matrix (n by n). For every matrix A, we will show that $A^{\mathrm{T}}A$ *has the same nullspace as* A. When the columns of A are linearly independent, this nullspace contains only the zero vector. Then $A^{\mathrm{T}}A$, which has this same nullspace, is invertible.

Let A be any matrix. If $\boldsymbol{x}$ is in its nullspace, then $A\boldsymbol{x} = \mathbf{0}$. Multiplying by A^{T} gives $A^{\mathrm{T}}A\boldsymbol{x} = \mathbf{0}$. So $\boldsymbol{x}$ is also in the nullspace of $A^{\mathrm{T}}A$.

Now start with the nullspace of $A^{\mathrm{T}}A$. From $A^{\mathrm{T}}A\boldsymbol{x} = \mathbf{0}$ we must prove that $A\boldsymbol{x} = \mathbf{0}$. We can't multiply by $(A^{\mathrm{T}})^{-1}$, which generally doesn't exist. A better way is to multiply by $\boldsymbol{x}^{\mathrm{T}}$:

$$(\boldsymbol{x}^{\mathrm{T}})A^{\mathrm{T}}A\boldsymbol{x} = (\boldsymbol{x}^{\mathrm{T}})\mathbf{0} \quad \text{or} \quad (A\boldsymbol{x})^{\mathrm{T}}(A\boldsymbol{x}) = 0 \quad \text{or} \quad \|A\boldsymbol{x}\|^2 = 0.$$

The vector $A\boldsymbol{x}$ has length zero. Therefore $A\boldsymbol{x} = \mathbf{0}$. Every vector $\boldsymbol{x}$ in one nullspace is in the other nullspace. If A has dependent columns, so does $A^{\mathrm{T}}A$. If A has independent columns, so does $A^{\mathrm{T}}A$. This is the good case:

When A has independent columns, $A^{\mathrm{T}}A$ is ***square, symmetric,*** and ***invertible.***

To repeat for emphasis: $A^{\mathrm{T}}A$ is (n by m) times (m by n). It is always square (n by n). It is always symmetric, because its transpose is $(A^{\mathrm{T}}A)^{\mathrm{T}} = A^{\mathrm{T}}(A^{\mathrm{T}})^{\mathrm{T}}$ which equals $A^{\mathrm{T}}A$. We also proved that $A^{\mathrm{T}}A$ is invertible—provided A has independent columns. Watch the difference between dependent and independent columns:

$$\begin{bmatrix} 1 & 1 & 0 \\ 2 & 2 & 0 \end{bmatrix} \underset{\text{dependent}}{\overset{A}{\begin{bmatrix} 1 & 2 \\ 1 & 2 \\ 0 & 0 \end{bmatrix}}} = \underset{\text{singular}}{\overset{A^{\mathrm{T}}A}{\begin{bmatrix} 2 & 4 \\ 4 & 8 \end{bmatrix}}} \qquad \begin{bmatrix} 1 & 1 & 0 \\ 2 & 2 & 1 \end{bmatrix} \underset{\text{indep.}}{\overset{A}{\begin{bmatrix} 1 & 2 \\ 1 & 2 \\ 0 & 1 \end{bmatrix}}} = \underset{\text{invertible}}{\overset{A^{\mathrm{T}}A}{\begin{bmatrix} 2 & 4 \\ 4 & 9 \end{bmatrix}}}$$

Very brief summary To find the projection $\boldsymbol{p} = \hat{x}_1\boldsymbol{a}_1 + \cdots + \hat{x}_n\boldsymbol{a}_n$, solve $A^{\mathrm{T}}A\hat{\boldsymbol{x}} = A^{\mathrm{T}}\boldsymbol{b}$. The projection is $A\hat{\boldsymbol{x}}$ and the error is $\boldsymbol{e} = \boldsymbol{b} - \boldsymbol{p} = \boldsymbol{b} - A\hat{\boldsymbol{x}}$. The projection matrix $P = A(A^{\mathrm{T}}A)^{-1}A^{\mathrm{T}}$ gives $\boldsymbol{p} = P\boldsymbol{b}$.

This matrix satisfies $P^2 = P$. ***The distance from b to the subspace is*** $\|\boldsymbol{e}\|$.

Problem Set 4.2

Questions 1–9 ask for projections onto lines. Also errors $\boldsymbol{e} = \boldsymbol{b} - \boldsymbol{p}$ and matrices P.

1 Project the vector $\boldsymbol{b}$ onto the line through $\boldsymbol{a}$. Check that $\boldsymbol{e}$ is perpendicular to $\boldsymbol{a}$:

$$\text{(a)} \quad \boldsymbol{b} = \begin{bmatrix} 1 \\ 2 \\ 2 \end{bmatrix} \quad \text{and} \quad \boldsymbol{a} = \begin{bmatrix} 1 \\ 1 \\ 1 \end{bmatrix} \qquad \text{(b)} \quad \boldsymbol{b} = \begin{bmatrix} 1 \\ 3 \\ 1 \end{bmatrix} \quad \text{and} \quad \boldsymbol{a} = \begin{bmatrix} -1 \\ -3 \\ -1 \end{bmatrix}.$$

2 *Draw* the projection of $\boldsymbol{b}$ onto $\boldsymbol{a}$ and also compute it from $\boldsymbol{p} = \hat{x}\boldsymbol{a}$:

$$\text{(a)} \quad \boldsymbol{b} = \begin{bmatrix} \cos\theta \\ \sin\theta \end{bmatrix} \quad \text{and} \quad \boldsymbol{a} = \begin{bmatrix} 1 \\ 0 \end{bmatrix} \qquad \text{(b)} \quad \boldsymbol{b} = \begin{bmatrix} 1 \\ 1 \end{bmatrix} \quad \text{and} \quad \boldsymbol{a} = \begin{bmatrix} 1 \\ -1 \end{bmatrix}.$$

3 In Problem 1, find the projection matrices $P_1 = \boldsymbol{a}\boldsymbol{a}^{\mathrm{T}}/\boldsymbol{a}^{\mathrm{T}}\boldsymbol{a}$ and similarly P_2 onto the lines through $\boldsymbol{a}$ in (a) and (b). Verify that $P_1^2 = P_1$. Multiply $P\boldsymbol{b}$ in each case to compute the projection $\boldsymbol{p}$.

4 Construct the projection matrices P_1 and P_2 onto the lines through the $\boldsymbol{a}$'s in Problem 2. Verify that $P_2^2 = P_2$. Explain why P_2^2 *should* equal P_2.

5 Compute the projection matrices $\boldsymbol{a}\boldsymbol{a}^{\mathrm{T}}/\boldsymbol{a}^{\mathrm{T}}\boldsymbol{a}$ onto the lines through $\boldsymbol{a}_1 = (-1, 2, 2)$ and $\boldsymbol{a}_2 = (2, 2, -1)$. Multiply those projection matrices and explain why their product $P_1 P_2$ is what it is.

6 Project $\boldsymbol{b} = (1, 0, 0)$ onto the lines through $\boldsymbol{a}_1$ and $\boldsymbol{a}_2$ in Problem 5 and also onto $\boldsymbol{a}_3 = (2, -1, 2)$. Add up the three projections $\boldsymbol{p}_1 + \boldsymbol{p}_2 + \boldsymbol{p}_3$.

7 Continuing Problems 5–6, find the projection matrix P_3 onto $\boldsymbol{a}_3 = (2, -1, 2)$. Verify that $P_1 + P_2 + P_3 = I$.

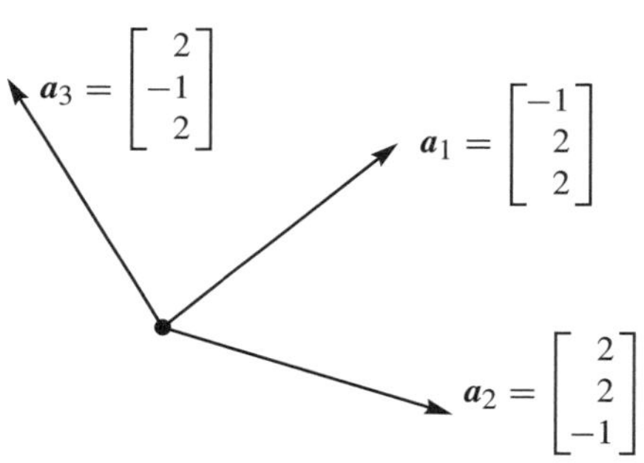

Questions 5-6-7

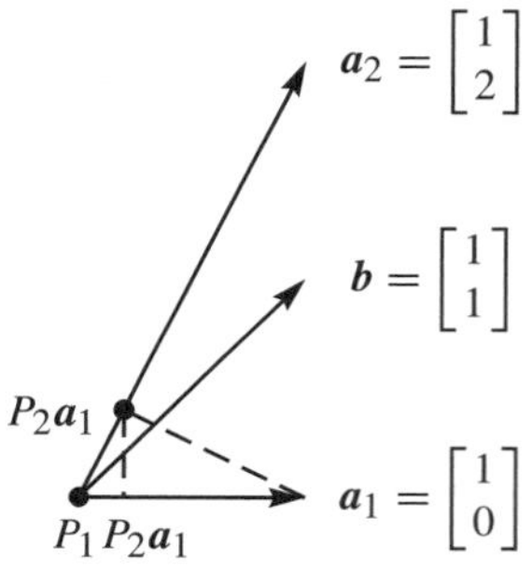

Questions 8-9-10

8 Project the vector $\boldsymbol{b} = (1, 1)$ onto the lines through $\boldsymbol{a}_1 = (1, 0)$ and $\boldsymbol{a}_2 = (1, 2)$. Draw the projections $\boldsymbol{p}_1$ and $\boldsymbol{p}_2$ onto the second figure and add $\boldsymbol{p}_1 + \boldsymbol{p}_2$. The projections do not add to $\boldsymbol{b}$ because the $\boldsymbol{a}$'s are not orthogonal.

9 Project $\boldsymbol{a}_1 = (1, 0)$ onto $\boldsymbol{a}_2 = (1, 2)$. Then project the result back onto $\boldsymbol{a}_1$. Draw these projections on a copy of the second figure. Multiply the projection matrices $P_1 P_2$: Is this a projection?

10 The projection of $\boldsymbol{b}$ onto the *plane* of $\boldsymbol{a}_1$ and $\boldsymbol{a}_2$ will equal $\boldsymbol{b}$. The projection matrix is $P =$ ______. Check $P = A(A^{\mathrm{T}}A)^{-1}A^{\mathrm{T}}$ for $A = \begin{bmatrix} \boldsymbol{a}_1 & \boldsymbol{a}_2 \end{bmatrix} = \begin{bmatrix} 1 & 1 \\ 0 & 2 \end{bmatrix}$.

Questions 11–20 ask for projections, and projection matrices, onto subspaces.

11 Project $\boldsymbol{b}$ onto the column space of A by solving $A^{\mathrm{T}}A\hat{\boldsymbol{x}} = A^{\mathrm{T}}\boldsymbol{b}$ and $\boldsymbol{p} = A\hat{\boldsymbol{x}}$:

$$\text{(a)}\quad A = \begin{bmatrix} 1 & 1 \\ 0 & 1 \\ 0 & 0 \end{bmatrix} \text{ and } \boldsymbol{b} = \begin{bmatrix} 2 \\ 3 \\ 4 \end{bmatrix} \qquad \text{(b)}\quad A = \begin{bmatrix} 1 & 1 \\ 1 & 1 \\ 0 & 1 \end{bmatrix} \text{ and } \boldsymbol{b} = \begin{bmatrix} 4 \\ 4 \\ 6 \end{bmatrix}.$$

Find $\boldsymbol{e} = \boldsymbol{b} - \boldsymbol{p}$. It should be perpendicular to the columns of A.

12 Compute the projection matrices P_1 and P_2 onto the column spaces in Problem 11. Verify that $P_1\boldsymbol{b}$ gives the first projection $\boldsymbol{p}_1$. Also verify $P_2^2 = P_2$.

13 Suppose A is the 4 by 4 identity matrix with its last column removed. A is 4 by 3. Project $\boldsymbol{b} = (1, 2, 3, 4)$ onto the column space of A. What shape is the projection matrix P and what is P?

14 Suppose $\boldsymbol{b}$ equals 2 times the first column of A. What is the projection of $\boldsymbol{b}$ onto the column space of A? Is $P = I$ in this case? Compute $\boldsymbol{p}$ and P when $\boldsymbol{b} = (0, 2, 4)$ and the columns of A are $(0, 1, 2)$ and $(1, 2, 0)$.

15 If A is doubled, then $P = 2A(4A^{\mathrm{T}}A)^{-1}2A^{\mathrm{T}}$. This is the same as $A(A^{\mathrm{T}}A)^{-1}A^{\mathrm{T}}$. The column space of $2A$ is the same as ______. Is $\hat{\boldsymbol{x}}$ the same for A and $2A$?

16 What linear combination of $(1, 2, -1)$ and $(1, 0, 1)$ is closest to $\boldsymbol{b} = (2, 1, 1)$?

17 (*Important*) If $P^2 = P$ show that $(I - P)^2 = I - P$. When P projects onto the column space of A, $I - P$ projects onto the ______.

18 (a) If P is the 2 by 2 projection matrix onto the line through $(1, 1)$, then $I - P$ is the projection matrix onto ______.

(b) If P is the 3 by 3 projection matrix onto the line through $(1, 1, 1)$, then $I - P$ is the projection matrix onto ______.

19 To find the projection matrix onto the plane $x - y - 2z = 0$, choose two vectors in that plane and make them the columns of A. The plane should be the column space. Then compute $P = A(A^{\mathrm{T}}A)^{-1}A^{\mathrm{T}}$.

20 To find the projection matrix P onto the same plane $x - y - 2z = 0$, write down a vector $\boldsymbol{e}$ that is perpendicular to that plane. Compute the projection $Q = \boldsymbol{e}\boldsymbol{e}^{\mathrm{T}}/\boldsymbol{e}^{\mathrm{T}}\boldsymbol{e}$ and then $P = I - Q$.

Questions 21–26 show that projection matrices satisfy $P^2 = P$ and $P^{\mathrm{T}} = P$.

21 Multiply the matrix $P = A(A^{\mathrm{T}}A)^{-1}A^{\mathrm{T}}$ by itself. Cancel to prove that $P^2 = P$. Explain why $P(P\boldsymbol{b})$ always equals $P\boldsymbol{b}$: The vector $P\boldsymbol{b}$ is in the column space so its projection is ______.

22 Prove that $P = A(A^{\mathrm{T}}A)^{-1}A^{\mathrm{T}}$ is symmetric by computing P^{T}. Remember that the inverse of a symmetric matrix is symmetric.

23 If A is square and invertible, the warning against splitting $(A^{\mathrm{T}}A)^{-1}$ does not apply: $P = AA^{-1}(A^{\mathrm{T}})^{-1}A^{\mathrm{T}} = I$. *When A is invertible, why is $P = I$?* ***What is the error*** $\boldsymbol{e}$?

24 The nullspace of A^{T} is _____ to the column space $\boldsymbol{R}(A)$. So if $A^{\mathrm{T}}\boldsymbol{b} = \boldsymbol{0}$, the projection of $\boldsymbol{b}$ onto $\boldsymbol{R}(A)$ should be $\boldsymbol{p} =$ _____. Check that $P = A(A^{\mathrm{T}}A)^{-1}A^{\mathrm{T}}$ gives this answer.

25 Explain why the projection matrix P onto an n-dimensional subspace has rank $r = n$. What is the column space of P?

26 If an m by m matrix has $A^2 = A$ and its rank is m, prove that $A = I$.

27 The important fact in Theorem **4G** is this: *If* $A^{\mathrm{T}}A\boldsymbol{x} = \boldsymbol{0}$ *then* $A\boldsymbol{x} = \boldsymbol{0}$. New proof: The vector $A\boldsymbol{x}$ is in the nullspace of _____. $A\boldsymbol{x}$ is always in the column space of _____. To be in both perpendicular spaces, $A\boldsymbol{x}$ must be zero.

28 The first four ***wavelets*** are in the columns of this wavelet matrix W:

$$W = \frac{1}{2}\begin{bmatrix} 1 & 1 & \sqrt{2} & 0 \\ 1 & 1 & -\sqrt{2} & 0 \\ 1 & -1 & 0 & \sqrt{2} \\ 1 & -1 & 0 & -\sqrt{2} \end{bmatrix}.$$

What is special about the columns of W? Find the inverse wavelet transform W^{-1}. What is the relation of W^{-1} to W?

4.3 Least-Squares Approximations

It often happens that $A\boldsymbol{x} = \boldsymbol{b}$ has no solution. The usual reason is: *too many equations*. The matrix has more rows than columns. There are more equations than unknowns (m is greater than n). The n columns span a small part of m-dimensional space. Unless all measurements are perfect, $\boldsymbol{b}$ is outside that column space. Elimination reaches an impossible equation and stops. But these are real problems and they need an answer.

To repeat: We cannot always get the error $\boldsymbol{e} = \boldsymbol{b} - A\boldsymbol{x}$ down to zero. When $\boldsymbol{e}$ is zero, $\boldsymbol{x}$ is an exact solution to $A\boldsymbol{x} = \boldsymbol{b}$. ***When the length of e is as small as possible, $\hat{\boldsymbol{x}}$ is a least-squares solution.*** Our goal in this section is to compute $\hat{\boldsymbol{x}}$ and use it.

The previous section emphasized $\boldsymbol{p}$ (the projection). This section emphasizes $\hat{\boldsymbol{x}}$ (the least-squares solution). They are connected by $\boldsymbol{p} = A\hat{\boldsymbol{x}}$. The fundamental equation is still $A^{\mathrm{T}}A\hat{\boldsymbol{x}} = A^{\mathrm{T}}\boldsymbol{b}$. When the original $A\boldsymbol{x} = \boldsymbol{b}$ has no solution, *multiply by* A^{T} *and solve* $A^{\mathrm{T}}A\hat{\boldsymbol{x}} = A^{\mathrm{T}}\boldsymbol{b}$.

Example 4.9 Find the closest straight line to three points $(0, 6)$, $(1, 0)$, and $(2, 0)$.

No straight line goes through those points. We are asking for two numbers C and D that satisfy three equations. The line is $b = C + Dt$. Here are the equations at $t = 0, 1, 2$ to match the given values $b = 6, 0, 0$:

$$\begin{array}{ll} \text{The first point is on the line if} & C + D \cdot 0 = 6 \\ \text{The second point is on the line if} & C + D \cdot 1 = 0 \\ \text{The third point is on the line if} & C + D \cdot 2 = 0. \end{array}$$

This 3 by 2 system has no solution. The right side $\boldsymbol{b}$ is not a combination of the columns of A:

$$A = \begin{bmatrix} 1 & 0 \\ 1 & 1 \\ 1 & 2 \end{bmatrix} \quad \boldsymbol{x} = \begin{bmatrix} C \\ D \end{bmatrix} \quad \boldsymbol{b} = \begin{bmatrix} 6 \\ 0 \\ 0 \end{bmatrix} \quad A\boldsymbol{x} = \boldsymbol{b} \text{ is unsolvable.}$$

The same numbers were in Example 4.8 in the last section. In practical problems, the data points are closer to a line than these points. But they don't match any $C + Dt$, and there easily could be $m = 100$ points instead of $m = 3$. The numbers 6, 0, 0 exaggerate the error so you can see it clearly.

Minimizing the Error

How do we make $\boldsymbol{e} = \boldsymbol{b} - A\boldsymbol{x}$ as small as possible? This is an important question with a beautiful answer. The best $\boldsymbol{x}$ (called $\hat{\boldsymbol{x}}$) can be found by geometry or algebra or calculus:

By geometry All vectors $A\boldsymbol{x}$ lie in the plane of the columns $(1, 1, 1)$ and $(0, 1, 2)$. In that plane, we are looking for the point closest to $\boldsymbol{b}$. *The nearest point is the projection* $\boldsymbol{p}$.

The best choice for $A\boldsymbol{x}$ is $\boldsymbol{p}$. The smallest possible error is $\boldsymbol{e} = \boldsymbol{b} - \boldsymbol{p}$.

By algebra Every vector $\boldsymbol{b}$ has a part in the column space of A and a perpendicular part in the left nullspace. The column space part is $\boldsymbol{p}$. The left nullspace part is $\boldsymbol{e}$. There is an equation we cannot solve ($A\boldsymbol{x} = \boldsymbol{b}$). There is an equation we do solve (by removing $\boldsymbol{e}$):

$$A\boldsymbol{x} = \boldsymbol{b} = \boldsymbol{p} + \boldsymbol{e} \quad \text{is impossible;} \qquad A\hat{\boldsymbol{x}} = \boldsymbol{p} \quad \text{is least squares.} \tag{4.14}$$

The solution to $A\boldsymbol{x} = \boldsymbol{p}$ makes the error as small as possible, because for any $\boldsymbol{x}$:

$$\|A\boldsymbol{x} - \boldsymbol{b}\|^2 = \|A\boldsymbol{x} - \boldsymbol{p}\|^2 + \|\boldsymbol{e}\|^2. \tag{4.15}$$

This is the law $c^2 = a^2 + b^2$ for a right triangle. The vector $A\boldsymbol{x} - \boldsymbol{p}$ in the column space is perpendicular to $\boldsymbol{e}$ in the left nullspace. We reduce $A\boldsymbol{x} - \boldsymbol{p}$ to zero by choosing $\boldsymbol{x}$ to be $\hat{\boldsymbol{x}}$. That leaves the smallest possible error (namely $\boldsymbol{e}$).

Notice what "smallest" means. The *squared length* of $Ax - b$ is being minimized:

The least-squares approximation makes $E = \|Ax - b\|^2$ as small as possible.

By calculus Most functions are minimized by calculus! The derivatives are zero at the minimum. The graph bottoms out and the slope in every direction is zero. Here the error function to be minimized is a *sum of squares*:

$$E = \|Ax - b\|^2 = (C + D \cdot 0 - 6)^2 + (C + D \cdot 1)^2 + (C + D \cdot 2)^2. \tag{4.16}$$

The unknowns are C and D. Those are the components of $\hat{x}$, and they determine the line. With two unknowns there are *two derivatives*—both zero at the minimum. They are called "partial derivatives" because the C derivative treats D as constant and the D derivative treats C as constant:

$$\frac{\partial E}{\partial C} = 2(C + D \cdot 0 - 6) \quad + 2(C + D \cdot 1) \quad + 2(C + D \cdot 2) \quad = 0$$

$$\frac{\partial E}{\partial D} = 2(C + D \cdot 0 - 6)(0) + 2(C + D \cdot 1)(1) + 2(C + D \cdot 2)(2) = 0.$$

$\partial E/\partial D$ contains the extra factors 0, 1, 2. Those are the numbers in E that multiply D. They appear because of the chain rule. (The derivative of $(4 + 5x)^2$ is 2 times $4 + 5x$ times an extra 5.) In the C derivative the corresponding factors are (1)(1)(1), because C is always multiplied by 1. It is no accident that 1, 1, 1 and 0, 1, 2 are the columns of A.

Now cancel 2 from every term and collect all C's and all D's:

$$\begin{aligned} &\text{The } C \text{ derivative } \frac{\partial E}{\partial C} \text{ is zero:} \quad 3C + 3D = 6 \\ &\text{The } D \text{ derivative } \frac{\partial E}{\partial D} \text{ is zero:} \quad 3C + 5D = 0. \end{aligned} \tag{4.17}$$

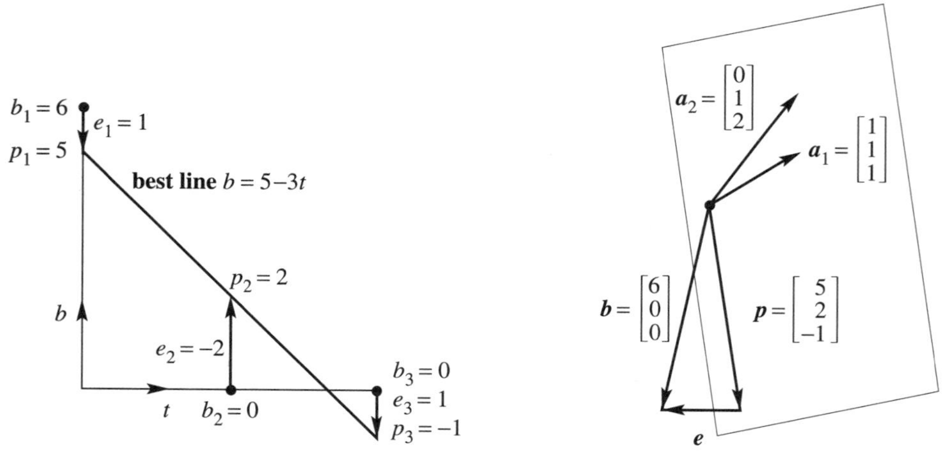

Figure 4.6 The closest line has heights $p = (5, 2, -1)$ with errors $e = (1, -2, 1)$. The equations $A^TA\hat{x} = A^Tb$ give $\hat{x} = (5, -3)$. The line is $b = 5 - 3t$ and the projection is $5a_1 - 3a_2$.

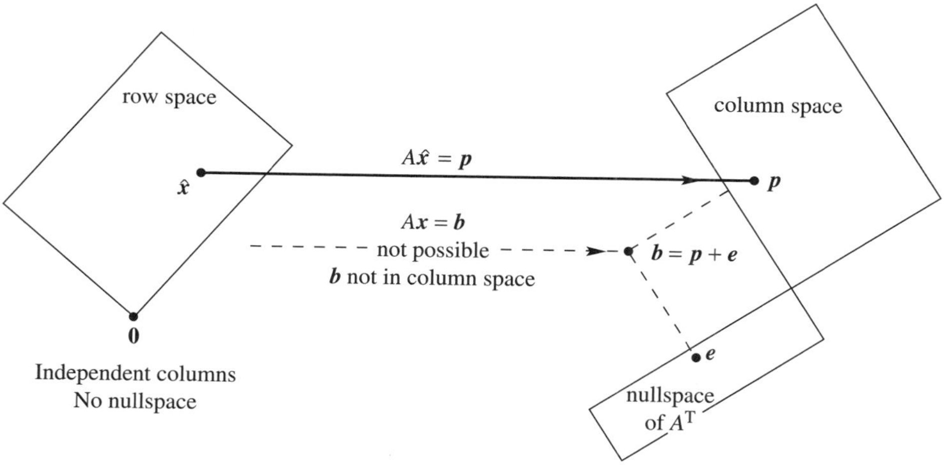

Figure 4.7 The projection p is closest to b, so $\|b - p\|^2 = \|b - A\hat{x}\|^2$ gives minimum error.

These equations are identical with $\boldsymbol{A^TA\hat{x} = A^Tb}$. The best C and D are the components of $\hat{x}$. The equations (4.17) from calculus are the same as the "normal equations" from linear algebra:

The partial derivatives of $\|Ax - b\|^2$ *are zero when* $A^TA\hat{x} = A^Tb$.

The solution is $C = 5$ and $D = -3$. Therefore $b = 5 - 3t$ is the best line—it comes closest to the three points. At $t = 0, 1, 2$ this line goes through 5, 2, -1. It could not go through 6, 0, 0. The errors are 1, -2, 1. This is the vector e!

Figure 4.6a shows the closest line to the three points. It misses by distances e_1, e_2, e_3. *Those are vertical distances*. The least-squares line is chosen to minimize $E = e_1^2 + e_2^2 + e_3^2$.

Figure 4.6b shows the same problem in another way (in 3-dimensional space). The vector b is not in the column space of A. That is why we could not put a line through the three points. The smallest possible error is the perpendicular vector e to the plane. This is $e = b - A\hat{x}$, the vector of errors $(1, -2, 1)$ in the three equations—and the distances from the best line. Behind both figures is the fundamental equation $A^TA\hat{x} = A^Tb$.

Notice that the errors 1, -2, 1 add to zero. The error $e = (e_1, e_2, e_3)$ is perpendicular to the first column $(1, 1, 1)$ in A. The dot product gives $e_1 + e_2 + e_3 = 0$.

The Big Picture

The key figure of this book shows the four subspaces and the true action of a matrix. The vector x on the left side of Figure 4.2 went to $b = Ax$ on the right side. In that figure x was split into $x_r + x_n$.

In this section the situation is just the opposite. There are *no* solutions to $Ax = b$. ***Instead of splitting up*** x ***we are splitting up*** b. Figure 4.7 shows the big picture for least squares. Instead of $Ax = b$ we solve $A\hat{x} = p$. The error $e = b - p$ is unavoidable.

Notice how the missing nullspace $\boldsymbol{N}(A)$ is very small—just one point. With independent columns, the only solution to $A\boldsymbol{x} = \boldsymbol{0}$ is $\boldsymbol{x} = \boldsymbol{0}$. Then $A^{\mathrm{T}}A$ is invertible. The equation $A^{\mathrm{T}}A\hat{\boldsymbol{x}} = A^{\mathrm{T}}\boldsymbol{b}$ fully determines the best vector $\hat{\boldsymbol{x}}$.

Section 7.2 will have the complete picture—all four subspaces included. Every $\boldsymbol{x}$ splits into $\boldsymbol{x}_r + \boldsymbol{x}_n$, and every $\boldsymbol{b}$ splits into $\boldsymbol{p} + \boldsymbol{e}$. The best solution is still $\hat{\boldsymbol{x}}$ (or $\hat{\boldsymbol{x}}_r$) in the row space. We can't help $\boldsymbol{e}$ and we don't want $\boldsymbol{x}_n$—this leaves $A\hat{\boldsymbol{x}} = \boldsymbol{p}$.

Fitting a Straight Line

This is the clearest application of least squares. It starts with m points in a plane—hopefully near a straight line. At times $t_1, \ldots, t_m$ those points are at heights $b_1, \ldots, b_m$. Figure 4.6a shows the best line $b = C + Dt$, which misses the points by distances $e_1, \ldots, e_m$. *Those are vertical distances.* No line is perfect, and the least-squares line is chosen to minimize $E = e_1^2 + \cdots + e_m^2$.

The first example in this section had three points. Now we allow m points (m can be large). The algebra still leads to the same two equations $A^{\mathrm{T}}A\hat{\boldsymbol{x}} = A^{\mathrm{T}}\boldsymbol{b}$. The components of $\hat{\boldsymbol{x}}$ are still C and D.

Figure 4.6b shows the same problem in another way (in m-dimensional space). A line goes exactly through the m points when we exactly solve $A\boldsymbol{x} = \boldsymbol{b}$. Generally we can't do it. There are only two unknowns C and D because A has $n = 2$ columns:

$$A\boldsymbol{x} = \boldsymbol{b} \quad \text{is} \quad \begin{array}{c} C + Dt_1 = b_1 \\ C + Dt_2 = b_2 \\ \vdots \\ C + Dt_m = b_m \end{array} \quad \text{and} \quad A = \begin{bmatrix} 1 & t_1 \\ 1 & t_2 \\ \vdots & \vdots \\ 1 & t_m \end{bmatrix}. \tag{4.18}$$

The column space is so thin that almost certainly $\boldsymbol{b}$ is outside of it. The components of $\boldsymbol{e}$ are the errors $e_1, \ldots, e_m$.

When $\boldsymbol{b}$ happens to lie in the column space, the points happen to lie on a line. In that case $\boldsymbol{b} = \boldsymbol{p}$. Then $A\boldsymbol{x} = \boldsymbol{b}$ is solvable and the errors are $\boldsymbol{e} = (0, \ldots, 0)$.

The closest line has heights $p_1, \ldots, p_m$ with errors $e_1, \ldots, e_m$.
The equations $A^{\mathrm{T}}A\hat{x} = A^{\mathrm{T}}b$ give $\hat{x} = (C, D)$. The errors are $e_i = b_i - C - Dt_i$.

Fitting points by straight lines is so important that we give the equations once and for all. Remember that $b = C + Dt$ exactly fits the data points if

$$\begin{array}{c} C + Dt_1 = b_1 \\ \vdots \\ C + Dt_m = b_m \end{array} \quad \text{or} \quad \begin{bmatrix} 1 & t_1 \\ \vdots & \vdots \\ 1 & t_m \end{bmatrix} \begin{bmatrix} C \\ D \end{bmatrix} = \begin{bmatrix} b_1 \\ \vdots \\ b_m \end{bmatrix}. \tag{4.19}$$

This is our equation $A\boldsymbol{x} = \boldsymbol{b}$. It is generally unsolvable, if A is rectangular. But there is one good feature. The columns of A are independent (unless all times t_i are the same). So we turn to least squares and solve $A^TA\hat{x} = A^Tb$. The "dot-product matrix" A^TA is 2 by 2:

$$A^TA = \begin{bmatrix} 1 & \cdots & 1 \\ t_1 & \cdots & t_m \end{bmatrix} \begin{bmatrix} 1 & t_1 \\ \vdots & \vdots \\ 1 & t_m \end{bmatrix} = \begin{bmatrix} m & \sum t_i \\ \sum t_i & \sum t_i^2 \end{bmatrix}. \tag{4.20}$$

On the right side of the equation is the 2 by 1 vector A^Tb:

$$A^Tb = \begin{bmatrix} 1 & \cdots & 1 \\ t_1 & \cdots & t_m \end{bmatrix} \begin{bmatrix} b_1 \\ \vdots \\ b_m \end{bmatrix} = \begin{bmatrix} \sum b_i \\ \sum t_i b_i \end{bmatrix}. \tag{4.21}$$

In a specific problem, all these numbers are given. The m equations reduce to two equations. A formula for C and D will be in equation (4.30), but solving the two equations directly is probably faster than substituting into the formula.

4H The line $C + Dt$ which minimizes $e_1^2 + \cdots + e_m^2$ is determined by $A^TA\hat{x} = A^Tb$:

$$\begin{bmatrix} m & \sum t_i \\ \sum t_i & \sum t_i^2 \end{bmatrix} \begin{bmatrix} C \\ D \end{bmatrix} = \begin{bmatrix} \sum b_i \\ \sum t_i b_i \end{bmatrix}. \tag{4.22}$$

The vertical errors at the m points on the line are the components of $e = b - p$.

As always, those equations come from geometry or linear algebra or calculus. The error vector $b - A\hat{x}$ is perpendicular to the columns of A (geometry). It is in the nullspace of A^T (linear algebra). The best $\hat{x} = (C, D)$ minimizes the total error E, the sum of squares:

$$E(x) = \|Ax - b\|^2 = (C + Dt_1 - b_1)^2 + \cdots + (C + Dt_m - b_m)^2.$$

When calculus sets the derivatives $\partial E/\partial C$ and $\partial E/\partial D$ to zero, it produces the two equations in (4.22).

Other least-squares problems have more than two unknowns. Fitting a parabola has $n = 3$ coefficients C, D, E (see below). In general x stands for n unknowns $x_1, \ldots, x_n$. The matrix A has n columns. The total error is a function $E(x) = \|Ax - b\|^2$ of n variables. Its derivatives give the n equations $A^TA\hat{x} = A^Tb$. The derivative of a square is linear—this is why the method of least squares is so popular.

Fitting by a Parabola

If we throw a ball, it would be crazy to fit the path by a straight line. A parabola $b = C + Dt + Et^2$ allows the ball to go up and come down again (b is the height at time t). The actual path is not a perfect parabola, but the whole theory of projectiles starts with that approximation.

When Galileo dropped a stone from the Leaning Tower of Pisa, it accelerated. The distance contains a quadratic term $\frac{1}{2}gt^2$. (Galileo's point was that the stone's mass is not involved.) Without that term we could never send a satellite into the right orbit. But even with a nonlinear function like t^2, the unknowns C, D, E appear linearly. Choosing the best parabola is still a problem in linear algebra.

Problem Fit $b_1, \ldots, b_m$ at times $t_1, \ldots, t_m$ by a parabola $b = C + Dt + Et^2$.

With $m > 3$ points, the equations to fit a parabola exactly are generally unsolvable:

$$\begin{aligned} C + Dt_1 + Et_1^2 &= b_1 \\ &\vdots \\ C + Dt_m + Et_m^2 &= b_m \end{aligned} \qquad \text{has the } m \text{ by 3 matrix} \qquad A = \begin{bmatrix} 1 & t_1 & t_1^2 \\ \vdots & \vdots & \vdots \\ 1 & t_m & t_m^2 \end{bmatrix}. \tag{4.23}$$

Least squares The best parabola chooses $\hat{x} = (C, D, E)$ to satisfy the three normal equations $A^{\mathrm{T}}A\hat{x} = A^{\mathrm{T}}\boldsymbol{b}$.

May we ask you to convert this to a problem of projection? The column space of A has dimension _____. The projection of $\boldsymbol{b}$ is $\boldsymbol{p} = A\hat{x}$, which combines the three columns using the coefficients _____, _____, _____. The error at the first data point is $e_1 = b_1 - C - Dt_1 - Et_1^2$. The total squared error is $E = e_1^2 +$ _____. If you prefer to minimize by calculus, take the partial derivatives of E with respect to _____, _____, _____. These three derivatives will be zero when $\hat{x} = (C, D, E)$ solves the 3 by 3 system of equations _____.

Example 4.10 For a parabola $b = C + Dt + Et^2$ to go through $b = 6, 0, 0$ when $t = 0, 1, 2$, the equations are

$$\begin{aligned} C + D \cdot 0 + E \cdot 0^2 &= 6 \\ C + D \cdot 1 + E \cdot 1^2 &= 0 \\ C + D \cdot 2 + E \cdot 2^2 &= 0. \end{aligned} \tag{4.24}$$

This is $A\boldsymbol{x} = \boldsymbol{b}$. We can solve it exactly. Three data points give three equations and a square matrix. The solution is $\boldsymbol{x} = (C, D, E) = (6, -9, 3)$. The parabola through the three points is $b = 6 - 9t + 3t^2$.

What does this mean for projection? The matrix has three columns, which span the whole space $\mathbf{R}^3$. The projection matrix is I. The projection of $\boldsymbol{b}$ is $\boldsymbol{b}$. The error is zero. We didn't need $A^{\mathrm{T}}A\hat{x} = A^{\mathrm{T}}\boldsymbol{b}$, because we solved $A\boldsymbol{x} = \boldsymbol{b}$. Of course we could multiply by A^{T}, but there is no reason to do it.

Figure 4.8a shows the parabola through the three points. It also shows a fourth point b_4 at time t_4. If that falls on the parabola, the new $A\boldsymbol{x} = \boldsymbol{b}$ (four equations) is still solvable. When the fourth point is not on the parabola, we turn to $A^{\mathrm{T}}A\hat{x} = A^{\mathrm{T}}\boldsymbol{b}$. Will the least-squares parabola stay the same, with all the error at the fourth point? Not likely!

An error vector $(0, 0, 0, e_4)$ would not be perpendicular to the column $(1, 1, 1, 1)$. Least squares balances out the four errors, and they add to zero.

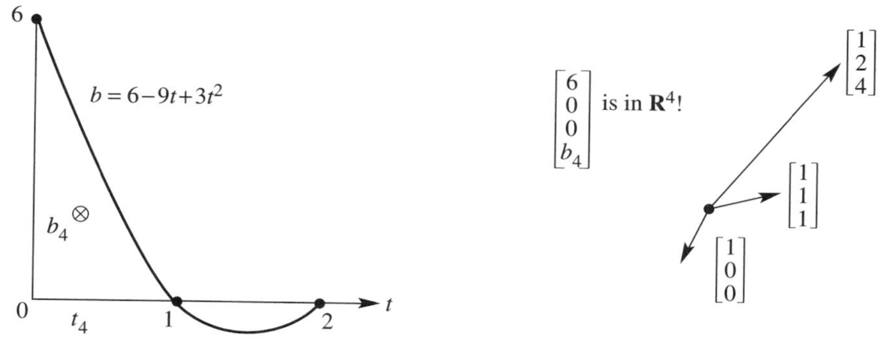

Figure 4.8 Exact fit of parabola through three points means $p = b$ and $e = 0$. Fourth point requires least squares.

Problem Set 4.3

Problems 1–10 use four data points to bring out the key ideas.

1 (Straight line $b = C + Dt$ through four points) With $b = 0, 8, 8, 20$ at times $t = 0, 1, 3, 4$, write down the four equations $Ax = b$ (unsolvable). Change the measurements to $p = 1, 5, 13, 17$ and find an exact solution to $A\hat{x} = p$.

2 With $b = 0, 8, 8, 20$ at $t = 0, 1, 3, 4$, set up and solve the normal equations $A^TA\hat{x} = A^Tb$. For the best straight line in Figure 4.9a, find its four heights and four errors. What is the minimum value of $E = e_1^2 + e_2^2 + e_3^2 + e_4^2$?

3 Compute $p = A\hat{x}$ for the same b and A using $A^TA\hat{x} = A^Tb$. Check that $e = b - p$ is perpendicular to both columns of A. What is the shortest distance $\|e\|$ from b to the column space?

4 (Use calculus) Write down $E = \|Ax - b\|^2$ as a sum of four squares involving C and D. Find the derivative equations $\partial E/\partial C = 0$ and $\partial E/\partial D = 0$. Divide by 2 to obtain the normal equations $A^TA\hat{x} = A^Tb$.

5 Find the height C of the best *horizontal line* to fit $b = (0, 8, 8, 20)$. An exact fit would solve the unsolvable equations $C = 0, C = 8, C = 8, C = 20$. Find the 4 by 1 matrix A in these equations and solve $A^TA\hat{x} = A^Tb$. Redraw Figure 4.9a to show the best height $\hat{x} = C$ and the four errors in e.

6 Project $b = (0, 8, 8, 20)$ onto the line through $a = (1, 1, 1, 1)$. Find $\hat{x} = a^Tb/a^Ta$ and the projection $p = \hat{x}a$. Redraw Figure 4.9b and check that $e = b - p$ is perpendicular to a. What is the shortest distance $\|e\|$ from b to the line in your figure?

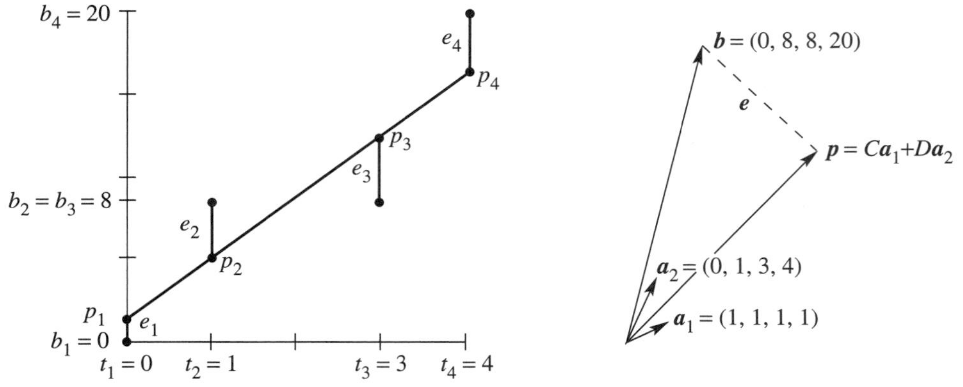

Figure 4.9 The closest line in 2 dimensions matches the projection in $\mathbf{R}^4$.

7 Find the closest line $b = Dt$, *through the origin*, to the same four points. An exact fit would solve $D \cdot 0 = 0$, $D \cdot 1 = 8$, $D \cdot 3 = 8$, $D \cdot 4 = 20$. Find the 4 by 1 matrix and solve $A^{\mathrm{T}}A\hat{x} = A^{\mathrm{T}}\boldsymbol{b}$. Redraw Figure 4.9a showing the best slope $\hat{x} = D$ and the four errors.

8 Project $\boldsymbol{b} = (0, 8, 8, 20)$ onto the line through $\boldsymbol{a} = (0, 1, 3, 4)$. Find $\hat{x}$ and $\boldsymbol{p} = \hat{x}\boldsymbol{a}$. The best C in Problems 5–6 and the best D in Problems 7–8 do not agree with the best (C, D) in Problems 1–4. That is because $(1, 1, 1, 1)$ and $(0, 1, 3, 4)$ are _____ perpendicular.

9 For the closest parabola $b = C + Dt + Et^2$ to the same four points, write down the unsolvable equations $A\boldsymbol{x} = \boldsymbol{b}$. Set up the three normal equations $A^{\mathrm{T}}A\hat{\boldsymbol{x}} = A^{\mathrm{T}}\boldsymbol{b}$ (solution not required). In Figure 4.9a you are now fitting a parabola—what is happening in Figure 4.9b?

10 For the closest cubic $b = C + Dt + Et^2 + Ft^3$ to the same four points, write down the four equations $A\boldsymbol{x} = \boldsymbol{b}$. Solve them by elimination. In Figure 4.9a this cubic now goes exactly through the points. Without computation write $\boldsymbol{p}$ and $\boldsymbol{e}$ in Figure 4.9b.

11 The average of the four times is $\hat{t} = \frac{1}{4}(0 + 1 + 3 + 4) = 2$. The average of the four b's is $\hat{b} = \frac{1}{4}(0 + 8 + 8 + 20) = 9$.

(a) Verify that the best line goes through the center point $(\hat{t}, \hat{b}) = (2, 9)$.

(b) Explain why $C + D\hat{t} = \hat{b}$ comes from the first of the two equations in (4.33).

Questions 12–16 introduce basic ideas of statistics—the foundation for least squares.

12 (Recommended) This problem projects $\boldsymbol{b} = (b_1, \ldots, b_m)$ onto the line through $\boldsymbol{a} = (1, \ldots, 1)$.

(a) Solve $\boldsymbol{a}^{\mathrm{T}}\boldsymbol{a}\hat{x} = \boldsymbol{a}^{\mathrm{T}}\boldsymbol{b}$ to show that $\hat{x}$ is the ***mean*** of the $\boldsymbol{b}$'s.

(b) Find the error vector $\boldsymbol{e}$ and the ***variance*** $\|\boldsymbol{e}\|^2$ and the ***standard deviation*** $\|\boldsymbol{e}\|$.

(c) Draw a graph with $\boldsymbol{b} = (1, 2, 6)$ fitted by a horizontal line. What are $\boldsymbol{p}$ and $\boldsymbol{e}$ on the graph? Check that $\boldsymbol{p}$ is perpendicular to $\boldsymbol{e}$ and find the matrix P.

13 First assumption behind least squares: Each measurement error has "expected value" zero. Multiply the eight error vectors $\boldsymbol{b} - A\boldsymbol{x} = (\pm 1, \pm 1, \pm 1)$ by $(A^{\mathrm{T}}A)^{-1}A^{\mathrm{T}}$ to show that the eight vectors $\hat{\boldsymbol{x}} - \boldsymbol{x}$ also average to zero. The expected value of $\hat{\boldsymbol{x}}$ is the correct $\boldsymbol{x}$.

14 Second assumption behind least squares: The measurement errors are independent and have the same variance σ^2. The average of $(\boldsymbol{b} - A\boldsymbol{x})(\boldsymbol{b} - A\boldsymbol{x})^{\mathrm{T}}$ is $\sigma^2 I$. Multiply on the left by $(A^{\mathrm{T}}A)^{-1}A^{\mathrm{T}}$ and multiply on the right by $A(A^{\mathrm{T}}A)^{-1}$ to show that the average of $(\hat{\boldsymbol{x}} - \boldsymbol{x})(\hat{\boldsymbol{x}} - \boldsymbol{x})^{\mathrm{T}}$ is $\sigma^2(A^{\mathrm{T}}A)^{-1}$. This is the "covariance matrix" for the error in $\hat{\boldsymbol{x}}$.

15 A doctor takes m readings $b_1, \ldots, b_m$ of your pulse rate. The least-squares solution to the m equations $x = b_1, x = b_2, \ldots, x = b_m$ is the average $\hat{x} = (b_1 + \cdots + b_m)/m$. The matrix A is a column of 1's. Problem 14 gives the expected error $(\hat{x} - x)^2$ as $\sigma^2(A^{\mathrm{T}}A)^{-1} =$ ______. By taking m measurements, the variance drops from σ^2 to σ^2/m.

16 If you know the average $\hat{x}_{99}$ of 99 numbers $b_1, \ldots, b_{99}$, how can you quickly find the average $\hat{x}_{100}$ with one more number b_{100}? The idea of *recursive* least squares is to avoid adding 100 numbers. What coefficient correctly gives $\hat{x}_{100}$ from b_{100} and $\hat{x}_{99}$?

$$\frac{1}{100}b_{100} + \underline{\qquad\quad}\,\hat{x}_{99} = \frac{1}{100}(b_1 + \cdots + b_{100}).$$

Questions 17–25 give more practice with $\hat{x}$ and p and e.

17 Write down three equations for the line $b = C + Dt$ to go through $b = 7$ at $t = -1$, $b = 7$ at $t = 1$, and $b = 21$ at $t = 2$. Find the least-squares solution $\hat{\boldsymbol{x}} = (C, D)$ and draw the closest line.

18 Find the projection $\boldsymbol{p} = A\hat{\boldsymbol{x}}$ in Problem 17. This gives the three ______ of the closest line. Show that the error vector is $\boldsymbol{e} = (2, -6, 4)$.

19 Suppose the measurements at $t = -1, 1, 2$ are the errors $2, -6, 4$ in Problem 18. Compute $\hat{\boldsymbol{x}}$ and the closest line. Explain the answer: $\boldsymbol{b} = (2, -6, 4)$ is perpendicular to ______ so the projection is $\boldsymbol{p} = \boldsymbol{0}$.

20 Suppose the measurements at $t = -1, 1, 2$ are $\boldsymbol{b} = (5, 13, 17)$. Compute $\hat{\boldsymbol{x}}$ and the closest line and $\boldsymbol{e}$. The error is $\boldsymbol{e} = \boldsymbol{0}$ because this $\boldsymbol{b}$ is ______.

21 Which of the four subspaces contains the error vector $\boldsymbol{e}$? Which contains $\boldsymbol{p}$? Which contains $\hat{\boldsymbol{x}}$? What is the nullspace of A?

22 Find the best line $C + Dt$ to fit $b = 4, 2, -1, 0, 0$ at times $t = -2, -1, 0, 1, 2$.

23 (Distance between lines) The points $P = (x, x, x)$ are on a line through $(1, 1, 1)$ and $Q = (y, 3y, -1)$ are on another line. Choose x and y to minimize the squared distance $\|P - Q\|^2$.

24 Is the error vector $\boldsymbol{e}$ orthogonal to $\boldsymbol{b}$ or $\boldsymbol{p}$ or $\boldsymbol{e}$ or $\hat{\boldsymbol{x}}$? Show that $\|\boldsymbol{e}\|^2$ equals $\boldsymbol{e}^{\mathrm{T}}\boldsymbol{b}$ which equals $\boldsymbol{b}^{\mathrm{T}}\boldsymbol{b} - \boldsymbol{b}^{\mathrm{T}}\boldsymbol{p}$. This is the smallest total error E.

25 The derivatives of $\|A\boldsymbol{x}\|^2$ with respect to the variables $x_1, \ldots, x_n$ fill the vector $2A^{\mathrm{T}}A\boldsymbol{x}$. The derivatives of $2\boldsymbol{b}^{\mathrm{T}}A\boldsymbol{x}$ fill the vector $2A^{\mathrm{T}}\boldsymbol{b}$. So the derivatives of $\|A\boldsymbol{x} - \boldsymbol{b}\|^2$ are zero when ______.

4.4 Orthogonal Bases and Gram-Schmidt

This section has two goals. The first is to see how orthogonal vectors make calculations simpler. Dot products are zero—this removes the work in $A^{\mathrm{T}}A$. The second goal is to ***construct orthogonal vectors***. We will pick combinations of the original vectors to produce right angles. Those original vectors are the columns of A. The orthogonal vectors will be the columns of a new matrix Q.

You know from Chapter 3 what a basis consists of—independent vectors that span the space. Numerically, we always compute with a basis. It gives a set of coordinate axes. The axes could meet at any angle (except 0° and 180°). But every time we visualize axes, they are perpendicular. *In our imagination, the coordinate axes are practically always orthogonal.* This simplifies the picture and it greatly simplifies the computations.

The vectors $\boldsymbol{q}_1, \ldots, \boldsymbol{q}_n$ are ***orthogonal*** when their dot products $\boldsymbol{q}_i \cdot \boldsymbol{q}_j$ are zero. More exactly $\boldsymbol{q}_i^{\mathrm{T}}\boldsymbol{q}_j = 0$ whenever $i \neq j$. With one more step—just divide each vector by its length—the vectors become orthogonal *unit* vectors. Their lengths are all 1. Then the basis is called *orthonormal.*

Definition The vectors $\boldsymbol{q}_1, \ldots, \boldsymbol{q}_n$ are ***orthonormal*** if

$$\boldsymbol{q}_i^{\mathrm{T}}\boldsymbol{q}_j = \begin{cases} 0 & \text{when } i \neq j \quad (\textbf{\textit{orthogonal}} \text{ vectors}) \\ 1 & \text{when } i = j \quad (\textbf{\textit{unit}} \text{ vectors: } \|\boldsymbol{q}_i\| = 1) \end{cases}$$

A matrix with orthonormal columns is assigned the special letter Q.

The matrix Q is easy to work with because $Q^{\mathrm{T}}Q = I$. This says in matrix language that the columns are orthonormal. It is equation (4.25) below, and Q is not required to be square. When the matrix *is* square, $Q^{\mathrm{T}}Q = I$ means that $Q^{\mathrm{T}} = Q^{-1}$.

4I A matrix Q with orthonormal columns satisfies $Q^{\mathrm{T}}Q = I$:

$$Q^{\mathrm{T}}Q = \begin{bmatrix} - \boldsymbol{q}_1^{\mathrm{T}} - \\ - \boldsymbol{q}_2^{\mathrm{T}} - \\ - \boldsymbol{q}_n^{\mathrm{T}} - \end{bmatrix} \begin{bmatrix} | & | & | \\ \boldsymbol{q}_1 & \boldsymbol{q}_2 & \boldsymbol{q}_n \\ | & | & | \end{bmatrix} = \begin{bmatrix} 1 & 0 & \cdots & 0 \\ 0 & 1 & \cdots & 0 \\ \vdots & \vdots & \ddots & \vdots \\ 0 & 0 & \cdots & 1 \end{bmatrix} = I. \qquad (4.25)$$

When row i of Q^{T} multiplies column j of Q, the dot product is $\boldsymbol{q}_i^{\mathrm{T}}\boldsymbol{q}_j$. Off the diagonal $(i \neq j)$ that is zero by orthogonality. On the diagonal $(i = j)$ the unit vectors give $\boldsymbol{q}_i^{\mathrm{T}}\boldsymbol{q}_i = \|\boldsymbol{q}_i\|^2 = 1$.

If the columns are only orthogonal (not unit vectors), then $Q^{\mathrm{T}}Q$ is a diagonal matrix (not the identity matrix). We wouldn't use the letter Q. But this matrix is almost as good. The important thing is orthogonality—then it is easy to produce unit vectors.

To repeat: $Q^{\mathrm{T}}Q = I$ even when Q is rectangular. In that case Q^{T} is only an inverse from the left. For square matrices we also have $QQ^{\mathrm{T}} = I$, so Q^{T} is the two-sided inverse of Q. The rows of a square Q are orthonormal like the columns. To invert the matrix we just transpose it. In this square case we call Q an ***orthogonal matrix***.†

Here are three examples of orthogonal matrices—rotation and permutation and reflection.

Example 4.11 (Rotation) Q rotates every vector in the plane through the angle θ:

$$Q = \begin{bmatrix} \cos\theta & -\sin\theta \\ \sin\theta & \cos\theta \end{bmatrix} \quad \text{and} \quad Q^{\mathrm{T}} = Q^{-1} = \begin{bmatrix} \cos\theta & \sin\theta \\ -\sin\theta & \cos\theta \end{bmatrix}.$$

The columns of Q are orthogonal (take their dot product). They are unit vectors because $\sin^2\theta + \cos^2\theta = 1$. Those columns give an ***orthonormal basis*** for the plane $\mathbf{R}^2$. The standard basis vectors $\boldsymbol{i}$ and $\boldsymbol{j}$ are rotated through θ (see Figure 4.10a).

Q^{-1} rotates vectors back through $-\theta$. It agrees with Q^{T}, because the cosine of $-\theta$ is the cosine of θ, and $\sin(-\theta) = -\sin\theta$. We have $Q^{\mathrm{T}}Q = I$ and $QQ^{\mathrm{T}} = I$.

Example 4.12 (Permutation) These matrices change the order of the components:

$$\begin{bmatrix} 0 & 1 & 0 \\ 0 & 0 & 1 \\ 1 & 0 & 0 \end{bmatrix} \begin{bmatrix} x \\ y \\ z \end{bmatrix} = \begin{bmatrix} y \\ z \\ x \end{bmatrix} \quad \text{and} \quad \begin{bmatrix} 0 & 1 \\ 1 & 0 \end{bmatrix} \begin{bmatrix} x \\ y \end{bmatrix} = \begin{bmatrix} y \\ x \end{bmatrix}.$$

All columns of these Q's are unit vectors (their lengths are obviously 1). They are also orthogonal (the 1's appear in different places). *The inverse of a permutation matrix is its transpose*. The inverse puts the components back into their original order:

$$\text{Inverse} = \text{transpose:} \quad \begin{bmatrix} 0 & 0 & 1 \\ 1 & 0 & 0 \\ 0 & 1 & 0 \end{bmatrix} \begin{bmatrix} y \\ z \\ x \end{bmatrix} = \begin{bmatrix} x \\ y \\ z \end{bmatrix} \quad \text{and} \quad \begin{bmatrix} 0 & 1 \\ 1 & 0 \end{bmatrix} \begin{bmatrix} y \\ x \end{bmatrix} = \begin{bmatrix} x \\ y \end{bmatrix}.$$

The 2 by 2 permutation is a reflection across the 45° line in Figure 4.10b.

Example 4.13 (Reflection) If $\boldsymbol{u}$ is any unit vector, set $Q = I - 2\boldsymbol{u}\boldsymbol{u}^{\mathrm{T}}$. Notice that $\boldsymbol{u}\boldsymbol{u}^{\mathrm{T}}$ is a matrix while $\boldsymbol{u}^{\mathrm{T}}\boldsymbol{u}$ is the number $\|\boldsymbol{u}\|^2 = 1$. Then Q^{T} and Q^{-1} both equal Q:

$$Q^{\mathrm{T}} = I - 2\boldsymbol{u}\boldsymbol{u}^{\mathrm{T}} = Q \quad \text{and} \quad Q^{\mathrm{T}}Q = I - 4\boldsymbol{u}\boldsymbol{u}^{\mathrm{T}} + 4\boldsymbol{u}\boldsymbol{u}^{\mathrm{T}}\boldsymbol{u}\boldsymbol{u}^{\mathrm{T}} = I. \tag{4.26}$$

†"Orthonormal matrix" would have been a better name for Q, but it's not used. Any matrix with orthonormal columns can be Q, but we only call it an *orthogonal matrix* when it is square.

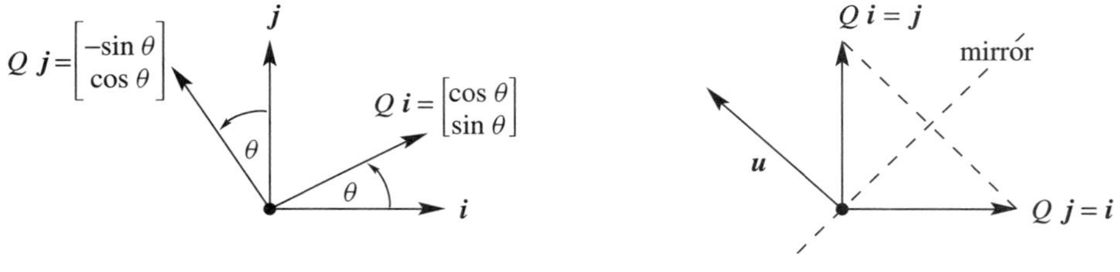

Figure 4.10 Rotation by $Q = \left[\begin{smallmatrix} c & -s \\ s & c \end{smallmatrix}\right]$ and reflection by $Q = \left[\begin{smallmatrix} 0 & 1 \\ 1 & 0 \end{smallmatrix}\right]$.

Reflection matrices $I - 2\boldsymbol{u}\boldsymbol{u}^{\mathrm{T}}$ are symmetric and also orthogonal. If you square them, you get the identity matrix. Reflecting twice through a mirror brings back the original, and $Q^2 = Q^{\mathrm{T}}Q = I$. Notice $\boldsymbol{u}^{\mathrm{T}}\boldsymbol{u} = 1$ near the end of equation (4.26).

As examples we choose the unit vectors $\boldsymbol{u}_1 = (1, 0)$ and then $\boldsymbol{u}_2 = (1/\sqrt{2}, -1/\sqrt{2})$. Compute $2\boldsymbol{u}\boldsymbol{u}^{\mathrm{T}}$ (column times row) and subtract from I:

$$Q_1 = I - 2\begin{bmatrix} 1 \\ 0 \end{bmatrix}\begin{bmatrix} 1 & 0 \end{bmatrix} = \begin{bmatrix} -1 & 0 \\ 0 & 1 \end{bmatrix} \quad \text{and} \quad Q_2 = I - 2\begin{bmatrix} .5 & -.5 \\ -.5 & .5 \end{bmatrix} = \begin{bmatrix} 0 & 1 \\ 1 & 0 \end{bmatrix}.$$

Q_1 reflects $\boldsymbol{u}_1 = (1, 0)$ across the y axis to $(-1, 0)$. Every vector (x, y) goes into its mirror image $(-x, y)$ across the y axis:

$$\text{Reflection from } Q_1\text{:} \quad \begin{bmatrix} -1 & 0 \\ 0 & 1 \end{bmatrix}\begin{bmatrix} x \\ y \end{bmatrix} = \begin{bmatrix} -x \\ y \end{bmatrix}.$$

Q_2 is reflection across the 45° line. Every (x, y) goes to (y, x)—this was the permutation in Example 4.12. A vector like (3, 3) doesn't move when you exchange 3 and 3—it is on the mirror line. Figure 4.10b shows the mirror.

Rotations preserve the length of a vector. So do reflections. So do permutations. So does multiplication by any orthogonal matrix—***lengths don't change***. This is a key property of Q:

4J If Q has orthonormal columns ($Q^{\mathrm{T}}Q = I$), it leaves lengths unchanged:

$$\|Q\boldsymbol{x}\| = \|\boldsymbol{x}\| \text{ for every vector } \boldsymbol{x}. \tag{4.27}$$

Q also preserves dot products and angles: $(Q\boldsymbol{x})^{\mathrm{T}}(Q\boldsymbol{y}) = \boldsymbol{x}^{\mathrm{T}}Q^{\mathrm{T}}Q\boldsymbol{y} = \boldsymbol{x}^{\mathrm{T}}\boldsymbol{y}$.

Proof $\|Q\boldsymbol{x}\|^2$ is the same as $\|\boldsymbol{x}\|^2$ because $(Q\boldsymbol{x})^{\mathrm{T}}(Q\boldsymbol{x}) = \boldsymbol{x}^{\mathrm{T}}Q^{\mathrm{T}}Q\boldsymbol{x} = \boldsymbol{x}^{\mathrm{T}}I\boldsymbol{x}$. Orthogonal matrices are excellent for computations—numbers can never grow too large when lengths are fixed. Good computer codes use Q's as much as possible, to be numerically stable.

Projections Using Orthogonal Bases: Q Replaces A

This chapter is about projections onto subspaces. We developed the equations for $\hat{\boldsymbol{x}}$ and $\boldsymbol{p}$ and P. When the columns of A were a basis for the subspace, all formulas involved $A^{\mathrm{T}}A$. (You have not forgotten $A^{\mathrm{T}}A\hat{\boldsymbol{x}} = A^{\mathrm{T}}\boldsymbol{b}$.) The entries of $A^{\mathrm{T}}A$ are the dot products $\boldsymbol{a}_i \cdot \boldsymbol{a}_j$.

Suppose the basis vectors are not only independent but orthonormal. The $\boldsymbol{a}$'s become $\boldsymbol{q}$'s. Their dot products are zero (except $\boldsymbol{q}_i \cdot \boldsymbol{q}_i = 1$). The new matrix Q has *orthonormal columns*: $A^{\mathrm{T}}A$ simplifies to $Q^{\mathrm{T}}Q = I$. Look at the improvements in $\hat{\boldsymbol{x}}$ and $\boldsymbol{p}$ and P. Instead of $Q^{\mathrm{T}}Q$ we print a blank for the identity matrix:

$$\underline{\qquad}\,\hat{\boldsymbol{x}} = Q^{\mathrm{T}}\boldsymbol{b} \quad \text{and} \quad \boldsymbol{p} = Q\hat{\boldsymbol{x}} \quad \text{and} \quad P = Q\,\underline{\qquad}\,Q^{\mathrm{T}}. \tag{4.28}$$

The least-squares solution of $Qx = b$ is $\hat{x} = Q^{\mathrm{T}}b$. The projection matrix simplifies to $P = QQ^{\mathrm{T}}$.

There are no matrices to invert, and Gaussian elimination is not needed. This is the point of an orthonormal basis. The components of $\hat{\boldsymbol{x}} = Q^{\mathrm{T}}\boldsymbol{b}$ are just dot products of $\boldsymbol{b}$ with the rows of Q^{T}, which are the $\boldsymbol{q}$'s. We have n separate 1-dimensional projections. The "coupling matrix" or "correlation matrix" $A^{\mathrm{T}}A$ is now $Q^{\mathrm{T}}Q = I$. There is no coupling. Here is a display of $\hat{\boldsymbol{x}} = Q^{\mathrm{T}}\boldsymbol{b}$ and $\boldsymbol{p} = Q\hat{\boldsymbol{x}}$:

$$\hat{\boldsymbol{x}} = \begin{bmatrix} - & \boldsymbol{q}_1^{\mathrm{T}} & - \\ & \vdots & \\ - & \boldsymbol{q}_n^{\mathrm{T}} & - \end{bmatrix} \begin{bmatrix} \\ \boldsymbol{b} \\ \\ \end{bmatrix} = \begin{bmatrix} \boldsymbol{q}_1^{\mathrm{T}}\boldsymbol{b} \\ \vdots \\ \boldsymbol{q}_n^{\mathrm{T}}\boldsymbol{b} \end{bmatrix}$$

$$\boldsymbol{p} = \begin{bmatrix} | & & | \\ \boldsymbol{q}_1 & \cdots & \boldsymbol{q}_n \\ | & & | \end{bmatrix} \begin{bmatrix} \boldsymbol{q}_1^{\mathrm{T}}\boldsymbol{b} \\ \vdots \\ \boldsymbol{q}_n^{\mathrm{T}}\boldsymbol{b} \end{bmatrix} = \boldsymbol{q}_1(\boldsymbol{q}_1^{\mathrm{T}}\boldsymbol{b}) + \cdots + \boldsymbol{q}_n(\boldsymbol{q}_n^{\mathrm{T}}\boldsymbol{b}).$$

When Q is a square matrix, the subspace is the whole space. Then Q^{T} is the same as Q^{-1}, and $\hat{\boldsymbol{x}} = Q^{\mathrm{T}}\boldsymbol{b}$ is the same as $\boldsymbol{x} = Q^{-1}\boldsymbol{b}$. The solution is exact! The projection of $\boldsymbol{b}$ onto the whole space is $\boldsymbol{b}$ itself. In this case $\boldsymbol{p} = \boldsymbol{b}$ and $P = QQ^{\mathrm{T}} = I$.

You may think that projection onto the whole space is not worth mentioning. But when $\boldsymbol{p} = \boldsymbol{b}$, our formula assembles $\boldsymbol{b}$ out of its 1-dimensional projections. If $\boldsymbol{q}_1, \ldots, \boldsymbol{q}_n$ is an orthonormal basis for the whole space, every $\boldsymbol{b}$ is the sum of its components along the $\boldsymbol{q}$'s:

$$\boldsymbol{b} = \boldsymbol{q}_1(\boldsymbol{q}_1^{\mathrm{T}}\boldsymbol{b}) + \boldsymbol{q}_2(\boldsymbol{q}_2^{\mathrm{T}}\boldsymbol{b}) + \cdots + \boldsymbol{q}_n(\boldsymbol{q}_n^{\mathrm{T}}\boldsymbol{b}). \tag{4.29}$$

That is $QQ^{\mathrm{T}} = I$. It is the foundation of Fourier series and all the great "transforms" of applied mathematics. They break vectors or functions into perpendicular pieces. Then by adding the pieces, the inverse transform puts the function back together.

Example 4.14 The columns of this matrix Q are orthonormal vectors $\boldsymbol{q}_1, \boldsymbol{q}_2, \boldsymbol{q}_3$:

$$Q = \frac{1}{3}\begin{bmatrix} -1 & 2 & 2 \\ 2 & -1 & 2 \\ 2 & 2 & -1 \end{bmatrix} \quad \text{has first column} \quad \boldsymbol{q}_1 = \begin{bmatrix} -\frac{1}{3} \\ \frac{2}{3} \\ \frac{2}{3} \end{bmatrix}.$$

The separate projections of $\boldsymbol{b} = (0, 0, 1)$ onto $\boldsymbol{q}_1$ and $\boldsymbol{q}_2$ and $\boldsymbol{q}_3$ are

$$\boldsymbol{q}_1(\boldsymbol{q}_1^{\mathrm{T}}\boldsymbol{b}) = \tfrac{2}{3}\boldsymbol{q}_1 \quad \text{and} \quad \boldsymbol{q}_2(\boldsymbol{q}_2^{\mathrm{T}}\boldsymbol{b}) = \tfrac{2}{3}\boldsymbol{q}_2 \quad \text{and} \quad \boldsymbol{q}_3(\boldsymbol{q}_3^{\mathrm{T}}\boldsymbol{b}) = -\tfrac{1}{3}\boldsymbol{q}_3.$$

The sum of the first two is the projection of $\boldsymbol{b}$ onto the *plane* of $\boldsymbol{q}_1$ and $\boldsymbol{q}_2$. The sum of all three is the projection of $\boldsymbol{b}$ onto the *whole space*—which is $\boldsymbol{b}$ itself:

$$\frac{2}{3}\boldsymbol{q}_1 + \frac{2}{3}\boldsymbol{q}_2 - \frac{1}{3}\boldsymbol{q}_3 = \frac{1}{9}\begin{bmatrix} -2+4-2 \\ 4-2-2 \\ 4+4+1 \end{bmatrix} = \begin{bmatrix} 0 \\ 0 \\ 1 \end{bmatrix} = \boldsymbol{b}.$$

Example 4.15 Fitting a straight line leads to $A\boldsymbol{x} = \boldsymbol{b}$ and least squares. The columns of A are orthogonal in one special case—when the measurement times add to zero. Suppose $b = 1, 2, 4$ at times $t = -2, 0, 2$. Those times add to zero. The dot product with $1, 1, 1$ is zero and A has orthogonal columns:

$$\begin{bmatrix} C + D(-2) = 1 \\ C + \quad D(0) = 2 \\ C + \quad D(2) = 4 \end{bmatrix} \qquad \text{or} \qquad A\boldsymbol{x} = \begin{bmatrix} 1 & -2 \\ 1 & 0 \\ 1 & 2 \end{bmatrix}\begin{bmatrix} C \\ D \end{bmatrix} = \begin{bmatrix} 1 \\ 2 \\ 4 \end{bmatrix}.$$

The measurements $1, 2, 4$ are not on a line. There is no exact C and D and $\boldsymbol{x}$. Look at the matrix $A^{\mathrm{T}}A$ in the least-squares equation for $\hat{\boldsymbol{x}}$:

$$A^{\mathrm{T}}A\hat{\boldsymbol{x}} = A^{\mathrm{T}}\boldsymbol{b} \qquad \text{is} \qquad \begin{bmatrix} 3 & 0 \\ 0 & 8 \end{bmatrix}\begin{bmatrix} \widehat{C} \\ \widehat{D} \end{bmatrix} = \begin{bmatrix} 7 \\ 6 \end{bmatrix}.$$

Main point: $A^{\mathrm{T}}A$ *is diagonal*. We can solve separately for $\widehat{C} = \frac{7}{3}$ and $\widehat{D} = \frac{6}{8}$. The zeros in $A^{\mathrm{T}}A$ are dot products of perpendicular columns in A. The denominators 3 and 8 are not 1 and 1, because the columns are not unit vectors. But a diagonal matrix is virtually as good as the identity matrix.

Orthogonal columns are so helpful that it is worth moving the time origin to produce them. To do that, subtract away the average time $\hat{t} = (t_1 + \cdots + t_m)/m$. Then the shifted measurement times $T_i = t_i - \hat{t}$ add to zero. With the columns now orthogonal, $A^{\mathrm{T}}A$ is diagonal. The best C and D (like $\frac{7}{3}$ and $\frac{6}{8}$ above) have direct formulas without inverse matrices:

$$\widehat{C} = \frac{b_1 + \cdots + b_m}{m} \qquad \text{and} \qquad \widehat{D} = \frac{b_1T_1 + \cdots + b_mT_m}{T_1^2 + \cdots + T_m^2}. \tag{4.30}$$

The best line is $\widehat{C} + \widehat{D}T$ or $\widehat{C} + \widehat{D}(t - \hat{t})$. The time shift is an example of the Gram-Schmidt process, which *orthogonalizes the columns in advance*. We now see how that process works for other matrices A. It changes A to Q.

The Gram-Schmidt Process

The point of this section is that "orthogonal is good." Projections and least-squares solutions normally need the inverse of $A^{\mathrm{T}}A$. When this matrix becomes $Q^{\mathrm{T}}Q = I$, the inverse is no problem. If the vectors are orthonormal, the one-dimensional projections are uncoupled and $\boldsymbol{p}$ is their sum. The best $\boldsymbol{x}$ is $Q^{\mathrm{T}}\boldsymbol{b}$ (n separate dot products with $\boldsymbol{b}$). For this to

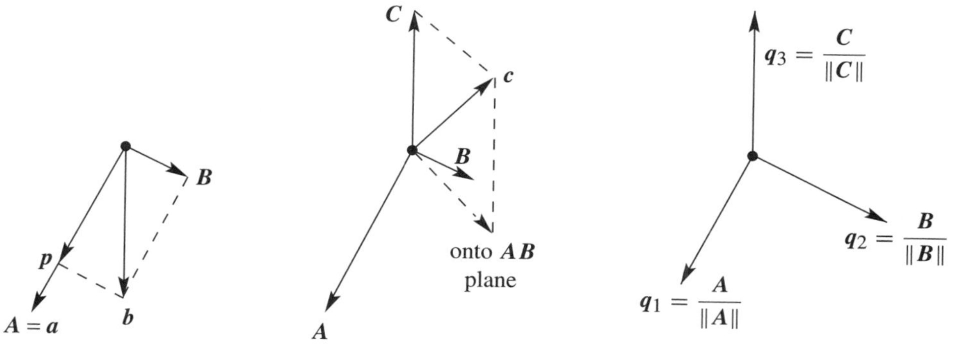

Figure 4.11 Subtract projections from $\boldsymbol{b}$ and $\boldsymbol{c}$ to find $\boldsymbol{B}$ and $\boldsymbol{C}$. Divide by $\|\boldsymbol{A}\|$, $\|\boldsymbol{B}\|$, and $\|\boldsymbol{C}\|$.

be true, we had to say "*If* the vectors are orthonormal." Now we find a way to *make* them orthonormal.

Start with three independent vectors $\boldsymbol{a}, \boldsymbol{b}, \boldsymbol{c}$. We intend to construct three orthogonal vectors $\boldsymbol{A}, \boldsymbol{B}, \boldsymbol{C}$. Then (at the end is easiest) we divide $\boldsymbol{A}, \boldsymbol{B}, \boldsymbol{C}$ by their lengths. That produces three orthonormal vectors $\boldsymbol{q}_1 = \boldsymbol{A}/\|\boldsymbol{A}\|$, $\boldsymbol{q}_2 = \boldsymbol{B}/\|\boldsymbol{B}\|$, $\boldsymbol{q}_3 = \boldsymbol{C}/\|\boldsymbol{C}\|$.

Gram-Schmidt Begin by choosing $\boldsymbol{A} = \boldsymbol{a}$. This gives the first direction. The next direction $\boldsymbol{B}$ must be perpendicular to $\boldsymbol{A}$. *Start with* $\boldsymbol{b}$ *and subtract its projection along* $\boldsymbol{A}$. This leaves the perpendicular part, which is the orthogonal vector $\boldsymbol{B}$:

$$\boldsymbol{B} = \boldsymbol{b} - \frac{\boldsymbol{A}^{\mathrm{T}}\boldsymbol{b}}{\boldsymbol{A}^{\mathrm{T}}\boldsymbol{A}}\boldsymbol{A}. \tag{4.31}$$

$\boldsymbol{A}$ and $\boldsymbol{B}$ are orthogonal in Figure 4.11. Take the dot product with $\boldsymbol{A}$ to verify that $\boldsymbol{A}^{\mathrm{T}}\boldsymbol{B} = \boldsymbol{A}^{\mathrm{T}}\boldsymbol{b} - \boldsymbol{A}^{\mathrm{T}}\boldsymbol{b} = 0$. This vector $\boldsymbol{B}$ is what we have called the error vector $\boldsymbol{e}$, perpendicular to $\boldsymbol{A}$. Notice that $\boldsymbol{B}$ in equation (4.31) is not zero (otherwise $\boldsymbol{a}$ and $\boldsymbol{b}$ would be dependent). The directions $\boldsymbol{A}$ and $\boldsymbol{B}$ are now set.

The third direction starts with $\boldsymbol{c}$. This is not a combination of $\boldsymbol{A}$ and $\boldsymbol{B}$ (because $\boldsymbol{c}$ is not a combination of $\boldsymbol{a}$ and $\boldsymbol{b}$). But most likely $\boldsymbol{c}$ is not perpendicular to $\boldsymbol{A}$ and $\boldsymbol{B}$. So subtract off its components in those two directions to get $\boldsymbol{C}$:

$$\boldsymbol{C} = \boldsymbol{c} - \frac{\boldsymbol{A}^{\mathrm{T}}\boldsymbol{c}}{\boldsymbol{A}^{\mathrm{T}}\boldsymbol{A}}\boldsymbol{A} - \frac{\boldsymbol{B}^{\mathrm{T}}\boldsymbol{c}}{\boldsymbol{B}^{\mathrm{T}}\boldsymbol{B}}\boldsymbol{B}. \tag{4.32}$$

This is the one and only idea of the Gram-Schmidt process. ***Subtract from every new vector its projections in the directions already set.*** That idea is repeated at every step.[†] If we also had a fourth vector $\boldsymbol{d}$, we would subtract its projections onto $\boldsymbol{A}$, $\boldsymbol{B}$, and $\boldsymbol{C}$. That gives $\boldsymbol{D}$. At the end, divide the orthogonal vectors $\boldsymbol{A}, \boldsymbol{B}, \boldsymbol{C}, \boldsymbol{D}$ by their lengths. The resulting vectors $\boldsymbol{q}_1, \boldsymbol{q}_2, \boldsymbol{q}_3, \boldsymbol{q}_4$ are orthonormal.

[†] We think Gram had the idea. We don't know where Schmidt came in.

Example 4.16 Suppose the independent vectors $\boldsymbol{a}, \boldsymbol{b}, \boldsymbol{c}$ are

$$\boldsymbol{a} = \begin{bmatrix} 1 \\ -1 \\ 0 \end{bmatrix} \quad \text{and} \quad \boldsymbol{b} = \begin{bmatrix} 2 \\ 0 \\ -2 \end{bmatrix} \quad \text{and} \quad \boldsymbol{c} = \begin{bmatrix} 3 \\ -3 \\ 3 \end{bmatrix}.$$

Then $\boldsymbol{A} = \boldsymbol{a}$ has $\boldsymbol{A}^{\mathrm{T}}\boldsymbol{A} = 2$. Subtract from $\boldsymbol{b}$ its projection along $\boldsymbol{A}$:

$$\boldsymbol{B} = \boldsymbol{b} - \frac{\boldsymbol{A}^{\mathrm{T}}\boldsymbol{b}}{\boldsymbol{A}^{\mathrm{T}}\boldsymbol{A}}\boldsymbol{A} = \boldsymbol{b} - \tfrac{2}{2}\boldsymbol{A} = \begin{bmatrix} 1 \\ 1 \\ -2 \end{bmatrix}.$$

Check: $\boldsymbol{A}^{\mathrm{T}}\boldsymbol{B} = 0$ as required. Now subtract two projections from $\boldsymbol{c}$:

$$\boldsymbol{C} = \boldsymbol{c} - \frac{\boldsymbol{A}^{\mathrm{T}}\boldsymbol{c}}{\boldsymbol{A}^{\mathrm{T}}\boldsymbol{A}}\boldsymbol{A} - \frac{\boldsymbol{B}^{\mathrm{T}}\boldsymbol{c}}{\boldsymbol{B}^{\mathrm{T}}\boldsymbol{B}}\boldsymbol{B} = \boldsymbol{c} - \tfrac{6}{2}\boldsymbol{A} + \tfrac{6}{6}\boldsymbol{B} = \begin{bmatrix} 1 \\ 1 \\ 1 \end{bmatrix}.$$

Check: $\boldsymbol{C}$ is perpendicular to $\boldsymbol{A}$ and $\boldsymbol{B}$. Finally convert $\boldsymbol{A}, \boldsymbol{B}, \boldsymbol{C}$ to unit vectors (length 1, orthonormal). Divide by the lengths $\sqrt{2}$ and $\sqrt{6}$ and $\sqrt{3}$. The orthonormal basis is

$$\boldsymbol{q}_1 = \frac{1}{\sqrt{2}}\begin{bmatrix} 1 \\ -1 \\ 0 \end{bmatrix} \quad \text{and} \quad \boldsymbol{q}_2 = \frac{1}{\sqrt{6}}\begin{bmatrix} 1 \\ 1 \\ -2 \end{bmatrix} \quad \text{and} \quad \boldsymbol{q}_3 = \frac{1}{\sqrt{3}}\begin{bmatrix} 1 \\ 1 \\ 1 \end{bmatrix}.$$

Usually $\boldsymbol{A}, \boldsymbol{B}, \boldsymbol{C}$ contain fractions. Almost always $\boldsymbol{q}_1, \boldsymbol{q}_2, \boldsymbol{q}_3$ contain square roots. These vectors had $m = 3$ components, but the steps of Gram-Schmidt are always the same. Subtract projections and then change to unit vectors.

The Factorization $A = QR$

We started with a matrix A, whose columns were $\boldsymbol{a}, \boldsymbol{b}, \boldsymbol{c}$. We ended with a matrix Q, whose columns are $\boldsymbol{q}_1, \boldsymbol{q}_2, \boldsymbol{q}_3$. How are those matrices related? Since the vectors $\boldsymbol{a}, \boldsymbol{b}, \boldsymbol{c}$ are combinations of the $\boldsymbol{q}$'s (and vice versa), there must be a third matrix connecting A to Q. Call it R.

The first step was $\boldsymbol{q}_1 = \boldsymbol{a}/\|\boldsymbol{a}\|$ (other vectors not involved). The second equation was (4.31), where $\boldsymbol{b}$ is a combination of $\boldsymbol{A}$ and $\boldsymbol{B}$. At that stage $\boldsymbol{C}$ and $\boldsymbol{q}_3$ were not involved. This non-involvement of later vectors is the key point of the process:

- The vectors $\boldsymbol{a}$ and $\boldsymbol{A}$ and $\boldsymbol{q}_1$ are along a single line.
- The vectors $\boldsymbol{a}, \boldsymbol{b}$ and $\boldsymbol{A}, \boldsymbol{B}$ and $\boldsymbol{q}_1, \boldsymbol{q}_2$ are in a single plane.
- The vectors $\boldsymbol{a}, \boldsymbol{b}, \boldsymbol{c}$ and $\boldsymbol{A}, \boldsymbol{B}, \boldsymbol{C}$ and $\boldsymbol{q}_1, \boldsymbol{q}_2, \boldsymbol{q}_3$ are in a subspace (dimension 3).

At every step, $\boldsymbol{a}_1, \ldots, \boldsymbol{a}_k$ are combinations of $\boldsymbol{q}_1, \ldots, \boldsymbol{q}_k$. Later $\boldsymbol{q}$'s are not involved. The connecting matrix R is a *triangular matrix*:

$$\begin{bmatrix} a & b & c \end{bmatrix} = \begin{bmatrix} q_1 & q_2 & q_3 \end{bmatrix} \begin{bmatrix} q_1^{\mathrm{T}}a & q_1^{\mathrm{T}}b & q_1^{\mathrm{T}}c \\ & q_2^{\mathrm{T}}b & q_2^{\mathrm{T}}c \\ & & q_3^{\mathrm{T}}c \end{bmatrix} \quad \text{or} \quad A = QR. \tag{4.33}$$

$A = QR$ is Gram-Schmidt in a nutshell. To understand those dot products in R, premultiply $A = QR$ by Q^{T}. Since $Q^{\mathrm{T}}Q = I$ this leaves $Q^{\mathrm{T}}A = R$. The entries of $Q^{\mathrm{T}}A$ are rows times columns. They are dot products of $\boldsymbol{q}_1, \boldsymbol{q}_2, \boldsymbol{q}_3$ with $\boldsymbol{a}, \boldsymbol{b}, \boldsymbol{c}$. That is what you see in R.

Look in particular at the second column of $A = QR$. The matrix multiplication is saying that $\boldsymbol{b} = \boldsymbol{q}_1(\boldsymbol{q}_1^{\mathrm{T}}\boldsymbol{b}) + \boldsymbol{q}_2(\boldsymbol{q}_2^{\mathrm{T}}\boldsymbol{b})$. This is exactly like equation (4.29), except that the third term $\boldsymbol{q}_3(\boldsymbol{q}_3^{\mathrm{T}}\boldsymbol{b})$ is not present. At the second stage of Gram-Schmidt, $\boldsymbol{q}_3$ is not involved! The dot product $\boldsymbol{q}_3^{\mathrm{T}}\boldsymbol{b}$ is zero in column 2 of R.

4K (Gram-Schmidt and $A = QR$) The Gram-Schmidt process starts with independent vectors $\boldsymbol{a}_1, \ldots, \boldsymbol{a}_n$. It constructs orthonormal vectors $\boldsymbol{q}_1, \ldots, \boldsymbol{q}_n$. The matrices with these columns satisfy $A = QR$. R is upper triangular because later $\boldsymbol{q}$'s are orthogonal to earlier $\boldsymbol{a}$'s and $R_{ij} = \boldsymbol{q}_i^{\mathrm{T}}\boldsymbol{a}_j = 0$.

Here are the $\boldsymbol{a}$'s and $\boldsymbol{q}$'s from the example. The i, j entry of $R = Q^{\mathrm{T}}A$ is the dot product of $\boldsymbol{q}_i$ with $\boldsymbol{a}_j$:

$$A = \begin{bmatrix} 1 & 2 & 3 \\ -1 & 0 & -3 \\ 0 & -2 & 3 \end{bmatrix} = \begin{bmatrix} 1/\sqrt{2} & 1/\sqrt{6} & 1/\sqrt{3} \\ -1/\sqrt{2} & 1/\sqrt{6} & 1/\sqrt{3} \\ 0 & -2/\sqrt{6} & 1/\sqrt{3} \end{bmatrix} \begin{bmatrix} \sqrt{2} & \sqrt{2} & \sqrt{18} \\ 0 & \sqrt{6} & -\sqrt{6} \\ 0 & 0 & \sqrt{3} \end{bmatrix} = QR.$$

The lengths of $\boldsymbol{A}, \boldsymbol{B}, \boldsymbol{C}$ are the numbers $\sqrt{2}, \sqrt{6}, \sqrt{3}$ on the diagonal of R. Because of the square roots, QR looks less beautiful than LU. Both factorizations are absolutely central to calculations in linear algebra.

Any m by n matrix A with independent columns can be factored into QR. The m by n matrix Q has orthonormal columns, and the square matrix R is upper triangular with positive diagonal. We must not forget why this is useful for least squares: ***$A^{\mathrm{T}}A$ equals $R^{\mathrm{T}}Q^{\mathrm{T}}QR = R^{\mathrm{T}}R$.*** The least-squares equation $A^{\mathrm{T}}A\hat{x} = A^{\mathrm{T}}\boldsymbol{b}$ simplifies to

$$R^{\mathrm{T}}R\hat{x} = R^{\mathrm{T}}Q^{\mathrm{T}}\boldsymbol{b} \quad \text{or} \quad R\hat{x} = Q^{\mathrm{T}}\boldsymbol{b}. \tag{4.34}$$

Instead of solving $A\boldsymbol{x} = \boldsymbol{b}$, which is impossible, we solve $R\hat{\boldsymbol{x}} = Q^{\mathrm{T}}\boldsymbol{b}$ by back substitution—which is very fast. The real cost is the mn^2 multiplications in the Gram-Schmidt process, which are needed to find Q and R.

Here is an informal code. It executes equations (4.35) and (4.36), for $k = 1$ then $k = 2$ and eventually $k = n$. Equation (4.35) divides orthogonal vectors by their lengths r_{kk} to get unit vectors. The orthogonal vectors are called $\boldsymbol{a}_1, \boldsymbol{a}_2, \boldsymbol{a}_3, \ldots$ like the original vectors, but equation (4.36) has put $\boldsymbol{A}, \boldsymbol{B}, \boldsymbol{C}, \ldots$ in their place. At the start, $k = 1$ and $\boldsymbol{a}_1 = \boldsymbol{A}$ has components a_{i1}:

$$r_{kk} = \left(\sum_{i=1}^{m} a_{ik}^2\right)^{1/2} \quad \text{and} \quad q_{ik} = \frac{a_{ik}}{r_{kk}} \quad i = 1, \ldots, m. \tag{4.35}$$

From each later $\boldsymbol{a}_j$ $(j = k+1, \ldots, n)$, subtract its projection on this new $\boldsymbol{q}_k$. The dot product $\boldsymbol{q}_k \cdot \boldsymbol{a}_j$ is r_{kj} in the triangular matrix R:

$$r_{kj} = \sum_{i=1}^{m} q_{ik} a_{ij} \quad \text{and} \quad a_{ij} = a_{ij} - q_{ik} r_{kj} \quad i = 1, \ldots, m. \tag{4.36}$$

Increase k to $k+1$ and return to (4.35). Starting from $\boldsymbol{a}, \boldsymbol{b}, \boldsymbol{c} = \boldsymbol{a}_1, \boldsymbol{a}_2, \boldsymbol{a}_3$ this code will

1 Construct $\boldsymbol{q}_1 = \boldsymbol{a}_1 / \|\boldsymbol{a}_1\|$.

2 Construct $\boldsymbol{B} = \boldsymbol{a}_2 - (\boldsymbol{q}_1^{\mathrm{T}} \boldsymbol{a}_2)\boldsymbol{q}_1$ and $\boldsymbol{C}^* = \boldsymbol{a}_3 - (\boldsymbol{q}_1^{\mathrm{T}} \boldsymbol{a}_3)\boldsymbol{q}_1$.

3 Construct $\boldsymbol{q}_2 = \boldsymbol{B} / \|\boldsymbol{B}\|$.

4 Construct $\boldsymbol{C} = \boldsymbol{C}^* - (\boldsymbol{q}_2^{\mathrm{T}} \boldsymbol{C}^*)\boldsymbol{q}_2$.

5 Construct $\boldsymbol{q}_3 = \boldsymbol{C} / \|\boldsymbol{C}\|$.

Equation (4.36) subtracts off projections as soon as the new vector $\boldsymbol{q}_k$ is found. This change to "one projection at a time" is called ***modified Gram-Schmidt***. It is numerically more stable than equation (4.32) which subtracts all projections at once.

Problem Set 4.4

Problems 1–12 are about orthogonal vectors and orthogonal matrices.

1 Are these pairs of vectors orthonormal or only orthogonal or only independent?

(a) $\begin{bmatrix} 1 \\ 0 \end{bmatrix}$ and $\begin{bmatrix} -1 \\ 1 \end{bmatrix}$ (b) $\begin{bmatrix} .6 \\ .8 \end{bmatrix}$ and $\begin{bmatrix} .4 \\ -.3 \end{bmatrix}$ (c) $\begin{bmatrix} \cos\theta \\ \sin\theta \end{bmatrix}$ and $\begin{bmatrix} -\sin\theta \\ \cos\theta \end{bmatrix}$.

Change the second vector when necessary to produce orthonormal vectors.

2 The vectors $(2, 2, -1)$ and $(-1, 2, 2)$ are orthogonal. Divide them by their lengths to find orthonormal vectors $\boldsymbol{q}_1$ and $\boldsymbol{q}_2$. Put those into the columns of Q and multiply $Q^{\mathrm{T}}Q$ and QQ^{T}.

3 (a) If A has three orthogonal columns each of length 4, what is $A^{\mathrm{T}}A$?

(b) If A has three orthogonal columns of lengths $1, 2, 3$, what is $A^{\mathrm{T}}A$?

4 Give an example of each of the following:

(a) A matrix Q that has orthonormal columns but $QQ^{\mathrm{T}} \neq I$.

(b) Two orthogonal vectors that are not linearly independent.

(c) An orthonormal basis for $\mathbf{R}^4$, where every component is $\frac{1}{2}$ or $-\frac{1}{2}$.

5 Find two orthogonal vectors that lie in the plane $x + y + 2z = 0$. Make them orthonormal.

6 If Q_1 and Q_2 are orthogonal matrices, show that their product $Q_1 Q_2$ is also an orthogonal matrix. (Use $Q^{\mathrm{T}} Q = I$.)

7 If the columns of Q are orthonormal, what is the least-squares solution $\hat{\boldsymbol{x}}$ to $Q\boldsymbol{x} = \boldsymbol{b}$? Give an example with $\boldsymbol{b} \neq 0$ but $\hat{\boldsymbol{x}} = \mathbf{0}$.

8 (a) Compute $P = QQ^{\mathrm{T}}$ when $\boldsymbol{q}_1 = (.8, .6, 0)$ and $\boldsymbol{q}_2 = (-.6, .8, 0)$. Verify that $P^2 = P$.

(b) Prove that always $(QQ^{\mathrm{T}})(QQ^{\mathrm{T}}) = QQ^{\mathrm{T}}$ by using $Q^{\mathrm{T}} Q = I$. $P = QQ^{\mathrm{T}}$ is the projection matrix onto the column space of Q.

9 Orthonormal vectors are automatically linearly independent.

(a) Vector proof: When $c_1\boldsymbol{q}_1 + c_2\boldsymbol{q}_2 + c_3\boldsymbol{q}_3 = \mathbf{0}$, what dot product leads to $c_1 = 0$? Similarly $c_2 = 0$ and $c_3 = 0$ and the $\boldsymbol{q}$'s are independent.

(b) Matrix proof: Show that $Q\boldsymbol{x} = \mathbf{0}$ leads to $\boldsymbol{x} = \mathbf{0}$. Since Q may be rectangular, you can use Q^{T} but not Q^{-1}.

10 (a) Find orthonormal vectors $\boldsymbol{q}_1$ and $\boldsymbol{q}_2$ in the plane of $\boldsymbol{a} = (1, 3, 4, 5, 7)$ and $\boldsymbol{b} = (-6, 6, 8, 0, 8)$.

(b) Which vector in this plane is closest to $(1, 0, 0, 0, 0)$?

11 If $\boldsymbol{q}_1$ and $\boldsymbol{q}_2$ are orthonormal vectors in $\mathbf{R}^5$, what combination ____ $\boldsymbol{q}_1 +$ ____ $\boldsymbol{q}_2$ is closest to a given vector $\boldsymbol{b}$?

12 If $\boldsymbol{a}_1, \boldsymbol{a}_2, \boldsymbol{a}_3$ is a basis for $\mathbf{R}^3$, any vector $\boldsymbol{b}$ can be written as

$$\boldsymbol{b} = x_1\boldsymbol{a}_1 + x_2\boldsymbol{a}_2 + x_3\boldsymbol{a}_3 \qquad \text{or} \qquad \begin{bmatrix} \boldsymbol{a}_1 & \boldsymbol{a}_2 & \boldsymbol{a}_3 \end{bmatrix} \begin{bmatrix} x_1 \\ x_2 \\ x_3 \end{bmatrix} = \boldsymbol{b}.$$

(a) Suppose the $\boldsymbol{a}$'s are orthonormal. Show that $x_1 = \boldsymbol{a}_1^{\mathrm{T}}\boldsymbol{b}$.

(b) Suppose the $\boldsymbol{a}$'s are orthogonal. Show that $x_1 = \boldsymbol{a}_1^{\mathrm{T}}\boldsymbol{b}/\boldsymbol{a}_1^{\mathrm{T}}\boldsymbol{a}_1$.

(c) If the $\boldsymbol{a}$'s are independent, x_1 is the first component of ____ times $\boldsymbol{b}$.

Problems 13–24 are about the Gram-Schmidt process and $A = QR$.

13 What multiple of $\boldsymbol{a} = \left[\begin{smallmatrix} 1 \\ 1 \end{smallmatrix}\right]$ should be subtracted from $\boldsymbol{b} = \left[\begin{smallmatrix} 4 \\ 0 \end{smallmatrix}\right]$ to make the result $\boldsymbol{B}$ orthogonal to $\boldsymbol{a}$? Sketch a figure to show $\boldsymbol{a}$, $\boldsymbol{b}$, and $\boldsymbol{B}$.

14 Complete the Gram-Schmidt process in Problem 13 by computing $q_1 = a/\|a\|$ and $q_2 = B/\|B\|$ and factoring into QR:

$$\begin{bmatrix} 1 & 4 \\ 1 & 0 \end{bmatrix} = \begin{bmatrix} q_1 & q_2 \end{bmatrix} \begin{bmatrix} \|a\| & ? \\ 0 & \|B\| \end{bmatrix}.$$

15 (a) Find orthonormal vectors q_1, q_2, q_3 such that q_1, q_2 span the column space of

$$A = \begin{bmatrix} 1 & 1 \\ 2 & -1 \\ -2 & 4 \end{bmatrix}.$$

(b) Which of the four fundamental subspaces contains q_3?

(c) Solve $Ax = (1, 2, 7)$ by least squares.

16 What multiple of $a = (4, 5, 2, 2)$ is closest to $b = (1, 2, 0, 0)$? Find orthonormal vectors q_1 and q_2 in the plane of a and b.

17 Find the projection of b onto the line through a:

$$a = \begin{bmatrix} 1 \\ 1 \\ 1 \end{bmatrix} \quad \text{and} \quad b = \begin{bmatrix} 1 \\ 3 \\ 5 \end{bmatrix} \quad \text{and} \quad p = ? \quad \text{and} \quad e = b - p = ?$$

Compute the orthonormal vectors $q_1 = a/\|a\|$ and $q_2 = e/\|e\|$.

18 If $A = QR$ then $A^{\mathrm{T}}A = R^{\mathrm{T}}R =$ ______ triangular times ______ triangular. *Gram-Schmidt on A corresponds to elimination on* $A^{\mathrm{T}}A$. Compare the pivots for $A^{\mathrm{T}}A$ with $\|a\|^2 = 3$ and $\|e\|^2 = 8$ in Problem 17:

$$A = \begin{bmatrix} 1 & 1 \\ 1 & 3 \\ 1 & 5 \end{bmatrix} \quad \text{and} \quad A^{\mathrm{T}}A = \begin{bmatrix} 3 & 9 \\ 9 & 35 \end{bmatrix}.$$

19 True or false (give an example in either case):

(a) The inverse of an orthogonal matrix is an orthogonal matrix.

(b) If Q (3 by 2) has orthonormal columns then $\|Qx\|$ always equals $\|x\|$.

20 Find an orthonormal basis for the column space of A:

$$A = \begin{bmatrix} 1 & -2 \\ 1 & 0 \\ 1 & 1 \\ 1 & 3 \end{bmatrix} \quad \text{and} \quad b = \begin{bmatrix} -4 \\ -3 \\ 3 \\ 0 \end{bmatrix}.$$

Then compute the projection of b onto that column space.

21 Find orthogonal vectors $\boldsymbol{A}, \boldsymbol{B}, \boldsymbol{C}$ by Gram-Schmidt from

$$\boldsymbol{a} = \begin{bmatrix} 1 \\ 1 \\ 2 \end{bmatrix} \quad \text{and} \quad \boldsymbol{b} = \begin{bmatrix} 1 \\ -1 \\ 0 \end{bmatrix} \quad \text{and} \quad \boldsymbol{c} = \begin{bmatrix} 1 \\ 0 \\ 4 \end{bmatrix}.$$

22 Find $\boldsymbol{q}_1, \boldsymbol{q}_2, \boldsymbol{q}_3$ (orthonormal) as combinations of $\boldsymbol{a}, \boldsymbol{b}, \boldsymbol{c}$ (independent columns). Then write A as QR:

$$A = \begin{bmatrix} 1 & 2 & 4 \\ 0 & 0 & 5 \\ 0 & 3 & 6 \end{bmatrix}.$$

23 (a) Find a basis for the subspace $\boldsymbol{S}$ in $\mathbf{R}^4$ spanned by all solutions of

$$x_1 + x_2 + x_3 - x_4 = 0.$$

(b) Find a basis for the orthogonal complement $\boldsymbol{S}^{\perp}$.

(c) Find $\boldsymbol{b}_1$ in $\boldsymbol{S}$ and $\boldsymbol{b}_2$ in $\boldsymbol{S}^{\perp}$ so that $\boldsymbol{b}_1 + \boldsymbol{b}_2 = \boldsymbol{b} = (1, 1, 1, 1)$.

24 If $ad - bc > 0$, the entries in $A = QR$ are

$$\begin{bmatrix} a & b \\ c & d \end{bmatrix} = \frac{\begin{bmatrix} a & -c \\ c & a \end{bmatrix}}{\sqrt{a^2 + c^2}} \frac{\begin{bmatrix} a^2 + c^2 & ab + cd \\ 0 & ad - bc \end{bmatrix}}{\sqrt{a^2 + c^2}}.$$

Write down $A = QR$ when $a, b, c, d = 2, 1, 1, 1$ and also $1, 1, 1, 1$. Which entry becomes zero when Gram-Schmidt breaks down?

Problems 25–28 are about the QR code in equations (11–12). It executes Gram-Schmidt.

25 Show why $\boldsymbol{C}$ (found via $\boldsymbol{C}^*$ in the steps after (4.36)) is equal to $\boldsymbol{C}$ in equation (4.32).

26 Equation (4.32) subtracts from $\boldsymbol{c}$ its components along $\boldsymbol{A}$ and $\boldsymbol{B}$. Why not subtract the components along $\boldsymbol{a}$ and along $\boldsymbol{b}$?

27 Write a working code and apply it to $\boldsymbol{a} = (2, 2, -1)$, $\boldsymbol{b} = (0, -3, 3)$, $\boldsymbol{c} = (1, 0, 0)$. What are the $\boldsymbol{q}$'s?

28 Where are the mn^2 multiplications in equations (4.35) and (4.36)?

Problems 29–32 involve orthogonal matrices that are special.

29 (a) Choose c so that Q is an orthogonal matrix:

$$Q = c \begin{bmatrix} 1 & -1 & -1 & -1 \\ -1 & 1 & -1 & -1 \\ -1 & -1 & 1 & -1 \\ -1 & -1 & -1 & 1 \end{bmatrix}.$$

(b) Change the first row and column to all 1's and fill in another orthogonal matrix.

30 Project $\boldsymbol{b} = (1, 1, 1, 1)$ onto the first column in Problem 29(a). Then project $\boldsymbol{b}$ onto the plane of the first two columns.

31 If $\boldsymbol{u}$ is a unit vector, then $Q = I - 2\boldsymbol{u}\boldsymbol{u}^{\mathrm{T}}$ is a reflection matrix (Example 4.13). Find Q from $\boldsymbol{u} = (0, 1)$ and also from $\boldsymbol{u} = (0, \sqrt{2}/2, \sqrt{2}/2)$. Draw figures to show reflections of (x, y) and (x, y, z).

32 $Q = I - 2\boldsymbol{u}\boldsymbol{u}^{\mathrm{T}}$ is a reflection matrix when $\boldsymbol{u}^{\mathrm{T}}\boldsymbol{u} = 1$.

(a) Prove that $Q\boldsymbol{u} = -\boldsymbol{u}$. The mirror is perpendicular to $\boldsymbol{u}$.

(b) Find $Q\boldsymbol{v}$ when $\boldsymbol{u}^{\mathrm{T}}\boldsymbol{v} = 0$. The mirror contains $\boldsymbol{v}$.

5

DETERMINANTS

5.1 The Properties of Determinants

The determinant of a square matrix is a single number. That number contains an amazing amount of information about the matrix. It tells immediately whether the matrix is invertible. ***The determinant is zero when the matrix has no inverse***. When A is invertible, the determinant of A^{-1} is $1/(\det A)$. If $\det A = 2$ then $\det A^{-1} = \frac{1}{2}$. In fact the determinant leads to a formula for every entry in A^{-1}.

This is one use for determinants—to find formulas for inverses and pivots and solutions $A^{-1}\boldsymbol{b}$. For a matrix of numbers, we seldom use those formulas. (Or rather, we use elimination as the quickest way to evaluate the answer.) For a matrix with entries a, b, c, d, its determinant shows how A^{-1} changes as A changes:

$$A = \begin{bmatrix} a & b \\ c & d \end{bmatrix} \quad \text{has inverse} \quad A^{-1} = \frac{1}{ad - bc} \begin{bmatrix} d & -b \\ -c & a \end{bmatrix}. \tag{5.1}$$

Multiply those matrices to get I. The determinant of A is $ad - bc$. When $\det A = 0$, we are asked to divide by zero and we can't—then A has no inverse. (The rows are parallel when $a/c = b/d$. This gives $ad = bc$ and a zero determinant.) Dependent rows lead to $\det A = 0$.

There is also a connection to the pivots, which are a and $d - (c/a)b$. The product of the two pivots is the determinant:

$$a\left(d - \frac{c}{a}b\right) = ad - bc \quad \text{which is} \quad \det A.$$

After a row exchange the pivots are c and $b - (a/c)d$. Those pivots multiply to give *minus* the determinant.

Looking ahead The determinant of an n by n matrix can be found in three ways:

1 Multiply the n pivots (times 1 or -1).

2 Add up $n!$ terms (times 1 or -1).

3 Combine n smaller determinants (times 1 or -1).

You see that plus or minus signs—the decisions between 1 and -1—play a very big part in determinants. That comes from the following rule:

The determinant changes sign when two rows* (or two columns) *are exchanged.

The identity matrix has determinant $+1$. Exchange two rows and $\det P = -1$. Exchange two more rows and the new permutation has $\det P = +1$. Half of all permutations are *even* ($\det P = 1$) and half are *odd* ($\det P = -1$). Starting from I, half of the P's involve an even number of exchanges and half require an odd number. In the 2 by 2 case, ad has a plus sign and bc has minus—coming from the row exchange:

$$\det\begin{bmatrix}1 & 0\\ 0 & 1\end{bmatrix} = 1 \quad \text{and} \quad \det\begin{bmatrix}0 & 1\\ 1 & 0\end{bmatrix} = -1.$$

The other essential rule is linearity—but a warning comes first. Linearity does not mean that $\det(A+B) = \det A + \det B$. ***This is absolutely false***. That kind of linearity is not even true when $A = I$ and $B = I$. The false rule would say that $\det 2I = 1 + 1 = 2$. The true rule is $\det 2I = 2^n$. Determinants are multiplied by t^n (not just by t) when matrices are multiplied by t. But we are getting ahead of ourselves. In the choice between defining the determinant by its properties or its formulas, we choose its properties—*sign reversal and linearity*. The properties are simple (Section 5.1). They prepare for the formulas (Section 5.2). Then come the applications, including these three:

(A1) Determinants give A^{-1} and $A^{-1}\boldsymbol{b}$ (by *Cramer's Rule*).

(A2) The *volume* of an n-dimensional box is $|\det A|$, when the edges of the box come from the rows of A.

(A3) The numbers λ for which $\det(A - \lambda I) = 0$ are the *eigenvalues* of A. This is the most important application and it fills Chapter 6.

The Properties of the Determinant

There are three basic properties (rules 1, 2, 3). By using them we can compute the determinant of any square matrix A. ***This number is written in two ways,* det *A* and |*A*|.** Notice: Brackets for the matrix, straight bars for its determinant. When A is a 2 by 2 matrix, the three properties lead to the answer we expect:

$$\text{The determinant of} \quad \begin{bmatrix}a & b\\ c & d\end{bmatrix} \quad \text{is} \quad \begin{vmatrix}a & b\\ c & d\end{vmatrix} = ad - bc.$$

We will check each rule against this 2 by 2 formula, but do not forget: The rules apply to any n by n matrix. When we prove that properties 4–10 follow from 1–3, the proof must apply to all square matrices.

1 ***The determinant is a linear function of the first row*** (when the other rows stay fixed). If the first row is multiplied by t, the determinant is multiplied by t. If first rows are added, determinants are added. This rule only applies when the other rows do not change! Notice how $[\,c \quad d\,]$ stays the same:

$$\textit{multiply row 1 by } t\text{:} \quad \begin{vmatrix} ta & tb \\ c & d \end{vmatrix} = t \begin{vmatrix} a & b \\ c & d \end{vmatrix}$$

$$\textit{add row 1 of } A \text{ to row 1 of } A'\text{:} \quad \begin{vmatrix} a+a' & b+b' \\ c & d \end{vmatrix} = \begin{vmatrix} a & b \\ c & d \end{vmatrix} + \begin{vmatrix} a' & b' \\ c & d \end{vmatrix}.$$

In the first case, both sides are $tad - tbc$. Then t factors out. In the second case, both sides are $ad + a'd - bc - b'c$. These rules still apply when A is n by n, and the last $n-1$ rows don't change. May we emphasize this with numbers:

$$\begin{vmatrix} t & 0 & 0 \\ 0 & 1 & 0 \\ 0 & 0 & 1 \end{vmatrix} = t \begin{vmatrix} 1 & 0 & 0 \\ 0 & 1 & 0 \\ 0 & 0 & 1 \end{vmatrix} \quad \text{and} \quad \begin{vmatrix} 1 & 2 & 3 \\ 0 & 1 & 0 \\ 0 & 0 & 1 \end{vmatrix} = \begin{vmatrix} 1 & 0 & 0 \\ 0 & 1 & 0 \\ 0 & 0 & 1 \end{vmatrix} + \begin{vmatrix} 0 & 2 & 3 \\ 0 & 1 & 0 \\ 0 & 0 & 1 \end{vmatrix}.$$

By itself, rule 1 does not say what any of those determinants are. It just says that the determinants must pass these two tests for linearity.

Combining multiplication and addition, we get any linear combination in the first row: t(row 1 of A) $+\, t'$(row 1 of A'). With this combined row, the determinant is t times $\det A$ plus t' times $\det A'$. Eventually the other rows have to get into the picture. Rule 2 does that.

2 ***The determinant changes sign when two rows are exchanged*** (sign reversal):

$$\text{Check:} \quad \begin{vmatrix} c & d \\ a & b \end{vmatrix} = - \begin{vmatrix} a & b \\ c & d \end{vmatrix} \quad \text{(both sides equal } bc - ad\text{)}.$$

Because of this rule, there is nothing special about row 1. The determinant is a *linear function of each row separately*. If row 2 is multiplied by t, so is $\det A$.

Proof Exchange rows 1 and 2. Then multiply the new row 1 by t. Then exchange back. The determinant is multiplied by -1 then t then -1, in other words by t—as we expected.

To operate on any row we can exchange it with the first row—then use rule 1 for the first row and exchange back. This still does not mean that $\det 2I = 2 \det I$. To obtain $2I$ we have to multiply *both* rows by 2, and the factor 2 comes out both times:

$$\begin{vmatrix} 2 & 0 \\ 0 & 2 \end{vmatrix} = 2^2 = 4 \quad \text{and} \quad \begin{vmatrix} t & & \\ & \ddots & \\ & & t \end{vmatrix} = t^n.$$

This is just like area and volume. Expand a rectangle by 2 and its area increases by 4. Expand an n-dimensional box by t and its volume increases by t^n. The connection is no accident—we will see how *determinants equal volumes*.

Property 3 (the easiest rule) matches the determinant of I with the volume of a unit cube.

3 ***The determinant of the n by n identity matrix is* 1.**

$$\begin{vmatrix} 1 & 0 \\ 0 & 1 \end{vmatrix} = 1 \qquad \text{and} \qquad \begin{vmatrix} 1 & & \\ & \ddots & \\ & & 1 \end{vmatrix} = 1.$$

Rules 1 and 2 left a scaling constant undecided. Rule 3 decides it by $\det I = 1$.

Pay special attention to rules 1–3. They completely determine the number $\det A$—but for a big matrix that fact is not obvious. We could stop here to find a formula for n by n determinants. It would be a little complicated—we prefer to go gradually. Instead we write down other properties which follow directly from the first three. These extra rules make determinants much easier to work with.

4 ***If two rows of A are equal, then* $\det A = 0$.**

$$\text{Check 2 by 2:} \quad \begin{vmatrix} a & b \\ a & b \end{vmatrix} = 0.$$

Rule 4 follows from rule 2. (Remember we must use the rules and not the 2 by 2 formula.) *Exchange the two equal rows.* The determinant D is supposed to change sign. But also D has to stay the same, because the matrix is not changed. The only number with $-D = D$ is $D = 0$—this must be the determinant. (Note: In Boolean algebra the reasoning fails, because $-1 = 1$. Then D is defined by rules 1, 3, 4.)

A matrix with two equal rows has no inverse. Rule 4 makes $\det A = 0$. But matrices can be singular and determinants can be zero without having equal rows! Rule 5 will be the key.

5 ***Subtracting a multiple of one row from another row leaves* $\det A$ *unchanged.***

$$\begin{vmatrix} a & b \\ c - la & d - lb \end{vmatrix} = \begin{vmatrix} a & b \\ c & d \end{vmatrix}.$$

Linearity splits the left side into the right side plus another term $-l\left[\begin{smallmatrix} a & b \\ a & b \end{smallmatrix}\right]$. This extra term is zero by rule 4. Therefore rule 5 is correct. Note how only one row changes while the others stay the same—as required by rule 1.

Conclusion *The determinant is not changed by the usual elimination steps*: $\det A$ *equals* $\det U$. If we can find determinants of triangular matrices U, we can find determinants of all matrices A. Every row exchange reverses the sign, so always $\det A = \pm \det U$. We have narrowed the problem to triangular matrices.

6 ***A matrix with a row of zeros has* $\det A = 0$.**

$$\begin{vmatrix} 0 & 0 \\ c & d \end{vmatrix} = 0 \qquad \text{and} \qquad \begin{vmatrix} a & b \\ 0 & 0 \end{vmatrix} = 0.$$

For an easy proof, add some other row to the zero row. The determinant is not changed (rule 5). But the matrix now has two equal rows. So $\det A = 0$ by rule 4.

7 ***If A is a triangular matrix then* det $A = a_{11}a_{22}\cdots a_{nn}$ *=product of diagonal entries.***

$$\begin{vmatrix} a & b \\ 0 & d \end{vmatrix} = ad \qquad \text{and also} \qquad \begin{vmatrix} a & 0 \\ c & d \end{vmatrix} = ad.$$

Suppose all diagonal entries of A are nonzero. Eliminate the off-diagonal entries by the usual steps. (If A is lower triangular, subtract multiples of each row from lower rows. If A is upper triangular, subtract from rows above.) By rule 5 the determinant is not changed—and now the matrix is diagonal:

$$\text{We must still prove that} \quad \begin{vmatrix} a_{11} & & & 0 \\ & a_{22} & & \\ & & \ddots & \\ 0 & & & a_{nn} \end{vmatrix} = a_{11}a_{22}\cdots a_{nn}.$$

For this we apply rules 1 and 3. Factor a_{11} from the first row. Then factor a_{22} from the second row. Eventually factor a_{nn} from the last row. The determinant is a_{11} times a_{22} times $\cdots$ times a_{nn} times $\det I$. Then rule 3 (used at last!) is $\det I = 1$.

What if a diagonal entry of the triangular matrix is zero? Then the matrix is singular. Elimination will produce a *zero row*. By rule 5 the determinant is unchanged, and by rule 6 a zero row means $\det A = 0$. Thus rule 7 is proved—the determinants of triangular matrices come directly from their main diagonals.

8 ***If A is singular then* det $A = 0$. *If A is invertible then* det $A \neq 0$.**

$$\begin{bmatrix} a & b \\ c & d \end{bmatrix} \quad \text{is singular if and only if} \quad ad - bc = 0.$$

Proof By elimination go from A to U. If A is singular then U has a zero row. The rules give $\det A = \det U = 0$. If A is invertible then U has nonzeros (the pivots) along its diagonal. The product of nonzero pivots (using rule 7) gives a nonzero determinant:

$$\det A = \pm \det U = \pm \text{ (product of the pivots)}.$$

This is the first formula for the determinant. A computer would use it to find $\det A$ from the pivots. The plus or minus sign depends on whether the number of row exchanges is even or odd. In other words, $+1$ or -1 is the determinant of the permutation matrix P that does the row exchanges. With no row exchanges, the number zero is even and $P = I$ and $\det A = \det U$. Note that always $\det L = 1$, because L is triangular with 1's on the diagonal. What we have is this:

$$\text{If} \quad PA = LU \quad \text{then} \quad \det P \det A = \det L \det U. \tag{5.2}$$

Again, $\det P = \pm 1$ and $\det A = \pm \det U$. Equation (5.2) is our first case of rule 9.

9 ***The determinant of AB is the product of the separate determinants: $|AB| = |A|\,|B|$.***

$$\begin{vmatrix} a & b \\ c & d \end{vmatrix} \begin{vmatrix} p & q \\ r & s \end{vmatrix} = \begin{vmatrix} ap+br & aq+bs \\ cp+dr & cq+ds \end{vmatrix}.$$

In particular—when the matrix B is A^{-1}—*the determinant of A^{-1} is $1/\det A$*:

$$AA^{-1} = I \quad \text{so} \quad (\det A)(\det A^{-1}) = \det I = 1.$$

This product rule is the most intricate so far. We could check the 2 by 2 case by algebra:

$$(ad-bc)(ps-qr) = (ap+br)(cq+ds) - (aq+bs)(cp+dr).$$

For the n by n case, here is a snappy proof that $|AB| = |A|\,|B|$. The idea is to consider the ratio $D(A) = |AB|/|B|$. If this ratio has properties 1–3—which we now check—it must equal the determinant $|A|$. (The case $|B| = 0$ is separate and easy, because AB is singular when B is singular. The rule $|AB| = |A|\,|B|$ becomes $0 = 0$.) Here are the three properties of the ratio $|AB|/|B|$:

Property 3 (Determinant of I): If $A = I$ then the ratio becomes $|B|/|B| = 1$.

Property 2 (Sign reversal): When two rows of A are exchanged, so are the same two rows of AB. Therefore $|AB|$ changes sign and so does the ratio $|AB|/|B|$.

Property 1 (Linearity): When row 1 of A is multiplied by t, so is row 1 of AB. This multiplies $|AB|$ by t and multiplies the ratio by t—as desired. Now suppose row 1 of A is added to row 1 of A' (the other rows staying the same throughout). Then row 1 of AB is added to row 1 of $A'B$. By rule 1, the determinants add. After dividing by $|B|$, the ratios add.

Conclusion This ratio $|AB|/|B|$ has the same three properties that define $|A|$. Therefore it equals $|A|$. This proves the product rule $|AB| = |A|\,|B|$.

10 ***The transpose A^{T} has the same determinant as A.***

Check: $\begin{vmatrix} a & b \\ c & d \end{vmatrix} = \begin{vmatrix} a & c \\ b & d \end{vmatrix}$ since both sides equal $ad - bc$.

The equation $|A^{\mathrm{T}}| = |A|$ becomes $0 = 0$ when A is singular (we know that A^{T} is also singular). Otherwise A has the usual factorization $PA = LU$. Transposing both sides gives $A^{\mathrm{T}}P^{\mathrm{T}} = U^{\mathrm{T}}L^{\mathrm{T}}$. The proof of $|A| = |A^{\mathrm{T}}|$ comes by using rule 9 for products and comparing:

$$\det P \det A = \det L \det U \quad \text{and} \quad \det A^{\mathrm{T}} \det P^{\mathrm{T}} = \det U^{\mathrm{T}} \det L^{\mathrm{T}}.$$

First, $\det L = \det L^{\mathrm{T}}$ (both have 1's on the diagonal). Second, $\det U = \det U^{\mathrm{T}}$ (transposing leaves the main diagonal unchanged, and triangular determinants only involve that diagonal). Third, $\det P = \det P^{\mathrm{T}}$ (permutations have $P^{\mathrm{T}} = P^{-1}$, so $|P|\,|P^{\mathrm{T}}| = 1$ by rule 9;

thus $|P|$ and $|P^{\mathrm{T}}|$ both equal 1 or both equal -1). Fourth and finally, the comparison proves that $\det A$ equals $\det A^{\mathrm{T}}$.

Important comment Rule 10 practically doubles our list of properties. Every rule for the rows can apply also to the columns (just by transposing, since $|A| = |A^{\mathrm{T}}|$). The determinant changes sign when two columns are exchanged. *A zero column or two equal columns will make the determinant zero.* If a column is multiplied by t, so is the determinant. The determinant is a linear function of each column separately.

It is time to stop. The list of properties is long enough. Next we find and use an explicit formula for the determinant.

Problem Set 5.1

Questions 1–12 are about the rules for determinants.

1 If a 4 by 4 matrix has $\det A = 2$, find $\det(2A)$ and $\det(-A)$ and $\det(A^2)$ and $\det(A^{-1})$.

2 If a 3 by 3 matrix has $\det A = -3$, find $\det\left(\frac{1}{2}A\right)$ and $\det(-A)$ and $\det(A^2)$ and $\det(A^{-1})$.

3 True or false, with reason or counterexample:

(a) The determinant of $I + A$ is $1 + \det A$.

(b) The determinant of ABC is $|A|\,|B|\,|C|$.

(c) The determinant of A^4 is $|A|^4$.

(d) The determinant of $4A$ is $4|A|$.

4 Which row exchanges show that these "reverse identity matrices" J_3 and J_4 have $|J_3| = -1$ but $|J_4| = +1$?

$$\det\begin{bmatrix} 0 & 0 & 1 \\ 0 & 1 & 0 \\ 1 & 0 & 0 \end{bmatrix} = -1 \qquad \text{but} \qquad \det\begin{bmatrix} 0 & 0 & 0 & 1 \\ 0 & 0 & 1 & 0 \\ 0 & 1 & 0 & 0 \\ 1 & 0 & 0 & 0 \end{bmatrix} = +1.$$

5 For $n = 5, 6, 7$, count the row exchanges to permute the reverse identity J_n to the identity matrix I_n. Propose a rule for every size n and predict whether J_{101} is even or odd.

6 Show how Rule 6 (determinant $= 0$ if a row is all zero) comes from Rule 1.

7 Prove from the product rule $|AB| = |A|\,|B|$ that an orthogonal matrix Q has determinant 1 or -1. Also prove that $|Q| = |Q^{-1}| = |Q^{\mathrm{T}}|$.

8 Find the determinants of rotations and reflections:

$$Q = \begin{bmatrix} \cos\theta & -\sin\theta \\ \sin\theta & \cos\theta \end{bmatrix} \quad \text{and} \quad Q = \begin{bmatrix} 1-2\cos^2\theta & -2\cos\theta\sin\theta \\ -2\cos\theta\sin\theta & 1-2\sin^2\theta \end{bmatrix}.$$

9 Prove that $|A^{\mathrm{T}}| = |A|$ by transposing $A = QR$. (R is triangular and Q is orthogonal; note Problem 7.) Why does $|R^{\mathrm{T}}| = |R|$?

10 If the entries in every row of A add to zero, prove that $\det A = 0$. If every row of A adds to one, prove that $\det(A - I) = 0$. Does this guarantee that $\det A = 1$?

11 Suppose that $CD = -DC$ and find the flaw in this reasoning: Taking determinants gives $|C|\,|D| = -|D|\,|C|$. Therefore $|C| = 0$ or $|D| = 0$. One or both of the matrices must be singular. (That is not true.)

12 The inverse of a 2 by 2 matrix seems to have determinant = 1:

$$\det A^{-1} = \det \frac{1}{ad-bc}\begin{bmatrix} d & -b \\ -c & a \end{bmatrix} = \frac{ad-bc}{ad-bc} = 1.$$

What is wrong with this calculation?

Questions 13–26 use the rules to compute specific determinants.

13 By applying row operations to produce an upper triangular U, compute

$$\det\begin{bmatrix} 1 & 2 & 3 & 0 \\ 2 & 6 & 6 & 1 \\ -1 & 0 & 0 & 3 \\ 0 & 2 & 0 & 5 \end{bmatrix} \quad \text{and} \quad \det\begin{bmatrix} 2 & -1 & 0 & 0 \\ -1 & 2 & -1 & 0 \\ 0 & -1 & 2 & -1 \\ 0 & 0 & -1 & 2 \end{bmatrix}.$$

14 Use row operations to show that the 3 by 3 "Vandermonde determinant" is

$$\det\begin{bmatrix} 1 & a & a^2 \\ 1 & b & b^2 \\ 1 & c & c^2 \end{bmatrix} = (b-a)(c-a)(c-b).$$

15 Find the determinants of a rank one matrix and a skew-symmetric matrix:

$$A = \begin{bmatrix} 1 \\ 2 \\ 3 \end{bmatrix}\begin{bmatrix} 1 & -4 & 5 \end{bmatrix} \quad \text{and} \quad K = \begin{bmatrix} 0 & 1 & 3 \\ -1 & 0 & 4 \\ -3 & -4 & 0 \end{bmatrix}.$$

16 A skew-symmetric matrix has $K^{\mathrm{T}} = -K$. Insert a, b, c for $1, 3, 4$ in Question 15 and show that $|K| = 0$. Write down a 4 by 4 example with $|K| = 1$.

17 Use row operations to simplify and compute these determinants:

$$\det \begin{bmatrix} 101 & 201 & 301 \\ 102 & 202 & 302 \\ 103 & 203 & 303 \end{bmatrix} \quad \text{and} \quad \det \begin{bmatrix} 1 & t & t^2 \\ t & 1 & t \\ t^2 & t & 1 \end{bmatrix}.$$

18 Find the determinants of U and U^{-1} and U^2:

$$U = \begin{bmatrix} 1 & 2 & 3 \\ 0 & 4 & 5 \\ 0 & 0 & 6 \end{bmatrix} \quad \text{and} \quad U = \begin{bmatrix} a & b \\ 0 & d \end{bmatrix}.$$

19 Suppose you do two row operations at once, going from

$$\begin{bmatrix} a & b \\ c & d \end{bmatrix} \quad \text{to} \quad \begin{bmatrix} a - Lc & b - Ld \\ c - la & d - lb \end{bmatrix}.$$

Find the second determinant. Does it equal $ad - bc$?

20 Add row 1 of A to row 2, then subtract row 2 from row 1. Then add row 1 to row 2 and multiply row 1 by -1 to reach B. Which rules show that

$$A = \begin{bmatrix} a & b \\ c & d \end{bmatrix} \quad \text{and} \quad B = \begin{bmatrix} c & d \\ a & b \end{bmatrix} \quad \text{have} \quad \det B = -\det A?$$

Those rules could replace Rule 2 in the definition of the determinant.

21 From $ad - bc$, find the determinants of A and A^{-1} and $A - \lambda I$:

$$A = \begin{bmatrix} 2 & 1 \\ 1 & 2 \end{bmatrix} \quad \text{and} \quad A^{-1} = \frac{1}{3}\begin{bmatrix} 2 & -1 \\ -1 & 2 \end{bmatrix} \quad \text{and} \quad A - \lambda I = \begin{bmatrix} 2-\lambda & 1 \\ 1 & 2-\lambda \end{bmatrix}.$$

Which two numbers λ lead to $\det(A - \lambda I) = 0$? Write down the matrix $A - \lambda I$ for each of those numbers λ—it should not be invertible.

22 From $A = \left[\begin{smallmatrix} 4 & 1 \\ 2 & 3 \end{smallmatrix}\right]$ find A^2 and A^{-1} and $A - \lambda I$ and their determinants. Which two numbers λ lead to $|A - \lambda I| = 0$?

23 Elimination reduces A to U. Then $A = LU$:

$$A = \begin{bmatrix} 3 & 3 & 4 \\ 6 & 8 & 7 \\ -3 & 5 & -9 \end{bmatrix} = \begin{bmatrix} 1 & 0 & 0 \\ 2 & 1 & 0 \\ -1 & 4 & 1 \end{bmatrix} \begin{bmatrix} 3 & 3 & 4 \\ 0 & 2 & -1 \\ 0 & 0 & -1 \end{bmatrix} = LU.$$

Find the determinants of L, U, A, $U^{-1}L^{-1}$, and $U^{-1}L^{-1}A$.

24 If the i, j entry of A is i times j, show that $\det A = 0$. (Exception when $A = [\,1\,]$.)

25 If the i, j entry of A is $i + j$, show that $\det A = 0$. (Exception when $n = 1$ or 2.)

26 Compute the determinants of these matrices by row operations:

$$A = \begin{bmatrix} 0 & a & 0 \\ 0 & 0 & b \\ c & 0 & 0 \end{bmatrix} \quad \text{and} \quad B = \begin{bmatrix} 0 & a & 0 & 0 \\ 0 & 0 & b & 0 \\ 0 & 0 & 0 & c \\ d & 0 & 0 & 0 \end{bmatrix} \quad \text{and} \quad C = \begin{bmatrix} a & a & a \\ a & b & b \\ a & b & c \end{bmatrix}.$$

27 True or false (give a reason if true or a 2 by 2 example if false):

(a) If A is not invertible then AB is not invertible.

(b) The determinant of A is the product of its pivots.

(c) The determinant of $A - B$ equals $\det A - \det B$.

(d) AB and BA have the same determinant.

28 (Calculus question) Show that the partial derivatives of $f(A) = \ln(\det A)$ give A^{-1}!

$$f(a, b, c, d) = \ln(ad - bc) \qquad \text{leads to} \qquad \begin{bmatrix} \partial f/\partial a & \partial f/\partial c \\ \partial f/\partial b & \partial f/\partial d \end{bmatrix} = A^{-1}.$$

5.2 Cramer's Rule, Inverses, and Volumes

This section is about the applications of determinants, first to $A\boldsymbol{x} = \boldsymbol{b}$ and A^{-1}. In the entries of A^{-1}, you will see $\det A$ in every denominator—we divide by it. (If it is zero then A^{-1} doesn't exist.) Each number in A^{-1} is a determinant divided by a determinant. So is every component of $\boldsymbol{x} = A^{-1}\boldsymbol{b}$.

Start with *Cramer's Rule* to find $\boldsymbol{x}$. A neat idea gives the solution immediately. Put $\boldsymbol{x}$ in column 1 of I and call that matrix Z_1. When you multiply A times Z_1, the first column becomes $A\boldsymbol{x}$ which is $\boldsymbol{b}$:

$$AZ_1 = \begin{bmatrix} & & \\ & A & \\ & & \end{bmatrix} \begin{bmatrix} x_1 & 0 & 0 \\ x_2 & 1 & 0 \\ x_3 & 0 & 1 \end{bmatrix} = \begin{bmatrix} b_1 & a_{12} & a_{13} \\ b_2 & a_{22} & a_{23} \\ b_3 & a_{32} & a_{33} \end{bmatrix} = B_1. \tag{5.3}$$

We multiplied A times Z_1 a column at a time. The first column is $A\boldsymbol{x}$, the other columns are just copied from A. *Now take determinants.* The rule is $(\det A)(\det Z_1) = \det B_1$. But Z_1 is triangular with determinant x_1:

$$(\det A)(x_1) = \det B_1 \qquad \text{or} \qquad x_1 = \frac{\det B_1}{\det A}. \tag{5.4}$$

This is the first component of $\boldsymbol{x}$ in Cramer's Rule! Changing the first column of A produced B_1.

To find x_2, put the vector $\boldsymbol{x}$ into the *second* column of the identity matrix. This new matrix is Z_2 and its determinant is x_2. Now multiply Z_2 by A:

$$AZ_2 = \begin{bmatrix} \boldsymbol{a}_1 & \boldsymbol{a}_2 & \boldsymbol{a}_3 \end{bmatrix} \begin{bmatrix} 1 & x_1 & 0 \\ 0 & x_2 & 0 \\ 0 & x_3 & 1 \end{bmatrix} = \begin{bmatrix} \boldsymbol{a}_1 & \boldsymbol{b} & \boldsymbol{a}_3 \end{bmatrix} = B_2. \tag{5.5}$$

In Z_2, the vector $\boldsymbol{x}$ replaced a column of I. So in B_2, the vector $\boldsymbol{b}$ replaced a column of A. Take determinants to find $(\det A)(\det Z_2) = \det B_2$. The determinant of Z_2 is x_2. This gives x_2 in Cramer's Rule:

5B (Cramer's Rule) If $\det A$ is not zero, then $A\boldsymbol{x} = \boldsymbol{b}$ has the unique solution

$$x_1 = \frac{\det B_1}{\det A}, \quad x_2 = \frac{\det B_2}{\det A}, \quad \ldots, \quad x_n = \frac{\det B_n}{\det A}.$$

The matrix B_j is obtained by replacing the jth column of A by the vector b.

The computer program for Cramer's rule only needs one formula line:

```
x(j) = det([A(:, 1:j−1) b A(:, j + 1:n)])/det(A)
```

To solve an n by n system, Cramer's Rule evaluates $n + 1$ determinants. When each one is the sum of $n!$ terms—applying the "big formula" with all permutations—this makes $(n + 1)!$ different terms. It would be crazy to solve equations that way. But we do finally have an explicit formula for the solution $\boldsymbol{x}$.

Example 5.1 Use Cramer's Rule (it needs four determinants) to solve

$$\begin{aligned} x_1 + x_2 + x_3 &= 1 \\ -2x_1 + x_2 &= 0 \\ -4x_1 + x_3 &= 0. \end{aligned}$$

The first determinant is $|A|$. It should not be zero. Then the right side $(1, 0, 0)$ goes into columns 1, 2, 3 to produce the matrices B_1, B_2, B_3:

$$|A| = \begin{vmatrix} 1 & 1 & 1 \\ -2 & 1 & 0 \\ -4 & 0 & 1 \end{vmatrix} = 7 \quad \text{and} \quad |B_1| = \begin{vmatrix} 1 & 1 & 1 \\ 0 & 1 & 0 \\ 0 & 0 & 1 \end{vmatrix} = 1 \quad \text{and}$$

$$|B_2| = \begin{vmatrix} 1 & 1 & 1 \\ -2 & 0 & 0 \\ -4 & 0 & 1 \end{vmatrix} = 2 \quad \text{and} \quad |B_3| = \begin{vmatrix} 1 & 1 & 1 \\ -2 & 1 & 0 \\ -4 & 0 & 0 \end{vmatrix} = 4.$$

Cramer's Rule takes ratios to find the components of $\boldsymbol{x}$. Always divide by $|A|$:

$$x_1 = \frac{|B_1|}{|A|} = \frac{1}{7} \quad \text{and} \quad x_2 = \frac{|B_2|}{|A|} = \frac{2}{7} \quad \text{and} \quad x_3 = \frac{|B_3|}{|A|} = \frac{4}{7}.$$

This example will be used again, to find the inverse matrix A^{-1}. We always substitute the x's back into the equations, to check the calculations.

A Formula for A^{-1}

In Example 1, the right side $\boldsymbol{b}$ was special. It was $(1, 0, 0)$, the first column of I. *The solution $\boldsymbol{x}$ must be the first column of A^{-1}:*

$$A\begin{bmatrix}1/7\\2/7\\4/7\end{bmatrix}=\begin{bmatrix}1\\0\\0\end{bmatrix} \quad \text{is the first column of} \quad A\begin{bmatrix}1/7 & \cdots\\2/7 & \cdots\\4/7 & \cdots\end{bmatrix}=\begin{bmatrix}1&0&0\\0&1&0\\0&0&1\end{bmatrix}. \tag{5.6}$$

The number 7 is the 3 by 3 determinant $|A|$. The numbers 1, 2, 4 are cofactors (one size smaller). Look back at $|B_1|$ to see how it reduces to a 2 by 2 determinant:

$$|B_1| = 1 \quad \text{is the cofactor} \quad C_{11} = \begin{vmatrix}1 & 0\\0 & 1\end{vmatrix}$$

$$|B_2| = 2 \quad \text{is the cofactor} \quad C_{12} = -\begin{vmatrix}-2 & 0\\-4 & 1\end{vmatrix}$$

$$|B_3| = 4 \quad \text{is the cofactor} \quad C_{13} = \begin{vmatrix}-2 & 1\\-4 & 0\end{vmatrix}.$$

Main point: ***The numerators in A^{-1} are cofactors***. When $\boldsymbol{b} = (1, 0, 0)$ replaces a column of A to produce B_j, the determinant is just 1 times a cofactor.

For the second column of A^{-1}, change $\boldsymbol{b}$ to $(0, 1, 0)$. Watch how *the determinants of B_1, B_2, B_3 are cofactors along row* 2—including the signs $(-)(+)(-)$:

$$\begin{vmatrix}0&1&1\\1&1&0\\0&0&1\end{vmatrix}=-1 \quad \text{and} \quad \begin{vmatrix}1&0&1\\-2&1&0\\-4&0&1\end{vmatrix}=5 \quad \text{and} \quad \begin{vmatrix}1&1&0\\-2&1&1\\-4&0&0\end{vmatrix}=-4.$$

With these numbers $-1, 5, -4$ in Cramer's Rule, the second column of A^{-1} is $-\frac{1}{7}, \frac{5}{7}, -\frac{4}{7}$.

For the third column of A^{-1}, the right side is $\boldsymbol{b} = (0, 0, 1)$. The three determinants become cofactors along the *third row*. The numbers are $-1, -2, 3$. We always divide by $|A| = 7$. Now we have all columns of A^{-1}:

$$A=\begin{bmatrix}1&1&1\\-2&1&0\\-4&0&1\end{bmatrix} \quad \text{times} \quad A^{-1}=\begin{bmatrix}1/7&-1/7&-1/7\\2/7&5/7&-2/7\\4/7&-4/7&3/7\end{bmatrix} \quad \text{equals} \quad I.$$

Summary We found A^{-1} by solving $AA^{-1} = I$. The three columns of I on the right led to the columns of A^{-1} on the left. After stating a short formula for A^{-1}, we will give a direct proof. Then you have two ways to approach A^{-1}—by putting columns of I into Cramer's Rule (above), or by cofactors (the lightning way using equation (5.10) below).

5C (Formula for A^{-1}) The i, j entry of A^{-1} is the cofactor C_{ji} divided by the determinant of A:

$$(A^{-1})_{ij} = \frac{C_{ji}}{\det A} \quad \text{or} \quad A^{-1} = \frac{C^{\mathrm{T}}}{\det A}. \tag{5.7}$$

The cofactors C_{ij} go into the "cofactor matrix" C. *Then this matrix is* ***transposed***. To compute the i, j entry of A^{-1}, cross out row j and column i of A. Multiply the determinant by $(-1)^{i+j}$ to get the cofactor, and divide by $\det A$.

Example 5.2 The matrix $A = \left[\begin{smallmatrix} a & b \\ c & d \end{smallmatrix}\right]$ has cofactor matrix $C = \left[\begin{smallmatrix} d & -c \\ -b & a \end{smallmatrix}\right]$. Look at A times the transpose of C:

$$AC^{\mathrm{T}} = \begin{bmatrix} a & b \\ c & d \end{bmatrix} \begin{bmatrix} d & -b \\ -c & a \end{bmatrix} = \begin{bmatrix} ad-bc & 0 \\ 0 & ad-bc \end{bmatrix}. \tag{5.8}$$

The matrix on the right is $\det A$ times I. So divide by $\det A$. Then A times $C^{\mathrm{T}}/\det A$ is I, which reveals A^{-1}:

$$A^{-1} \quad \text{is} \quad \frac{C^{\mathrm{T}}}{\det A} \quad \text{which is} \quad \frac{1}{ad-bc}\begin{bmatrix} d & -b \\ -c & a \end{bmatrix}. \tag{5.9}$$

This 2 by 2 example uses letters. The 3 by 3 example used numbers. Inverting a 4 by 4 matrix would need sixteen cofactors (each one is a 3 by 3 determinant). Elimination is faster—but now we know an explicit formula for A^{-1}.

Direct proof of the formula $A^{-1} = C^{\mathrm{T}}/\det A$ The idea is to multiply A times C^{T}:

$$\begin{bmatrix} a_{11} & a_{12} & a_{13} \\ a_{21} & a_{22} & a_{23} \\ a_{31} & a_{32} & a_{33} \end{bmatrix} \begin{bmatrix} C_{11} & C_{21} & C_{31} \\ C_{12} & C_{22} & C_{32} \\ C_{13} & C_{23} & C_{33} \end{bmatrix} = \begin{bmatrix} \det A & 0 & 0 \\ 0 & \det A & 0 \\ 0 & 0 & \det A \end{bmatrix}. \tag{5.10}$$

Row 1 of A times column 1 of the cofactors yields $\det A$ on the right:

$$\text{By cofactors of row 1:} \quad a_{11}C_{11} + a_{12}C_{12} + a_{13}C_{13} = \det A.$$

Similarly row 2 of A times column 2 of C^{T} yields $\det A$. The entries a_{2j} are multiplying cofactors C_{2j} as they should.

Reason for zeros off the main diagonal in equation (5.10). Rows of A are combined with cofactors from *different* rows. Row 2 of A times column 1 of C^{T} gives zero, but why?

$$a_{21}C_{11} + a_{22}C_{12} + a_{23}C_{13} = 0. \tag{5.11}$$

Answer: This is the determinant of a matrix A^* which has two equal rows. A^* is the same as A, except that its first row is a copy of its second row. So $\det A^* = 0$, which is equation (5.11). It is the expansion of $\det A^*$ along row 1, where A^* has the same cofactors C_{11}, C_{12}, C_{13} as A—because all rows agree after the first row. Thus the remarkable matrix multiplication (5.10) is correct.

Equation (5.10) immediately gives A^{-1}. We have $\det A$ times I on the right side:

$$AC^{\mathrm{T}} = (\det A)I \qquad \text{or} \qquad A^{-1} = \frac{C^{\mathrm{T}}}{\det A}.$$

The 2 by 2 case in equation (5.8) shows the zeros off the main diagonal of AC^{T}.

Example 5.3 A triangular matrix of 1's has determinant 1. The inverse matrix contains cofactors:

$$A = \begin{bmatrix} 1 & 0 & 0 & 0 \\ 1 & 1 & 0 & 0 \\ 1 & 1 & 1 & 0 \\ 1 & 1 & 1 & 1 \end{bmatrix} \quad \text{has inverse} \quad A^{-1} = \frac{C^{\mathrm{T}}}{1} = \begin{bmatrix} 1 & 0 & 0 & 0 \\ -1 & 1 & 0 & 0 \\ 0 & -1 & 1 & 0 \\ 0 & 0 & -1 & 1 \end{bmatrix}.$$

Cross out row 1 and column 1 of A to see the cofactor $C_{11} = 1$. Now cross out row 1 and column 2. The 3 by 3 submatrix is still triangular with determinant 1. But the cofactor C_{12} is -1 because of the sign $(-1)^{1+2}$. This number -1 goes into the $(2, 1)$ entry of A^{-1}—don't forget to transpose C!

Crossing out row 1 and column 3 leaves a matrix with two columns of 1's. Its determinant is $C_{13} = 0$. This is the $(3, 1)$ entry of A^{-1}. The $(4, 1)$ entry is zero for the same reason. The $(1, 4)$ entry of A^{-1} comes from the $(4, 1)$ cofactor—which has a column of zeros.

The inverse of a triangular matrix is triangular. Cofactors explain why.

Example 5.4 If all cofactors are nonzero, is A sure to be invertible? *No way.*

Example 5.5 Here is part of a direct computation of A^{-1} by cofactors:

$$A = \begin{bmatrix} 0 & 1 & 3 \\ 1 & 0 & 1 \\ 2 & 1 & 0 \end{bmatrix} \quad \text{and} \quad \begin{matrix} |A| = 5 \\ C_{12} = -(-2) \\ C_{22} = -6 \end{matrix} \quad \text{and} \quad A^{-1} = \begin{bmatrix} x & x & x \\ \frac{2}{5} & -\frac{6}{5} & x \\ x & x & x \end{bmatrix}.$$

6

EIGENVALUES AND EIGENVECTORS

6.1 Introduction to Eigenvalues

Linear equations $Ax = b$ come from steady state problems. Eigenvalues have their greatest importance in *dynamic problems*. The solution is changing with time—growing or decaying or oscillating. We can't find it by elimination. This chapter enters a new part of linear algebra. As for determinants, all matrices are square.

A good dynamical model is the sequence of matrices $A, A^2, A^3, A^4, \ldots$. Suppose you need the hundredth power A^{100}. The underlying equation multiplies by A at every step (and 99 steps is not many in scientific computing). The starting matrix A becomes unrecognizable after a few steps:

$$\underset{A}{\begin{bmatrix} .9 & .3 \\ .1 & .7 \end{bmatrix}} \quad \underset{A^2}{\begin{bmatrix} .84 & .48 \\ .16 & .52 \end{bmatrix}} \quad \underset{A^3}{\begin{bmatrix} .804 & .588 \\ .196 & .412 \end{bmatrix}} \quad \underset{A^4}{\begin{bmatrix} .7824 & .6528 \\ .2176 & .3472 \end{bmatrix}} \qquad \underset{A^{100}}{\begin{bmatrix} .75 & .75 \\ .25 & .25 \end{bmatrix}}$$

A^{100} was found by using the *eigenvalues* of A, not by multiplying 100 matrices.

To explain eigenvalues, we first explain eigenvectors. Almost all vectors, when multiplied by A, change direction. ***Certain exceptional vectors x are in the same direction as Ax. Those are the "eigenvectors."*** Multiply an eigenvector by A, and the vector Ax is a number λ times the original x.

The basic equation is $Ax = \lambda x$. The number λ is the ***"eigenvalue"***. It tells whether the special vector x is stretched or shrunk or reversed or left unchanged—when it is multiplied by A. We may find $\lambda = 2$ or $\frac{1}{2}$ or -1 or 1. The eigenvalue can be zero. The equation $Ax = 0x$ means that this eigenvector x is in the nullspace.

If A is the identity matrix, every vector is unchanged. All vectors are eigenvectors for $A = I$. The eigenvalue (the number lambda) is $\lambda = 1$. This is unusual to say the least. Most 2 by 2 matrices have *two* eigenvector directions and *two* eigenvalues. This section teaches how to compute the x's and λ's.

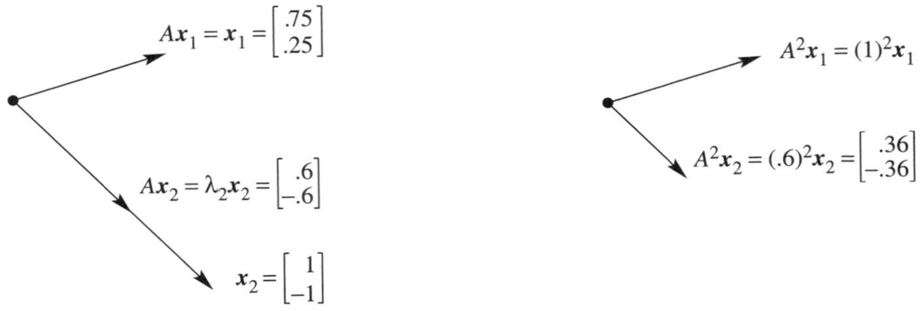

Figure 6.1 The eigenvectors keep their directions. The eigenvalues for A^2 are λ^2.

For the matrix above, here are eigenvectors $\boldsymbol{x}_1$ and $\boldsymbol{x}_2$. Multiplying those vectors by A gives $\boldsymbol{x}_1$ and $.6\boldsymbol{x}_2$. The eigenvalues λ are 1 and .6:

$$\boldsymbol{x}_1 = \begin{bmatrix} .75 \\ .25 \end{bmatrix} \quad \text{and} \quad A\boldsymbol{x}_1 = \begin{bmatrix} .9 & .3 \\ .1 & .7 \end{bmatrix}\begin{bmatrix} .75 \\ .25 \end{bmatrix} = \boldsymbol{x}_1 \quad (A\boldsymbol{x} = \boldsymbol{x} \text{ means that } \lambda_1 = 1)$$

$$\boldsymbol{x}_2 = \begin{bmatrix} 1 \\ -1 \end{bmatrix} \quad \text{and} \quad A\boldsymbol{x}_2 = \begin{bmatrix} .9 & .3 \\ .1 & .7 \end{bmatrix}\begin{bmatrix} 1 \\ -1 \end{bmatrix} = \begin{bmatrix} .6 \\ -.6 \end{bmatrix} \quad (\text{this is } .6\,\boldsymbol{x}_2 \text{ so } \lambda_2 = .6).$$

If we multiply $\boldsymbol{x}_1$ by A again, we still get $\boldsymbol{x}_1$. Every power of A will give $A^n\boldsymbol{x}_1 = \boldsymbol{x}_1$. Multiplying $\boldsymbol{x}_2$ by A gave $.6\boldsymbol{x}_2$, and if we multiply again we get $(.6)^2\boldsymbol{x}_2$. When A is squared, ***the eigenvectors $\boldsymbol{x}_1$ and $\boldsymbol{x}_2$ stay the same***. The λ's are now 1^2 and $(.6)^2$. ***The eigenvalues are squared!*** This pattern keeps going, because the eigenvectors stay in their own directions and never get mixed. The eigenvectors of A^{100} are the same $\boldsymbol{x}_1$ and $\boldsymbol{x}_2$. The eigenvalues of A^{100} are $1^{100} = 1$ and $(.6)^{100}$ = very small number.

Other vectors do change direction. But those other vectors are combinations of the two eigenvectors. The first column of A is

$$\begin{bmatrix} .9 \\ .1 \end{bmatrix} = \begin{bmatrix} .75 \\ .25 \end{bmatrix} + \begin{bmatrix} .15 \\ -.15 \end{bmatrix} = \boldsymbol{x}_1 + .15\,\boldsymbol{x}_2. \tag{6.1}$$

Multiplying by A gives the first column of A^2. Do it separately for $\boldsymbol{x}_1$ and $.15\,\boldsymbol{x}_2$:

$$A\begin{bmatrix} .9 \\ .1 \end{bmatrix} = \begin{bmatrix} .84 \\ .16 \end{bmatrix} \quad \text{is really} \quad A\boldsymbol{x}_1 + .15\,A\boldsymbol{x}_2 = \boldsymbol{x}_1 + (.15)(.6)\boldsymbol{x}_2.$$

Each eigenvector is multiplied by its eigenvalue, whenever A is applied. We didn't need these eigenvectors to find A^2. But it is the good way to do 99 multiplications:

$$A^{99}\begin{bmatrix} .9 \\ .1 \end{bmatrix} \quad \text{is really} \quad 1^{99}\boldsymbol{x}_1 + (.15)(.6)^{99}\boldsymbol{x}_2 = \begin{bmatrix} .75 \\ .25 \end{bmatrix} + \begin{bmatrix} \text{very} \\ \text{small} \\ \text{vector} \end{bmatrix}.$$

This is the first column of A^{100}. The second column is A^{99} times column 2 of A. But column 2 is $\boldsymbol{x}_1 - .45\,\boldsymbol{x}_2$. Multiply by A^{99} to get $\boldsymbol{x}_1 - (.45)(.6)^{99}\boldsymbol{x}_2$. Again $(.6)^{99}$ is extremely small. So both columns of A^{100} are practically equal to $\boldsymbol{x}_1 = (.75,\ .25)$.

The eigenvector $\boldsymbol{x}_1$ is a "steady state" that doesn't change (because $\lambda_1 = 1$). The eigenvector $\boldsymbol{x}_2$ is a "decaying mode" that virtually disappears (because $\lambda_2 = .6$). The higher the power of A, the closer its columns to the steady state.

We mention that this particular A is a ***Markov matrix***. Its entries are positive and every column adds to 1. Those facts guarantee that the largest eigenvalue is $\lambda = 1$ (as we found). The eigenvector for $\lambda = 1$ is the steady state—which all columns of A^k will approach.

Example 6.1 The projection matrix $P = \begin{bmatrix} .5 & .5 \\ .5 & .5 \end{bmatrix}$ has eigenvectors $\boldsymbol{x}_1 = (1, 1)$ and $\boldsymbol{x}_2 = (1, -1)$. For those vectors, $P\boldsymbol{x}_1 = \boldsymbol{x}_1$ and $P\boldsymbol{x}_2 = \boldsymbol{0}$. This matrix illustrates three things at once:

1 It is a Markov matrix, so $\lambda = 1$ is an eigenvalue.

2 It is a singular matrix, so $\lambda = 0$ is an eigenvalue.

3 It is a symmetric matrix, so $\boldsymbol{x}_1$ and $\boldsymbol{x}_2$ are perpendicular.

The only possible eigenvalues of a projection matrix are 0 and 1. The eigenvectors for $\lambda = 0$ (which means $P\boldsymbol{x} = 0\boldsymbol{x}$) fill up the nullspace. The eigenvectors for $\lambda = 1$ (which means $P\boldsymbol{x} = \boldsymbol{x}$) fill up the column space. The nullspace goes to zero. Vectors in the column space are projected onto themselves. An in-between vector like $\boldsymbol{v} = (3, 1)$ partly disappears and partly stays:

$$\boldsymbol{v} = \begin{bmatrix} 1 \\ -1 \end{bmatrix} + \begin{bmatrix} 2 \\ 2 \end{bmatrix} \quad \text{and} \quad P\boldsymbol{v} = \begin{bmatrix} 0 \\ 0 \end{bmatrix} + \begin{bmatrix} 2 \\ 2 \end{bmatrix} \quad \text{which is} \quad \begin{bmatrix} .5 & .5 \\ .5 & .5 \end{bmatrix}\begin{bmatrix} 3 \\ 1 \end{bmatrix}.$$

The projection P keeps the column space part and destroys the nullspace part, for every $\boldsymbol{v}$. To emphasize: *Special properties of a matrix lead to special eigenvalues and eigenvectors.* That is a major theme of this chapter. Here $\lambda = 0$ and 1. The next matrix is also special.

Example 6.2 The reflection matrix $R = \begin{bmatrix} 0 & 1 \\ 1 & 0 \end{bmatrix}$ has eigenvalues 1 and -1. The first eigenvector is $(1, 1)$—unchanged by R. The second eigenvector is $(1, -1)$—its signs are reversed by R. A matrix with no negative entries can still have a negative eigenvalue! The perpendicular eigenvectors are the same $\boldsymbol{x}_1$ and $\boldsymbol{x}_2$ that we found for projection. Behind that fact is a relation between the matrices R and P:

$$2P - I = R \quad \text{or} \quad 2\begin{bmatrix} .5 & .5 \\ .5 & .5 \end{bmatrix} - \begin{bmatrix} 1 & 0 \\ 0 & 1 \end{bmatrix} = \begin{bmatrix} 0 & 1 \\ 1 & 0 \end{bmatrix}. \tag{6.2}$$

Here is the point. If $P\boldsymbol{x} = \lambda\boldsymbol{x}$ then $2P\boldsymbol{x} = 2\lambda\boldsymbol{x}$. The eigenvalues are doubled when the matrix is doubled. Now subtract $I\boldsymbol{x} = \boldsymbol{x}$. The result is $(2P - I)\boldsymbol{x} = (2\lambda - 1)\boldsymbol{x}$. *When a matrix is shifted by I, each λ is shifted by* 1. The eigenvectors stay the same.

The eigenvalues are related exactly as the matrices are related:

$$2P - I = R \quad \text{so the eigenvalues of } R \text{ are} \quad \begin{array}{l} 2(1) - 1 = 1 \\ 2(0) - 1 = -1. \end{array}$$

Similarly the eigenvalues of R^2 are λ^2. In this case $R^2 = I$ and $(1)^2 = 1$ and $(-1)^2 = 1$.

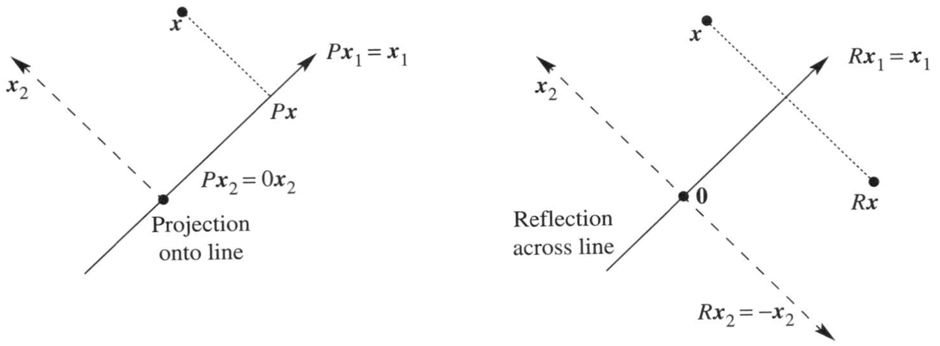

Figure 6.2 Projections have $\lambda = 1$ and 0. Reflections have $\lambda = 1$ and -1. *Perpendicular eigenvectors.*

This example is a *permutation* as well as a *reflection*. We used the letter R instead of P for two reasons. First, P was already taken by the projection. Second, other reflections have $\lambda = -1$, but many permutations don't. The eigenvectors with $\lambda = 1$ are their own reflection. The eigenvectors with $\lambda = -1$ go "through the mirror"—they are reversed to the other side. The next example is a rotation, and it brings bad news.

Example 6.3 ***The 90° rotation turns every vector so $Q = \left[\begin{smallmatrix} 0 & 1 \\ -1 & 0 \end{smallmatrix}\right]$ has no real eigenvectors or eigenvalues.***

This matrix Q rotates every vector in the xy plane. *No vector stays in the same direction* (except the zero vector which is useless). There cannot be an eigenvector, unless we go to ***imaginary numbers***. Which we do.

To see how i can help, look at Q^2. If Q is rotation through 90°, then Q^2 is rotation through 180°. In other words Q^2 is $-I$. Its eigenvalues are -1 and -1. (Certainly $-I\boldsymbol{x} = -1\boldsymbol{x}$.) If squaring Q is supposed to square its eigenvalues λ, we must have $\lambda^2 = -1$. *The eigenvalues of the* 90° *rotation matrix Q are $+i$ and $-i$.*

We meet the imaginary number i (with $i^2 = -1$) also in the eigenvectors of Q:

$$\begin{bmatrix} 0 & 1 \\ -1 & 0 \end{bmatrix}\begin{bmatrix} 1 \\ i \end{bmatrix} = i\begin{bmatrix} 1 \\ i \end{bmatrix} \quad \text{and} \quad \begin{bmatrix} 0 & 1 \\ -1 & 0 \end{bmatrix}\begin{bmatrix} i \\ 1 \end{bmatrix} = -i\begin{bmatrix} i \\ 1 \end{bmatrix}.$$

Somehow these complex vectors keep their direction as they are rotated. Don't ask how. This example makes the all-important point that real matrices can have complex eigenvalues. It also illustrates two special properties:

1 Q is a skew-symmetric matrix so each λ is pure imaginary.

2 Q is an orthogonal matrix so the absolute value of λ is 1.

A symmetric matrix ($A^{\mathrm{T}} = A$) can be compared to a real number. A skew-symmetric matrix ($A^{\mathrm{T}} = -A$) can be compared to an imaginary number. An orthogonal matrix ($A^{\mathrm{T}}A = I$) can be compared to a complex number with $|\lambda| = 1$. For the eigenvalues those are more than analogies—they are theorems to be proved. The eigenvectors for all these special matrices are perpendicular.

The Equation for the Eigenvalues

In those examples we could solve $Ax = \lambda x$ by trial and error. Now we use determinants and linear algebra. This is the key calculation in the whole chapter—***to compute the λ's and x's.***

First move λx to the left side. Write the equation $Ax = \lambda x$ as $(A - \lambda I)x = \mathbf{0}$. The matrix $A - \lambda I$ times the eigenvector x is the zero vector. The eigenvectors make up the nullspace of $A - \lambda I$! When we know an eigenvalue λ, we find an eigenvector by solving $(A - \lambda I)x = \mathbf{0}$.

Eigenvalues first. If $(A - \lambda I)x = \mathbf{0}$ has a nonzero solution, $A - \lambda I$ is not invertible. The matrix $A - \lambda I$ must be singular. ***The determinant of A − λI must be zero.*** This is how to recognize when λ is an eigenvalue:

6A The number λ is an eigenvalue of A if and only if

$$\det(A - \lambda I) = 0. \tag{6.3}$$

This "characteristic equation" involves only λ, not x. Then each root λ leads to x:

$$(A - \lambda I)x = \mathbf{0} \qquad \text{or} \qquad Ax = \lambda x. \tag{6.4}$$

Since $\det(A - \lambda I) = 0$ is an equation of the nth degree, there are n eigenvalues.

Example 6.4 $A = \begin{bmatrix} 1 & 2 \\ 2 & 4 \end{bmatrix}$ is already singular (zero determinant). Find its λ's and x's.

When A is singular, $\lambda = 0$ is one of the eigenvalues. The equation $Ax = 0x$ has solutions. They are the eigenvectors for $\lambda = 0$. But to find *all* λ's and x's, begin as always by subtracting λI from A:

$$\text{Subtract } \lambda \text{ from the diagonal to find} \quad A - \lambda I = \begin{bmatrix} 1-\lambda & 2 \\ 2 & 4-\lambda \end{bmatrix}.$$

Take the determinant of this matrix. From $1 - \lambda$ times $4 - \lambda$, the determinant involves $\lambda^2 - 5\lambda + 4$. The other term, not containing λ, is 2 times 2. Subtract as in $ad - bc$:

$$\begin{vmatrix} 1-\lambda & 2 \\ 2 & 4-\lambda \end{vmatrix} = (1-\lambda)(4-\lambda) - (2)(2) = \lambda^2 - 5\lambda. \tag{6.5}$$

This determinant $\lambda^2 - 5\lambda$ is zero when λ is an eigenvalue. Factoring into λ times $\lambda - 5$, the two roots are $\lambda = 0$ and $\lambda = 5$:

$$\det(A - \lambda I) = \lambda^2 - 5\lambda = 0 \quad \text{yields the eigenvalues} \quad \lambda_1 = 0 \quad \text{and} \quad \lambda_2 = 5.$$

Now find the eigenvectors. Solve $(A - \lambda I)\boldsymbol{x} = \mathbf{0}$ separately for $\lambda_1 = 0$ and $\lambda_2 = 5$:

$$(A - 0I)\boldsymbol{x} = \begin{bmatrix} 1 & 2 \\ 2 & 4 \end{bmatrix}\begin{bmatrix} y \\ z \end{bmatrix} = \begin{bmatrix} 0 \\ 0 \end{bmatrix} \text{ yields the eigenvector } \begin{bmatrix} y \\ z \end{bmatrix} = \begin{bmatrix} 2 \\ -1 \end{bmatrix} \text{ for } \lambda_1 = 0$$

$$(A - 5I)\boldsymbol{x} = \begin{bmatrix} -4 & 2 \\ 2 & -1 \end{bmatrix}\begin{bmatrix} y \\ z \end{bmatrix} = \begin{bmatrix} 0 \\ 0 \end{bmatrix} \text{ yields the eigenvector } \begin{bmatrix} y \\ z \end{bmatrix} = \begin{bmatrix} 1 \\ 2 \end{bmatrix} \text{ for } \lambda_2 = 5.$$

We need to emphasize: *There is nothing exceptional about* $\lambda = 0$. Like every other number, zero might be an eigenvalue and it might not. If A is singular, it is. The eigenvectors fill the nullspace: $A\boldsymbol{x} = 0\boldsymbol{x} = \mathbf{0}$. If A is invertible, zero is not an eigenvalue. We shift A by a multiple of I to *make it singular*. In the example, the shifted matrix $A - 5I$ was singular and 5 was the other eigenvalue.

Summary To solve the eigenvalue problem for an n by n matrix, follow these steps:

1 *Compute the determinant of* $A - \lambda I$. With λ subtracted along the diagonal, this determinant starts with λ^n or $-\lambda^n$. It is a polynomial in λ of degree n.

2 *Find the roots of this polynomial, so that* $\det(A - \lambda I) = 0$. The n roots are the n eigenvalues of A.

3 For each eigenvalue solve the system $(A - \lambda I)\boldsymbol{x} = \mathbf{0}$.

Since the determinant is zero, there are solutions other than $\boldsymbol{x} = \mathbf{0}$. Those $\boldsymbol{x}$'s are the eigenvectors.

A note on quick computations, when A is 2 by 2. The determinant of $A - \lambda I$ is a quadratic (starting with λ^2). By factoring or by the quadratic formula, we find its two roots (the eigenvalues). Then the eigenvectors come immediately from $A - \lambda I$. This matrix is singular, so both rows are multiples of a vector (a, b). *The eigenvector is any multiple of* $(b, -a)$. The example had $\lambda = 0$ and $\lambda = 5$:

$\lambda = 0$: rows of $A - 0I$ in the direction $(1, 2)$; eigenvector in the direction $(2, -1)$

$\lambda = 5$: rows of $A - 5I$ in the direction $(-4, 2)$; eigenvector in the direction $(2, 4)$.

Previously we wrote that last eigenvector as $(1, 2)$. *Both* $(1, 2)$ *and* $(2, 4)$ *are correct.* There is a whole line of eigenvectors—any nonzero multiple of $\boldsymbol{x}$ is as good as $\boldsymbol{x}$. Often we divide by the length, to make the eigenvector into a unit vector.

We end with a warning. Some 2 by 2 matrices have only *one* line of eigenvectors. Some n by n matrices don't have n independent eigenvectors. This can only happen when two eigenvalues are equal. (On the other hand $A = I$ has equal eigenvalues and plenty of eigenvectors.) Without n eigenvectors, we don't have a basis. We can't write every $\boldsymbol{v}$ as a combination of eigenvectors. In the language of the next section, we can't diagonalize the matrix.

Good News, Bad News

Bad news first: If you add a row of A to another row, the eigenvalues usually change. Elimination destroys the λ's. After elimination, the triangular U has *its* eigenvalues sitting along the diagonal—they are the pivots. But they are not the eigenvalues of A! Eigenvalues are changed when row 1 is added to row 2:

$$A = \begin{bmatrix} 1 & -1 \\ -1 & 1 \end{bmatrix} \quad \text{has } \lambda = 0 \text{ and } \lambda = 2; \qquad U = \begin{bmatrix} 1 & -1 \\ 0 & 0 \end{bmatrix} \quad \text{has } \lambda = 0 \text{ and } \lambda = 1.$$

Good news second: The *product* λ_1 times λ_2 and the *sum* $\lambda_1 + \lambda_2$ can be found quickly from the matrix A. Here the product is 0 times 2. That agrees with the determinant of A (which is 0). The sum of the eigenvalues is $0+2$. That agrees with the sum down the main diagonal of A (which is $1+1$). These quick checks always work:

6B ***The product of the n eigenvalues equals the determinant of A.***

6C ***The sum of the n eigenvalues equals the sum of the n diagonal entries.*** This number is called the ***trace*** of A:

$$\lambda_1 + \lambda_2 + \cdots + \lambda_n = \textbf{\textit{trace}} = a_{11} + a_{22} + \cdots + a_{nn}. \tag{6.6}$$

Those checks are very useful. They are proved in Problems 14–15 and again in the next section. They don't remove the pain of computing the λ's. But when the computation is wrong, they generally tell us so. For the correct λ's, go back to $\det(A - \lambda I) = 0$.

The determinant test makes the *product* of the λ's equal to the *product* of the pivots (assuming no row exchanges). But the sum of the λ's is not the sum of the pivots—as the example showed. The individual λ's have almost nothing to do with the individual pivots. In this new part of linear algebra, the key equation is really nonlinear: λ multiplies $\boldsymbol{x}$.

Problem Set 6.1

1 Find the eigenvalues and the eigenvectors of these two matrices:

$$A = \begin{bmatrix} 1 & 4 \\ 2 & 3 \end{bmatrix} \quad \text{and} \quad A + I = \begin{bmatrix} 2 & 4 \\ 2 & 4 \end{bmatrix}.$$

$A + I$ has the ______ eigenvectors as A. Its eigenvalues are ______ by 1.

2 Compute the eigenvalues and eigenvectors of A and A^{-1}:

$$A = \begin{bmatrix} 0 & 2 \\ 2 & 3 \end{bmatrix} \quad \text{and} \quad A^{-1} = \begin{bmatrix} -3/4 & 1/2 \\ 1/2 & 0 \end{bmatrix}.$$

A^{-1} has the ______ eigenvectors as A. When A has eigenvalues λ_1 and λ_2, its inverse has eigenvalues ______ .

3 Compute the eigenvalues and eigenvectors of A and A^2:

$$A = \begin{bmatrix} -1 & 3 \\ 2 & 0 \end{bmatrix} \quad \text{and} \quad A^2 = \begin{bmatrix} 7 & -3 \\ -2 & 6 \end{bmatrix}.$$

A^2 has the same _____ as A. When A has eigenvalues λ_1 and λ_2, A^2 has eigenvalues _____.

4 Find the eigenvalues of A and B and $A + B$:

$$A = \begin{bmatrix} 1 & 0 \\ 1 & 1 \end{bmatrix} \quad \text{and} \quad B = \begin{bmatrix} 1 & 1 \\ 0 & 1 \end{bmatrix} \quad \text{and} \quad A + B = \begin{bmatrix} 2 & 1 \\ 1 & 2 \end{bmatrix}.$$

Eigenvalues of $A + B$ (are equal to)(are not equal to) eigenvalues of A plus eigenvalues of B.

5 Find the eigenvalues of A and B and AB and BA:

$$A = \begin{bmatrix} 1 & 0 \\ 1 & 1 \end{bmatrix} \quad \text{and} \quad B = \begin{bmatrix} 1 & 1 \\ 0 & 1 \end{bmatrix} \quad \text{and} \quad AB = \begin{bmatrix} 1 & 1 \\ 1 & 2 \end{bmatrix} \quad \text{and} \quad BA = \begin{bmatrix} 2 & 1 \\ 1 & 1 \end{bmatrix}.$$

Eigenvalues of AB (are equal to)(are not equal to) eigenvalues of A times eigenvalues of B. Eigenvalues of AB (are equal to)(are not equal to) eigenvalues of BA.

6 Elimination produces $A = LU$. The eigenvalues of U are on its diagonal; they are the _____. The eigenvalues of L are on its diagonal; they are all _____. The eigenvalues of A are not the same as _____.

7 (a) If you know $\boldsymbol{x}$ is an eigenvector, the way to find λ is to _____.

(b) If you know λ is an eigenvalue, the way to find $\boldsymbol{x}$ is to _____.

8 What do you do to $A\boldsymbol{x} = \lambda\boldsymbol{x}$, in order to prove (a), (b), and (c)?

(a) λ^2 is an eigenvalue of A^2, as in Problem 3.

(b) λ^{-1} is an eigenvalue of A^{-1}, as in Problem 2.

(c) $\lambda + 1$ is an eigenvalue of $A + I$, as in Problem 1.

9 Find the eigenvalues and eigenvectors for both of these Markov matrices A and A^∞. Explain why A^{100} is close to A^∞:

$$A = \begin{bmatrix} .6 & .2 \\ .4 & .8 \end{bmatrix} \quad \text{and} \quad A^\infty = \begin{bmatrix} 1/3 & 1/3 \\ 2/3 & 2/3 \end{bmatrix}.$$

10 Find the eigenvalues and eigenvectors for the projection matrices P and P^{100}:

$$P = \begin{bmatrix} .2 & .4 & 0 \\ .4 & .8 & 0 \\ 0 & 0 & 1 \end{bmatrix}.$$

If two eigenvectors share the same λ, so do all their linear combinations. Find an eigenvector of P with no zero components.

11 From the unit vector $\boldsymbol{u} = \left(\frac{1}{6}, \frac{1}{6}, \frac{3}{6}, \frac{5}{6}\right)$ construct the rank one projection matrix $P = \boldsymbol{u}\boldsymbol{u}^{\mathrm{T}}$.

(a) Show that $P\boldsymbol{u} = \boldsymbol{u}$. Then $\boldsymbol{u}$ is an eigenvector with $\lambda = 1$.

(b) If $\boldsymbol{v}$ is perpendicular to $\boldsymbol{u}$ show that $P\boldsymbol{v} = \mathbf{0}$. Then $\lambda = 0$.

(c) Find three independent eigenvectors of P all with eigenvalue $\lambda = 0$.

12 Solve $\det(Q - \lambda I) = 0$ by the quadratic formula to reach $\lambda = \cos\theta \pm i \sin\theta$:

$$Q = \begin{bmatrix} \cos\theta & -\sin\theta \\ \sin\theta & \cos\theta \end{bmatrix} \quad \text{rotates the } xy \text{ plane by the angle } \theta.$$

Find the eigenvectors of Q by solving $(Q - \lambda I)\boldsymbol{x} = \mathbf{0}$. Use $i^2 = -1$.

13 Every permutation matrix leaves $\boldsymbol{x} = (1, 1, \ldots, 1)$ unchanged. Then $\lambda = 1$. Find two more λ's for these permutations:

$$P = \begin{bmatrix} 0 & 1 & 0 \\ 0 & 0 & 1 \\ 1 & 0 & 0 \end{bmatrix} \quad \text{and} \quad P = \begin{bmatrix} 0 & 0 & 1 \\ 0 & 1 & 0 \\ 1 & 0 & 0 \end{bmatrix}.$$

14 Prove that the determinant of A equals the product $\lambda_1\lambda_2\cdots\lambda_n$. Start with the polynomial $\det(A - \lambda I)$ separated into its n factors. Then set $\lambda =$ ______:

$$\det(A - \lambda I) = (\lambda_1 - \lambda)(\lambda_2 - \lambda)\cdots(\lambda_n - \lambda) \quad \text{so} \quad \det A = \_\_\_\_\_\_.$$

15 The sum of the diagonal entries (the *trace*) equals the sum of the eigenvalues:

$$A = \begin{bmatrix} a & b \\ c & d \end{bmatrix} \quad \text{has} \quad \det(A - \lambda I) = \lambda^2 - (a + d)\lambda + ad - bc = 0.$$

If A has $\lambda_1 = 3$ and $\lambda_2 = 4$ then $\det(A - \lambda I) =$ ______. The quadratic formula gives the eigenvalues $\lambda = (a + d + \sqrt{})/2$ and $\lambda =$ ______. Their sum is ______.

16 If A has $\lambda_1 = 4$ and $\lambda_2 = 5$ then $\det(A - \lambda I) = (\lambda - 4)(\lambda - 5) = \lambda^2 - 9\lambda + 20$. Find three matrices that have trace $a + d = 9$ and determinant 20 and $\lambda = 4, 5$.

17 A 3 by 3 matrix B is known to have eigenvalues 0, 1, 2. This information is enough to find three of these:

(a) the rank of B

(b) the determinant of $B^{\mathrm{T}}B$

(c) the eigenvalues of $B^{\mathrm{T}}B$

(d) the eigenvalues of $(B + I)^{-1}$.

18 Choose the second row of $A = \left[\begin{smallmatrix} 0 & 1 \\ * & * \end{smallmatrix}\right]$ so that A has eigenvalues 4 and 7.

19 Choose a, b, c, so that $\det(A - \lambda I) = 9\lambda - \lambda^3$. Then the eigenvalues are $-3, 0, 3$:

$$A = \begin{bmatrix} 0 & 1 & 0 \\ 0 & 0 & 1 \\ a & b & c \end{bmatrix}.$$

20 ***The eigenvalues of A equal the eigenvalues of A^{T}***. This is because $\det(A - \lambda I)$ equals $\det(A^{\mathrm{T}} - \lambda I)$. That is true because ______. Show by an example that the eigenvectors of A and A^{T} are *not* the same.

21 Construct any 3 by 3 Markov matrix M: positive entries down each column add to 1. If $\boldsymbol{e} = (1, 1, 1)$ verify that $M^{\mathrm{T}}\boldsymbol{e} = \boldsymbol{e}$. By Problem 20, $\lambda = 1$ is also an eigenvalue of M. Challenge: A 3 by 3 singular Markov matrix with trace $\frac{1}{2}$ has eigenvalues $\lambda =$ ______.

22 Find three 2 by 2 matrices that have $\lambda_1 = \lambda_2 = 0$. The trace is zero and the determinant is zero. The matrix A might not be 0 but check that $A^2 = 0$.

23 This matrix is singular with rank one. Find three λ's and three eigenvectors:

$$A = \begin{bmatrix} 1 \\ 2 \\ 1 \end{bmatrix} \begin{bmatrix} 2 & 1 & 2 \end{bmatrix} = \begin{bmatrix} 2 & 1 & 2 \\ 4 & 2 & 4 \\ 2 & 1 & 2 \end{bmatrix}.$$

24 Suppose A and B have the same eigenvalues $\lambda_1, \ldots, \lambda_n$ with the same independent eigenvectors $\boldsymbol{x}_1, \ldots, \boldsymbol{x}_n$. Then $A = B$. *Reason*: Any vector $\boldsymbol{x}$ is a combination $c_1\boldsymbol{x}_1 + \cdots + c_n\boldsymbol{x}_n$. What is $A\boldsymbol{x}$? What is $B\boldsymbol{x}$?

25 The block B has eigenvalues 1, 2 and C has eigenvalues 3, 4 and D has eigenvalues 5, 7. Find the eigenvalues of the 4 by 4 matrix A:

$$A = \begin{bmatrix} B & C \\ 0 & D \end{bmatrix} = \begin{bmatrix} 0 & 1 & 3 & 0 \\ -2 & 3 & 0 & 4 \\ 0 & 0 & 6 & 1 \\ 0 & 0 & 1 & 6 \end{bmatrix}.$$

26 Find the rank and the four eigenvalues of

$$A = \begin{bmatrix} 1 & 1 & 1 & 1 \\ 1 & 1 & 1 & 1 \\ 1 & 1 & 1 & 1 \\ 1 & 1 & 1 & 1 \end{bmatrix} \quad \text{and} \quad C = \begin{bmatrix} 1 & 0 & 1 & 0 \\ 0 & 1 & 0 & 1 \\ 1 & 0 & 1 & 0 \\ 0 & 1 & 0 & 1 \end{bmatrix}.$$

27 Subtract I from the previous A. Find the λ's and then the determinant:

$$B = A - I = \begin{bmatrix} 0 & 1 & 1 & 1 \\ 1 & 0 & 1 & 1 \\ 1 & 1 & 0 & 1 \\ 1 & 1 & 1 & 0 \end{bmatrix}.$$

When A (all ones) is 5 by 5, the eigenvalues of A and $B = A - I$ are ____ and ____.

28 (Review) Find the eigenvalues of A, B, and C:

$$A = \begin{bmatrix} 1 & 2 & 3 \\ 0 & 4 & 5 \\ 0 & 0 & 6 \end{bmatrix} \quad \text{and} \quad B = \begin{bmatrix} 0 & 0 & 1 \\ 0 & 2 & 0 \\ 3 & 0 & 0 \end{bmatrix} \quad \text{and} \quad C = \begin{bmatrix} 2 & 2 & 2 \\ 2 & 2 & 2 \\ 2 & 2 & 2 \end{bmatrix}.$$

29 When $a + b = c + d$ show that $(1, 1)$ is an eigenvector and find both eigenvalues of

$$A = \begin{bmatrix} a & b \\ c & d \end{bmatrix}.$$

30 When P exchanges rows 1 and 2 *and* columns 1 and 2, the eigenvalues don't change. Find eigenvectors of A and PAP for $\lambda = 11$:

$$A = \begin{bmatrix} 1 & 2 & 1 \\ 3 & 6 & 3 \\ 4 & 8 & 4 \end{bmatrix} \quad \text{and} \quad PAP = \begin{bmatrix} 6 & 3 & 3 \\ 2 & 1 & 1 \\ 8 & 4 & 4 \end{bmatrix}.$$

31 Suppose A has eigenvalues 0, 3, 5 with independent eigenvectors $\boldsymbol{u}$, $\boldsymbol{v}$, $\boldsymbol{w}$.

(a) Give a basis for the nullspace and a basis for the column space.

(b) Find a particular solution to $A\boldsymbol{x} = \boldsymbol{v} + \boldsymbol{w}$. Find all solutions.

(c) Show that $A\boldsymbol{x} = \boldsymbol{u}$ has no solution. (If it did then ____ would be in the column space.)

6.2 Diagonalizing a Matrix

When $\boldsymbol{x}$ is an eigenvector, multiplication by A is just multiplication by a single number: $A\boldsymbol{x} = \lambda\boldsymbol{x}$. All the difficulties of matrices are swept away. Instead of watching every interconnection in the system, we follow the eigenvectors separately. It is like having a *diagonal matrix*, with no off-diagonal interconnections. A diagonal matrix is easy to square and easy to invert, because it breaks into 1 by 1 matrices:

$$\begin{bmatrix} \lambda_1 & & \\ & \ddots & \\ & & \lambda_n \end{bmatrix}^2 = \begin{bmatrix} \lambda_1^2 & & \\ & \ddots & \\ & & \lambda_n^2 \end{bmatrix} \quad \text{and} \quad \begin{bmatrix} \lambda_1 & & \\ & \ddots & \\ & & \lambda_n \end{bmatrix}^{-1} = \begin{bmatrix} \lambda_1^{-1} & & \\ & \ddots & \\ & & \lambda_n^{-1} \end{bmatrix}.$$

The point of this section is very direct. ***The matrix A turns into this diagonal matrix Λ when we use the eigenvectors properly***. This is the matrix form of our key idea. We start right off with that one essential computation.

6D Diagonalization Suppose the n by n matrix A has n linearly independent eigenvectors. Put them into the columns of an ***eigenvector matrix*** S. Then $S^{-1}AS$ is the ***eigenvalue matrix*** Λ:

$$S^{-1}AS = \Lambda = \begin{bmatrix} \lambda_1 & & \\ & \ddots & \\ & & \lambda_n \end{bmatrix}. \tag{6.7}$$

The matrix A is "diagonalized." We use capital lambda for the eigenvalue matrix, because of the small λ's (the eigenvalues) on its diagonal.

Proof Multiply A times its eigenvectors, which are the columns of S. The first column of AS is $A\boldsymbol{x}_1$. That is $\lambda_1\boldsymbol{x}_1$:

$$AS = A\begin{bmatrix} & & \\ \boldsymbol{x}_1 & \cdots & \boldsymbol{x}_n \\ & & \end{bmatrix} = \begin{bmatrix} & & \\ \lambda_1\boldsymbol{x}_1 & \cdots & \lambda_n\boldsymbol{x}_n \\ & & \end{bmatrix}.$$

The trick is to split this matrix AS into S times Λ:

$$\begin{bmatrix} & & \\ \lambda_1\boldsymbol{x}_1 & \cdots & \lambda_n\boldsymbol{x}_n \\ & & \end{bmatrix} = \begin{bmatrix} & & \\ \boldsymbol{x}_1 & \cdots & \boldsymbol{x}_n \\ & & \end{bmatrix}\begin{bmatrix} \lambda_1 & & \\ & \ddots & \\ & & \lambda_n \end{bmatrix} = S\Lambda.$$

Keep those matrices in the right order! Then λ_1 multiplies the first column $\boldsymbol{x}_1$, as shown. The diagonalization is complete, and we can write $AS = S\Lambda$ in two good ways:

$$AS = S\Lambda \quad \text{is} \quad S^{-1}AS = \Lambda \quad \text{or} \quad A = S\Lambda S^{-1}. \tag{6.8}$$

The matrix S has an inverse, because its columns (the eigenvectors of A) were assumed to be linearly independent. Without n independent eigenvectors, we absolutely can't diagonalize.

The matrices A and Λ have the same eigenvalues $\lambda_1, \ldots, \lambda_n$. The eigenvectors are different. The job of the original eigenvectors was to diagonalize A—those eigenvectors of A went into S. The new eigenvectors, for the diagonal matrix Λ, are just the columns of I. By using Λ, we can solve differential equations or difference equations or even $A\boldsymbol{x} = \boldsymbol{b}$.

Example 6.5 The projection matrix $P = \left[\begin{smallmatrix} .5 & .5 \\ .5 & .5 \end{smallmatrix}\right]$ has eigenvalues 1 and 0. The eigenvectors $(1, 1)$ and $(-1, 1)$ go into S. From S comes S^{-1}. Then $S^{-1}PS$ is diagonal:

$$S^{-1}PS = \begin{bmatrix} .5 & .5 \\ -.5 & .5 \end{bmatrix}\begin{bmatrix} .5 & .5 \\ .5 & .5 \end{bmatrix}\begin{bmatrix} 1 & -1 \\ 1 & 1 \end{bmatrix} = \begin{bmatrix} 1 & 0 \\ 0 & 0 \end{bmatrix} = \Lambda.$$

The original projection satisfied $P^2 = P$. The new projection satisfies $\Lambda^2 = \Lambda$. The column space has swung around from $(1, 1)$ to $(1, 0)$. The nullspace has swung around from $(-1, 1)$ to $(0, 1)$. It is "just" a numerical simplification, but it makes all future computations easy.

Here are four small remarks about diagonalization, before the applications.

Remark 1 Suppose the numbers $\lambda_1, \ldots, \lambda_n$ are all different. Then it is automatic that the eigenvectors $\boldsymbol{x}_1, \ldots, \boldsymbol{x}_n$ are independent. See 6E below. Therefore *any matrix that has no repeated eigenvalues can be diagonalized.*

Remark 2 The eigenvector matrix S is not unique. We can multiply its columns (the eigenvectors) by any nonzero constants. Suppose we multiply by 5 and -1. Divide the rows of S^{-1} by 5 and -1 to find the new inverse:

$$S_{\text{new}}^{-1} P S_{\text{new}} = \begin{bmatrix} .1 & .1 \\ .5 & -.5 \end{bmatrix} \begin{bmatrix} .5 & .5 \\ .5 & .5 \end{bmatrix} \begin{bmatrix} 5 & 1 \\ 5 & -1 \end{bmatrix} = \begin{bmatrix} 1 & 0 \\ 0 & 0 \end{bmatrix} = \text{same } \Lambda.$$

The extreme case is $A = I$, when every vector is an eigenvector. Any invertible matrix S can be the eigenvector matrix. Then $S^{-1} I S = I$ (which is Λ).

Remark 3 To diagonalize A we *must* use an eigenvector matrix. From $S^{-1} A S = \Lambda$ we know that $AS = S\Lambda$. If the first column of S is y, the first column of AS is Ay. The first column of $S\Lambda$ is $\lambda_1 y$. For those to be equal, y must be an eigenvector.

The eigenvectors in S come in the same order as the eigenvalues in Λ. To reverse the order in S and Λ, put $(-1, 1)$ before $(1, 1)$:

$$\begin{bmatrix} -.5 & .5 \\ .5 & .5 \end{bmatrix} \begin{bmatrix} .5 & .5 \\ .5 & .5 \end{bmatrix} \begin{bmatrix} -1 & 1 \\ 1 & 1 \end{bmatrix} = \begin{bmatrix} 0 & 0 \\ 0 & 1 \end{bmatrix} = \text{new } \Lambda.$$

Remark 4 (repeated warning for repeated eigenvalues) Some matrices have too few eigenvectors. *Those matrices are not diagonalizable.* Here are two examples:

$$A = \begin{bmatrix} 1 & -1 \\ 1 & -1 \end{bmatrix} \quad \text{and} \quad A = \begin{bmatrix} 0 & 1 \\ 0 & 0 \end{bmatrix}.$$

Their eigenvalues happen to be 0 and 0. Nothing is special about zero—it is the repeated λ that counts. Look for eigenvectors of the second matrix.

$$A\boldsymbol{x} = 0\boldsymbol{x} \quad \text{means} \quad \begin{bmatrix} 0 & 1 \\ 0 & 0 \end{bmatrix} \begin{bmatrix} \boldsymbol{x} \end{bmatrix} = \begin{bmatrix} 0 \\ 0 \end{bmatrix}.$$

The eigenvector $\boldsymbol{x}$ is a multiple of $(1, 0)$. There is no second eigenvector, so A cannot be diagonalized. This matrix is the best example to test any statement about eigenvectors. In many true-false questions, this matrix leads to *false*.

Remember that there is no connection between invertibility and diagonalizability:

- ***Invertibility*** is concerned with the ***eigenvalues*** (zero or not).
- ***Diagonalizability*** is concerned with the ***eigenvectors*** (too few or enough).

Each eigenvalue has at least one eigenvector! If $(A-\lambda I)\boldsymbol{x} = \mathbf{0}$ leads you to $\boldsymbol{x} = \mathbf{0}$, then λ is *not* an eigenvalue. Look for a mistake in solving $\det(A-\lambda I) = 0$. If you have eigenvectors for n different λ's, those eigenvectors are independent and A is diagonalizable.

6E (Independent $\boldsymbol{x}$ from different λ) Eigenvectors $\boldsymbol{x}_1, \ldots, \boldsymbol{x}_j$ that correspond to distinct (all different) eigenvalues are linearly independent. An n by n matrix with n different eigenvalues (no repeated λ's) must be diagonalizable.

Proof Suppose $c_1\boldsymbol{x}_1 + c_2\boldsymbol{x}_2 = \mathbf{0}$. Multiply by A to find $c_1\lambda_1\boldsymbol{x}_1 + c_2\lambda_2\boldsymbol{x}_2 = \mathbf{0}$. Multiply by λ_2 to find $c_1\lambda_2\boldsymbol{x}_1 + c_2\lambda_2\boldsymbol{x}_2 = \mathbf{0}$. Now subtract one from the other:

$$\text{Subtraction leaves} \qquad (\lambda_1 - \lambda_2)c_1\boldsymbol{x}_1 = \mathbf{0}.$$

Since the λ's are different and $\boldsymbol{x}_1 \neq \mathbf{0}$, we are forced to the conclusion that $c_1 = 0$. Similarly $c_2 = 0$. No other combination gives $c_1\boldsymbol{x}_1 + c_2\boldsymbol{x}_2 = \mathbf{0}$, so the eigenvectors $\boldsymbol{x}_1$ and $\boldsymbol{x}_2$ must be independent.

This proof extends from two eigenvectors to j eigenvectors. Suppose $c_1\boldsymbol{x}_1 + \cdots + c_j\boldsymbol{x}_j = \mathbf{0}$. Multiply by A, then multiply separately by λ_j, and subtract. This removes $\boldsymbol{x}_j$. Now multiply by A and by λ_{j-1} and subtract. This removes $\boldsymbol{x}_{j-1}$. Eventually only a multiple of $\boldsymbol{x}_1$ is left:

$$(\lambda_1 - \lambda_2)\cdots(\lambda_1 - \lambda_j)c_1\boldsymbol{x}_1 = \mathbf{0} \quad \text{which forces} \quad c_1 = 0. \tag{6.9}$$

Similarly every $c_i = 0$. When the λ's are all different, the eigenvectors are independent.

With $j = n$ different eigenvalues, the full set of eigenvectors goes into the columns of S. Then A is diagonalized.

Example 6.6 The Markov matrix $A = \left[\begin{smallmatrix} .9 & .3 \\ .1 & .7 \end{smallmatrix}\right]$ in the last section had $\lambda_1 = 1$ and $\lambda_2 = .6$. Here is $S\Lambda S^{-1}$:

$$\begin{bmatrix} .9 & .3 \\ .1 & .7 \end{bmatrix} = \begin{bmatrix} .75 & 1 \\ .25 & -1 \end{bmatrix} \begin{bmatrix} 1 & 0 \\ 0 & .6 \end{bmatrix} \begin{bmatrix} 1 & 1 \\ .25 & -.75 \end{bmatrix} = S\Lambda S^{-1}.$$

The eigenvectors $(.75, .25)$ and $(1, -1)$ are in the columns of S. We know that they are also the eigenvectors of A^2. Therefore A^2 has the same S, and the eigenvalues in Λ are squared:

$$A^2 = S\Lambda S^{-1}S\Lambda S^{-1} = S\Lambda^2 S^{-1}.$$

Just keep going, and you see why the high powers A^k approach a "steady state":

$$A^k = S\Lambda^k S^{-1} = \begin{bmatrix} .75 & 1 \\ .25 & -1 \end{bmatrix} \begin{bmatrix} 1^k & 0 \\ 0 & (.6)^k \end{bmatrix} \begin{bmatrix} 1 & 1 \\ .25 & -.75 \end{bmatrix}.$$

As k gets larger, $(.6)^k$ gets smaller. In the limit it disappears completely. That limit is

$$A^\infty = \begin{bmatrix} .75 & 1 \\ .25 & -1 \end{bmatrix} \begin{bmatrix} 1 & 0 \\ 0 & 0 \end{bmatrix} \begin{bmatrix} 1 & 1 \\ .25 & -.75 \end{bmatrix} = \begin{bmatrix} .75 & .75 \\ .25 & .25 \end{bmatrix}.$$

The limit has the eigenvector $\boldsymbol{x}_1$ in both columns. We saw this steady state in the last section. Now we see it more quickly from powers of $A = S\Lambda S^{-1}$.

Eigenvalues of AB and $A+B$

The first guess about the eigenvalues of AB is not true. An eigenvalue λ of A times an eigenvalue β of B usually does *not* give an eigenvalue of AB. It is very tempting to think it should. *Here is a false proof*:

$$ABx = A\beta x = \beta Ax = \beta\lambda x. \tag{6.10}$$

It seems that β times λ is an eigenvalue. When x is an eigenvector for A and B, this proof is correct. ***The mistake is to expect that A and B automatically share the same eigenvector x.*** Usually they don't. Eigenvectors of A are not generally eigenvectors of B.

In this example A and B have zero eigenvalues, while 1 is an eigenvalue of AB:

$$A = \begin{bmatrix} 0 & 1 \\ 0 & 0 \end{bmatrix} \quad \text{and} \quad B = \begin{bmatrix} 0 & 0 \\ 1 & 0 \end{bmatrix}; \quad \text{then} \quad AB = \begin{bmatrix} 1 & 0 \\ 0 & 0 \end{bmatrix} \quad \text{and} \quad A+B = \begin{bmatrix} 0 & 1 \\ 1 & 0 \end{bmatrix}.$$

For the same reason, the eigenvalues of $A+B$ are generally not $\lambda+\beta$. Here $\lambda+\beta=0$ while $A+B$ has eigenvalues 1 and -1. (At least they add to zero.)

The false proof suggests what is true. Suppose x really is an eigenvector for both A and B. Then we do have $ABx = \lambda\beta x$. Sometimes all n eigenvectors are shared, and we *can* multiply eigenvalues. The test for A and B to share eigenvectors is important in quantum mechanics—time out to mention this application of linear algebra:

6F Commuting matrices share eigenvectors Suppose A and B can each be diagonalized. They share the same eigenvector matrix S if and only if $AB = BA$.

The uncertainty principle In quantum mechanics, the position matrix P and the momentum matrix Q do not commute. In fact $QP - PQ = I$ (these are infinite matrices). Then we cannot have $Px = \mathbf{0}$ at the same time as $Qx = \mathbf{0}$ (unless $x = \mathbf{0}$). If we knew the position exactly, we could not also know the momentum exactly. Problem 32 derives Heisenberg's uncertainty principle from the Schwarz inequality.

Fibonacci Numbers

We present a famous example, which leads to powers of matrices. The Fibonacci numbers start with $F_0 = 0$ and $F_1 = 1$. Then every new F is ***the sum of the two previous F's***:

$$\text{The sequence} \quad 0, 1, 1, 2, 3, 5, 8, 13, \ldots \quad \text{comes from} \quad F_{k+2} = F_{k+1} + F_k.$$

These numbers turn up in a fantastic variety of applications. Plants and trees grow in a spiral pattern, and a pear tree has 8 growths for every 3 turns. For a willow those numbers can be 13 and 5. The champion is a sunflower of Daniel O'Connell, whose seeds chose the almost unbelievable ratio $F_{12}/F_{13} = 144/233$. Our problem is more basic.

Problem: Find the Fibonacci number F_{100} The slow way is to apply the rule $F_{k+2} = F_{k+1} + F_k$ one step at a time. By adding $F_6 = 8$ to $F_7 = 13$ we reach $F_8 = 21$. Eventually we come to F_{100}. Linear algebra gives a better way.

The key is to begin with a matrix equation $\boldsymbol{u}_{k+1} = A\boldsymbol{u}_k$. That is a one-step rule for vectors, while Fibonacci gave a two-step rule for scalars. We match them by putting Fibonacci numbers into the vectors:

$$\text{Let} \quad \boldsymbol{u}_k = \begin{bmatrix} F_{k+1} \\ F_k \end{bmatrix}. \qquad \text{The rule} \quad \begin{matrix} F_{k+2} = F_{k+1} + F_k \\ F_{k+1} = F_{k+1} \end{matrix} \quad \text{becomes} \quad \boldsymbol{u}_{k+1} = \begin{bmatrix} 1 & 1 \\ 1 & 0 \end{bmatrix} \boldsymbol{u}_k. \tag{6.11}$$

Every step multiplies by $A = \left[\begin{smallmatrix} 1 & 1 \\ 1 & 0 \end{smallmatrix}\right]$. After 100 steps we reach $\boldsymbol{u}_{100} = A^{100}\boldsymbol{u}_0$:

$$\boldsymbol{u}_0 = \begin{bmatrix} 1 \\ 0 \end{bmatrix}, \quad \boldsymbol{u}_1 = \begin{bmatrix} 1 \\ 1 \end{bmatrix}, \quad \boldsymbol{u}_2 = \begin{bmatrix} 2 \\ 1 \end{bmatrix}, \quad \boldsymbol{u}_3 = \begin{bmatrix} 3 \\ 2 \end{bmatrix}, \quad \boldsymbol{u}_4 = \begin{bmatrix} 5 \\ 3 \end{bmatrix}, \quad \ldots .$$

The second component of $\boldsymbol{u}_{100}$ will be F_{100}. The Fibonacci numbers are in the powers of A—and we can compute A^{100} without multiplying 100 matrices.

This problem of A^{100} is just right for eigenvalues. Subtract λ from the diagonal of A:

$$A - \lambda I = \begin{bmatrix} 1-\lambda & 1 \\ 1 & -\lambda \end{bmatrix} \quad \text{leads to} \quad \det(A - \lambda I) = \lambda^2 - \lambda - 1.$$

The eigenvalues solve $\lambda^2 - \lambda - 1 = 0$. They come from the quadratic formula $\left(-b \pm \sqrt{b^2 - 4ac}\right)/2a$:

$$\lambda_1 = \frac{1+\sqrt{5}}{2} \approx 1.618 \qquad \text{and} \qquad \lambda_2 = \frac{1-\sqrt{5}}{2} \approx -.618.$$

These eigenvalues λ_1 and λ_2 lead to eigenvectors $\boldsymbol{x}_1$ and $\boldsymbol{x}_2$. This completes step 1:

$$\begin{bmatrix} 1-\lambda_1 & 1 \\ 1 & -\lambda_1 \end{bmatrix} \begin{bmatrix} \boldsymbol{x}_1 \end{bmatrix} = \begin{bmatrix} 0 \\ 0 \end{bmatrix} \quad \text{when} \quad \boldsymbol{x}_1 = \begin{bmatrix} \lambda_1 \\ 1 \end{bmatrix}$$

$$\begin{bmatrix} 1-\lambda_2 & 1 \\ 1 & -\lambda_2 \end{bmatrix} \begin{bmatrix} \boldsymbol{x}_2 \end{bmatrix} = \begin{bmatrix} 0 \\ 0 \end{bmatrix} \quad \text{when} \quad \boldsymbol{x}_2 = \begin{bmatrix} \lambda_2 \\ 1 \end{bmatrix}.$$

Step 2 finds the combination of those eigenvectors that gives $\boldsymbol{u}_0 = (1, 0)$:

$$\begin{bmatrix} 1 \\ 0 \end{bmatrix} = \frac{1}{\lambda_1 - \lambda_2}\left(\begin{bmatrix} \lambda_1 \\ 1 \end{bmatrix} - \begin{bmatrix} \lambda_2 \\ 1 \end{bmatrix}\right) \qquad \text{or} \qquad \boldsymbol{u}_0 = \frac{\boldsymbol{x}_1 - \boldsymbol{x}_2}{\lambda_1 - \lambda_2}. \tag{6.12}$$

The final step multiplies by A^{100} to find $\boldsymbol{u}_{100}$. The eigenvectors stay separate! They are multiplied by $(\lambda_1)^{100}$ and $(\lambda_2)^{100}$:

$$\boldsymbol{u}_{100} = \frac{(\lambda_1)^{100}\boldsymbol{x}_1 - (\lambda_2)^{100}\boldsymbol{x}_2}{\lambda_1 - \lambda_2}. \tag{6.13}$$

We want F_{100} = second component of $\boldsymbol{u}_{100}$. The second components of $\boldsymbol{x}_1$ and $\boldsymbol{x}_2$ are 1. Substitute the numbers λ_1 and λ_2 into equation (6.13), to find $\lambda_1 - \lambda_2 = \sqrt{5}$ and F_{100}:

$$F_{100} = \frac{1}{\sqrt{5}}\left[\left(\frac{1+\sqrt{5}}{2}\right)^{100} - \left(\frac{1-\sqrt{5}}{2}\right)^{100}\right] \approx 3.54 \cdot 10^{20}. \tag{6.14}$$

Is this a whole number? *Yes.* The fractions and square roots must disappear, because Fibonacci's rule $F_{k+2} = F_{k+1} + F_k$ stays with integers. The second term in (6.14) is less than $\frac{1}{2}$, so it must move the first term to the nearest whole number:

$$k\text{th Fibonacci number} \quad = \quad \text{nearest integer to} \quad \frac{1}{\sqrt{5}}\left(\frac{1+\sqrt{5}}{2}\right)^k. \tag{6.15}$$

The ratio of F_6 to F_5 is $8/5 = 1.6$. The ratio F_{101}/F_{100} must be very close to $(1+\sqrt{5})/2$. The Greeks called this number the *"golden mean."* For some reason a rectangle with sides 1.618 and 1 looks especially graceful.

Matrix Powers A^k

Fibonacci's example is a typical difference equation $\boldsymbol{u}_{k+1} = A\boldsymbol{u}_k$. ***Each step multiplies by A.*** The solution is $\boldsymbol{u}_k = A^k\boldsymbol{u}_0$. We want to make clear how diagonalizing the matrix gives a quick way to compute A^k.

The eigenvector matrix S produces $A = S\Lambda S^{-1}$. This is a factorization of the matrix, like $A = LU$ or $A = QR$. The new factorization is perfectly suited to computing powers, because ***every time S^{-1} multiplies S we get I***:

$$A^2 = S\Lambda S^{-1}S\Lambda S^{-1} = S\Lambda^2 S^{-1} \quad \text{and} \quad A^k = (S\Lambda S^{-1})\cdots(S\Lambda S^{-1}) = S\Lambda^k S^{-1}. \tag{6.16}$$

The eigenvector matrix for A^k is still S, and the eigenvalue matrix is Λ^k. We knew that. The eigenvectors don't change, and the eigenvalues are taken to the kth power. When A is diagonalized, $A^k\boldsymbol{u}_0$ is easy to compute:

1 Find the eigenvalues and n independent eigenvectors.

2 Write $\boldsymbol{u}_0$ as a combination $c_1\boldsymbol{x}_1 + \cdots + c_n\boldsymbol{x}_n$ of the eigenvectors.

3 Multiply each eigenvector $\boldsymbol{x}_i$ by $(\lambda_i)^k$.

4 Then $\boldsymbol{u}_k$ is the combination $c_1\lambda_1^k\boldsymbol{x}_1 + \cdots + c_n\lambda_n^k\boldsymbol{x}_n$.

In matrix language A^k is $(S\Lambda S^{-1})^k$ which is S times λ^k times S^{-1}. In vector language, the eigenvectors in S lead to the c's:

$$\boldsymbol{u}_0 = c_1\boldsymbol{x}_1 + \cdots + c_n\boldsymbol{x}_n = \begin{bmatrix} \boldsymbol{x}_1 & \cdots & \boldsymbol{x}_n \end{bmatrix}\begin{bmatrix} c_1 \\ \vdots \\ c_n \end{bmatrix}. \qquad \text{This says that} \quad \boldsymbol{u}_0 = S\boldsymbol{c}.$$

The coefficients $c_1, \ldots, c_n$ in Step 2 are $\boldsymbol{c} = S^{-1}\boldsymbol{u}_0$. Then Step 3 multiplies by Λ^k. Step 4 collects together the new combination $\sum c_i\lambda_i^k\boldsymbol{x}_i$. It is all in the matrices S and Λ^k and S^{-1}:

$$A^k\boldsymbol{u}_0 = S\Lambda^k S^{-1}\boldsymbol{u}_0 = S\Lambda^k\boldsymbol{c} = \begin{bmatrix} \boldsymbol{x}_1 & \cdots & \boldsymbol{x}_n \end{bmatrix}\begin{bmatrix} \lambda_1^k & & \\ & \ddots & \\ & & \lambda_n^k \end{bmatrix}\begin{bmatrix} c_1 \\ \vdots \\ c_n \end{bmatrix}. \tag{6.17}$$

This result is exactly $\boldsymbol{u}_k = c_1\lambda_1^k\boldsymbol{x}_1 + \cdots + c_n\lambda_n^k\boldsymbol{x}_n$. It is the solution to $\boldsymbol{u}_{k+1} = A\boldsymbol{u}_k$.

Example 6.7 Choose a matrix for which S and Λ and S^{-1} contain whole numbers:

$$A = \begin{bmatrix} 1 & 1 \\ 0 & 2 \end{bmatrix} \quad \text{has} \quad \lambda_1 = 1 \quad \text{and} \quad \boldsymbol{x}_1 = \begin{bmatrix} 1 \\ 0 \end{bmatrix}, \quad \lambda_2 = 2 \quad \text{and} \quad \boldsymbol{x}_2 = \begin{bmatrix} 1 \\ 1 \end{bmatrix}.$$

A is triangular, so its eigenvalues 1 and 2 are on the diagonal. A^k is also triangular, with 1 and 2^k on the diagonal. Those numbers stay separate in Λ^k. They are combined in A^k:

$$A^k = S\Lambda^k S^{-1} = \begin{bmatrix} 1 & 1 \\ 0 & 1 \end{bmatrix} \begin{bmatrix} 1^k & \\ & 2^k \end{bmatrix} \begin{bmatrix} 1 & -1 \\ 0 & 1 \end{bmatrix} = \begin{bmatrix} 1 & 2^k - 1 \\ 0 & 2^k \end{bmatrix}.$$

With $k = 1$ we get A. With $k = 0$ we get I. With $k = -1$ we get A^{-1}.

Note The zeroth power of every nonsingular matrix is $A^0 = I$. The product $S\Lambda^0 S^{-1}$ becomes SIS^{-1} which is I. Every λ to the zeroth power is 1. But the rule breaks down when $\lambda = 0$. Then 0^0 is not determined. We don't know A^0 when A is singular.

Nondiagonalizable Matrices (Optional)

Suppose $\lambda = 0$ is an eigenvalue of A. We discover that fact in two ways:

1 Eigenvectors (geometric) There are nonzero solutions to $A\boldsymbol{x} = 0\boldsymbol{x}$.

2 Eigenvalues (algebraic) A has a zero determinant and $\det(A - 0I) = 0$.

The number $\lambda = 0$ may be a simple eigenvalue or a multiple eigenvalue, and we want to know its ***multiplicity***. Most eigenvalues have multiplicity $M = 1$ (simple eigenvalues). Then there is a single line of eigenvectors, and $\det(A - \lambda I)$ does not have a double factor. For exceptional matrices, an eigenvalue (for example $\lambda = 0$) can be ***repeated***. Then there are two different ways to determine its multiplicity:

1 (Geometric multiplicity) Count the independent eigenvectors for $\lambda = 0$. This is the dimension of the nullspace of A. As always, the answer is $n - r$.

2 (Algebraic multiplicity) Count how often $\lambda = 0$ is a solution of $\det(A - \lambda I) = 0$. If the algebraic multiplicity is M, then λ^M is a factor of $\det(A - \lambda I)$.

The following matrix A is the standard example of trouble. Its eigenvalue $\lambda = 0$ is repeated. It is a double eigenvalue ($M = 2$) with only one eigenvector ($n - r = 1$). The geometric multiplicity can be below the algebraic multiplicity—it is never larger:

$$A = \begin{bmatrix} 0 & 1 \\ 0 & 0 \end{bmatrix} \quad \text{has rank} \quad r = 1 \quad \text{but} \quad \det(A - \lambda I) = \begin{vmatrix} -\lambda & 1 \\ 0 & -\lambda \end{vmatrix} = \lambda^2.$$

There "should" be two eigenvectors, because the equation $\lambda^2 = 0$ has a double root. The double factor λ^2 makes $M = 2$. But there is only one eigenvector $\boldsymbol{x} = (1, 0)$. ***This shortage of eigenvectors means that A is not diagonalizable.***

We have to emphasize: There is nothing special about $\lambda = 0$. It makes for easy computations, but these three matrices also have the same shortage of eigenvectors. Their repeated eigenvalue is $\lambda = 5$:

$$A = \begin{bmatrix} 5 & 1 \\ 0 & 5 \end{bmatrix} \quad \text{and} \quad A = \begin{bmatrix} 6 & -1 \\ 1 & 4 \end{bmatrix} \quad \text{and} \quad A = \begin{bmatrix} 7 & 2 \\ -2 & 3 \end{bmatrix}.$$

Those all have $\det(A - \lambda I) = (\lambda - 5)^2$. The algebraic multiplicity is $M = 2$. But $A - 5I$ has rank $r = 1$. The geometric multiplicity is $n - r = 1$. There is only one eigenvector, and these matrices are not diagonalizable.

Problem Set 6.2

Questions 1–8 are about the eigenvalue and eigenvector matrices.

1 Factor these two matrices into $A = S\Lambda S^{-1}$:

$$A = \begin{bmatrix} 1 & 2 \\ 0 & 3 \end{bmatrix} \quad \text{and} \quad A = \begin{bmatrix} 1 & 1 \\ 2 & 2 \end{bmatrix}.$$

2 If $A = S\Lambda S^{-1}$ then $A^3 = (\quad)(\quad)(\quad)$ and $A^{-1} = (\quad)(\quad)(\quad)$.

3 If A has $\lambda_1 = 2$ with eigenvector $\boldsymbol{x}_1 = \begin{bmatrix} 1 \\ 0 \end{bmatrix}$ and $\lambda_2 = 5$ with $\boldsymbol{x}_2 = \begin{bmatrix} 1 \\ 1 \end{bmatrix}$, use $S\Lambda S^{-1}$ to find A. No other matrix has the same λ's and $\boldsymbol{x}$'s.

4 Suppose $A = S\Lambda S^{-1}$. What is the eigenvalue matrix for $A + 2I$? What is the eigenvector matrix? Check that $A + 2I = (\quad)(\quad)(\quad)^{-1}$.

5 True or false: If the columns of S (eigenvectors of A) are linearly independent, then

(a) A is invertible (b) A is diagonalizable

(c) S is invertible (d) S is diagonalizable.

6 If the eigenvectors of A are the columns of I, then A is a ______ matrix. If the eigenvector matrix S is triangular, then S^{-1} is triangular. Prove that A is also triangular.

7 Describe all matrices S that diagonalize this matrix A:

$$A = \begin{bmatrix} 4 & 0 \\ 1 & 2 \end{bmatrix}.$$

Then describe all matrices that diagonalize A^{-1}.

8 Write down the most general matrix that has eigenvectors $\begin{bmatrix} 1 \\ 1 \end{bmatrix}$ and $\begin{bmatrix} 1 \\ -1 \end{bmatrix}$.

Questions 9–14 are about Fibonacci and Gibonacci numbers.

9 For the Fibonacci matrix $A = \begin{bmatrix} 1 & 1 \\ 1 & 0 \end{bmatrix}$, compute A^2 and A^3 and A^4. Then use the text and a calculator to find F_{20}.

10 Suppose each number G_{k+2} is the *average* of the two previous numbers G_{k+1} and G_k. Then $G_{k+2} = \frac{1}{2}(G_{k+1} + G_k)$:

$$\begin{aligned} G_{k+2} &= \tfrac{1}{2}G_{k+1} + \tfrac{1}{2}G_k \\ G_{k+1} &= G_{k+1} \end{aligned} \quad \text{is} \quad \begin{bmatrix} G_{k+2} \\ G_{k+1} \end{bmatrix} = \begin{bmatrix} & A & \end{bmatrix} \begin{bmatrix} G_{k+1} \\ G_k \end{bmatrix}.$$

(a) Find the eigenvalues and eigenvectors of A.

(b) Find the limit as $n \to \infty$ of the matrices $A^n = S\Lambda^n S^{-1}$.

(c) If $G_0 = 0$ and $G_1 = 1$ show that the Gibonacci numbers approach $\frac{2}{3}$.

11 Diagonalize the Fibonacci matrix by completing S^{-1}:

$$\begin{bmatrix} 1 & 1 \\ 1 & 0 \end{bmatrix} = \begin{bmatrix} \lambda_1 & \lambda_2 \\ 1 & 1 \end{bmatrix} \begin{bmatrix} \lambda_1 & 0 \\ 0 & \lambda_2 \end{bmatrix} \begin{bmatrix} & \\ & \end{bmatrix}.$$

Do the multiplication $S\Lambda^k S^{-1}\begin{bmatrix} 1 \\ 0 \end{bmatrix}$ to find its second component $F_k = (\lambda_1^k - \lambda_2^k)/(\lambda_1 - \lambda_2)$.

12 The numbers λ_1^k and λ_2^k satisfy the Fibonacci rule $F_{k+2} = F_{k+1} + F_k$:

$$\lambda_1^{k+2} = \lambda_1^{k+1} + \lambda_1^k \quad \text{and} \quad \lambda_2^{k+2} = \lambda_2^{k+1} + \lambda_2^k.$$

Prove this by using the original equation for the λ's. Then any combination of λ_1^k and λ_2^k satisfies the rule. The combination $F_k = (\lambda_1^k - \lambda_2^k)/(\lambda_1 - \lambda_2)$ gives the right start $F_0 = 0$ and $F_1 = 1$.

13 Suppose Fibonacci had started with $F_0 = 2$ and $F_1 = 1$. The rule $F_{k+2} = F_{k+1} + F_k$ is the same so the matrix A is the same. Its eigenvectors add to

$$\boldsymbol{x}_1 + \boldsymbol{x}_2 = \begin{bmatrix} \frac{1}{2}(1+\sqrt{5}) \\ 1 \end{bmatrix} + \begin{bmatrix} \frac{1}{2}(1-\sqrt{5}) \\ 1 \end{bmatrix} = \begin{bmatrix} 1 \\ 2 \end{bmatrix} = \begin{bmatrix} F_1 \\ F_0 \end{bmatrix}.$$

After 20 steps the second component of $A^{20}(\boldsymbol{x}_1 + \boldsymbol{x}_2)$ is $(\quad)^{20} + (\quad)^{20}$. Compute that number F_{20}.

14 Prove that every third Fibonacci number in $0, 1, 1, 2, 3, \ldots$ is even.

Questions 15–18 are about diagonalizability.

15 True or false: If the eigenvalues of A are 2, 2, 5 then the matrix is certainly

(a) invertible (b) diagonalizable (c) not diagonalizable.

16 True or false: If the only eigenvectors of A are multiples of $(1, 4)$ then A has

(a) no inverse (b) a repeated eigenvalue (c) no diagonalization $S\Lambda S^{-1}$.

17 Complete these matrices so that $\det A = 25$. Then check that $\lambda = 5$ is repeated—the determinant of $A - \lambda I$ is $(\lambda - 5)^2$. Find an eigenvector with $A\boldsymbol{x} = 5\boldsymbol{x}$. These matrices will not be diagonalizable because there is no second line of eigenvectors.

$$A = \begin{bmatrix} 8 & \\ & 2 \end{bmatrix} \quad \text{and} \quad A = \begin{bmatrix} 9 & 4 \\ & 1 \end{bmatrix} \quad \text{and} \quad A = \begin{bmatrix} 10 & 5 \\ -5 & \end{bmatrix}$$

18 The matrix $A = \left[\begin{smallmatrix} 3 & 1 \\ 0 & 3 \end{smallmatrix}\right]$ is not diagonalizable because the rank of $A - 3I$ is _____. Change one entry by .01 to make A diagonalizable. Which entries could you change?

Questions 19–23 are about powers of matrices.

19 $A^k = S\Lambda^k S^{-1}$ approaches the zero matrix as $k \to \infty$ if and only if every λ has absolute value less than _____. Which of these matrices has $A^k \to 0$?

$$A = \begin{bmatrix} .6 & .4 \\ .4 & .6 \end{bmatrix} \quad \text{and} \quad B = \begin{bmatrix} .6 & .9 \\ .1 & .6 \end{bmatrix}.$$

20 (Recommended) Find Λ and S to diagonalize A in Problem 19. What is the limit of Λ^k as $k \to \infty$? What is the limit of $S\Lambda^k S^{-1}$? In the columns of this limiting matrix you see the _____.

21 Find Λ and S to diagonalize B in Problem 19. What is $B^{10}\boldsymbol{u}_0$ for these $\boldsymbol{u}_0$?

$$\boldsymbol{u}_0 = \begin{bmatrix} 3 \\ 1 \end{bmatrix} \quad \text{and} \quad \boldsymbol{u}_0 = \begin{bmatrix} 3 \\ -1 \end{bmatrix} \quad \text{and} \quad \boldsymbol{u}_0 = \begin{bmatrix} 6 \\ 0 \end{bmatrix}.$$

22 Diagonalize A and compute $S\Lambda^k S^{-1}$ to prove this formula for A^k:

$$A = \begin{bmatrix} 2 & 1 \\ 1 & 2 \end{bmatrix} \quad \text{has} \quad A^k = \frac{1}{2}\begin{bmatrix} 1+3^k & 1-3^k \\ 1-3^k & 1+3^k \end{bmatrix}.$$

23 Diagonalize B and compute $S\Lambda^k S^{-1}$ to prove this formula for B^k:

$$B = \begin{bmatrix} 3 & 1 \\ 0 & 2 \end{bmatrix} \quad \text{has} \quad B^k = \begin{bmatrix} 3^k & 3^k - 2^k \\ 0 & 2^k \end{bmatrix}.$$

Questions 24–29 are new applications of $A = S\Lambda S^{-1}$.

24 Suppose that $A = S\Lambda S^{-1}$. Take determinants to prove that $\det A = \lambda_1\lambda_2\cdots\lambda_n =$ product of λ's. This quick proof only works when _____.

25 Show that trace AB = trace BA, by directly adding the diagonal entries of AB and BA:

$$A = \begin{bmatrix} a & b \\ c & d \end{bmatrix} \quad \text{and} \quad B = \begin{bmatrix} q & r \\ s & t \end{bmatrix}.$$

Choose A as S and B as ΛS^{-1}. Then $S\Lambda S^{-1}$ has the same trace as $\Lambda S^{-1}S$. The trace of A equals the trace of _____ which is _____.

26 $AB - BA = I$ is impossible since the left side has trace = ______. But find an elimination matrix so that $A = E$ and $B = E^{\mathrm{T}}$ give

$$AB - BA = \begin{bmatrix} -1 & 0 \\ 0 & 1 \end{bmatrix} \quad \text{which has trace zero.}$$

27 If $A = S\Lambda S^{-1}$, diagonalize the block matrix $B = \left[\begin{smallmatrix} A & 0 \\ 0 & 2A \end{smallmatrix}\right]$. Find its eigenvalue and eigenvector matrices.

28 Consider all 4 by 4 matrices A that are diagonalized by the same fixed eigenvector matrix S. Show that the A's form a subspace (cA and $A_1 + A_2$ have this same S). What is this subspace when $S = I$? What is its dimension?

29 Suppose $A^2 = A$. On the left side A multiplies each column of A. Which of our four subspaces contains eigenvectors with $\lambda = 1$? Which subspace contains eigenvectors with $\lambda = 0$? From the dimensions of those subspaces, A has a full set of independent eigenvectors and can be diagonalized.

30 (Recommended) Suppose $A\boldsymbol{x} = \lambda\boldsymbol{x}$. If $\lambda = 0$ then $\boldsymbol{x}$ is in the nullspace. If $\lambda \neq 0$ then $\boldsymbol{x}$ is in the column space. Those spaces have dimensions $(n - r) + r = n$. So why doesn't every square matrix have n linearly independent eigenvectors?

31 The eigenvalues of A are 1 and 9, the eigenvalues of B are -1 and 9:

$$A = \begin{bmatrix} 5 & 4 \\ 4 & 5 \end{bmatrix} \quad \text{and} \quad B = \begin{bmatrix} 4 & 5 \\ 5 & 4 \end{bmatrix}.$$

Find a matrix square root of A from $R = S\sqrt{\Lambda}\,S^{-1}$. Why is there no real matrix square root of B?

32 (Heisenberg's Uncertainty Principle) $AB - BA = I$ can happen for infinite matrices with $A = A^{\mathrm{T}}$ and $B = -B^{\mathrm{T}}$. Then

$$\boldsymbol{x}^{\mathrm{T}}\boldsymbol{x} = \boldsymbol{x}^{\mathrm{T}}AB\boldsymbol{x} - \boldsymbol{x}^{\mathrm{T}}BA\boldsymbol{x} \leq 2\|A\boldsymbol{x}\|\,\|B\boldsymbol{x}\|.$$

Explain that last step by using the Schwarz inequality. Then the inequality says that $\|A\boldsymbol{x}\|/\|\boldsymbol{x}\|$ times $\|B\boldsymbol{x}\|/\|\boldsymbol{x}\|$ is at least $\frac{1}{2}$. It is impossible to get the position error and momentum error both very small.

33 If A and B have the same λ's with the same independent eigenvectors, their factorizations into ______ are the same. So $A = B$.

34 Suppose the same S diagonalizes both A and B, so that $A = S\Lambda_1 S^{-1}$ and $B = S\Lambda_2 S^{-1}$. Prove that $AB = BA$.

35 If $A = S\Lambda S^{-1}$ show why the product $(A - \lambda_1 I)(A - \lambda_2 I) \cdots (A - \lambda_n I)$ is the zero matrix. The ***Cayley-Hamilton Theorem*** says that this product is zero for every matrix.

36 The matrix $A = \begin{bmatrix} -3 & 4 \\ -2 & 3 \end{bmatrix}$ has $\det(A - \lambda I) = \lambda^2 - 1$. Show from Problem 35 that $A^2 - I = 0$. Deduce that $A^{-1} = A$ and check that this is correct.

37 (a) When do the eigenvectors for $\lambda = 0$ span the nullspace $N(A)$?

(b) When do all the eigenvectors for $\lambda \neq 0$ span the column space $\boldsymbol{R}(A)$?

6.3 Symmetric Matrices

The eigenvalues of projection matrices are 1 and 0. The eigenvalues of reflection matrices are 1 and -1. Now we open up to all other ***symmetric matrices***. It is no exaggeration to say that these are the most important matrices the world will ever see—in the theory of linear algebra and also in the applications. We will come immediately to the two basic questions in this subject. Not only the questions, but also the answers.

You can guess the questions. The first is about the eigenvalues. The second is about the eigenvectors. ***What is special about $Ax = \lambda x$ when A is symmetric***? In matrix language, we are looking for special properties of Λ and S when $A = A^{\mathrm{T}}$.

The diagonalization $A = S\Lambda S^{-1}$ should reflect the fact that A is symmetric. We get some hint by transposing $S\Lambda S^{-1}$ to $(S^{-1})^{\mathrm{T}}\Lambda S^{\mathrm{T}}$. Since $A = A^{\mathrm{T}}$ those are the same. Possibly S^{-1} in the first form equals S^{T} in the second form. In that case $S^{\mathrm{T}}S = I$. We are near the answers and here they are:

1 A symmetric matrix has ***real eigenvalues***.

2 The ***eigenvectors*** can be chosen ***orthonormal***.

Those orthonormal eigenvectors go into the columns of S. There are n of them (independent because they are orthonormal). Every symmetric matrix can be diagonalized. ***Its eigenvector matrix S becomes an orthogonal matrix Q***. Orthogonal matrices have $Q^{-1} = Q^{\mathrm{T}}$—what we suspected about S is true. To remember it we write Q in place of S, when we choose orthonormal eigenvectors.

Why do we use the word "choose"? Because the eigenvectors do not *have* to be unit vectors. Their lengths are at our disposal. We will choose unit vectors—eigenvectors of length one, which are orthonormal and not just orthogonal. Then $A = S\Lambda S^{-1}$ is in its special and particular form for symmetric matrices:

6H (Spectral Theorem) Every symmetric matrix $A = A^{\mathrm{T}}$ has the factorization $Q\Lambda Q^{\mathrm{T}}$ with real diagonal Λ and orthogonal matrix Q:

$$A = Q\Lambda Q^{-1} = Q\Lambda Q^{\mathrm{T}} \quad \text{with} \quad Q^{-1} = Q^{\mathrm{T}}.$$

It is easy to see that $Q\Lambda Q^{\mathrm{T}}$ is symmetric. Take its transpose. You get $(Q^{\mathrm{T}})^{\mathrm{T}}\Lambda^{\mathrm{T}}Q^{\mathrm{T}}$, which is $Q\Lambda Q^{\mathrm{T}}$ again. So every matrix of this form is symmetric. The harder part is to prove that every symmetric matrix has real λ's and orthonormal $\boldsymbol{x}$'s. This is the *"spectral theorem"* in mathematics and the *"principal axis theorem"* in geometry and physics. We approach it in three steps:

1 By an example (which proves nothing, except that the spectral theorem might be true)

2 By calculating the 2 by 2 case (which convinces most fair-minded people)

3 By a proof when no eigenvalues are repeated (leaving only real diehards).

The diehards are worried about repeated eigenvalues. Are there still n orthonormal eigenvectors? Yes, there are. They go into the columns of S (which becomes Q). The last page before the problems outlines this fourth and final step.

We now take steps 1 and 2. In a sense they are optional. The 2 by 2 case is mostly for fun, since it is included in the final n by n case.

Example 6.8 Find the λ's and $\boldsymbol{x}$'s when $A = \begin{bmatrix} 1 & 2 \\ 2 & 4 \end{bmatrix}$ and $A - \lambda I = \begin{bmatrix} 1-\lambda & 2 \\ 2 & 4-\lambda \end{bmatrix}$.

Solution The equation $\det(A - \lambda I) = 0$ is $\lambda^2 - 5\lambda = 0$. The eigenvalues are 0 and 5 (*both real*). We can see them directly: $\lambda = 0$ is an eigenvalue because A is singular, and $\lambda = 5$ is the other eigenvalue so that $0 + 5$ agrees with $1 + 4$. This is the *trace* down the diagonal of A.

The eigenvectors are $(2, -1)$ and $(1, 2)$—orthogonal but not yet orthonormal. The eigenvector for $\lambda = 0$ is in the *nullspace* of A. The eigenvector for $\lambda = 5$ is in the *column space*. We ask ourselves, why are the nullspace and column space perpendicular? The Fundamental Theorem says that the nullspace is perpendicular to the ***row space***—not the column space. But our matrix is *symmetric*! Its row and column spaces are the same. So the eigenvectors $(2, -1)$ and $(1, 2)$ are perpendicular—which their dot product tells us anyway.

These eigenvectors have length $\sqrt{5}$. Divide them by $\sqrt{5}$ to get unit vectors. Put the unit vectors into the columns of S (which is Q). Then A is diagonalized:

$$Q^{-1}AQ = \frac{\begin{bmatrix} 2 & -1 \\ 1 & 2 \end{bmatrix}}{\sqrt{5}} \begin{bmatrix} 1 & 2 \\ 2 & 4 \end{bmatrix} \frac{\begin{bmatrix} 2 & 1 \\ -1 & 2 \end{bmatrix}}{\sqrt{5}} = \begin{bmatrix} 0 & 0 \\ 0 & 5 \end{bmatrix} = \Lambda.$$

Now comes the calculation for *any* 2 by 2 symmetric matrix. First, real eigenvalues. Second, perpendicular eigenvectors. The λ's come from

$$\det \begin{bmatrix} a-\lambda & b \\ b & c-\lambda \end{bmatrix} = \lambda^2 - (a+c)\lambda + (ac - b^2) = 0. \tag{6.18}$$

This quadratic factors into $(\lambda - \lambda_1)(\lambda - \lambda_2)$. The product $\lambda_1\lambda_2$ is the determinant $D = ac - b^2$. The sum $\lambda_1 + \lambda_2$ is the trace $T = a + c$. The quadratic formula could produce both λ's—but we only want to know they are real.

The test for real roots of $Ax^2 + Bx + C = 0$ is based on $B^2 - 4AC$. *This must not be negative*, or its square root in the quadratic formula would be imaginary. Our equation has different letters, $\lambda^2 - T\lambda + D = 0$, so the test is based on $T^2 - 4D$:

Real eigenvalues: $T^2 - 4D = (a+c)^2 - 4(ac - b^2)$ must not be negative. Rewrite that as $a^2 + 2ac + c^2 - 4ac + 4b^2$. Rewrite again as $(a-c)^2 + 4b^2$. This is not negative! So the roots of the quadratic (the eigenvalues) are certainly real.

Perpendicular eigenvectors: Compute $\boldsymbol{x}_1$ and $\boldsymbol{x}_2$ and their dot product:

$$(A - \lambda_1 I)\boldsymbol{x}_1 = \begin{bmatrix} a-\lambda_1 & b \\ b & c-\lambda_1 \end{bmatrix}\begin{bmatrix} \boldsymbol{x}_1 \end{bmatrix} = \boldsymbol{0} \quad \text{so} \quad \boldsymbol{x}_1 = \begin{bmatrix} b \\ \lambda_1 - a \end{bmatrix} \begin{matrix}\text{from}\\ \text{first}\\ \text{row}\end{matrix}$$

$$(A - \lambda_2 I)\boldsymbol{x}_2 = \begin{bmatrix} a-\lambda_2 & b \\ b & c-\lambda_2 \end{bmatrix}\begin{bmatrix} \boldsymbol{x}_2 \end{bmatrix} = \boldsymbol{0} \quad \text{so} \quad \boldsymbol{x}_2 = \begin{bmatrix} \lambda_2 - c \\ b \end{bmatrix} \begin{matrix}\text{from}\\ \text{second}\\ \text{row}\end{matrix} \tag{6.19}$$

When b is zero, A has perpendicular eigenvectors $(1, 0)$ and $(0, 1)$. Otherwise, take the dot product of $\boldsymbol{x}_1$ and $\boldsymbol{x}_2$ to prove they are perpendicular:

$$\boldsymbol{x}_1 \cdot \boldsymbol{x}_2 = b(\lambda_2 - c) + (\lambda_1 - a)b = b(\lambda_1 + \lambda_2 - a - c) = 0. \tag{6.20}$$

This is zero because $\lambda_1 + \lambda_2$ equals the trace $a + c$. Thus $\boldsymbol{x}_1 \cdot \boldsymbol{x}_2 = 0$.

Now comes the general n by n case, with real λ's and perpendicular eigenvectors.

6I Real Eigenvalues The eigenvalues of a real symmetric matrix are real.

Proof Suppose that $A\boldsymbol{x} = \lambda\boldsymbol{x}$. Until we know otherwise, λ might be a complex number. Then λ has the form $a + ib$ (a and b real). *Its complex conjugate is* $\bar{\lambda} = a - ib$. Similarly the components of $\boldsymbol{x}$ may be complex numbers, and switching the signs of their imaginary parts gives $\bar{\boldsymbol{x}}$. The good thing is that $\bar{\lambda}$ times $\bar{\boldsymbol{x}}$ is the conjugate of λ times $\boldsymbol{x}$. So take conjugates of $A\boldsymbol{x} = \lambda\boldsymbol{x}$, remembering that $A = \bar{A} = A^{\mathrm{T}}$ is real and symmetric:

$$A\boldsymbol{x} = \lambda\boldsymbol{x} \quad \text{leads to} \quad A\bar{\boldsymbol{x}} = \bar{\lambda}\bar{\boldsymbol{x}}. \qquad \text{Transpose to} \quad \bar{\boldsymbol{x}}^{\mathrm{T}}A = \bar{\boldsymbol{x}}^{\mathrm{T}}\bar{\lambda}. \tag{6.21}$$

Now take the dot product of the first equation with $\bar{\boldsymbol{x}}$ and the last equation with $\boldsymbol{x}$:

$$\bar{\boldsymbol{x}}^{\mathrm{T}}A\boldsymbol{x} = \bar{\boldsymbol{x}}^{\mathrm{T}}\lambda\boldsymbol{x} \qquad \text{and also} \qquad \bar{\boldsymbol{x}}^{\mathrm{T}}A\boldsymbol{x} = \bar{\boldsymbol{x}}^{\mathrm{T}}\bar{\lambda}\boldsymbol{x}. \tag{6.22}$$

The left sides are the same so the right sides are equal. One equation has λ, the other has $\bar{\lambda}$. They multiply $\bar{\boldsymbol{x}}^{\mathrm{T}}\boldsymbol{x}$ which is not zero—it is the squared length of the eigenvector. *Therefore λ must equal $\bar{\lambda}$*, and $a + ib$ equals $a - ib$. The imaginary part is $b = 0$ and the number $\lambda = a$ is real. Q.E.D.

The eigenvectors come from solving the real equation $(A - \lambda I)\boldsymbol{x} = \boldsymbol{0}$. So the $\boldsymbol{x}$'s are also real. The important fact is that they are perpendicular.

6J Orthogonal Eigenvectors Eigenvectors of a real symmetric matrix (when they correspond to different λ's) are always perpendicular.

When A has real eigenvalues and real orthogonal eigenvectors, then $A = A^{\mathrm{T}}$.

Proof Suppose $Ax = \lambda_1 x$ and $Ay = \lambda_2 y$ and $A = A^{\mathrm{T}}$. Take dot products with y and x:

$$(\lambda_1 x)^{\mathrm{T}} y = (Ax)^{\mathrm{T}} y = x^{\mathrm{T}} A^{\mathrm{T}} y = x^{\mathrm{T}} A y = x^{\mathrm{T}} \lambda_2 y. \tag{6.23}$$

The left side is $x^{\mathrm{T}}\lambda_1 y$, the right side is $x^{\mathrm{T}}\lambda_2 y$. Since $\lambda_1 \neq \lambda_2$, this proves that $x^{\mathrm{T}} y = 0$. Eigenvectors are perpendicular.

Example 6.9 Find the λ's and x's for this symmetric matrix with trace zero:

$$A = \begin{bmatrix} -3 & 4 \\ 4 & 3 \end{bmatrix} \quad \text{has} \quad \det(A - \lambda I) = \begin{vmatrix} -3-\lambda & 4 \\ 4 & 3-\lambda \end{vmatrix} = \lambda^2 - 25.$$

The roots of $\lambda^2 - 25 = 0$ are $\lambda_1 = 5$ and $\lambda_2 = -5$ (both real). The eigenvectors $x_1 = (1, 2)$ and $x_2 = (-2, 1)$ are perpendicular. They are not unit vectors, but they can be made into unit vectors. Divide by their lengths $\sqrt{5}$. The new x_1 and x_2 are the columns of Q, and Q^{-1} equals Q^{T}:

$$A = Q\Lambda Q^{-1} = \frac{\begin{bmatrix} 1 & -2 \\ 2 & 1 \end{bmatrix}}{\sqrt{5}} \begin{bmatrix} 5 & 0 \\ 0 & -5 \end{bmatrix} \frac{\begin{bmatrix} 1 & 2 \\ -2 & 1 \end{bmatrix}}{\sqrt{5}}.$$

This illustrates the rule to remember. Every 2 by 2 symmetric matrix looks like

$$A = Q\Lambda Q^{-1} = \begin{bmatrix} x_1 & x_2 \end{bmatrix} \begin{bmatrix} \lambda_1 & \\ & \lambda_2 \end{bmatrix} \begin{bmatrix} x_1^{\mathrm{T}} \\ x_2^{\mathrm{T}} \end{bmatrix}. \tag{6.24}$$

One more step. ***The columns x_1 and x_2 times the rows $\lambda_1 x_1^{\mathrm{T}}$ and $\lambda_2 x_2^{\mathrm{T}}$ produce A***:

$$A = \lambda_1 x_1 x_1^{\mathrm{T}} + \lambda_2 x_2 x_2^{\mathrm{T}}. \tag{6.25}$$

This is the great factorization $Q\Lambda Q^{\mathrm{T}}$, written in terms of λ's and x's. When the symmetric matrix is n by n, there are n columns in Q multiplying n rows in Q^{T}. The n pieces are $\lambda_i x_i x_i^{\mathrm{T}}$. Those are matrices!

$$A = \begin{bmatrix} -3 & 4 \\ 4 & 3 \end{bmatrix} = 5 \begin{bmatrix} 1/5 & 2/5 \\ 2/5 & 4/5 \end{bmatrix} - 5 \begin{bmatrix} 4/5 & -2/5 \\ -2/5 & 1/5 \end{bmatrix}. \tag{6.26}$$

On the right, each $x_i x_i^{\mathrm{T}}$ is a ***projection matrix***. It is like uu^{T} in Chapter 4. The spectral theorem for symmetric matrices says that A is a combination of projection matrices:

$$A = \lambda_1 P_1 + \cdots + \lambda_n P_n \qquad \lambda_i = \text{eigenvalue}, \quad P_i = \text{projection onto eigenvector}.$$

Complex Eigenvalues of Real Matrices

Equation (6.21) went from $Ax = \lambda x$ to $A\bar{x} = \bar{\lambda}\bar{x}$. In the end, λ and x were real. Those two equations were the same. But a *non*symmetric matrix can easily produce λ and x that are complex. In this case, $A\bar{x} = \bar{\lambda}\bar{x}$ is different from $Ax = \lambda x$. It gives us a new eigenvalue (which is $\bar{\lambda}$) and a new eigenvector (which is $\bar{x}$):

For real matrices, complex λ's and x's come in "conjugate pairs."

$$\textit{If} \quad Ax = \lambda x \quad \textit{then} \quad A\bar{x} = \bar{\lambda}\bar{x}.$$

Example 6.10 The rotation $\left[\begin{smallmatrix} \cos\theta & -\sin\theta \\ \sin\theta & \cos\theta \end{smallmatrix}\right]$ has $\lambda_1 = \cos\theta + i\sin\theta$ and $\lambda_2 = \cos\theta - i\sin\theta$. Those eigenvalues are conjugate to each other. They are λ and $\bar{\lambda}$, because the imaginary part $\sin\theta$ switches sign. The eigenvectors must be x and $\bar{x}$ (all this is true because the matrix is real):

$$\begin{aligned} \begin{bmatrix} \cos\theta & -\sin\theta \\ \sin\theta & \cos\theta \end{bmatrix} \begin{bmatrix} 1 \\ -i \end{bmatrix} &= (\cos\theta + i\sin\theta) \begin{bmatrix} 1 \\ -i \end{bmatrix} \\ \begin{bmatrix} \cos\theta & -\sin\theta \\ \sin\theta & \cos\theta \end{bmatrix} \begin{bmatrix} 1 \\ i \end{bmatrix} &= (\cos\theta - i\sin\theta) \begin{bmatrix} 1 \\ i \end{bmatrix}. \end{aligned} \tag{6.27}$$

One is $Ax = \lambda x$, the other is $A\bar{x} = \bar{\lambda}\bar{x}$. One eigenvector has $-i$, the other has $+i$. For this real matrix the eigenvalues and eigenvectors are not real. But they are conjugates.

By Euler's formula, $\cos\theta + i\sin\theta$ is the same as $e^{i\theta}$. Its absolute value is $|\lambda| = 1$, because $\cos^2\theta + \sin^2\theta = 1$. This fact $|\lambda| = 1$ holds for the eigenvalues of every orthogonal matrix—including this rotation.

We apologize that a touch of complex numbers slipped in. They are unavoidable even when the matrix is real.

6.4 Positive Definite Matrices

This section concentrates on ***symmetric matrices that have positive eigenvalues***. If symmetry makes a matrix important, this extra property (all $\lambda > 0$) makes it special. When we say special, we don't mean rare. Symmetric matrices with positive eigenvalues enter all kinds of applications of linear algebra.

The first problem is to recognize these matrices. You may say, just find the eigenvalues and test $\lambda > 0$. That is exactly what we want to avoid. Calculating eigenvalues is work. When the λ's are needed, we can compute them. But if we just want to know that they are positive, there are faster ways. Here are the two goals of this section:

1 To find quick tests that guarantee positive eigenvalues.

2 To explain the applications.

The matrices are symmetric to start with, so the λ's are automatically real numbers.

An important case is 2 by 2. ***When does $A = \left[\begin{smallmatrix} a & b \\ b & c \end{smallmatrix}\right]$ have $\lambda_1 > 0$ and $\lambda_2 > 0$?***

6L ***The eigenvalues of $A = A^{\mathrm{T}}$ are positive if and only if $a > 0$ and $ac - b^2 > 0$.***

This test is passed by $\begin{bmatrix} 4 & 5 \\ 5 & 7 \end{bmatrix}$ and failed by $\begin{bmatrix} 4 & 5 \\ 5 & 6 \end{bmatrix}$. Also failed by $\begin{bmatrix} -1 & 0 \\ 0 & -1 \end{bmatrix}$ even though the determinant is 1.

Proof without computing the λ's Suppose $\lambda_1 > 0$ and $\lambda_2 > 0$. Their product $\lambda_1\lambda_2$ equals the determinant $ac - b^2$. That must be positive. Therefore ac is also positive. Then a and c have the same sign. That sign has to be positive, because $\lambda_1 + \lambda_2$ equals the trace $a + c$. Proved so far: Positive λ's require $ac - b^2 > 0$ and $a > 0$ (and also $c > 0$).

The statement was "if and only if," so there is another half to prove. Start with $a > 0$ and $ac - b^2 > 0$. This also ensures $c > 0$. Since $\lambda_1\lambda_2$ equals the determinant $ac - b^2$, the λ's are both positive or both negative. Since $\lambda_1 + \lambda_2$ equals the trace $a + c > 0$, the λ's must be positive. End of proof.

We think of a and $ac - b^2$ as a 1 by 1 and a 2 by 2 determinant. Here is another form of the test. Instead of determinants, it checks for *positive pivots*.

6M ***The eigenvalues of $A = A^{\mathrm{T}}$ are positive if and only if the pivots are positive:***

$$a > 0 \qquad \text{and} \qquad \frac{ac - b^2}{a} > 0.$$

A new proof is unnecessary. The ratio of positive numbers is certainly positive:

$$a > 0 \quad \text{and} \quad ac - b^2 > 0 \qquad \text{if and only if} \qquad a > 0 \quad \text{and} \quad \frac{ac - b^2}{a} > 0.$$

The point is to recognize that last ratio as the second pivot of A:

$$\begin{bmatrix} a & b \\ b & c \end{bmatrix} \quad \underset{\text{The multiplier is } b/a}{\xrightarrow{\text{The first pivot is } a}} \quad \begin{bmatrix} a & b \\ 0 & c - \frac{b}{a}b \end{bmatrix} \quad \begin{matrix} \text{The second pivot is} \\ c - \dfrac{b^2}{a} = \dfrac{ac - b^2}{a}. \end{matrix}$$

This doesn't add information, to change from a and $ac - b^2$ to their ratio (the pivot). But it connects two big parts of linear algebra. ***Positive eigenvalues mean positive pivots and vice versa*** (for symmetric matrices!). If that holds for n by n symmetric matrices, and it does, then we have a quick test for $\lambda > 0$. The pivots are a lot faster to compute than the eigenvalues. It is very satisfying to see pivots and determinants and eigenvalues and even least squares come together in this course.

Example 6.11 This matrix has $a = 1$ (positive). But $ac - b^2 = (1)(3) - (2)^2$ is negative:

$$\begin{bmatrix} 1 & 2 \\ 2 & 3 \end{bmatrix} \quad \text{has a negative eigenvalue and a negative pivot.}$$

The pivots are 1 and -1. The eigenvalues also multiply to give -1. One eigenvalue is negative (we don't want its formula, just its sign).

Next comes a totally different way to look at symmetric matrices with positive eigenvalues. From $A\boldsymbol{x} = \lambda\boldsymbol{x}$, multiply by $\boldsymbol{x}^{\mathrm{T}}$ to get $\boldsymbol{x}^{\mathrm{T}}A\boldsymbol{x} = \lambda\boldsymbol{x}^{\mathrm{T}}\boldsymbol{x}$. The right side is a positive λ times a positive $\boldsymbol{x}^{\mathrm{T}}\boldsymbol{x} = \|\boldsymbol{x}\|^2$. So the left side $\boldsymbol{x}^{\mathrm{T}}A\boldsymbol{x}$ is positive when $\boldsymbol{x}$ is an eigenvector.

The new idea is that ***this number*** $\boldsymbol{x}^{\mathrm{T}}A\boldsymbol{x}$ ***is positive for all vectors*** $\boldsymbol{x}$, not just the eigenvectors. (Of course $\boldsymbol{x}^{\mathrm{T}}A\boldsymbol{x} = 0$ for the trivial vector $\boldsymbol{x} = \boldsymbol{0}$.)

There is a name for matrices with this property $\boldsymbol{x}^{\mathrm{T}}A\boldsymbol{x} > 0$. They are called ***positive definite***. We will prove that these are exactly the matrices whose eigenvalues and pivots are all positive.

Definition The matrix A is ***positive definite*** if $\boldsymbol{x}^{\mathrm{T}}A\boldsymbol{x} > 0$ for every nonzero vector:

$$\boldsymbol{x}^{\mathrm{T}}A\boldsymbol{x} = \begin{bmatrix} x & y \end{bmatrix}\begin{bmatrix} a & b \\ b & c \end{bmatrix}\begin{bmatrix} x \\ y \end{bmatrix} = ax^2 + 2bxy + cy^2 > 0.$$

Multiplying 1 by 2 times 2 by 2 times 2 by 1 produces a single number. It is $\boldsymbol{x}^{\mathrm{T}}A\boldsymbol{x}$. The four entries a, b, b, c give the four parts of that number. From a and c on the diagonal come the pure squares ax^2 and cy^2. From b and b off the diagonal come the cross terms bxy and byx (the same). Adding those four parts gives $\boldsymbol{x}^{\mathrm{T}}A\boldsymbol{x} = ax^2 + 2bxy + cy^2$.

We could have written x_1 and x_2 for the components of $\boldsymbol{x}$. They will be used often, so we avoid subscripts. The number $\boldsymbol{x}^{\mathrm{T}}A\boldsymbol{x}$ is a quadratic function of x and y:

$$f(x, y) = ax^2 + 2bxy + cy^2 \quad \text{is "second degree."}$$

The rest of this book has been linear (mostly $A\boldsymbol{x}$). Now the degree has gone from 1 to 2. Where the first derivatives of $A\boldsymbol{x}$ are constant, it is the *second* derivatives of $ax^2 + 2bxy + cy^2$ that are constant. Those second derivatives are $2a$, $2b$, $2b$, $2c$. They go into the matrix $2A$!

$$\frac{\partial f}{\partial x} = 2ax + 2by \qquad\qquad \begin{bmatrix} \dfrac{\partial^2 f}{\partial x^2} & \dfrac{\partial^2 f}{\partial y \partial x} \\[2ex] \dfrac{\partial^2 f}{\partial x \partial y} & \dfrac{\partial^2 f}{\partial y^2} \end{bmatrix} = \begin{bmatrix} 2a & 2b \\ 2b & 2c \end{bmatrix}.$$
$$\frac{\partial f}{\partial y} = 2bx + 2cy \qquad \text{and}$$

This is the 2 by 2 version of what everybody knows for 1 by 1. There the function is ax^2, its slope is $2ax$, and its second derivative is $2a$. Now the function is $\boldsymbol{x}^{\mathrm{T}}A\boldsymbol{x}$, its first derivatives are in the vector $2A\boldsymbol{x}$, and its second derivatives are in the matrix $2A$. Third derivatives are zero.

Where does calculus use second derivatives? They give the *bending* of the graph. When f'' is positive, the curve bends up from the tangent line. The parabola $y = ax^2$ is convex up or concave down according to $a > 0$ or $a < 0$. The point $x = 0$ is a minimum point of $y = x^2$ and a maximum point of $y = -x^2$. To decide minimum versus maximum for a two-variable function $f(x, y)$, we need to look at a matrix.

A is positive definite when $f = \boldsymbol{x}^{\mathrm{T}}A\boldsymbol{x}$ ***has a minimum at*** $x = y = 0$. At other points f is positive, at the origin f is zero. The statement "A is a positive definite matrix" is the 2 by 2 version of "a is a positive number."

Example 6.12 This matrix is positive definite. The function $f(x, y)$ is positive:

$$A = \begin{bmatrix} 1 & 2 \\ 2 & 7 \end{bmatrix} \quad \text{has pivots 1 and 3.}$$

The function is $x^T A x = x^2 + 4xy + 7y^2$. It is positive because it is a sum of squares:

$$x^2 + 4xy + 7y^2 = (x + 2y)^2 + 3y^2.$$

The pivots 1 and 3 multiply those squares. This is no accident! We prove below, by the algebra of "completing the square," that this always happens. So when the pivots are positive, the sum $f(x, y)$ is guaranteed to be positive.

Comparing Examples 6.11 and 6.12, the only difference is that a_{22} changed from 3 to 7. The borderline is when $a_{22} = 4$. Above 4, the matrix is positive definite. At $a_{22} = 4$, the borderline matrix is only "semidefinite." Then (> 0) changes to (≥ 0):

$$\begin{bmatrix} 1 & 2 \\ 2 & 4 \end{bmatrix} \quad \text{has pivots 1 and } \underline{\qquad}.$$

It has eigenvalues 5 and 0. It has $a > 0$ but $ac - b^2 = 0$.

We will summarize this section so far. We have four ways to recognize a positive definite matrix:

6N When a 2 by 2 symmetric matrix has one of these four properties, it has them all:

1 The eigenvalues are positive.

2 The 1 by 1 and 2 by 2 determinants are positive: $a > 0$ and $ac - b^2 > 0$.

3 The pivots are positive: $a > 0$ and $(ac - b^2)/a > 0$.

4 $x^T A x = ax^2 + 2bxy + cy^2$ is positive except at $(0, 0)$.

When A has any one (therefore all) of these four properties, it is a ***positive definite matrix***.

Note We deal only with symmetric matrices. The cross derivative $\partial^2 f/\partial x \partial y$ always equals $\partial^2 f/\partial y \partial x$. For $f(x, y, z)$ the nine second derivatives fill a symmetric 3 by 3 matrix. It is positive definite when the three pivots (and the three eigenvalues, and the three determinants) are positive.

Example 6.13 Is $f(x, y) = x^2 + 8xy + 3y^2$ everywhere positive—except at $(0, 0)$?

Solution The second derivatives are $f_{xx} = 2$ and $f_{xy} = f_{yx} = 8$ and $f_{yy} = 6$—all positive. But the test is *not* positive derivatives. We look for positive *definiteness*. The answer is *no*, this function is not always positive. By trial and error we locate a point $x = 1,\ y = -1$ where $f(1, -1) = 1 - 8 + 3 = -4$. Better to do linear algebra, and apply the exact tests to the matrix that produced $f(x, y)$:

$$x^2 + 8xy + 3y^2 = \begin{bmatrix} x & y \end{bmatrix} \begin{bmatrix} 1 & 4 \\ 4 & 3 \end{bmatrix} \begin{bmatrix} x \\ y \end{bmatrix}. \tag{6.28}$$

The matrix has $ac - b^2 = 3 - 16$. The pivots are 1 and -13. The eigenvalues are $\underline{\qquad}$ (we don't need them). The matrix is not positive definite.

Note how $8xy$ comes from $a_{12} = 4$ above the diagonal and $a_{21} = 4$ symmetrically below. Please do that matrix multiplication on the right side of (6.28) to see the function appear.

Main point The sign of b is not the essential thing. The cross derivative $\partial^2 f/\partial x \partial y$ can be positive or negative—it is b^2 that enters the tests. The *size* of b, compared to a and c, decides whether the matrix is positive definite and the function has a minimum.

Example 6.14 For which numbers c is $x^2 + 8xy + cy^2$ always positive (or zero)?

Solution The matrix is $A = \begin{bmatrix} 1 & 4 \\ 4 & c \end{bmatrix}$. Again $a = 1$ passes the first test. The second test has $ac - b^2 = c - 16$. For a positive definite matrix we need $c > 16$.

At the "semidefinite" borderline, $\begin{bmatrix} 1 & 4 \\ 4 & 16 \end{bmatrix}$ has $\lambda = 17$ and 0, determinants 1 and 0, pivots 1 and ______. The function $x^2 + 8xy + 16y^2$ is $(x + 4y)^2$. Its graph does not go below zero, but it stays equal to zero along the line $x + 4y = 0$. This is close to positive definite, but each test just misses.

Instead of two squares (see the next example), $(x + 4y)^2$ has only one square. The function can't be negative, but it is zero when $x = 4$ and $y = -1$: A minimum but not a strict minimum.

Example 6.15 When A is positive definite, write $f(x, y)$ as a sum of two squares.

Solution This is called "completing the square." The part $ax^2 + 2bxy$ is completed to the first square $a\left(x + \frac{b}{a}y\right)^2$. Multiplying that out, ax^2 and $2bxy$ are correct—but we have to add in $a\left(\frac{b}{a}y\right)^2$. To stay even, this added amount b^2y^2/a has to be subtracted off again:

$$ax^2 + 2bxy + cy^2 = a\left(x + \frac{b}{a}y\right)^2 + \left(\frac{ac - b^2}{a}\right)y^2.$$

After that gentle touch of algebra, the situation is clearer. There are two perfect squares (never negative). They are multiplied by two numbers, which could be positive or negative. *Those numbers a and $(ac - b^2)/a$ are the pivots!* So positive pivots give a sum of squares and a positive definite matrix. Think back to the factorization $A = LU$ and also $A = LDL^{\mathrm{T}}$:

$$\begin{bmatrix} a & b \\ b & c \end{bmatrix} = \begin{bmatrix} 1 & 0 \\ b/a & 1 \end{bmatrix} \begin{bmatrix} a & b \\ 0 & (ac - b^2)/a \end{bmatrix} \qquad \text{(this is } LU\text{)}$$

$$= \begin{bmatrix} 1 & 0 \\ b/a & 1 \end{bmatrix} \begin{bmatrix} a & \\ & (ac - b^2)/a \end{bmatrix} \begin{bmatrix} 1 & b/a \\ 0 & 1 \end{bmatrix} \qquad \text{(this is } LDL^{\mathrm{T}}\text{).} \qquad (6.29)$$

To complete the square, we dealt with a and b and fixed the rest later. *Elimination does exactly the same.* It deals with the first column, and fixes the rest later. We can work with the function $f(x, y)$ or the matrix. The numbers that come out are identical.

Outside the squares are the pivots. Inside $\left(x + \frac{b}{a}y\right)^2$ are the numbers 1 and $\frac{b}{a}$—which are in L. ***Every positive definite symmetric matrix factors into $A = LDL^{\mathrm{T}}$ with positive pivots.***

Important to compare $A = LDL^{\mathrm{T}}$ with $A = Q\Lambda Q^{\mathrm{T}}$. One is based on pivots (in D), the other is based on eigenvalues (in Λ). *Please* do not think that the pivots equal the eigenvalues. Their signs are the same, but the numbers are entirely different.

Positive Definite Matrices: n by n

For a 2 by 2 matrix, the "positive definite test" uses *eigenvalues* or *determinants* or *pivots*. All those numbers must be positive. We hope and expect that the same tests carry over to larger matrices. They do.

6O When an n by n symmetric matrix has one of these four properties, it has them all:

1 All the ***eigenvalues*** are positive.

2 All the ***upper left determinants*** are positive.

3 All the ***pivots*** are positive.

4 $\boldsymbol{x}^{\mathrm{T}}A\boldsymbol{x}$ is positive except at $\boldsymbol{x} = \mathbf{0}$. The matrix is ***positive definite***.

The upper left determinants are 1 by 1, 2 by 2, ..., n by n. The last one is the determinant of A. This remarkable theorem ties together the whole linear algebra course—at least for symmetric matrices. We believe that two examples are more helpful than a proof.

Example 6.16 Test the matrices A and A^* for positive definiteness:

$$A = \begin{bmatrix} 2 & -1 & 0 \\ -1 & 2 & -1 \\ 0 & -1 & 2 \end{bmatrix} \quad \text{and} \quad A^* = \begin{bmatrix} 2 & -1 & b \\ -1 & 2 & -1 \\ b & -1 & 2 \end{bmatrix}.$$

Solution A is an old friend (or enemy). Its pivots are 2 and $\frac{3}{2}$ and $\frac{4}{3}$, all positive. Its upper left determinants are 2 and 3 and 4, all positive. After some calculation, its eigenvalues are $2-\sqrt{2}$ and 2 and $2+\sqrt{2}$, all positive. We can write $\boldsymbol{x}^{\mathrm{T}}A\boldsymbol{x}$ as a sum of *three* squares (since $n = 3$). Using $A = LDL^{\mathrm{T}}$ the pivots appear outside the squares and the multipliers are inside:

$$\begin{aligned} \boldsymbol{x}^{\mathrm{T}}A\boldsymbol{x} &= 2\left(x_1^2 - x_1x_2 + x_2^2 - x_2x_3 + x_3^2\right) \\ &= 2\left(x_1 - \tfrac{1}{2}x_2\right)^2 + \tfrac{3}{2}\left(x_2 - \tfrac{2}{3}x_3\right)^2 + \tfrac{4}{3}x_3^2. \end{aligned}$$

Go to the second matrix A^*. *The determinant test is easiest.* The 1 by 1 determinant is 2 and the 2 by 2 determinant is 3. The 3 by 3 determinant comes from A^* itself:

$$\det A^* = 4 + 2b - 2b^2 \quad \text{must be positive.}$$

At $b = -1$ and $b = 2$ we get $\det A^* = 0$. In those cases A^* is positive *semi*definite (no inverse, zero eigenvalue, $\boldsymbol{x}^{\mathrm{T}}A^*\boldsymbol{x} \geq 0$). Between $b = -1$ and $b = 2$ the matrix is positive definite. The corner entry $b = 0$ in the first example was safely between.

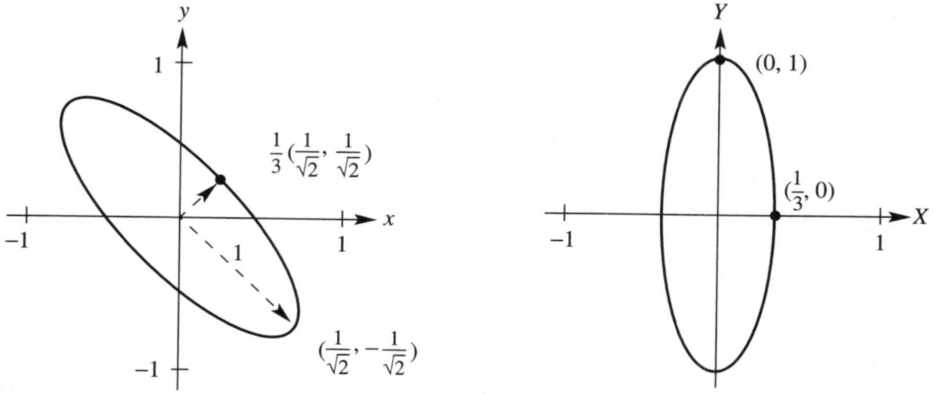

Figure 6.3 The tilted ellipse $5x^2 + 8xy + 5y^2 = 1$. Lined up it is $9X^2 + Y^2 = 1$.

The Ellipse $ax^2 + 2bxy + cy^2 = 1$

Think of a tilted ellipse centered at (0, 0), as in Figure 6.3a. Turn it to line up with the coordinate axes. That is Figure 6.3b. These two pictures show the geometry behind $A = Q\Lambda Q^{-1}$:

1 The tilted ellipse is associated with A. Its equation is $\boldsymbol{x}^{\mathrm{T}}A\boldsymbol{x} = 1$.

2 The lined-up ellipse is associated with Λ. Its equation is $\boldsymbol{X}^{\mathrm{T}}\Lambda\boldsymbol{X} = 1$.

3 The rotation that lines up the ellipse is Q.

Example 6.17 Find the axes of the tilted ellipse $5x^2 + 8xy + 5y^2 = 1$.

Solution Start with the positive definite matrix that matches this function:

$$\text{The function is} \quad \begin{bmatrix} x & y \end{bmatrix} \begin{bmatrix} 5 & 4 \\ 4 & 5 \end{bmatrix} \begin{bmatrix} x \\ y \end{bmatrix}. \qquad \text{The matrix is} \quad A = \begin{bmatrix} 5 & 4 \\ 4 & 5 \end{bmatrix}.$$

The eigenvalues of A are $\lambda_1 = 9$ and $\lambda_2 = 1$. The eigenvectors are $\begin{bmatrix} 1 \\ 1 \end{bmatrix}$ and $\begin{bmatrix} 1 \\ -1 \end{bmatrix}$. To make them unit vectors, divide by $\sqrt{2}$. Then $A = Q\Lambda Q^{\mathrm{T}}$ is

$$\begin{bmatrix} 5 & 4 \\ 4 & 5 \end{bmatrix} = \frac{\begin{bmatrix} 1 & 1 \\ 1 & -1 \end{bmatrix}}{\sqrt{2}} \begin{bmatrix} 9 & 0 \\ 0 & 1 \end{bmatrix} \frac{\begin{bmatrix} 1 & 1 \\ 1 & -1 \end{bmatrix}}{\sqrt{2}}.$$

Now multiply both sides by $[\, x \quad y \,]$ on the left and $\begin{bmatrix} x \\ y \end{bmatrix}$ on the right. The matrix $\begin{bmatrix} 1 & 1 \\ 1 & -1 \end{bmatrix}$ yields $x + y$ and $x - y$. The whole multiplication gives

$$5x^2 + 8xy + 5y^2 = 9\left(\frac{x+y}{\sqrt{2}}\right)^2 + 1\left(\frac{x-y}{\sqrt{2}}\right)^2. \tag{6.30}$$

The function is again a sum of two squares. But this is different from completing the square. The coefficients are not the pivots 5 and 9/5 from D, they are the eigenvalues 9 and 1 from Λ. Inside *these* squares are the eigenvectors $(1, 1)$ and $(1, -1)$. Previously L was inside the squares and the pivots were outside. Now Q is inside and the λ's are outside.

The axes of the tilted ellipse point along the eigenvectors. This explains why $A = Q\Lambda Q^{\mathrm{T}}$ is called the "principal axis theorem"—it displays the axes. Not only the axis directions (from the eigenvectors) but also the axis lengths (from the eigenvalues). To see it all, use capital letters for the new coordinates that line up the ellipse:

$$\frac{x+y}{\sqrt{2}} = X \qquad \text{and} \qquad \frac{x-y}{\sqrt{2}} = Y.$$

The ellipse becomes $9X^2 + Y^2 = 1$. The largest value of X^2 is $\frac{1}{9}$. The point at the end of the shorter axis has $X = \frac{1}{3}$ and $Y = 0$. Notice: The *bigger* eigenvalue λ_1 gives the *shorter* axis, of half-length $1/\sqrt{\lambda_1} = \frac{1}{3}$. The point at the end of the major axis has $X = 0$ and $Y = 1$. The smaller eigenvalue $\lambda_2 = 1$ gives the greater length $1/\sqrt{\lambda_2} = 1$. Those are really *half*-lengths because we start from $(0, 0)$.

In the xy system, the axes are along the eigenvectors of A. In the XY system, the axes are along the eigenvectors of Λ—the coordinate axes. Everything comes from the diagonalization $A = Q\Lambda Q^{\mathrm{T}}$.

6P Suppose $A = Q\Lambda Q^{\mathrm{T}}$ is positive definite. Then $\boldsymbol{x}^{\mathrm{T}}A\boldsymbol{x} = 1$ yields an ellipse:

$$\begin{bmatrix} x & y \end{bmatrix} Q\Lambda Q^{\mathrm{T}} \begin{bmatrix} x \\ y \end{bmatrix} = \begin{bmatrix} X & Y \end{bmatrix} \Lambda \begin{bmatrix} X \\ Y \end{bmatrix} = \lambda_1 X^2 + \lambda_2 Y^2 = 1.$$

The half-lengths of the axes are $1/\sqrt{\lambda_1}$ (when $X = 1$ and $Y = 0$) and $1/\sqrt{\lambda_2}$.

Note that A must be positive definite. If an eigenvalue is negative (exchange the 4's with the 5's in A), we don't have an ellipse. The sum of squares becomes a *difference of squares*: $9X^2 - Y^2 = 1$. This is a *hyperbola*.

Problem Set 6.4

Problems 1–13 are about tests for positive definiteness.

1 Which of A_1, A_2, A_3, A_4 has two positive eigenvalues? Use the test, don't compute the λ's:

$$A_1 = \begin{bmatrix} 5 & 6 \\ 6 & 7 \end{bmatrix} \quad A_2 = \begin{bmatrix} -1 & -2 \\ -2 & -5 \end{bmatrix} \quad A_3 = \begin{bmatrix} 1 & 10 \\ 10 & 100 \end{bmatrix} \quad A_4 = \begin{bmatrix} 1 & 10 \\ 10 & 101 \end{bmatrix}.$$

Explain why $c > 0$ (instead of $a > 0$) combined with $ac - b^2 > 0$ is also a complete test for $\left[\begin{smallmatrix} a & b \\ b & c \end{smallmatrix}\right]$ to have positive eigenvalues.

2 For which numbers b and c are these matrices positive definite?

$$A = \begin{bmatrix} 1 & b \\ b & 9 \end{bmatrix} \qquad \text{and} \qquad A = \begin{bmatrix} 2 & 4 \\ 4 & c \end{bmatrix}.$$

Factor each A into LU and then into LDL^{T}.

3 What is the quadratic $f = ax^2 + 2bxy + cy^2$ for these matrices? Complete the square to write f as a sum of one or two squares $d_1(\ \ )^2 + d_2(\ \ )^2$.

$$A = \begin{bmatrix} 1 & 2 \\ 2 & 7 \end{bmatrix} \quad \text{and} \quad A = \begin{bmatrix} 1 & 2 \\ 2 & 4 \end{bmatrix}.$$

4 Show that $f(x, y) = x^2 + 4xy + 3y^2$ does not have a minimum at $(0, 0)$ even though it has positive coefficients. Write f as a *difference* of squares and find a point (x, y) where f is negative.

5 The function $f(x, y) = 2xy$ certainly has a saddle point and not a minimum at $(0, 0)$. What symmetric matrix A produces this f? What are its eigenvalues?

6 Test to see if $A^{\mathrm{T}}A$ is positive definite:

$$A = \begin{bmatrix} 1 & 2 \\ 0 & 3 \end{bmatrix} \quad \text{and} \quad A = \begin{bmatrix} 1 & 1 \\ 1 & 2 \\ 2 & 1 \end{bmatrix} \quad \text{and} \quad A = \begin{bmatrix} 1 & 1 & 2 \\ 1 & 2 & 1 \end{bmatrix}.$$

7 (Important) If A has independent columns then $A^{\mathrm{T}}A$ is square, symmetric and invertible (Section 4.2). ***Show why $\boldsymbol{x}^{\mathrm{T}}A^{\mathrm{T}}A\boldsymbol{x}$ is positive except when $\boldsymbol{x} = \boldsymbol{0}$.*** Then $A^{\mathrm{T}}A$ is more than invertible, it is positive definite.

8 The function $f(x, y) = 3(x + 2y)^2 + 4y^2$ is positive except at $(0, 0)$. What is the matrix A, so that $f = [x \ \ y]A[x \ \ y]^{\mathrm{T}}$? Check that the pivots of A are 3 and 4.

9 Find the 3 by 3 matrix A and its pivots, rank, eigenvalues, and determinant:

$$\begin{bmatrix} x_1 & x_2 & x_3 \end{bmatrix} \begin{bmatrix} & & \\ & A & \\ & & \end{bmatrix} \begin{bmatrix} x_1 \\ x_2 \\ x_3 \end{bmatrix} = 4(x_1 - x_2 + 2x_3)^2.$$

10 Which 3 by 3 symmetric matrices A produce these functions $f = \boldsymbol{x}^{\mathrm{T}}A\boldsymbol{x}$? Why is the first matrix positive definite but not the second one?

(a) $f = 2(x_1^2 + x_2^2 + x_3^2 - x_1x_2 - x_2x_3)$

(b) $f = 2(x_1^2 + x_2^2 + x_3^2 - x_1x_2 - x_1x_3 - x_2x_3)$.

11 Compute the three upper left determinants to establish positive definiteness. Verify that their ratios give the second and third pivots.

$$A = \begin{bmatrix} 2 & 2 & 0 \\ 2 & 5 & 3 \\ 0 & 3 & 8 \end{bmatrix}.$$

12 For what numbers c and d are A and B positive definite? Test the 3 determinants:

$$A = \begin{bmatrix} c & 1 & 1 \\ 1 & c & 1 \\ 1 & 1 & c \end{bmatrix} \quad \text{and} \quad B = \begin{bmatrix} 1 & 2 & 3 \\ 2 & d & 4 \\ 3 & 4 & 5 \end{bmatrix}.$$

13 Find a matrix with $a > 0$ and $c > 0$ and $a + c > 2b$ that has a negative eigenvalue.

Problems 14–20 are about applications of the tests.

14 If A is positive definite then A^{-1} is positive definite. *First proof*: The eigenvalues of A^{-1} are positive because ______. *Second proof* (2 by 2): The entries of

$$A^{-1} = \frac{1}{ac - b^2} \begin{bmatrix} c & -b \\ -b & a \end{bmatrix} \quad \text{pass the test} \quad \_\_\_\_\_\_.$$

15 If A and B are positive definite, show that $A + B$ is also positive definite. Pivots and eigenvalues are not convenient for this; better to prove $\boldsymbol{x}^{\mathrm{T}}(A + B)\boldsymbol{x} > 0$ from the positive definiteness of A and B.

16 For a *block* positive definite matrix, the upper left block A must be positive definite:

$$\begin{bmatrix} \boldsymbol{x}^{\mathrm{T}} & \boldsymbol{y}^{\mathrm{T}} \end{bmatrix} \begin{bmatrix} A & B \\ B^{\mathrm{T}} & C \end{bmatrix} \begin{bmatrix} \boldsymbol{x} \\ \boldsymbol{y} \end{bmatrix} \quad \text{reduces to} \quad \boldsymbol{x}^{\mathrm{T}} A \boldsymbol{x} \quad \text{when} \quad \boldsymbol{y} = \_\_\_\_\_\_.$$

The complete block test is that A and $C - B^{\mathrm{T}} A^{-1} B$ must be positive definite.

17 A positive definite matrix cannot have a zero (or worse, a negative number) on its diagonal. Show that this matrix is not positive definite:

$$\begin{bmatrix} x_1 & x_2 & x_3 \end{bmatrix} \begin{bmatrix} 4 & 1 & 1 \\ 1 & 0 & 2 \\ 1 & 2 & 5 \end{bmatrix} \begin{bmatrix} x_1 \\ x_2 \\ x_3 \end{bmatrix} \quad \text{is not positive when} \quad (x_1, x_2, x_3) = (\quad , \quad , \quad).$$

18 The first entry a_{11} of a symmetric matrix A cannot be smaller than all the eigenvalues. If it were, then $A - a_{11} I$ would have ______ eigenvalues but it has a ______ on the main diagonal. Similarly no diagonal entry can be larger than all the eigenvalues.

19 If $\boldsymbol{x}$ is an eigenvector of A then $\boldsymbol{x}^{\mathrm{T}} A \boldsymbol{x} =$ ______. Prove that λ is positive when A is positive definite.

20 Give a quick reason why each of these statements is true:

(a) Every positive definite matrix is invertible.

(b) The only positive definite permutation matrix is $P = I$.

(c) The only positive definite projection matrix is $P = I$.

(d) A diagonal matrix with positive diagonal entries is positive definite.

(e) A symmetric matrix with a positive determinant might not be positive definite!

Problems 21–24 use the eigenvalues; Problems 25–27 are based on pivots.

21 For which s and t do these matrices have positive eigenvalues (therefore positive definite)?

$$A = \begin{bmatrix} s & -4 & -4 \\ -4 & s & -4 \\ -4 & -4 & s \end{bmatrix} \quad \text{and} \quad B = \begin{bmatrix} t & 3 & 0 \\ 3 & t & 4 \\ 0 & 4 & t \end{bmatrix}.$$

22 From $A = Q\Lambda Q^{\mathrm{T}}$ compute the positive definite symmetric square root $Q\Lambda^{1/2}Q^{\mathrm{T}}$ of each matrix. Check that this square root gives $R^2 = A$:

$$A = \begin{bmatrix} 5 & 4 \\ 4 & 5 \end{bmatrix} \quad \text{and} \quad A = \begin{bmatrix} 10 & 6 \\ 6 & 10 \end{bmatrix}.$$

23 You may have seen the equation for an ellipse as $\left(\frac{x}{a}\right)^2 + \left(\frac{y}{b}\right)^2 = 1$. What are a and b when the equation is written as $\lambda_1 x^2 + \lambda_2 y^2 = 1$? The ellipse $9x^2 + 16y^2 = 1$ has axes with half-lengths $a =$ ____ and $b =$ ____.

24 Draw the tilted ellipse $x^2 + xy + y^2 = 1$ and find the half-lengths of its axes from the eigenvalues of the corresponding A.

25 With positive pivots in D, the factorization $A = LDL^{\mathrm{T}}$ becomes $L\sqrt{D}\sqrt{D}L^{\mathrm{T}}$. (Square roots of the pivots give $D = \sqrt{D}\sqrt{D}$.) Then $C = L\sqrt{D}$ yields the ***Cholesky factorization*** $A = CC^{\mathrm{T}}$:

$$\text{From} \quad C = \begin{bmatrix} 3 & 0 \\ 1 & 2 \end{bmatrix} \quad \text{find } A. \qquad \text{From} \quad A = \begin{bmatrix} 4 & 8 \\ 8 & 25 \end{bmatrix} \quad \text{find } C.$$

26 In the Cholesky factorization $A = CC^{\mathrm{T}}$, with $C = L\sqrt{D}$, the ____ of the pivots are on the diagonal of C. Find C (lower triangular) for

$$A = \begin{bmatrix} 9 & 0 & 0 \\ 0 & 1 & 2 \\ 0 & 2 & 8 \end{bmatrix} \quad \text{and} \quad A = \begin{bmatrix} 1 & 1 & 1 \\ 1 & 2 & 2 \\ 1 & 2 & 7 \end{bmatrix}.$$

27 The symmetric factorization $A = LDL^{\mathrm{T}}$ means that $\boldsymbol{x}^{\mathrm{T}}A\boldsymbol{x} = \boldsymbol{x}^{\mathrm{T}}LDL^{\mathrm{T}}\boldsymbol{x}$. This is

$$\begin{bmatrix} x & y \end{bmatrix} \begin{bmatrix} a & b \\ b & c \end{bmatrix} \begin{bmatrix} x \\ y \end{bmatrix} = \begin{bmatrix} x & y \end{bmatrix} \begin{bmatrix} 1 & 0 \\ b/a & 1 \end{bmatrix} \begin{bmatrix} a & 0 \\ 0 & (ac - b^2)/a \end{bmatrix} \begin{bmatrix} 1 & b/a \\ 0 & 1 \end{bmatrix} \begin{bmatrix} x \\ y \end{bmatrix}.$$

Multiplication produces $ax^2 + 2bxy + cy^2 = a\left(x + \frac{b}{a}y\right)^2 +$ ____ y^2. The second pivot completes the square. Test with $a = 2,\ b = 4,\ c = 10$.

28 Without multiplying $A = \begin{bmatrix} \cos\theta & -\sin\theta \\ \sin\theta & \cos\theta \end{bmatrix} \begin{bmatrix} 2 & 0 \\ 0 & 5 \end{bmatrix} \begin{bmatrix} \cos\theta & \sin\theta \\ -\sin\theta & \cos\theta \end{bmatrix}$, find

(a) the determinant of A (b) the eigenvalues of A

(c) the eigenvectors of A (d) a reason why A is positive definite.

29 For $f_1(x, y) = \frac{1}{4}x^4 + x^2y + y^2$ and $f_2(x, y) = x^3 + xy - x$ find the second derivative matrices A_1 and A_2:

$$A = \begin{bmatrix} \partial^2 f/\partial x^2 & \partial^2 f/\partial x \partial y \\ \partial^2 f/\partial y \partial x & \partial^2 f/\partial y^2 \end{bmatrix}.$$

A_1 is positive definite so f_1 is concave up (= convex). Find the minimum point of f_1 and the saddle point of f_2 (where first derivatives are zero).

30 The graph of $z = x^2 + y^2$ is a bowl opening upward. The graph of $z = x^2 - y^2$ is a saddle. The graph of $z = -x^2 - y^2$ is a bowl opening downward. What is a test for $z = ax^2 + 2bxy + cy^2$ to have a saddle at (0, 0)?

31 Which values of c give a bowl and which give a saddle for the graph of the function $z = 4x^2 + 12xy + cy^2$? Describe the graph at the borderline value of c.

6.5 Stability and Preconditioning

Up to now, our approach to $A\boldsymbol{x} = \boldsymbol{b}$ has been "direct." We accepted A as it came. We attacked it with Gaussian elimination. This section is about ***"iterative"*** methods, which replace A by a simpler matrix S. The difference $T = S - A$ is moved over to the right side of the equation. The problem becomes easier to solve, with S instead of A. But there is a price—*the simpler system has to be solved over and over.*

An iterative method is easy to invent. Just split A into $S - T$. Then $A\boldsymbol{x} = \boldsymbol{b}$ is the same as

$$S\boldsymbol{x} = T\boldsymbol{x} + \boldsymbol{b}. \tag{6.31}$$

The novelty is to solve (6.31) iteratively. Each guess $\boldsymbol{x}_k$ leads to the next $\boldsymbol{x}_{k+1}$:

$$S\boldsymbol{x}_{k+1} = T\boldsymbol{x}_k + \boldsymbol{b}. \tag{6.32}$$

Start with any $\boldsymbol{x}_0$. Then solve $S\boldsymbol{x}_1 = T\boldsymbol{x}_0 + \boldsymbol{b}$. Continue to the second iteration $S\boldsymbol{x}_2 = T\boldsymbol{x}_1 + \boldsymbol{b}$. A hundred iterations are very common—maybe more. Stop when (and if!) the new vector $\boldsymbol{x}_{k+1}$ is sufficiently close to $\boldsymbol{x}_k$—or when the residual $\boldsymbol{b} - A\boldsymbol{x}_k$ is near zero. We can choose the stopping test. Our hope is to get near the true solution, more quickly than by elimination. When the sequence $\boldsymbol{x}_k$ converges, its limit $\boldsymbol{x} = \boldsymbol{x}_\infty$ does solve equation (6.31). The proof is to let $k \to \infty$ in equation (6.32).

The two goals of the splitting $A = S - T$ are ***speed per step*** and ***fast convergence of the $\boldsymbol{x}_k$***. The speed of each step depends on S and the speed of convergence depends on $S^{-1}T$:

1 Equation (6.32) should be easy to solve for $\boldsymbol{x}_{k+1}$. The ***"preconditioner"*** S could be diagonal or triangular. When its LU factorization is known, each iteration step is fast.

2 The difference $\boldsymbol{x} - \boldsymbol{x}_k$ (this is the error $\boldsymbol{e}_k$) should go quickly to zero. Subtracting equation (6.32) from (6.31) cancels $\boldsymbol{b}$, and it leaves the ***error equation***:

$$S\boldsymbol{e}_{k+1} = T\boldsymbol{e}_k \quad \text{which means} \quad \boldsymbol{e}_{k+1} = S^{-1}T\boldsymbol{e}_k. \tag{6.33}$$

At every step the error is multiplied by $S^{-1}T$. If $S^{-1}T$ is small, its powers go quickly to zero. But what is "small"?

The extreme splitting is $S = A$ and $T = 0$. Then the first step of the iteration is the original $A\boldsymbol{x} = \boldsymbol{b}$. Convergence is perfect and $S^{-1}T$ is zero. But the cost of that step is what we wanted to avoid. The choice of S is a battle between speed per step (a simple S) and fast convergence (S close to A). Here are some popular choices:

(J) $S =$ diagonal part of A (the iteration is called *Jacobi's method*)
(GS) $S =$ lower triangular part of A (*Gauss-Seidel method*)
(SOR) $S =$ combination of Jacobi and Gauss-Seidel (*successive overrelaxation*)
(ILU) $S =$ approximate L times approximate U (*incomplete LU method*).

Our first question is pure linear algebra: ***When do the $\boldsymbol{x}_k$'s converge to $\boldsymbol{x}$?*** The answer uncovers the number $|\lambda|_{\max}$ that controls convergence. It is the largest eigenvalue of the iteration matrix $S^{-1}T$.

The Spectral Radius Controls Convergence

Equation (6.33) is $\boldsymbol{e}_{k+1} = S^{-1}T\boldsymbol{e}_k$. Every iteration step multiplies the error by the same matrix $B = S^{-1}T$. The error after k steps is $\boldsymbol{e}_k = B^k\boldsymbol{e}_0$. ***The error approaches zero if the powers of $B = S^{-1}T$ approach zero.*** It is beautiful to see how the eigenvalues of B—the largest eigenvalue in particular—control the matrix powers B^k.

Convergence The powers B^k approach zero if and only if every eigenvalue of B satisfies $|\lambda| < 1$. ***The rate of convergence is controlled by the spectral radius $|\lambda|_{\max}$.***

The test for convergence is $|\lambda|_{\max} < 1$. Real eigenvalues must lie between -1 and 1. Complex eigenvalues $\lambda = a + ib$ must lie inside the unit circle in the complex plane. In that case the absolute value $|\lambda|$ is the square root of $a^2 + b^2$. In every case the spectral radius is the largest distance from the origin 0 to the eigenvalues $\lambda_1, \ldots, \lambda_n$. Those are eigenvalues of the iteration matrix $B = S^{-1}T$.

To see why $|\lambda|_{\max} < 1$ is necessary, suppose the starting error $\boldsymbol{e}_0$ happens to be an eigenvector of B. After one step the error is $B\boldsymbol{e}_0 = \lambda\boldsymbol{e}_0$. After k steps the error is $B^k\boldsymbol{e}_0 = \lambda^k\boldsymbol{e}_0$. If we start with an eigenvector, we continue with that eigenvector—and it grows or decays with the powers λ^k. *This factor λ^k goes to zero when* $|\lambda| < 1$. Since this condition is required of every eigenvalue, we need $|\lambda|_{\max} < 1$.

To see why $|\lambda|_{\max} < 1$ is sufficient for the error to approach zero, suppose $\boldsymbol{e}_0$ is a combination of eigenvectors:

$$\boldsymbol{e}_0 = c_1\boldsymbol{x}_1 + \cdots + c_n\boldsymbol{x}_n \quad \text{leads to} \quad \boldsymbol{e}_k = c_1\lambda_1^k\boldsymbol{x}_1 + \cdots + c_n\lambda_n^k\boldsymbol{x}_n. \tag{6.34}$$

This is the point of eigenvectors! They grow independently, each one controlled by its eigenvalue. When we multiply by B, the eigenvector $\boldsymbol{x}_i$ is multiplied by λ_i. If all $|\lambda_i| < 1$ then equation (6.34) ensures that $\boldsymbol{e}_k$ goes to zero.

Examples $B = \begin{bmatrix} .6 & .5 \\ .6 & .5 \end{bmatrix}$ has $\lambda_{\max} = 1.1$, $\quad B' = \begin{bmatrix} .6 & 1.1 \\ 0 & .5 \end{bmatrix}$ has $\lambda_{\max} = .6$.
B^2 is 1.1 times B. Then B^3 is $(1.1)^2$ times B. The powers of B blow up. Contrast with the powers of B'. The matrix $(B')^k$ has $(.6)^k$ and $(.5)^k$ on its diagonal. The off-diagonal entries also involve $(.6)^k$, which sets the speed of convergence.

Note There is a technical difficulty when B does not have n independent eigenvectors. (To produce this effect in B', change .5 to .6 .) The starting error $\boldsymbol{e}_0$ may not be a combination of eigenvectors—there are too few for a basis. Then diagonalization is impossible and equation (6.34) is not correct. We turn to the *Jordan form*:

$$B = SJS^{-1} \qquad \text{and} \qquad B^k = SJ^kS^{-1}. \tag{6.35}$$

The Jordan forms J and J^k are made of "blocks" with one repeated eigenvalue:

$$\text{The powers of a 2 by 2 block are} \qquad \begin{bmatrix} \lambda & 1 \\ 0 & \lambda \end{bmatrix}^k = \begin{bmatrix} \lambda^k & k\lambda^{k-1} \\ 0 & \lambda^k \end{bmatrix}.$$

If $|\lambda| < 1$ then these powers approach zero. The extra factor k from a double eigenvalue is overwhelmed by the decreasing factor λ^{k-1}. This applies to all Jordan blocks. A larger block has $k^2\lambda^{k-2}$ in J^k, which also approaches zero when $|\lambda| < 1$.

If all $|\lambda| < 1$ then $J^k \to 0$ and $B^k \to 0$. This proves our Theorem: ***Convergence requires*** $|\boldsymbol{\lambda}|_{\max} < \mathbf{1}$.

We emphasize that the same requirement $|\lambda|_{\max} < 1$ is the condition for ***stability*** of a difference equation $\boldsymbol{u}_{k+1} = A\boldsymbol{u}_k$:

$$\boldsymbol{u}_k = A^k\boldsymbol{u}_0 \to \mathbf{0} \quad \text{if and only if all} \quad |\lambda(A)| < 1.$$

7

LINEAR TRANSFORMATIONS

7.1 The Idea of a Linear Transformation

When a matrix A multiplies a vector $\boldsymbol{v}$, it produces another vector $A\boldsymbol{v}$. You could think of the matrix as "transforming" the first vector $\boldsymbol{v}$ into the second vector $A\boldsymbol{v}$. ***In goes $\boldsymbol{v}$, out comes $A\boldsymbol{v}$***. This transformation follows the same idea as a function. In goes a number x, out comes $f(x)$. For one vector $\boldsymbol{v}$ or one number x, we multiply by the matrix or we evaluate the function. The deeper goal is to see the complete picture—all $\boldsymbol{v}$'s and all $A\boldsymbol{v}$'s.

Start again with a matrix A. It transforms $\boldsymbol{v}$ to $A\boldsymbol{v}$. It transforms $\boldsymbol{w}$ to $A\boldsymbol{w}$. Then we *know* what happens to $\boldsymbol{u} = \boldsymbol{v} + \boldsymbol{w}$. There is no doubt about $A\boldsymbol{u}$, it has to equal $A\boldsymbol{v} + A\boldsymbol{w}$. Matrix multiplication gives a ***linear transformation***:

Definition A transformation T assigns an output $T(\boldsymbol{v})$ to each input vector $\boldsymbol{v}$. The transformation is *linear* if it meets these requirements for all $\boldsymbol{v}$ and $\boldsymbol{w}$:

(a) $T(\boldsymbol{v} + \boldsymbol{w}) = T(\boldsymbol{v}) + T(\boldsymbol{w})$

(b) $T(c\boldsymbol{v}) = cT(\boldsymbol{v})$ for all c.

If the input is $\boldsymbol{v} = \mathbf{0}$, the output must be $T(\boldsymbol{v}) = \mathbf{0}$. Very often requirements (a) and (b) are combined into one:

$$\textbf{\textit{Linearity:}} \quad T(c\boldsymbol{v} + d\boldsymbol{w}) = cT(\boldsymbol{v}) + dT(\boldsymbol{w}).$$

A linear transformation is highly restricted. Suppose we add $\boldsymbol{u}_0$ to every vector. Then $T(\boldsymbol{v}) = \boldsymbol{v} + \boldsymbol{u}_0$ and $T(\boldsymbol{w}) = \boldsymbol{w} + \boldsymbol{u}_0$. This isn't good, or at least it isn't linear. Applying T to $\boldsymbol{v} + \boldsymbol{w}$ gives $\boldsymbol{v} + \boldsymbol{w} + \boldsymbol{u}_0$. That is not the same as $T(\boldsymbol{v}) + T(\boldsymbol{w})$:

$$T(\boldsymbol{v}) + T(\boldsymbol{w}) = \boldsymbol{v} + \boldsymbol{u}_0 + \boldsymbol{w} + \boldsymbol{u}_0$$

does not equal

$$T(\boldsymbol{v} + \boldsymbol{w}) = \boldsymbol{v} + \boldsymbol{w} + \boldsymbol{u}_0.$$

The exception is when $\boldsymbol{u}_0 = \mathbf{0}$. The transformation reduces to $T(\boldsymbol{v}) = \boldsymbol{v}$. This is the ***identity transformation*** (nothing moves, as in multiplication by I). That is certainly linear. In this case the input space $\mathbf{V}$ is the same as the output space $\mathbf{W}$.

The transformation $T(\boldsymbol{v}) = \boldsymbol{v} + \boldsymbol{u}_0$ may not be linear but it is *"affine."* Straight lines stay straight, but lines through the origin don't stay through the origin. *Affine means "linear plus shift."* Its output can be $A\boldsymbol{v} + \boldsymbol{u}_0$, for any matrix A.

Computer graphics works with affine transformations. We focus on the $A\boldsymbol{v}$ part—which is linear.

Example 7.1 Choose a fixed vector $\boldsymbol{a} = (1, 3, 4)$, and let $T(\boldsymbol{v})$ be the dot product $\boldsymbol{a} \cdot \boldsymbol{v}$:

The input is $\boldsymbol{v} = (v_1, v_2, v_3)$. The output is $T(\boldsymbol{v}) = \boldsymbol{a} \cdot \boldsymbol{v} = v_1 + 3v_2 + 4v_3$.

This is linear. The inputs $\boldsymbol{v}$ come from three-dimensional space, so $\mathbf{V} = \mathbf{R}^3$. The outputs are just numbers, so we can say $\mathbf{W} = \mathbf{R}^1$. We are multiplying by the row matrix $A = [\,1 \;\; 3 \;\; 4\,]$. Then $T(\boldsymbol{v}) = A\boldsymbol{v}$.

You will get good at recognizing which transformations are linear. If the output involves squares or products or lengths, v_1^2 or $v_1 v_2$ or $\|\boldsymbol{v}\|$, then T is not linear.

Example 7.2 $T(\boldsymbol{v}) = \|\boldsymbol{v}\|$ is not linear. Requirement (a) for linearity would be $\|\boldsymbol{v}+\boldsymbol{w}\| = \|\boldsymbol{v}\| + \|\boldsymbol{w}\|$. Requirement (b) would be $\|c\boldsymbol{v}\| = c\|\boldsymbol{v}\|$. The first is false (the sides of a triangle satisfy an *inequality* $\|\boldsymbol{v} + \boldsymbol{w}\| \leq \|\boldsymbol{v}\| + \|\boldsymbol{w}\|$). The second is also false, when we choose $c = -1$. The length $\| -\boldsymbol{v}\|$ is not $-\|\boldsymbol{v}\|$.

Example 7.3 (More important) T is the transformation that *rotates every vector by* 30°. The domain is the xy plane (where the input vector $\boldsymbol{v}$ is). The range is also the xy plane (where the rotated vector $T(\boldsymbol{v})$ is). We described T without mentioning a matrix: just rotate the plane.

Is rotation linear? *Yes it is.* We can rotate two vectors and add the results. Or we can first add the vectors and then rotate. One result $T(\boldsymbol{v}) + T(\boldsymbol{w})$ is the same as the other result $T(\boldsymbol{v} + \boldsymbol{w})$. The whole plane is turning together, in this linear transformation.

Note Transformations have a language of their own. Where there is no matrix, we can't talk about a column space. But the idea can be rescued and used. The column space consisted of all outputs $A\boldsymbol{v}$. The nullspace consisted of all inputs for which $A\boldsymbol{v} = \mathbf{0}$. Translate those into "range" and "kernel":

Range of T = set of all outputs $T(\boldsymbol{v})$: corresponds to column space

Kernel of T = set of all inputs for which $T(\boldsymbol{v}) = \mathbf{0}$: corresponds to nullspace.

The range is a subspace of the output space $\mathbf{W}$. The kernel is a subspace of the input space $\mathbf{V}$. When T is multiplication by a matrix, $T(\boldsymbol{v}) = A\boldsymbol{v}$, you can translate back to column space and nullspace. We won't always say range and kernel when these other words are available.

For an m by n matrix, the nullspace is a subspace of $\mathbf{V} = \mathbf{R}^n$. The column space is a subspace of ______. The range might or might not be the whole output space $\mathbf{W}$.

Examples of Transformations (mostly linear)

Example 7.4 Project every 3-dimensional vector down onto the xy plane. The range is that plane, which contains every $T(\boldsymbol{v})$. The kernel is the z axis (which projects down to zero). This projection is linear.

Example 7.5 Project every 3-dimensional vector onto the horizontal plane $z = 1$. The vector $\boldsymbol{v} = (x, y, z)$ is transformed to $T(\boldsymbol{v}) = (x, y, 1)$. This transformation is not linear. Why not?

Example 7.6 Multiply every 3-dimensional vector by a 3 by 3 matrix A. This is definitely a linear transformation!

$$T(\boldsymbol{v}+\boldsymbol{w}) = A(\boldsymbol{v}+\boldsymbol{w}) \quad \text{which equals} \quad A\boldsymbol{v} + A\boldsymbol{w} = T(\boldsymbol{v}) + T(\boldsymbol{w}).$$

Example 7.7 Suppose $T(\boldsymbol{v})$ is multiplication by an *invertible matrix*. The kernel is the zero vector; the range $\mathbf{W}$ equals the domain $\mathbf{V}$. Another linear transformation is multiplication by A^{-1}. This is the ***inverse transformation*** T^{-1}, which brings every vector $T(\boldsymbol{v})$ back to $\boldsymbol{v}$:

$$T^{-1}(T(\boldsymbol{v})) = \boldsymbol{v} \quad \text{matches the matrix multiplication} \quad A^{-1}(A\boldsymbol{v}) = \boldsymbol{v}.$$

We are reaching an unavoidable question. ***Are all linear transformations produced by matrices?*** Each m by n matrix does produce a linear transformation from $\mathbf{R}^n$ to $\mathbf{R}^m$. The rule is $T(\boldsymbol{v}) = A\boldsymbol{v}$. Our question is the converse. When a linear T is described as a "rotation" or "projection" or ..., is there always a matrix hiding behind it?

The answer is *yes*. This is an approach to linear algebra that doesn't start with matrices. The next section shows that we still end up with matrices.

Linear Transformations of the Plane

It is more interesting to *see* a transformation than to define it. When a 2 by 2 matrix A multiplies all vectors in $\mathbf{R}^2$, we can watch how it acts. Start with a "house" in the xy plane. It has eleven endpoints. Those eleven vectors $\boldsymbol{v}$ are transformed into eleven vectors $A\boldsymbol{v}$. Straight lines between $\boldsymbol{v}$'s become straight lines between the transformed vectors $A\boldsymbol{v}$. (The transformation is linear!) Therefore T (house) is some kind of a house—possibly stretched or rotated or otherwise unlivable.

This part of the book is visual not theoretical. We will show six houses and the matrices that produce them. The columns of H are the eleven circled points of the first house. (H is 2 by 12, so plot2d connects the last circle to the first.) The 11 points in the house matrix are transformed by $A * H$ to produce the other houses:

$$H = \begin{bmatrix} -6 & -6 & -7 & 0 & 7 & 6 & 6 & -3 & -3 & 0 & 0 & -6 \\ -7 & 2 & 1 & 8 & 1 & 2 & -7 & -7 & -2 & -2 & -7 & -7 \end{bmatrix}.$$

Problem Set 7.1

1 A linear transformation must leave the zero vector fixed: $T(\mathbf{0}) = \mathbf{0}$. Prove this from $T(\boldsymbol{v}+\boldsymbol{w}) = T(\boldsymbol{v}) + T(\boldsymbol{w})$ by choosing $\boldsymbol{w} =$ _____. Prove it also from requirement (b) by choosing $c =$ _____.

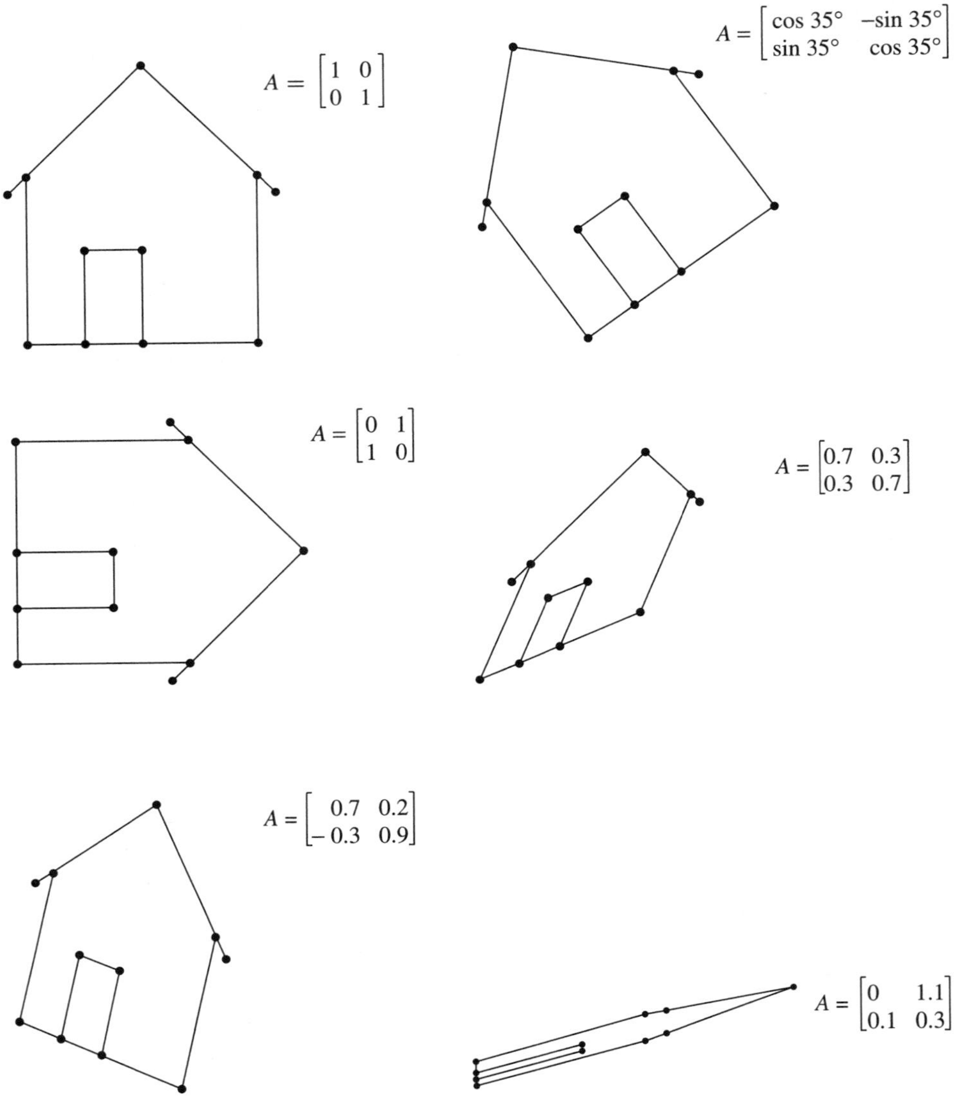

Figure 7.1 Linear transformations of a house drawn by plot2d(A*H).

2 Requirement (b) gives $T(c\boldsymbol{v}) = cT(\boldsymbol{v})$ and also $T(d\boldsymbol{w}) = dT(\boldsymbol{w})$. Then by addition, requirement (a) gives $T(\quad) = (\quad)$. What is $T(c\boldsymbol{v} + d\boldsymbol{w} + e\boldsymbol{u})$?

3 Which of these transformations is not linear? The input is $\boldsymbol{v} = (v_1, v_2)$:

(a) $T(\boldsymbol{v}) = (v_2, v_1)$ (b) $T(\boldsymbol{v}) = (v_1, v_1)$ (c) $T(\boldsymbol{v}) = (0, v_1)$

(d) $T(\boldsymbol{v}) = (0, 1)$.

4 If S and T are linear transformations, is $S(T(\boldsymbol{v}))$ linear or quadratic?

(a) (Special case) If $S(\boldsymbol{v}) = \boldsymbol{v}$ and $T(\boldsymbol{v}) = \boldsymbol{v}$, then $S(T(\boldsymbol{v})) = \boldsymbol{v}$ or $\boldsymbol{v}^2$?

(b) (General case) $S(\boldsymbol{w}_1+\boldsymbol{w}_2) = S(\boldsymbol{w}_1)+S(\boldsymbol{w}_2)$ and $T(\boldsymbol{v}_1+\boldsymbol{v}_2) = T(\boldsymbol{v}_1)+T(\boldsymbol{v}_2)$ combine into

$$S(T(\boldsymbol{v}_1 + \boldsymbol{v}_2)) = S(\underline{\qquad}) = \underline{\qquad} + \underline{\qquad}.$$

5 Suppose $T(\boldsymbol{v}) = \boldsymbol{v}$ except that $T(0, v_2) = (0, 0)$. Show that this transformation satisfies $T(c\boldsymbol{v}) = cT(\boldsymbol{v})$ but not $T(\boldsymbol{v} + \boldsymbol{w}) = T(\boldsymbol{v}) + T(\boldsymbol{w})$.

6 Which of these transformations satisfy $T(\boldsymbol{v} + \boldsymbol{w}) = T(\boldsymbol{v}) + T(\boldsymbol{w})$ and which satisfy $T(c\boldsymbol{v}) = cT(\boldsymbol{v})$?

(a) $T(\boldsymbol{v}) = \boldsymbol{v}/\|\boldsymbol{v}\|$ (b) $T(\boldsymbol{v}) = v_1+v_2+v_3$ (c) $T(\boldsymbol{v}) = (v_1, 2v_2, 3v_3)$

(d) $T(\boldsymbol{v}) =$ largest component of $\boldsymbol{v}$.

7 For these transformations of $\mathbf{V} = \mathbf{R}^2$ to $\mathbf{W} = \mathbf{R}^2$, find $T(T(\boldsymbol{v}))$. Is this transformation T^2 linear?

(a) $T(\boldsymbol{v}) = -\boldsymbol{v}$ (b) $T(\boldsymbol{v}) = \boldsymbol{v} + (1, 1)$

(c) $T(\boldsymbol{v}) = 90°$ rotation $= (-v_2, v_1)$

(d) $T(\boldsymbol{v}) =$ projection $= \left(\frac{v_1+v_2}{2}, \frac{v_1+v_2}{2}\right)$.

8 Find the range and kernel (like the column space and nullspace) of T:

(a) $T(v_1, v_2) = (v_2, v_1)$ (b) $T(v_1, v_2, v_3) = (v_1, v_2)$

(c) $T(v_1, v_2) = (0, 0)$ (d) $T(v_1, v_2) = (v_1, v_1)$.

9 The "cyclic" transformation T is defined by $T(v_1, v_2, v_3) = (v_2, v_3, v_1)$. What is $T(T(\boldsymbol{v}))$? What is $T^3(\boldsymbol{v})$? What is $T^{100}(\boldsymbol{v})$? Apply T three times and 100 times to $\boldsymbol{v}$.

10 A linear transformation from $\mathbf{V}$ to $\mathbf{W}$ has an *inverse* from $\mathbf{W}$ to $\mathbf{V}$ when the range is all of $\mathbf{W}$ and the kernel contains only $\boldsymbol{v} = \mathbf{0}$. Why are these transformations not invertible?

(a) $T(v_1, v_2) = (v_2, v_2)$ $\mathbf{W} = \mathbf{R}^2$

(b) $T(v_1, v_2) = (v_1, v_2, v_1 + v_2)$ $\mathbf{W} = \mathbf{R}^3$

(c) $T(v_1, v_2) = v_1$ $\mathbf{W} = \mathbf{R}^1$

11 If $T(\boldsymbol{v}) = A\boldsymbol{v}$ and A is m by n, then T is "multiplication by A."

(a) What are the input and output spaces $\mathbf{V}$ and $\mathbf{W}$?

(b) Why is range of T = column space of A?

(c) Why is kernel of T = nullspace of A?

12 Suppose a linear T transforms $(1, 1)$ to $(2, 2)$ and $(2, 0)$ to $(0, 0)$. Find $T(\boldsymbol{v})$ when

(a) $\boldsymbol{v} = (2, 2)$ (b) $\boldsymbol{v} = (3, 1)$ (c) $\boldsymbol{v} = (-1, 1)$ (d) $\boldsymbol{v} = (a, b)$.

Problems 13–20 may be harder. The input space V contains all 2 by 2 matrices M.

13 M is any 2 by 2 matrix and $A = \begin{bmatrix} 1 & 2 \\ 3 & 4 \end{bmatrix}$. The transformation T is defined by $T(M) = AM$. What rules of matrix multiplication show that T is linear?

14 Suppose $A = \begin{bmatrix} 1 & 2 \\ 3 & 5 \end{bmatrix}$. Show that the range of T is the whole matrix space **V** and the kernel is the zero matrix:

(1) If $AM = 0$ prove that M must be the zero matrix.

(2) Find a solution to $AM = B$ for any 2 by 2 matrix B.

15 Suppose $A = \begin{bmatrix} 1 & 2 \\ 3 & 6 \end{bmatrix}$. Show that the identity matrix I is not in the range of T. Find a nonzero matrix M such that $T(M) = AM$ is zero.

16 Suppose T transposes every matrix M. Try to find a matrix A which gives $AM = M^{\mathrm{T}}$ for every M. Show that no matrix A will do it. *To professors:* Is this a linear transformation that doesn't come from a matrix?

17 The transformation T that transposes every matrix is definitely linear. Which of these extra properties are true?

(a) $T^2 =$ identity transformation.

(b) The kernel of T is the zero matrix.

(c) Every matrix is in the range of T.

(d) $T(M) = -M$ is impossible.

18 Suppose $T(M) = \begin{bmatrix} 1 & 0 \\ 0 & 0 \end{bmatrix} \begin{bmatrix} M \end{bmatrix} \begin{bmatrix} 0 & 0 \\ 0 & 1 \end{bmatrix}$. Find a matrix with $T(M) \neq 0$. Describe all matrices with $T(M) = 0$ (the kernel of T) and all output matrices $T(M)$ (the range of T).

19 If $A \neq 0$ and $B \neq 0$ then there is a matrix M such that $AMB \neq 0$. Show by example that $M = I$ might fail. For your example find an M that succeeds.

20 If A and B are invertible and $T(M) = AMB$, find $T^{-1}(M)$ in the form $(\quad)M(\quad)$.

Questions 21–27 are about house transformations by matrices. The output is T (house).

21 How can you tell from the picture of T (house) that A is

(a) a diagonal matrix?

(b) a rank-one matrix?

(c) a lower triangular matrix?

22 Draw a picture of T (house) for these matrices:

$$D = \begin{bmatrix} 2 & 0 \\ 0 & 1 \end{bmatrix} \quad \text{and} \quad A = \begin{bmatrix} .7 & .7 \\ .3 & .3 \end{bmatrix} \quad \text{and} \quad U = \begin{bmatrix} 1 & 1 \\ 0 & 1 \end{bmatrix}.$$

23 What are the conditions on $A = \left[\begin{smallmatrix} a & b \\ c & d \end{smallmatrix}\right]$ to ensure that T (house) will

(a) sit straight up?

(b) expand the house by 3 in all directions?

(c) rotate the house with no change in its shape?

24 What are the conditions on $\det A = ad - bc$ to ensure that T (house) will

(a) be squashed onto a line?

(b) keep its endpoints in clockwise order (not reflected)?

(c) have the same area as the original house?

If one side of the house stays in place, how do you know that $A = I$?

25 Describe T (house) when $T(\boldsymbol{v}) = -\boldsymbol{v} + (1, 0)$. This T is "affine."

26 Change the house matrix H to add a chimney.

27 This MATLAB program creates a vector of 50 angles called theta, and then draws the unit circle and T (circle) = ellipse. You can change A.

```
A = [2 1;1 2]
theta = [0:2 * pi/50:2 * pi];
circle = [cos(theta); sin(theta)];
ellipse = A * circle;
axis([−4 4 −4 4]); axis('square')
plot(circle(1,:), circle(2,:), ellipse(1,:), ellipse(2,:))
```

28 Add two eyes and a smile to the circle in Problem 27. (If one eye is dark and the other is light, you can tell when the face is reflected across the y axis.) Multiply by matrices A to get new faces.

29 The first house is drawn by this program plot2d(H). Circles from o and lines from –:

```
x = H(1,:)'; y = H(2,:)';
axis([−10 10 −10 10]), axis('square')
plot(x,y,'o',x,y,'−');
```

Test plot2d(A' * H) and plot2d(A' * A * H) with the matrices in Figure 7.1.

30 Without a computer describe the houses $A * H$ for these matrices A:

$$\begin{bmatrix} 1 & 0 \\ 0 & .1 \end{bmatrix} \quad \text{and} \quad \begin{bmatrix} .5 & .5 \\ .5 & .5 \end{bmatrix} \quad \text{and} \quad \begin{bmatrix} .5 & .5 \\ -.5 & .5 \end{bmatrix} \quad \text{and} \quad \begin{bmatrix} 1 & 1 \\ 1 & 0 \end{bmatrix}.$$

7.2 Choice of Basis: Similarity and SVD

This section is about diagonalizing a matrix (any matrix). The example we know best is $S^{-1}AS = \Lambda$, for square matrices only. By placing the eigenvectors in S we produce the eigenvalues in Λ. You can look at this diagonalization in two ways, as a ***factorization*** of A or as a ***good choice of basis***:

- ***Factorization of the matrix:*** $A = S\Lambda S^{-1}$
- ***Choice of eigenvector basis: The matrix becomes*** Λ.

Chapter 6 emphasized the first way. Matrix multiplication gave $AS = S\Lambda$ (each column is just $A\boldsymbol{x} = \lambda\boldsymbol{x}$). This chapter is emphasizing the second way. We change from the standard basis, where the matrix is A, to a better basis. There have to be n independent eigenvectors, which are the input basis and also the output basis. Then output equals Λ times input.

When A is symmetric, its eigenvectors can be chosen orthonormal. The matrix with those columns is called Q. The diagonal matrix Λ is $Q^{-1}AQ$ which is also $Q^{\mathrm{T}}AQ$. The $S\Lambda S^{-1}$ factorization of a symmetric matrix becomes $A = Q\Lambda Q^{\mathrm{T}}$.

Nothing is new in those paragraphs. But this section moves to something entirely new and very important. You know the Fundamental Theorem of Linear Algebra, which is true for every matrix. It involves four subspaces—the row space, the column space, and the two nullspaces. By the row operations of elimination, we produced bases for those subspaces. But those bases are not the best! They are not orthonormal and the matrix did not become diagonal. Now we choose the best bases.

We will explain the new result in two ways, first the bases and then the factorization. Every m by n matrix is allowed.

7C There are orthonormal bases $\boldsymbol{v}_1, \ldots, \boldsymbol{v}_r$ for the row space and $\boldsymbol{u}_1, \ldots, \boldsymbol{u}_r$ for the column space such that $A\boldsymbol{v}_i = \sigma_i \boldsymbol{u}_i$. We can also ensure that $\sigma_i > 0$.

7D Singular Value Decomposition Every m by n matrix can be factored into $A = U\Sigma V^{\mathrm{T}}$, where U and V are orthogonal matrices and Σ is diagonal:

$$A = U\Sigma V^{\mathrm{T}} = \underset{m \text{ by } m}{\begin{bmatrix} \boldsymbol{u}_1 & \cdots & \boldsymbol{u}_r & \cdots & \boldsymbol{u}_m \end{bmatrix}} \underset{m \text{ by } n}{\begin{bmatrix} \sigma_1 & & \\ & \ddots & \\ & & \sigma_r \end{bmatrix}} \underset{n \text{ by } n}{\begin{bmatrix} \boldsymbol{v}_1 & \cdots & \boldsymbol{v}_r & \cdots & \boldsymbol{v}_n \end{bmatrix}^{\mathrm{T}}}.$$

The matrix Σ has the "singular values" $\sigma_1, \ldots, \sigma_r$ on its diagonal and is otherwise zero.

Compare with the symmetric case $A = Q\Lambda Q^{\mathrm{T}}$. The orthogonal matrices U and V are no longer the same Q. The input basis is *not the same* as the output basis. The input basis starts with $\boldsymbol{v}_1, \ldots, \boldsymbol{v}_r$ from the row space. It finishes with any orthonormal basis $\boldsymbol{v}_{r+1}, \ldots, \boldsymbol{v}_n$ for the nullspace. Similarly the output basis starts with the good $\boldsymbol{u}_1, \ldots, \boldsymbol{u}_r$ in the column space and ends with any orthonormal $\boldsymbol{u}_{r+1}, \ldots, \boldsymbol{u}_m$ in the left nullspace.

You are seeing again the four dimensions r and $n-r$ and r and $m-r$. The orthogonality of row space to nullspace is here too: $\boldsymbol{v}_1, \ldots, \boldsymbol{v}_r$ are automatically orthogonal to $\boldsymbol{v}_{r+1}, \ldots, \boldsymbol{v}_n$. We are taking the final step in the Fundamental Theorem of Linear Algebra: ***To choose bases that make the matrix diagonal.***

Important point: The singular values σ_i are not eigenvalues of A. In fact σ_i^2 is an eigenvalue of $A^{\mathrm{T}}A$ (and also of AA^{T}). Those matrices are symmetric. Their orthogonal eigenvectors are the $\boldsymbol{v}$'s and $\boldsymbol{u}$'s.

This is an extra long section of the book. We don't expect the first course to reach this far. But the **SVD** is absolutely a high point of linear algebra—the dimensions are right and the orthogonality is right and now the bases are right. The Fundamental Theorem of Linear Algebra is complete.

Similarity: $M^{-1}AM$ and $S^{-1}AS$

We begin with a square matrix and one basis. The input space **V** is $\mathbf{R}^n$ and the output space **W** is also $\mathbf{R}^n$. The basis vectors are the columns of I. The matrix with this basis is n by n, and we call it A. The linear transformation is just "multiplication by A."

Most of this book has been about one fundamental problem—*to make the matrix simple*. We made it triangular in Chapter 2 (by elimination), and we made it diagonal in Chapter 6 (by eigenvectors). Now a change in the matrix comes from a *change of basis*.

Here are the main facts in advance. When you change the basis for **V**, the matrix changes from A to AM. Because **V** is the input space, the matrix M goes on the right (to come first). When you change the basis for **W**, the new matrix is $M^{-1}A$. We are dealing with the output space so M^{-1} is on the left (to come last). ***If you change both bases in the same way, the new matrix is $M^{-1}AM$.*** The good basis vectors are the eigenvectors, which go into the columns of $M = S$. The matrix becomes $S^{-1}AS = \Lambda$.

7E When the basis consists of the eigenvectors $\boldsymbol{x}_1, \ldots, \boldsymbol{x}_n$, the matrix for T becomes Λ.

Reason To find column 1 of the matrix, input the first basis vector $\boldsymbol{x}_1$. The transformation multiplies by A. The output is $A\boldsymbol{x}_1 = \lambda_1\boldsymbol{x}_1$. This is λ_1 times the first basis vector plus zero times the other basis vectors. Therefore the first column of the matrix is $(\lambda_1, 0, \ldots, 0)$. In the eigenvector basis, the matrix is diagonal.

Example 7.8 Find the *eigenvector basis* for projection onto the 135° line $y = -x$. The standard vectors $(1, 0)$ and $(0, 1)$ are projected in Figure 7.2. In the standard basis

$$A = \begin{bmatrix} .5 & -.5 \\ -.5 & .5 \end{bmatrix}.$$

Solution The eigenvectors for this projection are $\boldsymbol{x}_1 = (1, -1)$ and $\boldsymbol{x}_2 = (1, 1)$. The first is on the projection line and the second is perpendicular (Figure 7.2). Their projections are $\boldsymbol{x}_1$ and $\mathbf{0}$. The eigenvalues are $\lambda_1 = 1$ and $\lambda_2 = 0$. In the eigenvector basis the projection matrix is

$$\Lambda = \begin{bmatrix} 1 & 0 \\ 0 & 0 \end{bmatrix}.$$

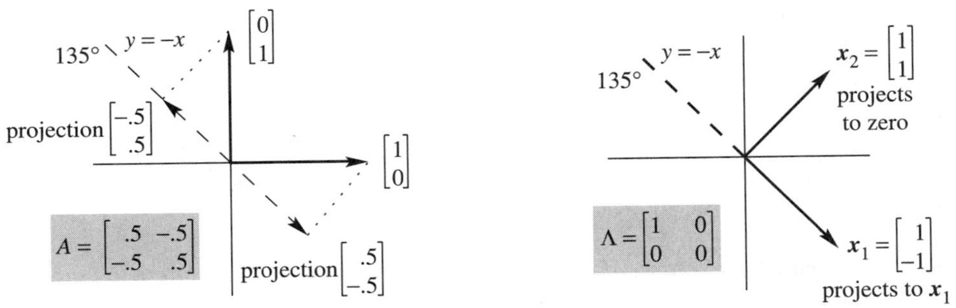

Figure 7.2 Projection on the 135° line $y = -x$. Standard basis vs. eigenvector basis.

What if you choose a different basis $\boldsymbol{v}_1 = (2,0)$ and $\boldsymbol{v}_2 = (1,1)$? There are two ways to find the new matrix B, and the main point of this page is to show you both ways:

First way Project $\boldsymbol{v}_1$ to reach $(1,-1)$. This is $\boldsymbol{v}_1 - \boldsymbol{v}_2$. The first column of B contains these coefficients 1 and -1. Project $\boldsymbol{v}_2$ to reach $(0,0)$. This is $0\boldsymbol{v}_1 + 0\boldsymbol{v}_2$. Column 2 contains 0 and 0. With basis $\boldsymbol{v}_1 = (2,0)$ and $\boldsymbol{v}_2 = (1,1)$ the matrix from Figure 7.3 is

$$B = \begin{bmatrix} 1 & 0 \\ -1 & 0 \end{bmatrix}.$$

Second way Find the matrix B in three steps. Change from the $\boldsymbol{v}$'s to the standard basis, using M. Project in that standard basis, using A. Change back to the $\boldsymbol{v}$'s with M^{-1}:

$$B_{\boldsymbol{v}\text{'s to }\boldsymbol{v}\text{'s}} = M^{-1}_{\text{standard to }\boldsymbol{v}\text{'s}} \; A_{\text{standard}} \; M_{\boldsymbol{v}\text{'s to standard}}$$

The change of basis matrix M has the $\boldsymbol{v}$'s in its columns. Then $M^{-1}AM$ is B:

$$B = \begin{bmatrix} .5 & -.5 \\ 0 & 1 \end{bmatrix} \begin{bmatrix} .5 & -.5 \\ -.5 & .5 \end{bmatrix} \begin{bmatrix} 2 & 1 \\ 0 & 1 \end{bmatrix} = \begin{bmatrix} 1 & 0 \\ -1 & 0 \end{bmatrix}.$$

Conclusions The matrix B is the same both ways. The second way shows that $B = M^{-1}AM$. Then B is *similar* to A. B and A represent the same transformation T—in this case a projection. M and M^{-1} only represent the identity transformation I. (They are not identity matrices! Their input and output bases are different.) The product of matrices $M^{-1}AM$ copies the product of transformations ITI.

7F In one basis $\boldsymbol{w}_1, \ldots, \boldsymbol{w}_n$ the transformation T has matrix A. In another basis $\boldsymbol{v}_1, \ldots, \boldsymbol{v}_n$ the same transformation has matrix B. The identity transformation I from $\boldsymbol{v}$'s to $\boldsymbol{w}$'s has the matrix M (*change of basis matrix*). Then B is similar to A:

$$T_{\boldsymbol{v}\text{ to }\boldsymbol{v}} = I_{\boldsymbol{w}\text{ to }\boldsymbol{v}} \, T_{\boldsymbol{w}\text{ to }\boldsymbol{w}} \, I_{\boldsymbol{v}\text{ to }\boldsymbol{w}} \quad \text{leads to} \quad B = M^{-1}AM.$$

It can be shown that A and B have the same eigenvalues and the same determinant. A is invertible when B is invertible (and when T is invertible).

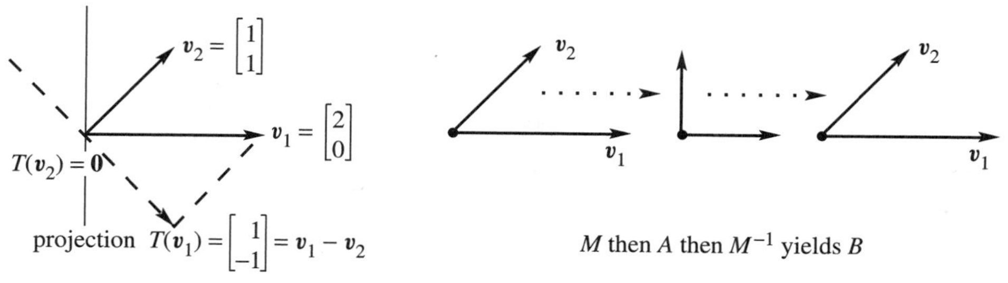

Figure 7.3 Projection matrix with a new basis. First way: Stay with $\boldsymbol{v}$'s. Second way: Go to the standard basis and back by $M^{-1}AM$.

Suppose the $\boldsymbol{v}$'s are the eigenvectors and the $\boldsymbol{w}$'s are the standard basis. The change of basis matrix M is S. Its columns are the eigenvectors written in the standard basis! Then the similar matrix $B = M^{-1}AM$ is the diagonal matrix $\Lambda = S^{-1}AS$.

Example 7.9 T reflects every vector $\boldsymbol{v}$ across the straight line at angle θ. The output $T(\boldsymbol{v})$ is the mirror image of $\boldsymbol{v}$ on the other side of the line. Find the matrix A in the standard basis and the matrix Λ in the eigenvector basis.

Solution The eigenvector $\boldsymbol{v}_1 = (\cos\theta, \sin\theta)$ is on the line. It is reflected to itself so $\lambda_1 = 1$. The eigenvector $\boldsymbol{v}_2 = (-\sin\theta, \cos\theta)$ is perpendicular to the line. Its reflection is $-\boldsymbol{v}_2$ on the other side. In this basis the matrix is

$$B = \Lambda = \begin{bmatrix} 1 & 0 \\ 0 & -1 \end{bmatrix}.$$

Now use the standard basis $(1, 0)$ and $(0, 1)$. Find A by going to the $\boldsymbol{v}$'s and back. The change of basis matrix is $M = S$. Its columns contain the $\boldsymbol{v}$'s. Then A is $MBM^{-1} = S\Lambda S^{-1}$:

$$A = \begin{bmatrix} \cos\theta & -\sin\theta \\ \sin\theta & \cos\theta \end{bmatrix} \begin{bmatrix} 1 & 0 \\ 0 & -1 \end{bmatrix} \begin{bmatrix} \cos\theta & \sin\theta \\ -\sin\theta & \cos\theta \end{bmatrix} = \begin{bmatrix} \cos^2\theta - \sin^2\theta & 2\sin\theta\cos\theta \\ 2\sin\theta\cos\theta & \sin^2\theta - \cos^2\theta \end{bmatrix}. \tag{7.1}$$

With the identities for $\cos 2\theta$ and $\sin 2\theta$ we can recognize the reflection matrix of Section 6.1:

$$A = \begin{bmatrix} \cos 2\theta & \sin 2\theta \\ \sin 2\theta & -\cos 2\theta \end{bmatrix}. \tag{7.2}$$

This matrix has $A^2 = I$. Two reflections bring back the original.

The Singular Value Decomposition (SVD)

Now comes a highlight of linear algebra. A is any m by n matrix. It can be square or rectangular, and we will diagonalize it. Its row space is r-dimensional (inside $\mathbf{R}^n$) and its

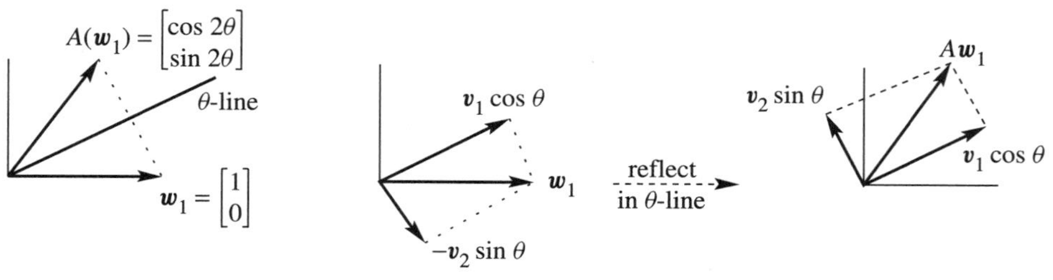

Figure 7.4 Column 1 of the reflection matrix: one step (standard basis) or three steps (to $\boldsymbol{v}$'s and back).

column space is r-dimensional (inside $\mathbf{R}^m$). We are going to choose orthonormal bases for those spaces. The row space basis will be $\boldsymbol{v}_1, \ldots, \boldsymbol{v}_r$ and the column space basis will be $\boldsymbol{u}_1, \ldots, \boldsymbol{u}_r$.

Start with a 2 by 2 matrix: $m = n = 2$. Let its rank be $r = 2$, so it is invertible. Its row space is the plane $\mathbf{R}^2$ and its column space is also the plane $\mathbf{R}^2$. We want $\boldsymbol{v}_1$ and $\boldsymbol{v}_2$ to be perpendicular unit vectors, and we want $\boldsymbol{u}_1$ and $\boldsymbol{u}_2$ to be perpendicular unit vectors. As a specific example, we work with

$$A = \begin{bmatrix} 2 & 2 \\ -1 & 1 \end{bmatrix}.$$

First point Why not choose the standard basis? *Because then the matrix is not diagonal.*

Second point Why not choose the eigenvector basis? *Because that basis is not orthonormal.*

We are aiming for orthonormal bases that also diagonalize A. The two bases will be different—one basis cannot do it. When the inputs are $\boldsymbol{v}_1$ and $\boldsymbol{v}_2$, the outputs are $A\boldsymbol{v}_1$ and $A\boldsymbol{v}_2$. We want those to line up with $\boldsymbol{u}_1$ and $\boldsymbol{u}_2$. The basis vectors have to give $A\boldsymbol{v}_1 = \sigma_1\boldsymbol{u}_1$ and also $A\boldsymbol{v}_2 = \sigma_2\boldsymbol{u}_2$. With those vectors as columns you can see what we are asking for:

$$A\begin{bmatrix} \boldsymbol{v}_1 & \boldsymbol{v}_2 \end{bmatrix} = \begin{bmatrix} \sigma_1\boldsymbol{u}_1 & \sigma_2\boldsymbol{u}_2 \end{bmatrix} = \begin{bmatrix} \boldsymbol{u}_1 & \boldsymbol{u}_2 \end{bmatrix}\begin{bmatrix} \sigma_1 & \\ & \sigma_2 \end{bmatrix}. \tag{7.3}$$

In matrix notation that is $AV = U\Sigma$. The diagonal matrix Σ is like Λ (capital sigma versus capital lambda). One contains the *singular values* σ_1, σ_2 and the other contains the eigenvalues λ_1, λ_2.

The difference comes from U and V. When they both equal S, we have $AS = S\Lambda$ which means $S^{-1}AS = \Lambda$. The matrix is diagonalized but the eigenvectors are not generally orthonormal. The new requirement is that *U and V must be orthogonal matrices.* The basis vectors in their columns must be orthonormal:

$$V^{\mathrm{T}}V = \begin{bmatrix} - \boldsymbol{v}_1^{\mathrm{T}} - \\ - \boldsymbol{v}_2^{\mathrm{T}} - \end{bmatrix}\begin{bmatrix} \boldsymbol{v}_1 & \boldsymbol{v}_2 \end{bmatrix} = \begin{bmatrix} 1 & 0 \\ 0 & 1 \end{bmatrix}.$$

Thus $V^{\mathrm{T}}V = I$ which means $V^{\mathrm{T}} = V^{-1}$. Similarly $U^{\mathrm{T}}U = I$ and $U^{\mathrm{T}} = U^{-1}$.

7G The ***Singular Value Decomposition*** (SVD) has $AV = U\Sigma$ with orthogonal matrices U and V. Then

$$A = U\Sigma V^{-1} = U\Sigma V^{\mathrm{T}}. \tag{7.4}$$

This is the new factorization: ***orthogonal*** times ***diagonal*** times ***orthogonal***.

We have two matrices U and V instead of one matrix S. But there is a neat way to get U out of the picture and see V by itself: *Multiply A^{T} times A.*

$$A^{\mathrm{T}}A = (U\Sigma V^{\mathrm{T}})^{\mathrm{T}}(U\Sigma V^{\mathrm{T}}) = V\Sigma^{\mathrm{T}}U^{\mathrm{T}}U\Sigma V^{\mathrm{T}}. \tag{7.5}$$

$U^{\mathrm{T}}U$ disappears because it equals I. Then Σ^{T} is next to Σ. Multiplying those diagonal matrices gives σ_1^2 and σ_2^2. That leaves an ordinary factorization of the symmetric matrix $A^{\mathrm{T}}A$:

$$A^{\mathrm{T}}A = V\begin{bmatrix}\sigma_1^2 & 0\\ 0 & \sigma_2^2\end{bmatrix}V^{\mathrm{T}}. \tag{7.6}$$

In Chapter 6 we would have called this $Q\Lambda Q^{\mathrm{T}}$. The symmetric matrix was A itself. Now the symmetric matrix is $A^{\mathrm{T}}A$! *And the columns of V are its eigenvectors.*

This tells us how to find V. We are ready to complete the example.

Example 7.10 Find the singular value decomposition of $A = \left[\begin{smallmatrix}2 & 2\\ -1 & 1\end{smallmatrix}\right]$.

Solution Compute $A^{\mathrm{T}}A$ and its eigenvectors. Then make them unit vectors:

$$A^{\mathrm{T}}A = \begin{bmatrix}5 & 3\\ 3 & 5\end{bmatrix} \quad \text{has eigenvectors} \quad \boldsymbol{v}_1 = \begin{bmatrix}-1/\sqrt{2}\\ 1/\sqrt{2}\end{bmatrix} \quad \text{and} \quad \boldsymbol{v}_2 = \begin{bmatrix}1/\sqrt{2}\\ 1/\sqrt{2}\end{bmatrix}.$$

The eigenvalues of $A^{\mathrm{T}}A$ are 2 and 8. The $\boldsymbol{v}$'s are perpendicular, because eigenvectors of every symmetric matrix are perpendicular—and $A^{\mathrm{T}}A$ is automatically symmetric.

What about $\boldsymbol{u}_1$ and $\boldsymbol{u}_2$? They are quick to find, because $A\boldsymbol{v}_1$ is in the direction of $\boldsymbol{u}_1$ and $A\boldsymbol{v}_2$ is in the direction of $\boldsymbol{u}_2$:

$$A\boldsymbol{v}_1 = \begin{bmatrix}2 & 2\\ -1 & 1\end{bmatrix}\begin{bmatrix}-1/\sqrt{2}\\ 1/\sqrt{2}\end{bmatrix} = \begin{bmatrix}0\\ \sqrt{2}\end{bmatrix}. \qquad \text{The unit vector is} \quad \boldsymbol{u}_1 = \begin{bmatrix}0\\ 1\end{bmatrix}.$$

$$A\boldsymbol{v}_2 = \begin{bmatrix}2 & 2\\ -1 & 1\end{bmatrix}\begin{bmatrix}1/\sqrt{2}\\ 1/\sqrt{2}\end{bmatrix} = \begin{bmatrix}2\sqrt{2}\\ 0\end{bmatrix}. \qquad \text{The unit vector is} \quad \boldsymbol{u}_2 = \begin{bmatrix}1\\ 0\end{bmatrix}.$$

Since the eigenvalues of $A^{\mathrm{T}}A$ are $\sigma_1^2 = 2$ and $\sigma_2^2 = 8$, the singular values of A are their square roots. Thus $\sigma_1 = \sqrt{2}$. In the display above, $A\boldsymbol{v}_1$ has that factor $\sqrt{2}$. In fact $A\boldsymbol{v}_1 = \sigma_1\boldsymbol{u}_1$ exactly as required. Similarly $\sigma_2 = \sqrt{8} = 2\sqrt{2}$. This factor is also in the display to give $A\boldsymbol{v}_2 = \sigma_2\boldsymbol{u}_2$. We have completed the SVD:

$$A = U\Sigma V^{\mathrm{T}} \quad \text{is} \quad \begin{bmatrix}2 & 2\\ -1 & 1\end{bmatrix} = \begin{bmatrix}0 & 1\\ 1 & 0\end{bmatrix}\begin{bmatrix}\sqrt{2} & \\ & 2\sqrt{2}\end{bmatrix}\begin{bmatrix}-1/\sqrt{2} & 1/\sqrt{2}\\ 1/\sqrt{2} & 1/\sqrt{2}\end{bmatrix}. \tag{7.7}$$

This matrix, and every invertible 2 by 2 matrix, ***transforms the unit circle to an ellipse***. You can see that in Figure 7.5.

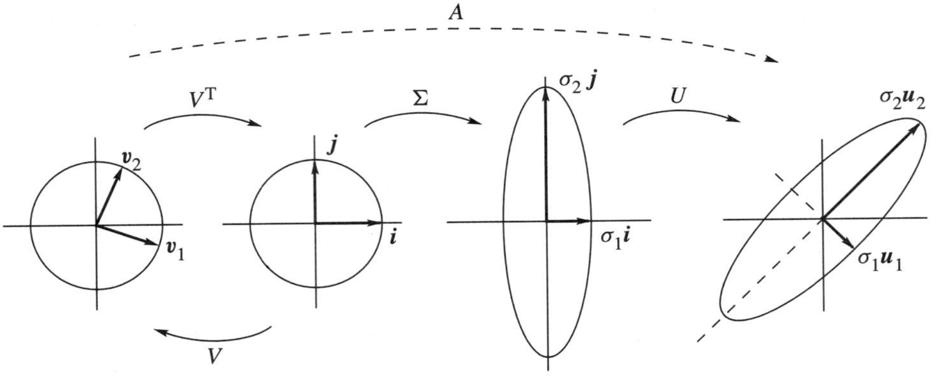

Figure 7.5 U and V are rotations and reflections. Σ is a stretching matrix.

One final point about that example. We found the $\boldsymbol{u}$'s from the $\boldsymbol{v}$'s. Could we find the $\boldsymbol{u}$'s directly? *Yes*, by multiplying AA^{T} instead of $A^{\mathrm{T}}A$:

$$AA^{\mathrm{T}} = (U\Sigma V^{\mathrm{T}})(V\Sigma^{\mathrm{T}}U^{\mathrm{T}}) = U\Sigma\Sigma^{\mathrm{T}}U^{\mathrm{T}}. \tag{7.8}$$

This time it is $V^{\mathrm{T}}V = I$ that disappears. Multiplying $\Sigma\Sigma^{\mathrm{T}}$ gives σ_1^2 and σ_2^2 as before. We have an ordinary factorization of the symmetric matrix AA^{T}. *The columns of U are the eigenvectors of AA^{T}.*

Example 7.11 Compute the eigenvectors $\boldsymbol{u}_1$ and $\boldsymbol{u}_2$ directly from AA^{T}. The eigenvalues are again $\sigma_1^2 = 2$ and $\sigma_2^2 = 8$. The singular values are still their square roots:

$$AA^{\mathrm{T}} = \begin{bmatrix} 2 & 2 \\ -1 & 1 \end{bmatrix} \begin{bmatrix} 2 & -1 \\ 2 & 1 \end{bmatrix} = \begin{bmatrix} 8 & 0 \\ 0 & 2 \end{bmatrix}.$$

This matrix happens to be diagonal. Its eigenvectors are $(1, 0)$ and $(0, 1)$. This agrees with $\boldsymbol{u}_1$ and $\boldsymbol{u}_2$ found earlier, but *in the opposite order.* Why should we take $\boldsymbol{u}_1$ to be $(0, 1)$ instead of $(1, 0)$? Because we have to follow the order of the eigenvalues.

We originally chose $\sigma_1^2 = 2$. The eigenvectors $\boldsymbol{v}_1$ (for $A^{\mathrm{T}}A$) and $\boldsymbol{u}_1$ (for AA^{T}) have to stay with that choice. If you want the σ's in decreasing order, that is also possible (and generally preferred). Then $\sigma_1^2 = 8$ and $\sigma_2^2 = 2$. This exchanges $\boldsymbol{v}_1$ and $\boldsymbol{v}_2$ in V, and it exchanges $\boldsymbol{u}_1$ and $\boldsymbol{u}_2$ in U. The new SVD is still correct:

$$A = U\Sigma V^{\mathrm{T}} \quad \text{is now} \quad \begin{bmatrix} 2 & 2 \\ -1 & 1 \end{bmatrix} = \begin{bmatrix} 1 & 0 \\ 0 & 1 \end{bmatrix} \begin{bmatrix} 2\sqrt{2} & 0 \\ 0 & \sqrt{2} \end{bmatrix} \begin{bmatrix} 1/\sqrt{2} & -1/\sqrt{2} \\ 1/\sqrt{2} & 1/\sqrt{2} \end{bmatrix}^{\mathrm{T}}.$$

The other small bit of freedom is to multiply an eigenvector by -1. The result is still a unit eigenvector. If we do this to $\boldsymbol{v}_1$ we must also do it to $\boldsymbol{u}_1$—because $A\boldsymbol{v}_1 = \sigma_1\boldsymbol{u}_1$. It is the signs of the eigenvectors that keep σ_1 positive, and it is the order of the eigenvectors that puts σ_1 first in Σ.

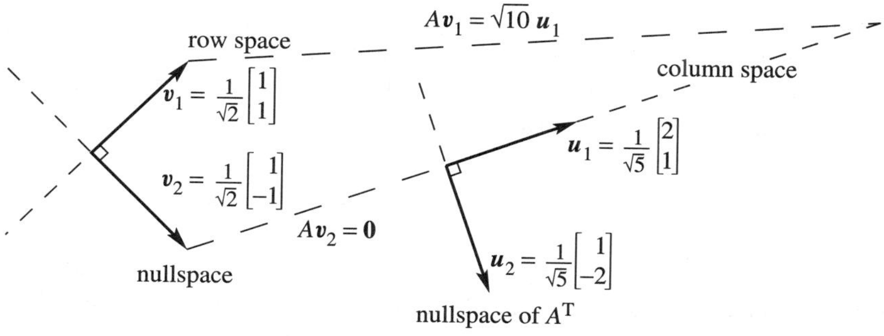

Figure 7.6 The SVD chooses orthonormal bases so that $A\boldsymbol{v}_i = \sigma_i \boldsymbol{u}_i$.

Example 7.12 Find the SVD of the singular matrix $A = \left[\begin{smallmatrix}2&2\\1&1\end{smallmatrix}\right]$. The rank is $r = 1$. The row space has only one basis vector $\boldsymbol{v}_1$. The column space has only one basis vector $\boldsymbol{u}_1$. We can see those vectors in A, and make them into unit vectors:

$$\boldsymbol{v}_1 = \text{multiple of row } \begin{bmatrix}1\\1\end{bmatrix} = \frac{1}{\sqrt{2}}\begin{bmatrix}1\\1\end{bmatrix}$$

$$\boldsymbol{u}_1 = \text{multiple of column } \begin{bmatrix}2\\1\end{bmatrix} = \frac{1}{\sqrt{5}}\begin{bmatrix}2\\1\end{bmatrix}.$$

Then $A\boldsymbol{v}_1$ must equal $\sigma_1\boldsymbol{u}_1$. It does, with singular value $\sigma_1 = \sqrt{10}$. The SVD could stop there (it usually doesn't):

$$\begin{bmatrix}2&2\\1&1\end{bmatrix} = \begin{bmatrix}2/\sqrt{5}\\1/\sqrt{5}\end{bmatrix}\begin{bmatrix}\sqrt{10}\end{bmatrix}\begin{bmatrix}1/\sqrt{2} & 1/\sqrt{2}\end{bmatrix}.$$

It is customary for U and V to be square. The matrices need a second column. The vector $\boldsymbol{v}_2$ must be orthogonal to $\boldsymbol{v}_1$, and $\boldsymbol{u}_2$ must be orthogonal to $\boldsymbol{u}_1$:

$$\boldsymbol{v}_2 = \frac{1}{\sqrt{2}}\begin{bmatrix}1\\-1\end{bmatrix} \quad\text{and}\quad \boldsymbol{u}_2 = \frac{1}{\sqrt{5}}\begin{bmatrix}1\\-2\end{bmatrix}.$$

The vector $\boldsymbol{v}_2$ is in the nullspace. It is perpendicular to $\boldsymbol{v}_1$ in the row space. Multiply by A to get $A\boldsymbol{v}_2 = \boldsymbol{0}$. We could say that the second singular value is $\sigma_2 = 0$, but this is against the rules. Singular values are like pivots—only the r nonzeros are counted.

If A is 2 by 2 then all three matrices U, Σ, V are 2 by 2 in the true SVD:

$$\begin{bmatrix}2&2\\1&1\end{bmatrix} = U\Sigma V^{\mathrm{T}} = \frac{1}{\sqrt{5}}\begin{bmatrix}2&1\\1&-2\end{bmatrix}\begin{bmatrix}\sqrt{10}&0\\0&0\end{bmatrix}\frac{1}{\sqrt{2}}\begin{bmatrix}1&1\\1&-1\end{bmatrix}. \qquad (7.9)$$

The matrices U and V contain orthonormal bases for all four fundamental subspaces:

first	r	columns of V :	*row space* of A
last	$n - r$	columns of V :	*nullspace* of A
first	r	columns of U :	*column space* of A
last	$m - r$	columns of U :	*nullspace* of A^{T}.

The first columns $\boldsymbol{v}_1, \ldots, \boldsymbol{v}_r$ and $\boldsymbol{u}_1, \ldots, \boldsymbol{u}_r$ are the hardest to choose, because $A\boldsymbol{v}_i$ has to fall in the direction of $\boldsymbol{u}_i$. The last $\boldsymbol{v}$'s and $\boldsymbol{u}$'s (in the nullspaces) are easier. As long as those are orthonormal, the SVD will be correct. The $\boldsymbol{v}$'s are eigenvectors of $A^{\mathrm{T}}A$ and the $\boldsymbol{u}$'s are eigenvectors of AA^{T}. This example has

$$A^{\mathrm{T}}A = \begin{bmatrix} 5 & 5 \\ 5 & 5 \end{bmatrix} \quad \text{and} \quad AA^{\mathrm{T}} = \begin{bmatrix} 8 & 4 \\ 4 & 2 \end{bmatrix}.$$

Those matrices have the same eigenvalues 10 and 0. The first has eigenvectors $\boldsymbol{v}_1$ and $\boldsymbol{v}_2$, the second has eigenvectors $\boldsymbol{u}_1$ and $\boldsymbol{u}_2$. Multiplication shows that $A\boldsymbol{v}_1 = \sqrt{10}\,\boldsymbol{u}_1$ and $A\boldsymbol{v}_2 = \boldsymbol{0}$. It always happens that $A\boldsymbol{v}_i = \sigma_i \boldsymbol{u}_i$, and we now explain why.

Starting from $A^{\mathrm{T}}A\boldsymbol{v}_i = \sigma_i^2 \boldsymbol{v}_i$, the two key steps are to multiply by $\boldsymbol{v}_i^{\mathrm{T}}$ and by A:

$$\boldsymbol{v}_i^{\mathrm{T}}A^{\mathrm{T}}A\boldsymbol{v}_i = \sigma_i^2 \boldsymbol{v}_i^{\mathrm{T}}\boldsymbol{v}_i \quad \text{gives} \quad \|A\boldsymbol{v}_i\|^2 = \sigma_i^2 \quad \text{so that} \quad \|A\boldsymbol{v}_i\| = \sigma_i \tag{7.10}$$

$$AA^{\mathrm{T}}A\boldsymbol{v}_i = \sigma_i^2 A\boldsymbol{v}_i \quad \text{gives} \quad \boldsymbol{u}_i = A\boldsymbol{v}_i/\sigma_i \quad \text{as a unit eigenvector of} \quad AA^{\mathrm{T}}. \tag{7.11}$$

Equation (7.10) used the small trick of placing parentheses in $(\boldsymbol{v}_i^{\mathrm{T}}A^{\mathrm{T}})(A\boldsymbol{v}_i)$. This is a vector times its transpose, giving $\|A\boldsymbol{v}_i\|^2$. Equation (7.11) placed the parentheses in $(AA^{\mathrm{T}})(A\boldsymbol{v}_i)$. This shows that $A\boldsymbol{v}_i$ is an eigenvector of AA^{T}. We divide it by its length σ_i to get the unit vector $\boldsymbol{u}_i = A\boldsymbol{v}_i/\sigma_i$. This is the equation $A\boldsymbol{v}_i = \sigma_i\boldsymbol{u}_i$, which says that A is diagonalized by these outstanding bases.

We will give you our opinion directly. The SVD is the climax of this linear algebra course. We think of it as the final step in the Fundamental Theorem. First come the *dimensions* of the four subspaces. Then their *orthogonality*. Then the *bases which diagonalize A*. It is all in the formula $A = U\Sigma V^{\mathrm{T}}$. Applications are coming—they are certainly important!—but you have made it to the top.

Polar Decomposition and SVD Applications

Every complex number has the polar form $re^{i\theta}$. A nonnegative number r multiplies a number on the unit circle. (Remember that $|e^{i\theta}| = |\cos\theta + i\sin\theta| = 1$.) Thinking of these numbers as 1 by 1 matrices, $r \ge 0$ corresponds to a *positive semidefinite matrix* (call it H) and $e^{i\theta}$ corresponds to an *orthogonal matrix* Q. The SVD extends this $re^{i\theta}$ factorization to matrices (even m by n with rectangular Q).

7H Every real square matrix can be factored into $A = QH$, where Q is ***orthogonal*** and H is ***symmetric positive semidefinite***. If A is invertible then H is positive definite.

For the proof we just insert $V^{\mathrm{T}}V = I$ into the middle of the SVD:

$$A = U\Sigma V^{\mathrm{T}} = (UV^{\mathrm{T}})(V\Sigma V^{\mathrm{T}}) = (Q)(H). \tag{7.12}$$

The first factor UV^{T} is Q. The product of orthogonal matrices is orthogonal. The second factor $V\Sigma V^{\mathrm{T}}$ is H. It is positive semidefinite because its eigenvalues are in Σ. If A is invertible then H is also invertible—it is symmetric positive definite. ***H is the square root of*** $\boldsymbol{A^{\mathrm{T}}A}$. Equation (7.5) says that $H^2 = V\Sigma^2V^{\mathrm{T}} = A^{\mathrm{T}}A$.

There is also a polar decomposition $A = KQ$ in the reverse order. Q is the same but now $K = U\Sigma U^{\mathrm{T}}$. This is the square root of AA^{T} by equation (7.8).

Example 7.13 Find the polar decomposition $A = QH$ from the SVD in Example 7.11:

$$A = \begin{bmatrix} 2 & 2 \\ -1 & 1 \end{bmatrix} = \begin{bmatrix} 0 & 1 \\ 1 & 0 \end{bmatrix} \begin{bmatrix} \sqrt{2} & \\ & 2\sqrt{2} \end{bmatrix} \begin{bmatrix} -1/\sqrt{2} & 1/\sqrt{2} \\ 1/\sqrt{2} & 1/\sqrt{2} \end{bmatrix} = U\Sigma V^{\mathrm{T}}.$$

Solution The orthogonal part is $Q = UV^{\mathrm{T}}$. The positive definite part is $H = V\Sigma V^{\mathrm{T}} = Q^{-1}A$:

$$Q = \begin{bmatrix} 0 & 1 \\ 1 & 0 \end{bmatrix} \begin{bmatrix} -1/\sqrt{2} & 1/\sqrt{2} \\ 1/\sqrt{2} & 1/\sqrt{2} \end{bmatrix} = \begin{bmatrix} 1/\sqrt{2} & 1/\sqrt{2} \\ -1/\sqrt{2} & 1/\sqrt{2} \end{bmatrix}$$

$$H = \begin{bmatrix} 1/\sqrt{2} & -1/\sqrt{2} \\ 1/\sqrt{2} & 1/\sqrt{2} \end{bmatrix} \begin{bmatrix} 2 & 2 \\ -1 & 1 \end{bmatrix} = \begin{bmatrix} 3/\sqrt{2} & 1/\sqrt{2} \\ 1/\sqrt{2} & 3/\sqrt{2} \end{bmatrix}.$$

In mechanics, the polar decomposition separates the rotation (in Q) from the stretching (in H). The eigenvalues of H are the singular values of A; they give the stretching factors. The eigenvectors of H are the eigenvectors of $A^{\mathrm{T}}A$; they give the stretching directions (*the principal axes*).

The polar decomposition just splits the key equation $A\boldsymbol{v}_i = \sigma_i \boldsymbol{u}_i$ into two steps. The "H" part multiplies $\boldsymbol{v}_i$ by σ_i. The "Q" part swings the $\boldsymbol{v}$ direction around to the $\boldsymbol{u}$ direction. The other order $A = KQ$ swings $\boldsymbol{v}$'s to $\boldsymbol{u}$'s first (with the same Q). Then K multiplies $\boldsymbol{u}_i$ by σ_i to complete the job of A.

The Pseudoinverse

By choosing good bases, the action of A has become clear. It multiplies $\boldsymbol{v}_i$ in the row space to give $\sigma_i \boldsymbol{u}_i$ in the column space. The inverse matrix must do the opposite! If $A\boldsymbol{v} = \sigma\boldsymbol{u}$ then $A^{-1}\boldsymbol{u} = \boldsymbol{v}/\sigma$. The singular values of A^{-1} are $1/\sigma$, just as the eigenvalues of A^{-1} are $1/\lambda$. The bases are reversed. The $\boldsymbol{u}$'s are in the row space of A^{-1}, the $\boldsymbol{v}$'s are now in the column space.

Until this moment we would have added the words "*if A^{-1} exists*". Now we don't. A matrix that multiplies $\boldsymbol{u}_i$ to produce $\boldsymbol{v}_i/\sigma_i$ *does* exist. It is denoted by A^+:

$$A^+ = V\Sigma^+U^{\mathrm{T}} = \underset{n \text{ by } n}{\begin{bmatrix} & & & & \\ \boldsymbol{v}_1 & \cdots & \boldsymbol{v}_r & \cdots & \boldsymbol{v}_n \\ & & & & \end{bmatrix}} \underset{n \text{ by } m}{\begin{bmatrix} \sigma_1^{-1} & & \\ & \ddots & \\ & & \sigma_r^{-1} \end{bmatrix}} \underset{m \text{ by } m}{\begin{bmatrix} & & & & \\ \boldsymbol{u}_1 & \cdots & \boldsymbol{u}_r & \cdots & \boldsymbol{u}_m \\ & & & & \end{bmatrix}^{\mathrm{T}}}. \tag{7.13}$$

A^+ is the ***pseudoinverse*** of A. It is an n by m matrix. If A^{-1} exists (we said it again), then A^+ is the same as A^{-1}. In that case $m = n = r$ and we are inverting $U\Sigma V^{\mathrm{T}}$ to get

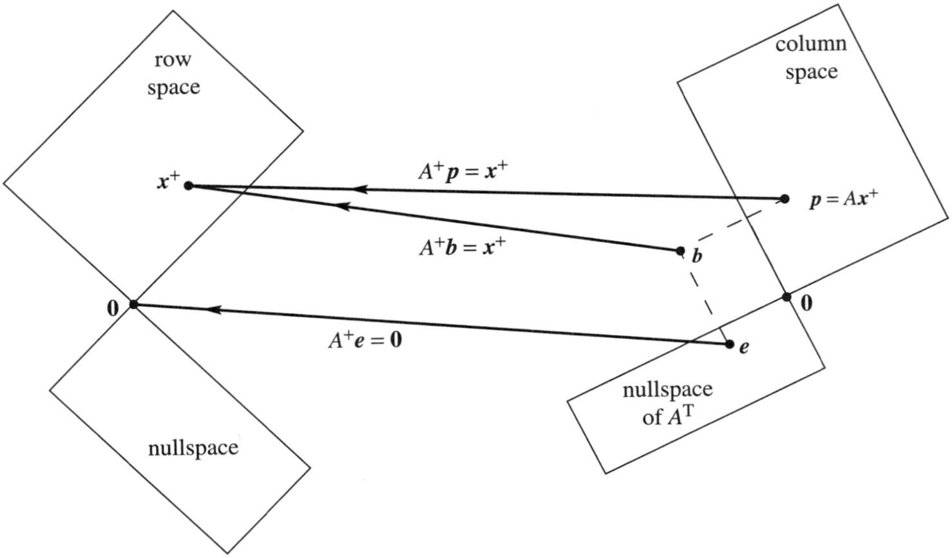

Figure 7.7 A is invertible from row space to column space. A^+ inverts it.

$V\Sigma^{-1}U^T$. The new symbol A^+ is needed when $r < m$ or $r < n$. Then A has no two-sided inverse but it has a *pseudo*inverse A^+ with these properties:

$$A^+\boldsymbol{u}_i = \frac{1}{\sigma_i}\boldsymbol{v}_i \quad \text{for } i \le r \qquad \text{and} \qquad A^+\boldsymbol{u}_i = \boldsymbol{0} \quad \text{for } i > r.$$

When we know what happens to each basis vector $\boldsymbol{u}_i$, we know A^+. The vectors $\boldsymbol{u}_1, \ldots, \boldsymbol{u}_r$ in the column space of A go back to the row space. The other vectors $\boldsymbol{u}_{r+1}, \ldots, \boldsymbol{u}_m$ are in the left nullspace, and A^+ in Figure 7.7 sends them to zero.

Example 7.14 The pseudoinverse of $A = \left[\begin{smallmatrix} 2 & 2 \\ -1 & 1 \end{smallmatrix}\right]$ is $A^+ = A^{-1}$, because A is invertible. Inverting $U\Sigma V^T$ is immediate from Example 7.11 because $U^{-1} = U^T$ and Σ is diagonal and $(V^T)^{-1} = V$:

$$A^+ = A^{-1} = V\Sigma^{-1}U^T = \begin{bmatrix} 1/4 & -1/2 \\ 1/4 & 1/2 \end{bmatrix}.$$

Example 7.15 Find the pseudoinverse of $A = \left[\begin{smallmatrix} 2 & 2 \\ 1 & 1 \end{smallmatrix}\right]$. This matrix is not invertible. The rank is 1. The singular value is $\sqrt{10}$ from Example 7.12. That is inverted in Σ^+:

$$A^+ = V\Sigma^+U^T = \frac{1}{\sqrt{2}}\begin{bmatrix} 1 & 1 \\ 1 & -1 \end{bmatrix}\begin{bmatrix} 1/\sqrt{10} & 0 \\ 0 & 0 \end{bmatrix}\frac{1}{\sqrt{5}}\begin{bmatrix} 2 & 1 \\ 1 & -2 \end{bmatrix} = \frac{1}{10}\begin{bmatrix} 2 & 1 \\ 2 & 1 \end{bmatrix}.$$

A^+ also has rank 1. Its column space is the row space of $A = \left[\begin{smallmatrix} 2 & 2 \\ 1 & 1 \end{smallmatrix}\right]$. When A takes $(1, 1)$ in the row space to $(4, 2)$ in the column space, A^+ does the reverse. Every rank one matrix is a column times a row. With unit vectors $\boldsymbol{u}$ and $\boldsymbol{v}$, that is $A = \sigma\boldsymbol{u}\boldsymbol{v}^T$. Then the best inverse we have is $A^+ = \frac{1}{\sigma}\boldsymbol{v}\boldsymbol{u}^T$.

The product AA^+ is $\boldsymbol{u}\boldsymbol{u}^{\mathrm{T}}$, the projection onto the line through $\boldsymbol{u}$. The product A^+A is $\boldsymbol{v}\boldsymbol{v}^{\mathrm{T}}$, the projection onto the line through $\boldsymbol{v}$. For all matrices, AA^+ and A^+A are the projections onto the column space and row space.

Problem 19 will show how $\boldsymbol{x}^+ = A^+\boldsymbol{b}$ is the ***shortest least squares solution to*** $A\boldsymbol{x} = \boldsymbol{b}$. Any other vector that solves the normal equation $A^{\mathrm{T}}A\hat{\boldsymbol{x}} = A^{\mathrm{T}}\boldsymbol{b}$ is longer than $\boldsymbol{x}^+$.

Problem Set 7.2

Problems 1–6 compute and use the SVD of a particular matrix (not invertible).

1 Compute $A^{\mathrm{T}}A$ and its eigenvalues and unit eigenvectors $\boldsymbol{v}_1$ and $\boldsymbol{v}_2$:

$$A = \begin{bmatrix} 1 & 2 \\ 3 & 6 \end{bmatrix}.$$

What is the only singular value σ_1? The rank of A is $r = 1$.

2 (a) Compute AA^{T} and its eigenvalues and unit eigenvectors $\boldsymbol{u}_1$ and $\boldsymbol{u}_2$.

(b) Verify from Problem 1 that $A\boldsymbol{v}_1 = \sigma_1\boldsymbol{u}_1$. Put numbers into the SVD:

$$\begin{bmatrix} 1 & 2 \\ 3 & 6 \end{bmatrix} = \begin{bmatrix} \boldsymbol{u}_1 & \boldsymbol{u}_2 \end{bmatrix} \begin{bmatrix} \sigma_1 & \\ & 0 \end{bmatrix} \begin{bmatrix} \boldsymbol{v}_1 & \boldsymbol{v}_2 \end{bmatrix}^{\mathrm{T}}.$$

3 From the $\boldsymbol{u}$'s and $\boldsymbol{v}$'s write down orthonormal bases for the four fundamental subspaces of this matrix A.

4 Draw a picture like Figure 7.5 to show the three steps of the SVD for this A.

5 From U, V, and Σ find the orthogonal matrix $Q = UV^{\mathrm{T}}$ and the symmetric matrix $H = V\Sigma V^{\mathrm{T}}$. Verify the polar decomposition $A = QH$. This H is only semidefinite because ______.

6 Compute the pseudoinverse $A^+ = V\Sigma^+U^{\mathrm{T}}$. The diagonal matrix Σ^+ contains $1/\sigma_1$. Rename the four subspaces (for A) in Figure 7.7 as four subspaces for A^+. Compute A^+A and AA^+.

Problems 7–11 are about the SVD of an invertible matrix.

7 Compute $A^{\mathrm{T}}A$ and its eigenvalues and unit eigenvectors $\boldsymbol{v}_1$ and $\boldsymbol{v}_2$. What are the singular values σ_1 and σ_2 for this matrix A?

$$A = \begin{bmatrix} 3 & 3 \\ -1 & 1 \end{bmatrix}.$$

8 AA^{T} has the same eigenvalues σ_1^2 and σ_2^2 as $A^{\mathrm{T}}A$. Find unit eigenvectors $\boldsymbol{u}_1$ and $\boldsymbol{u}_2$. Put numbers into the SVD:

$$A = \begin{bmatrix} 3 & 3 \\ -1 & 1 \end{bmatrix} = \begin{bmatrix} \boldsymbol{u}_1 & \boldsymbol{u}_2 \end{bmatrix} \begin{bmatrix} \sigma_1 & \\ & \sigma_2 \end{bmatrix} \begin{bmatrix} \boldsymbol{v}_1 & \boldsymbol{v}_2 \end{bmatrix}^{\mathrm{T}}.$$

9 In Problem 8, multiply columns times rows to show that $A = \sigma_1 \boldsymbol{u}_1 \boldsymbol{v}_1^{\mathrm{T}} + \sigma_2 \boldsymbol{u}_2 \boldsymbol{v}_2^{\mathrm{T}}$. Prove from $A = U \Sigma V^{\mathrm{T}}$ that every matrix of rank r is the sum of r matrices of rank one.

10 From U, V, and Σ find the orthogonal matrix $Q = UV^{\mathrm{T}}$ and the symmetric matrix $K = U \Sigma U^{\mathrm{T}}$. Verify the polar decomposition in the reverse order $A = KQ$.

11 The pseudoinverse of this A is the same as ______ because ______.

Problems 12–13 compute and use the SVD of a 1 by 3 rectangular matrix.

12 Compute $A^{\mathrm{T}}A$ and AA^{T} and their eigenvalues and unit eigenvectors when the matrix is $A = [\,3 \;\; 4 \;\; 0\,]$. What are the singular values of A?

13 Put numbers into the singular value decomposition of A:

$$A = \begin{bmatrix} 3 & 4 & 0 \end{bmatrix} = \begin{bmatrix} \boldsymbol{u}_1 \end{bmatrix} \begin{bmatrix} \sigma_1 & 0 & 0 \end{bmatrix} \begin{bmatrix} \boldsymbol{v}_1 & \boldsymbol{v}_2 & \boldsymbol{v}_3 \end{bmatrix}^{\mathrm{T}}.$$

Put numbers into the pseudoinverse of A. *Compute AA^+ and A^+A*:

$$A^+ = \begin{bmatrix} \\ \\ \\ \end{bmatrix} = \begin{bmatrix} \boldsymbol{v}_1 & \boldsymbol{v}_2 & \boldsymbol{v}_3 \end{bmatrix} \begin{bmatrix} 1/\sigma_1 \\ 0 \\ 0 \end{bmatrix} \begin{bmatrix} \boldsymbol{u}_1 \end{bmatrix}^{\mathrm{T}}.$$

14 What is the only 2 by 3 matrix that has no pivots and no singular values? What is Σ for that matrix? A^+ is the zero matrix, but what shape?

15 If $\det A = 0$ how do you know that $\det A^+ = 0$?

16 When are the factors in $U \Sigma V^{\mathrm{T}}$ the same as in $Q \Lambda Q^{\mathrm{T}}$? The eigenvalues λ_i must be positive, to equal the σ_i. Then A must be ______ and positive ______.

Questions 17–20 bring out the main properties of A^+ and $x^+ = A^+ b$.

17 In Example 6 all matrices have rank one. The vector $\boldsymbol{b}$ is (b_1, b_2).

$$A = \begin{bmatrix} 2 & 2 \\ 1 & 1 \end{bmatrix}, \quad A^+ = \begin{bmatrix} .2 & .1 \\ .2 & .1 \end{bmatrix}, \quad AA^+ = \begin{bmatrix} .8 & .4 \\ .4 & .2 \end{bmatrix}, \quad A^{\mathrm{T}}A = \begin{bmatrix} 5 & 5 \\ 5 & 5 \end{bmatrix}.$$

18 (a) The equation $A^{\mathrm{T}}A\hat{\boldsymbol{x}} = A^{\mathrm{T}}\boldsymbol{b}$ has many solutions because $A^{\mathrm{T}}A$ is ______.

(b) Verify that $\boldsymbol{x}^+ = A^+\boldsymbol{b} = (.2b_1 + .1b_2,\ .2b_1 + .1b_2)$ does solve $A^{\mathrm{T}}A\boldsymbol{x}^+ = A^{\mathrm{T}}\boldsymbol{b}$.

(c) AA^+ projects onto the column space of A. Therefore ______ projects onto the nullspace of A^{T}. Then $A^{\mathrm{T}}(AA^+ - I)\boldsymbol{b} = \boldsymbol{0}$. Then $A^{\mathrm{T}}A\boldsymbol{x}^+ = A^{\mathrm{T}}\boldsymbol{b}$ and $\hat{\boldsymbol{x}}$ can be $\boldsymbol{x}^+$.

19 *The vector x^+ is the shortest possible solution to* $A^{\mathrm{T}}A\hat{x} = A^{\mathrm{T}}\boldsymbol{b}$. The difference $\hat{x} - \boldsymbol{x}^+$ is in the nullspace of $A^{\mathrm{T}}A$. This is also the nullspace of A (see **4C**). Explain how it follows that

$$\|\hat{x}\|^2 = \|\boldsymbol{x}^+\|^2 + \|\hat{x} - \boldsymbol{x}^+\|^2.$$

Any other solution $\hat{x}$ has greater length than $\boldsymbol{x}^+$.

20 Every $\boldsymbol{b}$ in $\mathbf{R}^m$ is $\boldsymbol{p} + \boldsymbol{e}$. This is the column space part plus the left nullspace part. Every $\boldsymbol{x}$ in $\mathbf{R}^n$ is $\boldsymbol{x}_r + \boldsymbol{x}_n$ = (row space part) + (nullspace part). Then

$$AA^+\boldsymbol{p} = \_\_\_\_, \qquad AA^+\boldsymbol{e} = \_\_\_\_, \qquad A^+A\boldsymbol{x}_r = \_\_\_\_, \qquad A^+A\boldsymbol{x}_n = \_\_\_\_.$$

21 Find A^+ and A^+A and AA^+ for the 2 by 1 matrix whose SVD is

$$A = \begin{bmatrix} 3 \\ 4 \end{bmatrix} = \begin{bmatrix} .6 & -.8 \\ .8 & .6 \end{bmatrix} \begin{bmatrix} 5 \\ 0 \end{bmatrix} \begin{bmatrix} 1 \end{bmatrix}.$$

Questions 21–23 are about factorizations of 2 by 2 matrices.

22 A general 2 by 2 matrix A is determined by *four* numbers. If triangular, it is determined by *three*. If diagonal, by *two*. If a rotation, by *one*. Check that the total count is four for each factorization of A:

$$LU, \qquad LDU, \qquad QR, \qquad U\Sigma V^{\mathrm{T}}, \qquad S\Lambda S^{-1}.$$

23 Following Problem 22, check that LDL^{T} and $Q\Lambda Q^{\mathrm{T}}$ are determined by *three* numbers. This is correct because the matrix A is ____.

24 A new factorization! Factor $\left[\begin{smallmatrix} a & b \\ c & d \end{smallmatrix}\right]$ into $A = EH$, where E is lower triangular with 1's on the diagonal and H is symmetric. When is this impossible?

Part II

Geodesy

8

LEVELING NETWORKS

8.1 Heights by Least Squares

Our first example in geodesy is ***leveling***—the determination of *heights*. The problem is to find the heights $x_1, \dots, x_n$ at n specified points. What we actually measure is *differences of heights*. The height at point i is measured from point j, to give a value b_{ij} (probably not exact) for the height difference:

$$x_i - x_j = b_{ij} - \text{error}. \tag{8.1}$$

These differences are measured for certain pairs i, j. From the measurements b_{ij} we are to estimate the actual heights.

Suppose first that there are no errors in the measurements. Then we expect to solve the equations exactly. But if you look at the equations for $n = 3$ points and $m = 3$ measurements, you will see a difficulty:

$$\begin{aligned} x_1 - x_2 &= b_{12} \\ x_2 - x_3 &= b_{23} \\ x_3 - x_1 &= b_{31}. \end{aligned} \tag{8.2}$$

This system of equations is ***singular***. Its coefficient matrix is

$$A = \begin{bmatrix} 1 & -1 & 0 \\ 0 & 1 & -1 \\ -1 & 0 & 1 \end{bmatrix}.$$

The rows of A add to the zero row. The matrix is not invertible. The determinant of A has to be zero (we refuse to compute determinants). When we add the three equations, the result on the left side is zero:

$$0 = b_{12} + b_{23} + b_{31}. \tag{8.3}$$

A singular system of equations has two possibilities, no solution or too many:

1 There is ***no solution***. The measurements b_{12}, b_{23}, b_{31} do not add to zero and the three equations are inconsistent.

2 The equations are consistent but the solution x_1, x_2, x_3 is ***not unique***. There are infinitely many solutions when the consistency condition in equation (8.3) is met.

For measurements with errors, we expect to be in case 1: no solution. For exact measurements we must be in case 2: many solutions. This is our situation, and the reason is clear:

> We cannot determine absolute heights purely from height differences. One or more of the heights x_j must be *postulated* (given a priori). Each ***fixed height*** is removed from the list of unknowns.

Suppose the third height is fixed at $x_3 = H$. Our equations become

$$\begin{aligned} x_1 - x_2 &= b_{12} \\ x_2 &= b_{23} + H \\ -x_1 \quad &= b_{31} - H. \end{aligned} \tag{8.4}$$

Now we have three equations and only two unknowns. Notice that they add to the same consistency equation $0 = b_{12} + b_{23} + b_{31}$. There are still two possibilities but the second is different because the matrix is different:

1 There is ***no solution*** (the measurements are not consistent).

2 There is ***exactly one solution*** (the consistency equation holds).

In the language of linear algebra, we have a 3 by 2 matrix. The third column corresponding to x_3 has been removed. The two remaining columns are ***independent***:

$$A_{\text{reduced}} = \begin{bmatrix} 1 & -1 \\ 0 & 1 \\ -1 & 0 \end{bmatrix}.$$

The rank is 2 (full column rank). *There is either no solution or one solution.* The nullspace of A contains only the zero vector.

Remark 8.1 Our problem is closely parallel to computing voltages (instead of heights) in an electrical circuit. The consistency equation $0 = b_{12} + b_{23} + b_{31}$ is Kirchhoff's Voltage Law: The differences around a loop add to zero. The fixed height $x_3 = H$ is like a fixed voltage, which allows the other voltages to be uniquely determined. Postulating $x_3 = 0$ is "grounding a node."

Section 8.4 will develop further the analogy between heights in leveling networks and voltages in electrical networks. This viewpoint is important. It places geodesy into the basic framework of applied mathematics.

For exact measurements, consistency will hold. We can solve two of the equations for x_1 and x_2, and the third equation automatically follows. This is the nice case but in practice it almost never happens.

For measurements with errors, we expect the three equations in (8.4) to be inconsistent. They cannot be solved. We look for a "best" solution, which makes an agreed measure E of overall system error as small as possible. The solution is best for that error measure E. For ***least squares***, we minimize the sum of squares from the m equations:

$$E^2 = r_1^2 + r_2^2 + r_3^2 = (b_{12} - x_1 + x_2)^2 + (b_{23} + H - x_2)^2 + (b_{31} - H + x_1)^2.$$

This will be our starting point: ordinary least squares. It is not our finishing point. Usually it is *not* agreed that this E is the error measure that should be minimized. Other error measures give other "best" solutions and here are three of the most important:

$$E^2_{\text{weighted}} = \frac{r_1^2}{\sigma_1^2} + \frac{r_2^2}{\sigma_2^2} + \frac{r_3^2}{\sigma_3^2} \qquad \text{(weighted } l^2 \text{ norm)}$$

$$E_{\text{sum}} = |r_1| + |r_2| + |r_3| \qquad (l^1 \text{ norm})$$

$$E_{\max} = \text{maximum of } \{|r_1|, |r_2|, |r_3|\} \qquad (l^\infty \text{ norm}).$$

Most of our attention will go to ***weighted least squares***. We must explain why the particular weights are chosen in E^2_{weighted} and how they affect the estimated solution $\hat{x}_1, \hat{x}_2$.

The quantities $\sigma_1^2, \sigma_2^2, \sigma_3^2$ are ***variances***. They measure the reliabilities of the three measurements. More reliable measurements have smaller variances and larger weights (because the weight $1/\sigma^2$ is the reciprocal of the variance). Equations that are weighted more heavily are solved more exactly when we minimize the overall error E^2_{weighted}.

Remark 8.2 The next chapter gives a detailed discussion of variances and covariances. This link to statistics is essential. The useful output from our problem should be the height estimates $\hat{x}_i$ and also *an indication of their reliability*. ***We want to know the variances of the output errors $\hat{x}_i - x_i$, given the variances of the input measurement errors r_i.*** It will be proved that the output variances are *smallest* when the weights are reciprocals of the input variances. That is the reason for the weights $1/\sigma^2$.

More generally, the optimum weight matrix is the inverse of the covariance matrix.

Remark 8.3 The error measures $E_{\text{sum}} = \sum |r_i|$ and $E_{\max} = |r_i|_{\max}$ are not quadratic (and not even differentiable) because of the corners in the absolute value function. Minimization gives piecewise linear instead of linear equations. Thus E_{sum} and $E_{\max}$ lead to *linear programming*, in which a subset of the equations $A\boldsymbol{x} = \boldsymbol{b}$ holds exactly. The difficult problem is to find that subset. The simplex method is quite efficient as a direct method. Iterative methods use weighted least squares at each linear step, with weights taken from the preceding iteration (this is called downweighting in geodesy).

$E_{\max}$ is almost never used but E_{sum} is very helpful. It is more robust than least squares, and less willing to conform to wild measurements (outliers). In practice there are almost always gross errors among the measurements b_{ij}. Observations get wrongly identified; numbers are wrongly reproduced. Some geodesists estimate that 5% of their data are infected. A least-squares fit will smooth over these errors too successfully. By minimizing E_{sum} instead of E, the gross errors can be identified in the residual $\boldsymbol{b} - A\hat{\boldsymbol{x}}$. Those incorrect observations are removed from the data before computing the final estimate $\hat{\boldsymbol{x}}$ that minimizes E^2_{weighted}.

Completion of the Example

We return to the three equations for two unknowns:

$$\begin{aligned} x_1 - x_2 &= b_{12} \\ x_2 &= b_{23} + H \\ -x_1 \qquad &= b_{31} - H. \end{aligned} \tag{8.5}$$

This is our system $A\boldsymbol{x} = \boldsymbol{b}$. It will be solved by least squares, and also by weighted least squares. We will choose the weights that are appropriate in this leveling problem. The example is small enough to carry through in full detail.

The matrix A and the right side $\boldsymbol{b}$ are

$$A = \begin{bmatrix} 1 & -1 \\ 0 & 1 \\ -1 & 0 \end{bmatrix} \qquad \text{and} \qquad \boldsymbol{b} = \begin{bmatrix} b_{12} \\ b_{23} + H \\ b_{31} - H \end{bmatrix}.$$

For ordinary least squares, with unit weights, the *normal equation* is $A^{\mathrm{T}}A\hat{\boldsymbol{x}} = A^{\mathrm{T}}\boldsymbol{b}$. Its solution is the estimate $\hat{\boldsymbol{x}} = (\hat{x}_1, \hat{x}_2)$ of the unknown heights at the first two observation sites. The third height was postulated as $x_3 = H$.

Multiply $A\hat{\boldsymbol{x}} = \boldsymbol{b}$ by the 2 by 3 matrix A^{T} to find the equation $A^{\mathrm{T}}A\hat{\boldsymbol{x}} = A^{\mathrm{T}}\boldsymbol{b}$:

$$\begin{bmatrix} 2 & -1 \\ -1 & 2 \end{bmatrix} \begin{bmatrix} \hat{x}_1 \\ \hat{x}_2 \end{bmatrix} = \begin{bmatrix} b_{12} - b_{31} + H \\ b_{23} - b_{12} + H \end{bmatrix}. \tag{8.6}$$

This matrix $A^{\mathrm{T}}A$ has the properties we expect: It is symmetric and it is invertible. The third column of A was removed by postulating $x_3 = H$, leaving two independent columns. The inverse of $A^{\mathrm{T}}A$ would not be computed in large problems, but here it is easy to do:

$$\begin{bmatrix} \hat{x}_1 \\ \hat{x}_2 \end{bmatrix} = \frac{1}{3} \begin{bmatrix} 2 & 1 \\ 1 & 2 \end{bmatrix} \begin{bmatrix} b_{12} - b_{31} + H \\ b_{23} - b_{12} + H \end{bmatrix}. \tag{8.7}$$

This gives the unweighted least-squares estimates

$$\begin{aligned} \hat{x}_1 &= \tfrac{1}{3}\big(b_{12} + b_{23} - 2b_{31}\big) + H \\ \hat{x}_2 &= \tfrac{1}{3}\big(-b_{12} - b_{31} + 2b_{23}\big) + H. \end{aligned} \tag{8.8}$$

Note how all heights are raised by the same amount H. By fixing $x_3 = H$, we set the "arbitrary constant." If heights were measured from a different sea level, all components of $\hat{\boldsymbol{x}}$ would go up or down together.

Notice also the possibility that the original equations are consistent: $b_{12} + b_{23} + b_{31} = 0$. In that case the estimate $\hat{\boldsymbol{x}}$ is also the genuine solution $\boldsymbol{x}$. It is the unique solution to $A\boldsymbol{x} = \boldsymbol{b}$. Replacing $b_{12} + b_{23} + b_{31}$ by zero gives the exact solution *when it exists*:

$$\begin{aligned} x_1 &= -b_{31} + H \\ x_2 &= b_{23} + H \\ x_3 &= H. \end{aligned} \tag{8.9}$$

Again all heights move together with H. But least squares is telling us that (8.8) is a better estimate than (8.9) when the equations are inconsistent.

Weighted Least Squares

Change from the unweighted to the weighted error measure:

$$E^2 = r_1^2 + r_2^2 + r_3^2 \qquad \text{becomes} \qquad E^2_{\text{weighted}} = \frac{r_1^2}{\sigma_1^2} + \frac{r_2^2}{\sigma_2^2} + \frac{r_3^2}{\sigma_3^2}.$$

The variances $\sigma_1^2, \sigma_2^2, \sigma_3^2$ represent the spread of the measurement errors around their mean. For the leveling problem there is a very useful empirical rule: ***The variance is proportional to the distance between observation points***. Thus we choose

$$\begin{aligned}
\sigma_1^2 &= \sigma_0^2 l_1 = \sigma_0^2 \times \text{distance between sites 1 and 2}\\
\sigma_2^2 &= \sigma_0^2 l_2 = \sigma_0^2 \times \text{distance between sites 2 and 3}\\
\sigma_3^2 &= \sigma_0^2 l_3 = \sigma_0^2 \times \text{distance between sites 3 and 1.}
\end{aligned}$$

The factor σ_0^2 is the ***variance of unit weight***. In Chapter 9 (where variances and covariances are properly introduced), this factor σ_0^2 plays a valuable role. It allows us to rescale a covariance matrix, when our estimates of this matrix proves (from the actual data) to be unrealistic. In a statistically perfect world, our a priori covariance matrix would be correct and the scaling factor would be $\sigma_0^2 = 1$.

Chapter 9 will also give a specific formula (9.64) to estimate σ_0^2.

We still need to postulate one height: $x_3 = H$. Our three measurements still have errors r_1, r_2, r_3 (and we know their statistics: mean zero and variances $\sigma_0^2 l_1, \sigma_0^2 l_2, \sigma_0^2 l_3$):

$$\begin{aligned}
x_1 - x_2 &= b_{12} && - r_1\\
x_2 &= b_{23} + H && - r_2\\
-x_1 \qquad &= b_{31} - H && - r_3.
\end{aligned} \tag{8.10}$$

The best solution $\hat{x}_1, \hat{x}_2$ minimizes E^2_{weighted}. The variances $\sigma_0^2 l_1, \sigma_0^2 l_2, \sigma_0^2 l_3$ are in the denominators. When we take derivatives of E^2_{weighted} with respect to x_1 and x_2, these numbers $\sigma_0^2 l_1, \sigma_0^2 l_2, \sigma_0^2 l_3$ will appear in the linear equations. Calculus gives two equations from the x_1 derivative and the x_2 derivative of the weighted sum of squares:

$$\begin{aligned}
\frac{\partial}{\partial x_1}: &\quad (x_1 - x_2 - b_{12})/(\sigma_0^2 l_1) - (-x_1 - b_{31} + H)/(\sigma_0^2 l_3) = 0\\
\frac{\partial}{\partial x_2}: &\quad -(x_1 - x_2 - b_{12})/(\sigma_0^2 l_1) + (x_2 - b_{23} - H)/(\sigma_0^2 l_2) = 0.
\end{aligned} \tag{8.11}$$

It is slightly inconvenient to have these fractions. Matrix notation will be better. The numbers $1/\sigma_i^2 = 1/(\sigma_0^2 l_i)$ will go into a weighting matrix C. Our example has a diagonal weighting matrix, because the three errors are assumed independent:

$$C = \begin{bmatrix} (\sigma_0^2 l_1)^{-1} & & \\ & (\sigma_0^2 l_2)^{-1} & \\ & & (\sigma_0^2 l_3)^{-1} \end{bmatrix} = \begin{bmatrix} c_1 & & \\ & c_2 & \\ & & c_3 \end{bmatrix}.$$

The next section will do the algebra for a general weighting matrix C:

When $Ax = b$ is weighted by C, the normal equations become $A^{\mathrm{T}}CA\hat{x} = A^{\mathrm{T}}Cb$.

This is exactly equation (8.11). We will use c_i instead of $1/(\sigma_0^2 l_i)$, but you will see the same coefficients from (8.11) in the following matrix $A^{\mathrm{T}}CA$:

$$A^{\mathrm{T}}CA = \begin{bmatrix} 1 & 0 & -1 \\ -1 & 1 & 0 \end{bmatrix} \begin{bmatrix} c_1 & & \\ & c_2 & \\ & & c_3 \end{bmatrix} \begin{bmatrix} 1 & -1 \\ 0 & 1 \\ -1 & 0 \end{bmatrix} = \begin{bmatrix} c_1 + c_3 & -c_1 \\ -c_1 & c_1 + c_2 \end{bmatrix}. \tag{8.12}$$

This is symmetric and invertible and positive definite. With $c_1 = c_2 = c_3 = 1$ it is the unweighted matrix $A^{\mathrm{T}}A$ computed above. Its determinant is $\Delta = c_1c_2 + c_1c_3 + c_2c_3$. The right side $A^{\mathrm{T}}Cb$ is also straightforward:

$$\begin{bmatrix} 1 & 0 & -1 \\ -1 & 1 & 0 \end{bmatrix} \begin{bmatrix} c_1 & & \\ & c_2 & \\ & & c_3 \end{bmatrix} \begin{bmatrix} b_{12} \\ b_{23} + H \\ b_{31} - H \end{bmatrix} = \begin{bmatrix} c_1b_{12} - c_3b_{31} + c_3H \\ c_2b_{23} - c_1b_{12} + c_2H \end{bmatrix}. \tag{8.13}$$

For the sake of curiosity we explicitly invert $A^{\mathrm{T}}CA$ and solve $A^{\mathrm{T}}CA\hat{x} = A^{\mathrm{T}}Cb$:

$$\begin{aligned} \begin{bmatrix} \hat{x}_1 \\ \hat{x}_2 \end{bmatrix} &= \frac{1}{\Delta} \begin{bmatrix} c_1 + c_2 & c_1 \\ c_1 & c_1 + c_3 \end{bmatrix} \begin{bmatrix} c_1b_{12} - c_3b_{31} + c_3H \\ c_2b_{23} - c_1b_{12} + c_2H \end{bmatrix} \\ &= \frac{1}{\Delta} \begin{bmatrix} c_1c_2(b_{12} + b_{23}) - (c_1c_3 + c_2c_3)b_{31} \\ c_1c_3(-b_{12} - b_{31}) + (c_1c_2 + c_2c_3)b_{23} \end{bmatrix} + \begin{bmatrix} H \\ H \end{bmatrix}. \end{aligned}$$

Again all estimated heights move up or down with H. The unit weights $c_1 = c_2 = c_3 = 1$ bring back the unweighted $\hat{x}$ computed earlier. These pencil and paper calculations are never to be repeated (in this book!). They show explicitly how the weights $c_i = 1/(\sigma_0^2 l_i)$ enter into the estimates. The next section derives the key equation $A^{\mathrm{T}}CA\hat{x} = A^{\mathrm{T}}Cb$.

8.2 Weighted Least Squares

The previous section gave an example of *weighted least squares*. The weighting matrix C was diagonal, because the observation errors were not correlated. The matrix becomes $C = I$ (or really $C = \sigma^2 I$) when the errors all have the same variance. This includes the *i. i. d.* case of independent and identically distributed errors. When errors are not independent, geodesy must deal with any symmetric positive definite matrix $C = \Sigma^{-1}$. *This is the inverse of the covariance matrix* Σ.

This section extends the normal equations $A^{\mathrm{T}}CA\hat{x} = A^{\mathrm{T}}Cb$ and the basic theory to allow for C. That matrix changes the way we measure the errors $r = b - Ax$. The squared length $r^{\mathrm{T}}r$ becomes $r^{\mathrm{T}}Cr$, including the weights:

$$\|r\|^2 = r^{\mathrm{T}}r \qquad \text{changes to} \qquad \|r\|_C^2 = r^{\mathrm{T}}Cr.$$

When lengths change, so do inner products: $a^{\mathrm{T}}b$ becomes $a^{\mathrm{T}}Cb$. Angles change too. Two vectors a and b are now perpendicular when $a^{\mathrm{T}}Cb = 0$.

In this sense, the best combination $A\hat{x}$ of the columns of A is still a perpendicular projection of $\boldsymbol{b}$. *The fundamental equation of least squares still requires that the error* $\boldsymbol{r} = \boldsymbol{b} - A\hat{\boldsymbol{x}}$ *shall be perpendicular to all columns of* A. The inner products $\boldsymbol{a}^{\mathrm{T}}C\boldsymbol{r}$, between columns of A and the error $\boldsymbol{r}$, are all zero. Those columns are the rows of A^{T}, so

$$A^{\mathrm{T}}C\boldsymbol{r} = \boldsymbol{0} \quad \text{or} \quad A^{\mathrm{T}}C(\boldsymbol{b} - A\hat{\boldsymbol{x}}) = \boldsymbol{0} \quad \text{or} \quad A^{\mathrm{T}}CA\hat{\boldsymbol{x}} = A^{\mathrm{T}}C\boldsymbol{b}. \tag{8.14}$$

This is the ***weighted normal equation***.

Remember the other source of this equation, which is minimization. The vector $\hat{\boldsymbol{x}}$ is chosen to make $E^2_{\text{weighted}} = \|\boldsymbol{r}\|^2_C$ as small as possible:

$$\text{Minimize} \quad \|\boldsymbol{r}\|^2_C = (\boldsymbol{b} - A\boldsymbol{x})^{\mathrm{T}}C(\boldsymbol{b} - A\boldsymbol{x}). \tag{8.15}$$

Expand that expression into $\boldsymbol{x}^{\mathrm{T}}(A^{\mathrm{T}}CA)\boldsymbol{x} - 2\boldsymbol{x}^{\mathrm{T}}(A^{\mathrm{T}}C\boldsymbol{b}) + \boldsymbol{b}^{\mathrm{T}}C\boldsymbol{b} = \boldsymbol{x}^{\mathrm{T}}K\boldsymbol{x} - 2\boldsymbol{x}^{\mathrm{T}}\boldsymbol{c} +$ constant. Now we have a pure calculus problem. We can set derivatives to zero. That gives n equations for the n components of $\hat{\boldsymbol{x}}$. The equations will be linear because $\|\boldsymbol{r}\|^2_C$ is quadratic. Matrix algebra will find those minimizing equations, once and for all. They are $K\hat{\boldsymbol{x}} = \boldsymbol{c}$ with $K = A^{\mathrm{T}}CA$ and $\boldsymbol{c} = A^{\mathrm{T}}C\boldsymbol{b}$:

Theorem When K is symmetric positive definite, the quadratic $Q(\boldsymbol{x}) = \boldsymbol{x}^{\mathrm{T}}K\boldsymbol{x} - 2\boldsymbol{x}^{\mathrm{T}}\boldsymbol{c}$ is minimized at the point where $K\hat{\boldsymbol{x}} = \boldsymbol{c}$. The minimum value of Q, at that point $\hat{\boldsymbol{x}} = K^{-1}\boldsymbol{c}$, is $Q(\hat{\boldsymbol{x}}) = -\boldsymbol{c}^{\mathrm{T}}K^{-1}\boldsymbol{c}$.

Proof Compare $Q(\hat{\boldsymbol{x}})$ with all other $Q(\boldsymbol{x})$, to show that $Q(\hat{\boldsymbol{x}})$ is smallest:

$$\begin{aligned} Q(\boldsymbol{x}) - Q(\hat{\boldsymbol{x}}) &= \boldsymbol{x}^{\mathrm{T}}K\boldsymbol{x} - 2\boldsymbol{x}^{\mathrm{T}}\boldsymbol{c} - \hat{\boldsymbol{x}}^{\mathrm{T}}K\hat{\boldsymbol{x}} + 2\hat{\boldsymbol{x}}^{\mathrm{T}}\boldsymbol{c} \\ &= \boldsymbol{x}^{\mathrm{T}}K\boldsymbol{x} - 2\boldsymbol{x}^{\mathrm{T}}K\hat{\boldsymbol{x}} + \hat{\boldsymbol{x}}^{\mathrm{T}}K\hat{\boldsymbol{x}} \qquad \text{(substitute } K\hat{\boldsymbol{x}} \text{ for } \boldsymbol{c}\text{)} \\ &= (\boldsymbol{x} - \hat{\boldsymbol{x}})^{\mathrm{T}}K(\boldsymbol{x} - \hat{\boldsymbol{x}}). \end{aligned}$$

Since K is positive definite, this difference is never negative. $Q(\hat{\boldsymbol{x}})$ is the smallest possible value. At that point $\hat{\boldsymbol{x}} = K^{-1}\boldsymbol{c}$, the minimum of Q is

$$Q_{\min} = (K^{-1}\boldsymbol{c})^{\mathrm{T}}K(K^{-1}\boldsymbol{c}) - 2(K^{-1}\boldsymbol{c})^{\mathrm{T}}\boldsymbol{c} = -\boldsymbol{c}^{\mathrm{T}}K^{-1}\boldsymbol{c}. \tag{8.16}$$

Corollary The minimum of the weighted error $\|\boldsymbol{r}\|^2_C$ is attained when $(A^{\mathrm{T}}CA)\hat{\boldsymbol{x}} = A^{\mathrm{T}}C\boldsymbol{b}$. The minimum value is

$$\begin{aligned} \|\boldsymbol{r}\|^2_C &= \text{minimum of} \quad \boldsymbol{x}^{\mathrm{T}}K\boldsymbol{x} - 2\boldsymbol{x}^{\mathrm{T}}\boldsymbol{c} + \boldsymbol{b}^{\mathrm{T}}C\boldsymbol{b} \\ &= -\boldsymbol{c}^{\mathrm{T}}K^{-1}\boldsymbol{c} + \boldsymbol{b}^{\mathrm{T}}C\boldsymbol{b} \\ &= -\boldsymbol{b}^{\mathrm{T}}CA(A^{\mathrm{T}}CA)^{-1}A^{\mathrm{T}}C\boldsymbol{b} + \boldsymbol{b}^{\mathrm{T}}C\boldsymbol{b}. \end{aligned}$$

If A was a square invertible matrix, this whole error would reduce to zero! The solution $\hat{\boldsymbol{x}} = K^{-1}\boldsymbol{c}$ would be exact. The inverse of $K = A^{\mathrm{T}}CA$ could be split into $A^{-1}C^{-1}(A^{\mathrm{T}})^{-1}$, and everything cancels. ***But this splitting is not legal for a rectangular matrix A.*** Our only assumption is that A has independent columns, which makes K positive definite (and invertible). The product $A^{\mathrm{T}}CA$ is invertible but not its separate factors.

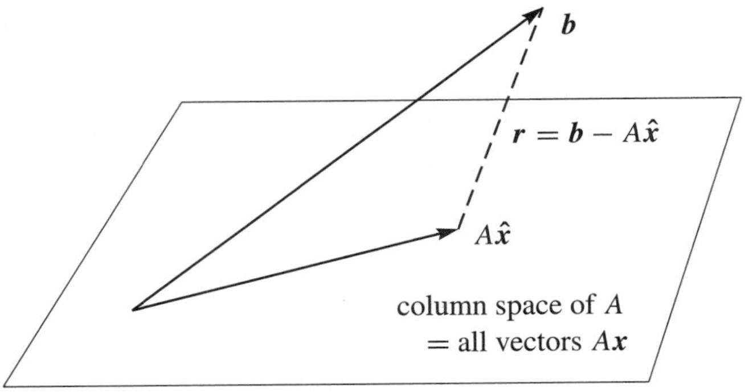

Figure 8.1 The projection is perpendicular in the C-inner product: $\boldsymbol{a}^{\mathrm{T}}C\boldsymbol{r} = 0$ for every column $\boldsymbol{a}$ of A. Then $A^{\mathrm{T}}C\boldsymbol{r} = \mathbf{0}$ is the weighted normal equation. The C-right triangle has $\|\boldsymbol{r}\|_C^2 + \|A\hat{\boldsymbol{x}}\|_C^2 = \|\boldsymbol{b}\|_C^2$, which is $\boldsymbol{b}^{\mathrm{T}}C\boldsymbol{b}$.

Figure 8.1 shows the projection geometrically. Our eyes use the eye-inner product. ($C = I$ is C = eye in MATLAB.) So visually $\boldsymbol{r}$ does not look perpendicular to its projection. But the angle is right in the C-inner product, which gives the key equation $A^{\mathrm{T}}C\boldsymbol{r} = \mathbf{0}$. This is $A^{\mathrm{T}}CA\hat{\boldsymbol{x}} = A^{\mathrm{T}}C\boldsymbol{b}$.

Remark 8.4 *On notation.* Since the time of Gauss the method of least squares has been named *adjustment theory* in geodesy and other applied sciences. The traditional notation defines the ***residual r*** by

$$A\boldsymbol{x} = \boldsymbol{b} + \boldsymbol{r}. \tag{8.17}$$

In agreement with statistics and numerical linear algebra we define the residual with ***opposite sign***: $\boldsymbol{r} = \boldsymbol{b} - A\boldsymbol{x}$. Gauss used the notation P for the weights (Latin: 'pondus'). For various reasons we have chosen to change this notation to C.

8.3 Leveling Networks and Graphs

For a closer look at C we need the basic ideas of statistics (variance and covariance). That discussion will come in Chapter 9. Here we take a much closer look at the rectangular matrix A. This has a special and important form for networks, and the leveling problem becomes a basic example in applied mathematics. The matrix A, whose entries are all 1's and -1's and 0's, is the ***incidence matrix for a graph***.

A graph consists of ***nodes*** and ***edges***. There are n nodes (the points where the heights x_i are to be determined). There is an edge between node i and node j if we measure the height difference $x_i - x_j$. This makes m edges, with $m > n$. We will show how the equations that govern a leveling network fall into the general framework of applied mathematics and also into the special pattern involving the incidence matrix of a graph.

Figure 8.2 shows a graph with four nodes and six edges. This is a ***directed graph***, because edge directions are assigned by the arrows. The height difference along edge 1

$$A = \begin{bmatrix} -1 & 1 & 0 & 0 \\ -1 & 0 & 1 & 0 \\ 0 & -1 & 1 & 0 \\ -1 & 0 & 0 & 1 \\ 0 & -1 & 0 & 1 \\ 0 & 0 & -1 & 1 \end{bmatrix}$$

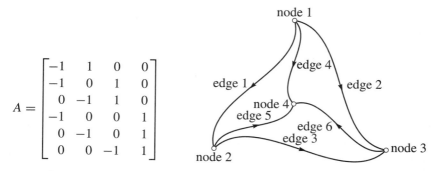

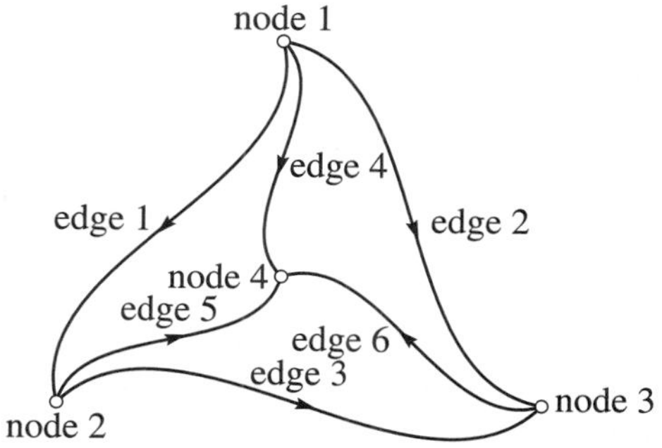

Figure 8.2 A graph and its 6 by 4 edge-node incidence matrix.

is $d_1 = x_2 - x_1$. Our actual measurement of this difference is b_{12}. The arrow does not imply that node 2 is higher than node 1, it just specifies the difference as $x_2 - x_1$, rather than $x_1 - x_2$. (The directions of the arrows are arbitrary, but fixed.) Now we introduce the ***incidence matrix*** or difference matrix or connection matrix of the graph.

The incidence matrix A has a row for every edge and a column for every node. In our example the matrix is 6 by 4. Each row has two nonzero entries, $+1$ and -1, to show which node the arrow enters and which node it leaves. Thus the nonzero entries in row 1 (for edge 1) are $+1$ in column 2, and -1 in column 1:

$$A = \begin{bmatrix} -1 & 1 & 0 & 0 \\ -1 & 0 & 1 & 0 \\ 0 & -1 & 1 & 0 \\ -1 & 0 & 0 & 1 \\ 0 & -1 & 0 & 1 \\ 0 & 0 & -1 & 1 \end{bmatrix} . \quad \begin{matrix} \leftarrow \text{edge 1} \\ \\ \\ \\ \\ \\ \end{matrix}$$

This matrix contains all information about the graph. This particular example is a "complete graph," with no edges missing. (A complete graph has all $m = \frac{1}{2}n(n+1)$ edges. If an edge is removed from the graph, a row is removed from the matrix.) We could allow, but we don't, double edges between nodes and an edge from a node to itself. A double edge would just mean that the height difference $x_i - x_j$ was measured twice.

The incidence matrix is more than a passive record of the edge connections in the graph. The matrix A is also active; *it computes differences*. When we apply A to a vector $\boldsymbol{x} = (x_1, x_2, x_3, x_4)$ of heights, the output $A\boldsymbol{x}$ is a set of six ***height differences***:

$$A\boldsymbol{x} = \begin{bmatrix} -1 & 1 & 0 & 0 \\ -1 & 0 & 1 & 0 \\ 0 & -1 & 1 & 0 \\ -1 & 0 & 0 & 1 \\ 0 & -1 & 0 & 1 \\ 0 & 0 & -1 & 1 \end{bmatrix} \begin{bmatrix} x_1 \\ x_2 \\ x_3 \\ x_4 \end{bmatrix} = \begin{bmatrix} x_2 - x_1 \\ x_3 - x_1 \\ x_3 - x_2 \\ x_4 - x_1 \\ x_4 - x_2 \\ x_4 - x_3 \end{bmatrix} . \tag{8.18}$$

This is important. These differences are measured by $b_{12}, b_{13}, \dots, b_{34}$. These measurements involve errors. The six equations to be solved by weighted least squares are

$$\begin{aligned} x_2 - x_1 &= b_{12} \\ &\vdots \\ x_4 - x_3 &= b_{34} \end{aligned} \qquad \text{or in matrix form} \qquad A\boldsymbol{x} = \boldsymbol{b}. \tag{8.19}$$

The six equations in four unknowns are probably inconsistent (because of the measurement errors $r_1, \dots, r_6$). We do not expect an exact solution; there are more equations than unknowns. We form the (weighted) normal equations to arrive at the best estimate $\hat{\boldsymbol{x}}$. But there is an issue to be dealt with first:

> The four columns of the matrix add to the zero column. Since A has linearly *dependent* columns, the matrices $A^{\mathrm{T}}A$ and $A^{\mathrm{T}}CA$ will not be invertible. Action must be taken. One or more heights must be fixed!

In linear algebra, this question is about the "column space" of the matrix. Assuming that the graph is connected (it doesn't separate into two parts with no edges between them) there is only *one* relation between the columns: they add to the zero column. The nullspace is one-dimensional, containing the vector $(1, 1, 1, 1)$. The rank is $n - 1$. If we remove any *one* column, the new matrix A has full rank and the new $A^{\mathrm{T}}A$ is invertible. If we fix *one* height (like $x_3 = H$ in the previous section), all other heights can be estimated.

We are free to fix k heights, not just one. Then k columns of A are removed. The resulting matrix has full rank $n - k$. The normal equations will yield the $n - k$ unknown heights. Those will be the best (weighted) estimates $\hat{\boldsymbol{x}}$ from the m observations.

Remark 8.5 We plan to insert a special Section 8.4 about the incidence matrix of a graph. The dimensions of the four subspaces will give a linear algebra proof of Euler's famous formula, which is the grandfather of topology:

$$n - m + l = (\text{\# of nodes}) - (\text{\# of edges}) + (\text{\# of loops}) = 1. \tag{8.20}$$

Figure 8.2 has $l = 3$ independent loops. Euler's alternating sum is $4 - 6 + 3 = 1$. This formula will be seen as equivalent to the fundamental dimension theorem of linear algebra:

$$(\text{dimension of column space}) + (\text{dimension of nullspace}) = n. \tag{8.21}$$

Summary

Leveling networks are described by directed graphs and incidence matrices A. The graph has a node for each height x_i; there are k fixed heights and $n - k$ unknown heights. There are m measurements b_{ij} of height differences. Those correspond to the edges of the graph, each with a direction arrow. The graph changes to a *network* when we assign numbers $c_1, \dots, c_m$ to the edges.

Each number c_i is the weight of an observation. Statistically c_i is $1/\sigma_i^2$, the reciprocal of the variance when we measure a height difference. For the leveling problem $c_i = 1/(\sigma_0^2 l_i)$ is proportional to the inverse length of that edge. These numbers go into the diagonal matrix C which is m by m. It reflects characteristics about the edges while the incidence matrix A describes the connections in the network. In leveling networks the vector $\boldsymbol{x}$ denotes heights, and $A\boldsymbol{x}$ denotes differences of heights.

All nodes are assigned a *height* $\hat{x}_1, \ldots, \hat{x}_n$. The difference of heights along a loop is the sum $(\hat{x}_2 - \hat{x}_1) + (\hat{x}_3 - \hat{x}_2) + (\hat{x}_1 - \hat{x}_3) = 0$ in which everything cancels.

Loop law: ***Components of*** $A\hat{\boldsymbol{x}}$ ***add to zero around every loop.***

This is the equivalent in geodesy of Kirchhoff's Voltage Law for circuits.

For edge i the *weighted error* y_i equals c_i times the residual $r_i = (\boldsymbol{b} - A\hat{\boldsymbol{x}})_i$. Together for all edges this is a vector equation: $\boldsymbol{y} = C\boldsymbol{r}$. The weighted normal equation is $A^{\mathrm{T}}\boldsymbol{y} = \boldsymbol{0}$ or $A^{\mathrm{T}}C(\boldsymbol{b} - A\hat{\boldsymbol{x}}) = \boldsymbol{0}$. This is the equivalent of Kirchhoff's Current Law at each node:

Node law: $A^{\mathrm{T}}C\hat{\boldsymbol{r}} = \boldsymbol{0}$ ***at each node.***

The node law secures what in statistics is called ***unbiasedness***. We get unbiased estimates $\hat{x}_i$ for the unknown heights. The expected (average) value of $\hat{x}_i$ is correct.

In statistical applications we are furnished with observational equations of the type $A\boldsymbol{x} = \boldsymbol{b} - \boldsymbol{r}$, where $\boldsymbol{b}$ is the observed difference of height and $\boldsymbol{r}$ a residual vector. The norm of $\boldsymbol{r}$ is to be minimized: $E = \|\boldsymbol{r}\|^2$ is $\boldsymbol{r}^{\mathrm{T}}\boldsymbol{r}$ or more generally $\boldsymbol{r}^{\mathrm{T}}C\boldsymbol{r}$. The major part of a height difference $(A\boldsymbol{x})_i$ is determined by the observed height difference b_i while the remaining part is given by the residual $r_i = b_i - (A\boldsymbol{x})_i$. Hence $\boldsymbol{y} = C\boldsymbol{r}$ evaluates to

$$\boldsymbol{y} = C\boldsymbol{b} - CA\boldsymbol{x} \qquad \text{or} \qquad C^{-1}\boldsymbol{y} + A\boldsymbol{x} = \boldsymbol{b}.$$

This expression links the heights $\boldsymbol{x}$ with the *weighted errors* $\boldsymbol{y}$. We no longer try to solve $A\boldsymbol{x} = \boldsymbol{b}$ which has more equations than unknowns. There is a new term $C^{-1}\boldsymbol{y}$, the weighted errors. The special case in which $A\boldsymbol{x} = \boldsymbol{b}$ has a solution is also the one in which all errors are zero.

We summarize the basic equations for equilibrium:

$$\begin{aligned} C^{-1}\boldsymbol{y} + A\boldsymbol{x} &= \boldsymbol{b} \\ A^{\mathrm{T}}\boldsymbol{y} \qquad &= \boldsymbol{0}. \end{aligned} \tag{8.22}$$

This system is linear and symmetric. The unknowns are the weighted errors $\boldsymbol{y}$ and the heights $\boldsymbol{x}$. We may write it in block form as

$$\begin{bmatrix} C^{-1} & A \\ A^{\mathrm{T}} & 0 \end{bmatrix} \begin{bmatrix} \boldsymbol{y} \\ \boldsymbol{x} \end{bmatrix} = \begin{bmatrix} \boldsymbol{b} \\ \boldsymbol{0} \end{bmatrix}. \tag{8.23}$$

We may even use elimination on this block matrix. The pivot is C^{-1}, the factor multiplying row 1 is $A^{\mathrm{T}}C$, and addition eliminates A^{T} below the pivot. The result is

$$\begin{bmatrix} C^{-1} & A \\ 0 & A^{\mathrm{T}}CA \end{bmatrix} \begin{bmatrix} \boldsymbol{y} \\ \boldsymbol{x} \end{bmatrix} = \begin{bmatrix} \boldsymbol{b} \\ A^{\mathrm{T}}C\boldsymbol{b} \end{bmatrix}. \tag{8.24}$$

The equation to be solved for $\boldsymbol{x}$ is in the bottom row:

$$A^{\mathrm{T}}CA\boldsymbol{x} = A^{\mathrm{T}}C\boldsymbol{b}. \tag{8.25}$$

Equation (8.25) is the ***normal equation***. The error vector is ***normal*** to the columns of A (in the inner product determined by C). We get this equation by substituting $\boldsymbol{y} = C(\boldsymbol{b} - A\boldsymbol{x})$

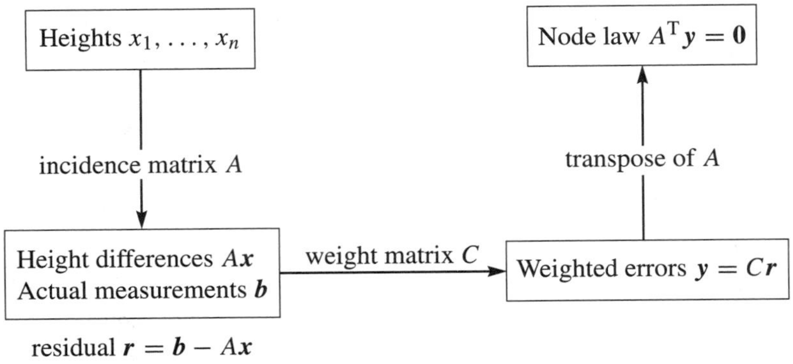

Figure 8.3 Description of a leveling network.

into $A^{\mathrm{T}}\boldsymbol{y} = \boldsymbol{0}$. The weighted errors $\boldsymbol{y}$ are eliminated and this yields equation (8.25) for the heights $\boldsymbol{x}$.

It is important that at least one height x_j is given beforehand. When $x_j = H$, the jth height is fixed and the jth column in the original incidence matrix is removed. We eliminate the nullspace by this maneuver. The resulting matrix is what we finally understand by A; it is m by $n-1$ and its columns are linearly independent. The square matrix $A^{\mathrm{T}}CA$ which is the key for the solution of equation (8.25) for $\boldsymbol{x}$ is an invertible matrix of order $n-1$ and with full rank

$$\underset{(n-1)\text{ by }m}{A^{\mathrm{T}}} \quad \underset{m\text{ by }m}{C} \quad \underset{m\text{ by }(n-1)}{A} \quad = \quad \underset{(n-1)\text{ by }(n-1)}{A^{\mathrm{T}}CA}.$$

In practical leveling networks the height is fixed at *several* nodes. Let their number be k and consider the following procedure:

- Enumerate all nodes from 1 to n, whether the node has a fixed height or not.
- Write down the m by n incidence matrix A and the m by m weight matrix C.
- Delete the k columns in A belonging to fixed nodes.
- Bring those k columns, multiplied by the k fixed heights, to the right side and include them into $\boldsymbol{b}$. The m measurement equations $A\boldsymbol{x} = \boldsymbol{b}$ in the $n-k$ unknown heights have $\boldsymbol{b} = \boldsymbol{b}_{\text{measured}} - (k \text{ columns})(k \text{ fixed heights})$.
- Calculate $A^{\mathrm{T}}CA$ and $A^{\mathrm{T}}C\boldsymbol{b}$ and solve the system $A^{\mathrm{T}}CA\hat{\boldsymbol{x}} = A^{\mathrm{T}}C\boldsymbol{b}$.

This procedure is turned into the M-file lev.

Example 8.1 For the leveling network depicted in Figure 8.4 as a directed graph, the following points have fixed heights:

$$H_A = 10.021\,\text{m}$$
$$H_B = 10.321\,\text{m}$$
$$H_C = 11.002\,\text{m}.$$

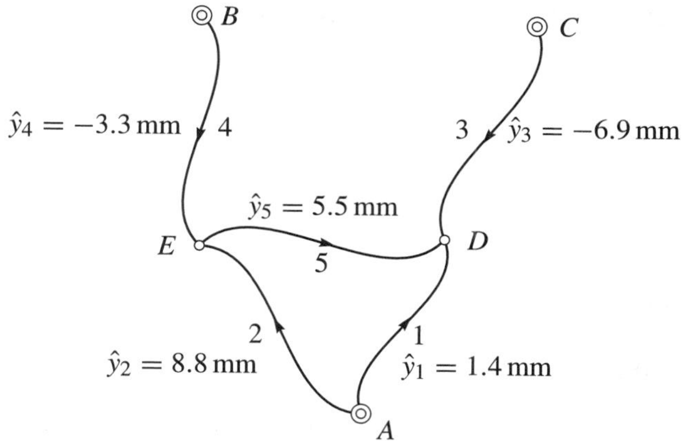

Figure 8.4 Directed graph for a geometric leveling network.

The observations of height differences and the lengths of the leveling lines l_i are

$$\begin{aligned} b_1 &= 1.978\,\text{m} & l_1 &= 1.02\,\text{km} \\ b_2 &= 0.732\,\text{m} & l_2 &= 0.97\,\text{km} \\ b_3 &= 0.988\,\text{m} & l_3 &= 1.11\,\text{km} \\ b_4 &= 0.420\,\text{m} & l_4 &= 1.07\,\text{km} \\ b_5 &= 1.258\,\text{m} & l_5 &= 0.89\,\text{km}. \end{aligned}$$

The edge-node incidence matrix is 5 by 5. According to the given rules it looks like

$$\begin{bmatrix} -1 & 0 & 0 & 1 & 0 \\ -1 & 0 & 0 & 0 & 1 \\ 0 & 0 & -1 & 1 & 0 \\ 0 & -1 & 0 & 0 & 1 \\ \underbrace{0 \quad 0 \quad 0}_{A_1} & & & \underbrace{1 \quad -1}_{A_2} & \end{bmatrix}.$$

This is the incidence matrix in case all points are free, that is, no height is fixed a priori. But on the contrary we want to keep the height of nodes A, B, and C fixed at given values. This means that the three first columns of A must be deleted and the right side is modified accordingly. The modified observation equations $A\boldsymbol{x} = \boldsymbol{b} - \boldsymbol{r}$ are

$$\begin{bmatrix} 1 & 0 \\ 0 & 1 \\ 1 & 0 \\ 0 & 1 \\ 1 & -1 \end{bmatrix} \begin{bmatrix} H_D \\ H_E \end{bmatrix} = \begin{bmatrix} 1.978 + 10.021 \\ 0.732 + 10.021 \\ 0.988 + 11.002 \\ 0.420 + 10.321 \\ 1.258 \end{bmatrix} - \begin{bmatrix} r_1 \\ r_2 \\ r_3 \\ r_4 \\ r_5 \end{bmatrix}.$$

The weighted matrix for the two unknowns (at D and E) is with $\sigma_0^2 = 1$:

$$A^{\mathrm{T}}CA = \begin{bmatrix} 1 & 0 & 1 & 0 & 1 \\ 0 & 1 & 0 & 1 & -1 \end{bmatrix} \begin{bmatrix} 0.980 & 0 & 0 & 0 & 0 \\ 0 & 1.031 & 0 & 0 & 0 \\ 0 & 0 & 0.901 & 0 & 0 \\ 0 & 0 & 0 & 0.935 & 0 \\ 0 & 0 & 0 & 0 & 1.124 \end{bmatrix} \begin{bmatrix} 1 & 0 \\ 0 & 1 \\ 1 & 0 \\ 0 & 1 \\ 1 & -1 \end{bmatrix}$$

$$= \begin{bmatrix} 3.005\,0 & -1.124\,0 \\ -1.124\,0 & 3.090\,0 \end{bmatrix}.$$

And accordingly the right side is

$$A^{\mathrm{T}}C\boldsymbol{b} = \begin{bmatrix} 1 & 0 & 1 & 0 & 1 \\ 0 & 1 & 0 & 1 & -1 \end{bmatrix} \begin{bmatrix} 0.980 & 0 & 0 & 0 & 0 \\ 0 & 1.031 & 0 & 0 & 0 \\ 0 & 0 & 0.901 & 0 & 0 \\ 0 & 0 & 0 & 0.935 & 0 \\ 0 & 0 & 0 & 0 & 1.124 \end{bmatrix} \begin{bmatrix} 11.999 \\ 10.753 \\ 11.990 \\ 10.741 \\ 1.258 \end{bmatrix}$$

$$= \begin{bmatrix} 23.976\,0 \\ 19.715\,2 \end{bmatrix}.$$

Now the normal equations are $A^{\mathrm{T}}CA\hat{\boldsymbol{x}} = A^{\mathrm{T}}C\boldsymbol{b}$:

$$\begin{bmatrix} 3.005\,0 & -1.124\,0 \\ -1.124\,0 & 3.090\,0 \end{bmatrix} \begin{bmatrix} H_D \\ H_E \end{bmatrix} = \begin{bmatrix} 23.976\,0 \\ 19.715\,2 \end{bmatrix}.$$

The solution is

$$\widehat{H}_D = 11.997\,6\,\mathrm{m}; \qquad \widehat{H}_E = 10.744\,5\,\mathrm{m};$$

$$\boldsymbol{p} = A\hat{\boldsymbol{x}} = \begin{bmatrix} 1 & 0 \\ 0 & 1 \\ 1 & 0 \\ 0 & 1 \\ 1 & -1 \end{bmatrix} \begin{bmatrix} 11.997\,6 \\ 10.744\,5 \end{bmatrix} = \begin{bmatrix} 11.997\,6 \\ 10.744\,5 \\ 11.997\,6 \\ 10.744\,5 \\ 1.253\,1 \end{bmatrix};$$

$$\hat{\boldsymbol{r}} = \boldsymbol{b} - \boldsymbol{p} = \begin{bmatrix} 0.001\,4 \\ 0.008\,5 \\ -0.007\,6 \\ -0.003\,5 \\ 0.004\,9 \end{bmatrix}.$$

8.4 Graphs and Incidence Matrices

Any time you have a connected system, with each part depending on other parts, you have a matrix. Linear algebra deals with interacting systems, provided the laws that govern them are linear. One special model appears so often, and has become so basic and useful, that we always put it first. The model consists of ***nodes connected by edges***. It appears in the leveling problem, and now we want to go beyond that problem—to see the linear algebra of graphs.

This section is *entirely optional*. It relates the leveling problem to the problem of voltages and currents in an electrical network. The incidence matrix A appears in Kirchhoff's Current Law, the weights $c_i = 1/(\sigma_0^2 l_i)$ become conductances ($= 1/$resistances), and the height observations b_{ij} turn into batteries!

A graph of the usual kind displays a function like $f(x)$. Graphs of a different kind (m edges connecting n nodes) lead to matrices. This section is about the incidence matrix A of a graph—and especially about the four subspaces that come with it.

For any m by n matrix there are two subspaces in $\mathbf{R}^n$ and two in $\mathbf{R}^m$. They are the column spaces and nullspaces of A and A^{T}. Their *dimensions* are related by the most important theorem in linear algebra. The second part of that theorem is the *orthogonality* of the subspaces. Our goal is to show how examples from graphs illuminate the Fundamental Theorem of Linear Algebra.

We review the four subspaces (for any matrix). Then we construct a ***directed graph*** and its ***incidence matrix***. The dimensions will be easy to discover. But we want the subspaces themselves—this is where orthogonality helps. It is essential to connect the subspaces to the graph they come from. By specializing to incidence matrices, the laws of linear algebra become Kirchhoff's laws. Please don't be put off by the words "current" and "potential" and "Kirchhoff." These rectangular matrices are the best.

Every entry of an incidence matrix is 0 or 1 or -1. This continues to hold during elimination. All pivots and multipliers are ± 1. Therefore both factors in $A = LU$ also contain $0, 1, -1$. Remarkably this persists for the nullspaces of A and A^{T}. All four subspaces have basis vectors with these exceptionally simple components. The matrices are not concocted for a textbook, they come from a model that is absolutely essential in pure and applied mathematics.

Review of the Four Subspaces

Start with an m by n matrix. Its columns are vectors in $\mathbf{R}^m$. Their linear combinations produce the *column space*, a subspace of $\mathbf{R}^m$. Those combinations are exactly the matrix-vector products Ax. So if we regard A as a linear transformation (taking $\boldsymbol{x}$ to $A\boldsymbol{x}$), the column space is its *range*. Call this subspace $\boldsymbol{R}(A)$.

The rows of A have n components. They are vectors in $\mathbf{R}^n$ (or they would be, if they were column vectors). Their linear combinations produce the *row space*. To avoid any inconvenience with rows, we transpose the matrix. The row space becomes $\boldsymbol{R}(A^{\mathrm{T}})$, the column space of A^{T}.

The central questions of linear algebra come from these two ways of looking at the same numbers, by columns and by rows.

The *nullspace* $\boldsymbol{N}(A)$ contains every $\boldsymbol{x}$ that satisfies $A\boldsymbol{x} = \mathbf{0}$—this is a subspace of $\mathbf{R}^n$. The *"left" nullspace* contains all solutions to $A^{\mathrm{T}}\boldsymbol{y} = \mathbf{0}$. Now $\boldsymbol{y}$ has m components, and

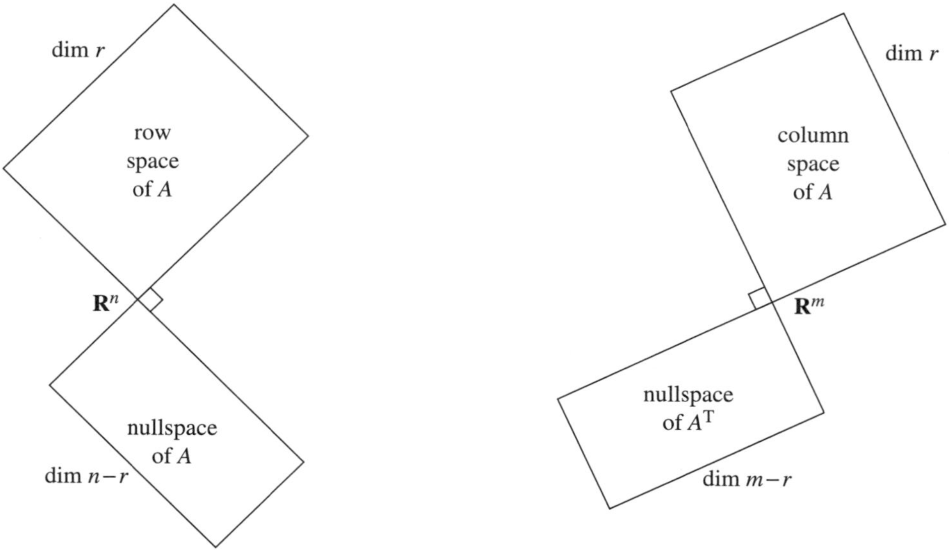

Figure 8.5 The four subspaces with their dimensions and orthogonality.

$\boldsymbol{N}(A^{\mathrm{T}})$ is a subspace of $\mathbf{R}^m$. Written as $\boldsymbol{y}^{\mathrm{T}}A = \mathbf{0}^{\mathrm{T}}$, we are combining rows of A to produce the zero row. The four subspaces are illustrated by Figure 8.5, which shows $\mathbf{R}^n$ on one side and $\mathbf{R}^m$ on the other. The link between them is A.

The information in that figure is crucial. First come the dimensions, which obey the two central laws of linear algebra:

$$\dim \boldsymbol{R}(A) = \dim \boldsymbol{R}(A^{\mathrm{T}}) \quad \text{and} \quad \dim \boldsymbol{R}(A) + \dim \boldsymbol{N}(A) = n.$$

When the row space has dimension r, the nullspace has dimension $n-r$. Elimination leaves these two spaces unchanged, and it changes A into its echelon form U. The dimension count is easy for U. There are r rows and columns with pivots. There are $n - r$ columns without pivots, and those lead to vectors in the nullspace.

The pivot rows are a basis for the row space. The pivot columns are a basis for the column space *of the echelon form U*—we are seeing $\boldsymbol{R}(U)$ and not $\boldsymbol{R}(A)$. The columns are changed by elimination, but their dependence or independence is not changed. The following matrix A comes from a graph, and its echelon form is U:

$$A = \begin{bmatrix} -1 & 1 & 0 & 0 \\ -1 & 0 & 1 & 0 \\ 0 & -1 & 1 & 0 \\ -1 & 0 & 0 & 1 \\ 0 & -1 & 0 & 1 \\ 0 & 0 & -1 & 1 \end{bmatrix} \quad \text{goes to} \quad U = \begin{bmatrix} -1 & 1 & 0 & 0 \\ 0 & -1 & 1 & 0 \\ 0 & 0 & -1 & 1 \\ 0 & 0 & 0 & 0 \\ 0 & 0 & 0 & 0 \\ 0 & 0 & 0 & 0 \end{bmatrix}.$$

The key is that $A\boldsymbol{x} = \mathbf{0}$ exactly when $U\boldsymbol{x} = \mathbf{0}$. Columns of A are dependent exactly when corresponding columns of U are dependent. The nullspace stays the same, and the *dimension* of the column space is unchanged. In this example the nullspace is the line through $\boldsymbol{x} = (1, 1, 1, 1)$. The column spaces of A and U have dimension $r = 3$.

Figure 8.5 shows more—the subspaces are orthogonal. ***The nullspace is perpendicular to the row space***. This comes directly from the m equations $A\boldsymbol{x} = \mathbf{0}$. The first equation says that $\boldsymbol{x}$ is orthogonal to the first row of A (to produce the first zero). The last equation says that $\boldsymbol{x}$ is orthogonal to the last row of A (to produce the last zero). For A and U above, $\boldsymbol{x} = (1, 1, 1, 1)$ is perpendicular to all rows and thus to the whole row space.

This review of the subspaces applies to any matrix A—only the example was special. Now we concentrate on that example. It is the incidence matrix for a particular graph, and we look to the graph for the meaning of each subspace.

Directed Graphs and Incidence Matrices

Figure 8.2 displays a *graph*. It has $m = 6$ edges and $n = 4$ nodes. The incidence matrix tells which nodes are connected by which edges. It also tells the directions of the arrows (this is a *directed* graph). The entries -1 and $+1$ in the first row of A give a record of the first edge. Row numbers are edge numbers, column numbers are node numbers.

Edge 2 goes from node 1 to node 3, so $a_{21} = -1$ and $a_{23} = +1$. Each row shows which node the edge leaves (by -1), and which node it enters (by $+1$). You can write down A by looking at the graph.

The graph in Figure 8.6 has the same four nodes but only three edges. Its incidence matrix is 3 by 4. The first graph is *complete*—every pair of nodes is connected by an edge. The second graph is a ***tree***—the graph has ***no closed loops***. Those graphs are the two extremes, with the maximum number of edges $m = \frac{1}{2}n(n-1)$ and the minimum number $m = n - 1$. We are assuming that the graph is *connected*, and it makes no fundamental difference which way the arrows go. On each edge, flow with the arrow is "positive." Flow in the opposite direction counts as negative. The flow might be a current or a signal or a force—or a measurement of the difference in height!

The rows of B match the nonzero rows of U—the echelon form found earlier. ***Elimination reduces every graph to a tree.*** The first step subtracts row 1 from row 2, which

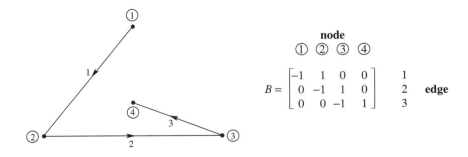

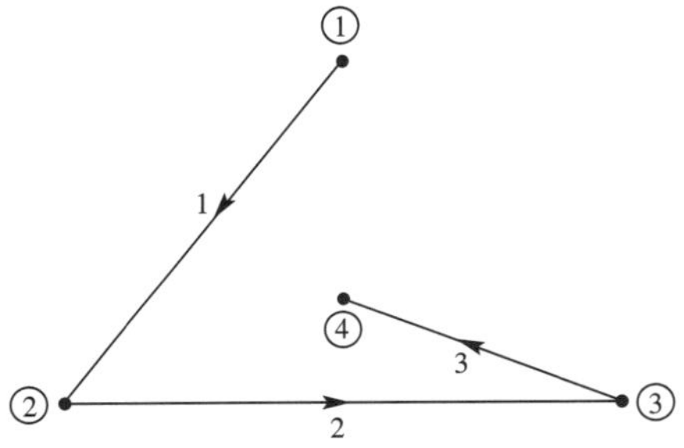

node
① ② ③ ④

$$B = \begin{bmatrix} -1 & 1 & 0 & 0 \\ 0 & -1 & 1 & 0 \\ 0 & 0 & -1 & 1 \end{bmatrix} \begin{matrix} 1 \\ 2 \\ 3 \end{matrix} \quad \textbf{edge}$$

Figure 8.6 Tree with 4 nodes.

creates a copy of row 3. The second step creates a zero row—the loop is gone:

$$\begin{matrix} -1 & 1 & 0 & 0 \\ -1 & 0 & 1 & 0 \\ 0 & -1 & 1 & 0 \end{matrix} \quad \rightarrow \quad \begin{matrix} -1 & 1 & 0 & 0 \\ 0 & -1 & 1 & 0 \\ 0 & -1 & 1 & 0 \end{matrix} \quad \rightarrow \quad \begin{matrix} -1 & 1 & 0 & 0 \\ 0 & -1 & 1 & 0 \\ 0 & 0 & 0 & 0 \end{matrix} .$$

Those steps are typical. When two edges share a node, elimination produces the "shortcut edge" without that node. If we already have a copy of this shortcut edge, elimination removes it. When the dust clears we have a tree. If we renumber the edges, we may reach a different tree. The row space has many bases and the complete graph has $n^{n-2} = 4^2 = 16$ trees.

An idea suggests itself: ***Rows are dependent when edges form a loop***. Independent rows come from trees.

The next step is to look at $A\boldsymbol{x}$, which is a vector of differences:

$$A\boldsymbol{x} = \begin{bmatrix} -1 & 1 & 0 & 0 \\ -1 & 0 & 1 & 0 \\ 0 & -1 & 1 & 0 \\ -1 & 0 & 0 & 1 \\ 0 & -1 & 0 & 1 \\ 0 & 0 & -1 & 1 \end{bmatrix} \begin{bmatrix} x_1 \\ x_2 \\ x_3 \\ x_4 \end{bmatrix} = \begin{bmatrix} x_2 - x_1 \\ x_3 - x_1 \\ x_3 - x_2 \\ x_4 - x_1 \\ x_4 - x_2 \\ x_4 - x_3 \end{bmatrix} . \tag{8.26}$$

The unknowns x_1, x_2, x_3, x_4 represent ***potentials*** at the nodes. Then $A\boldsymbol{x}$ gives the ***potential differences*** across the edges. It is these differences that cause flows. We now examine the meaning of each subspace.

1 The *nullspace* contains the solutions to $A\boldsymbol{x} = \boldsymbol{0}$. All six potential differences are zero. This means: *All four potentials are equal.* Every $\boldsymbol{x}$ in the nullspace is a constant vector (c, c, c, c). The nullspace of A is a line in $\mathbf{R}^n$—its dimension is $n - r = 1$.

The second incidence matrix B has the same nullspace for the same reason. Another way to explain $\boldsymbol{x} = (1, 1, 1, 1)$: The columns add up to the zero column.

We can raise or lower all potentials by the same amount c, without changing the differences. There is an "arbitrary constant" in the potentials. Compare this with the same statement for functions. We can raise or lower $F(x)$ by the same amount C, without changing its derivative. There is an arbitrary constant C in the integral.

Calculus adds "$+C$" to indefinite integrals. Graph theory adds (c, c, c, c) to potentials. Linear algebra adds any vector $\boldsymbol{x}_n$ in the nullspace to one particular solution of $A\boldsymbol{x} = \boldsymbol{b}$.

The "$+C$" disappears in calculus when the integral starts at a known point $x = a$. Similarly the nullspace disappears when we set $x_4 = 0$. The unknown x_4 is removed and so are the fourth columns of A and B. Electrical engineers would say that node 4 has been "grounded."

2 The *row space* contains all combinations of the six rows. Its dimension is certainly not six. The equation $r + (n - r) = n$ must be $3 + 1 = 4$. The rank is $r = 3$, as we also saw from elimination. Since each row of A adds to zero, this must be true for every vector $\boldsymbol{v}$ in the row space.

3 The *column space* contains all combinations of the four columns. We expect three independent columns, since there were three independent rows. The first three columns are independent (so are any three). But the four columns add to the zero vector, which says again that $(1, 1, 1, 1)$ is in the nullspace. *How can we tell if a particular vector* $\boldsymbol{b}$ *is in the column space, and* $A\boldsymbol{x} = \boldsymbol{b}$ *has an exact solution*?

Kirchhoff's answer $A\boldsymbol{x}$ is the vector of differences in equation (8.26). If we add differences around a closed loop in the graph, the cancellation leaves zero. Around the big triangle formed by edges $1, 3, -2$ (the arrow goes backward on edge 2) the differences are

$$(x_2 - x_1) + (x_3 - x_2) - (x_3 - x_1) = 0.$$

This is the ***voltage law***: ***The components of*** $A\boldsymbol{x}$ ***add to zero around every loop.*** When $\boldsymbol{b}$ is in the column space, it equals $A\boldsymbol{x}$ for some $\boldsymbol{x}$. Therefore $\boldsymbol{b}$ obeys the law:

$$b_1 + b_3 - b_2 = 0.$$

By testing each loop, we decide whether $\boldsymbol{b}$ is in the column space. $A\boldsymbol{x} = \boldsymbol{b}$ can be solved exactly when the components of $\boldsymbol{b}$ satisfy all the same dependencies as the rows of A. Then elimination leads to $0 = 0$, and $A\boldsymbol{x} = \boldsymbol{b}$ is consistent.

4 The *left nullspace* contains the solutions to $A^{\mathrm{T}}\boldsymbol{y} = \boldsymbol{0}$. Its dimension is $m - r = 6 - 3$:

$$A^{\mathrm{T}}\boldsymbol{y} = \begin{bmatrix} -1 & -1 & 0 & -1 & 0 & 0 \\ 1 & 0 & -1 & 0 & -1 & 0 \\ 0 & 1 & 1 & 0 & 0 & -1 \\ 0 & 0 & 0 & 1 & 1 & 1 \end{bmatrix} \begin{bmatrix} y_1 \\ y_2 \\ y_3 \\ y_4 \\ y_5 \\ y_6 \end{bmatrix} = \begin{bmatrix} 0 \\ 0 \\ 0 \\ 0 \end{bmatrix}. \qquad (8.27)$$

The true number of equations is $r = 3$ and not $n = 4$. Reason: The four equations above add to $0 = 0$. The fourth equation follows automatically from the first three.

The first equation says that $-y_1 - y_2 - y_4 = 0$. ***The net flow into node 1 is zero.*** The fourth equation says that $y_4 + y_5 + y_6 = 0$. ***The net flow into node 4 is zero.*** The word "net" means "flow into the node minus flow out." That is zero at every node, and for currents the law has a name:

Kirchhoff's Current Law: ***Flow in equals flow out at each node.***

This law deserves first place among the equations of applied mathematics. It expresses *"conservation"* and *"continuity"* and *"balance."* Nothing is lost, nothing is gained. When currents or forces are in equilibrium, the equation to solve is $A^{\mathrm{T}}\boldsymbol{y} = \boldsymbol{0}$.

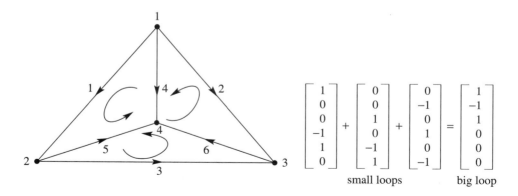

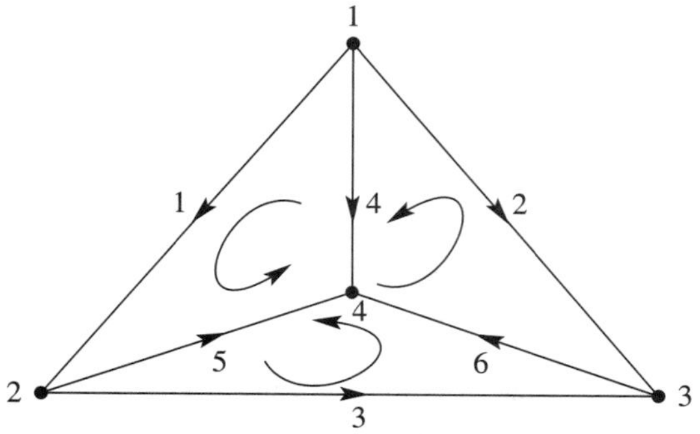

$$\begin{bmatrix} 1 \\ 0 \\ 0 \\ -1 \\ 1 \\ 0 \end{bmatrix} + \begin{bmatrix} 0 \\ 0 \\ 1 \\ 0 \\ -1 \\ 1 \end{bmatrix} + \begin{bmatrix} 0 \\ -1 \\ 0 \\ 1 \\ 0 \\ -1 \end{bmatrix} = \begin{bmatrix} 1 \\ -1 \\ 1 \\ 0 \\ 0 \\ 0 \end{bmatrix}$$

small loops big loop

Notice the beautiful fact that the matrix in this balance equation is the transpose of the incidence matrix A.

What are the actual solutions to $A^{\mathrm{T}}\boldsymbol{y} = \boldsymbol{0}$? The currents must balance themselves. The easiest way is to **flow around a loop**. If a unit of current goes around the big triangle (forward on edge 1, forward on 3, backward on 2), the current vector is $\boldsymbol{y} = (1, -1, 1, 0, 0, 0)$. This satisfies $A^{\mathrm{T}}\boldsymbol{y} = \boldsymbol{0}$. Every loop current yields a solution $\boldsymbol{y}$, because flow in equals flow out at every node. A smaller loop goes forward on edge 1, forward on 5, back on 4. Then $\boldsymbol{y} = (1, 0, 0, -1, 1, 0)$ is also in the left nullspace.

We expect three independent $\boldsymbol{y}$'s, since $6-3=3$. The three small loops in the graph are independent. The big triangle seems to give a fourth $\boldsymbol{y}$, but it is the sum of flows around the small loops.

Summary The incidence matrix A comes from a connected graph with n nodes and m edges. Here are the four fundamental subspaces:

1 The constant vectors $(c, c, \dots, c)$ make up the nullspace of A.

2 There are $r = n - 1$ independent rows, using edges from any tree.

3 ***Voltage law***: The components of $A\boldsymbol{x}$ add to zero around every loop.

4 ***Current law***: $A^{\mathrm{T}}\boldsymbol{y} = \boldsymbol{0}$ is solved by loop currents. $N(A^{\mathrm{T}})$ has dimension $m - r$. ***There are $m - r = m - n + 1$ independent loops in the graph.***

For a graph in a plane, the small loops are independent. Then linear algebra yields *Euler's formula*:

$$(\textit{number of nodes}) - (\textit{number of edges}) + (\textit{number of small loops}) \\ = n - m + (m - n + 1) = 1.$$

A single triangle has (3 nodes) − (3 edges) + (1 loop). For the graph in our example, nodes − edges + loops becomes $4-6+3$. On a 10-node tree Euler's count is $10-9+0$. All planar graphs lead to the same answer 1.

Networks and $A^{\mathrm{T}}CA$

The current $\boldsymbol{y}$ along an edge is the product of two numbers. One number is the difference between the potentials $\boldsymbol{x}$ at the ends of the edge. This difference is $A\boldsymbol{x}$ and it drives the flow. The other number is the ***"conductance"*** c—which measures how easily flow gets through.

In physics and engineering, c is decided by the material. It is high for metal and low for plastics. For a superconductor, c is nearly infinite (for electrical current). If we consider elastic stretching, c might be low for metal and higher for plastics. In economics, c measures the capacity of an edge or its cost.

To summarize, the graph is known from its "connectivity matrix" A. This tells the connections between nodes and edges. A ***network*** goes further, and assigns a conductance c to each edge. These numbers $c_1, \ldots, c_m$ go into the "conductance matrix" C—which is diagonal.

For a network of resistors, the conductance is $c = 1/(\text{resistance})$. In addition to Kirchhoff's laws for the whole system of currents, we have Ohm's law for each particular current. Ohm's law connects the current y_1 on edge 1 to the potential difference $x_2 - x_1$ between the nodes:

$$\text{current along edge} = (\text{conductance}) \text{ times } (\text{potential difference}).$$

Ohm's law for all m currents is $\boldsymbol{y} = -CA\boldsymbol{x}$. The vector $A\boldsymbol{x}$ gives the potential differences, and C multiplies by the conductances. Combining Ohm's law with Kirchhoff's current law $A^{\mathrm{T}}\boldsymbol{y} = \boldsymbol{0}$, we get $A^{\mathrm{T}}CA\boldsymbol{x} = \boldsymbol{0}$. This is *almost* the central equation for network flows. The only thing wrong is the zero on the right side! The network needs power from outside—a voltage source or a current source—to make something happen.

Note about applied mathematics Every new application has its own form of Ohm's law. For elastic structures $\boldsymbol{y} = CA\boldsymbol{x}$ is Hooke's law. The stress $\boldsymbol{y}$ is (elasticity C) times (stretching $A\boldsymbol{x}$). For heat conduction, $A\boldsymbol{x}$ is a temperature gradient and C is the conductivity. For fluid flows $A\boldsymbol{x}$ is a pressure gradient. There is a similar law for least-square regression in statistics—and this is at the heart of our book. C is the inverse of the covariance matrix Σ_b.

The textbook *Introduction to Applied Mathematics* (Wellesley-Cambridge Press) is practically built on "$A^{\mathrm{T}}CA$." This is the key to equilibrium. In geodesy, the measurement of *lengths* will also fit this pattern. That is a 2-D or 3-D problem, where heights are 1-D. In fact the matrix A for lengths appears in the earlier book as the matrix for structures (trusses). Applied mathematics is more organized than it looks.

We end by an example with a current source. Kirchhoff's law changes from $A^{\mathrm{T}}\boldsymbol{y} = \boldsymbol{0}$ to $A^{\mathrm{T}}\boldsymbol{y} = \boldsymbol{f}$, to balance the source $\boldsymbol{f}$ from outside. *Flow into each node still equals flow out.* Figure 8.7 shows the current source going into node 1. The source comes out at node 4 to keep the balance (in = out). The problem is: ***Find the currents on the six edges.***

Example 8.2 All conductances are $c = 1$, so that $C = I$. A current y_4 travels directly from node 1 to node 4. Other current goes the long way from node 1 to node 2 to node 4 (this is $y_1 = y_5$). Current also goes from node 1 to node 3 to node 4 (this is $y_2 = y_6$). We

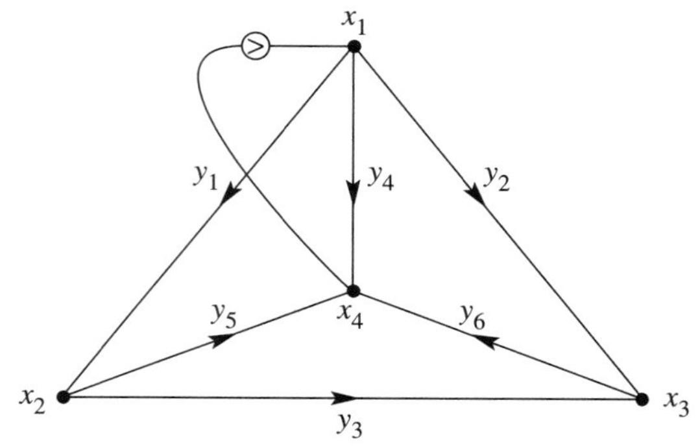

Figure 8.7 A network of conductors with a current source.

can find the six currents by using special rules for symmetry, or we can do it right by using $A^{\mathrm{T}}CA$. Since $C = I$, this matrix is

$$A^{\mathrm{T}}CA = A^{\mathrm{T}}A = \begin{bmatrix} -1 & -1 & 0 & -1 & 0 & 0 \\ 1 & 0 & -1 & 0 & -1 & 0 \\ 0 & 1 & 1 & 0 & 0 & -1 \\ 0 & 0 & 0 & 1 & 1 & 1 \end{bmatrix} \begin{bmatrix} -1 & 1 & 0 & 0 \\ -1 & 0 & 1 & 0 \\ 0 & -1 & 1 & 0 \\ -1 & 0 & 0 & 1 \\ 0 & -1 & 0 & 1 \\ 0 & 0 & -1 & 1 \end{bmatrix}$$

$$= \begin{bmatrix} 3 & -1 & -1 & -1 \\ -1 & 3 & -1 & -1 \\ -1 & -1 & 3 & -1 \\ -1 & -1 & -1 & 3 \end{bmatrix}.$$

That last matrix is not invertible! We cannot solve for all four potentials because (1, 1, 1, 1) is in the nullspace. One node has to be grounded. Setting $x_4 = 0$ removes the fourth row and column, and this leaves a 3 by 3 invertible matrix. Now we solve $A^{\mathrm{T}}CA\boldsymbol{x} = \boldsymbol{f}$ for the unknown potentials x_1, x_2, x_3. The right side $\boldsymbol{f}$ shows the source strength S into node 1:

$$\begin{bmatrix} 3 & -1 & -1 \\ -1 & 3 & -1 \\ -1 & -1 & 3 \end{bmatrix} \begin{bmatrix} x_1 \\ x_2 \\ x_3 \end{bmatrix} = \begin{bmatrix} S \\ 0 \\ 0 \end{bmatrix} \quad \text{gives} \quad \begin{bmatrix} x_1 \\ x_2 \\ x_3 \end{bmatrix} = \begin{bmatrix} S/2 \\ S/4 \\ S/4 \end{bmatrix}.$$

From these potentials, Ohm's law $\boldsymbol{y} = -CA\boldsymbol{x}$ yields the six currents. Remember $C = I$:

$$\begin{bmatrix} y_1 \\ y_2 \\ y_3 \\ y_4 \\ y_5 \\ y_6 \end{bmatrix} = - \begin{bmatrix} -1 & 1 & 0 & 0 \\ -1 & 0 & 1 & 0 \\ 0 & -1 & 1 & 0 \\ -1 & 0 & 0 & 1 \\ 0 & -1 & 0 & 1 \\ 0 & 0 & -1 & 1 \end{bmatrix} \begin{bmatrix} S/2 \\ S/4 \\ S/4 \\ 0 \end{bmatrix} = \begin{bmatrix} S/4 \\ S/4 \\ 0 \\ S/2 \\ S/4 \\ S/4 \end{bmatrix}.$$

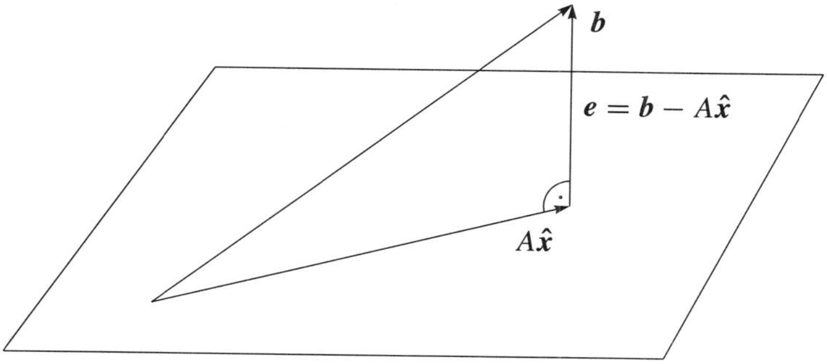

Figure 8.8 Projection of $\boldsymbol{b}$ on A's column space.

Half the current goes directly on edge 4. That is $y_4 = S/2$. The rest travels along other edges. No current crosses from node 2 to node 3. Symmetry indicated $y_3 = 0$ and now the solution proves it.

A computer system for circuit analysis forms the matrix $A^{\mathrm{T}}CA$. Then it solves $A^{\mathrm{T}}CA\boldsymbol{x} = \boldsymbol{f}$. This example shows how large networks are studied. The same matrix $A^{\mathrm{T}}A$ appears in least squares. Nature distributes the currents to minimize the heat loss, where statistics chooses $\hat{\boldsymbol{x}}$ to minimize the error.

The Dual Problem

All good minimization problems have (hidden in the shadows behind them) a *dual problem*. This dual is a new problem that looks very different from the primal, but it uses the same inputs. And in a subtle way it leads to the same solution. The route to that solution is often through *Lagrange multipliers*, which are introduced to handle the constraints (also called conditions) in an optimization problem.

We describe first the dual to the direct unweighted problem of minimizing $\|\boldsymbol{b}-A\boldsymbol{x}\|^2$. This is the fundamental objective of least squares, and it is *not* constrained: every vector $\boldsymbol{x}$ is admitted to the competition. Geometrically, this primal problem projects $\boldsymbol{b}$ onto the subspace of all vectors $A\boldsymbol{x}$ (the column space of A). The residual $\boldsymbol{e} = \boldsymbol{b} - A\hat{\boldsymbol{x}}$ is normal to the subspace, and this produces the normal equations!

$$A^{\mathrm{T}}\boldsymbol{e} = \boldsymbol{0} \qquad \text{or} \qquad A^{\mathrm{T}}A\hat{\boldsymbol{x}} = A^{\mathrm{T}}\boldsymbol{b}.$$

Figure 8.8 shows the triangle of vectors $\boldsymbol{b} = A\hat{\boldsymbol{x}}$ (projection) $+\, \boldsymbol{b} - A\hat{\boldsymbol{x}}$ (error) .

The dual problem simply projects the same vector $\boldsymbol{b}$ *onto the perpendicular space.* That subspace, perpendicular to the column space of A, is perfectly described by the Fundamental Theorem of Linear Algebra in Section 4.1. *It is the nullspace of* A^{T}. Its vectors will be called $\boldsymbol{y}$. Thus the dual problem is to *find the vector* $\boldsymbol{y}$ *in the nullspace of* A^{T} *that is closest to* $\boldsymbol{b}$:

$$\text{Dual: } \textit{Minimize } \tfrac{1}{2}\|\boldsymbol{b}-\boldsymbol{y}\|^2 \quad \textit{subject to} \quad A^{\mathrm{T}}\boldsymbol{y} = \boldsymbol{0}.$$

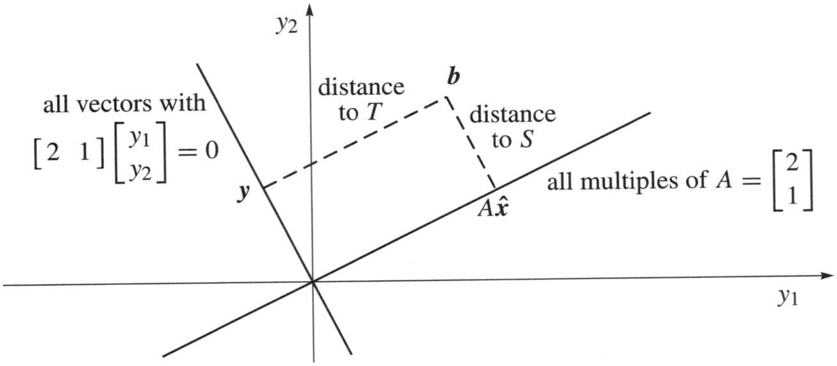

Figure 8.9 Projection of $\boldsymbol{b}$ onto column space of A and nullspace of A^{T}.

The constraint is $A^{\mathrm{T}}\boldsymbol{y} = \boldsymbol{0}$. This represents n equations in the m unknowns $y_1, \ldots, y_m$. We suppose as usual that A is m by n, with $m > n$, and that its n columns are independent. The rank is n (which makes $A^{\mathrm{T}}A$ invertible).

Figure 8.9 shows the geometric solution to the dual problem as well as the primal. *The dual solution is exactly* $\boldsymbol{y} = \boldsymbol{e}$. It is the other leg of the triangle. If $\boldsymbol{e}$ is computed first, it reveals the primal leg $A\hat{\boldsymbol{x}}$ as $\boldsymbol{b} - \boldsymbol{e}$. The "primal-dual" solution is the whole triangle, and it includes both answers $\boldsymbol{x} = \hat{\boldsymbol{x}}$ and $\boldsymbol{y} = \boldsymbol{e}$.

How is the dual problem actually solved? We are minimizing but only over a subspace of $\boldsymbol{y}$'s. The constraint $A^{\mathrm{T}}\boldsymbol{y} = \boldsymbol{0}$ (n linear equations) must be respected. It needs to be built into the function before we can use calculus and set derivatives to zero. Constraints are built in by introducing new unknowns $x_1, \ldots, x_n$ called Lagrange multipliers—one for each constraint. It is the Lagrangian function L whose derivatives we actually set to zero:

$$L(\boldsymbol{x}, \boldsymbol{y}) = \tfrac{1}{2}\|\boldsymbol{b} - \boldsymbol{y}\|^2 + \boldsymbol{x}^{\mathrm{T}}(A^{\mathrm{T}}\boldsymbol{y}). \tag{8.28}$$

This is a function of $m + n$ variables, $\boldsymbol{x}$'s and $\boldsymbol{y}$'s. It is linear in the $\boldsymbol{x}$'s. The derivatives $\partial L/\partial x_j$ just recover the n constraints:

$$\begin{aligned} \frac{\partial L}{\partial x_j} &= \text{coefficient of } x_j \text{ in (8.28)} \\ &= j\text{th component of } A^{\mathrm{T}}\boldsymbol{y}. \end{aligned}$$

The n equations $\partial L/\partial x_j = 0$ are exactly the constraints $A^{\mathrm{T}}\boldsymbol{y} = \boldsymbol{0}$. The other m equations come from the derivatives with respect to $\boldsymbol{y}$:

$$\begin{aligned} L &= \tfrac{1}{2}(b_1 - y_1)^2 + \cdots + \tfrac{1}{2}(b_m - y_m)^2 + (A\boldsymbol{x})^{\mathrm{T}}\boldsymbol{y} \\ \frac{\partial L}{\partial y_j} &= y_j - b_j + (j\text{th component of } Ax). \end{aligned}$$

Setting these derivatives to zero gives the vector equation $\boldsymbol{y} + A\boldsymbol{x} = \boldsymbol{b}$. Again we see our triangle. Note that the primal could be solved directly (no constraints and therefore no Lagrange multipliers). The minimizing equation was $A^{\mathrm{T}}A\hat{x} = A^{\mathrm{T}}\boldsymbol{b}$. In contrast, the dual has dragged in $\boldsymbol{x}$ as a set of Lagrange multipliers, and produced a larger system of equations:

$$\begin{aligned} \boldsymbol{y} + A\boldsymbol{x} &= \boldsymbol{b} \\ A^{\mathrm{T}}\boldsymbol{y} &= \boldsymbol{0}. \end{aligned} \tag{8.29}$$

If the problem involves a covariance matrix Σ, it multiplies $\boldsymbol{y}$ in the first equation above—because the dual problem is minimizing $(\boldsymbol{b} - \boldsymbol{y})^{\mathrm{T}}\Sigma(\boldsymbol{b} - \boldsymbol{y})$.

Substituting $\boldsymbol{y} = \boldsymbol{b} - A\boldsymbol{x}$ from the first equation into the second, we again find $A^{\mathrm{T}}A\boldsymbol{x} = A^{\mathrm{T}}\boldsymbol{b}$. The dual problem leads to the same normal equation! Thus the Lagrange multipliers $\boldsymbol{x} = \hat{\boldsymbol{x}}$ are computed first. In an important sense, they give *the derivative of the minimum value with respect to the constraints*. This topic is further developed in the first author's textbook *Introduction to Applied Mathematics* (Wellesley-Cambridge Press, 1986) and in all serious texts on optimization.

The Solution to $A^{\mathrm{T}}\boldsymbol{y} = \boldsymbol{0}$

There is an alternative to Lagrange multipliers. We can try to solve directly the constraint equations $A^{\mathrm{T}}\boldsymbol{y} = \boldsymbol{0}$. These are n equations in m unknowns. So there will be $m - n$ degrees of freedom in the solutions. Suppose we write B for a matrix whose $l = m - n$ columns are a basis for this solution subspace. Then B is m by l.

This approach has the advantage that the number of unknowns is *reduced* by n (using the n constraints) instead of *increased* by n (introducing Lagrange multipliers for those constraints). It has the disadvantage that we have to construct the basis matrix B. For the leveling problem, which is so closely related to the example of a network of resistors, we can give an explicit description of the solutions to $A^{\mathrm{T}}\boldsymbol{y} = \boldsymbol{0}$. This solution space has $l = m - n$ nice basis vectors. But even then, with the tempting possibility of fewer unknowns, the dual problem of "adjustment by conditions" is very seldom used!

Comment The situation is similar in the *finite element method*. There the unknowns $\boldsymbol{x}$ and $\boldsymbol{y}$ are the displacements and stresses. The primal method (*displacement method*) is chosen 99.9% of the time. The dual method (*stress method*) is much less popular—because it requires a basis matrix B of solutions to the equilibrium equations $A^{\mathrm{T}}\boldsymbol{y} = \boldsymbol{0}$. Notice that in the finite element method, the original problem is expressed by differential equations. Then approximation by piecewise polynomials reduces to a discrete (finite matrix) formulation. In leveling, the problem is discrete in the first place.

Now we identify the matrix B for a network of resistors (the electrical circuit example) or of baselines (the leveling example). The columns of B are solutions to $A^{\mathrm{T}}\boldsymbol{y} = \boldsymbol{0}$. This is Kirchhoff's Current Law: flow in = flow out at each node. It is clearly satisfied by *flow around a loop*. A loop is a closed path of network edges (baselines). At each node along the loop, one edge enters and another edge leaves. A unit current $\boldsymbol{y}$ around this loop is described by a vector whose entries are $+1$ or -1 or zero:

$$y_i = \begin{cases} +1 & \text{if the loop follows the arrow on edge } i \\ -1 & \text{if the loop goes against the arrow on edge } i \\ 0 & \text{if the loop does not include edge } i. \end{cases}$$

There are l independent loops in the network, and they give l independent columns of B. There remains the task of identifying l independent loops. For leveling this is easy; just take the small loops displayed in the network. Then large loops involving many baselines will be combinations of these small basic loops. Example 8.5 describes the general code findloop.

The general solution to $A^{\mathrm{T}}\boldsymbol{y} = \mathbf{0}$ is a combination of these loop vectors $\boldsymbol{y}$ in the columns of B. Their combinations are all vectors $\boldsymbol{y} = B\boldsymbol{\lambda}$. The $l = m - n$ unknowns are the coefficients $\lambda_1, \ldots, \lambda_l$ and we now have an unconstrained minimization—since these vectors $\boldsymbol{y} = B\boldsymbol{\lambda}$ satisfy the constraints. Instead of minimizing $\frac{1}{2}\|\boldsymbol{b} - \boldsymbol{y}\|^2$ with constraints $A^{\mathrm{T}}\boldsymbol{y} = \mathbf{0}$, we solve the equivalent dual problem:

$$\text{Minimize} \quad \tfrac{1}{2}\|\boldsymbol{b} - B\boldsymbol{\lambda}\|^2 \quad \text{over all vectors } \boldsymbol{\lambda}. \tag{8.30}$$

The minimizing equation is the dual normal equation

$$B^{\mathrm{T}}B\boldsymbol{\lambda} = B^{\mathrm{T}}\boldsymbol{b}. \tag{8.31}$$

If there is a covariance matrix Σ in the primal problem (which led to $A^{\mathrm{T}}\Sigma^{-1}A$), then this matrix appears in $B^{\mathrm{T}}\Sigma B$. The dual energy or *complementary energy* to be minimized is $\frac{1}{2}(\boldsymbol{b} - B\boldsymbol{\lambda})^{\mathrm{T}}\Sigma(\boldsymbol{b} - B\boldsymbol{\lambda})$. The weights are Σ not Σ^{-1}. The dual normal equations become $B^{\mathrm{T}}\Sigma B\boldsymbol{\lambda} = B^{\mathrm{T}}\Sigma\boldsymbol{b}$.

Now we consider the specific meaning of the unknowns $y_1, \ldots, y_m$ (and the conditioned unknowns $\lambda_1, \ldots, \lambda_l$) for *geodesy*.

Example 8.3 *A dual formulation of the leveling problem*: ***adjustment by conditions***. The idea is to make the loop law (Kirchhoff's Voltage Law) fundamental instead of the current law. The sum of height differences around loops is zero, and known heights contribute to $\boldsymbol{c}$:

$$B^{\mathrm{T}}(\boldsymbol{b} + \boldsymbol{r}) = \boldsymbol{c} \qquad \text{or} \qquad B^{\mathrm{T}}\boldsymbol{r} = \boldsymbol{w} = \boldsymbol{c} - B^{\mathrm{T}}\boldsymbol{b}. \tag{8.32}$$

With $\boldsymbol{w} = \mathbf{0}$ this is the loop law. In order to avoid confusion with the $A^{\mathrm{T}}CA$ formulation we have written B which has dimensions m by l; here l denotes the number of loops and m the number of observations. Equation (8.32) is the ***condition equation***.

The least-squares condition still is $\boldsymbol{r}^{\mathrm{T}}C\boldsymbol{r} = \min$, but this is a ***constrained*** minimization. The components of the vector $\boldsymbol{r}$ of height differences are subject to the conditions $B^{\mathrm{T}}\boldsymbol{r} = \boldsymbol{w}$ expressed by equation (8.32). We introduce the Lagrange multipliers $\boldsymbol{\lambda} = (\lambda_1, \lambda_2, \ldots, \lambda_l)$, one multiplier for each constraint. Instead of minimizing $\boldsymbol{r}^{\mathrm{T}}C\boldsymbol{r}$ we work with the extended function $L = \frac{1}{2}\boldsymbol{r}^{\mathrm{T}}C\boldsymbol{r} + \boldsymbol{\lambda}^{\mathrm{T}}(B^{\mathrm{T}}\boldsymbol{r} - \boldsymbol{w})$. The $m + l$ unknowns, namely $\boldsymbol{r}$ and $\boldsymbol{\lambda}$, are determined so that the $m + l$ partial derivatives of L are zero:

$$\frac{\partial L}{\partial \boldsymbol{r}} = C\boldsymbol{r} + B\boldsymbol{\lambda} = \mathbf{0} \qquad \text{and} \qquad \frac{\partial L}{\partial \boldsymbol{\lambda}} = B^{\mathrm{T}}\boldsymbol{r} - \boldsymbol{w} = \mathbf{0}. \tag{8.33}$$

Note that the derivative with respect to $\boldsymbol{\lambda}$ brings back the original constraint. Solving the first equation (with $C = \Sigma^{-1}$) gives

$$\boldsymbol{r} = -C^{-1}B\boldsymbol{\lambda} = -\Sigma B\boldsymbol{\lambda}. \tag{8.34}$$

This is known as the ***correlate equation***. Substitute into $B^{\mathrm{T}}\boldsymbol{r} = \boldsymbol{w}$ to reach

$$-B^{\mathrm{T}}\Sigma B\boldsymbol{\lambda} = \boldsymbol{w} \tag{8.35}$$

which are the dual ***normal equations***. These l equations yield $\boldsymbol{\lambda} = -(B^{\mathrm{T}}\Sigma B)^{-1}\boldsymbol{w}$. From the correlate equation (8.34), the estimated residuals are

$$\hat{\boldsymbol{r}} = \Sigma B(B^{\mathrm{T}}\Sigma B)^{-1}\boldsymbol{w}. \tag{8.36}$$

With $\boldsymbol{w} = \boldsymbol{c} - B^{\mathrm{T}}\boldsymbol{b}$ we get the estimated observations $\hat{\boldsymbol{b}} = A\hat{\boldsymbol{x}}$:

$$\hat{\boldsymbol{b}} = \boldsymbol{b} - \hat{\boldsymbol{r}} = \left(I - \Sigma B(B^{\mathrm{T}}\Sigma B)^{-1}B^{\mathrm{T}}\right)\boldsymbol{b} - \Sigma B(B^{\mathrm{T}}\Sigma B)^{-1}\boldsymbol{c}. \tag{8.37}$$

One possible expression for the estimate of the weighted sum of squares of residuals is

$$\hat{\boldsymbol{r}}^{\mathrm{T}}C\hat{\boldsymbol{r}} = \boldsymbol{\lambda}^{\mathrm{T}}B^{\mathrm{T}}\Sigma B\boldsymbol{\lambda}.$$

For completeness we bring the expression for the estimate of the variance of unit weight. This quantity will be defined in Chapter 9:

$$\hat{\sigma}_0 = \sqrt{\frac{\boldsymbol{\lambda}^{\mathrm{T}}B^{\mathrm{T}}\Sigma B\boldsymbol{\lambda}}{l}}. \tag{8.38}$$

In most geodetic problems the number of redundant observations $l = m - n$ is much smaller than m. Rough calculations reveal that $l/m = 0.1$–0.2 in traverses, and $l/m > 0.4$ in plane networks. This implies that the number l of normal equations for the dual least-squares formulation is much smaller than the number n for the standard formulation. This fact was important in earlier times when the solution of the normal equations was the great obstacle in a least-squares procedure. Today this is of less importance. It is far more difficult to program a computer to set up the conditions than just to read the observation equations. So today most software uses the standard formulation as described in the rest of this book.

Example 8.4 We shall illustrate this dual theory by means of the data in Example 8.1.

There are $m = 5$ observations, $n = 2$ unknowns, hence $l = m - n = 3$ conditions. Two of these conditions are of a special type as they involve points with fixed heights. Only the last condition (sum of height differences around a loop) is a typical one:

$$\begin{aligned} w_1 &= (H_A - H_C) + (b_1 - b_3) \\ w_2 &= (H_A - H_B) + (b_2 - b_4) \\ w_3 &= -b_1 + b_2 + b_5. \end{aligned}$$

The loops appear in the rows of B^{T}, and $\boldsymbol{w}$ is given by the problem:

$$B^{\mathrm{T}} = \begin{bmatrix} 1 & 0 & -1 & 0 & 0 \\ 0 & 1 & 0 & -1 & 0 \\ -1 & 1 & 0 & 0 & 1 \end{bmatrix} \qquad \text{and} \qquad \boldsymbol{w} = \begin{bmatrix} 0.009 \\ 0.012 \\ 0.012 \end{bmatrix}.$$

With $\sigma_0^2 = 1$ the covariance matrix comes from the distances l_i between points:

$$\Sigma = C^{-1} = \begin{bmatrix} 1.02 & & & & \\ & 0.97 & & & \\ & & 1.11 & & \\ & & & 1.07 & \\ & & & & 0.89 \end{bmatrix}.$$

The solution of the dual normal equations (with minus sign) is

$$\boldsymbol{\lambda} = \begin{bmatrix} -0.006\,9 \\ -0.003\,3 \\ -0.005\,5 \end{bmatrix}.$$

Then the residuals and estimated observations are given as

$$\hat{\boldsymbol{r}} = \begin{bmatrix} 0.001\,4 \\ 0.008\,5 \\ -0.007\,6 \\ -0.003\,5 \\ 0.004\,9 \end{bmatrix} \qquad \text{and} \qquad \hat{\boldsymbol{b}} = \boldsymbol{b} - \hat{\boldsymbol{r}} = \begin{bmatrix} 1.976\,6 \\ 0.723\,5 \\ 0.995\,6 \\ 0.423\,5 \\ 1.253\,1 \end{bmatrix}.$$

The only ordinary loop condition is satisfied, too:

$$-\hat{b}_1 + \hat{b}_2 + \hat{b}_5 = 0.$$

Example 8.5 In that example it was easy to recognize the independent loops from the figure. For larger networks it is not always too simple to find a full set of loops. So we look for an automatic way of creating independent loops.

Our starting point will be a list of edges. Each edge is described by the "from" node (which the edge leaves) and the "to" node (which it enters). The leveled height differences are known for all edges, and they should sum to zero around every loop.

Each edge forms a row of the matrix A. We have just learned that the loops are the special solutions to the equation $A^{\mathrm{T}}\boldsymbol{y} = \boldsymbol{0}$. So for given A we need a code to compute a complete set of special solutions—these are vectors with $y_i = \pm 1$ when an edge is in the loop and otherwise $y_i = 0$. This is done by the collection of M-files: looplist, findloop, findnode, plu, and ref. (The file findloop is a renamed version of the Teaching Code called null described in Section 3.2.) The M-file ref computes the reduced echelon form of A^{T}, and findloop puts the special solutions of $A^{\mathrm{T}}\boldsymbol{y} = \boldsymbol{0}$ into the nullspace matrix B.

The output is a list of edges for each loop, and the loop sums that should be zero. By the M-file findnode we produce the sequence of nodes in each loop. *Recall that a network has many complete sets of loops.* Our particular set is determined by the order of the edges and nodes in A. Repeated edges make a special case and they create no difficulties.

A modified version (we have three differences instead of one) of the same code can be used to combine GPS baselines to form loops of 3-dimensional vectors! The M-script vectors contains a set of GPS vectors, the M-file v_loops does the computation and the result can be read in vectors.log.

On calculational accuracy The observed height differences b_i are rounded to mm. It is natural in all subsequent calculations to keep this accuracy. To the normal equations and their solution is even added one extra digit (and also in the standard deviation of the two heights).

The weight matrix is indicated with 3 significant digits. In well conditioned problems the weights are not so decisive for the result, see Section 11.6. Besides, the determination of weights is not always a well defined matter.

The ***node law*** $A^{\mathrm{T}}C\hat{\boldsymbol{r}} = \boldsymbol{0}$ in Example 8.1 is fulfilled at the free nodes, but the fixed nodes cause a modification. The node law as derived in Section 8.3 works perfectly well for a network without fixed nodes. As point of departure we arrange the observation equations so that the free nodes are at the bottom of $\boldsymbol{x}$. (The columns of the *original* incidence matrix A are interchanged accordingly.) Columns in A referring to fixed nodes are contained in A_1 and the fixed heights are denoted $\boldsymbol{x}^0$:

$$A\boldsymbol{x} = \begin{bmatrix} A_1 & A_2 \end{bmatrix} \begin{bmatrix} \boldsymbol{x}^0 \\ \boldsymbol{x} \end{bmatrix} = \boldsymbol{b} - \boldsymbol{r}.$$

The unknown part $\boldsymbol{x}$ has

$$A_2\boldsymbol{x} = (\boldsymbol{b} - A_1\boldsymbol{x}^0) - \boldsymbol{r}. \tag{8.39}$$

As usual, we form the normal equations by multiplication to the left by $A^{\mathrm{T}}C$:

$$\begin{bmatrix} A_1^{\mathrm{T}} \\ A_2^{\mathrm{T}} \end{bmatrix} CA_2\boldsymbol{x} = \begin{bmatrix} A_1^{\mathrm{T}} \\ A_2^{\mathrm{T}} \end{bmatrix} C(\boldsymbol{b} - A_1\boldsymbol{x}^0) - \begin{bmatrix} A_1^{\mathrm{T}} \\ A_2^{\mathrm{T}} \end{bmatrix} C\boldsymbol{r}.$$

The lower equation is just the set of normal equations which we already solved. The upper equation is

$$A_1^{\mathrm{T}}CA_2\boldsymbol{x} = A_1^{\mathrm{T}}C(\boldsymbol{b} - A_1\boldsymbol{x}^0) - A_1^{\mathrm{T}}C\boldsymbol{r}.$$

Rearranged this becomes

$$\begin{aligned} A_1^{\mathrm{T}}C\boldsymbol{r} &= -A_1^{\mathrm{T}}CA_2\boldsymbol{x} + A_1^{\mathrm{T}}C(\boldsymbol{b} - A_1\boldsymbol{x}^0) \\ &= A_1^{\mathrm{T}}C(-A_2\boldsymbol{x} + \boldsymbol{b} - A_1\boldsymbol{x}^0). \end{aligned}$$

Remembering $\boldsymbol{y} = C\boldsymbol{r}$, we get

$$\boldsymbol{y} = \underset{m \text{ by } m}{C}\Big(\big(\underset{m \text{ by } 1}{\boldsymbol{b}} - \underset{m \text{ by } (n-f)}{A_2}\ \underset{(n-f) \text{ by } 1}{\boldsymbol{x}}\big) - \underset{m \text{ by } f}{A_1}\ \underset{f \text{ by } 1}{\boldsymbol{x}^0}\Big). \tag{8.40}$$

The oriented sum of y's along any path connecting two points with fixed heights equals the weight times the difference of observed height differences and fixed heights.

Problem Set 8.4

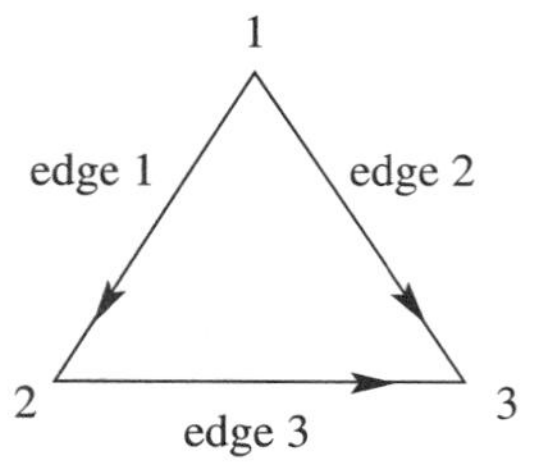

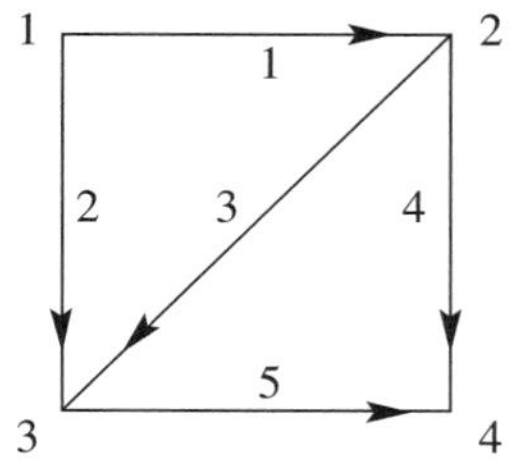

1 Write down the 3 by 3 incidence matrix A for the triangle graph. The first row has -1 in column 1 and $+1$ in column 2. What vectors (x_1, x_2, x_3) are in its nullspace? How do you know that $(1, 0, 0)$ is not in its row space?

2 Write down A^T for the triangle graph. Find a vector $\boldsymbol{y}$ in its nullspace. The components of $\boldsymbol{y}$ are currents on the edges—how much current is going around the triangle?

3 Eliminate x_1 and x_2 from the third equation to find the echelon matrix U. What tree corresponds to the two nonzero rows of U?

$$\begin{aligned} -x_1 + x_2 &= b_1 \\ -x_1 + x_3 &= b_2 \\ -x_2 + x_3 &= b_3. \end{aligned}$$

4 Choose a vector (b_1, b_2, b_3) for which $A\boldsymbol{x} = \boldsymbol{b}$ can be solved, and another vector $\boldsymbol{b}$ that allows no solution. How are those $\boldsymbol{b}$'s related to $\boldsymbol{y} = (1, -1, 1)$?

5 Choose a vector (f_1, f_2, f_3) for which $A^T\boldsymbol{y} = \boldsymbol{f}$ can be solved, and a vector $\boldsymbol{f}$ that allows no solution. How are those $\boldsymbol{f}$'s related to $\boldsymbol{x} = (1, 1, 1)$? The equation $A^T\boldsymbol{y} = \boldsymbol{f}$ is Kirchhoff's ______ law.

6 Multiply matrices to find A^TA. Choose a vector $\boldsymbol{f}$ for which $A^TA\boldsymbol{x} = \boldsymbol{f}$ can be solved, and solve for $\boldsymbol{x}$. Put those potentials $\boldsymbol{x}$ and the currents $\boldsymbol{y} = -A\boldsymbol{x}$ and current sources $\boldsymbol{f}$ onto the triangle graph. Conductances are 1 because $C = I$.

7 With conductances $c_1 = 1$ and $c_2 = c_3 = 2$, multiply matrices to find A^TCA. For $\boldsymbol{f} = (1, 0, -1)$ find a solution to $A^TCA\boldsymbol{x} = \boldsymbol{f}$. Write the potentials $\boldsymbol{x}$ and currents $\boldsymbol{y} = -CA\boldsymbol{x}$ on the triangle graph, when the current source $\boldsymbol{f}$ goes into node 1 and out from node 3.

8 Write down the 5 by 4 incidence matrix A for the square graph with two loops. Find one solution to $A\boldsymbol{x} = \boldsymbol{0}$ and two solutions to $A^T\boldsymbol{y} = \boldsymbol{0}$.

9 Find two requirements on the b's for the five differences $x_2 - x_1$, $x_3 - x_1$, $x_3 - x_2$, $x_4 - x_2$, $x_4 - x_3$ to equal b_1, b_2, b_3, b_4, b_5. You have found Kirchhoff's ______ law around the two ______ in the graph.

10 Reduce A to its echelon form U. The three nonzero rows give the incidence matrix for what graph? You found one tree in the square graph—find the other seven trees.

11 Multiply matrices to find $A^{\mathrm{T}}A$ and guess how its entries come from the graph:

(a) The diagonal of $A^{\mathrm{T}}A$ tells how many ______ into each node.

(b) The off-diagonals -1 or 0 tell which pairs of nodes are ______.

12 Why is each statement true about $A^{\mathrm{T}}A$? *Answer for* $A^{\mathrm{T}}A$ *not* A.

(a) Its nullspace contains $(1, 1, 1, 1)$. Its rank is $n - 1$.

(b) It is positive semidefinite but not positive definite.

(c) Its four eigenvalues are real and their signs are ______.

13 With conductances $c_1 = c_2 = 2$ and $c_3 = c_4 = c_5 = 3$, multiply the matrices $A^{\mathrm{T}}CA$. Find a solution to $A^{\mathrm{T}}CA\boldsymbol{x} = \boldsymbol{f} = (1, 0, 0, -1)$. Write these potentials $\boldsymbol{x}$ and currents $\boldsymbol{y} = -CA\boldsymbol{x}$ on the nodes and edges of the square graph.

14 The matrix $A^{\mathrm{T}}CA$ is not invertible. What vectors $\boldsymbol{x}$ are in its nullspace? Why does $A^{\mathrm{T}}CA\boldsymbol{x} = \boldsymbol{f}$ have a solution if and only if $f_1 + f_2 + f_3 + f_4 = 0$?

15 A connected graph with 7 nodes and 7 edges has how many loops?

16 For the graph with 4 nodes, 6 edges, and 3 loops, add a new node. If you connect it to one old node, Euler's formula becomes $(\quad) - (\quad) + (\quad) = 1$. If you connect it to two old nodes, Euler's formula becomes $(\quad) - (\quad) + (\quad) = 1$.

17 Suppose A is a 12 by 9 incidence matrix from a connected (but unknown) graph.

(a) How many columns of A are independent?

(b) What condition on $\boldsymbol{f}$ makes it possible to solve $A^{\mathrm{T}}\boldsymbol{y} = \boldsymbol{f}$?

(c) The diagonal entries of $A^{\mathrm{T}}A$ give the number of edges into each node. What is the sum of those diagonal entries?

18 Why does a complete graph with $n = 6$ nodes have $m = 15$ edges? A tree connecting 6 nodes has ______ edges.

8.5 One-Dimensional Distance Networks

There are a few other *linear* least-squares problems in geodesy. The following section touches upon two of them. Chapter 13 will discuss other cases.

Example 8.6 The simplest linear situation in surveying is the following: Between 4 points A, B, C, and D situated on a straight line we have measured the distances AB, BC, CD, AC, AD, and BD. The six measurements are

$$(b_1, b_2, \dots, b_6) = (3.17, 1.12, 2.25, 4.31, 6.51, 3.36).$$

A similar problem concerning directions is the so-called station adjustment which is treated in Section 13.3.

As unknowns we choose $AB = x_1$ and $BC = x_2$ and $CD = x_3$. We have $m = 6$ and $n = 3$. The weights are set to unity so $C = I$. The observation equations then are

$$\begin{aligned} x_1 &= b_1 - r_1 \\ x_2 &= b_2 - r_2 \\ x_3 &= b_3 - r_3 \\ x_1 + x_2 &= b_4 - r_4 \\ x_1 + x_2 + x_3 &= b_5 - r_5 \\ x_2 + x_3 &= b_6 - r_6. \end{aligned}$$

In matrix form this is

$$\begin{bmatrix} 1 & 0 & 0 \\ 0 & 1 & 0 \\ 0 & 0 & 1 \\ 1 & 1 & 0 \\ 1 & 1 & 1 \\ 0 & 1 & 1 \end{bmatrix} \begin{bmatrix} x_1 \\ x_2 \\ x_3 \end{bmatrix} = \begin{bmatrix} 3.17 \\ 1.12 \\ 2.25 \\ 4.31 \\ 6.51 \\ 3.36 \end{bmatrix} - \begin{bmatrix} r_1 \\ r_2 \\ r_3 \\ r_4 \\ r_5 \\ r_6 \end{bmatrix}.$$

The normal equations become

$$\underbrace{\begin{bmatrix} 3 & 2 & 1 \\ 2 & 4 & 2 \\ 1 & 2 & 3 \end{bmatrix}}_{A^{\mathrm{T}}A} \begin{bmatrix} x_1 \\ x_2 \\ x_3 \end{bmatrix} = \begin{bmatrix} 13.99 \\ 15.30 \\ 12.12 \end{bmatrix}.$$

The solution is

$$\hat{x} = \begin{bmatrix} 3.170 \\ 1.122 \\ 2.235 \end{bmatrix}.$$

Example 8.7 Next we shall modify the situation described in Example 8.6 by introducing an extra unknown z, the *zero constant*. It is due to the fact that the zero mark of the

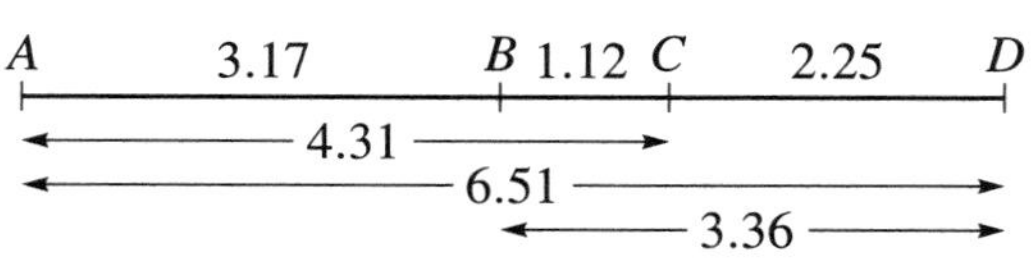

measuring tape is possibly not situated exactly at the physical end point. So we want to determine z along with the earlier 3 unknowns.

The new version of the observation equations looks like

$$\begin{bmatrix} 1 & 0 & 0 & -1 \\ 0 & 1 & 0 & -1 \\ 0 & 0 & 1 & -1 \\ 1 & 1 & 0 & -1 \\ 1 & 1 & 1 & -1 \\ 0 & 1 & 1 & -1 \end{bmatrix} \begin{bmatrix} x_1 \\ x_2 \\ x_3 \\ z \end{bmatrix} = \begin{bmatrix} 3.17 \\ 1.12 \\ 2.25 \\ 4.31 \\ 6.51 \\ 3.36 \end{bmatrix} - \begin{bmatrix} r_1 \\ r_2 \\ r_3 \\ r_4 \\ r_5 \\ r_6 \end{bmatrix}.$$

The modified normal equations are

$$\begin{bmatrix} 3 & 2 & 1 & -3 \\ 2 & 4 & 2 & -4 \\ 1 & 2 & 3 & -3 \\ -3 & -4 & -3 & 6 \end{bmatrix} \begin{bmatrix} x_1 \\ x_2 \\ x_3 \\ z \end{bmatrix} = \begin{bmatrix} 13.99 \\ 15.30 \\ 12.12 \\ -20.72 \end{bmatrix}.$$

The solution is

$$\boldsymbol{x} = \begin{bmatrix} 3.1625 \\ 1.1150 \\ 2.2275 \end{bmatrix} \qquad \text{and} \qquad z = -0.0150.$$

Example 8.8 A yet more comprehensive model also allows for a change of scale k of the measuring tape. The modified observation equations are

$$\begin{aligned} k x_1 + z &= b_1 - r_1 \\ k x_2 + z &= b_2 - r_2 \\ k x_3 + z &= b_3 - r_3 \\ k x_1 + k x_2 + z &= b_4 - r_4 \\ k x_1 + k x_2 + k x_3 + z &= b_5 - r_5 \\ k x_2 + k x_3 + z &= b_6 - r_6. \end{aligned}$$

This problem is not linear. Products $k x_1, k x_2, k x_3$ of unknowns appear. How to solve this type of problem is the subject of Chapter 10.

9

RANDOM VARIABLES AND COVARIANCE MATRICES

9.1 The Normal Distribution and χ^2

A great deal of practical geodesy consists in taking observations. The ***observed values*** are real numbers. But they are not exact; they include "random" errors. In order to analyze these values we take them to be ***continuous random variables***. In distinction to a real variable, a random variable is furnished with a probability distribution. Often, the various values of X are not equally probable.

The probability of hitting a particular real number $x = n$ is zero. Instead we introduce a ***probability density*** $p(x)$. The chance that a random X falls between a and b is found by integrating the density $p(x)$:

$$\text{Prob}(a \le X \le b) = \int_a^b p(x)\,dx. \tag{9.1}$$

Roughly speaking, $p(x)\,dx$ is the chance of falling between x and $x + dx$. Certainly $p(x) \ge 0$. If a and b are the extreme limits $-\infty$ and ∞, including all possible outcomes, the probability is necessarily one:

$$\text{Prob}(-\infty < X < +\infty) = \int_{-\infty}^{\infty} p(x)\,dx = 1. \tag{9.2}$$

The most important density function is the ***normal distribution*** which is defined as

$$p(x) = \frac{1}{\sigma\sqrt{2\pi}} e^{-(x-\mu)^2/2\sigma^2}. \tag{9.3}$$

Equation (9.3) involves two parameters. They are the mean value μ and the standard deviation σ (the standard deviation measures the spread around the mean value). Often the two parameters are given the "standardized" values $\mu = 0$ and $\sigma = 1$. Any normal distribution (9.3) can be standardized by the substitution $y = (x - \mu)/\sigma$. Then $p(x)$ becomes symmetric around $x = 0$ and the variance is $\sigma^2 = 1$:

$$p(x) = \frac{1}{\sqrt{2\pi}} e^{-x^2/2}. \tag{9.4}$$

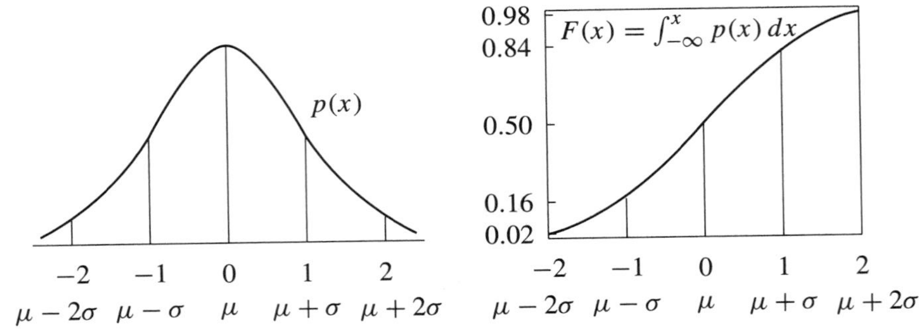

Figure 9.1 The normal distribution $p(x)$ (bell-shaped) and its cumulative density $F(x)$: $p(x)$ is given in (9.3)–(9.4) and there is no explicit formula for its integral $F(x)$.

The factor $\sqrt{2\pi}$ is included to make $\int p(x)\,dx = 1$.

The ***bell-shaped graph*** of $p(x)$ in Figure 9.1 is symmetric around the middle point $x = \mu$. The width of the graph is governed by the second parameter σ—which stretches the x axis and shrinks the y axis (leaving total area equal to 1). The axes are labeled to show the standard case $\mu = 0,\ \sigma = 1$ and also the graph of $p(x)$ for any other μ and σ.

We now give a name to the integral of $p(x)$. The limits will be $-\infty$ and x, so the integral $F(x)$ measures the ***probability that a random sample is below*** x:

$$\text{Prob}(X \le x) = \int_{-\infty}^{x} p(x)\,dx = \textbf{\textit{cumulative density function}}\ F(x). \tag{9.5}$$

$F(x)$ accumulates the probabilities given by $p(x)$, so $dF(x)/dx = p(x)$. The reverse of the integral is the derivative. The total probability is $F(\infty) = 1$. This integral from $-\infty$ to ∞ covers all outcomes.

Figure 9.1b shows the integral of the bell-shaped normal distribution. The middle point $x = \mu$ has $F = \frac{1}{2}$. By symmetry, there is a 50-50 chance of an outcome below the mean. The cumulative density $F(x)$ is near 0.16 at $\mu - \sigma$ and near 0.84 at $\mu + \sigma$. The chance of falling in the interval $[\mu - \sigma, \mu + \sigma]$ is $0.84 - 0.16 = 0.68$. Thus 68% of the outcomes are less than one deviation σ from the center μ.

Moving out to $\mu - 1.96\sigma$ and $\mu + 1.96\sigma$, 95% of the area is in between. ***With 95% confidence X is less than two deviations from the mean.*** Only one outcome in 20 is further out (less than one in 40 on each side). This 95% confidence interval is often taken as an indicator. When an observation is outside this interval, when it is more than 2σ away

Table 9.1 Cumulative density function $F(x)$ for the normal distribution

x	0.0	0.5	1.0	1.5	2.0	2.5	3.0
$F(x)$	0.5	0.6915	0.8413	0.9332	0.9773	0.9938	0.9987

Table 9.2 One-dimensional normal distribution: Confidence intervals

Range of x	Probability
$\|x-\mu\| \le \sigma$	0.682 7
$\|x-\mu\| \le 2\sigma$	0.954 4
$\|x-\mu\| \le 3\sigma$	0.997 3

Probability	Range of x
0.50	$\|x-\mu\| \le 0.674\sigma$
0.95	$\|x-\mu\| \le 1.960\sigma$
0.99	$\|x-\mu\| \le 2.576\sigma$

from μ, we may accept that this happened with probability below 5% or we may question whether the estimated values for σ and μ are correct.

Geodesy frequently sets the standard error further out, at $\mu + 3\sigma$. The cumulative density at that point is $F = 0.998\,7$. Thus the probability of a (one dimensional) sample further than *three* standard deviations from the mean (above or below) is $2(1-F) = 0.002\,7$. Tables 9.1 and 9.2 show cumulative probabilities at important breakpoints.

The integral of $p(x)$ from a to b is the probability that a random sample X will lie in this interval:

$$\text{Prob}(a \le X \le b) = \text{Prob}(X \le b) - \text{Prob}(X \le a) = F(b) - F(a).$$

For a random variable X which is normally distributed with $\mu = 1$ and $\sigma = 2$ we have

$$\begin{aligned}\text{Prob}(2 \le X \le 3) &= \text{Prob}(2 < X < 3) = F\left(\tfrac{3-1}{2}\right) - F\left(\tfrac{2-1}{2}\right)\\ &= F(1) - F\left(\tfrac{1}{2}\right) = 0.8413 - 0.6915 = 0.1498.\end{aligned}$$

If $x_1, \ldots, x_n$ are independent random variables each with mean μ and variance σ_0^2, then $\hat{x} = \sum x_i/n$ is still normally distributed with mean μ. The variance is σ_0^2/n.

Two-dimensional Normal Distribution

In two dimensions we have probability densities $p(x, y)$. Again the integral over all possibilities, $-\infty < x < \infty$ and $-\infty < y < \infty$, is 1. The normal distribution is always of the greatest importance, and we write the standardized form first. The mean is $\mu = (0, 0)$ and the 2 by 2 covariance matrix is $\Sigma = I$:

$$p(x, y) = \frac{1}{2\pi} e^{-(x^2+y^2)/2}. \tag{9.6}$$

The M-file twod generated the 100 points (x, y) in Figure 9.2 in accordance with this two-dimensional distribution. (You can obtain your own different set of 100 points.) The figure shows the circle of radius $2\sigma = 2$ around the mean. In the two-dimensional normal distribution, only 86.47% of the sample points are expected to be in this circle. This is because $\iint_{x^2+y^2<4} p(x, y)\,dx\,dy = .864\,7$. We count 16 points outside the circle in this particular sample.

Table 9.6 will show the corresponding numbers for the circles of radius σ and 3σ. The squared distance $x^2 + y^2$ is governed by a χ_2^2 distribution, which has the simple form

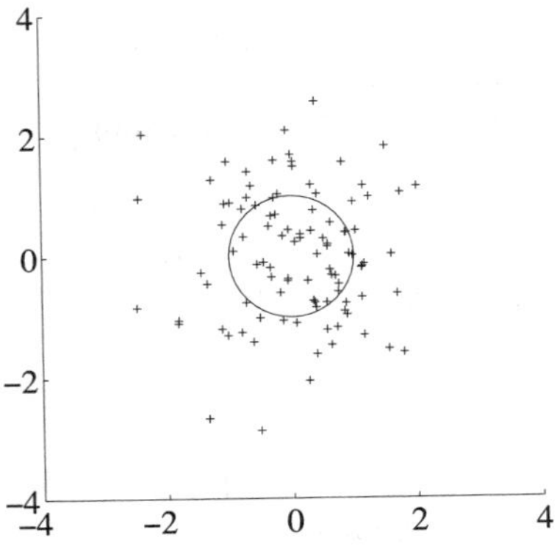

Figure 9.2 The two-dimensional normal distribution and the circle of radius 2σ.

$p_2(s) = \frac{1}{2}e^{-s/2}$. The table also shows the (non-integer) multiples of σ that will give 90%, 95%, and 99% *confidence* intervals around the mean μ (assuming we know μ).

We get the ***marginal*** distribution of y, when we integrate over all x:

$$M(y) = \int_{-\infty}^{\infty} p(x, y)\,dx. \tag{9.7}$$

In this instance $M(y)$ is simply the one-dimensional normal distribution for y. That is because $p(x, y)$ in (9.6) separates into a product $p(x)p(y)$. The variables x and y are uncorrelated; the matrix covariance Σ is diagonal. But certainly we meet many situations in which correlation is present and Σ is not diagonal. This highly important idea is defined and developed in Section 9.3. We give here the normal distribution when the covariance matrix Σ has diagonal entries σ_x^2 and σ_y^2 and off-diagonal entry $\Sigma_{12} = \Sigma_{21} = \sigma_{xy}$:

$$p(x, y) = \frac{1}{2\pi\sqrt{|\det \Sigma|}} e^{-[x\ \ y]\Sigma^{-1}[x\ \ y]^{\mathrm{T}}/2}. \tag{9.8}$$

If the mean is moved from $(0, 0)$ to (μ_x, μ_y), then replace x and y by $x - \mu_x$ and $y - \mu_y$.

χ^2 Distribution

Next we investigate the probability distribution for a random variable defined as the *sum of squares* of the Gaussian random variables $x_1, \ldots, x_n$:

$$\chi_n^2 = x_1^2 + x_2^2 + \cdots + x_n^2.$$

We assume that the x_i are independent and normally distributed with $\mu = 0$ and $\sigma = 1$. The random variable χ_n has a geometrical interpretation: χ_n is the distance from the origin to a point with coordinates $(x_1, \ldots, x_n)$.

In Example 9.1 we derive the density function for $x = \chi_1^2$, but already now we want to present the general expression for every n:

$$p_n(x) = \frac{1}{2^{n/2}\Gamma(n/2)} x^{(n/2)-1} e^{-x/2}, \qquad x \geq 0. \tag{9.9}$$

For small values of n this χ_n^2 distribution is strongly unsymmetric. Its mean value by (9.22) is $\mu = n$:

$$\begin{aligned} E\{\chi_n^2\} &= \frac{1}{2^{n/2}\Gamma(n/2)} \int_0^\infty x e^{-x/2} x^{(n/2)-1}\, dx \\ &= \frac{1}{2^{n/2}\Gamma(n/2)} \int_0^\infty e^{-x/2} x^{n/2}\, dx \\ &= \frac{1}{2^{n/2}\Gamma(n/2)} \Gamma(1+n/2)\, 2^{1+(n/2)} \\ &= \frac{(n/2)\Gamma(n/2)\, 2^{1+(n/2)}}{2^{n/2}\Gamma(n/2)} = n. \end{aligned} \tag{9.10}$$

Similarly we calculate the variance as $2n$:

$$\begin{aligned} \sigma^2(\chi_n^2) &= \frac{1}{2^{n/2}\Gamma(n/2)} \int_0^\infty x^2 e^{-x/2} x^{(n/2)-1}\, dx - (\text{mean})^2 \\ &= \frac{1}{2^{n/2}\Gamma(n/2)} \Gamma(2+n/2)\, 2^{2+(n/2)} - n^2 \\ &= \frac{(1+n/2)(n/2)\Gamma(n/2)\, 2^{2+(n/2)}}{2^{n/2}\Gamma(n/2)} - n^2 \\ &= (2+n)n - n^2 = 2n. \end{aligned} \tag{9.11}$$

For $n \to \infty$, the distribution of χ_n^2/n tends to the normal distribution with mean 1 and variance $\sigma^2 = 2/n$. Figure 9.3 shows $p_n(x)$ for $n = 1, 2, 3, 4$, and 10.

Example 9.1 Derivation of equation (9.9). Start with the distribution of x_1^2:

$$\begin{aligned} \text{Prob}(x < \chi_1^2 < x + dx) &= \text{Prob}(\sqrt{x} < x_1 < \sqrt{x+dx}) + \text{Prob}(-\sqrt{x+dx} < x_1 < -\sqrt{x}) \\ &= 2\,\text{Prob}(\sqrt{x} < x_1 < \sqrt{x+dx}). \end{aligned}$$

We expand $\sqrt{x+dx}$ into a Taylor series $\sqrt{x} + dx/2\sqrt{x} + \cdots$ to obtain

$$\begin{aligned} 2\,\text{Prob}(\sqrt{x} < x_1 < \sqrt{x+dx}) &\approx 2\,\text{Prob}\big(\sqrt{x} < x_1 < \sqrt{x} + \tfrac{dx}{2\sqrt{x}}\big) \\ &\approx \tfrac{2}{\sqrt{2\pi}} e^{-x/2} \tfrac{dx}{2\sqrt{x}} = \tfrac{1}{\sqrt{2\pi x}} e^{-x/2} dx. \end{aligned}$$

This informal argument agrees with (9.9) since $\Gamma\big(\frac{1}{2}\big) = \sqrt{\pi}$:

$$p_1(x) = \tfrac{1}{\sqrt{2\pi x}} e^{-x/2} \qquad \text{for } x > 0. \tag{9.12}$$

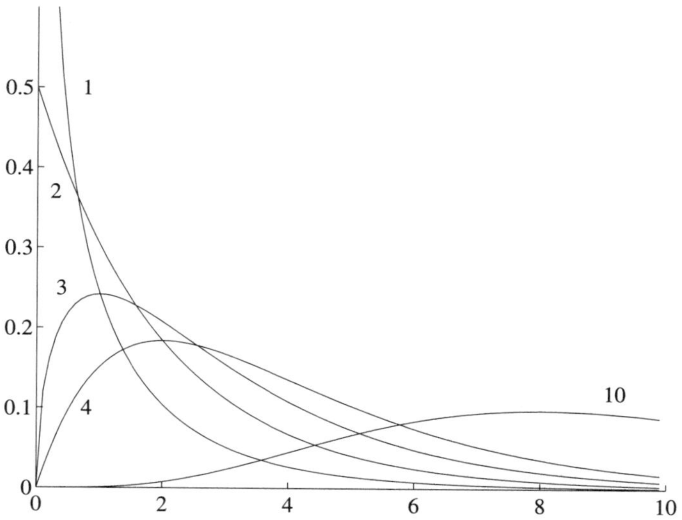

Figure 9.3 Probability density $p_n(\chi^2)$ with $n = 1, 2, 3, 4, 10$ degrees of freedom.

Alternatively one could start from the cumulative distribution for χ_1^2:

$$\text{Prob}(\chi_1^2 < x) = \text{Prob}(-\sqrt{x} < \chi_1 < \sqrt{x}) = 2F(\sqrt{x}) - 1.$$

Then formally $p_1(x)$ is the derivative:

$$p_1(x) = \frac{d}{dx}\big(2F(\sqrt{x}) - 1\big) = \tfrac{1}{\sqrt{x}}\tfrac{1}{\sqrt{2\pi}}e^{-x/2}.$$

We intend to prove by induction the general expression

$$p_n(x) = \frac{1}{2^{n/2}\Gamma(n/2)}x^{(n-2)/2}e^{-x/2} \qquad \text{for } x \geq 0. \tag{9.13}$$

We start by assuming that equation (9.13) is valid for the indices $1, 2, \ldots, n$ and want to prove it for $n+1$. According to the definition, $\chi_n^2 = x_1^2 + \cdots + x_n^2$ with independent terms. Consequently $\chi_{n+1}^2 = \chi_n^2 + \chi_1^2$. Again χ_n^2 and χ_1^2 are independent. Now we need the convolution theorem for probability densities: ***The probability density function of the sum of two independent random variables is the convolution of their probability density functions:***

$$\begin{aligned} p_{n+1}(x) &= \int_0^x p_n(y)p_1(x-y)\,dy = c_n \int_0^x y^{(n/2)-1}e^{-y/2}\frac{e^{-(x-y)/2}}{\sqrt{x-y}}\,dy \\ &= c_n\,e^{-x/2}\int_0^x \frac{y^{(n/2)-1}}{\sqrt{x-y}}\,dy \\ &= c_n\,e^{-x/2}x^{(n-1)/2}\int_0^1 \big(\tfrac{y}{x}\big)^{(n/2)-1}\big(1-\tfrac{y}{x}\big)^{-1/2}\,d\big(\tfrac{y}{x}\big) \\ &= C_n\,x^{(n-1)/2}e^{-x/2}. \end{aligned} \tag{9.14}$$

We determine the constant C_n so that the total probability is 1:

$$\int_0^\infty p_{n+1}(x)\,dx = C_n \int_0^\infty x^{(n-1)/2} e^{-x/2}\,dx = 1.$$

Replacing $x/2$ by t, the integral becomes the gamma function:

$$C_n\, 2^{(n+1)/2}\Gamma\big(\tfrac{n+1}{2}\big) = 1.$$

Substituting for C_n, equation (9.14) becomes

$$p_{n+1}(x) = \frac{1}{2^{(n+1)/2}\Gamma\big(\frac{n+1}{2}\big)} x^{(n-1)/2} e^{-x/2}. \tag{9.15}$$

Compared with (9.13), all indices have been incremented by exactly one. Therefore (9.13) is valid for all positive integers n.

In many applications the cumulative function $K_n(x)$ is more useful than $p_n(x)$ itself:

$$K_n(x) = \text{Prob}(\chi_n^2 \le x) = \int_0^x p_n(t)\,dt. \tag{9.16}$$

$K_n(x)$ gives the probability that the square sum of n normally distributed random variables x_i is smaller than a given number x. This function is tabulated for various values of n. Without the factor $1/\big(2^{n/2}\Gamma(n/2)\big)$, $K_n(x)$ is called the incomplete gamma function.

Often one encounters distribution functions closely related to χ^2. For example if $X = y_1^2 + \cdots + y_n^2$ with $E\{y_i\} = 0$ and $E\{y_i^2\} = \sigma^2$ then $X = \sigma^2\chi_n^2$. We will also meet random variables of the form

$$\begin{aligned}
\tfrac{1}{n}X &= \tfrac{1}{n}(y_1^2 + \cdots + y_n^2)\\
\sqrt{X} &= \sqrt{y_1^2 + \cdots + y_n^2}\\
\sqrt{\tfrac{1}{n}X} &= \sqrt{\tfrac{1}{n}(y_1^2 + \cdots + y_n^2)}.
\end{aligned}$$

All these variables have density distributions related to $p_n(x)$. We gather them in Table 9.3.

Statistical test theory assumes that we can present two clear-cut alternative hypotheses. In geodesy this is seldom possible. So instead the concept of confidence regions is much more used, and that brings the χ^2 distribution into play.

In particular, the random variable $(n-1)\hat{\sigma}^2/\sigma_0^2$ using the sample variance $\hat{\sigma}^2 = \frac{1}{n-1}\sum(x_i - \hat{x})^2$ follows a χ^2 distribution ***with n* − 1 *degrees of freedom***.

Remember $K_n(x)$ is a probability. When this probability is fixed at some specific value $K_n(x) = P\%$ the corresponding value for x is known as the $P\%$-fractile. For the χ^2 distribution with n degrees of freedom the $P\%$-fractile is represented by the symbol $\chi^2_{n,P}$. Values of $\chi^2_{n,P}$ are given in Table 9.4.

By means of the χ^2 distribution and the estimated variance $\hat{\sigma}^2$ we can derive confidence intervals for the (theoretical) variance σ_0^2. Estimates are usually denoted by a hat (ˆ).

Table 9.3 Random variables and their probability densities

Random variable	Probability density when y_i are $N(0,\sigma^2)$
$X=\sum_{i=1}^{n} y_i^2$	$\frac{1}{\sigma^2}p_n\Big(\frac{y}{\sigma}\Big)=\frac{1}{2^{n/2}\Gamma(n/2)\sigma^n}\,y^{(n/2)-1}e^{-y/(2\sigma^2)}$
$\frac{1}{n}X=\frac{1}{n}\sum_{i=1}^{n} y_i^2$	$\frac{n}{\sigma^2}p_n\Big(\frac{ny}{\sigma}\Big)=\frac{(n/2)^{n/2}}{\Gamma(n/2)\sigma^n}\,y^{(n/2)-1}e^{-ny/(2\sigma^2)}$
$\sqrt{X}=\sqrt{\sum_{i=1}^{n} y_i^2}$	$\frac{2y}{\sigma^2}p_n\Big(\frac{y^2}{\sigma^2}\Big)=\frac{2}{2^{n/2}\Gamma(n/2)\sigma^n}\,y^{n-1}e^{-y^2/(2\sigma^2)}$
$\sqrt{\frac{1}{n}X}=\sqrt{\frac{1}{n}\sum_{i=1}^{n} y_i^2}$	$\frac{2ny}{\sigma^2}p_n\Big(\frac{ny^2}{\sigma^2}\Big)=\frac{2(n/2)^{n/2}}{\Gamma(n/2)\sigma^n}\,y^{n-1}e^{-ny^2/(2\sigma^2)}$

Let the P%-fractile of the χ^2 distribution with $n-1$ degrees of freedom be called $\chi^2_{n-1,P}$. Then

$$\mathrm{Prob}\bigg(\chi^2_{n-1,Q}<\frac{(n-1)\hat\sigma^2}{\sigma_0^2}<\chi^2_{n-1,P}\bigg)=P-Q.$$

We solve this equation for σ_0^2 and obtain the $(P-Q)$%-confidence interval for σ_0^2:

$$\frac{n-1}{\chi^2_{n-1,P}}\hat\sigma^2<\sigma_0^2<\frac{n-1}{\chi^2_{n-1,Q}}\hat\sigma^2. \tag{9.17}$$

If we have u unknowns instead of 1, the number $n-1$ changes to $n-u$.

For the two-sided confidence interval $Q>0$ and $P<100$, we put $Q=100-P$. Then equation (9.17) becomes

$$\frac{n-u}{\chi^2_{n-u,P}}\hat\sigma^2<\sigma_0^2<\frac{n-u}{\chi^2_{n-u,(100-P)}}\hat\sigma^2. \tag{9.18}$$

If the confidence is selected as 90% then often this is split into two equal halves and one calculates the probability to get a lower and an upper limit.

Suppose an empirical standard deviation $\hat\sigma$ is determined on the basis of f degrees of freedom. We define a ***confidence interval*** (of confidence $1-\alpha$) for σ_0 through the probability conditions

$$\mathrm{Prob}(a\le\sigma_0<b)=1-\alpha \quad\text{and}\quad \mathrm{Prob}(\sigma_0<a)=\mathrm{Prob}(\sigma_0>b)=\alpha/2$$

and we seek the limits a and b. Assume the original observations are normally distributed, so that $\chi_f^2=f\hat\sigma^2/\sigma_0^2$ and

$$\mathrm{Prob}\big(\chi_f^2\le\chi^2_{f,\alpha/2}\big)=\alpha/2 \quad\text{and}\quad \mathrm{Prob}\big(\chi_f^2\le\chi^2_{f,1-(\alpha/2)}\big)=1-(\alpha/2).$$

Table 9.4 Probabilities for one-sided P%-fractiles of the χ^2 distribution

$n-u$	1%	5%	10%	50%	90%	95%	99%
1	0.00016	0.0039	0.016	0.5	2.7	3.8	6.6
2	0.020	0.103	0.211	1.4	4.6	6.0	9.2
3	0.115	0.352	0.584	2.4	6.3	7.8	11.3
4	0.30	0.71	1.06	3.4	7.8	9.5	13.3
5	0.55	1.15	1.61	4.4	9.2	11.1	15.1
6	0.87	1.64	2.20	5.3	10.6	12.6	16.8
8	1.65	2.73	3.49	7.3	13.4	15.5	20.1
10	2.56	3.94	4.87	9.3	16.0	18.3	23.2
20	8.3	10.9	12.4	19.3	28.4	31.4	37.6
30	15.0	18.5	20.6	29.3	40.3	43.8	50.9
40	22.2	26.5	29.1	39.3	51.8	55.8	63.7
50	29.7	34.8	37.7	49.3	63.2	67.5	76.2
100	70.1	77.9	82.4	99.3	118.5	124.3	135.8

Those combine into

$$\operatorname{Prob}(a \leq \sigma_0 \leq b) = 1 - \alpha$$

with

$$a = \sqrt{\frac{f}{\chi^2_{f,1-(\alpha/2)}}}\,\hat{\sigma} = \gamma_a \hat{\sigma} \qquad \text{and} \qquad b = \sqrt{\frac{f}{\chi^2_{f,\alpha/2}}}\,\hat{\sigma} = \gamma_b \hat{\sigma}.$$

The factors γ_a and γ_b obviously depend on the confidence probability $1-\alpha$ and the number of degrees of freedom. The factors are shown in Table 9.5 for $1 - \alpha = 0.95$. This uses Table 9.4 which can be found in Abramowitz & Stegun (1972), Table 26.8. For $f = 5$ we get $0.672\,\hat{\sigma} < \sigma_0 < 2.09\,\hat{\sigma}$.

A common geodetic least-squares problem is the estimation of coordinates of control points for a network. Simultaneously we get a covariance matrix for the estimates. This leads to the construction of the confidence ellipse, see Section 9.8. We now ask what is the

Table 9.5 Limiting factors γ_a and γ_b for confidence interval with probability 0.95

f	1	2	3	4	5	10	20	50	100
γ_a	0.446	0.521	0.567	0.599	0.624	0.699	0.765	0.837	0.879
γ_b	31.9	6.28	3.73	2.87	2.45	1.76	1.44	1.24	1.16

Table 9.6 χ^2 with $n = 2$: Probability $K_2(c^2)$ that $x^2 + y^2 \le c^2$

c	$K_2(c^2)$
1	0.393 5
2	0.864 7
3	0.988 9

$K_2(c^2)$	c
0.90	2.146
0.95	2.448
0.99	3.035

probability that the estimated point will lie within this confidence ellipse. $K_2(c^2)$ tells the probability for the following inequality:

$$\frac{y_1^2}{\lambda_1^2} + \frac{y_2^2}{\lambda_2^2} < c^2.$$

The integral of $p_2(x) = \frac{1}{2\Gamma(1)} x^0 e^{-x/2} = \frac{1}{2} e^{-x/2}$ gives the cumulative probability K_2:

$$K_2(c^2) = \text{Prob}\big(\chi_2^2 < c^2\big) = \tfrac{1}{2} \int_0^{c^2} e^{-x/2}\, dx = 1 - e^{-c^2/2}. \tag{9.19}$$

We indicate some values in Table 9.6.

F Distribution

The *ratio* of sums of squares is often encountered in geodetic practice. Let $x_1, \dots, x_m$ and $y_1, \dots, y_n$ be normally distributed independent random variables. We define κ as the dimensionless quantity

$$\kappa = \frac{\frac{1}{m} \sum_{i=1}^{m} x_i^2}{\frac{1}{n} \sum_{j=1}^{n} y_j^2} = \frac{s/m}{t/n}. \tag{9.20}$$

We want the probability $F_{mn}(x)$ that this fraction κ is smaller than a given constant x.

The χ^2 probability densities for the individual sums s and t are

$$p_m(x) = \frac{1}{2^{m/2}\Gamma(m/2)} x^{(m-2)/2} e^{-x/2} \quad \text{and} \quad p_n(x) = \frac{1}{2^{n/2}\Gamma(n/2)} x^{(n-2)/2} e^{-x/2}.$$

The combined probability density follows from convolution

$$p_{mn}(\kappa) = \frac{1}{2^{(m+n)/2}\Gamma(m/2)\Gamma(n/2)} x^{(m-2)/2} y^{(n-2)/2} e^{-(x+y)/2}.$$

After some calculations we get the final formula for this *F distribution*:

$$F_{mn}(x) = \frac{\Gamma\big((m+n)/2\big)}{\Gamma(m/2)\Gamma(n/2)} \Big(\frac{m}{n}\Big)^{m/2} \frac{x^{(m-2)/2}}{\big(1 + (m/n)x\big)^{(m+n)/2}}. \tag{9.21}$$

9.2 Mean, Variance, and Standard Deviation

We now find the mean μ of any distribution $p(x)$. The symmetry of the normal distribution guarantees that the built-in number μ is the mean. In general μ is the "expected value." ***To find μ, multiply outcome x by probability $p(x)$ and integrate:***

$$\textbf{\textit{mean}} = \mu = E\{X\} = \int_{-\infty}^{\infty} x p(x)\, dx. \tag{9.22}$$

This is usually different from the point $x = m$ where $F(m) = \frac{1}{2}$. That point is the median, since there are equal (50%) probabilities of $x \leq m$ and $x \geq m$.

Together with the mean we introduce the ***variance***. It is always written σ^2 and in the normal distribution it measures the width of the curve. The variance σ^2 is the expected value of $(X - \text{mean})^2 = (X - \mu)^2$. Multiply outcome times probability and integrate:

$$\sigma^2 = \int_{-\infty}^{\infty} (x - \mu)^2 p(x)\, dx = \int_{-\infty}^{\infty} x^2 p(x)\, dx - \mu^2. \tag{9.23}$$

The ***standard deviation*** (written σ) is the square root of σ^2.

In practice we may repeat an observation many times. Usually, economy will dictate some limit. Each independent observation produces an outcome X. The average of the outcomes from N observations is $\widehat{X}$ (called "X hat"):

$$\widehat{X} = \frac{X_1 + X_2 + \cdots + X_N}{N} = \textbf{\textit{average outcome}}. \tag{9.24}$$

Frequently all we know about the random variable X is its mean μ and variance σ^2. It is amazing how much information this gives about the average $\widehat{X}$:

Law of Averages: $\widehat{X}$ is almost sure to approach μ as $N \to \infty$.

Central Limit Theorem: The sum of a large number of independent identically distributed random variables with finite means and variances, standardized to have mean 0 and variance 1, is approximately normally distributed.

No matter what the probability for X, the probabilities for $\widehat{X}$ move toward the normal bell-shaped curve. The standard deviation of the average is close to $\sigma/\sqrt{N}$. In the Law of Averages, "almost sure" means that the chance of $\widehat{X}$ *not* approaching μ is zero.

The quantity σ/μ is referred to as the ***relative accuracy*** or the ***variation coefficient***. The inverse quantity μ/σ is the ***signal-to-noise ratio***.

Finally we derive some useful computational rules for means and variances. Let h be a real function. Then $Y = h(X)$ is a random variable and we have

$$E\{Y\} = E\{h(X)\} = \int_{-\infty}^{\infty} h(x) p(x)\, dx. \tag{9.25}$$

If a and b are real numbers and $h(x) = ax + b$ this yields

$$E\{aX + b\} = aE\{X\} + b. \tag{9.26}$$

Evaluating (9.25) for $h(x) = \big(x - E\{X\}\big)^2 = (x-\mu)^2$ gives

$$\sigma^2\{X\} = \int_{-\infty}^{\infty} \big(x - E\{X\}\big)^2 p(x)\,dx. \tag{9.27}$$

When we introduce a linear substitution $aX + b$, we get

$$\begin{aligned}\sigma^2\{aX+b\} &= \int_{-\infty}^{\infty} \big(ax + b - E\{aX+b\}\big)^2 p(x)\,dx \\ &= \int_{-\infty}^{\infty} \big(ax + b - aE\{X\} - b\big)^2 p(x)\,dx \\ &= a^2 \int_{-\infty}^{\infty} \big(x - E\{X\}\big)^2 p(x)\,dx = a^2\sigma^2\{X\}.\end{aligned}$$

The shift by b leaves σ^2 unchanged. The scaling by a multiplies that variance by a^2:

$$\sigma^2\{aX+b\} = a^2\sigma^2\{X\}. \tag{9.28}$$

Notice that in the formulation of mean value and standard deviation, the order of observations is of no consequence. Since none of the observations are time-dependent, the time at which an element is observed within the sequence cannot affect the outcome.

A *time-ordered sequence* of random variables is called a *discrete random process*.

9.3 Covariance

Until now we have studied only independent random variables. Next we make the step to *two* random variables X_1 and X_2, possibly correlated. Consider the probability that X_1 falls between x_1 and x_1+dx_1 and in the same observation X_2 falls between x_2 and x_2+dx_2. This produces the joint probability density $p(x_1,x_2)\,dx_1dx_2$. Again its integral is 1:

$$\text{Prob}(-\infty < X_1, X_2 < +\infty) = \int_{-\infty}^{\infty}\int_{-\infty}^{\infty} p(x_1,x_2)\,dx_1dx_2 = 1. \tag{9.29}$$

The mean value μ_1 of X_1, and the mean value of μ_2 of X_2, are

$$\begin{aligned}\mu_1 &= \int_{-\infty}^{\infty}\int_{-\infty}^{\infty} x_1 p(x_1,x_2)\,dx_1dx_2 \\ \mu_2 &= \int_{-\infty}^{\infty}\int_{-\infty}^{\infty} x_2 p(x_1,x_2)\,dx_1dx_2.\end{aligned}$$

The mean value of a function $\varphi(x_1,x_2)$ is

$$E\{\varphi(X_1,X_2)\} = \int_{-\infty}^{\infty}\int_{-\infty}^{\infty} \varphi(x_1,x_2)p(x_1,x_2)\,dx_1dx_2. \tag{9.30}$$

For the important case $\varphi(X_1,X_2) = X_1 + X_2$ we get

$$E\{X_1+X_2\} = \int_{-\infty}^{\infty}\int_{-\infty}^{\infty} (x_1+x_2)p(x_1,x_2)\,dx_1dx_2 = E\{X_1\} + E\{X_2\}. \tag{9.31}$$

The ***covariance*** between X_1 and X_2 is defined as

$$\sigma_{12} = \text{cov}\{X_1, X_2\} = \int_{-\infty}^{\infty}\int_{-\infty}^{\infty}(x_1 - \mu_1)(x_2 - \mu_2)p(x_1, x_2)\,dx_1dx_2 \tag{9.32}$$

$$= E\{X_1X_2\} - \mu_1\mu_2. \tag{9.33}$$

In case $X_2 = X_1$ we get the variance σ_1^2 of the random variable x_1. The covariance of a variable with itself is its variance. Thus by convention σ_{11} is written σ_1^2.

Let X_1 denote the weight and X_2 the height of a number of individuals. An observed value of X_2 tends to be small if the corresponding value of X_1 is small and conversely. The random variables X_1 and X_2, weight and height, are said to be *dependent*. They do not satisfy the following requirement: Two random variables with probability densities $p_1(x_1)$ and $p_2(x_2)$ and joint probability density $p(x_1, x_2)$ are *independent* if

$$p(x_1, x_2) = p_1(x_1)p_2(x_2) \qquad \text{for all } x_1 \text{ and } x_2. \tag{9.34}$$

Independence is an exceedingly important property. In general we do not know $p(x_1, x_2)$ from the separate probabilities.

Finally we generalize to n random variables. We observe $\boldsymbol{X} = (X_1, X_2, \ldots, X_n)$. The ***vector of means*** is $\boldsymbol{\mu} = (\mu_1, \mu_2, \ldots, \mu_n)$ where $\mu_i = E\{X_i\}$. The most important quantity in geodetic statistics is the expectation of the matrix $(\boldsymbol{X} - \boldsymbol{\mu})(\boldsymbol{X} - \boldsymbol{\mu})^{\mathrm{T}}$ which contains all the products $(X_i - \mu_i)(X_j - \mu_j)$. The expectations of all these products enter the covariance matrix Σ_X:

$$\Sigma_X = E\{(\boldsymbol{X} - \boldsymbol{\mu})(\boldsymbol{X} - \boldsymbol{\mu})^{\mathrm{T}}\} = E\{\boldsymbol{X}\boldsymbol{X}^{\mathrm{T}}\} - \boldsymbol{\mu}\boldsymbol{\mu}^{\mathrm{T}}. \tag{9.35}$$

This ***covariance matrix*** contains every $\sigma_{ij} = E\{(X_i - \mu_i)(X_j - \mu_j)\}$ and is symmetric:

$$\Sigma_X = \begin{bmatrix} \sigma_1^2 & \sigma_{12} & \cdots & \sigma_{1n} \\ \sigma_{21} & \sigma_2^2 & \cdots & \sigma_{2n} \\ \vdots & \vdots & \ddots & \vdots \\ \sigma_{n1} & \sigma_{n2} & \cdots & \sigma_n^2 \end{bmatrix}. \tag{9.36}$$

Covariance is a measure of the stochastic dependence between two parameters X_i and X_j. Clearly σ_{ij} measures a coupling between the errors of X_i and the errors of X_j. One can speak about ***stochastic dependence or independence***. If X_i and X_j have a tendency to deviate either both positively or both negatively, then σ_{ij} will be positive. This does not imply that a positive $X_i - \mu_i$ cannot occur together with a negative $X_j - \mu_j$—but it is less likely. Similarly σ_{ij} will be negative if a positive $X_i - \mu_i$ prefers to be coupled to a negative $X_j - \mu_j$. It is important to note the difference between two random variables being independent and being uncorrelated. They are independent if and only if $p(x_1, x_2) = p_1(x_1)p_2(x_2)$. They are uncorrelated if $\sigma_{ij} = 0$. Two independent random variables X_i and X_j are uncorrelated.

The covariance matrix Σ_X is ***positive definite*** (or at least semidefinite). For proof, note that any linear combination $l = c_1(X_1 - \mu_1) + \cdots + c_n(X_n - \mu_n)$ has

$$0 \le E\{l^2\} = \sum\sum c_i c_j \sigma_{ij} = \begin{bmatrix} c_1 & \ldots & c_n \end{bmatrix} \Sigma_X \begin{bmatrix} c_1 \\ \vdots \\ c_n \end{bmatrix}. \tag{9.37}$$

Thus $\boldsymbol{c}^{\mathrm{T}} \Sigma_X \boldsymbol{c}$ is nonnegative for any vector $\boldsymbol{c}$, which makes Σ_X semidefinite.

The ***correlation coefficient*** ρ_{ij} is a standardized covariance, never exceeding 1:

$$\rho_{ij} = \sigma_{ij} \Big/ \sqrt{\sigma_i^2 \sigma_j^2} = \sigma_{ij}/\sigma_i\sigma_j \quad \text{and} \quad |\rho_{ij}| \le 1. \tag{9.38}$$

Example 9.2 Covariance matrices must be *nonnegative definite*. In practice roundoff errors and other error sources may cause the matrix to become indefinite. One or more small negative eigenvalues can appear. Whenever one encounters such a matrix it must be repaired in some way.

A good procedure is to build up a modified matrix using only the positive eigenvalues and their eigenvectors. The following MATLAB code will do the job by discarding any $\lambda_i < 0$:

M-file: repair

```
function R = repair(S)
%   Repair of indefinite covariance matrix
[v,lambda] = eig(S);
[n,n] = size(S);
R = zeros(n,n);
for i = 1:n
   if lambda(i,i) > 0
      s = lambda(i,i) * v(:,i) * v(:,i)';
   end;
   R = R + s;
end;
```

9.4 Inverse Covariances as Weights

Suppose we observe the position of a satellite, many times. Those observations $b_1, \ldots, b_m$ have errors and the errors may not be independent. We want to know (for example) the velocity of the satellite. The question in this section is how to use the a priori information we have about the accuracy and the degree of independence of the observations. That information is contained in the covariance matrix Σ_b.

As usual in least squares, we are fitting many observations (positions at different times) by a few parameters (initial position and velocity). A perfect fit is not expected. We minimize the lack-of-fit expressed by the residual $\boldsymbol{r} = \boldsymbol{b} - A\boldsymbol{x}$. When we choose the vector $\hat{\boldsymbol{x}}$ that makes $\boldsymbol{r}$ as small as possible, we are allowed to define what "small" means. A very basic choice of the quantity to minimize is the ***weighted*** length $\boldsymbol{r}^{\mathrm{T}}C\boldsymbol{r}$, and the problem is to choose appropriately the weights in C.

We want to prove that choosing the weight matrix C inversely proportional to the covariance matrix Σ_b leads to the *best linear unbiased estimate* (BLUE) of the unknown parameters. The estimate is $\hat{\boldsymbol{x}}$ and we will find *its* covariance.

First a few comments about the choice of C. All we know about the errors, except for their mean $E\{\boldsymbol{r}\} = E\{\boldsymbol{b} - A\boldsymbol{x}\} = \boldsymbol{0}$, is contained in the matrix Σ_b. This is the covariance matrix $E\{\boldsymbol{b}\boldsymbol{b}^{\mathrm{T}}\}$ or entrywise $E\{b_i b_j\}$ of the observation errors. What we require of any rule $\hat{\boldsymbol{x}} = L\boldsymbol{b}$, which estimates the true but unknown parameters $\boldsymbol{x}$ from the measurements $\boldsymbol{b}$, is that it be *linear* and *unbiased*. It is certainly linear if L is a matrix. It is unbiased if the expected error $\boldsymbol{x} - \hat{\boldsymbol{x}}$—the error in the estimate, not in the observations—is also zero:

$$E\{\boldsymbol{x} - \hat{\boldsymbol{x}}\} = E\{\boldsymbol{x} - L\boldsymbol{b}\} = E\{\boldsymbol{x} - LA\boldsymbol{x} - L\boldsymbol{r}\} = E\{(I - LA)\boldsymbol{x}\} = \boldsymbol{0}. \tag{9.39}$$

Thus L is unbiased if it is a left inverse of the rectangular matrix A: $LA = I$.* Under this restriction, Gauss picked out the best L (we call it L_0) and the best C in the following way: ***The matrix C should be σ_0^2 times the inverse of the covariance matrix Σ_b.***

His rule leads to $L_0 = (A^{\mathrm{T}}\Sigma_b^{-1}A)^{-1}A^{\mathrm{T}}\Sigma_b^{-1}$, which does satisfy $L_0 A = I$. For completeness we include a proof that this choice is optimal.

The best linear unbiased estimate (BLUE) is the one with $C = \sigma_0^2\Sigma_b^{-1}$. The covariance matrix Σ_x is as small as possible. The optimal estimate $\hat{x}$ and the optimal matrix L_0 are

$$\hat{\boldsymbol{x}} = (A^{\mathrm{T}}\Sigma_b^{-1}A)^{-1}A^{\mathrm{T}}\Sigma_b^{-1}\boldsymbol{b} = L_0\boldsymbol{b}. \tag{9.40}$$

This choice minimizes the expected error in the estimate, measured by the covariance matrix $\Sigma_x = E\{(\boldsymbol{x} - \hat{\boldsymbol{x}})(\boldsymbol{x} - \hat{\boldsymbol{x}})^{\mathrm{T}}\}$. Formula (9.42) shows this smallest (best) matrix Σ_x.

Proof The matrix to be minimized is

$$\begin{aligned}\Sigma_x &= E\{(\boldsymbol{x} - \hat{\boldsymbol{x}})(\boldsymbol{x} - \hat{\boldsymbol{x}})^{\mathrm{T}}\} = E\{(\boldsymbol{x} - L\boldsymbol{b})(\boldsymbol{x} - L\boldsymbol{b})^{\mathrm{T}}\}\\ &= E\{(\boldsymbol{x} - LA\boldsymbol{x} - L\boldsymbol{r})(\boldsymbol{x} - LA\boldsymbol{x} - L\boldsymbol{r})^{\mathrm{T}}\}.\end{aligned}$$

Since $\boldsymbol{x} = LA\boldsymbol{x}$ and L is linear, this is just

$$\Sigma_x = E\{(L\boldsymbol{r})(L\boldsymbol{r})^{\mathrm{T}}\} = LE\{\boldsymbol{r}\boldsymbol{r}^{\mathrm{T}}\}L^{\mathrm{T}} = L\Sigma_b L^{\mathrm{T}}.$$

Thus it is $L\Sigma_b L^{\mathrm{T}}$ that we minimize, subject to $LA = I$. To show that L_0 is the optimal choice, write any L as $L_0 + (L - L_0)$ and substitute into $\Sigma_x = L\Sigma_b L^{\mathrm{T}}$:

$$\Sigma_x = L_0\Sigma_b L_0^{\mathrm{T}} + (L - L_0)\Sigma_b L_0^{\mathrm{T}} + L_0\Sigma_b(L - L_0)^{\mathrm{T}} + (L - L_0)\Sigma_b(L - L_0)^{\mathrm{T}}. \tag{9.41}$$

The middle terms are transposes of one another, and they are zero:

$$(L - L_0)\Sigma_b L_0^{\mathrm{T}} = (L - L_0)\Sigma_b\Sigma_b^{-1}A(A^{\mathrm{T}}\Sigma_b^{-1}A)^{-1} = 0$$

*$LA = I$ just means: Whenever $\boldsymbol{b}$ can be fitted exactly by some $\boldsymbol{x}$ (so that $A\boldsymbol{x} = \boldsymbol{b}$) our choice $\hat{\boldsymbol{x}}$ should be $\boldsymbol{x}$: $\hat{\boldsymbol{x}} = L\boldsymbol{b} = LA\boldsymbol{x} = \boldsymbol{x}$. If the data lies on a straight line, then the best fit should be that line.

because $\Sigma_b \Sigma_b^{-1}$ is the identity and $(L - L_0)A = I - I = 0$. Furthermore the last term in (9.41) is symmetric and at least positive semidefinite. It is smallest when $L = L_0$, which is therefore the minimizing choice. The proof is complete.

The matrix Σ_x for this optimal estimate $\boldsymbol{x}$ comes out neatly when we simplify

$$\Sigma_x = L_0 \Sigma_b L_0^{\mathrm{T}} = (A^{\mathrm{T}} \Sigma_b^{-1} A)^{-1} A^{\mathrm{T}} \Sigma_b^{-1} \Sigma_b \Sigma_b^{-1} A (A^{\mathrm{T}} \Sigma_b^{-1} A)^{-1}.$$

Cancelling Σ_b and $A^{\mathrm{T}} \Sigma_b^{-1} A$ with their inverses gives the key formula for Σ_x:

When $\hat{x}$ is computed from the rule $A^{\mathrm{T}} \Sigma_b^{-1} A \hat{x} = A^{\mathrm{T}} \Sigma_b^{-1} b$, its covariance matrix is

$$\Sigma_x = (A^{\mathrm{T}} \Sigma_b^{-1} A)^{-1}. \tag{9.42}$$

This Σ_x gives the expected error in $\hat{\boldsymbol{x}}$ just as Σ_b gave the expected error in $\boldsymbol{b}$. These matrices average over all experiments—they do not depend on the particular measurement $\boldsymbol{b}$ that led to the particular estimate $\hat{\boldsymbol{x}}$. We emphasize that $\Sigma_x = E\{(\boldsymbol{x} - \hat{\boldsymbol{x}})(\boldsymbol{x} - \hat{\boldsymbol{x}})^{\mathrm{T}}\}$ is the fundamental matrix in filtering theory. Its inverse $A^{\mathrm{T}} \Sigma_b^{-1} A$ is the ***information matrix***, and it is exactly the triple product $A^{\mathrm{T}} C A / \sigma_0^2$. It measures the information content of the experiment. It goes up as the variance Σ_b goes down, since $C = \sigma_0^2 \Sigma_b^{-1}$. It also goes up as the experiment continues; every new row in A makes it larger.

In the most important case, with independent errors and unit variances and $C = I$, we are back to "white noise" and ordinary least squares. Then the information matrix is $A^{\mathrm{T}} A$. Its inverse is Σ_x.

Remark 9.1 We can always obtain $C = I$ from $C = \sigma_0^2 \Sigma_b^{-1}$ by a change of variables. Factor the matrix $\sigma_0^2 \Sigma_b^{-1}$ into $W^{\mathrm{T}} W$ and introduce $\bar{\boldsymbol{r}} = W\boldsymbol{r}$. (Often W is denoted $\sigma_0 \Sigma_b^{-1/2}$, the matrix square root of Σ_b^{-1}.) These standardized errors $\bar{\boldsymbol{r}} = W(A\boldsymbol{x} - \boldsymbol{b})$ still have mean zero, and their covariance matrix is the identity:

$$E\{(W\boldsymbol{r})(W\boldsymbol{r})^{\mathrm{T}}\} = W E\{\boldsymbol{r}\boldsymbol{r}^{\mathrm{T}}\} W^{\mathrm{T}} = W \Sigma_b W^{\mathrm{T}} = I.$$

The weighting matrix W returns us to white noise—a unit covariance problem, with simpler theory and simpler computations.†

Remark 9.2 If one of the variances is zero, say $\sigma_1^2 = 0$, then the first measurement is exact. The first row and column of Σ_b are zero, and Σ_b is not positive definite or even invertible. (The weighting matrix has $(\Sigma_b^{-1})_{11} = \infty$.) This just means that the first equation in $A\boldsymbol{x} = \boldsymbol{b}$ should be given infinite weight and be solved exactly. If this were true of all m measurements we would have to solve all m equations $A\boldsymbol{x} = \boldsymbol{b}$, without the help of least squares; but with exact measurements that would be possible.

To repeat we have the following relation between the covariance matrix and weight matrix:

$$C \Sigma_b = \sigma_0^2 I. \tag{9.43}$$

†Throughout mathematics there is this choice between a change of variable or a change in the definition of length. We introduce $\boldsymbol{X} = W\boldsymbol{x}$ or we measure the original $\boldsymbol{x}$ by $\|\boldsymbol{x}\|^2 = \boldsymbol{x}^{\mathrm{T}} \Sigma_b^{-1} \boldsymbol{x}$. It is really Hobson's choice; it makes no difference.

If the observations are uncorrelated this is simply

$$c_1\sigma_1^2 = c_2\sigma_2^2 = \cdots = c_n\sigma_n^2 = 1 \cdot \sigma_0^2. \tag{9.44}$$

Later we shall examine the quantity σ_0^2 more closely. For the moment just regard it as a "variance with unit weight."

For experienced readers Of course, the weights or the variances must be decided before any least-squares estimation. In practice this happens in the following way. If we want observation X_2 with variance σ_2^2 to have weight $c_2 = 1$, then the weight c_1 is given as σ_2^2/σ_1^2.

Most of the time one starts with uncorrelated observations. (A typical correlated observation is the *difference* of original observations. This is particularly the case with GPS observations or angles derived from theodolite direction observations.) Uncorrelated observations imply that all covariances σ_{ij} $(i \neq j)$ are zero. Then the covariance matrix of the observations $\boldsymbol{b}$ is diagonal:

$$\Sigma_b = \operatorname{diag}(\sigma_1^2, \ldots, \sigma_n^2) = \begin{bmatrix} \sigma_1^2 & & \\ & \ddots & \\ & & \sigma_n^2 \end{bmatrix}. \tag{9.45}$$

It is suitable to work with weights deviating as little as possible from 1. In geodesy we often introduce unit weight for

- a distance observed once and of length 1 km
- a direction as observed in one set
- a geometric or trigonometric leveling observed once and of length 1 km.

Example 9.3 Let m observations of the same variable x be given as $b_1, \ldots, b_m$. We want to prove that the *weighted mean* $\hat{x}$ and its variance are given as

$$\hat{x} = \frac{\boldsymbol{e}^{\mathrm{T}}C\boldsymbol{b}}{\boldsymbol{e}^{\mathrm{T}}C\boldsymbol{e}} \qquad \text{and} \qquad \hat{\sigma}_{\hat{x}}^2 = \frac{1}{(m-1)}\frac{\hat{\boldsymbol{r}}^{\mathrm{T}}C\hat{\boldsymbol{r}}}{\boldsymbol{e}^{\mathrm{T}}C\boldsymbol{e}}.$$

Here we introduced $\hat{\boldsymbol{r}} = \boldsymbol{b} - A\hat{\boldsymbol{x}}$ and the vector $\boldsymbol{e} = (1, 1, \ldots, 1)$. The m observation equations $x_i = b_i - r_i$ can be written in the matrix form $A\boldsymbol{x} = \boldsymbol{b} - \boldsymbol{r}$:

$$\begin{bmatrix} 1 \\ 1 \\ \vdots \\ 1 \end{bmatrix} x = \begin{bmatrix} b_1 \\ b_2 \\ \vdots \\ b_m \end{bmatrix} - \begin{bmatrix} r_1 \\ r_2 \\ \vdots \\ r_m \end{bmatrix}.$$

The normal equations $A^{\mathrm{T}}CA\boldsymbol{x} = A^{\mathrm{T}}C\boldsymbol{b}$ are

$$\begin{bmatrix} 1 & \ldots & 1 \end{bmatrix} \begin{bmatrix} c_1 & & \\ & \ddots & \\ & & c_m \end{bmatrix} \begin{bmatrix} 1 \\ \vdots \\ 1 \end{bmatrix} \hat{x} = \begin{bmatrix} 1 & \ldots & 1 \end{bmatrix} \begin{bmatrix} c_1 & & \\ & \ddots & \\ & & c_m \end{bmatrix} \begin{bmatrix} b_1 \\ \vdots \\ b_m \end{bmatrix}.$$

This is $\sum_{i=1}^m c_i\hat{x} = \sum_{i=1}^m c_i b_i$ and the solution is

$$\hat{x} = \sum_{i=1}^{m} c_i b_i \Big/ \sum_{i=1}^{m} c_i = \boldsymbol{e}^{\mathrm{T}} C \boldsymbol{b} / \boldsymbol{e}^{\mathrm{T}} C \boldsymbol{e}.$$

According to Table 9.7 (on page 336) we have $\hat{\boldsymbol{r}}^{\mathrm{T}} C \hat{\boldsymbol{r}} = \boldsymbol{b}^{\mathrm{T}} C \boldsymbol{b} - \boldsymbol{b}^{\mathrm{T}} C A \hat{x}$ which yields

$$\hat{\boldsymbol{r}}^{\mathrm{T}} C \hat{\boldsymbol{r}} = \sum c_i b_i^2 - \hat{x} \sum c_i b_i = \sum c_i b_i^2 + \hat{x}\Big(\hat{x} \sum c_i - 2 \sum c_i b_i\Big)$$

or

$$\hat{\boldsymbol{r}}^{\mathrm{T}} C \hat{\boldsymbol{r}} = \sum c_i (\hat{x} - b_i)^2.$$

Then (9.60) and (9.64) yield the variance for the weighted mean value $\hat{x}$:

$$\hat{\sigma}_{\hat{x}}^2 = \frac{\hat{\boldsymbol{r}}^{\mathrm{T}} C \hat{\boldsymbol{r}}}{m-1} \Big/ \sum_{i=1}^{m} c_i = \frac{1}{m-1} \frac{\hat{\boldsymbol{r}}^{\mathrm{T}} C \hat{\boldsymbol{r}}}{\boldsymbol{e}^{\mathrm{T}} C \boldsymbol{e}}.$$

Example 9.4 Finally we make a preliminary inspection of the concept of *correlation* and its eventual influence. Suppose we have observed a given distance twice with the results $b_1 = 100\,\mathrm{m}$ and $b_2 = 102\,\mathrm{m}$. The least-squares estimate for the length is $\hat{x}$. If $C = I$ then $\hat{x} = 101\,\mathrm{m}$ is the average. In case C is diagonal we get $b_1 < \hat{x} < b_2$. When $c_1 \to \infty$ we approach $\hat{x} = b_1$ and when $c_2 \to \infty$ we obtain $\hat{x} = b_2$. The least-squares result for diagonal C (no correlation) always lies between the smallest and the largest observation.

If the two observations are correlated the circumstances change drastically. We still have $A^{\mathrm{T}} = [\,1 \quad 1\,]$ for the two observations. Suppose the inverse covariance matrix is

$$\begin{bmatrix} 5 & 2 \\ 2 & 1 \end{bmatrix}^{-1} = \begin{bmatrix} 1 & -2 \\ -2 & 5 \end{bmatrix} = C.$$

Now $A^{\mathrm{T}} C A = 2$ and $A^{\mathrm{T}} C \boldsymbol{b} = 206$. Therefore $\hat{x} = 103 > b_2$. A strong positive correlation combined with a large weight for b_2 results in $\hat{x} > b_2$ which cannot happen for independent observations. In Section 11.6 we shall show how arbitrary covariance matrices may lead to arbitrary least-squares results. You may run the M-file corrdemo.

9.5 Estimation of Mean and Variance

In practice we often want to estimate unknown quantities on the basis of one or more series of observations. Let $X_1, X_2, \ldots, X_n$ be a set of independent random variables and let

$$Y = \varphi(X_1, X_2, \ldots, X_n) \tag{9.46}$$

be defined for all given sets of data—and for any value of n. If $E\{Y\} \to \eta$ and $\sigma^2\{Y\} \to 0$ for $n \to \infty$ we say that (9.46) is an unbiased ***estimate*** of the constant η.

Note especially that the sample mean $\widehat{X}$ is an unbiased estimate of μ:

$$E\{\widehat{X}\} = \mu \qquad \text{and} \qquad \sigma^2\{\widehat{X}\} = \frac{\sigma^2}{n}. \tag{9.47}$$

The average $\widehat{X}$ is an unbiased estimate of the mean μ.

Next we look for an estimate of σ^2. Start by assuming that μ is known. We consider the function

$$\hat{\sigma}^2 = \frac{\sum(X_i - \mu)^2}{n}. \tag{9.48}$$

According to the definition of variance we have $E\{(X_i - \mu)^2\} = \sigma^2$; hence $E\{\hat{\sigma}^2\} = \sigma^2$.

Generally the mean μ is not known. In this case we start with the identity

$$X_i - \mu = (X_i - \widehat{X}) + (\widehat{X} - \mu)$$

and get

$$\sum(X_i - \mu)^2 = \sum(X_i - \widehat{X})^2 + n(\widehat{X} - \mu)^2.$$

The double products vanish because $\sum(X_i - \widehat{X}) = 0$. We take the mean value on both sides and get

$$n\sigma^2 = E\{\sum(X_i - \widehat{X})^2\} + \sigma^2.$$

Rearrange this into

$$E\{\sum(X_i - \widehat{X})^2\} = (n-1)\sigma^2. \tag{9.49}$$

This says that

$$\hat{s}^2 = \frac{\sum(X_i - \widehat{X})^2}{n-1} \qquad \text{has} \qquad E\{\hat{s}^2\} = \sigma^2. \tag{9.50}$$

Since $\sigma\{\widehat{X}\} = \sigma / \sqrt{n}$, we also get an estimate of the standard deviation of the average:

$$\hat{\sigma}(\widehat{X}) = \frac{\hat{s}}{\sqrt{n}} = \sqrt{\frac{\sum(X_i - \widehat{X})^2}{n(n-1)}}. \tag{9.51}$$

These estimates both are biased due to the following fact: Let y be an unbiased estimate of η and φ a nonlinear function of y. Then the estimate $\varphi(y)$ for $\varphi(\eta)$ is likely to be biased because generally $E\{\varphi(y)\} \neq \varphi(E\{y\})$ for nonlinear φ.

Example 9.5 We want to prove that $\hat{s}$ is a biased estimate for σ.

Let a random variable X have a Gamma distribution G with shape parameter α and scale parameter β: $X \sim G(\alpha, \beta)$. Then $E\{X\} = \alpha/\beta$ and $\sigma^2(X) = \alpha/\beta^2$ and

$$p(x) = \frac{x^{\alpha-1}}{\Gamma(\alpha)\beta^\alpha} e^{-x/\beta} \qquad \text{for } x > 0.$$

The expected value of the square root of X is

$$\begin{aligned} E\{\sqrt{X}\} &= \int_0^\infty x^{1/2} \frac{x^{\alpha-1}}{\Gamma(\alpha)\beta^\alpha} e^{-x/\beta}\, dx \\ &= \frac{1}{\Gamma(\alpha)\beta^\alpha} \int_0^\infty x^{\alpha-(1/2)} e^{-x/\beta}\, dx \\ &= \frac{\Gamma\big(\alpha+\frac{1}{2}\big)}{\Gamma(\alpha)} \beta^{1/2}. \end{aligned} \tag{9.52}$$

For $\hat{s}^2$ as defined in (9.50) we have $\hat{s}^2 \sim G\big(\frac{n-1}{2}, \frac{2}{n-1}\sigma^2\big)$. By means of (9.52) we have

$$E\{\hat{s}\} = \frac{\Gamma\big(\frac{n}{2}\big)}{\Gamma\big(\frac{n-1}{2}\big)} \sqrt{\frac{2}{n-1}}\, \sigma$$

and from (9.50) we have $\sqrt{E\{\hat{s}^2\}} = \sigma$. Now we postulate that $E\{\hat{s}\} < \sqrt{E\{\hat{s}^2\}}$ or

$$\frac{\Gamma\big(\frac{n}{2}\big)}{\Gamma\big(\frac{n-1}{2}\big)} < \sqrt{\tfrac{n-1}{2}}.$$

For $n = 3, 4, 5, 6$ the left side is $\sqrt{\pi}/2,\ 2/\sqrt{\pi},\ 3\sqrt{\pi}/4$, and $8/3\sqrt{\pi}$. These values are certainly smaller than $\sqrt{(n-1)/2}$. Thus $\hat{s}$ is a biased estimate of σ. Consequently, $\hat{\sigma}(\widehat{X})$ in (9.51) is biased, too.

By introducing matrix notation we may rewrite some earlier results. Let $\boldsymbol{c}$ be the vector $(c_1, c_2, \ldots, c_n)$ and $\boldsymbol{e} = (1, 1, \ldots, 1)$. Then the mean is

$$\text{Unweighted} \quad \widehat{X} = \frac{\boldsymbol{e}^{\mathrm{T}} \boldsymbol{X}}{n} \quad \text{and weighted} \quad \widehat{X} = \frac{\boldsymbol{c}^{\mathrm{T}} \boldsymbol{X}}{\boldsymbol{c}^{\mathrm{T}} \boldsymbol{e}}. \tag{9.53}$$

The weighted version of (9.48) involves the diagonal matrix $C = \operatorname{diag}(c_1, \ldots, c_n)$:

$$\hat{\sigma}^2 = \frac{(\boldsymbol{X} - \mu \boldsymbol{e})^{\mathrm{T}} C (\boldsymbol{X} - \mu \boldsymbol{e})}{\boldsymbol{c}^{\mathrm{T}} \boldsymbol{e}} = \frac{\boldsymbol{X}^{\mathrm{T}} C \boldsymbol{X}}{\boldsymbol{c}^{\mathrm{T}} \boldsymbol{e}} - 2\mu \frac{\boldsymbol{c}^{\mathrm{T}} \boldsymbol{X}}{\boldsymbol{c}^{\mathrm{T}} \boldsymbol{e}} + \mu^2. \tag{9.54}$$

9.6 Propagation of Means and Covariances

In practice we are often more interested in linear functions of the random variables than the random variables themselves. The most often encountered examples are covered by the linear transformation

$$\boldsymbol{Y} = B\boldsymbol{X} + \boldsymbol{k},$$

where B is a known m by n matrix and $\boldsymbol{k}$ (also known) is m by 1. The linearity of the expectation operator E yields

$$E\{\boldsymbol{Y}\} = E\{B\boldsymbol{X} + \boldsymbol{k}\} = B\, E\{\boldsymbol{X}\} + \boldsymbol{k}. \tag{9.55}$$

Then the covariance for Y comes from the covariance for X. This formula is highly important, and we will use it often!

Law of Covariance Propagation If $Y = BX + \boldsymbol{k}$ and $\boldsymbol{k}$ is known then

$$\Sigma_Y = B\Sigma_X B^{\mathrm{T}}. \tag{9.56}$$

The proof is direct. The shift by $\boldsymbol{k}$ cancels itself, and the matrix B comes outside by linearity:

$$\begin{aligned}\Sigma_Y &= E\{(Y - E\{Y\})(Y - E\{Y\})^{\mathrm{T}}\}\\ &= E\{(BX + \boldsymbol{k} - BE\{X\} - \boldsymbol{k})(BX + \boldsymbol{k} - BE\{X\} - \boldsymbol{k})^{\mathrm{T}}\}\\ &= E\{(BX - BE\{X\})(BX - BE\{X\})^{\mathrm{T}}\}\\ &= BE\{(X - E\{X\})(X - E\{X\})^{\mathrm{T}}\}B^{\mathrm{T}} = B\,\Sigma_X\,B^{\mathrm{T}}.\end{aligned}$$

In geodesy, the law $\Sigma_Y = B\Sigma_X B^{\mathrm{T}}$ appears in many applications. Sometimes B gives a change of coordinates (in this case B is square). For least squares, X contains the n observations and Y contains the m fitted parameters (in this case B is rectangular, with $m > n$). We will frequently quote this ***law of covariance propagation***. In the special case that all X_i are independent, Σ_X is diagonal. Furthermore, if B is 1 by n and hence $Y = b_1X_1 + \cdots + b_nX_n$, one obtains the ***law of error propagation***:

$$\sigma_Y^2 = b_1^2\sigma_1^2 + \cdots + b_n^2\sigma_n^2. \tag{9.57}$$

This gives the variance for a combination of n independent measurements.

Note that if the vector $\boldsymbol{k}$ is also subject to error, independently of X and with mean zero and covariance matrix Σ_k, then this matrix would be added to Σ_Y. The law becomes

$$\Sigma_Y = B\Sigma_X B^{\mathrm{T}} + \Sigma_k.$$

This law is at the heart of the Kalman filter in Chapter 17.

Example 9.6 In leveling it is common practice to attribute the variance σ_0^2 to a leveling observation of length l. According to equation (9.57), in leveling lines of lengths $2l, \ldots, nl$ we have variances $2\sigma_0^2, \ldots, n\sigma_0^2$. Consequently the weights are $1/2, \ldots, 1/n$. For a leveling line observed under these conditions the *weight is reciprocal to the distance*.

Example 9.7 We have two independent random variables x and y with zero mean and variance 1. Then the linear expression $z = ax + by + c$ has a normal distribution with mean c and variance $a^2 + b^2$: $\Sigma_z = [\,a \;\; b\,][\,1\,]\big[\begin{smallmatrix}a\\ b\end{smallmatrix}\big] = a^2 + b^2$.

Example 9.8 A basic example to surveying is the coordinate transformation from polar to Cartesian coordinates:

$$x = r\cos\theta \qquad \text{and} \qquad y = r\sin\theta.$$

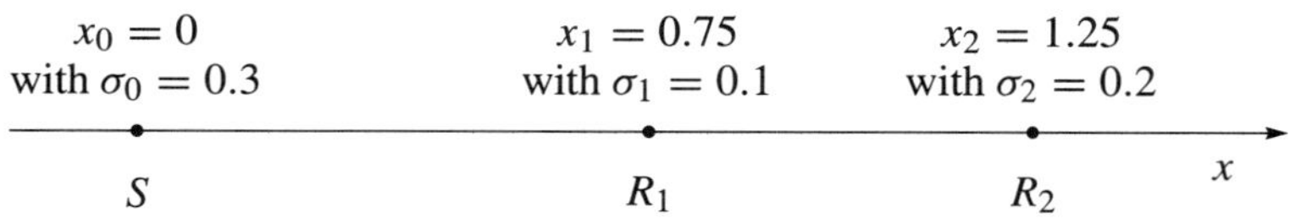

Figure 9.4 Pseudoranges from the satellite S to the receivers R_1 and R_2.

From given values of (r, θ) the coordinates (x, y) are found immediately. For given variances $(\sigma_r, \sigma_\theta)$, the propagation law yields σ_x and σ_y by linearization. The matrix B becomes the ***Jacobian matrix*** J containing the partial derivatives:

$$J = \begin{bmatrix} \partial x/\partial r & \partial x/\partial \theta \\ \partial y/\partial r & \partial y/\partial \theta \end{bmatrix} = \begin{bmatrix} \cos\theta & -r\sin\theta \\ \sin\theta & r\cos\theta \end{bmatrix}.$$

The law of propagation of variances yields the covariance matrix Σ_x for $\boldsymbol{x} = (x, y)$:

$$\Sigma_{\boldsymbol{x}} = J \begin{bmatrix} \sigma_r^2 & 0 \\ 0 & \sigma_\theta^2 \end{bmatrix} J^{\mathrm{T}} = \begin{bmatrix} \sigma_r^2\cos^2\theta + r^2\sigma_\theta^2\sin^2\theta & (\sigma_r^2 - r^2\sigma_\theta^2)\sin\theta\cos\theta \\ (\sigma_r^2 - r^2\sigma_\theta^2)\sin\theta\cos\theta & \sigma_r^2\sin^2\theta + r^2\sigma_\theta^2\cos^2\theta \end{bmatrix}.$$

Insertion of the actual values yields the covariance matrix $\Sigma_{\boldsymbol{x}}$. The x and y coordinates are uncorrelated when $\theta = 0$ or $\theta = \pi/4$. Then we have a polar point determination in the direction of one of the coordinate axes. Also $\sigma_{xy} = 0$ when the special condition $\sigma_r^2 = r^2\sigma_\theta^2$ prevails. Then the standard deviation σ_r of the distance measurement equals the perpendicular error $r\sigma_\theta$. This leads to circular confidence ellipses.

Example 9.9 (Single differences of pseudoranges) One of the simplest and best illustrations of the covariance propagation law is the process of taking differences. The difference of two receiver positions is the baseline vector between them. So we are solving for this difference $(x_2 - x_1)$ rather than the individual coordinates x_1 and x_2. Of course the difference operator is linear, given by a simple 1 by 2 matrix $[\,-1 \;\; 1\,]$. The hope is that the difference is more stable than the separate positions. The covariance matrix should be smaller.

This example is in *one dimension.* The satellite and the two receivers in Figure 9.4 are *collinear,* which never happens in practice. A later example (Section 14.5) will have two satellites and two receivers and work with *double differences.*

Physically, taking the difference cancels the sources of error that are common to both receivers (the satellite clock error and almost all of the ionospheric delay). *This is the basis for Differential GPS.* There are many applications of DGPS, coming in Part III of this book. If one of the receiver positions is well established (the home base), then an accurate baseline gives an accurate position for the second receiver (the rover). Differential GPS allows us to undo the dithering of the satellite clock. Our goal here is to see how the difference (the baseline) can be more accurate than either endpoint—and to calculate its covariance by the propagation law.

We can use the MATLAB command randn(1,100) to produce 100 explicit sample points. Then the covariance matrices can be illustrated by Figure 9.5, and by the very simple M-file oned. We owe the example and also its M-file to Mike Bevis. First come the formulas.

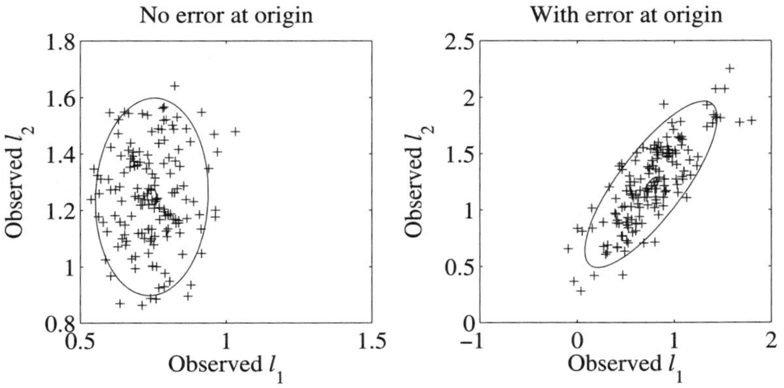

Figure 9.5 The one-dimensional normal distribution function and the 2σ-curve of confidence.

The positions of the satellite and the receivers are $x_0 = 0$, $x_1 = 0.75$, $x_2 = 1.25$. They are subject to independent random errors, normally distributed with mean zero and standard deviations $\sigma_0 = 0.3$, $\sigma_1 = 0.1$, $\sigma_2 = 0.2$. We chose σ_0 largest to make the figures clear. The measurement of x_1 (its pseudorange) has errors from the satellite and that first receiver. A vector of n samples comes from

$$\rho_1 = 0.75 \ * \text{ones}(1, n) + \sigma_1 \ * \ \text{randn}(1, n) - \sigma_0 * \ \text{randn}(1, n).$$

A similar command gives n samples of the pseudorange ρ_2. But since randn produces new numbers every time, we must define del $= \sigma_0 * \text{randn}(1, n)$ and subtract that *same* satellite error from both ρ_1 and ρ_2. This is the whole point(!), that the difference eliminates the satellite error. The M-file achieves this by $\sigma_0 = 0$, or we can work directly with $\rho_2 - \rho_1$.

Now come the covariance matrices. We use the important fact that the mean (the average) of n samples has variance (and covariance) reduced by $1/n$. Thus from the data we compute the scalars

$$m_1 = \text{mean}(\rho_1) \qquad \text{and} \qquad m_2 = \text{mean}(\rho_2)$$

$$V = \text{cov}(\rho_1, \rho_2) \qquad \text{and} \qquad V_m = V/n = \text{covariance of the mean.}$$

The first plot in Figure 9.5 shows the n sample values of (ρ_1, ρ_2), centered about the mean (m_1, m_2). The small inner ellipse is controlled by V_m and the outer ellipse by V. On average 86% of the sample points should lie in that outer ellipse with nsigma set to 2.

The key point is that ρ_1 is correlated with ρ_2. Therefore the *ellipses are tilted*. The same satellite error enters both ρ_1 and ρ_2, with the same sign. The correlation is positive. The errors are large because of $\sigma = 0.3$. The theoretical covariance matrix Σ and the sample covariance matrix V are

$$\Sigma = \begin{bmatrix} \sigma_1^2 & \sigma_{12} \\ \sigma_{12} & \sigma_2^2 \end{bmatrix} \qquad \text{and} \qquad V_{ij} = (\rho_i - m_i)^{\mathrm{T}}(\rho_j - m_j).$$

Now take differences $d = \rho_2 - \rho_1$. The linear transformation is $T = [-1 \;\; 1]$. The propagation law $T\Sigma T^{\mathrm{T}}$ says that the covariance (or just variance) of the scalar d should be

$$\text{covariance} = \Sigma_d = \begin{bmatrix} -1 & +1 \end{bmatrix} \Sigma \begin{bmatrix} -1 \\ +1 \end{bmatrix} = \sigma_1^2 - 2\sigma_{12} + \sigma_2^2.$$

Notice how the effect of $\sigma_0 = 0.3$ has disappeared from Σ_d. For the sample value of the covariance of d we compute around the sample mean:

$$m_d = \mathrm{mean}(d) \qquad \text{and} \qquad V_d = \mathrm{cov}(d) = \sum_1^n \big(d(i) - m_d\big)^2.$$

The MATLAB output compares the predicted Σ_d with the actual V_d.

One more plot is of great interest. It shows the measured distances d_1 and d_2 to the receivers, *without the satellite error* that they share:

$$d_1 = 0.75 * \mathrm{ones}(1, n) + \sigma_1 * \mathrm{randn}(1, n)$$

$$d_2 = 1.25 * \mathrm{ones}(1, n) + \sigma_2 * \mathrm{randn}(1, n).$$

Figure 9.5b shows the n points (d_1, d_2), with the sample mean m_{d_1,d_2} and the sample covariance V_{d_1,d_2}. *The error ellipse is now aligned with the axes.* Taking differences has *decorrelated* the errors in d_1 from the errors in d_2. The theoretical covariance matrix for (d_1, d_2) is *diagonal*:

$$\begin{bmatrix} \sigma_1^2 & 0 \\ 0 & \sigma_2^2 \end{bmatrix}.$$

This example verifies experimentally the propagation law. We used that law for $\Sigma_d = T\Sigma T^{\mathrm{T}}$. The samples of $d_2 - d_1$ give the experimental value of this covariance. The probability is very high that it is close to Σ_d.

As a final curiosity, consider *two* satellites and receivers all on the x axis. For satellites 1 and 2 we can use single differences d^1 and d^2 as above. But the double difference $d^1 - d^2$ from all four measurements is *automatically zero*. The measurements ρ_1^2 and ρ_2^2 to the second satellite both include the distance s between satellites, and everything cancels in the double difference:

$$\begin{aligned} d^1 - d^2 &= \big(\rho_1^1 - \rho_2^1\big) - \big(\rho_1^2 - \rho_2^2\big) \\ &= \big(\rho_1^1 - \rho_2^1\big) - \big(\rho_1^1 + s - \rho_2^1 - s\big) = 0. \end{aligned}$$

In two or three dimensions this will not occur! Length is nonlinear; we have differences of square roots. Satellites and receivers are not collinear. Double differencing is the fundamental tool in Chapter 15 on processing of GPS data.

Example 9.10 (Variance of local east-north-up coordinates from GPS) The relationship between increments of geocentric coordinates (x, y, z) and of the local east-north-up coordinates (e, n, u) is given by latitude and longitude:

$$\begin{bmatrix} e \\ n \\ u \end{bmatrix} = \begin{bmatrix} -\sin\lambda & \cos\lambda & 0 \\ -\sin\varphi\cos\lambda & -\sin\varphi\sin\lambda & \cos\varphi \\ \cos\varphi\cos\lambda & \cos\varphi\sin\lambda & \sin\varphi \end{bmatrix} \begin{bmatrix} x \\ y \\ z \end{bmatrix} = F \begin{bmatrix} x \\ y \\ z \end{bmatrix}. \tag{9.58}$$

We want to compute the covariance matrix for e, n, u. It is based on a given covariance matrix Σ_{xyz} for the geocentric coordinate increments in x, y, z. The law of propagation of variances gives

$$\Sigma_{enu} = F\Sigma_{xyz}F^{\mathrm{T}}. \tag{9.59}$$

A covariance matrix from practice, with units of mm^2, is

$$\Sigma_{xyz} = \begin{bmatrix} 25 & -7.970 & 18.220 \\ -7.970 & 4 & -6.360 \\ 18.220 & -6.360 & 16 \end{bmatrix}.$$

The position is $\varphi = 55^\circ\ 54'$, $\lambda = 12^\circ\ 29'$, so we get

$$F = \begin{bmatrix} -0.216 & 0.976 & 0 \\ -0.808 & -0.179 & 0.561 \\ 0.547 & 0.121 & 0.824 \end{bmatrix}.$$

Hence the covariance matrix for the coordinate increments Δe, Δn, Δn becomes

$$\Sigma_{enu} = \begin{bmatrix} 8.34 & 3.96 & -14.90 \\ 3.96 & 3.95 & -8.24 \\ -14.90 & -8.24 & 32.52 \end{bmatrix}.$$

The standard deviations are $\sigma_e = 2.9$ mm, $\sigma_n = 2.0$ mm and $\sigma_u = 5.7$ mm. These numbers are in good agreement with experience. The vertical deviation σ_u is about twice as large as the in-plane deviations.

9.7 Estimating the Variance of Unit Weight

The least-squares solution of a full rank problem comes from $A^{\mathrm{T}}\Sigma_b^{-1}A\hat{x} = A^{\mathrm{T}}\Sigma_b^{-1}\boldsymbol{b}$. The right side has covariance $A^{\mathrm{T}}\Sigma_b^{-1}\Sigma_b\Sigma_b^{-1}A = A^{\mathrm{T}}\Sigma_b^{-1}A = P$. The left side $P\hat{x}$ has covariance $P\Sigma_{\hat{x}}P$ (again from the propagation law). Therefore $\Sigma_{\hat{x}} = P^{-1}$:

$$\Sigma_{\hat{x}} = (A^{\mathrm{T}}\Sigma_b^{-1}A)^{-1} = \sigma_0^2(A^{\mathrm{T}}CA)^{-1}. \tag{9.60}$$

The covariance $\Sigma_{\hat{x}}$ is the inverse matrix from the normal equations. It is useful to separate the scalar factor σ_0^2 (***the variance of unit weight***) so that $\Sigma_b = \sigma_0^2C^{-1}$. The Cholesky decomposition of $A^{\mathrm{T}}CA$ gives

$$(A^{\mathrm{T}}CA)^{-1} = (R^{\mathrm{T}}R)^{-1} = R^{-1}R^{-\mathrm{T}}. \tag{9.61}$$

The upper triangular R is computed by the MATLAB command R=chol(A' * C * A).

The variance propagation law is valid for all linear functions of the unknowns $\hat{x}$. Each row in the j by n matrix F defines a linear function of $\hat{x}$. By the propagation law the covariance matrix for $F\hat{x}$ is $F(A^{\mathrm{T}}\Sigma_b^{-1}A)^{-1}F^{\mathrm{T}}$. In the special case $f = Ax$, the best estimators are $\hat{f} = A\hat{x} = \hat{p}$. Sometimes they are called the *estimated observations* or *fitted values*. Their covariance matrix is

$$\Sigma_{\hat{p}} = A(A^{\mathrm{T}}\Sigma_b^{-1}A)^{-1}A^{\mathrm{T}}. \tag{9.62}$$

This is the *a posteriori covariance matrix* for the observations. It is the covariance for the projection $\hat{p}$ of b. The *a priori covariance matrix*, of course, is given as $\Sigma_b = \sigma_0^2 C^{-1}$.

If A is n by n, the factors in $(A^{\mathrm{T}}CA)^{-1}$ can be inverted and we get $\Sigma_b = \Sigma_p$; but this is no longer a least-squares problem. The equation $Ax = b$ can be solved exactly.

For assessing gross errors in geodetic practice we use the ***standardized residuals*** $\tilde{r}$:

$$\hat{r} = A\hat{x} - b \qquad \text{and} \qquad \tilde{r} = \big(\mathrm{diag}(\Sigma_{\hat{x}})\big)^{-1/2}\hat{r}. \tag{9.63}$$

This achieves $\sigma^2(\tilde{r}_1) = \cdots = \sigma^2(\tilde{r}_n) = 1$. Björck (1996) uses $\tilde{r} = \mathrm{diag}(A\Sigma_{\hat{x}}A^{\mathrm{T}})$ instead.

In most cases the variance of unit weight σ_0^2 is unknown. Statistical theory gives the following unbiased estimate of σ_0^2 when A is m by n:

$$\hat{\sigma}_0^2 = \frac{\hat{r}^{\mathrm{T}}C\hat{r}}{m-n} = \frac{\|b - A\hat{x}\|^2}{m-n}. \tag{9.64}$$

We shall outline a proof in case $\Sigma_b = \sigma_0^2 I$. If we want a valid proof for $\Sigma_b \neq \sigma_0^2 I$ we have to decorrelate the variables by the transformation $b' = Wb$, $A' = WA$. This is described in Section 11.8 on weight normalization, where $C = W^{\mathrm{T}}W$.

The minimum value for the sum of squares is given by

$$\hat{r}^{\mathrm{T}}\hat{r} = (b - \hat{p})^{\mathrm{T}}(b - \hat{p}) = (b - Pb)^{\mathrm{T}}(b - Pb) = b^{\mathrm{T}}(I - P)(I - P)b = b^{\mathrm{T}}(I - P)b. \tag{9.65}$$

Here we exploited that P and $I - P$ are projections: $P = P^{\mathrm{T}} = P^2$. The expectation $b^{\mathrm{T}}(I - P)b$ is a quadratic form, therefore with a χ^2 distribution. The number of degrees of freedom is the rank of $I - P$. When P projects onto the n-dimensional column space of A, its rank is n and the rank of $I - P$ is $m - n$. Hence

$$b^{\mathrm{T}}(I - P)b \sim \sigma_0^2\chi_{m-n}^2.$$

To prove (9.64) we compute the mean value, using equation (9.10) for $E\{\chi^2\}$:

$$E\big\{b^{\mathrm{T}}(I - P)b\big\} = \sigma_0^2 E\{\chi_{m-n}^2\} = \sigma_0^2(m - n).$$

In words, ***the number of degrees of freedom in a sum of squares is diminished by the number of estimated parameters.*** In 1889 the Danish actuary T. N. Thiele presented a reasoning based on the only assumption that the observations are independent and normally distributed. He found that in the most simple linear model—the canonical model—n observations have *unknown mean values* and the remaining $m - n$ have *known mean values*. According to Thiele estimation in this model is evident. Next he showed that any linear model can be turned into a canonical form by an orthogonal transformation. The estimators in this model, by the inverse transformation, can be expressed by means of the original observations. This fundamental idea of Thiele was not understood by his contemporaries. Only in the 1930's the canonical form of the linear normal model was rediscovered.

A comprehensive historical exposition is given in Seal (1967).

Example 9.11 We apply equation (9.64) to Example 8.1 where $\hat{\boldsymbol{r}}^{\mathrm{T}} C \hat{\boldsymbol{r}} = 0.000\,167$ and $m - n = 5 - 2$:

$$\hat{\sigma}_0^2 = \frac{\hat{\boldsymbol{r}}^{\mathrm{T}} C \hat{\boldsymbol{r}}}{m-n} = \frac{0.000\,167}{3} = 0.000\,055\,\mathrm{m}^2/\mathrm{km} \quad \text{and} \quad \hat{\sigma}_0 = 0.007\,5\,\mathrm{m}/\sqrt{\mathrm{km}}.$$

This variance is valid for the weight $c = 1$. According to definition (9.44) this corresponds to a leveling line of length 1 km. The variance is called the ***variance of unit weight***.

The covariance matrix, in units of m^2, is

$$\Sigma_{\hat{x}} = \hat{\sigma}_0^2 (A^{\mathrm{T}} C A)^{-1} = \begin{bmatrix} 0.214\,3 & 0.077\,9 \\ 0.077\,9 & 0.208\,4 \end{bmatrix} \times 10^{-4}.$$

Finally the variances of the estimated heights H_D and H_E are the diagonal entries of $\Sigma_{\hat{x}}$:

$$\hat{\sigma}_{H_D}^2 = 0.000\,021\,\mathrm{m}^2, \qquad \hat{\sigma}_{H_D} = 0.004\,6\,\mathrm{m};$$
$$\hat{\sigma}_{H_E}^2 = 0.000\,021\,\mathrm{m}^2, \qquad \hat{\sigma}_{H_E} = 0.004\,6\,\mathrm{m}.$$

The standard deviation $\hat{\sigma}_0 = 7\,\mathrm{mm}/\sqrt{\mathrm{km}}$ for 1 kilometer of leveling does characterize the network as a fairly good geometric leveling.

Example 9.12 In Example 8.6 the standard deviation of unit weight is $\hat{\sigma}_0 = \|\hat{\boldsymbol{r}}\|/\sqrt{3}$. This is 0.016 8. The standard deviation of each component of the solution vector $\hat{\boldsymbol{x}}$ is 0.011 9 because the inverse coefficient matrix for the normals is

$$(A^{\mathrm{T}} A)^{-1} = \begin{bmatrix} \frac{1}{2} & -\frac{1}{4} & 0 \\ -\frac{1}{4} & \frac{1}{2} & -\frac{1}{4} \\ 0 & -\frac{1}{4} & \frac{1}{2} \end{bmatrix}.$$

Solving the Normal Equations and Estimating $\hat{\boldsymbol{r}}^{\mathrm{T}} C \hat{\boldsymbol{r}}$

Now we turn to the computational aspects of solving a least-squares problem. This means to solve the normal equations and to find expressions for various covariance matrices and error quantities. Specifically to calculate the weighted sum of squared residuals: $\hat{\boldsymbol{r}}^{\mathrm{T}} C \hat{\boldsymbol{r}}$. The topic is treated in more detail in Chapter 11. Here we restrict ourselves to using two standard methods: the QR factorization and the Cholesky method.

QR Factorization For any m by n matrix A there is an m by m matrix Q with orthonormal columns such that

$$Q^{-1} A = Q^{\mathrm{T}} A = \begin{bmatrix} R \\ 0 \end{bmatrix}. \tag{9.66}$$

The upper triangular matrix R has nonnegative diagonal entries. This ***QR decomposition*** of A is established in Section 4.4; it is the matrix statement of the Gram-Schmidt process. If A has full rank n, then R equals the Cholesky factor of $A^{\mathrm{T}} A$:

$$A^{\mathrm{T}} A = \begin{bmatrix} R^{\mathrm{T}} & 0 \end{bmatrix} Q^{\mathrm{T}} Q \begin{bmatrix} R \\ 0 \end{bmatrix} = R^{\mathrm{T}} R.$$

Table 9.7 Formulas for weighted least squares

Observation equations (A is m by n)	$A\boldsymbol{x} = \boldsymbol{b} - \boldsymbol{r}$
Covariance matrix for observations	Σ_b (notice $\sigma_0^2 = 1$ a priori)
Weight matrix (m by m)	$C = \Sigma_b^{-1}$
Normal equations (n by n)	$A^{\mathrm{T}}CA\hat{\boldsymbol{x}} = A^{\mathrm{T}}C\boldsymbol{b}$
Solution of normal equations	$\hat{\boldsymbol{x}} = (A^{\mathrm{T}}CA)^{-1}A^{\mathrm{T}}C\boldsymbol{b}$
Estimated observations	$\hat{\boldsymbol{p}} = A\hat{\boldsymbol{x}} = A(A^{\mathrm{T}}CA)^{-1}A^{\mathrm{T}}C\boldsymbol{b}$
Estimated residuals	$\hat{\boldsymbol{r}} = \boldsymbol{b} - \hat{\boldsymbol{p}} = \boldsymbol{b} - A\hat{\boldsymbol{x}}$
Estimated weighted sum of squares	$\hat{\boldsymbol{r}}^{\mathrm{T}}C\hat{\boldsymbol{r}} = \boldsymbol{b}^{\mathrm{T}}C\boldsymbol{b} - \boldsymbol{b}^{\mathrm{T}}CA\hat{\boldsymbol{x}}$
Estimated variance of unit weight	$\hat{\sigma}_0^2 = \hat{\boldsymbol{r}}^{\mathrm{T}}C\hat{\boldsymbol{r}}/(m-n)$
Covariance matrix for unknowns	$\Sigma_{\hat{x}} = \hat{\sigma}_0^2(A^{\mathrm{T}}CA)^{-1} = (A^{\mathrm{T}}\Sigma_b^{-1}A)^{-1}$
Covariance matrix for observations	$\Sigma_{A\hat{x}} = \hat{\sigma}_0^2 A(A^{\mathrm{T}}CA)^{-1}A^{\mathrm{T}}$
Covariance matrix for residuals	$\Sigma_{\hat{r}} = \hat{\sigma}_0^2\big(C^{-1} - A(A^{\mathrm{T}}CA)^{-1}A^{\mathrm{T}}\big)$

Since Q is orthogonal it preserves lengths so that $\boldsymbol{c} = Q^{\mathrm{T}}\boldsymbol{b}$ has the same length as $\boldsymbol{b}$:

$$\|\boldsymbol{b} - A\boldsymbol{x}\|^2 = \|Q^{\mathrm{T}}\boldsymbol{b} - Q^{\mathrm{T}}A\boldsymbol{x}\|^2 = \left\|\begin{bmatrix}\boldsymbol{c}_1\\ \boldsymbol{c}_2\end{bmatrix} - \begin{bmatrix}R\\ 0\end{bmatrix}\boldsymbol{x}\right\|^2 = \|\boldsymbol{c}_2\|^2 + \|R\boldsymbol{x} - \boldsymbol{c}_1\|^2. \quad (9.67)$$

The residual norm is minimized by the least-squares solution $\hat{\boldsymbol{x}} = R^{-1}\boldsymbol{c}_1$. The minimum residual equals the norm of $\boldsymbol{c}_2$.

Cholesky Factorization Often the Cholesky factorization applies not to the normal equations $A^{\mathrm{T}}CA$ themselves but rather to a system augmented with a column and a row:

$$\begin{bmatrix} A^{\mathrm{T}}CA & A^{\mathrm{T}}C\boldsymbol{b} \\ (A^{\mathrm{T}}C\boldsymbol{b})^{\mathrm{T}} & \boldsymbol{b}^{\mathrm{T}}C\boldsymbol{b} \end{bmatrix}. \quad (9.68)$$

The lower triangular Cholesky factor of the augmented matrix is

$$\widetilde{L} = \begin{bmatrix} \underset{n \text{ by } n}{L} & \underset{n \text{ by } 1}{\boldsymbol{0}} \\ \underset{1 \text{ by } n}{z^{\mathrm{T}}} & \underset{1 \text{ by } 1}{s} \end{bmatrix}.$$

We look closely at the lower right entry s and compute the product

$$\widetilde{L}\widetilde{L}^{\mathrm{T}} = \begin{bmatrix} LL^{\mathrm{T}} & Lz \\ z^{\mathrm{T}}L^{\mathrm{T}} & z^{\mathrm{T}}z + s^2 \end{bmatrix}.$$

Comparing this product with (9.68) we get

$$z^{\mathrm{T}}z + s^2 = \boldsymbol{b}^{\mathrm{T}}C\boldsymbol{b}. \quad (9.69)$$

The Cholesky factorization finds an n by n lower triangular matrix L such that $LL^{\mathrm{T}} = A^{\mathrm{T}}CA$. Given this "square root" L, the normal equations can be solved quickly and stably via two triangular systems:

$$Lz = A^{\mathrm{T}}C\boldsymbol{b} \qquad \text{and} \qquad L^{\mathrm{T}}\hat{\boldsymbol{x}} = z.$$

These two systems are identical to $A^{\mathrm{T}}CA\hat{\boldsymbol{x}} = A^{\mathrm{T}}C\boldsymbol{b}$. (Multiply the second by L and substitute the first.) From these systems also follows $z = L^{-1}A^{\mathrm{T}}C\boldsymbol{b}$ and consequently

$$z^{\mathrm{T}}z = \boldsymbol{b}^{\mathrm{T}}CAL^{-\mathrm{T}}L^{-1}A^{\mathrm{T}}C\boldsymbol{b} = \boldsymbol{b}^{\mathrm{T}}CA\underbrace{(A^{\mathrm{T}}CA)^{-1}A^{\mathrm{T}}C\boldsymbol{b}}_{\hat{\boldsymbol{x}}}. \tag{9.70}$$

Insertion into (9.69) reveals that $\hat{\boldsymbol{r}}^{\mathrm{T}}C\hat{\boldsymbol{r}}$ can be found directly from s:

$$s^2 = \boldsymbol{b}^{\mathrm{T}}C\boldsymbol{b} - z^{\mathrm{T}}z = \boldsymbol{b}^{\mathrm{T}}C\boldsymbol{b} - \boldsymbol{b}^{\mathrm{T}}CA\hat{\boldsymbol{x}} = -\boldsymbol{b}^{\mathrm{T}}C\hat{\boldsymbol{r}} = \hat{\boldsymbol{r}}^{\mathrm{T}}C\hat{\boldsymbol{r}} - \underbrace{\hat{\boldsymbol{x}}^{\mathrm{T}}A^{\mathrm{T}}C\hat{\boldsymbol{r}}}_{0} = \hat{\boldsymbol{r}}^{\mathrm{T}}C\hat{\boldsymbol{r}}. \tag{9.71}$$

We repeat the contrast between the QR and the Cholesky approach to the normal equations. One works very stably with $A = QR$ (orthogonalization). The other works more simply but a little more dangerously with $A^{\mathrm{T}}CA = LL^{\mathrm{T}}$.

The QR decomposition solves the normals as $\hat{\boldsymbol{x}} = R^{-1}Q^{\mathrm{T}}\boldsymbol{b}$ and $\hat{\boldsymbol{r}}^{\mathrm{T}}C\hat{\boldsymbol{r}} = \|\boldsymbol{c}_2\|^2$. The Cholesky method solves two triangular systems: $Lz = A^{\mathrm{T}}C\boldsymbol{b}$ and $L^{\mathrm{T}}\hat{\boldsymbol{x}} = z$. Then finally $\hat{\boldsymbol{r}}^{\mathrm{T}}C\hat{\boldsymbol{r}} = s^2$. All relevant formulas are surveyed in Table 9.7.

9.8 Confidence Ellipses

Let the estimate $(\widehat{X}_1, \widehat{X}_2)$ be the coordinates of one particular network point. These estimated coordinates have the covariance matrix

$$\Sigma = \begin{bmatrix} \sigma_1^2 & \sigma_{12} \\ \sigma_{21} & \sigma_2^2 \end{bmatrix} = \sigma_0^2 \begin{bmatrix} s_1^2 & s_{12} \\ s_{21} & s_2^2 \end{bmatrix} = \sigma_0^2 Q. \tag{9.72}$$

This 2 by 2 matrix is positive definite; so its inverse exists and is likewise positive definite. We introduce a local coordinate system with origin at $(\widehat{X}_1, \widehat{X}_2)$ and with axes parallel to the original ones. A statistician would write $\boldsymbol{x} \sim N_2(\mathbf{0}, \sigma_0^2 Q)$ which means that (x_1, x_2) has a two-dimensional normal distribution with zero mean and covariance matrix $\sigma_0^2 Q$.

In the local system (x_1, x_2) is a point on the curve described by the ***quadratic form***

$$\boldsymbol{x}^{\mathrm{T}}\Sigma^{-1}\boldsymbol{x} = c^2. \tag{9.73}$$

This curve is an ellipse because Σ^{-1} is positive definite. It is the ***confidence ellipse*** of the point—or ***error ellipse*** if it is conceived as a pure geometric quantity. It corresponds to the M-file errell.

We denote a unit vector in the direction φ by $\boldsymbol{\xi} = (\cos\varphi, \sin\varphi)$. The expression

$$\boldsymbol{\xi}^{\mathrm{T}}\boldsymbol{x} = \cos\varphi\, x_1 + \sin\varphi\, x_2$$

is the ***point error in the direction of*** $\boldsymbol{\xi}$. It is the projection of $\boldsymbol{x}$ in this direction $\boldsymbol{\xi}$.

From the law of error propagation (9.56), the variance of $\boldsymbol{\xi}^{\mathrm{T}}\boldsymbol{x}$ is

$$\sigma^2(\boldsymbol{\xi}^{\mathrm{T}}\boldsymbol{x}) = \boldsymbol{\xi}^{\mathrm{T}}\Sigma\boldsymbol{\xi} = \sigma_1^2\cos^2\varphi + 2\sigma_{12}\cos\varphi\sin\varphi + \sigma_2^2\sin^2\varphi. \tag{9.74}$$

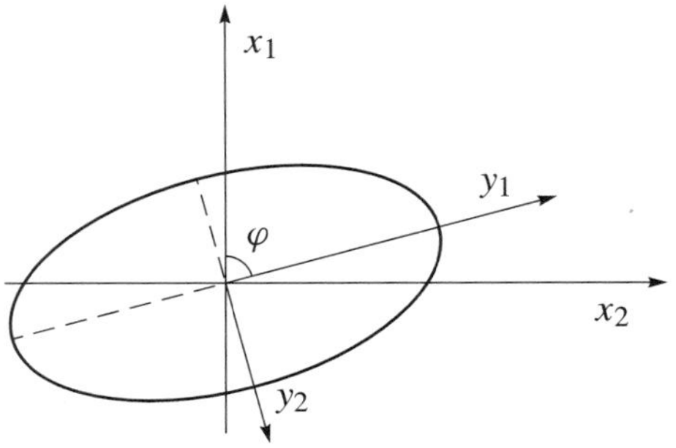

Figure 9.6 The principal axes of the confidence ellipse.

The maximum and minimum values are in the directions of the axes of the confidence ellipse. They are found as solutions to the equation $d\sigma^2/d\varphi = 0$:

$$-(\sigma_1^2 - \sigma_2^2)\sin 2\varphi + 2\sigma_{12}\cos 2\varphi = 0.$$

If $\sigma_1^2 = \sigma_2^2$ then the ellipse is oriented in the direction $\varphi = 45°$ (50 gon). If furthermore $\sigma_{12} = 0$ then the axes are indeterminate and $\Sigma = \sigma^2 I$ and the ellipse is a circle. In all other cases the axis direction φ is determined through

$$\tan 2\varphi = \frac{2\sigma_{12}}{\sigma_1^2 - \sigma_2^2}. \tag{9.75}$$

Now we rotate the x_1x_2 system by this angle φ around the point $(\widehat{X}_1, \widehat{X}_2)$ to a y_1y_2 system. The y_1 axis is collinear with the major axis of the confidence ellipse; the off-diagonal entries of the covariance matrix now vanish; hence y_1 and y_2 are independent. The eigenvectors of Σ are diagonalizing the matrix (and its inverse). Since y_1 is in the direction of the major axis we have $\lambda_1 > \lambda_2$ and the equation of the ellipse is

$$\begin{bmatrix} y_1 & y_2 \end{bmatrix} \begin{bmatrix} \lambda_1 & 0 \\ 0 & \lambda_2 \end{bmatrix}^{-1} \begin{bmatrix} y_1 \\ y_2 \end{bmatrix} = c^2 \tag{9.76}$$

or

$$\frac{y_1^2}{\left(c\sqrt{\lambda_1}\right)^2} + \frac{y_2^2}{\left(c\sqrt{\lambda_2}\right)^2} = 1. \tag{9.77}$$

Notice that λ_1 and λ_2 are the eigenvalues of the inverse matrix Σ^{-1}. They are given by

$$\left.\begin{matrix} \lambda_1 \\ \lambda_2 \end{matrix}\right\} = \tfrac{1}{2}\left(\sigma_1^2 + \sigma_2^2 \pm \sqrt{(\sigma_1^2 + \sigma_2^2)^2 - 4(\sigma_1^2\sigma_2^2 - \sigma_{12}^2)}\right). \tag{9.78}$$

By this we have the explicit expressions for determining the confidence ellipse:

1 The semi major axis is $a = c\sqrt{\lambda_1}$

2 The semi minor axis is $b = c\sqrt{\lambda_2}$

3 The major axis is rotated φ away from the x_1 axis.

The Support Function

The confidence ellipse is fully described. Next we shall demonstrate a geometric interpretation of the variance σ^2 in (9.74) in the direction $\boldsymbol{\xi}$. For convenience we rotate the ellipse to the y_1y_2 principal axes, and use ψ for the angle from the y_1 axis. The equation of the ellipse is

$$\frac{y_1^2}{a^2} + \frac{y_2^2}{b^2} = 1. \tag{9.79}$$

The ***support function*** $p(\psi)$ is the maximum of $y_1 \cos\psi + y_2 \sin\psi$ over the ellipse. This is *not* the distance from $(0,0)$ to the boundary along the ψ-line. If the tangent perpendicular to that direction touches the ellipse at $\boldsymbol{y}_\psi$, then the support $p(\psi)$ measures the projection of $\boldsymbol{y}_\psi$ in the direction ψ.

Figure 9.7 shows the support distance $p(0)$, measured to the vertical tangent perpendicular to $\psi = 0$. It also shows the distance $p(\frac{\pi}{2})$ to the horizontal tangent. All convex sets have support functions p. For an ellipse this is a fourth-order curve and not easily drawable, see Figure 9.7. The M-file support calculates and plots the support function of any ellipse, given the positive definite matrix Σ.

Confidence ellipses close to circular shape are close to their support curves. But for flat ellipses the difference is large except in the four small sectors around the end points.

We can connect $p(\psi)$ to σ^2 in (9.74). The length of the projection of the tangent point $\boldsymbol{y}_\psi = (y_1, y_2)$ is

$$p(\psi) = |y_1 \cos\psi + y_2 \sin\psi|.$$

We square this expression and get

$$p^2(\psi) = y_1^2 \cos^2\psi + 2y_1y_2 \cos\psi \sin\psi + y_2^2 \sin^2\psi. \tag{9.80}$$

The equation for the tangent to the ellipse at $\boldsymbol{y}_\psi$ is

$$-\frac{y_1}{a^2}\sin\psi + \frac{y_2}{b^2}\cos\psi = 0.$$

We square and multiply by a^2b^2:

$$\frac{b^2y_1^2}{a^2}\sin^2\psi + \frac{a^2y_2^2}{b^2}\cos^2\psi - 2y_1y_2\sin\psi\cos\psi = 0.$$

Next we add this to (9.80) in order that the mixed products cancel:

$$\begin{aligned} p^2(\psi) &= \Big(y_1^2\cos^2\psi + \frac{b^2}{a^2}y_1^2\sin^2\psi\Big) + \Big(y_2^2\sin^2\psi + \frac{a^2}{b^2}y_2^2\cos^2\psi\Big) \\ &= \Big(\frac{y_1^2}{a^2} + \frac{y_2^2}{b^2}\Big)\Big(a^2\cos^2\psi + b^2\sin^2\psi\Big). \end{aligned}$$

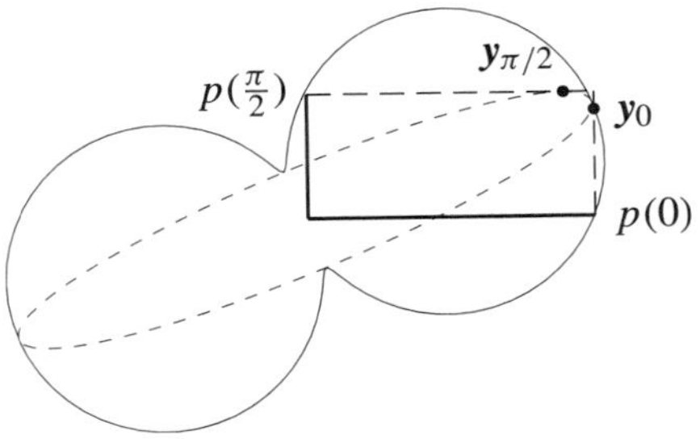

Figure 9.7 The support function at $\varphi = 0$ and $\varphi = \pi/2$: ellipse for $\Sigma = \left[\begin{smallmatrix} 2 & -4 \\ -4 & 10 \end{smallmatrix}\right]$.

Using (9.79) for the special point y_ψ we get

$$p^2(\psi) = a^2 \cos^2 \psi + b^2 \sin^2 \psi = \begin{bmatrix} \cos\psi & \sin\psi \end{bmatrix} \begin{bmatrix} c^2\lambda_1 & 0 \\ 0 & c^2\lambda_2 \end{bmatrix} \begin{bmatrix} \cos\psi \\ \sin\psi \end{bmatrix}.$$

In the original x_1x_2 system (where the angle is φ) this is

$$p^2(\psi) = \begin{bmatrix} \cos\varphi & \sin\varphi \end{bmatrix} \begin{bmatrix} \sigma_1^2 & \sigma_{12} \\ \sigma_{12} & \sigma_2^2 \end{bmatrix} \begin{bmatrix} \cos\varphi \\ \sin\varphi \end{bmatrix}. \tag{9.81}$$

This proves that the support function is actually the standard deviation σ in (9.74).

The preceding considerations were all based on a geometrical view of the quadratic form $x^{\mathrm{T}}\Sigma^{-1}x = $ constant. This view shall now be complemented by a statistical aspect.

The degree of confidence α for the ellipse (9.73) is given by the probability that the ellipse contains the true position (X_1, X_2). There are two situations: the covariance matrix Σ can be known or it may need rescaling by an unknown factor. Either $\sigma_0^2 = 1$ or else the variance is estimated by $\hat{\sigma}_0^2 = r^{\mathrm{T}}\Sigma^{-1}r/(m-n)$.

Known covariance matrix The random variables $y_1/c_\alpha\sqrt{\lambda_1}$ and $y_2/c_\alpha\sqrt{\lambda_2}$ are independent and normally distributed with mean value 0 and variance 1. Statistics tells us that the sum of squares $\left(y_1/c_\alpha\sqrt{\lambda_1}\right)^2 + \left(y_2/c_\alpha\sqrt{\lambda_2}\right)^2$ is χ^2-distributed with 2 degrees of freedom. The probability that

$$\frac{y_1^2}{\left(c_\alpha\sqrt{\lambda_1}\right)^2} + \frac{y_2^2}{\left(c_\alpha\sqrt{\lambda_2}\right)^2} \le 1 \tag{9.82}$$

is $K(c_\alpha^2) = 1 - e^{-c_\alpha^2/2}$, cf. (9.19) and Table 9.6. Specifically this means that if the confidence ellipse is a circle with radius 10 cm, then every 10th sample point falls outside a circle with radius 21.5 cm. Every 20th point falls outside a circle with radius 24.5 cm.

Unknown covariance matrix We insert (9.72) into the quadratic form (9.73) and get

$$x^{\mathrm{T}}\Sigma^{-1}x = \frac{x^{\mathrm{T}}Q^{-1}x}{\hat{\sigma}_0^2}. \tag{9.83}$$

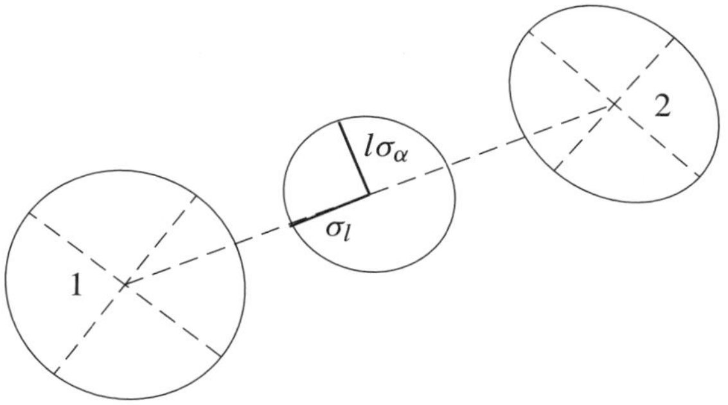

Figure 9.8 Confidence ellipses of points 1 and 2 and their relative confidence ellipse. The figure illustrates the geometrical meaning of σ_l and σ_α.

The numerator and denominator are independent with the distributions for χ_2^2 and χ_{m-n}^2. Hence the left side has the $F_{2,m-n}$ distribution given by (9.20). Note that Q is known a priori.

For $m-n$ sufficiently large, say 50, we can act as if $\hat{\sigma}_0^2$ were known; see Abramowitz & Stegun (1972), Table 26.9. We can choose α and c_α^2 such that

$$\text{Prob}\big(\boldsymbol{x}^{\mathrm{T}}\Sigma^{-1}\boldsymbol{x}/2 \le c_\alpha^2\big) = \text{Prob}\big(F_{2,m-n} \le c_\alpha^2\big) = \alpha,$$

This describes the ellipse in which $\boldsymbol{x}$ must lie with the prescribed probability α.

Relative Confidence Ellipses

A very important step in geodesy and GPS is to work with differences. We saw a one-dimensional example in Section 9.1 and in the M-file oned. The separate errors from a satellite to two receivers partly disappeared in the difference $x_2 - x_1$. The large tilted error ellipse (with off-diagonal correlation from the shared satellite error) became a smaller upright ellipse in Figure 9.5 from a diagonal covariance matrix. This is the ***relative*** confidence ellipse. We now extend this idea to two dimensions.

Points 1 and 2 have coordinates (x_1, y_1) and (x_2, y_2). The vector $\boldsymbol{d}$ between the points has components

$$dx = x_2 - x_1 \qquad \text{and} \qquad dy = y_2 - y_1.$$

This vector comes from multiplying the full set of four unknowns x_1, y_1, x_2, y_2 by the matrix

$$L = \begin{bmatrix} -1 & 0 & +1 & 0 \\ 0 & -1 & 0 & +1 \end{bmatrix} = \begin{bmatrix} -I & I \end{bmatrix}.$$

Then the propagation law says that the covariance matrix of the vector $\boldsymbol{d} = (dx, dy)$ is $\Sigma_d = L\Sigma L^{\mathrm{T}}$:

$$\Sigma_d = \begin{bmatrix} -I & I \end{bmatrix} \begin{bmatrix} \Sigma_1 & \Sigma_{12} \\ \Sigma_{12}^{\mathrm{T}} & \Sigma_2 \end{bmatrix} \begin{bmatrix} -I \\ I \end{bmatrix} = \begin{bmatrix} \Sigma_1 - \Sigma_{12} - \Sigma_{12}^{\mathrm{T}} + \Sigma_2 \end{bmatrix}. \tag{9.84}$$

Each of these blocks is 2 by 2. The diagonal blocks Σ_1 and Σ_2 contain covariances for point 1 and point 2 separately. They correspond to ellipses 1 and 2 in Figure 9.8, and the matrices are

$$\Sigma_1 = \begin{bmatrix} \sigma_{x_1}^2 & \sigma_{x_1 y_1} \\ \sigma_{x_1 y_1} & \sigma_{y_1}^2 \end{bmatrix} \quad \text{and} \quad \Sigma_2 = \begin{bmatrix} \sigma_{x_2}^2 & \sigma_{x_2 y_2} \\ \sigma_{x_2 y_2} & \sigma_{y_2}^2 \end{bmatrix}.$$

The off-diagonal blocks Σ_{12} and Σ_{12}^{T} in the full 4 by 4 symmetric matrix Σ contain covariances $\sigma_{x_1 x_2}, \sigma_{x_1 y_2}, \sigma_{x_2 y_1}, \sigma_{x_2 y_2}$ ***between*** points 1 and 2. Then the combination (9.84) of the four blocks is the relative covariance matrix Σ_d:

$$\Sigma_d = \begin{bmatrix} \sigma_{x_1}^2 - 2\sigma_{x_1 x_2} + \sigma_{x_2}^2 & \sigma_{x_1 y_1} - \sigma_{x_1 y_2} - \sigma_{x_2 y_1} + \sigma_{x_2 y_2} \\ \sigma_{x_1 y_1} - \sigma_{x_1 y_2} - \sigma_{x_2 y_1} + \sigma_{x_2 y_2} & \sigma_{y_1}^2 - 2\sigma_{y_1 y_2} + \sigma_{y_2}^2 \end{bmatrix}. \tag{9.85}$$

This covariance matrix Σ_d produces a ***relative confidence ellipse*** with the following remarkable and simple properties. The standard deviation σ_l of the length of the vector, $l^2 = (dx)^2 + (dy)^2$, can be found in Figure 9.8. The angle α has tangent dy/dx. The quantities σ_l and $l\sigma_\alpha$ are determined by the support function in the vector direction and the perpendicular direction. This geometrical construction makes the relative confidence ellipse useful. Of course points 1 and 2 can be any two points in the network.

The relative ellipse is often smaller than the separate ellipses 1 and 2. Typically this is the case when the covariances $\sigma_{x_1 x_2}$ and $\sigma_{y_1 y_2}$ are positive. Then errors common to the two points are removed in their difference $\boldsymbol{d} = (x_2 - x_1, y_2 - y_1)$.

The M-files relellip and ellaxes compute the relative confidence ellipse.

10

NONLINEAR PROBLEMS

10.1 Getting Around Nonlinearity

This section introduces two new items: a plane coordinate system in which many geodetic computations take place, and the concept of nonlinearity which we remedy by applying the process of linearization. The plane coordinate system involves *two* unknowns for the position of each nodal point. Leveling networks involved only one unknown, the height.

Basically, geodesy is about computing coordinates of selected points on or close to the surface of the Earth. This goal is achieved by making appropriate observations between these points. The most common forms are slope distances, zenith distances and horizontal directions. We take more observations than necessary. Here the principle of least squares enters the scene. It yields a unique result. Computing variances also furnishes us with statements about the accuracy.

The accuracy measures are mostly for the benefit of the geodesist and the coordinates are the product delivered to the eventual customer. After each discussion of the problem and its linearization and its solution, we will add an important paragraph about variances.

From ancient times it has been a tradition to characterize the position of a point by its latitude φ and longitude λ. We have inherited this tradition and even today most geodetic xy coordinate systems are oriented with x axis positive to the North, y axis positive to the East. Bearings α are reckoned ***counter clockwise*** from the positive x axis, see Figure 10.1. So in the usual rotation matrices we must substitute the rotational angle θ by $-\theta$.

Until now all observation equations have been linear. They are characterized by having constant entries in the matrix A. Only very few observation equations are that simple. Even a distance observation results in a nonlinear equation:

$$f_{i,\text{obsv}} = \sqrt{(X_k - X_j)^2 + (Y_k - Y_j)^2} + r_i. \tag{10.1}$$

Here and in the following we denote the coordinates of points by capitals while *small x's and y's denote coordinate increments*.

The principle of least squares leads directly to the normal equations in case of linear observation equations. Nonlinear observation equations first have to be linearized. This can only happen if we know *preliminary values* for the unknowns and subsequently solve

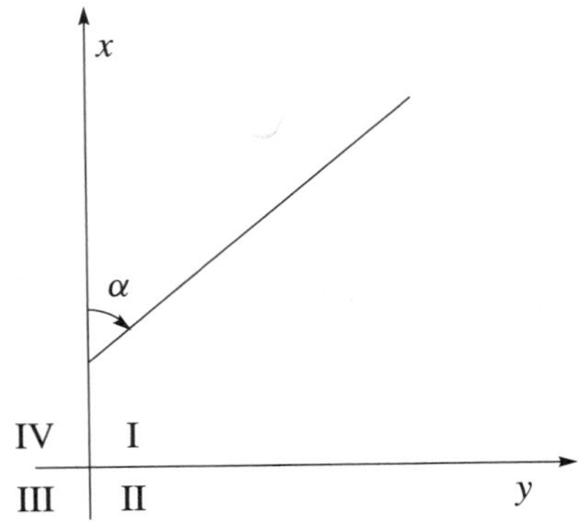

Figure 10.1 The coordinate system.

for the coordinate *increments*. When those increments are small, the linearization is justified. The procedure is to be repeated—to be ***iterated***. The estimates hopefully move the solution(point) closer and closer to the true solution. More about uniqueness in Section 10.1.

Thus we start by linearizing around the basepoint $\boldsymbol{X}^0 = (X_j^0, X_k^0, Y_j^0, Y_k^0)$. We assume that these ***preliminary values*** are known. They will be denoted by the superscript ^{0}. In vector form the linearized observation equation is

$$f(\boldsymbol{X}) = f(\boldsymbol{X}^0) + \left[\boldsymbol{X} - \boldsymbol{X}^0\right]^{\mathrm{T}} \operatorname{grad} f \,\big|_{X=X^0} + \text{second order terms.} \tag{10.2}$$

The ***increments*** of the coordinates are $\boldsymbol{X} - \boldsymbol{X}^0 = (x_j, x_k, y_j, y_k)$. In the calculations we restrict ourselves to a first order approximation (a linearization):

$$f(\boldsymbol{X}) - f(\boldsymbol{X}^0) = \left[\boldsymbol{X} - \boldsymbol{X}^0\right]^{\mathrm{T}} \operatorname{grad} f \,\big|_{X=X^0}\,. \tag{10.3}$$

The gradient of a function has the geometric meaning, independent of the coordinate system in use, of giving the derivative of f in all directions. Equation (10.3) can be written

$$\text{change in } f = (\text{change in position}) \times (\text{gradient of } f). \tag{10.4}$$

In Figure 10.2 we show the meaning of this equation when f is a function of one or more variables.

Now (10.2) can be written as

$$f_i = \sqrt{(X_k^0 - X_j^0)^2 + (Y_k^0 - Y_j^0)^2} + \begin{bmatrix} x_j & x_k & y_j & y_k \end{bmatrix} \begin{bmatrix} -\cos\alpha_{jk}^0 \\ \cos\alpha_{jk}^0 \\ -\sin\alpha_{jk}^0 \\ \sin\alpha_{jk}^0 \end{bmatrix} + r_i. \tag{10.5}$$

A typical derivative of the square root is

$$\frac{\partial f_i}{\partial Y_k}\bigg|_{X=X^0} = \tfrac{1}{2} \cdot \frac{2(Y_k - Y_j)}{f_i}\bigg|_{X=X^0} = \frac{Y_k^0 - Y_j^0}{f_i^0} = \sin\alpha_{jk}^0.$$

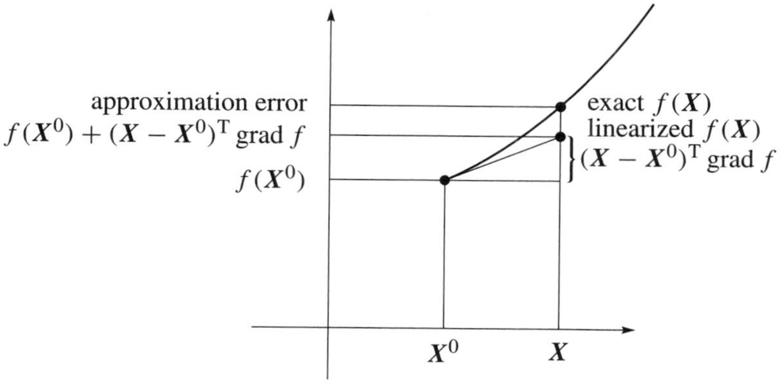

Figure 10.2 The geometrical meaning of the gradient and the linearization process. The figure can be read literally as if $\boldsymbol{X}$ is a single variable (it is actually multi-dimensional).

This is the last component in (10.5), and f_i^0 denotes $\sqrt{(X_k^0 - X_j^0)^2 + (Y_k^0 - Y_j^0)^2}$.

The unknowns (x_j, x_k, y_j, y_k) remain in the same sequence as indicated in $\boldsymbol{x}$. Equation (10.5) must be arranged according to the pattern in the linear observation equation $A\boldsymbol{x} = \boldsymbol{b} - \boldsymbol{r}$:

$$\begin{bmatrix} \cdots & -\cos\alpha_{jk}^0 & \cos\alpha_{jk}^0 & -\sin\alpha_{jk}^0 & \sin\alpha_{jk}^0 & \cdots \end{bmatrix} \begin{bmatrix} \vdots \\ x_j \\ x_k \\ y_j \\ y_k \\ \vdots \end{bmatrix} = b_i - r_i, \tag{10.6}$$

where $b_i = f_i - f_i^0$. Thus we have described the ith row of the system $A\boldsymbol{x} = \boldsymbol{b} - \boldsymbol{r}$.

Until now we have presupposed that the unknowns really represent unknown coordinates. If one or more coordinates are to be kept fixed, we omit the columns in A corresponding to those unknowns. They never obtain any increment. Likewise these unknowns are deleted from $\boldsymbol{x}$. This procedure is quite analogous to the one described in Section 8.3.

Example 10.1 (Distance measurement) We observe 3 distances in order to locate point P:

$$f_1 = 100.01 \quad \text{and} \quad f_2 = 100.02 \quad \text{and} \quad f_3 = 100.03$$

with weights $C = \operatorname{diag}(1, 1, 1) = I$. Figure 10.3 indicates all preliminary values for the coordinates. As preliminary coordinates for P we use $(X_P^0, Y_P^0) = (170.71, 170.71)$. These values are calculated using simple geometrical relations.

The first observation equation is

$$\begin{bmatrix} -\cos\alpha_{P,001}^0 & -\sin\alpha_{P,001}^0 \end{bmatrix} \begin{bmatrix} x_P \\ y_P \end{bmatrix} = f_1 - \sqrt{(X_{001}^0 - X_P^0)^2 + (Y_{001}^0 - Y_P^0)^2} - r_1$$

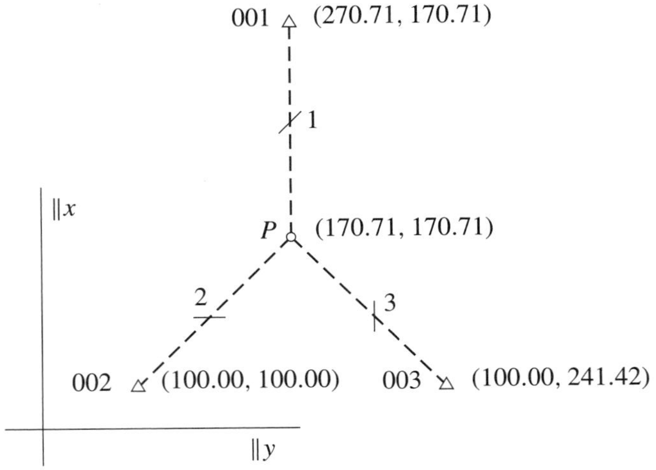

Figure 10.3 Determining the coordinates of a point by resection.

or

$$\begin{bmatrix} -1 & 0 \end{bmatrix}\begin{bmatrix} x_P \\ y_P \end{bmatrix} = f_1 - 100.000 - r_1.$$

Similarly the second observation equation is

$$\begin{bmatrix} 0.707\,1 & 0.707\,1 \end{bmatrix}\begin{bmatrix} x_P \\ y_P \end{bmatrix} = f_2 - 99.999 - r_2,$$

and finally the third observation equation

$$\begin{bmatrix} 0.707\,1 & -0.707\,1 \end{bmatrix}\begin{bmatrix} x_P \\ y_P \end{bmatrix} = f_3 - 99.999 - r_3.$$

Now all three equations together

$$\begin{bmatrix} -1 & 0 \\ 0.707\,1 & 0.707\,1 \\ 0.707\,1 & -0.707\,1 \end{bmatrix}\begin{bmatrix} x_P \\ y_P \end{bmatrix} = \begin{bmatrix} 100.01 - 100.000 \\ 100.02 - 99.999 \\ 100.03 - 99.999 \end{bmatrix} - \begin{bmatrix} r_1 \\ r_2 \\ r_3 \end{bmatrix}.$$

The normal equations $A^{\mathrm{T}}A\hat{\boldsymbol{x}} = A^{\mathrm{T}}\boldsymbol{b}$ are

$$\begin{bmatrix} 2 & 0 \\ 0 & 1 \end{bmatrix}\begin{bmatrix} \hat{x}_P \\ \hat{y}_P \end{bmatrix} = \begin{bmatrix} 0.026\,7 \\ -0.007\,1 \end{bmatrix}.$$

As the coefficient matrix became diagonal (why is $A^{\mathrm{T}}A$ diagonal in this case?) the solution is easily found as $(\hat{x}_P, \hat{y}_P) = (0.013\,4, -0.007\,1)$. The final coordinates are $(\widehat{X}_P, \widehat{Y}_P) = (X_P^0 + \hat{x}_P, Y_P^0 + \hat{y}_P) = (170.723\,\mathrm{m}, 170.703\,\mathrm{m})$.

In this simple example the residuals are most easily calculated as

$$\hat{\boldsymbol{r}} = \boldsymbol{b} - A\hat{\boldsymbol{x}} = \boldsymbol{b} - \boldsymbol{p} = \begin{bmatrix} 0.01 \\ 0.02 \\ 0.03 \end{bmatrix} - \begin{bmatrix} -0.0127 \\ 0.0040 \\ 0.0140 \end{bmatrix} = \begin{bmatrix} 0.0234 \\ 0.0165 \\ 0.0165 \end{bmatrix}.$$

Finally $\hat{\sigma}_0 = \sqrt{\hat{\boldsymbol{r}}^{\mathrm{T}}\hat{\boldsymbol{r}}/(3-2)} = 0.033\,\mathrm{m}$. The covariance matrix is

$$\Sigma_{\hat{x}} = \hat{\sigma}_0^2(A^{\mathrm{T}}A)^{-1} = \begin{bmatrix} 0.000\,514 & 0 \\ 0 & 0.001\,029 \end{bmatrix}$$

and the standard deviations on the coordinates are $\sigma_{\widehat{X}_P} = \sqrt{0.000\,514} = 0.023\,\mathrm{m}$ and $\sigma_{\widehat{Y}_P} = \sqrt{0.001\,029} = 0.033\,\mathrm{m}$.

If we had ***chosen another basepoint***—changing Y_P^0 to 170.69, say—we would have had the following calculations:

$$\boldsymbol{b} = \begin{bmatrix} 0.010 \\ 0.035 \\ 0.017 \end{bmatrix} \qquad \text{and} \qquad A^{\mathrm{T}}\boldsymbol{b} = \begin{bmatrix} 0.0267 \\ 0.0129 \end{bmatrix},$$

whereupon $(\hat{x}_P, \hat{y}_P) = (0.013, 0.013)$ and finally $(\widehat{X}_P, \widehat{Y}_P) = (170.723, 170.703)$ i.e. the same result as earlier.

The error calculation runs as follows

$$\hat{\boldsymbol{r}} = \begin{bmatrix} 0.010 \\ 0.035 \\ 0.017 \end{bmatrix} - A\begin{bmatrix} 0.013 \\ 0.013 \end{bmatrix} = \begin{bmatrix} 0.023 \\ 0.017 \\ 0.017 \end{bmatrix}.$$

The sum of squares is $\hat{\boldsymbol{r}}^{\mathrm{T}}\hat{\boldsymbol{r}} = 0.001\,094$ and the standard deviation for the unit weight is $\hat{\sigma}_0 = 0.033\,\mathrm{m}$.

The numerical differences in this example come into existence because we use two different basepoints and also because of common ***rounding errors***. The former can be handled by using a reasonable procedure in which the basepoint is calculated again after every iteration; this is discussed in the next section.

Iterative Improvements

The general least-squares procedure is as follows: If necessary the observation equations are linearized. As a result we have the gradient matrix A and the right side vector $\boldsymbol{b}$. Moreover the weights are given by the matrix C. Next the normal equations are solved. Their solution $\hat{\boldsymbol{x}}^{(0)} = \hat{\boldsymbol{x}}$ is added to the vector of preliminary values $\boldsymbol{X}^0$: $\hat{\boldsymbol{x}}^{(1)} = \boldsymbol{X}^{(0)} + \hat{\boldsymbol{x}}^{(0)}$. The observation and normal equations are repeated but now with $\hat{\boldsymbol{x}}^{(1)}$ as basepoint. Thereupon the solution of the normals is $\hat{\boldsymbol{x}}^{(1)}$. This is added to $\hat{\boldsymbol{x}}^{(1)}$ and we get $\hat{\boldsymbol{x}}^{(2)} = \hat{\boldsymbol{x}}^{(1)} + \hat{\boldsymbol{x}}^{(1)}$. It is not required that the gradient matrix A is known very accurately. In fact one may use the same A in several iterations.

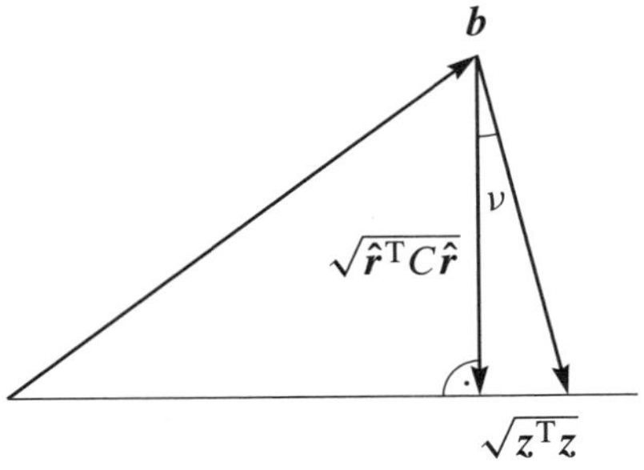

Figure 10.4 The exponent of numerical precision ν.

It is absolutely essential that the residuals r be computed with a higher precision than that of the rest of the computation. This is a general principle in all equation solving: Calculation of the residual is the critical computation and must be done with the most accuracy. It is rarely necessary to achieve such high accuracy in any other part of the algorithm, see Forsythe & Moler (1967), Section 13.

Note the useful relation between the error ϵ in x and the residual r:

$$\epsilon = A^{-1}b - x = A^{-1}(b - Ax) = A^{-1}r$$

or in norms

$$\|\epsilon\| \leq \|A^{-1}\| \|r\|. \tag{10.7}$$

So far we have not discussed how to obtain the first guess $X^{(0)}$. The geodesist knows best how to find a close first guess. In the geodetic terminology this is named the computation of ***preliminary coordinates***.

Usually the error is squared in each iteration until we reach the level of the rounding errors. But what happens if $X^{(0)}$ is far from the solution? The answer depends on f. Computational experience shows that $\hat{x}^{(k)}$ often will get close to a solution X, but it will move around this point in a certain distance and in many iteration steps. Sometimes it is impossible to guess which solution we finally will obtain.

Finally there are some instances in which $\hat{x}^{(k)}$ does not converge at all. Instead the solution will reach certain limit points. This happens in case of gross errors in the data.

How can we be sure that the solution is not spoiled by the effect of nonlinearity? If we start the procedure by using good preliminary values for the unknowns and continue the iterative process until the right side of the normal equations is sufficiently small then we can be rather sure that we have found a satisfactory solution of the least-squares problem.

Since 1967 geodesists at the former Danish Geodetic Institute have computed a quantity ν based on the Cholesky reduction and defined as

$$\nu = -\log \sqrt{\frac{z^{\mathrm{T}} z}{\hat{r}^{\mathrm{T}} C \hat{r}}}, \tag{10.8}$$

where z is the right side of the Cholesky reduced normal equations, see (9.70); ν is called the ***exponent of numerical precision*** and it is an estimate for the number of numerically significant digits in the least-squares problem. Geometrically ν is a measure for the angle at $\boldsymbol{b}$ under which the distance between the computed and the true solution is viewed, see Krarup (1982). This angle cannot be calculated, but ν can (see Figure 10.4).

10.2 Geodetic Observation Equations

Distance

Section 10.1 showed in all detail how to linearize an observation of distance. But computationally it can be an advantage to modify the distance observation s_i by dividing it by the value s_i^0 which is computed from the preliminary coordinates. Then the entries in A stemming from a distance observation become identical to those originating from an observation of a direction.

Hence the observation is the measured distance s_i divided by the distance s_i^0 calculated from the preliminary coordinates:

$$(s_i^0)^2 = (X_k^0 - X_j^0)^2 + (Y_k^0 - Y_j^0)^2.$$

By calculating the gradient of s_i/s_i^0, inserting the preliminary values, ***choosing*** the sequence of the unknowns as (x_j, x_k, y_j, y_k), the gradient becomes

$$\operatorname{grad}\frac{s_i}{s_i^0} = \Bigl(-\frac{X_k - X_j}{s_i s_i^0}, \frac{X_k - X_j}{s_i s_i^0}, -\frac{Y_k - Y_j}{s_i s_i^0}, \frac{Y_k - Y_j}{s_i s_i^0}\Bigr). \tag{10.9}$$

By use of the preliminary coordinates the linearized observation equation then is

$$\begin{bmatrix} \cdots & -\dfrac{\cos\alpha_{jk}^0}{s_i^0} & \dfrac{\cos\alpha_{jk}^0}{s_i^0} & -\dfrac{\sin\alpha_{jk}^0}{s_i^0} & \dfrac{\sin\alpha_{jk}^0}{s_i^0} & \cdots \end{bmatrix} \begin{bmatrix} \vdots \\ x_j \\ x_k \\ y_j \\ y_k \\ \vdots \end{bmatrix} = \frac{s_i - s_i^0}{s_i^0} - r_i = b_i - r_i. \tag{10.10}$$

The division with s_{jk}^0 formally can be looked upon as the result of ***observing the logarithm of*** s_{jk}, rather than s_{jk} itself. The subsequent differentiation of $\ln s_{jk}$ automatically leads to division by s_{jk}.

The a priori variance of one single observation of distance is

$$\sigma_s^2 = \sigma_a^2 + \sigma_c^2. \tag{10.11}$$

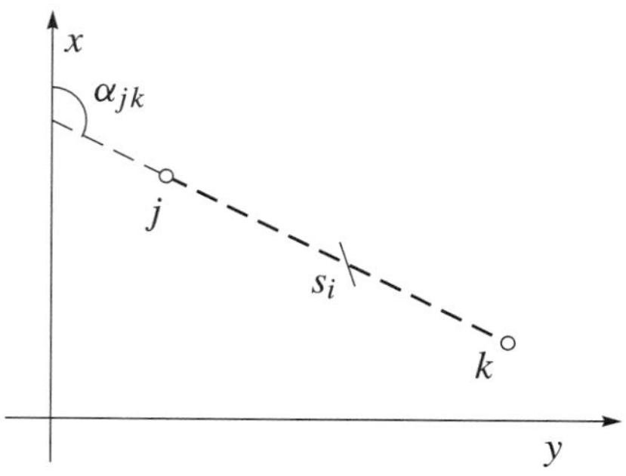

Figure 10.5 Observation of distance.

Here σ_a^2 is the variance of the distance measurement, and σ_c^2 is the collected variance of centricity of both instrument and reflector.

If there are n observations, the variance becomes

$$\sigma_s^2 = \frac{\sigma_a^2 + \sigma_c^2}{n} \tag{10.12}$$

following the law of error propagation. It has been argued that the centricity contribution shall be divided by n. To our knowledge the expression shown is in good accordance with common sense and practical experience.

Finally the ***weight for an observation of distance*** is set to $c_s = \sigma_0^2/\sigma_s^2$ where σ_0^2 denotes the variance of unit weight.

Example 10.2 (Observation of distance—revisited) In Example 10.1 we now modify the observed distance by dividing it by the value resulting from the preliminary coordinates. Thus we use observation equations of the type (10.10). The weight matrix is set to $C \equiv I$ and the preliminary coordinates of P are fixed to $(X_P^0, Y_P^0) = (170.00, 170.00)$.

The first observation equation looks like

$$\begin{bmatrix} -\dfrac{\cos\alpha^0_{P,001}}{s_1^0} & -\dfrac{\sin\alpha^0_{P,001}}{s_1^0} \end{bmatrix} \begin{bmatrix} x_P \\ y_P \end{bmatrix} = \frac{s_1 - s_1^0}{s_1^0} - r_1$$

or

$$\begin{bmatrix} -0.0099 & -0.0001 \end{bmatrix} \begin{bmatrix} x_P \\ y_P \end{bmatrix} = -0.0070 - r_1.$$

The second and third observation equations are similar. The three observation equations together are

$$\begin{bmatrix} -0.0099 & -0.0001 \\ 0.0071 & 0.0071 \\ 0.0070 & -0.0070 \end{bmatrix} \begin{bmatrix} x_P \\ y_P \end{bmatrix} = \begin{bmatrix} -0.0070 \\ 0.0103 \\ 0.0003 \end{bmatrix} - \begin{bmatrix} r_1 \\ r_2 \\ r_3 \end{bmatrix}.$$

The normal equations are formed and solved; the result is $(\hat{x}_P, \hat{y}_P) = (0.724, 0.699)$ or $(\widehat{X}_P, \widehat{Y}_P) = (X_P^0 + \hat{x}_P, Y_P^0 + \hat{y}_P) = (170.724\,\text{m}, 170.699\,\text{m})$.

The present solution deviates 1 and 4 mm from the one found in Example 10.1. So we compute a second iteration with $(X_P^0, Y_P^0) = (170.724\,\text{m}, 170.699\,\text{m})$ as basepoint. Here are the observation equations:

$$\begin{bmatrix} -0.0100 & 0.0000 \\ 0.00707 & 0.00707 \\ 0.00707 & -0.00707 \end{bmatrix} \begin{bmatrix} x_P \\ y_P \end{bmatrix} = \begin{bmatrix} 0.00024 \\ 0.00019 \\ 0.00013 \end{bmatrix} - \begin{bmatrix} r_1 \\ r_2 \\ r_3 \end{bmatrix}.$$

The solution now is $(\hat{x}_P, \hat{y}_P) = (-0.0006, 0.0039)$ or $(\widehat{X}_P, \widehat{Y}_P) = (X_P^0 + \hat{x}_P, Y_P^0 + \hat{y}_P) = (170.723\,\text{m}, 170.703\,\text{m})$. This result agrees well with Example 10.1.

One conclusion is that preliminary coordinates deviating up to 1 m from the final ones may result in changes of coordinates at mm level.

Quasi-Distance

When using electronic equipment for distance measurements it is difficult to avoid introducing systematic errors for various reasons: unknown frequency and wrong index of refraction, inadequate height reduction, incorrect map scale correction, a control network with systematically wrong scale, et cetera. For one or more of these reasons it is reasonable to introduce a change of scale μ in the distance observation as described by the following equation

$$s_i = (1 + \mu)\sqrt{(X_k - X_j)^2 + (Y_k - Y_j)^2}. \tag{10.13}$$

After having chosen the sequence of unknowns as $(x_j, x_k, y_j, y_k, \mu)$ the gradient of the left side is

$$\operatorname{grad} s_i = \Big(-(1 + \mu)\frac{X_k - X_j}{s_i},\ (1 + \mu)\frac{X_k - X_j}{s_i},\ -(1 + \mu)\frac{Y_k - Y_j}{s_i},\ (1 + \mu)\frac{Y_k - Y_j}{s_i},\ s_i \Big).$$

After inserting the preliminary values, the linearized observation equations are

$$\begin{bmatrix} \cdots & -(1+\mu)\cos\alpha_{jk}^0 & (1+\mu)\cos\alpha_{jk}^0 & -(1+\mu)\sin\alpha_{jk}^0 & (1+\mu)\sin\alpha_{jk}^0 & s_i^0 & \cdots \end{bmatrix} \begin{bmatrix} \vdots \\ x_j \\ x_k \\ y_j \\ y_k \\ \mu \\ \vdots \end{bmatrix} = b_i - r_i, \tag{10.14}$$

where $b_i = s_i - s_i^0$. The determination of weights occurs according to the variance given in (10.12).

Pseudodistance

Distance observations also can be influenced by an error which is independent of the distance itself. Such a distance is known as a ***pseudodistance***. It fits the distance calculated from the coordinates of the terminal points except for an additive constant Z. In the realm of electronic distance measurements it is called an ***addition constant*** or today more often a ***reflector constant***.

The observation is the distance between the points j and k plus the reflector constant:

$$s_{jk} = \sqrt{(X_k - X_j)^2 + (Y_k - Y_j)^2} + Z. \tag{10.15}$$

Let us choose the sequence of unknowns as (x_j, x_k, y_j, y_k, z). Then the gradient of the left side in (10.15) becomes

$$\operatorname{grad} s_{jk} = \Big(-\frac{X_k - X_j}{s_{jk}}, \frac{X_k - X_j}{s_{jk}}, -\frac{Y_k - Y_j}{s_{jk}}, \frac{Y_k - Y_j}{s_{jk}}, +1\Big). \tag{10.16}$$

The linearized observation equation is

$$\begin{bmatrix} -\cos\alpha_{jk}^0 & \cos\alpha_{jk}^0 & -\sin\alpha_{jk}^0 & \sin\alpha_{jk}^0 & +1 & \cdots \end{bmatrix} \begin{bmatrix} x_j \\ x_k \\ y_j \\ y_k \\ z \\ \vdots \end{bmatrix} = s_{jk} - s_{jk}^0 - r_i = b_i - r_i.$$

The weight of the observation is given by the expression (10.12).

It is advantageous to divide the observation equation by s_{jk}^0 so it becomes identical to (10.10), apart from the constant Z.

The constant Z, which is a characteristic for a given distance measuring instrument and reflector, is usually best determined through a least-squares procedure for a whole control network. All distance observations are entered as pseudodistances. This procedure is strongly recommended for small networks especially as encountered in connection with deformation measurements.

If one tries to determine Z in combination with measurements of a control network there will be too much common measurement noise to do it safely and the determination thus becomes too uncertain for any purpose. An alternative is to determine Z by measuring the distances between 3 points: 1, 2, and 3 on a straight line. Now Z can be found as

$$\widehat{Z} = s_{13} - (s_{12} + s_{23}).$$

Modern instruments have Z-values of a few mm. Since the standard deviations of average distances in control networks are greater, one might often neglect the constant Z.

Ratios of Distances

You may take the view that the precise unit of length is unknown. This is due to lack of information about the frequency of the carrier or the index of refraction n or the unit length in an existing network which our new network has to be merged into.

This view implies that only ***ratios between distances with a common terminal*** can be estimated. A ratio of distances only contributes to determination of the shape of a network, not its scale. This scale information has to be furnished otherwise.

The observation consists of the *ratio* of distances from point j to the points k and l:

$$\frac{s_{jk}}{s_{jl}} = \frac{\sqrt{(X_k - X_j)^2 + (Y_k - Y_j)^2}}{\sqrt{(X_l - X_j)^2 + (Y_l - Y_j)^2}}. \tag{10.17}$$

We choose the sequence of unknowns as $(x_j, x_k, x_l, y_j, y_k, y_l)$ after which the gradient of the left side in (10.17) is

$$\operatorname{grad}\frac{s_{jk}}{s_{jl}} = \Big(f\Big[\frac{X_l - X_j}{s_{jl}^2} - \frac{X_k - X_j}{s_{jk}^2}\Big],\ f\frac{X_k - X_j}{s_{jk}^2},\ -f\frac{X_l - X_j}{s_{jl}^2},$$
$$f\Big[\frac{Y_l - Y_j}{s_{jl}^2} - \frac{Y_k - Y_j}{s_{jk}^2}\Big],\ f\frac{Y_k - Y_j}{s_{jk}^2},\ -f\frac{Y_l - Y_j}{s_{jl}^2}\Big),$$

where we have put $f = s_{jk}/s_{jl}$. The linear observation equation now is

$$\begin{bmatrix} \cdots & f^0\Big(\frac{X_l^0 - X_j^0}{(s_{jl}^0)^2} - \frac{X_k^0 - X_j^0}{(s_{jk}^0)^2}\Big) & f^0\frac{X_k^0 - X_j^0}{(s_{jk}^0)^2} & -f^0\frac{X_l^0 - X_j^0}{(s_{jl}^0)^2} \\ & f^0\Big(\frac{Y_l^0 - Y_j^0}{(s_{jl}^0)^2} - \frac{Y_k^0 - Y_j^0}{(s_{jk}^0)^2}\Big) & f^0\frac{Y_k^0 - Y_j^0}{(s_{jk}^0)^2} & -f^0\frac{Y_l^0 - Y_j^0}{(s_{jl}^0)^2} & \cdots \end{bmatrix} \begin{bmatrix} \vdots \\ x_j \\ x_k \\ x_l \\ y_j \\ y_k \\ y_l \\ \vdots \end{bmatrix} = b_i - r_i$$

where $b_i = s_{jk}/s_{jl} - s_{jk}^0/s_{jl}^0$.

Let the variance of the two distances be $\sigma_{s_{jk}}^2$ and $\sigma_{s_{jl}}^2$. We obtain the correct weight for the ratio of the two distances by using the law of propagation of variances:

$$\sigma_f^2 = \frac{f^2}{s_{jk}^2}\sigma_{s_{jk}}^2 + \frac{f^2}{s_{jl}^2}\sigma_{s_{jl}}^2 - 2\frac{f^2}{s_{jk}s_{jl}}\sigma_{s_{jk}s_{jl}}.$$

The covariance between the two measured distances typically is positive. If we ignore the covariance $\sigma_{s_{jk}s_{jl}}$ the weight $c_f = \sigma_0^2/\sigma_f^2$ is systematically underestimated.

Observation of Coordinate and Height

The observation equation of a measured coordinate, for instance from GPS, or a measured height, for instance from a tide gauge, reads

$$+1\, x_i = X_{i,\text{obsv}} - X_i^0 - r_i. \tag{10.18}$$

If the observation is to be fixed to the measured value then the weight has to be 10–100 times larger than the other weights. If it shall enter at equal terms with the other observations then a weight of normal magnitude is appropriate. The actual situation will indicate the proper weight to be applied.

Azimuth

An azimuth observation determines the angle to the meridian. Astronomical azimuths are related to the geographical meridian while the magnetic azimuth relates to the magnetic meridian. In this case we have to know the angle between the magnetic and the geographical meridian.

Astronomical azimuths typically are determined with a standard deviation less than 0.1 mgon. Magnetic azimuths cannot be determined better than 0.1 gon.

Let the observed azimuth from point j to point k be denoted A_{jk}. Then the observation is

$$A_{jk} = \arctan \frac{Y_k - Y_j}{X_k - X_j}. \tag{10.19}$$

We choose the sequence of unknowns as (x_j, x_k, y_j, y_k) and the gradient of the left side of (10.19) becomes

$$\operatorname{grad} A_{jk} = \Big(-\frac{Y_k - Y_j}{(s_{jk})^2}, \frac{Y_k - Y_j}{(s_{jk})^2}, -\frac{X_k - X_j}{(s_{jk})^2}, \frac{X_k - X_j}{(s_{jk})^2} \Big). \tag{10.20}$$

The linearized version of (10.19) now is

$$\begin{bmatrix} \cdots & -\frac{\sin \alpha_{jk}^0}{s_{jk}^0} & \frac{\sin \alpha_{jk}^0}{s_{jk}^0} & -\frac{\cos \alpha_{jk}^0}{s_{jk}^0} & \frac{\cos \alpha_{jk}^0}{s_{jk}^0} & \cdots \end{bmatrix} \begin{bmatrix} \vdots \\ x_j \\ x_k \\ y_j \\ y_k \\ \vdots \end{bmatrix} = A_{jk} - A_{jk}^0 - r_i.$$

Horizontal Direction

The result of observations of horizontal directions at a point j are reduced sets (mean sets) in which the first direction has the reduced value 0 gon. The observation for the bearing

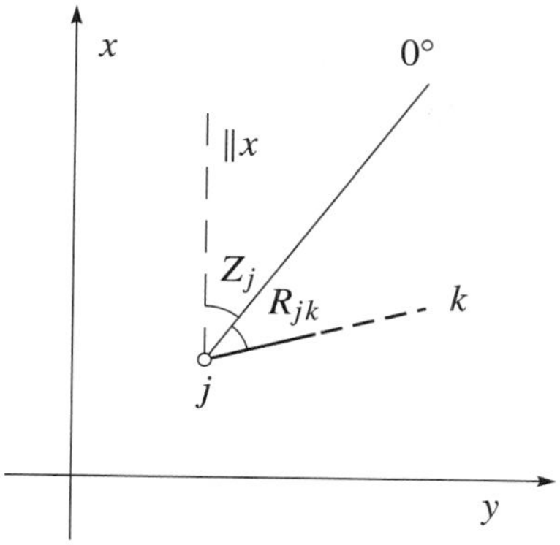

Figure 10.6 Observation of a horizontal direction.

A_{jk} of the side jk is then according to Figure 10.6

$$A_{jk} - Z_j = R_{jk}, \qquad k = 1, \dots, r, \tag{10.21}$$

where R_{jk} denotes the observed bearing for the same side. The constant Z_j is the bearing for the direction 0 gon of the horizontal circle of the theodolite. It is named the ***orientation unknown***.

We stress that Z_j is a characteristic quantity for each separate setup of the theodolite. If we again place the theodolite in the same point we get a new value of Z_j!

The bearing is

$$A_{jk} = \arctan \frac{Y_k - Y_j}{X_k - X_j} \qquad \text{and the derivative} \qquad \frac{d(\arctan x)}{dx} = \frac{1}{1+x^2}.$$

It is easy to calculate the gradient of the left side of the observation equation when the sequence of unknowns is $(x_j, x_k, y_j, y_k, z_j)$:

$$\operatorname{grad}(A_{jk} - Z_j) = \left(\frac{Y_k - Y_j}{(s_{jk})^2}, -\frac{Y_k - Y_j}{(s_{jk})^2}, -\frac{X_k - X_j}{(s_{jk})^2}, \frac{X_k - X_j}{(s_{jk})^2}, -1 \right). \tag{10.22}$$

We introduce preliminary values $Z_j = Z_j^0 + z_j$, $A_{jk}^0 = \arctan \frac{Y_k^0 - Y_j^0}{X_k^0 - X_j^0}$, and finally the linearized observation equation

$$\begin{bmatrix} \cdots & \dfrac{\sin \alpha_{jk}^0}{s_{jk}^0} & -\dfrac{\sin \alpha_{jk}^0}{s_{jk}^0} & \dfrac{\cos \alpha_{jk}^0}{s_{jk}^0} & -\dfrac{\cos \alpha_{jk}^0}{s_{jk}^0} & -1 & \cdots \end{bmatrix} \begin{bmatrix} \vdots \\ x_j \\ x_k \\ y_j \\ y_k \\ z_j \\ \vdots \end{bmatrix}$$

$$= R_{jk} + Z_j^0 - A_{jk}^0 - r_i. \tag{10.23}$$

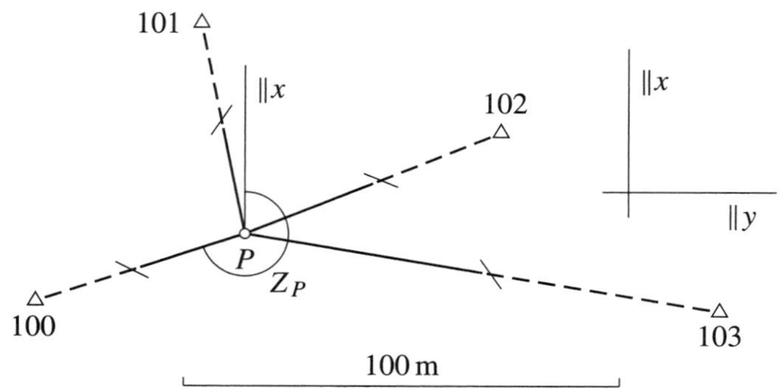

Figure 10.7 Free stationing.

The a priori variance of a bearing is

$$\sigma_r^2 = \frac{(\sigma_c\,\omega/s_{jk})^2 + \sigma_t^2}{n}, \tag{10.24}$$

where n denotes the number of observations (the number of sets if σ_r is to refer to a bearing measured with n sets) and σ_t^2 is the variance of a direction measured with 1 set and σ_c is the common centricity contribution from theodolite and signal.

Accordingly the ***weight of a direction observation is*** $c_r = \sigma_0^2/\sigma_r^2$ where σ_0^2 is the variance of unit weight.

For computational reasons it is an advantage to eliminate the orientation unknowns from the normal equations. The procedure is described in Section 11.7.

Example 10.3 To illustrate the interaction between different observation types we take an example from daily geodetic practice, namely a free stationing. The data were collected by a group of students making their first experiences with a total station.

The given points are situated as shown in Figure 10.7 and with postulated coordinates and observations:

Station	X [m]	Y [m]
100	60.10	−926.73
101	126.16	−888.81
102	105.38	−819.88
103	57.60	−769.61

Station	observed station i	reduced mean direction $v_{P,i}$ [gon]	horisontal distance $s_{P,i}$ [m]
P	100	0.000	51.086
	101	107.548	50.771
	102	191.521	65.249
	103	231.709	110.362

The a priori variances are fixed according to

1 distances, cf. (10.12)

$$\sigma_s^2 = \frac{\sigma_a^2 + \sigma_c^2}{n} = \frac{a^2 + (bs)^2 + \sigma_c^2}{n};$$

2 directions, cf. (10.24)

$$\sigma_r^2 = \frac{(\sigma_c\,\omega/s)^2 + \sigma_t^2}{n}.$$

We use the following values: $a = 5\,\text{mm}$, $b = 3 \times 10^{-5}$, $\sigma_c = 5\,\text{mm}$, $\sigma_t = 1\,\text{mgon}$, $n = 2$, and $\omega = 1/\sin(0.001\text{ gon}) = 63\,662$. We seek the coordinates of point P and their standard deviations.

First we have to calculate a set of preliminary coordinates for P. One solution is $(X_P^0, Y_P^0) = (76.40, -878.35)$. The preliminary value for Z_P is fixed as

$$Z_P^0 = \arctan \frac{Y_{100} - Y_P^0}{X_{100} - X_P^0} = 279.312\,\text{gon}$$

and we get the linearized observation equations for the four distances and the four directions, cf. (10.10) and (10.23):

$$-\frac{X_{100} - X_P^0}{(s_{P,100}^0)^2} x_P - \frac{Y_{100} - Y_P^0}{(s_{P,100}^0)^2} y_P + 0 \cdot z_P = \frac{s_{P,100} - s_{P,100}^0}{s_{P,100}^0} - r_1$$

$$\vdots$$

$$-\frac{X_{103} - X_P^0}{(s_{P,103}^0)^2} x_P - \frac{Y_{103} - Y_P^0}{(s_{P,103}^0)^2} y_P + 0 \cdot z_P = \frac{s_{P,103} - s_{P,103}^0}{s_{P,103}^0} - r_4$$

$$\frac{Y_{100} - Y_P^0}{(s_{P,100}^0)^2} x_P - \frac{X_{100} - X_P^0}{(s_{P,100}^0)^2} y_P - 1 \cdot z_P = Z_P^0 - \arctan \frac{Y_{100} - Y_P^0}{X_{100} - X_P^0} + v_{P,100} - r_5$$

$$\vdots$$

$$\frac{Y_{103} - Y_P^0}{(s_{P,103}^0)^2} x_P - \frac{X_{103} - X_P^0}{(s_{P,103}^0)^2} y_P - 1 \cdot z_P = Z_P^0 - \arctan \frac{Y_{103} - Y_P^0}{X_{103} - X_P^0} + v_{P,103} - r_8.$$

We arrange the unknowns in the following sequence

$$\boldsymbol{x} = \begin{bmatrix} x_P \\ y_P \\ z_P \end{bmatrix},$$

and the coefficient matrix A is

$$A = \begin{bmatrix} -\frac{X_{100} - X_P^0}{(s_{P,100}^0)^2} & -\frac{Y_{100} - Y_P^0}{(s_{P,100}^0)^2} & 0 \\ \vdots & & \\ \frac{Y_{100} - Y_P^0}{(s_{P,100}^0)^2} & -\frac{X_{100} - X_P^0}{(s_{P,100}^0)^2} & -1 \\ \vdots & & \end{bmatrix} = \begin{bmatrix} 0.0062 & -0.0185 & 0 \\ -0.0193 & -0.0041 & 0 \\ -0.0068 & 0.0137 & 0 \\ 0.0015 & 0.0089 & 0 \\ -0.0185 & -0.0062 & -1 \\ -0.0041 & 0.0193 & -1 \\ 0.0137 & 0.0068 & -1 \\ 0.0089 & -0.0015 & -1 \end{bmatrix}$$

$$\boldsymbol{b}^{\mathrm{T}} = \begin{bmatrix} 6.65 & -15.05 & -1.35 & 0.80 & 0.00 & 7.90 & 19.98 & 19.21 \end{bmatrix} \times 10^{-4}.$$

Note that the last four components in $\boldsymbol{b}$ are in units of radians. The diagonal entries of W are defined as the inverse of the standard deviation of the observations

$$\operatorname{diag}(W) = \begin{bmatrix} 0.195 & 0.195 & 0.193 & 0.181 & 0.224 & 0.224 & 0.285 & 0.463 \end{bmatrix}.$$

The *weight normalized* observation equations $WA\boldsymbol{x} = W\boldsymbol{b} - W\boldsymbol{r}$ are

$$\begin{bmatrix} 0.001\,218 & -0.003\,615 & 0 \\ -0.003\,764 & -0.000\,791 & 0 \\ -0.001\,314 & 0.002\,651 & 0 \\ 0.000\,279 & 0.001\,616 & 0 \\ -0.004\,153 & -0.001\,399 & -0.224 \\ -0.000\,909 & 0.004\,324 & -0.224 \\ 0.003\,914 & 0.001\,940 & -0.285 \\ 0.004\,134 & 0.000\,715 & -0.463 \end{bmatrix} \begin{bmatrix} x_P \\ y_P \\ z_P \end{bmatrix} = \begin{bmatrix} 0.000\,130 \\ -0.000\,293 \\ -0.000\,026 \\ 0.000\,014 \\ 0.000\,000 \\ 0.000\,177 \\ 0.000\,569 \\ 0.000\,889 \end{bmatrix} - W\boldsymbol{r}.$$

Accordingly the weighted normal equations $(WA)^{\mathrm{T}}(WA)\boldsymbol{x} = (WA)^{\mathrm{T}}W\boldsymbol{b}$. With $C = W^{\mathrm{T}}W$ this is $A^{\mathrm{T}}CA\boldsymbol{x} = A^{\mathrm{T}}C\boldsymbol{b}$:

$$\begin{bmatrix} 0.000\,067\,9 & 0.000\,002\,1 & -0.001\,895\,6 \\ 0.000\,002\,1 & 0.000\,048\,3 & -0.000\,877\,2 \\ -0.001\,895\,6 & -0.000\,877\,2 & 0.395\,946\,0 \end{bmatrix} \begin{bmatrix} x_P \\ y_P \\ z_P \end{bmatrix} = \begin{bmatrix} 0.000\,007\,0 \\ 0.000\,000\,9 \\ -0.000\,613\,7 \end{bmatrix}.$$

The normal equations are solved by Cholesky reduction:

$$\begin{aligned} & l_{11} = \sqrt{a_{11}} \\ & l_{21} = a_{12}/l_{11} \qquad l_{22} = \sqrt{a_{22} - l_{21}^2} \\ & l_{31} = a_{13}/l_{11} \qquad l_{32} = (a_{23} - l_{21}l_{31})/l_{22} \qquad l_{33} = \sqrt{a_{33} - l_{31}^2 - l_{32}^2} \end{aligned}$$

$$\begin{aligned} z_1 &= b_1/l_{11} \\ z_2 &= (b_2 - l_{21}z_1)/l_{22} \\ z_3 &= (b_3 - l_{31}z_1 - l_{32}z_2)/l_{33} \end{aligned}$$

or

$$\begin{aligned} & l_{11} = 0.008\,242 \\ & l_{21} = 0.000\,250 \qquad l_{22} = 0.006\,942 \\ & l_{31} = -0.229\,989 \qquad l_{32} = -0.118\,063 \qquad l_{33} = 0.573\,683 \end{aligned}$$

$$\begin{aligned} z_1 &= 0.000\,855 \\ z_2 &= 0.000\,106 \\ z_3 &= -0.000\,705 \\ \sum z_i^2 &= 1.239\,347 \times 10^{-6}. \end{aligned}$$

The back solution is

$$x_3 = z_3/l_{33}$$
$$x_2 = (z_2 - l_{32}x_3)/l_{22}$$
$$x_1 = (z_1 - l_{21}x_2 - l_{31}x_3)/l_{11}$$

The result is

$$\hat{x}_P = x_1 = 0.069\,6\,\text{m}$$
$$\hat{y}_P = x_2 = -0.005\,6\,\text{m}$$
$$\hat{z}_P = x_3 = -0.001\,229 \approx -78\,\text{mgon}$$

This results in the following coordinates for P:

X_P^0:	76.400 m	Y_P^0:	−878.350 m
$+\hat{x}_P$:	0.070 m	$+\hat{y}_P$:	−0.006 m
$\widehat{X}_P$:	76.470 m	$\widehat{Y}_P$:	−878.356 m

and the following value for the orientation unknown

$$\widehat{Z}_P = Z_P^0 + \hat{z}_P = 379.312 - 0.078 = 379.234\,\text{gon}.$$

The a posteriori variance calculation is done according to the survey on page 336 and (9.70)

$$\boldsymbol{r}^{\mathrm{T}}C\hat{\boldsymbol{r}} = \boldsymbol{b}^{\mathrm{T}}C\boldsymbol{b} - (A^{\mathrm{T}}C\boldsymbol{b})^{\mathrm{T}}\hat{\boldsymbol{x}} = \boldsymbol{b}^{\mathrm{T}}C\boldsymbol{b} - \boldsymbol{z}^{\mathrm{T}}\boldsymbol{z}$$
$$= 1.250\,470\,3 \times 10^{-6} - 1.239\,346\,9 \times 10^{-6} = 1.112\,3 \times 10^{-8}.$$

Finally $\hat{\sigma}_0^2$ is likewise calculated according to the same survey:

$$\hat{\sigma}_0^2 = \frac{\hat{\boldsymbol{r}}^{\mathrm{T}}C\hat{\boldsymbol{r}}}{m-n} = \frac{1.112\,3 \times 10^{-8}}{8-3} = 0.222\,5 \times 10^{-8}$$
$$\hat{\sigma}_0 = 0.47 \times 10^{-4}.$$

Next, we want to determine the confidence ellipse in P. We start by calculating the inverse normal equation matrix

$$(A^{\mathrm{T}}CA)^{-1} = \left[\begin{array}{cc|c} 1.701\,8 & 0.078\,5 & 0.008\,3 \\ 0.078\,5 & 2.162\,7 & 0.005\,2 \\ \hline 0.008\,3 & 0.005\,2 & 0.000\,30 \end{array}\right] \times 10^4.$$

The part concerning the coordinates $\widehat{X}_P$ and $\widehat{Y}_P$ is the upper left 2 by 2 block matrix $(A^{\mathrm{T}}CA)^{-1}_{11}$. Hence the covariance matrix $\Sigma_{\hat{x}}$ for the coordinates is

$$\Sigma_{\hat{x}} = \hat{\sigma}_0^2\,(A^{\mathrm{T}}CA)^{-1}_{11} = \begin{bmatrix} \hat{\sigma}_1^2 & \hat{\sigma}_{12} \\ \hat{\sigma}_{21} & \hat{\sigma}_2^2 \end{bmatrix} = \begin{bmatrix} 0.378\,6 & 0.017\,5 \\ 0.017\,5 & 0.481\,1 \end{bmatrix} \times 10^{-4}.$$

The bearing φ for the major axis of the confidence ellipse is calculated according to (9.75)

$$\tan 2\hat{\varphi} = \frac{2\hat{\sigma}_{12}}{\hat{\sigma}_1^2 - \hat{\sigma}_2^2} = \frac{2 \cdot 0.017\,5}{0.378\,6 - 0.481\,1} \qquad \text{and} \qquad \hat{\varphi} = 89.5\,\text{gon}.$$

The square root of the eigenvalues of $\Sigma_{\hat{x}}$ are

$$\sqrt{\lambda} = \begin{cases} \sqrt{\lambda_1} = 7.0\,\text{mm} = \hat{a} \\ \sqrt{\lambda_2} = 6.1\,\text{mm} = \hat{b} \end{cases}$$

which denote the semi major and the semi minor axis of the confidence ellipse of point P.

Trigonometric Leveling

An often used expression for the trigonometric leveling observation h looks like

$$h = s\cos z + \frac{1-k}{2R}s^2\sin^2 z + i - r. \tag{10.25}$$

Here s denotes the slope distance between the total station and the reflector, z is the observed zenith distance, i is the height of instrument above the one mark, and r is the height of the reflector above the other mark, see Figure 10.8. Futhermore $R \approx 6\,370\,\text{km}$ denotes the mean radius of the Earth and the coefficient of refraction is for mid-latitudes about $k = 0.13$ and for instance in Greenland 0.17–0.19. Under other skies k is even negative.

Height differences calculated according to (10.25) are usually applied on equal terms with height differences as determined by geometrical leveling. Only we have to determine the weight for the trigonometrical leveling observation.

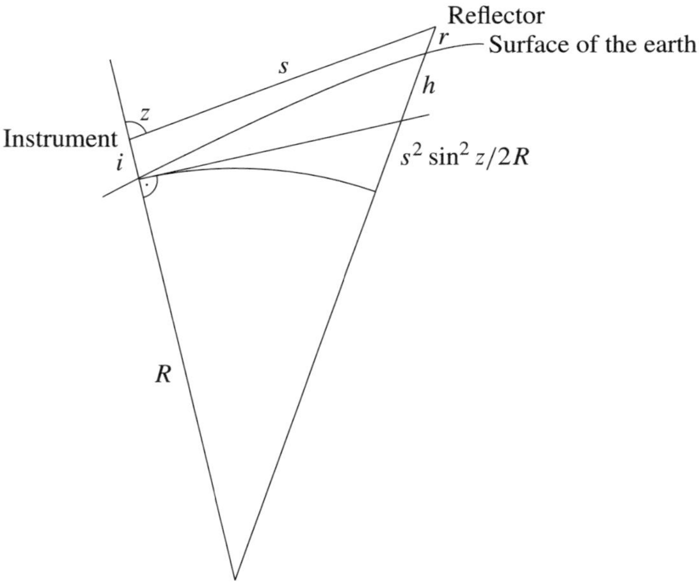

Figure 10.8 Trigonometric leveling

We apply the law of error propagation (9.57) and the weight relation (9.44) for determining the weight. We use (9.57) on (10.25) and get

$$\sigma_h^2 = \Big(\cos z + \frac{1-k}{R} s \sin^2 z\Big)^2 \sigma_s^2 + \Big(-s \sin z + \frac{1-k}{2R} s^2 \sin 2z\Big)^2 \sigma_z^2 + \frac{s^4 \sin^4 z}{4R^2}\sigma_k^2 + \sigma_i^2 + \sigma_r^2 \qquad (10.26)$$

where we regard R as a constant. It is reasonable to put $\sin z \approx 1$ and $s/R \approx 0$ to get the simplified expression

$$\sigma_h^2 = \cos^2 z\,\sigma_s^2 + s^2\sigma_z^2 + \frac{s^4}{4R^2}\sigma_k^2 + \sigma_i^2 + \sigma_r^2. \qquad (10.27)$$

We shall use this to investigate different situations:

1 The most frequent use of trigonometric leveling is by the total station in a free stationing. The height of the individual points is measured by placing the reflector at them. Realistic values for the standard deviations are $\sigma_r = 2$ mm, $\sigma_z = 1$ mgon, and $z = 95$ gon. As we only deal with differences of height, i is eliminated and consequently $\sigma_i = 0$. Let $\sigma_k = 0.04$, $s = 200$ m and $\sigma_s = 5$ mm; then the variance of a height difference h between two points at most 400 m apart is

$$\sigma_h^2 = \Big(0.006 \cdot 25 + \frac{200\,000^2}{63\,662^2} + \frac{200\,000^4 \cdot 0.04^2}{4(6\,370 \cdot 10^6)^2} + 2^2\Big)$$
$$= 2\big(0.15 + 9.8 + 0.016 + 4\big) = 28 \text{ mm}^2.$$

The standard deviation of a height difference between points up to 400 m apart is, under the given presumptions, 5 mm which is equivalent to $\sigma_h = 8\sqrt{L}$ mm where the length L of the leveled way is in units of km.

Is it possible to obtain a smaller variance by using sights of 50 m, for example, but in turn get 10 set-ups per km? The result is $20(0.15 + 0.7 + 4) = 100 \text{ mm}^2$. So the decisive fact is whether σ_r is set to 1 or 2 mm.

These considerations reveal that trigonometric leveling qualitatively fully competes with geometric leveling.

2 With the same assumptions, but with a longer distance $s = 1$ km the calculations are

$$\sigma_h^2 = 0.006 \cdot 100 + \frac{1\,000\,000^2}{63\,662^2} + \frac{1\,000\,000^4 \cdot 0.04^2}{4(6\,370 \cdot 10^6)^2} + 2^2$$
$$= 0.6 + 246.7 + 10 + 4 = 261.3 \text{ mm}^2.$$

Now we see that the important term is $s^2\sigma_z^2$ and why the fixing of weight has to happen according to

$$c = \frac{\sigma_0^2}{\sigma_h^2} = \frac{1}{s^2} \cdot \frac{\sigma_0^2}{\sigma_z^2}.$$

So ***trigonometric leveling is weighted inversely proportional to the square of the length of sight***.

The expression for the variance shows that trigonometric leveling with short sights, and a possibly less accurate observation of the zenith distance, gives fully reliable results. With longer sights even a good observation of the zenith distance cannot compensate for the worsening refraction influence. This is one reason why trigonometric leveling with long sight lengths will never obtain a wide acceptance in flat and hilly landscapes. In mountainous areas it can be adopted.

10.3 Three-Dimensional Model

At the beginning of this chapter we introduced a plane coordinate system X, Y and developed the whole theory in this plane.

Because of the introduction of GPS, many geodesists want to solve their least-squares problems in a three-dimensional coordinate system X, Y, Z. This is the natural setting for a combination of classical geodetic operations and the *global positioning system*. So we shall develop most of the observational types again, in a 3-dimensional setting.

First of all a few words about the whole idea. The classical geodetic observation practice is characterized by a lot of small reductions and corrections to the observations performed at the surface of the Earth. This happens in order that the least-squares computations can easily be done in a horizontal 2-dimensional model at mean sea level. One can perceive this reduction practice as unsatisfactory today. With excellent computational means such a procedure can be looked upon as outdated and less rigorous than a 3-dimensional model.

So we start by introducing the relation between Cartesian (X, Y, Z) coordinates and geographical (φ, λ, h) coordinates

$$\begin{bmatrix} X \\ Y \\ Z \end{bmatrix} = \begin{bmatrix} (N_\varphi + h)\cos\varphi\cos\lambda \\ (N_\varphi + h)\cos\varphi\sin\lambda \\ \big((1-f)^2 N_\varphi + h\big)\sin\varphi \end{bmatrix}. \tag{10.28}$$

The *reference ellipsoid* is the surface given by $X^2 + Y^2 + Z^2 = 1$. Recall the radius of curvature in the prime vertical (which is the vertical plane normal to the astronomical meridian)

$$N_\varphi = \frac{a}{\sqrt{1 - f(2-f)\sin^2\varphi}}. \tag{10.29}$$

The unit vector $\boldsymbol{u}$ normal to the surface is

$$\boldsymbol{u} = \begin{bmatrix} \cos\varphi\cos\lambda \\ \cos\varphi\sin\lambda \\ \sin\varphi \end{bmatrix}. \tag{10.30}$$

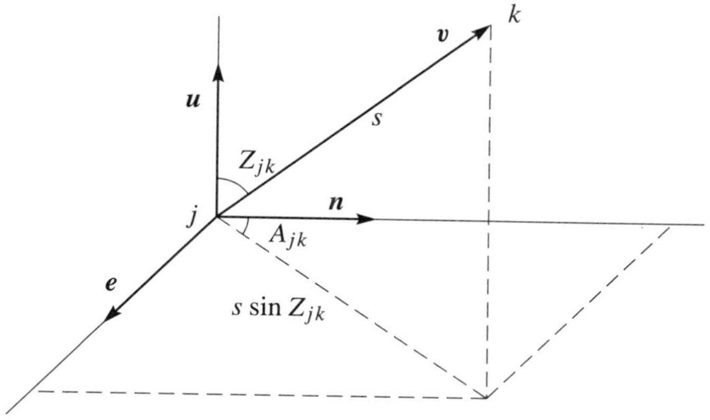

Figure 10.9 Observation of distance s, zenith distance Z_{jk} and azimuth A_{jk} between points j and k.

The tangent $\boldsymbol{n}$ and binormal $\boldsymbol{e}$ unit vectors are derivatives of $\boldsymbol{u}$ ($\boldsymbol{n}$ alludes to northing, $\boldsymbol{e}$ to easting, and $\boldsymbol{u}$ to up, i.e. the normal direction to the surface):

$$\boldsymbol{n} = \frac{\partial \boldsymbol{u}}{\partial \varphi} = \begin{bmatrix} -\sin\varphi\cos\lambda \\ -\sin\varphi\sin\lambda \\ \cos\varphi \end{bmatrix} \quad \text{and} \quad \boldsymbol{e} = \frac{1}{\cos\varphi}\frac{\partial \boldsymbol{u}}{\partial \lambda} = \begin{bmatrix} -\sin\lambda \\ \cos\lambda \\ 0 \end{bmatrix}. \tag{10.31}$$

Verify that $\boldsymbol{e} = \boldsymbol{n} \times \boldsymbol{u}$.

The unit vectors $\boldsymbol{n}, \boldsymbol{e}, \boldsymbol{u}$ provide the natural coordinate frame at a point on the reference ellipsoid, see Figure 10.9. For reasons of reference we collect the three unit vectors $\boldsymbol{e}, \boldsymbol{n}, \boldsymbol{u}$ into the orthogonal matrix

$$R = \begin{bmatrix} \boldsymbol{e} & \boldsymbol{n} & \boldsymbol{u} \end{bmatrix} = \begin{bmatrix} -\sin\lambda & -\sin\varphi\cos\lambda & \cos\varphi\cos\lambda \\ \cos\lambda & -\sin\varphi\sin\lambda & \cos\varphi\sin\lambda \\ 0 & \cos\varphi & \sin\varphi \end{bmatrix}. \tag{10.32}$$

Now let the difference vector $\boldsymbol{v}$ between two points j and k on the surface of the Earth—not necessarily on the ellipsoid—be

$$\boldsymbol{v} = \begin{bmatrix} X_k - X_j \\ Y_k - Y_j \\ Z_k - Z_j \end{bmatrix}. \tag{10.33}$$

Immediately we get from Figure 10.9 the $\boldsymbol{n}, \boldsymbol{e}, \boldsymbol{u}$ components of $\boldsymbol{v}$

$$\begin{aligned} \boldsymbol{v}^{\mathrm{T}}\boldsymbol{n} &= s\sin Z_{jk}\cos A_{jk} \\ \boldsymbol{v}^{\mathrm{T}}\boldsymbol{e} &= s\sin Z_{jk}\sin A_{jk} \\ \boldsymbol{v}^{\mathrm{T}}\boldsymbol{u} &= s\cos Z_{jk}. \end{aligned}$$

Theoretically, in 3-dimensional geodesy we have to consider 5 unknowns at each point: X, Y, Z and two parameters defining the ***direction of the plumb line***. A convenient choice is the astronomical latitude Φ and longitude Λ.

Above we already have used the geodetic latitude φ, and longitude λ, and ellipsoidal height h. They are point coordinates equivalent to X, Y, Z. The astronomical coordinates are to be looked upon as direction parameters defining the direction of the plumb line.

Horizontal Direction As usual the bearing A_{jk} is decribed by

$$A_{jk}=\arctan\frac{\boldsymbol{v}^{\mathrm{T}}\boldsymbol{e}}{\boldsymbol{v}^{\mathrm{T}}\boldsymbol{n}}=\arctan\frac{-(X_k-X_j)\sin\lambda+(Y_k-Y_j)\cos\lambda}{-(X_k-X_j)\sin\varphi\cos\lambda-(Y_k-Y_j)\sin\varphi\sin\lambda+(Z_k-Z_j)\cos\varphi}.$$

We introduce the earlier relation between the orientation unknown Z_j and the observed bearing R_{jk}:

$$A_{jk} = \mathrm{Z}_j + R_{jk}. \tag{10.34}$$

There is a little conflict between the notation for the coordinate Z_j (in italic) and the orientation unknown Z_j (in Roman type). As earlier we use the following sequence of the nine unknowns

$$\boldsymbol{x} = \begin{bmatrix} \dots & x_j & y_j & z_j & \delta\Phi_j & \delta\Lambda_j & x_k & y_k & z_k & \mathrm{z}_j & \dots \end{bmatrix}^{\mathrm{T}}.$$

We introduce preliminary values $X_j = X_j^0 + x_j,\ Y_j = Y_j^0 + y_j, \dots, \Phi_j = \Phi_j^0 + \delta\Phi_j, \dots,$ $\mathrm{Z}_j = \mathrm{Z}_j^0 + \mathrm{z}_j$.

The difficult step is now to perform the calculation for $\operatorname{grad}(A_{jk} - \mathrm{Z}_j)$ correctly. There are nine coefficients to be calculated. We start with

$$\frac{\partial A_{jk}}{\partial X_j} = \frac{1}{1+\tan^2 A_{jk}}\frac{\boldsymbol{v}^{\mathrm{T}}\boldsymbol{n}\sin\lambda - \sin\varphi\cos\lambda\,\boldsymbol{v}^{\mathrm{T}}\boldsymbol{e}}{(\boldsymbol{v}^{\mathrm{T}}\boldsymbol{n})^2}$$

$$= \cos^2 A_{jk}\frac{\sin\lambda - \sin\varphi\cos\lambda\tan A_{jk}}{s\sin Z_{jk}\cos A_{jk}} = \frac{\sin\lambda\cos A_{jk} - \sin\varphi\cos\lambda\sin A_{jk}}{s\sin Z_{jk}}$$

$$\frac{\partial A_{jk}}{\partial Y_j} = \cos^2 A_{jk}\frac{-\cos\lambda - \sin\varphi\sin\lambda\tan A_{jk}}{s\sin Z_{jk}\cos A_{jk}} = \frac{-\cos\lambda\cos A_{jk} - \sin\varphi\sin\lambda\sin A_{jk}}{s\sin Z_{jk}}$$

$$\frac{\partial A_{jk}}{\partial Z_j} = \cos^2 A_{jk}\frac{\boldsymbol{v}^{\mathrm{T}}\boldsymbol{e}\cos\varphi}{s\sin Z_{jk}\cos A_{jk}} = \frac{\cos\varphi\sin A_{jk}}{s\sin Z_{jk}}$$

Then the derivatives with respect to the astronomical latitude Φ and the astronomical longitude Λ:

$$\frac{\partial A_{jk}}{\partial\Phi} = \cos^2 A_{jk}\frac{(\boldsymbol{v}^{\mathrm{T}}\boldsymbol{n})\boldsymbol{v}^{\mathrm{T}}\frac{\partial\boldsymbol{e}}{\partial\Phi} - \boldsymbol{v}^{\mathrm{T}}\frac{\partial\boldsymbol{n}}{\partial\Phi}\boldsymbol{v}^{\mathrm{T}}\boldsymbol{e}}{(\boldsymbol{v}^{\mathrm{T}}\boldsymbol{n})^2} = \cos^2 A_{jk}\frac{\boldsymbol{v}^{\mathrm{T}}\boldsymbol{u}\boldsymbol{v}^{\mathrm{T}}\boldsymbol{e}}{(\boldsymbol{v}^{\mathrm{T}}\boldsymbol{n})^2} = \cot Z_{jk}\sin A_{jk}$$

$$\frac{\partial A_{jk}}{\partial\Lambda} = \cos^2 A_{jk}\frac{(\boldsymbol{v}^{\mathrm{T}}\boldsymbol{n})\boldsymbol{v}^{\mathrm{T}}\frac{\partial\boldsymbol{e}}{\partial\Lambda} - \boldsymbol{v}^{\mathrm{T}}\frac{\partial\boldsymbol{n}}{\partial\Lambda}\boldsymbol{v}^{\mathrm{T}}\boldsymbol{e}}{(\boldsymbol{v}^{\mathrm{T}}\boldsymbol{n})^2} = \sin\varphi - \cos A_{jk}\cos\varphi\cot Z_{jk}.$$

Now the gradient is

$$\operatorname{grad} A_{jk} = \Big(\dots, \frac{\sin\lambda_j \cos A_{jk} - \sin\varphi_j \cos\lambda_j \sin A_{jk}}{s \sin Z_{jk}},$$
$$\frac{-\cos\lambda_j \cos A_{jk} - \sin\varphi_j \sin\lambda_j \sin A_{jk}}{s\sin Z_{jk}}, \frac{\cos\varphi_j \sin A_{jk}}{s \sin Z_{jk}}, \cot Z_{jk}\sin A_{jk},$$
$$\sin\varphi_j - \cos\varphi_j\cos A_{jk}\cot Z_{jk}, -\frac{\sin\lambda_j \cos A_{jk} - \sin\varphi_j \cos\lambda_j \sin A_{jk}}{s \sin Z_{jk}},$$
$$-\frac{-\cos\lambda_j \cos A_{jk} - \sin\varphi_j \sin\lambda_j \sin A_{jk}}{s\sin Z_{jk}}, -\frac{\cos\varphi_j \sin A_{jk}}{s \sin Z_{jk}}, -1, \dots\Big).$$

The linearized observation equation then is

$$\operatorname{grad} A_{jk}\, \boldsymbol{x} = R_{jk} + Z_j^0 - A_{jk}^0 - r_{jk} \tag{10.35}$$

where

$$A_{jk}^0 = \arctan\frac{-(X_k^0 - X_j^0)\sin\lambda_j^0 + (Y_k^0 - Y_j^0)\cos\lambda_j}{-(X_k^0 - X_j^0)\sin\varphi_j^0\cos\lambda_j^0 - (Y_k^0 - Y_j^0)\sin\varphi_j^0\sin\lambda_j^0 + (Z_k^0 - Z_j^0)\cos\varphi_j^0}.$$

Zenith distance The zenith distance Z_{jk} is expressed as

$$Z_{jk} = \arccos\frac{\boldsymbol{v}^{\mathrm{T}}\boldsymbol{u}}{s} = \arccos\frac{(X_k - X_j)\cos\varphi\cos\lambda + (Y_k - Y_j)\cos\varphi\sin\lambda + (Z_k - Z_j)\sin\varphi}{\sqrt{(X_k - X_j)^2 + (Y_k - Y_j)^2 + (Z_k - Z_j)^2}}.$$

Again we linearize the observation equation; the unknowns are arranged in the following order

$$\boldsymbol{x} = \begin{bmatrix} \dots & x_j & y_j & z_j & \delta\Phi_j & \delta\Lambda_j & x_k & y_k & z_k & \dots \end{bmatrix}^{\mathrm{T}}.$$

It is not trivial to calculate grad Z_{jk}. We start by introducing the notation $Z_{jk} = \arccos\frac{t}{s}$ and get

$$\begin{aligned}\frac{\partial Z_{jk}}{\partial X_j} &= \frac{1}{\sqrt{1 - (\frac{t}{s})^2}}\frac{s\frac{\partial t}{\partial X_j} - \frac{\partial s}{\partial X_j}t}{s^2} = \frac{1}{\sqrt{1 - \cos^2 Z_{jk}}}\frac{s(-\cos\varphi\cos\lambda) - \frac{-(X_k - X_j)}{s}t}{s^2}\\ &= \frac{1}{\sin Z_{jk}}\frac{-s\cos\varphi\cos\lambda + (X_k - X_j)\cos Z_{jk}}{s^2}\\ &= \frac{(X_k - X_j)\cos Z_{jk} - s\cos\varphi\cos\lambda}{s^2\sin Z_{jk}}\end{aligned}$$

and similarly for $\partial Z_{jk}/\partial X_k,\ \partial Z_{jk}/\partial Y_j, \dots, \partial Z_{jk}/\partial Z_k$. The next derivation is

$$\frac{\partial Z_{jk}}{\partial\Phi} = \frac{1}{\sin Z_{jk}}\frac{\boldsymbol{v}^{\mathrm{T}}\frac{\partial\boldsymbol{u}}{\partial\Phi}}{s} = \frac{1}{\sin Z_{jk}}\frac{\boldsymbol{v}^{\mathrm{T}}\boldsymbol{n}}{s} = \cos A_{jk}.$$

Finally

$$\frac{\partial Z_{jk}}{\partial \Lambda} = \frac{1}{\sin Z_{jk}} \frac{\boldsymbol{v}^{\mathrm{T}} \frac{\partial \boldsymbol{u}}{\partial \Lambda}}{s} = \frac{\cos\varphi\, \boldsymbol{v}^{\mathrm{T}} \boldsymbol{e}}{s \sin Z_{jk}} = \frac{\cos\varphi\, s \sin Z_{jk} \sin A_{jk}}{s \sin Z_{jk}} = \cos\varphi \sin A_{jk}.$$

Consequently

$$\begin{aligned}\operatorname{grad} Z_{jk} = \Bigg(\ldots, &\frac{(X_k - X_j)\cos Z_{jk} - s\cos\varphi_j \cos\lambda_j}{s^2 \sin Z_{jk}},\\ &\frac{(Y_k - Y_j)\cos Z_{jk} - s\cos\varphi_j \sin\lambda_j}{s^2 \sin Z_{jk}}, \frac{(Z_k - Z_j)\cos Z_{jk} - s\sin\varphi_j}{s^2 \sin Z_{jk}},\\ &-\cos A_{jk}, -\cos\varphi_j \sin A_{jk}, -\frac{(X_k - X_j)\cos Z_{jk} - s\cos\varphi_j \cos\lambda_j}{s^2 \sin Z_{jk}},\\ -&\frac{(Y_k - Y_j)\cos Z_{jk} - s\cos\varphi_j \sin\lambda_j}{s^2 \sin Z_{jk}}, -\frac{(Z_k - Z_j)\cos Z_{jk} - s\sin\varphi_j}{s^2 \sin Z_{jk}}, \ldots\Bigg).\end{aligned}$$

The linearized observation equation for a zenith distance is

$$\operatorname{grad} Z_{jk}\, \boldsymbol{x} = Z_{jk,\mathrm{obsv}} - Z_{jk}^0 - r_{jk} \tag{10.36}$$

where

$$Z_{jk}^0 = \arccos \frac{(X_k^0 - X_j^0)\cos\varphi_j^0 \cos\lambda_j^0 + (Y_k^0 - Y_j^0)\cos\varphi_j^0 \sin\lambda_j^0 + (Z_k^0 - Z_j^0)\sin\varphi_j^0}{\sqrt{(X_k^0 - X_j^0)^2 + (Y_k^0 - Y_j^0)^2 + (Z_k^0 - Z_j^0)^2}}.$$

Slope Distance This is an easy case, similar to the 2-dimensional case. Only an extra term is added:

$$s_{jk} = \sqrt{\boldsymbol{v}^{\mathrm{T}} \boldsymbol{v}} = \sqrt{(X_k - X_j)^2 + (Y_k - Y_j)^2 + (Z_k - Z_j)^2}. \tag{10.37}$$

Distance observations are independent of the plumb line direction; consequently there are no unknowns for the corresponding parameters:

$$\boldsymbol{x} = \begin{bmatrix} \ldots & x_j & y_j & z_j & x_k & y_k & z_k & \ldots \end{bmatrix}^{\mathrm{T}}$$

and we get

$$\begin{aligned}\operatorname{grad} s_{jk} = \Bigg(\ldots, -\frac{(X_k - X_j)}{s}, -\frac{(Y_k - Y_j)}{s}, -\frac{(Z_k - Z_j)}{s},\\ \frac{(X_k - X_j)}{s}, \frac{(Y_k - Y_j)}{s}, \frac{(Z_k - Z_j)}{s}, \ldots\Bigg).\end{aligned}$$

The linearized observation equation is

$$\operatorname{grad} s_{jk}\, \boldsymbol{x} = s_{jk,\mathrm{obsv}} - s_{jk}^0 - r_{jk} \tag{10.38}$$

where

$$s_{jk}^0 = \sqrt{(X_k^0 - X_j^0)^2 + (Y_k^0 - Y_j^0)^2 + (Z_k^0 - Z_j^0)^2}.$$

Horizontal Distance This observation is just a special case of the combined slope distance, horizontal direction, and zenith distance with $Z_{jk} = \pi/2$.

Leveled Height Difference Heights are conceptually related to the ellipsoid. So the most direct and easiest way to deal with height differences $h_{jk} = h_k - h_j$ is to calculate each term from the well-known formulas (10.28). We repeat them for convenience

$$\begin{bmatrix} X \\ Y \\ Z \end{bmatrix} = \begin{bmatrix} (N_\varphi + h)\cos\varphi\cos\lambda \\ (N_\varphi + h)\cos\varphi\sin\lambda \\ \big((1-f)^2 N_\varphi + h\big)\sin\varphi \end{bmatrix}.$$

The solution of these nonlinear equations shall be presented according to an elegant development proposed by Clyde Goad. Suppose (X, Y, Z) are given and we want to solve for (φ, λ, h). By dividing the two upper equations we immediately obtain

$$\lambda = \arctan\frac{Y}{X}. \tag{10.39}$$

Of course, special precaution has to be taken if $X = 0$; in MATLAB code this happens by the statement lambda=atan2(Y,X). This function works well as long as either $X \neq 0$ or $Y \neq 0$. When $X = Y = 0$, λ is undefined and thus can take on any arbitrary value. Next we introduce the distance P from the point in question to the spin axis

$$P = \sqrt{X^2 + Y^2} = (N_\varphi + h)\cos\varphi \tag{10.40}$$

$$Z = \big((1-f)^2 N_\varphi + h\big)\sin\varphi \tag{10.41}$$

which is to be solved for (φ, h). We start with the values

$$r = \sqrt{P^2 + Z^2}$$

$$\varphi^0 \approx \arcsin(Z/r)$$

$$h^0 \approx r - a(1 - f\sin^2\varphi^0)$$

and by means of equations (10.40) and (10.41) we get (P^0, Z^0). The case $r = 0$ must be handled specially. Linearizing these equations we obtain

$$Z(\varphi, h) = Z(\varphi^0 + \Delta\varphi, h^0 + \Delta h) = Z(\varphi^0, h^0) + \frac{\partial Z}{\partial\varphi}\Delta\varphi + \frac{\partial Z}{\partial h}\Delta h + \cdots$$

$$P(\varphi, h) = P(\varphi^0 + \Delta\varphi, h^0 + \Delta h) = P(\varphi^0, h^0) + \frac{\partial P}{\partial\varphi}\Delta\varphi + \frac{\partial P}{\partial h}\Delta h + \cdots.$$

Neglecting higher order terms and rearranging we get a matrix form

$$\begin{bmatrix} Z - Z^0 \\ P - P^0 \end{bmatrix} = \begin{bmatrix} \Delta Z \\ \Delta P \end{bmatrix} \approx \begin{bmatrix} \dfrac{\partial Z}{\partial\varphi} & \dfrac{\partial Z}{\partial h} \\ \dfrac{\partial P}{\partial\varphi} & \dfrac{\partial P}{\partial h} \end{bmatrix} \begin{bmatrix} \Delta\varphi \\ \Delta h \end{bmatrix}$$

$$\approx \begin{bmatrix} (N_\varphi + h)\cos\varphi & \sin\varphi \\ -(N_\varphi + h)\sin\varphi & \cos\varphi \end{bmatrix} \begin{bmatrix} \Delta\varphi \\ \Delta h \end{bmatrix} = \begin{bmatrix} \cos\varphi & \sin\varphi \\ -\sin\varphi & \cos\varphi \end{bmatrix} \begin{bmatrix} (N_\varphi + h)\Delta\varphi \\ \Delta h \end{bmatrix}.$$

As a valid approximation we have set $f = 0$ in the partial derivatives. We invert (i.e., we transpose) the orthogonal rotation matrix to get

$$\begin{bmatrix} \cos\varphi & -\sin\varphi \\ \sin\varphi & \cos\varphi \end{bmatrix} \begin{bmatrix} \Delta Z \\ \Delta P \end{bmatrix} = \begin{bmatrix} (N_\varphi + h)\Delta\varphi \\ \Delta h \end{bmatrix}.$$

Finally

$$\Delta h = \sin\varphi \Delta Z + \cos\varphi \Delta P \tag{10.42}$$

$$\Delta\varphi = \frac{\cos\varphi \Delta Z - \sin\varphi \Delta P}{N_\varphi + h}. \tag{10.43}$$

The iterative procedure is continued until an adequate precision is achieved.

The M-file frgeod calculates Cartesian coordinates (X, Y, Z) given geodetic coordinates (φ, λ, h) according to (10.28). The M-file togeod calculates geodetic coordinates (φ, λ, h) given the Cartesian coordinates (X, Y, Z) according to the above derivation. Both functions are most useful when processing GPS vectors, cf. Section 14.4.

We prepare for the linearized observation equation for a height difference and rewrite equation (10.42): $\cos\varphi \Delta P + \sin\varphi \Delta Z = \Delta h$ or

$$\cos\varphi\left(\frac{\partial P}{\partial X}(x_k - x_j) + \frac{\partial P}{\partial Y}(y_k - y_j)\right) + \sin\varphi(z_k - z_j) = h_k - h_j.$$

The partial derivatives become

$$\frac{\partial P}{\partial X} = \frac{X}{P} = \cos\lambda \qquad \text{and} \qquad \frac{\partial P}{\partial Y} = \frac{Y}{P} = \sin\lambda.$$

The linearized observation equation for a height difference is

$$\operatorname{grad} h_{jk}\, \boldsymbol{x} = h_{jk,\text{obsv}} - h_{jk}^0 - r_{jk}. \tag{10.44}$$

If the unknowns are arranged as follows

$$\boldsymbol{x} = \begin{bmatrix} \dots & x_j & y_j & z_j & x_k & y_k & z_k & \dots \end{bmatrix}^{\mathrm{T}}$$

then

$$\operatorname{grad} h_{jk} = (\dots, -\cos\varphi\cos\lambda, -\cos\varphi\sin\lambda, -\sin\varphi, \cos\varphi\cos\lambda, \cos\varphi\sin\lambda, \sin\varphi, \dots).$$

One recognizes that $\operatorname{grad} h_{jk}\, \boldsymbol{x}$ is the dot product between the third column of R as defined in (10.32) with vector $\boldsymbol{v}$ as described in (10.33).

A preliminary value for the observed height difference h_{jk}^0 is calculated by the M-file togeod at both sites j and k.

11

LINEAR ALGEBRA FOR WEIGHTED LEAST SQUARES

11.1 Gram-Schmidt on A and Cholesky on $A^{\mathrm{T}}A$

The linear system is $Ax = b$. The matrix A is m by n and its n columns are independent. Then the n by n matrix $A^{\mathrm{T}}A$ is invertible and symmetric and positive definite.

The method of least squares yields an orthogonal projection of a point $\boldsymbol{b}$ onto the subspace $\boldsymbol{R}(A)$. This "column space" is spanned by the columns $\boldsymbol{a}_1, \dots, \boldsymbol{a}_n$ of A. This is a linear subspace of m-dimensional space $\mathbf{R}^m$, but its axes (columns of A) are generally not orthogonal. The projection of $\boldsymbol{b}$ is the combination $A\hat{\boldsymbol{x}}$, where $\hat{\boldsymbol{x}}$ satisfies the (weighted) normal equation $A^{\mathrm{T}}CA\hat{\boldsymbol{x}} = A^{\mathrm{T}}C\boldsymbol{b}$.

In least squares we write $\boldsymbol{b}$ as the combination

$$\boldsymbol{b} = \hat{x}_1\boldsymbol{a}_1 + \hat{x}_2\boldsymbol{a}_2 + \cdots + \hat{x}_n\boldsymbol{a}_n + \text{error } \boldsymbol{e}. \tag{11.1}$$

Let the inner product between any two vectors $\boldsymbol{a}$ and $\boldsymbol{b}$ be defined as $(\boldsymbol{a}, \boldsymbol{b}) = \boldsymbol{a}^{\mathrm{T}}C\boldsymbol{b}$. The symmetric positive definite matrix C is the weight matrix.

We want to demonstrate by explicit algorithms that Cholesky's elimination method (on $A^{\mathrm{T}}CA$) is equivalent to a Gram-Schmidt orthogonalization procedure on the columns of A.

Let the orthonormalized columns be $\boldsymbol{q}_i$. These are orthonormal in the "C-inner product" so that $(\boldsymbol{q}_i, \boldsymbol{q}_j) = \boldsymbol{q}_i^{\mathrm{T}}C\boldsymbol{q}_j = \delta_{ij}$. The vectors $\boldsymbol{q}_i$ are collected as columns of the matrix Q. Therefore we have $Q^{\mathrm{T}}CQ = I$. We define the matrix R by $Q^{-1}A$, so that

$$A = QR.$$

The matrix R is upper triangular with nonnegative diagonal entries. This follows from the order in which the Gram-Schmidt orthogonalization is executed—one vector at a time.

It is numerically safer to work with orthogonal and triangular matrices, Q and R. But we always modify the Gram-Schmidt algorithm, *to subtract one projection at a time*. And in most applications it is safe enough (and faster) to work directly with $A^{\mathrm{T}}CA$.

Calculation of R from Q and A

From the order of the Gram-Schmidt process, each new A_i is a combination of $\boldsymbol{a}_i$ and the vectors $A_1, \ldots, A_{i-1}$ that are already set. When the A's are normalized to unit vectors, each new $\boldsymbol{q}_i$ is a combination of $\boldsymbol{a}_i$ and $\boldsymbol{q}_1, \ldots, \boldsymbol{q}_{i-1}$. Therefore $\boldsymbol{a}_i$ is a combination of $\boldsymbol{q}_1, \ldots, \boldsymbol{q}_i$. ***An early a does not involve a later q!***

We can express this as a growing triangle:

$$\boldsymbol{a}_1 = (\boldsymbol{q}_1, \boldsymbol{a}_1)\boldsymbol{q}_1 \tag{11.2}$$

$$\boldsymbol{a}_2 = (\boldsymbol{q}_1, \boldsymbol{a}_2)\boldsymbol{q}_1 + (\boldsymbol{q}_2, \boldsymbol{a}_2)\boldsymbol{q}_2 \tag{11.3}$$

$$\boldsymbol{a}_3 = (\boldsymbol{q}_1, \boldsymbol{a}_3)\boldsymbol{q}_1 + (\boldsymbol{q}_2, \boldsymbol{a}_3)\boldsymbol{q}_2 + (\boldsymbol{q}_3, \boldsymbol{a}_3)\boldsymbol{q}_3. \tag{11.4}$$

In matrix form this is exactly $A = QR$. The entry r_{ij} is the inner product of $\boldsymbol{q}_i$ with $\boldsymbol{a}_j$. To see this, multiply to the left by $\boldsymbol{q}_1^{\mathrm{T}}$. Remember that $\boldsymbol{q}_i^{\mathrm{T}}\boldsymbol{q}_j = 0$ unless $i = j$:

$$r_{11} = (\boldsymbol{q}_1, \boldsymbol{a}_1)$$
$$r_{12} = (\boldsymbol{q}_1, \boldsymbol{a}_2)$$
$$r_{13} = (\boldsymbol{q}_1, \boldsymbol{a}_3).$$

Furthermore, a multiplication by $\boldsymbol{q}_2^{\mathrm{T}}$ yields

$$r_{22} = (\boldsymbol{q}_2, \boldsymbol{a}_2)$$
$$r_{23} = (\boldsymbol{q}_2, \boldsymbol{a}_3).$$

Finally, a multiplication by $\boldsymbol{q}_3^{\mathrm{T}}$ gives

$$r_{33} = (\boldsymbol{q}_3, \boldsymbol{a}_3).$$

In the following step we use the expressions (11.2), (11.3), and (11.4) to form the inner products, and we arrange the system into a recursive solution:

$$(\boldsymbol{a}_1, \boldsymbol{a}_1) = (\boldsymbol{q}_1, \boldsymbol{a}_1)^2$$
$$r_{11} = (\boldsymbol{q}_1, \boldsymbol{a}_1) = \sqrt{(\boldsymbol{a}_1, \boldsymbol{a}_1)}$$

$$(\boldsymbol{a}_1, \boldsymbol{a}_2) = (\boldsymbol{q}_1, \boldsymbol{a}_1)(\boldsymbol{q}_1, \boldsymbol{a}_2)$$
$$r_{12} = (\boldsymbol{q}_1, \boldsymbol{a}_2) = \frac{(\boldsymbol{a}_1, \boldsymbol{a}_2)}{(\boldsymbol{q}_1, \boldsymbol{a}_1)} = \frac{(\boldsymbol{a}_1, \boldsymbol{a}_2)}{\sqrt{(\boldsymbol{a}_1, \boldsymbol{a}_1)}}$$

$$(\boldsymbol{a}_1, \boldsymbol{a}_3) = (\boldsymbol{q}_1, \boldsymbol{a}_1)(\boldsymbol{q}_1, \boldsymbol{a}_3)$$
$$r_{13} = (\boldsymbol{q}_1, \boldsymbol{a}_3) = \frac{(\boldsymbol{a}_1, \boldsymbol{a}_3)}{\sqrt{(\boldsymbol{a}_1, \boldsymbol{a}_1)}}$$

$$(\boldsymbol{a}_2, \boldsymbol{a}_2) = (\boldsymbol{q}_1, \boldsymbol{a}_2)^2 + (\boldsymbol{q}_2, \boldsymbol{a}_2)^2$$
$$r_{22} = (\boldsymbol{q}_2, \boldsymbol{a}_2) = \sqrt{(\boldsymbol{a}_2, \boldsymbol{a}_2) - (\boldsymbol{q}_1, \boldsymbol{a}_2)^2}$$

$$(\boldsymbol{a}_2, \boldsymbol{a}_3) = (\boldsymbol{q}_1, \boldsymbol{a}_2)(\boldsymbol{q}_1, \boldsymbol{a}_3) + (\boldsymbol{q}_2, \boldsymbol{a}_2)(\boldsymbol{q}_2, \boldsymbol{a}_3)$$
$$r_{23} = (\boldsymbol{q}_2, \boldsymbol{a}_3) = \frac{(\boldsymbol{a}_2, \boldsymbol{a}_3) - (\boldsymbol{q}_1, \boldsymbol{a}_2)(\boldsymbol{q}_1, \boldsymbol{a}_3)}{(\boldsymbol{q}_2, \boldsymbol{a}_2)} = \frac{(\boldsymbol{a}_2, \boldsymbol{a}_3) - (\boldsymbol{q}_1, \boldsymbol{a}_2)(\boldsymbol{q}_1, \boldsymbol{a}_3)}{\sqrt{(\boldsymbol{a}_2, \boldsymbol{a}_2) - (\boldsymbol{q}_1, \boldsymbol{a}_2)^2}}$$

$$(a_3, a_3) = (q_1, a_3)^2 + (q_2, a_3)^2 + (q_3, a_3)^2$$
$$r_{33} = (q_3, a_3) = \sqrt{(a_3, a_3) - (q_2, a_3)^2 - (q_1, a_3)^2}.$$

It is useful to recognize that the upper triangular matrix R in $A = QR$ is also the Cholesky factor of the matrix $A^{\mathrm{T}}A$. If $A = QR$ then $A^{\mathrm{T}}A = (QR)^{\mathrm{T}}QR$. This equals $R^{\mathrm{T}}Q^{\mathrm{T}}QR$ which is $R^{\mathrm{T}}R$. Therefore R is the upper triangular Cholesky factor for $A^{\mathrm{T}}A$:

$$A^{\mathrm{T}}A \text{ factors into } R^{\mathrm{T}}R = (\text{ lower triangular })(\text{ upper triangular }).$$

In actual calculations this offers two ways to compute R:

1 Apply Gram-Schmidt to A.

2 Work with the coefficient matrix $A^{\mathrm{T}}A$.

The first method is slightly slower. For full matrices Gram-Schmidt needs about mn^2 separate multiplications. This method is more stable numerically (we mean modified Gram-Schmidt of course). The errors are roughly proportional to the condition number $c(A)$.

The second method (direct solution of the normal equations) is faster. For full matrices it takes $\frac{1}{2}mn^2$ multiplications and additions to compute the n^2 entries of $A^{\mathrm{T}}A$ (which is symmetric!). Then elimination requires about $\frac{1}{6}n^3$—again halved by symmetry. This method works directly with the normal equations and ***it is by far the most frequent choice in practice***, although numerically it is not quite as stable.

Example 11.1 We use this algorithm on Example 4.16 with $C = I$

$$A = \begin{bmatrix} 1 & 2 & 3 \\ -1 & 0 & -3 \\ 0 & -2 & 3 \end{bmatrix}$$

The result of the MATLAB command [Q,R]=qr(A) is

$$Q = \begin{bmatrix} -0.7071 & -0.4082 & 0.5774 \\ 0.7071 & -0.4082 & 0.5774 \\ 0 & 0.8165 & 0.5774 \end{bmatrix} \approx \begin{bmatrix} -\frac{1}{\sqrt{2}} & -\frac{1}{\sqrt{6}} & \frac{1}{\sqrt{3}} \\ \frac{1}{\sqrt{2}} & -\frac{1}{\sqrt{6}} & \frac{1}{\sqrt{3}} \\ 0 & \frac{2}{\sqrt{6}} & \frac{1}{\sqrt{3}} \end{bmatrix}$$

$$R = \begin{bmatrix} -1.4142 & -1.4142 & -4.2426 \\ 0 & -2.4495 & 2.4495 \\ 0 & 0 & 1.7321 \end{bmatrix} \approx \begin{bmatrix} -\sqrt{2} & -\sqrt{2} & -\sqrt{18} \\ 0 & -\sqrt{6} & \sqrt{6} \\ 0 & 0 & \sqrt{3} \end{bmatrix}.$$

Example 4.16 showed how these numbers arise in Gram-Schmidt.

Now suppose that C is not necessarily I. The normal equation matrix is $A^{\mathrm{T}}CA$. We still have $A = QR$, but now the columns q_i are orthogonal in the C-inner product:

$$q_i^{\mathrm{T}}Cq_j = \begin{cases} 1 & \text{if } i = j \\ 0 & \text{otherwise} \end{cases} \qquad \text{and} \qquad Q^{\mathrm{T}}CQ = I.$$

The matrix R is still a Cholesky factor!

$$N = A^{\mathrm{T}}CA = (QR)^{\mathrm{T}}C(QR) = R^{\mathrm{T}}(QCQ)R = R^{\mathrm{T}}R.$$

Notice that $R = R^{-\mathrm{T}}N$ and $N^{-1} = R^{-1}R^{-\mathrm{T}}$. Further we augment A by the right side $\boldsymbol{b}$: $[\,A \quad \boldsymbol{b}\,]$. Set $R^{\mathrm{T}}\boldsymbol{z} = A^{\mathrm{T}}C\boldsymbol{b}$ and $R\boldsymbol{x} = \boldsymbol{z}$. Then

$$N\boldsymbol{x} = R^{\mathrm{T}}R\boldsymbol{x} = R^{\mathrm{T}}\boldsymbol{z} = A^{\mathrm{T}}C\boldsymbol{b}.$$

Calculational procedure

1 Use Gram-Schmidt on A to obtain R

2 Calculate $\boldsymbol{z} = R^{-\mathrm{T}}A^{\mathrm{T}}C\boldsymbol{b}$

3 Solve $R\hat{\boldsymbol{x}} = \boldsymbol{z}$ by back substitution.

Thus the normal equations are solved.

A more sophisticated procedure is to augment the normal matrix by $\boldsymbol{b}$ to $\tilde{N}$:

$$\tilde{N} = \begin{bmatrix} A^{\mathrm{T}} \\ \boldsymbol{b}^{\mathrm{T}} \end{bmatrix} C \begin{bmatrix} A & \boldsymbol{b} \end{bmatrix} = \begin{bmatrix} A^{\mathrm{T}}CA & A^{\mathrm{T}}C\boldsymbol{b} \\ \boldsymbol{b}^{\mathrm{T}}CA & \boldsymbol{b}^{\mathrm{T}}C\boldsymbol{b} \end{bmatrix} \begin{bmatrix} R^{\mathrm{T}}R & R^{\mathrm{T}}Q^{\mathrm{T}}C\boldsymbol{b} \\ \boldsymbol{b}^{\mathrm{T}}CQR & \boldsymbol{b}^{\mathrm{T}}C\boldsymbol{b} \end{bmatrix}. \tag{11.5}$$

Simultaneously we augment the matrix R:

$$\tilde{R} = \begin{bmatrix} R & \boldsymbol{z} \\ \boldsymbol{0} & s \end{bmatrix}. \tag{11.6}$$

Then

$$\tilde{R}^{\mathrm{T}}\tilde{R} = \begin{bmatrix} R^{\mathrm{T}} & 0 \\ \boldsymbol{z}^{\mathrm{T}} & s \end{bmatrix} \begin{bmatrix} R & \boldsymbol{z} \\ \boldsymbol{0} & s \end{bmatrix} \begin{bmatrix} R^{\mathrm{T}}R & R^{\mathrm{T}}\boldsymbol{z} \\ \boldsymbol{z}^{\mathrm{T}}R & \boldsymbol{z}^{\mathrm{T}}\boldsymbol{z} + s^2 \end{bmatrix}. \tag{11.7}$$

Comparing with (11.5) we get

$$\boldsymbol{z}^{\mathrm{T}}\boldsymbol{z} + s^2 = \boldsymbol{b}^{\mathrm{T}}C\boldsymbol{b}.$$

Repeating Step 2 above we have $\boldsymbol{z} = R^{-1}A^{\mathrm{T}}C\boldsymbol{b}$ and

$$\boldsymbol{z}^{\mathrm{T}}\boldsymbol{z} = \boldsymbol{b}^{\mathrm{T}}CAR^{-1}R^{-\mathrm{T}}A^{\mathrm{T}}C\boldsymbol{b} = \boldsymbol{b}^{\mathrm{T}}CA\hat{\boldsymbol{x}} \quad \text{and} \quad s^2 = \boldsymbol{b}^{\mathrm{T}}C\boldsymbol{b} - \boldsymbol{b}CA\hat{\boldsymbol{x}} = \hat{\boldsymbol{r}}^{\mathrm{T}}C\hat{\boldsymbol{r}}.$$

If we put $\boldsymbol{b}^{\mathrm{T}}C\boldsymbol{b}$ in the lower left entry of $\tilde{R}$, then after the solution we recover $\hat{\boldsymbol{r}}^{\mathrm{T}}C\hat{\boldsymbol{r}}$ at the very same place. This square sum of residuals is valuable for estimating a lot of a posteriori variances. The residuals are defined as $A\hat{\boldsymbol{x}} = \boldsymbol{b} - \hat{\boldsymbol{r}}$.

11.2 Cholesky's Method in the Least-Squares Setting

We shall recast the least-squares problem and again solve it by means of orthogonal projections. We consider the present procedure as the natural one from a geometrical point of view. The interesting fact to discover is that we do not need to know the orthonormalized columns $\boldsymbol{q}_i$ of A explicitly. The idea is to make a conceptual short cut which offers a solution to the least-squares problem directly from the method of Cholesky. It is considered an essential contribution and not just a nice variation of the traditional procedure!

Let the observations be collected in the vector $\boldsymbol{b}$ and let $\boldsymbol{A}_0$ be described in parameter form as the subspace of all $A\boldsymbol{x}$ (the column space) shifted by $\boldsymbol{t}$:

$$\boldsymbol{y} = A\boldsymbol{x} + \boldsymbol{t}.$$

The least-squares problem is to determine that point $\boldsymbol{p} \in \boldsymbol{A}_0$ which is closest to $\boldsymbol{b}$. It is also the projection of $\boldsymbol{b}$ onto $\boldsymbol{A}_0$. For practical reasons we first translate by $-\boldsymbol{t}$; hence the problem is to find the projection of $\boldsymbol{l} = \boldsymbol{b} - \boldsymbol{t}$ onto the linear subspace of all $\boldsymbol{y} = A\boldsymbol{x}$. The resulting $\hat{\boldsymbol{x}}$ remains unchanged under this translation; but we must add $\boldsymbol{t}$ to the solution $\boldsymbol{p} = A\hat{\boldsymbol{x}}$.

To repeat: We want to find $\hat{\boldsymbol{x}}$ such that

$$\boldsymbol{l} - A\hat{\boldsymbol{x}} \perp A\boldsymbol{x} \qquad \text{for all } \boldsymbol{x}. \tag{11.8}$$

In other words, we want to split $\boldsymbol{l} = \boldsymbol{p} + \boldsymbol{r}$ into $A\hat{\boldsymbol{x}}$ + perpendicular error.

If the columns $\boldsymbol{a}_i$ of A are the orthonormalized vectors $\boldsymbol{q}_i$ we simply have

$$\boldsymbol{p} = \sum_{i=1}^{n} (\boldsymbol{q}_i, \boldsymbol{l})\boldsymbol{q}_i \qquad \text{and} \qquad \boldsymbol{r} = \boldsymbol{l} - \boldsymbol{p}. \tag{11.9}$$

The inner product $(\boldsymbol{a}, \boldsymbol{b})$ is defined as $(\boldsymbol{a}, \boldsymbol{b}) = \boldsymbol{a}^{\mathrm{T}} C \boldsymbol{b}$.

If the columns are not orthonormalized, we can make them so: Normalize the first column $\boldsymbol{a}_1$. Suppose the first i columns are orthonormalized. Then we simply write the next column as $\boldsymbol{a}_{i+1} = \boldsymbol{p} + \boldsymbol{r}$ where $\boldsymbol{p}$ is in the linear subspace spanned by the first i columns—orthonormalized or not—and $\boldsymbol{r}$ is perpendicular to this subspace. Thus we can use the Gram-Schmidt procedure as described in an earlier section and the orthonormal version of $\boldsymbol{a}_{i+1}$ is $\boldsymbol{p}/\|\boldsymbol{p}\| = \boldsymbol{q}_{i+1}$.

Now we present the recursive procedure for solving the least-squares problem given by the observation equation matrix A and the observations—the right side $\boldsymbol{b}$—via the Cholesky method:

$$\begin{aligned}
\boldsymbol{a}_1 &= (\boldsymbol{q}_1, \boldsymbol{a}_1)\boldsymbol{q}_1 \\
(\boldsymbol{a}_1, \boldsymbol{a}_1) &= (\boldsymbol{q}_1, \boldsymbol{a}_1)^2 \\
(\boldsymbol{q}_1, \boldsymbol{a}_1) &= \sqrt{(\boldsymbol{a}_1, \boldsymbol{a}_1)}
\end{aligned}$$

$$\begin{aligned}
\boldsymbol{a}_2 &= (\boldsymbol{q}_1, \boldsymbol{a}_2)\boldsymbol{q}_1 + (\boldsymbol{q}_2, \boldsymbol{a}_2)\boldsymbol{q}_2 \\
(\boldsymbol{a}_1, \boldsymbol{a}_2) &= (\boldsymbol{q}_1, \boldsymbol{a}_1)(\boldsymbol{q}_1, \boldsymbol{a}_2) \\
(\boldsymbol{q}_1, \boldsymbol{a}_2) &= \frac{(\boldsymbol{a}_1, \boldsymbol{a}_2)}{(\boldsymbol{q}_1, \boldsymbol{a}_1)} \\
(\boldsymbol{a}_2, \boldsymbol{a}_2) &= (\boldsymbol{q}_1, \boldsymbol{a}_2)^2 + (\boldsymbol{q}_2, \boldsymbol{a}_2)^2 \\
(\boldsymbol{q}_2, \boldsymbol{a}_2) &= \sqrt{(\boldsymbol{a}_2, \boldsymbol{a}_2) - (\boldsymbol{q}_1, \boldsymbol{a}_2)^2}
\end{aligned}$$

$$\begin{aligned}
\boldsymbol{a}_3 &= (\boldsymbol{q}_1, \boldsymbol{a}_3)\boldsymbol{q}_1 + (\boldsymbol{q}_2, \boldsymbol{a}_3)\boldsymbol{q}_2 + (\boldsymbol{q}_3, \boldsymbol{a}_3)\boldsymbol{q}_3 \\
(\boldsymbol{a}_1, \boldsymbol{a}_3) &= (\boldsymbol{q}_1, \boldsymbol{a}_1)(\boldsymbol{q}_1, \boldsymbol{a}_3) \\
(\boldsymbol{q}_1, \boldsymbol{a}_3) &= \frac{(\boldsymbol{a}_1, \boldsymbol{a}_3)}{(\boldsymbol{q}_1, \boldsymbol{a}_1)}
\end{aligned}$$

$$
\begin{aligned}
(\boldsymbol{a}_2, \boldsymbol{a}_3) &= (\boldsymbol{q}_1, \boldsymbol{a}_2)(\boldsymbol{q}_1, \boldsymbol{a}_3) + (\boldsymbol{q}_2, \boldsymbol{a}_2)(\boldsymbol{q}_2, \boldsymbol{a}_3) \\
(\boldsymbol{q}_2, \boldsymbol{a}_3) &= \frac{(\boldsymbol{a}_2, \boldsymbol{a}_3) - (\boldsymbol{q}_1, \boldsymbol{a}_2)(\boldsymbol{q}_1, \boldsymbol{a}_3)}{(\boldsymbol{q}_2, \boldsymbol{a}_2)} \\
(\boldsymbol{a}_3, \boldsymbol{a}_3) &= (\boldsymbol{q}_1, \boldsymbol{a}_3)^2 + (\boldsymbol{q}_2, \boldsymbol{a}_3)^2 + (\boldsymbol{q}_3, \boldsymbol{a}_3)^2 \\
(\boldsymbol{q}_3, \boldsymbol{a}_3) &= \sqrt{(\boldsymbol{a}_3, \boldsymbol{a}_3) - (\boldsymbol{q}_2, \boldsymbol{a}_3)^2 - (\boldsymbol{q}_1, \boldsymbol{a}_3)^2}.
\end{aligned}
$$

We compare the present Cholesky method with the QR factorization on page 370. In both cases we have

$$
\begin{aligned}
r_{11} &= \sqrt{(\boldsymbol{a}_1, \boldsymbol{a}_1)} \quad r_{12} = \frac{(\boldsymbol{a}_1, \boldsymbol{a}_2)}{\sqrt{(\boldsymbol{a}_1, \boldsymbol{a}_1)}} \quad && r_{13} = \frac{(\boldsymbol{a}_1, \boldsymbol{a}_3)}{\sqrt{(\boldsymbol{a}_1, \boldsymbol{a}_1)}} \\
r_{22} &= \sqrt{(\boldsymbol{a}_2, \boldsymbol{a}_2) - (\boldsymbol{q}_1, \boldsymbol{a}_2)^2} && r_{23} = \frac{(\boldsymbol{a}_2, \boldsymbol{a}_3) - (\boldsymbol{q}_1, \boldsymbol{a}_2)(\boldsymbol{q}_1, \boldsymbol{a}_3)}{\sqrt{(\boldsymbol{a}_2, \boldsymbol{a}_2) - (\boldsymbol{q}_1, \boldsymbol{a}_2)^2}} \\
& && r_{33} = \sqrt{(\boldsymbol{a}_3, \boldsymbol{a}_3) - (\boldsymbol{q}_2, \boldsymbol{a}_3)^2 - (\boldsymbol{q}_1, \boldsymbol{a}_3)^2}.
\end{aligned}
$$

By this we have established that the upper triangular matrix R in $A = QR$ is also the Cholesky factor of the matrix $A^{\mathrm{T}}A$ as proved directly on page 370.

The interesting fact is that we do not need to find the orthonormal columns $\boldsymbol{q}_i$ explicitly. So we have combined the Gram-Schmidt process and the normal equations and solve the latter by the Cholesky method.

We emphasize that (11.9) is equivalent to the normal equations.

As we proceed through this book we will recognize that ***matrix inversion only is needed for covariance information***, not for equation solving. In both cases we prefer procedures based on the Cholesky factorization.

Example 11.2 We describe a least-squares problem by the A matrix given in Example 11.1, except we delete the last column:

$$
A = \begin{bmatrix} 1 & 2 \\ -1 & 0 \\ 0 & -2 \end{bmatrix}.
$$

The right side is $\boldsymbol{b} = (1, -2, 2)$ and the weight matrix $C = I$. From the above procedure follows

$$
R = \begin{bmatrix} (\boldsymbol{q}_1, \boldsymbol{a}_1) & (\boldsymbol{q}_1, \boldsymbol{a}_2) \\ 0 & (\boldsymbol{q}_2, \boldsymbol{a}_2) \end{bmatrix} = \begin{bmatrix} \sqrt{2} & \sqrt{2} \\ 0 & \sqrt{6} \end{bmatrix}.
$$

Then

$$
R^{-1} = \frac{1}{\sqrt{12}} \begin{bmatrix} \sqrt{6} & -\sqrt{2} \\ 0 & \sqrt{2} \end{bmatrix} \quad \text{and} \quad A^{\mathrm{T}}\boldsymbol{b} = \begin{bmatrix} 3 \\ -2 \end{bmatrix}.
$$

Next

$$
z = R^{-\mathrm{T}} A^{\mathrm{T}} \boldsymbol{b} = \begin{bmatrix} \frac{3}{\sqrt{2}} \\ -\frac{5}{\sqrt{6}} \end{bmatrix}
$$

and the solution is

$$\hat{x} = R^{-1}z = \begin{bmatrix} \frac{7}{3} \\ -\frac{5}{6} \end{bmatrix}.$$

Finally the projection $\boldsymbol{p}$ of $\boldsymbol{b}$ is

$$\boldsymbol{p} = A\hat{x} = \frac{1}{3}\begin{bmatrix} 2 \\ -7 \\ 5 \end{bmatrix}.$$

According to (11.8) we also have

$$\boldsymbol{p} = (\boldsymbol{q}_1, \boldsymbol{b})\boldsymbol{q}_1 + (\boldsymbol{q}_2, \boldsymbol{b})\boldsymbol{q}_2 = \frac{3}{\sqrt{2}}\begin{bmatrix} \frac{1}{\sqrt{2}} \\ -\frac{1}{\sqrt{2}} \\ 0 \end{bmatrix} - \frac{5}{\sqrt{6}}\begin{bmatrix} \frac{1}{\sqrt{6}} \\ \frac{1}{\sqrt{6}} \\ \frac{2}{\sqrt{6}} \end{bmatrix} = \frac{1}{3}\begin{bmatrix} 2 \\ -7 \\ 5 \end{bmatrix}.$$

The last expression is based on the explicit knowledge of $\boldsymbol{q}_i$ whereas all other derivations in this example are not. The computationally intensive part is obtaining the inverse of R (or solving the triangular system for z).

11.3 SVD: The Canonical Form for Geodesy

In this section we want to introduce new coordinate systems in which the least squares problem become simpler and thus more lucid. After the coordinate transformation we say that we have obtained the "canonical form" of the least squares problem.

Let the linearized observation equations be

$$A\boldsymbol{x} = \boldsymbol{b} \tag{11.10}$$

where $\boldsymbol{x}$ is an n-dimensional vector decribing the corrections to the unknowns (coordinates), and $\boldsymbol{b}$ is the m-dimensional vector of observations. They are presumed uncorrelated and with equal weight. In other words, the covariance matrix for the observations is $\Sigma_b = I$.

The singular value decomposition demonstrates that it always is possible to find an orthogonal matrix V (in the row space of A or the coordinate space) and another U (in the column space of A or the observation space) so that

$$\boldsymbol{y} = V\boldsymbol{x} \qquad \text{and} \qquad \boldsymbol{c} = U\boldsymbol{b}. \tag{11.11}$$

Consequently equation (11.10) can be written

$$B\boldsymbol{y} = \boldsymbol{c} \tag{11.12}$$

where

$$B = UAV^{\mathrm{T}}. \tag{11.13}$$

The SVD chooses U and V so that B is of the form

$$B = \overset{n}{\begin{bmatrix} D \\ \hline \Theta \end{bmatrix}} \begin{matrix} \}\, n \\ \}\, m-n \end{matrix} \tag{11.14}$$

with the diagonal matrix D

$$D = \begin{bmatrix} \sigma_1 & & & \\ & \sigma_2 & & \\ & & \ddots & \\ & & & \sigma_n \end{bmatrix}. \tag{11.15}$$

The matrix Θ is an $(m-n)$ by n zero matrix. Furthermore, all the singular values σ's are ordered decreasingly

$$\sigma_1 \geq \sigma_2 \geq \sigma_3 \geq \cdots \geq \sigma_n \geq 0.$$

Evidently we have

$$B^{\mathrm{T}}B = VA^{\mathrm{T}}U^{\mathrm{T}}UAV^{\mathrm{T}} = VA^{\mathrm{T}}AV^{\mathrm{T}} = D^2 \tag{11.16}$$

and

$$BB^{\mathrm{T}} = UAV^{\mathrm{T}}VA^{\mathrm{T}}U^{\mathrm{T}} = UAA^{\mathrm{T}}U^{\mathrm{T}} = \left[\begin{array}{c|c} D^2 & \Theta \\ \hline \Theta & \Theta \end{array}\right]. \tag{11.17}$$

Remember that an orthogonal matrix Q has $Q^{\mathrm{T}}Q = QQ^{\mathrm{T}} = I$. Now put

$$V^{\mathrm{T}} = (\varphi_1, \varphi_2, \ldots, \varphi_n), \tag{11.18}$$

and

$$U^{\mathrm{T}} = (\psi_1, \psi_2, \ldots, \psi_n, \rho_1, \rho_2, \ldots, \rho_{m-n}), \tag{11.19}$$

where $\{\varphi_i\}$ denotes an orthonormal set of n-dimensional vectors, and $\{\psi_i\} \cup \{\rho_i\}$ is another set of m-dimensional vectors. The transposed version of equation (11.11) shows that $\boldsymbol{y}$ can be looked upon as the coefficients in the expansion of the vector $\boldsymbol{x}$ on the set $\{\varphi_i\}$, and correspondingly $\boldsymbol{c}$ can be conceived as coefficients in the expansion of $\boldsymbol{b}$ on $\{\psi_i\}$ and $\{\rho_i\}$. We call $\{\varphi_i\}$, $\{\psi_i\}$ and $\{\rho_i\}$ the *first*, the *second*, and the *third* set of ***canonical vectors*** even though they are not determined uniquely.

A statistical interpretation of the canonical form of the least-squares problem is given by Scheffé (1959), Chapter I. The space spanned by the $\{\rho_i\}$ vectors is called the ***error space***, and the space spanned by the $\{\varphi_i\}$ is called the ***estimation space***.

On the basis of (11.13) and (11.14) it is possible to show that the following relation between the first two sets of canonical vectors is valid:

$$A\varphi_i = \sigma_i\psi_i, \qquad i = 1, 2, \ldots, n. \tag{11.20}$$

Similarly, from the transformed version of (11.13) and (11.14) follows

$$A^{\mathrm{T}}\psi_i = \sigma_i\varphi_i, \qquad i = 1, 2, \ldots, n \tag{11.21}$$

$$A^{\mathrm{T}}\rho_i = 0, \qquad i = 1, 2, \ldots, m-n. \tag{11.22}$$

As the orthogonal transformation U of $\boldsymbol{b}$ leaves the observations $\boldsymbol{c}$ in (11.12) weight normalized we can solve the least-squares problem via the normal equations

$$B^{\mathrm{T}}B\boldsymbol{y} = B^{\mathrm{T}}\boldsymbol{c} \tag{11.23}$$

or

$$\sigma_i^2 y_i = \sigma_i c_i, \quad i = 1, 2, \ldots, n \qquad \text{or finally} \qquad y_j = \frac{c_j}{\sigma_j}, \quad j = 1, 2, \ldots, r$$

where r denotes the number of nonzero σ_j's and the Latin subscript j denotes the j'th component of these vectors.

These facts lead to the following important consequences:

1 About the *first* set of canonical vectors: The components of $\boldsymbol{x}$ which are best determined are those in directions defined by φ_i with large σ_i; entries in the direction of φ_i with $\sigma_i = 0$ are totally undetermined.

In order that this result be interpreted correctly the corrections of the coordinates must be measured in approximately the same unit. Usually one wants that all entries of $\boldsymbol{x}$ are determined with equal accuracy.

2 The *second* set reveals the observations which should have been performed with larger weight. It is so because we expand the individual observations—i.e. the unit vectors in the observation space—in $\{\psi_i\}$ and $\{\rho_i\}$ and subsequently choose those with dominating coefficients and corresponding to small values of $\sigma_i \neq 0$. These coefficients can be found by inspection of the matrix U. This is a consequence of the property $U^{\mathrm{T}}U = I$ which again can be taken as the one which yields the expansion of unit vectors into the sets $\{\psi_i\}$ and $\{\rho_i\}$.

3 For the *third* set we realize that the entries of the observations in directions determined by $\{\rho_i\}$ do not give any new information about the coordinates. This is also valid for ψ_i with $i > r$, i.e. the eigenvectors corresponding to eigenvalues $\sigma_i = 0$. Maybe it is relevant to define the ***redundancy*** for the i'th observation in the following manner

$$\mathrm{red}_i = \sqrt{\sum_{j=r+1}^{m} u_{ji}^2}. \tag{11.24}$$

11.4 The Condition Number

Geodetic network analysis is based on the fact that we can build the left side of the normals $A^{\mathrm{T}}CA$ when we know the topology of the network, described by A. We assume known weights C of the observations. Hence we may calculate the covariance matrix $\Sigma_x = (A^{\mathrm{T}}CA)^{-1}$. The actual observations only enter into the right side. So much network analysis can be performed without taking a single measurement.

One good measure for comparing various networks is the *condition number*. In the next section we define the condition number and demonstrate its use in connection with simple 1- and 2-dimensional geodetic networks.

The norm and condition number of a matrix play a vital role in the numerical calculations and design problems of geodesy. Norm inequalities for vectors and matrices often are used for *estimating the influence of roundoff errors and observational errors*. The condition number also can be introduced in the optimum *design of networks*. This number helps to decide if one network design is better than another.

The following matrix is typical for many least-squares problems:

$$A = \begin{bmatrix} 2 & -1 & & & \\ -1 & 2 & -1 & & \\ & & \ddots & & \\ & & -1 & 2 & -1 \\ & & & -1 & 2 \end{bmatrix}.$$

It is the coefficient matrix in the normal equation for

1 a regular traverse along the x-axis with postulated abscissas at both terminals. Only distances are measured and with equal weights.

2 a leveling line with postulated heights at both terminals. All observations are of equal weight.

3 A^2 is the coefficient matrix for a regular traverse along the x-axis with postulated y values at the terminals. All angles are supposed to be measured with equal weights.

The eigenvalues of A are

$$\lambda_i = 4\sin^2\frac{i\pi}{2(n+1)}, \qquad i = 1, 2, \dots, n. \tag{11.25}$$

The condition number of a positive definite matrix is $\lambda_{\max}/\lambda_{\min}$. For this example we find

$$c(A) = \frac{\lambda_{\max}}{\lambda_{\min}} = \frac{4\sin^2\frac{n\pi}{2(n+1)}}{4\sin^2\frac{\pi}{2(n+1)}} \approx \left(\frac{2(n+1)}{\pi}\right)^2 \approx 0.4n^2.$$

Finally $c(A^2) = c(A)^2 \approx 0.2n^4$.

This can be extended to a 2-dimensional leveling network, covering a rectangular area subdivided into m by n squares. All differences of height between neighboring points are observed with equal weight. The eigenvalues for this normal equation coefficient matrix are

$$\lambda_{jk} = 4\left(\sin^2\frac{j\pi}{2(m+1)} + \sin^2\frac{k\pi}{2(n+1)}\right), \qquad \begin{cases} j = 1, 2, \dots, m, \\ k = 1, 2, \dots, n. \end{cases}$$

Let $p = \max(m, n)$ and we get the following estimate of the condition of the network:

$$c = \frac{\cos^2\frac{\pi}{2(m+1)} + \cos^2\frac{\pi}{2(n+1)}}{\sin^2\frac{\pi}{2(p+1)}} \approx \frac{8(p+1)^2}{\pi^2} < p^2, \quad 0 < \tfrac{m}{n} < \text{constant}, \quad p > 10.$$

Comparing this condition number with the one from the leveling line we see that the condition number is doubled for the rectangular network. At the same time the number of points increased by the square of p. We may conclude: *If the difference of height between two points has to be determined as accurately as possible, the network cannot be too "narrow." A genuinely 2-dimensional network has better condition number*. Of course, the observational work increases tremendously compared to the accuracy. But this just confirms that accuracy costs money.

After this summary of elementary network analysis we shall derive the results in detail.

11.5 Regularly Spaced Networks

1-Dimensional Networks

We start by considering 1-dimensional networks as shown in the Figure 11.1 below. A straight line $P_1 P_n$ is subdivided into smaller segments by the points $P_2, P_3, \dots, P_{n-1}$. The length of each line segment between two consecutive points P_{i-1} and P_i is measured and the logarithm of the measured quantity is given with weight c_i. Thus the observation equations are

$$\ln(x_i - x_{i-1}) = s_i, \qquad \text{weight } c_i$$

where x_i is the abscissa of point P_i. The linearized, and weight normalized observation equations are

$$\sqrt{c_i}\,\frac{dx_i - dx_{i-1}}{x_i^0 - x_{i-1}^0} = \sqrt{c_i}\,ds_i, \qquad i = 2, 3, \dots, n.$$

We set

$$a_i = \frac{c_i}{(x_i^0 - x_{i-1}^0)^2};$$

then the $(n-2)$ by $(n-2)$ coefficient matrix of the normal equations is

$$N_d = \begin{bmatrix} a_2 + a_3 & -a_3 & & & & \\ -a_3 & a_3 + a_4 & -a_4 & & & \\ & -a_4 & a_4 + a_5 & -a_5 & & \\ & & & \ddots & & \\ & & & & -a_{n-1} & a_{n-1} + a_n \end{bmatrix}. \tag{11.26}$$

We have omitted the unknowns x_1^0 and x_n^0 corresponding to fixing the network at the points P_1 and P_n.

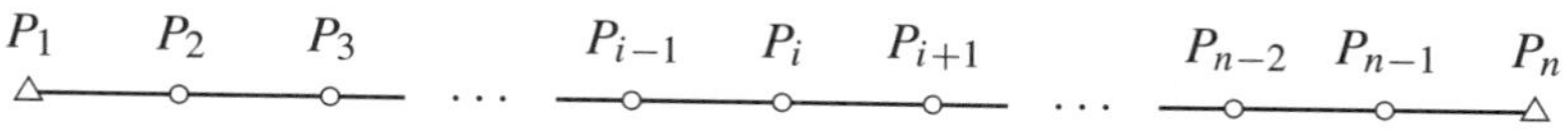

Figure 11.1 Regular traverse.

The matrix N_d is now transformed to $N_p = DN_dD$ with the transforming diagonal matrix

$$D = \operatorname{diag}\bigl(-1, 1, -1, \dots, (-1)^{n-2}\bigr).$$

The result is

$$N_p = \begin{bmatrix} a_2+a_3 & a_3 & & & & \\ a_3 & a_3+a_4 & a_4 & & & \\ & a_4 & a_4+a_5 & a_5 & & \\ & & \ddots & \ddots & \ddots & \\ & & & a_{n-2} & a_{n-2}+a_{n-1} & a_{n-1} \\ & & & & a_{n-1} & a_{n-1}+a_n \end{bmatrix}. \tag{11.27}$$

Evidently, all entries of N_p are positive. Since the matrix is positive definite all its principal minors are positive; therefore the theory of ***oscillating matrices*** applies to N_p, see Gantmacher (1959), Chapter III, § 9.

If we transfer this theory for matrix (11.27) onto matrix (11.26) we get:

Theorem 11.1 Let the eigenvalues of N_d be ordered as

$$0 < \lambda_1 < \lambda_2 < \dots < \lambda_{n-2}.$$

Let the eigenvector for λ_1 be φ_1; the components of this eigenvector are nonzero and have the same sign. The eigenvector φ_2 corresponding to λ_2 has a sequence of components with one change of sign. Generally, the eigenvector φ_k corresponding to eigenvalue λ_k has exactly $k-1$ changes of sign in its components.

For $a_i = k,\ i = 2, 3, \dots, n$, we find the explicit expressions for the eigenvalues

$$\lambda_i = 4k \sin^2 \frac{i\pi}{2(n-1)}, \qquad i = 1, 2, \dots, n-2 \tag{11.28}$$

and non-normalized eigenvectors

$$\varphi_j(i) = \sin \frac{ij\pi}{n-1}, \qquad i, j = 1, 2, \dots, n-2. \tag{11.29}$$

The condition number is

$$c(N_d) = \left(\frac{\sin \frac{(n-2)\pi}{2(n-1)}}{\sin \frac{\pi}{2(n-1)}} \right)^2 \approx \left(\frac{2(n-1)}{\pi} \right)^2.$$

From theorems in Gantmacher & Krein (1960), p. 127–129 the following inequalities can be set up

$$\min(a_i + a_{i+1}) - 2\max(a_i)\cos\frac{\pi}{n-1} \le \lambda_1 \le \max(a_i + a_{i+1}) - 2\min(a_i)\cos\frac{\pi}{n-1}$$

$$\min(a_i + a_{i+1}) + 2\min(a_i)\cos\frac{\pi}{n-1} \le \lambda_n \le \max(a_i + a_{i+1}) + 2\max(a_i)\cos\frac{\pi}{n-1}.$$

For $k = \min(a_i + a_{i+1}) = \max(a_i + a_{i+1}) = 2\min(a_i) = 2\max(a_i)$ we get

$$\lambda_1 = 2k \sin^2 \frac{\pi}{2(n-1)}$$
$$\lambda_n = 2k \cos^2 \frac{\pi}{2(n-1)} = 2k \sin^2 \frac{n\pi}{2(n-1)}.$$

Example 11.3 In order to illustrate the above theory we perform a small calculational experiment in MATLAB. Let

$$N_d = \begin{bmatrix} 2 & -1 & 0 & 0 \\ -1 & 2 & -1 & 0 \\ 0 & -1 & 2 & -1 \\ 0 & 0 & -1 & 2 \end{bmatrix} \quad \text{and} \quad D = \begin{bmatrix} -1 & 0 & 0 & 0 \\ 0 & 1 & 0 & 0 \\ 0 & 0 & -1 & 0 \\ 0 & 0 & 0 & 1 \end{bmatrix}.$$

Then

$$N_p = \begin{bmatrix} 2 & 1 & 0 & 0 \\ 1 & 2 & 1 & 0 \\ 0 & 1 & 2 & 1 \\ 0 & 0 & 1 & 2 \end{bmatrix}$$

which obviously has positive entries. The eigenvectors are the columns of

$$\varphi = \begin{bmatrix} 0.588 & 0.951 & 0.951 & 0.588 \\ 0.951 & 0.588 & -0.588 & -0.951 \\ 0.951 & -0.588 & -0.588 & 0.951 \\ 0.588 & -0.951 & 0.951 & -0.588 \end{bmatrix}$$

and the eigenvalues are arranged in increasing order

$$\lambda = \begin{bmatrix} 0.382 & 1.382 & 2.618 & 3.618 \end{bmatrix}.$$

Now note that φ_1 has no change of sign, φ_2 has one change of sign, φ_3 has two changes of sign, and finally φ_4 has three changes of sign.

Another situation is the following: let the length of the single sub-intervals be known but suppose the line $P_1 P_n$ has infinitesimal breaks at points $P_2, P_3, \dots, P_{n-1}$ and further suppose that the angles β_i at P_i between $P_i P_{i-1}$ and $P_i P_{i+1}$ for $i = 2, 3, \dots, n-1$ were measured. These angles are likely to be close to π. The linearized observation equations are

$$\sqrt{c_i}\left(\frac{dy_i - dy_{i-1}}{x_i^0 - x_{i-1}^0} - \frac{dy_{i+1} - dy_i}{x_{i+1}^0 - x_i^0}\right) = \sqrt{c_i}\, d\beta_i, \qquad i = 2, 3, \dots, n.$$

We put

$$a_i = \frac{\sqrt{c_i}}{x_i^0 - x_{i-1}^0} \quad \text{and} \quad b_i = \frac{\sqrt{c_i}}{x_{i+1}^0 - x_i^0}.$$

Then the upper left 3 by 3 entries of the normal equation matrix N_a becomes

$$\begin{bmatrix} (a_2+b_2)^2+a_3^2 & -b_2(a_2+b_2)-a_3(a_3+b_3) & a_3b_3 \\ -b_2(a_2+b_2)-a_3(a_3+b_3) & b_2^2+(a_3+b_3)^2+a_4^2 & -b_3(a_3+b_3)-a_4(a_4+b_4) \\ a_3b_3 & -b_3(a_3+b_3)-a_4(a_4+b_4) & b_3^2+(a_4+b_4)^2+a_5^2 \end{bmatrix}.$$

The matrix N_a can be transformed in a manner similar to N_d to obtain an oscillating matrix. Theorem 11.1 is valid for the eigenvalues and eigenvectors of N_a.

In order to obtain explicit results let us set $a_i^2 = b_i^2 = k$. Then the eigenvectors are those given in (11.29) and the eigenvalues are

$$\lambda_i = 16k \sin^4 \frac{i\pi}{2(n-1)}$$

and the condition number is

$$c(N_a) = \left(\frac{\sin \frac{(n-2)\pi}{2(n-1)}}{\sin \frac{\pi}{2(n-1)}} \right)^4 \approx \left(\frac{2(n-1)}{\pi} \right)^4.$$

We repeat the derivations and this time we combine both distance and angle observations. We restrict all side lengths to be 1. That leads to simpler expressions. Again we introduce unknowns $x_2, x_3, \ldots, x_{n-1}$, and $y_2, y_3, \ldots, y_{n-1}$, see Figure 11.1. The distance observations $s_{i,i+1}$ only depend on the x_i's and are of a simple linear type

$$s_{i,i+1} + r_{i,i+1} = -x_i + x_{i+1}.$$

The angular observations β_i depend only on the y_i's:

$$\beta_i = \arctan(y_i - y_{i+1}) - \arctan(y_{i-1} - y_i).$$

A linearization yields the following

$$\beta_i + r_i = y_i - y_{i+1} - (y_{i-1} - y_i) = -y_{i-1} + 2y_i - y_{i+1}.$$

We gather all the linearized observation equations in matrix form

$$\begin{bmatrix} 1 & & & & & & & & \\ -1 & 1 & & & & & & & \\ & \ddots & \ddots & & & & & & \\ & & -1 & 1 & & & & & \\ & & & -1 & & & & & \\ & & & & -1 & & & & \\ & & & & 2 & -1 & & & \\ & & & & -1 & 2 & -1 & & \\ & & & & & \ddots & \ddots & \ddots & \\ & & & & & & -1 & 2 & -1 \\ & & & & & & & -1 & 2 \end{bmatrix} \begin{bmatrix} x_2 \\ x_3 \\ \vdots \\ x_{n-2} \\ x_{n-1} \\ y_2 \\ y_3 \\ y_4 \\ \vdots \\ y_{n-3} \\ y_{n-2} \\ y_{n-1} \end{bmatrix} = \begin{bmatrix} A_1 & 0 \\ 0 & A_2 \end{bmatrix} \boldsymbol{x} = \boldsymbol{b} - \boldsymbol{r}. \tag{11.30}$$

The number of observations is $2n-1$, the number of unknowns is $2n-4$, hence 3 redundant observations. We assume all observations of equal weight: $C = I$. The problem evidently splits into two: one for distances and one for angles, as the distances only depend on x and the angles on y. For the distances we get

$$A_1^{\mathrm{T}} A_1 = \begin{bmatrix} 2 & -1 & & & \\ -1 & 2 & -1 & & \\ & \ddots & \ddots & \ddots & \\ & & -1 & 2 & -1 \\ & & & -1 & 2 \end{bmatrix}.$$

This $(n-2)$ by $(n-2)$ matrix is simple and we immediately quote the explicit expression for the inverse matrix

$$(A_1^{\mathrm{T}} A_1)_{ij}^{-1} = \frac{j(n-1-i)}{n-1} \qquad \text{for } i \le j \text{ and else symmetric.}$$

Especially note the variance at the mid-point $i = j = (n-1)/2$, which is

$$\sigma_{\hat{x}}^2 = \frac{n-1}{2} \frac{n-1-\frac{n-1}{2}}{n-1} \sigma_s^2 = \frac{n-1}{4} \sigma_s^2.$$

This result is in good agreement with the one from conventional least-squares.

The eigenvalues already are given in (11.28):

$$\lambda_i = 4 \sin^2 \frac{i\pi}{2(n-1)}, \qquad i = 1, 2, \ldots, n-2.$$

The condition number is $c(A_1^{\mathrm{T}} A_1) = \lambda_{\max}/\lambda_{\min} \approx 4(n-1)^2 \pi^{-2}$, and the norm is

$$\|A_1^{\mathrm{T}} A_1\|_2 = 4 \sin^2 \frac{(n-2)\pi}{2(n-1)}.$$

Likewise for the angular observations

$$A_2^{\mathrm{T}} A_2 = \begin{bmatrix} 6 & -4 & 1 & & & & \\ -4 & 6 & -4 & 1 & & & \\ 1 & -4 & 6 & -4 & 1 & & \\ & \ddots & \ddots & \ddots & \ddots & \ddots & \\ & & 1 & -4 & 6 & -4 & 1 \\ & & & 1 & -4 & 6 & -4 \\ & & & & 1 & -4 & 6 \end{bmatrix}.$$

The inverse of this $(n-2)$ by $(n-2)$ matrix is

$$(A_2^{\mathrm{T}} A_2)_{ij}^{-1} = \frac{i(i+1)(n-1-j)(n-j)}{6n(n^2-1)} \Big(2i(n-1-j) + (3n-1)(j-i) + n + 1 \Big).$$

This result is due to Torben Krarup who derived it for the present problem. He used a Gauss-like elimination technique.

Historically geodesists have focused on the diagonal term $i = j = (n-1)/2$:

$$(A_2^{\mathrm{T}} A_2)^{-1}_{\frac{n-1}{2},\frac{n-1}{2}} = \frac{\big(\frac{n-1}{2}\big)\big(\frac{n-1}{2}+1\big)\big(n-1-\frac{n-1}{2}\big)\big(n-\frac{n-1}{2}\big)}{6n(n^2-1)} \times \Big[(n-1)\big(n-1-\tfrac{n-1}{2}\big)+n+1\Big]$$

$$= \frac{n^2-1}{192n}\Big[(n-1)^2+2(n+1)\Big] = \frac{(n^2-1)(n^2+3)}{192n}.$$

This expression demonstrates that the standard deviation of the mid-point in a regular traverse grows like $n^{3/2}$.

2-Dimensional Networks

An adequate description of adjustment problems in 2-dimensions is obtained by introducing the *Kronecker product*. Let $A = (a_{ij})$ and $B = (b_{ij})$ be m by n and p by q matrices, respectively. Then the Kronecker product

$$A \otimes B = (a_{ij} B) \tag{11.31}$$

is a mp by nq matrix expressible as a partitioned matrix with $a_{ij}B$ as the (i, j)th partition, $i = 1, \dots, m$; $j = 1, \dots, n$.

The formal rules for operating with Kronecker products are as follows:

$$0 \otimes A = A \otimes 0 = 0 \tag{i}$$

$$(A_1 + A_2) \otimes B = (A_1 \otimes B) + (A_2 \otimes B) \tag{ii}$$

$$A \otimes (B_1 + B_2) = A \otimes B_1 + A \otimes B_2 \tag{iii}$$

$$\alpha A \otimes \beta B = \alpha\beta(A \otimes B) \tag{iv}$$

$$A_1 A_2 \otimes B_1 B_2 = (A_1 \otimes B_1)(A_2 \otimes B_2) \tag{v}$$

$$(A \otimes B)^{-1} = A^{-1} \otimes B^{-1} \quad \text{if the inverses exist.} \tag{vi}$$

$$(A \otimes B)^{\mathrm{T}} = A^{\mathrm{T}} \otimes B^{\mathrm{T}} \tag{vii}$$

Consider a regular leveling network consisting of m by n points in a rectangular mesh. Suppose the difference of height between neighboring points is observed with variance σ^2 and the individual differences are uncorrelated. Without loss of generality we set $\sigma^2 = 1$.

The height of point $P_{r,s}$ is denoted $h_{r,s}$ and the observed differences of height $d_{rp,sq}$ must be indexed with four variables. By means of the Kronecker product we denote the "horizontal" observations by

$$(I_m \otimes H_n)\boldsymbol{h} = \boldsymbol{d}_1$$

where the subscripts indicate the dimension of the square matrices. Similarly the "vertical" observations are

$$(K_m \otimes I_n)\boldsymbol{h} = \boldsymbol{d}_2$$

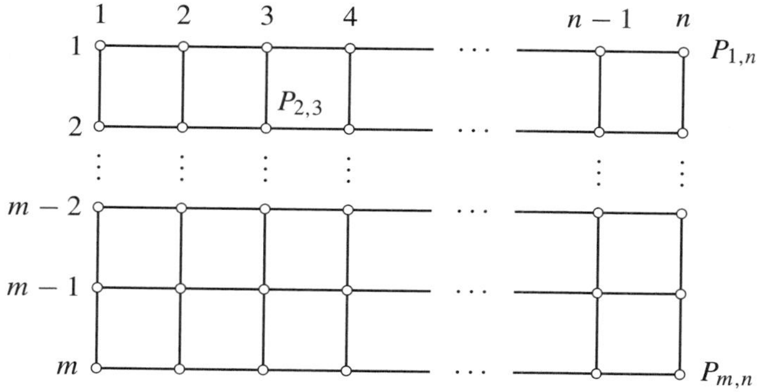

where $\boldsymbol{h}$ is an mn vector containing the heights of all "free" points, and $\boldsymbol{d}_1$ and $\boldsymbol{d}_2$ are $mn - m$ and $mn - n$ dimensional vectors containing the observed differences of height. Furthermore H_n is the $(n-1)$ by n dimensional matrix:

$$H_n = \begin{bmatrix} -1 & 1 & & & & & \\ & -1 & 1 & & & & \\ & & -1 & 1 & & & \\ & & & \ddots & \ddots & & \\ & & & & -1 & 1 & \\ & & & & & -1 & 1 \end{bmatrix}. \tag{11.32}$$

As usual the normal equations are made by multiplying to the left by the transposed coefficient matrix; the "horizontal" contribution is

$$(I_m \otimes H_n)^{\mathrm{T}}(I_m \otimes H_n) = (I_m \otimes H_n^{\mathrm{T}})(I_m \otimes H_n) = I_m \otimes (H_n^{\mathrm{T}} H_n)$$

according to the computational rules on page 384. The "vertical" contribution likewise is $(H_m^{\mathrm{T}} H_m) \otimes I_n$. The total normal equation matrix is

$$N = I_m \otimes (H_n^{\mathrm{T}} H_n) + (H_m^{\mathrm{T}} H_m) \otimes I_n.$$

Again we look for the eigenvalues and eigenvectors of N. We define the following eigenvalue problems

$$H_n^{\mathrm{T}} H_n \boldsymbol{\psi}_n = \lambda \boldsymbol{\psi}_n$$
$$H_m^{\mathrm{T}} H_m \boldsymbol{\varphi}_m = \mu \boldsymbol{\varphi}_m$$

and

$$\boldsymbol{\Phi} = \boldsymbol{\varphi}_m \otimes \boldsymbol{\psi}_n.$$

The total eigenvalue problem then is

$$\begin{aligned} N\boldsymbol{\Phi} &= \big(I_m \otimes (H_n^{\mathrm{T}} H_n) + (H_m^{\mathrm{T}} H_m) \otimes I_n\big)(\boldsymbol{\varphi}_m \otimes \boldsymbol{\psi}_n) \\ &= I_m \boldsymbol{\varphi}_m \otimes (H_n^{\mathrm{T}} H_n)\boldsymbol{\psi}_n + (H_m^{\mathrm{T}} H_m)\boldsymbol{\varphi}_m \otimes I_n \boldsymbol{\psi}_n \\ &= \boldsymbol{\varphi}_m \otimes \lambda \boldsymbol{\psi}_n + \mu \boldsymbol{\varphi}_m \otimes \boldsymbol{\psi}_n = (\lambda + \mu)\boldsymbol{\varphi}_m \otimes \boldsymbol{\psi}_n = (\lambda + \mu)\boldsymbol{\Phi}. \end{aligned}$$

Note that λ and μ are constants. The above derivation clearly shows that the eigenvalues for the total problem are given as the sum of the "horizontal" and the "vertical" problems. The explicit expressions for the eigenvalues can be shown to be

$$\lambda_i = 4\sin^2\frac{(i-1)\pi}{2m}, \qquad i = 1,\dots,m$$
$$\mu_j = 4\sin^2\frac{(j-1)\pi}{2n}, \qquad j = 1,\dots,n$$

and thus

$$\Lambda_{ij} = 4\Big(\sin^2\frac{i\pi}{2m} + \sin^2\frac{j\pi}{2n}\Big). \tag{11.33}$$

Accordingly a *network fixed along all boundaries* has the following sets of eigenvalues

$$\lambda_i = 4\sin^2\frac{i\pi}{2(m-2)}, \qquad i = 2,\dots,m-1$$
$$\mu_j = 4\sin^2\frac{j\pi}{2(n-2)}, \qquad j = 2,\dots,n-1$$

and all eigenvalues for the total problem

$$\Lambda_{ij} = 4\Big(\sin^2\frac{i\pi}{2(m-2)} + \sin^2\frac{j\pi}{2(n-2)}\Big). \tag{11.34}$$

We shall estimate the condition number for this problem. Let $p = \max\{m,n\}$ and let m/n be bounded, then

$$c(N) = \frac{4\Big(\sin^2\frac{(m-2)\pi}{2(m-2)} + \sin^2\frac{(n-2)\pi}{2(n-2)}\Big)}{4\Big(\sin^2\frac{2\pi}{2(m-2)} + \sin^2\frac{2\pi}{2(n-2)}\Big)} \approx \frac{4p^2}{\pi^2}.$$

Example 11.4 We want to describe in detail the situation for $m = 3$ and $n = 2$. The horizontal observations $(I_m \otimes H_n)\boldsymbol{x} = \boldsymbol{b}_1$ are

$$I_3 \otimes H_1\boldsymbol{x} = \begin{bmatrix} 1 & & \\ & 1 & \\ & & 1 \end{bmatrix} \otimes \begin{bmatrix} -1 & 1 \end{bmatrix} \begin{bmatrix} x_{1,1} \\ x_{2,1} \\ x_{3,1} \\ x_{1,2} \\ x_{2,2} \\ x_{3,2} \end{bmatrix} = \dots$$

or

$$\left[\begin{array}{cc|cc|cc} -1 & 1 & 0 & 0 & 0 & 0 \\ 0 & 0 & -1 & 1 & 0 & 0 \\ 0 & 0 & 0 & 0 & -1 & 1 \end{array}\right] \boldsymbol{x} = \dots.$$

Likewise the vertical observations $(H_2 \otimes I_2)\boldsymbol{x} = \boldsymbol{b}_2$:

$$\left[\begin{array}{cc|cc|cc} -1 & 0 & 1 & 0 & 0 & 0 \\ 0 & -1 & 0 & 1 & 0 & 0 \\ \hline 0 & 0 & -1 & 0 & 1 & 0 \\ 0 & 0 & 0 & -1 & 0 & 1 \end{array}\right] \begin{bmatrix} x_{1,1} \\ x_{2,1} \\ x_{3,1} \\ x_{1,2} \\ x_{2,2} \\ x_{3,2} \end{bmatrix} = \cdots .$$

The horizontal contribution to the normals is

$$(I_m \otimes H_{n-1})^{\mathrm{T}}(I_m \otimes H_{n-1}) = (I_m \otimes H_{n-1}^{\mathrm{T}})(I_m \otimes H_{n-1}) = I_m \otimes H_{n-1}^{\mathrm{T}} H_{n-1}$$

and the vertical contribution

$$(H_{m-1} \otimes I_n)^{\mathrm{T}}(H_{m-1} \otimes I_n) = H_{m-1}^{\mathrm{T}} H_{m-1} \otimes I_n$$

In general the total normal equations are

$$N = I_m \otimes H_{m-1}^{\mathrm{T}} H_{m-1} + H_{m-1}^{\mathrm{T}} H_{m-1} \otimes I_N$$

In our special case

$$N = I \otimes H_1^{\mathrm{T}} H_1 + H_2^{\mathrm{T}} H_2 \otimes I$$

$$= \left[\begin{array}{cc|cc|cc} 1 & -1 & 0 & 0 & 0 & 0 \\ -1 & 1 & 0 & 0 & 0 & 0 \\ \hline 0 & 0 & 1 & -1 & 0 & 0 \\ 0 & 0 & -1 & 1 & 0 & 0 \\ \hline 0 & 0 & 0 & 0 & 1 & -1 \\ 0 & 0 & 0 & 0 & -1 & 1 \end{array}\right] + \left[\begin{array}{ccc|ccc} 1 & -1 & 0 & 0 & 0 & 0 \\ -1 & 2 & -1 & 0 & 0 & 0 \\ 0 & -1 & 1 & 0 & 0 & 0 \\ \hline 0 & 0 & 0 & 1 & -1 & 0 \\ 0 & 0 & 0 & -1 & 2 & -1 \\ 0 & 0 & 0 & 0 & -1 & 1 \end{array}\right]$$

$$= \begin{bmatrix} 2 & -2 & 0 & 0 & 0 & 0 \\ -2 & 3 & -1 & 0 & 0 & 0 \\ 0 & -1 & 2 & -1 & 0 & 0 \\ 0 & 0 & -1 & 2 & -1 & 0 \\ 0 & 0 & 0 & -1 & 3 & -2 \\ 0 & 0 & 0 & 0 & -2 & 2 \end{bmatrix}.$$

Example 11.5 *Linear regression* at unit times $t = 1, 2, \ldots, m$ has observation equations $Ax = \boldsymbol{b} - \boldsymbol{r}$:

$$tx_1 + x_2 = b_t - r_t.$$

With $C \equiv I$ the matrix in the normal equations is

$$A^{\mathrm{T}}A = \begin{bmatrix} 1 & 2 & \cdots & m \\ 1 & 1 & \cdots & 1 \end{bmatrix} \begin{bmatrix} 1 & 1 \\ 2 & 1 \\ \vdots & \vdots \\ m & 1 \end{bmatrix} = \begin{bmatrix} \frac{1}{6}m(m+1)(2m+1) & \frac{1}{2}m(m+1) \\ \frac{1}{2}m(m+1) & m \end{bmatrix}.$$

The determinant is $(m^4 - m^3)/12$. The eigenvalues are

$$\lambda = \frac{m\big[(m+1)(2m+1)+6\big] \pm m\sqrt{(m+1)^2(2m+1)^2 + 12(m^2+3m+5)}}{12}.$$

This expression can hardly be reduced so we make an asymptotic expansion. The condition number $c(A^{\mathrm{T}}A) = \lambda_{\max}/\lambda_{\min}$ is approximately $\frac{16}{27}m^2 + \frac{24}{27}m \approx 0.59m^2 + 0.89m$. Here are exact values:

m	3	4	5	10
$c(A^{\mathrm{T}}A)$	46.14	55.78	69.99	187.12

Example 11.6 One might ask: What happens to the condition number under the elimination process? After some elimination steps the situation looks like

$$A = \begin{bmatrix} A_0 & B \\ B^{\mathrm{T}} & C \end{bmatrix}$$

with

$$A^{-1} = \begin{bmatrix} A_0^{-1} + A_0^{-1}B(C - B^{\mathrm{T}}A_0^{-1}B)^{-1}B^{\mathrm{T}}A_0^{-1} & -A_0^{-1}B(C - B^{\mathrm{T}}A_0^{-1}B)^{-1} \\ -(C - B^{\mathrm{T}}A_0^{-1}B)^{-1}B^{\mathrm{T}}A_0^{-1} & (C - B^{\mathrm{T}}A_0^{-1}B)^{-1} \end{bmatrix}.$$

It is important to recognize that the (2,2) entry $(C - B^{\mathrm{T}}A_0^{-1}B)^{-1}$ is symmetric. Its condition cannot be greater than $c(A^{-1})$, and as the condition number for a matrix and its inverse are equal we have $c\big((C - B^{\mathrm{T}}A_0^{-1}B)^{-1}\big) \le c(A)$. Now $C - B^{\mathrm{T}}A_0^{-1}B$ results from elimination of the first $\dim(A_0)$ rows of the A matrix and therefore the answer is: ***The condition number is not increasing during the elimination***, for symmetric positive definite matrices.

The condition number for the least-squares problem of linear regression and the distance measurements in the regular traverse is $O(p^2)$, p being the number of points. Many geodetic problems are of this type. A distinct other type is connected to observation of directions. This type has condition numbers proportional to p^4, much worse than the former.

The mathematical explanation is that the first type is connected to differential equations of second order while the latter type is connected to equations of the fourth order. Two-dimensional networks with combined distance and direction observations typically have condition numbers proportional to $p \ln p$.

Example 11.7 Table 11.1 indicates the error propagation in various geodetic networks. They are from Meissl (1981) and references therein. The variance changes depending on translation of origin, rotation of axes, and scale of the local coordinate system.

The observations are assumed uncorrelated with standard deviation σ. The least-squares problem is solved by using minimal constraints. The error propagation is measured by the largest point variance $\sigma_P^2 = (\sigma_x^2 + \sigma_y^2)/2$.

The boundary conditions are most important in case of direction observations.

Table 11.1 Asymptotic behavior of point variance σ_P^2

Square shaped networks	
Direction observations: unknown translation, rotation, and scale.[a]	
Error propagation depends on the boundary conditions:	
Free boundary: σ_P^2 is proportional to	n^2
Distances are observed between all neighboring boundary points	n
Azimuths are observed between all neighboring boundary points	n
Distances and azimuths are both observed along the boundary	n
All boundary points are fixed	$\ln n$
Pure distance observations: unknown translation and rotation	$\ln n$
Azimuth observations: unknown translation and scale	$\ln n$
Distance and azimuth observations: unknown translation	$\ln n$
GPS networks: known translation, rotation, and scale	constant
Oblong shaped networks	
Traverse	n^3
Chain of triangles	n^3

[a]This is also the photogrammetric case of observation of model coordinates.

A posteriori covariance In the beginning of the 1970's intensive research was made to establish continuous analogues of actual discrete networks. The attempt was successful as far as absolute observation types were concerned: levelling, distances, and azimuths.

A useful outcome of the continuous model is the equivalent of the covariance matrix, namely a ***covariance function***. Such covariance functions are indispensable for network design and other advanced topics.

The derivation of these covariance functions can be found in Borre (1989). A simple example is the covariance function for a regular, triangularly shaped network covering the whole plane, see Borre (1989), equation (3.77). The network has no boundary. The distance along all sides of the network is observed with weight a (per unit area) and the azimuth with weight b.

To get an elegant formula we introduce the variables $A = (9a + 3b)/4$ and $B = (3a + 9b)/4$. Then the a posteriori covariance function with distance and azimuth observations is nearly

$$G(P, Q) = \frac{A+B}{4AB\pi}\left\{\begin{bmatrix}\ln r & 0\\ 0 & \ln r\end{bmatrix} - \frac{A-B}{2(A+B)}\begin{bmatrix}\cos 2\varphi & \sin 2\varphi\\ \sin 2\varphi & -\cos 2\varphi\end{bmatrix}\right\} \tag{11.35}$$

where (r, φ) denotes the polar coordinates for the difference vector between any two points P and Q. The distance r has to be measured in units of the side length.

The formula for $G(P, Q)$ demonstrates the "Taylor-von Kármán structure." The radial error is a function of $\ln r$ while transverse error is a function of φ alone.

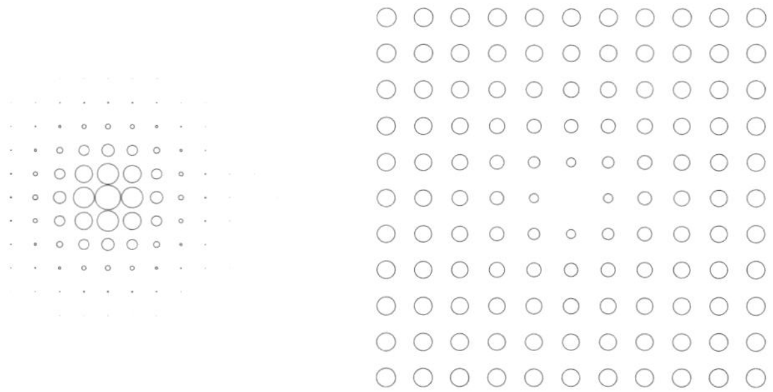

Figure 11.2 Confidence ellipses from the a priori covariance $\sigma_0^2 e^{-a^2 r^2}$ and the a posteriori covariance $G(P, Q)$ in (11.35). The 11 by 11 ellipses are calculated by fixing the midpoint P. The ellipses continue throughout the whole plane.

Hence $G(P, Q)$ acts as a rational covariance function for the plane. Now we shall state some applications of $G(P, Q)$.

Example 11.8 In an equilateral triangular network we have observed all distances and azimuths with uniform weights $a = b = 2/(\sqrt{3}\,\sigma^2)$ for the whole network. Equal weights yield a ***homogeneous*** network. We have $A = B = 3a$ in (11.35):

$$G(P, Q) = \frac{1}{2\sqrt{3}\pi}\begin{bmatrix} \ln r & 0 \\ 0 & \ln r \end{bmatrix}\sigma^2.$$

The covariance function is independent of φ; this phenomenon is called ***isotropy***.

Example 11.9 Consider the same network, but only with distance observations. Now $b = 0$ and $A = 9a/4$, $B = 3a/4$, and $a = 2/(\sqrt{3}\,\sigma^2)$. The covariance function is

$$G(P, Q) = \frac{4}{3\sqrt{3}\pi}\left\{\begin{bmatrix} \ln r & 0 \\ 0 & \ln r \end{bmatrix} - \frac{1}{4}\begin{bmatrix} \cos 2\varphi & \sin 2\varphi \\ \sin 2\varphi & -\cos 2\varphi \end{bmatrix}\right\}\sigma^2.$$

By other methods a better approximation is found in Bartelme & Meissl (1974) for the term $\ln r$ in the first matrix, namely $\ln r + 0.599\ldots$.

Example 11.10 We estimate the variance between the abscissae of points P and Q in Example 11.9. Formal use of the fundamental solution yields

$$\sigma^2(r_{PQ}) = \frac{4}{3\sqrt{3}\pi}\Big(\ln r - \frac{1}{4}\cos 2\varphi\Big)\sigma^2.$$

Such formulas are good approximations for what Baarda calls criterion matrices, but the continuous model yields singular expressions at $r = 0$.

Figure 11.2 illustrates the important difference between a priori and a posteriori covariances. We plot confidence ellipses based on the expressions for autocorrelation for a Gauss-Markov process and (11.35) for covariances on an infinite regular network. As long as the shape of the network does not become pathological the present figures reflect the error situation quite well.

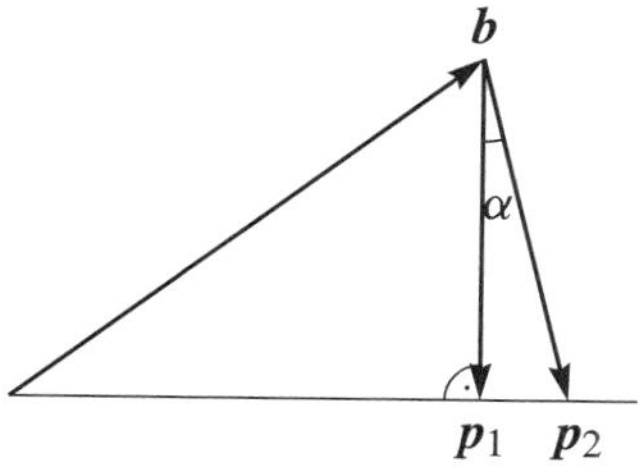

Figure 11.4 The dependence of the projection on the norm. The figure is drawn in C_1-norm.

line segment from $(3, -1)$ to $(5/11, 5/11)$ on the third row. *Letting all weights vary, we evidently can obtain solutions anywhere in the interior of the triangle.*

The M-file dw reflects what goes on.

This approach to changes of weight is excellent for a small number of observations. For a more qualitative knowledge about a larger problem, we shall turn to a more powerful tool.

Now we ask for more quantitative results: If the weights are changed, how much can the projection $\boldsymbol{p} = P\boldsymbol{b}$ move in the column space of A? A measure for this movement is

$$\tan\alpha = \frac{\|\boldsymbol{p}_2 - \boldsymbol{p}_1\|_{C_1}}{\|\boldsymbol{b} - \boldsymbol{p}_1\|_{C_1}}. \tag{11.42}$$

The norm is weighted by $\|\boldsymbol{x}\|_{C_1} = \|C_1\boldsymbol{x}\|$ and α is the angle at $\boldsymbol{b}$ which spans $\overline{\boldsymbol{p}_1\boldsymbol{p}_2}$. Note that Figure 11.4 is drawn in C_1-norm. The angle at $\boldsymbol{p}_2$ is C_2-orthogonal; but in the C_1-norm the angle is $\frac{\pi}{2} - \alpha$. The change from C_1 to C_2-norm leads to a change of the angle from $\frac{\pi}{2}$ to $\frac{\pi}{2} - \alpha$.

The square root W of a matrix C is the positive definite matrix that satisfies $W^2 = C$. Such a matrix certainly exists, is unique, and nonsingular. We define $D = W_1^{-1}C_2W_1^{-1}$, whose condition number $c(D)$—the ratio between the largest and smallest eigenvalue—can be related to the angle α:

$$2|\tan\alpha| \le \sqrt{c(D)} - \frac{1}{\sqrt{c(D)}}. \tag{11.43}$$

Let us repeat: A least-squares problem is given by the coefficient matrix A, the weight C_0, and the observations $\boldsymbol{b}$. We consider all weight matrices C such that the eigenvalues of $W_0^{-1}CW_0^{-1}$ lie in the closed interval $[s, t]$. In the C_0-norm we always find

$$|\tan\alpha| \le \tfrac{1}{2}\left(\sqrt{\tfrac{t}{s}} - \sqrt{\tfrac{s}{t}}\right). \tag{11.44}$$

The angle α measures the displacement of the least-squares result as seen from $\boldsymbol{b}$. At least one matrix C exists for which the equality sign is valid. Furthermore, the ratio between the norms of the residual vectors corresponding to C_1 and C_2 is bounded by

$$\frac{1}{\sqrt{\lambda_{\max}}} \le \frac{\|\boldsymbol{b} - \boldsymbol{p}_1\|_{C_1}}{\|\boldsymbol{b} - \boldsymbol{p}_2\|_{C_2}} \le \sqrt{\lambda_{\max}} \tag{11.45}$$

where $\lambda_{\max}$ is the largest eigenvalue of the matrix $W_2^{-1}C_1W_2^{-1}$.

The main results about the changes of weight in a least-squares problem are quoted from Krarup (1972). Once again the condition number is prominent. We are saying that the effect of ignoring a possible correlation between the observations can be dangerous if the condition number is large.

Example 11.12 We introduce an n by n covariance matrix with strong correlation:

$$\Sigma_b = \begin{bmatrix} 2 & -1 & & & \\ -1 & 2 & -1 & & \\ & \ddots & \ddots & \ddots & \\ & & -1 & 2 & -1 \\ & & & -1 & 2 \end{bmatrix}.$$

Its inverse has the special form

$$D = C_2 = \Sigma_b^{-1} = \begin{cases} \frac{i(n+1-j)}{n+1} & \text{for } i \le j \\ \frac{j(n+1-i)}{n+1} & \text{for } i \ge j. \end{cases}$$

The eigenvalues are $4\sin^2 \frac{i\pi}{2(n+1)}$ so the condition is $c(D) \approx \frac{4(n+1)^2}{\pi^2} < n^2$. By (11.44)

$$|\tan\alpha| \le \tfrac{1}{2}\left(\sqrt{c(D)} - \frac{1}{\sqrt{c(D)}}\right) \approx \tfrac{1}{\pi}(n+1) \to \infty.$$

A strong correlation can take us arbitrarily far from the solution corresponding to $C_1 = I$.

Example 11.13 The correlation is weaker if $D = C_2 = I + t\Sigma_b^{-1}$ and $t \approx 10^{-2}$:

$$c(D) = \frac{1 + t\,4\sin^2 \frac{n\pi}{2(n+1)}}{1 + t\,4\sin^2 \frac{\pi}{2(n+1)}} \approx (1+4t)\left(1 - 4t\frac{\pi^2}{4n^2}\right) \approx 1 + 4t.$$

$$|\tan\alpha| \le \tfrac{1}{2}\left(\sqrt{1+4t} - \frac{1}{\sqrt{1+4t}}\right) = \frac{2t}{\sqrt{1+4t}}.$$

Thus $|\tan\alpha| < 2t$.

11.7 Elimination of Unknowns

A Classical Procedure

Now we want to describe how to eliminate an unknown from a least-squares problem. Some unknowns may be of very little importance. They are introduced into a least-squares problem in order to treat correlation in a proper way; but otherwise these unknowns are of no interest. Sometimes such unknowns are called *nuisance* parameters, like the orientation unknowns in direction observations made by theodolite.

To begin with we will assume that all observation equations have weight 1; otherwise they would be normalized by the square root of the actual weight.

The actual procedure is easily described by means of a concrete example:

$$\begin{bmatrix} 1 & 0 \\ 1 & 1 \\ 1 & 3 \\ 1 & 4 \end{bmatrix} \begin{bmatrix} c \\ d \end{bmatrix} = \begin{bmatrix} 0 \\ 8 \\ 8 \\ 20 \end{bmatrix} - \boldsymbol{r}. \tag{11.46}$$

The normal equations are

$$\begin{bmatrix} 4 & 8 \\ 8 & 26 \end{bmatrix} \begin{bmatrix} c \\ d \end{bmatrix} = \begin{bmatrix} 36 \\ 112 \end{bmatrix} \tag{11.47}$$

and the solution is

$$\begin{bmatrix} c \\ d \end{bmatrix} = \begin{bmatrix} 1 \\ 4 \end{bmatrix}.$$

If we solve the equations according to the method of Cholesky the computations run as follows:

1 Triangular decomposition of the left side

$$l_{11} = \sqrt{4} = 2$$
$$l_{21} = 8/2 = 4 \qquad l_{22} = \sqrt{26 - 16} = \sqrt{10}$$

2 Forward elimination on the right side

$$z_1 = 36/2 = 18$$
$$z_2 = (112 - 4 \cdot 18)/\sqrt{10} = 40/\sqrt{10}$$

3 Back solution

$$d = 40/(\sqrt{10} \cdot \sqrt{10}) = 4$$
$$c = (18 - 4 \cdot 4)/2 = 1.$$

Starting from this example we shall study the following trick: We augment the existing observation equations with a fictitious equation. It is the sum of the given equations and it is assigned the weight $c_{\text{fict}} = -\frac{1}{t} = -\frac{1}{4}$. (The sum of the identical entries in the first column of A is t.) This new equation is

$$4c + 8d = 36 \qquad \text{with weight} \quad -\tfrac{1}{4}. \tag{11.48}$$

The augmented normal equation system has a singular matrix $A^{\mathrm{T}}CA$:

$$\begin{bmatrix} 1 & 1 & 1 & 1 & 4 \\ 0 & 1 & 3 & 4 & 8 \end{bmatrix} \begin{bmatrix} 1 & & & & \\ & 1 & & & \\ & & 1 & & \\ & & & 1 & \\ & & & & -\frac{1}{4} \end{bmatrix} \begin{bmatrix} 1 & 0 \\ 1 & 1 \\ 1 & 3 \\ 1 & 4 \\ 4 & 8 \end{bmatrix} = \begin{bmatrix} 0 & 0 \\ 0 & 10 \end{bmatrix}.$$

The system $A^{\mathrm{T}}CAx = A^{\mathrm{T}}Cb$ is still solvable (always):

$$\begin{bmatrix} 0 & 0 \\ 0 & 10 \end{bmatrix} \begin{bmatrix} c \\ d \end{bmatrix} = \begin{bmatrix} 0 \\ 40 \end{bmatrix} \tag{11.49}$$

which yields $d = 4$. By insertion into (11.48) we get $c = 1$. At first sight the singular normal equations (11.49) have a surprising structure which we want to illustrate.

Comparing the two systems of normal equations (11.47) and (11.49) it becomes evident that the unknown c has been eliminated. Our earlier standard method of elimination was elementary row operations. We demonstrate the method on the actual numbers:

$$\begin{aligned} 4c + 8d &= 36 \\ 8c + 26d &= 112 \end{aligned} \quad \longrightarrow \quad \begin{aligned} 4c + 8d &= 36 \\ 10d &= 40, \end{aligned} \tag{11.50}$$

and this reveals nothing new.

Earlier we assumed that all entries are identical in the column of A corresponding to the unknown to be eliminated. In practice these entries often are 1.

If the weights c_i of the single observations are varying then the weight of the fictitious equation has to be changed to $c_{\text{fict}} = -1 / \sum_{i=1}^{m} c_i$ and the summation in A has to be performed as a weighted summation.

The variance σ_c^2 of the eliminated unknown c is calculated as follows: In order to calculate the inverse matrix of the normal equations we use the same row operations as above on the unit matrix:

$$\begin{bmatrix} 1 & 0 \\ 0 & 1 \end{bmatrix} \quad \longrightarrow \quad \begin{bmatrix} 1 & 0 \\ -2 & 1 \end{bmatrix}.$$

We substitute the first column as right side in (11.50) and obtain the solution

$$v_1 = \begin{bmatrix} \frac{13}{20} \\ -\frac{1}{5} \end{bmatrix}$$

and substituting the second column as right side yields the following solution

$$v_2 = \begin{bmatrix} -\frac{1}{5} \\ \frac{1}{10} \end{bmatrix}$$

or

$$(A^{\mathrm{T}}CA)^{-1} = \begin{bmatrix} v_1 & v_2 \end{bmatrix} = \frac{1}{20} \begin{bmatrix} 13 & -4 \\ -4 & 2 \end{bmatrix}.$$

This result, of course, is also obtained when using ordinary methods for the inversion of the coefficient matrix (11.47).

In order to calculate the variance factor $\hat{\sigma}_0^2$ we have to use (9.71)

$$\hat{r}^{\mathrm{T}}C\hat{r} = b^{\mathrm{T}}Cb - z^{\mathrm{T}}z = b^{\mathrm{T}}Cb - \sum z_i^2 = 64 + 64 + 400 - (324 + 160) = 44.$$

An identical result is produced by calculating

$$p = A\hat{x} = \begin{bmatrix} 1 \\ 5 \\ 13 \\ 17 \end{bmatrix} \qquad \text{and then} \qquad \hat{r} = b - p = \begin{bmatrix} -1 \\ 3 \\ -5 \\ 3 \end{bmatrix}.$$

Subsequently, $\hat{\sigma}_0^2 = 44/(4-2) = 22$. Note that n is not to be reduced because of the elimination. Implicitly there still are n unknowns. Finally

$$\hat{\sigma}_c^2 = \frac{22 \cdot 26}{40} = 14.3 \qquad \text{and} \qquad \hat{\sigma}_d^2 = \frac{22 \cdot 4}{40} = 2.2.$$

We summarize the method: In a least-squares problem we want to eliminate an unknown with constant coefficients. This is done by augmenting the original problem with a fictitious observation equation which results as the sum of all observation equations in which this unknown appears. The new equation is given the weight $-1/$(sum of coefficients of the unknown). Next, the remaining $n-1$ normal equations are solved in usual manner. Subsequently, the eliminated unknown can be calculated from the new observation equation by insertion of the solution. The variance of the unknown is calculated from the normal equations by using unit vectors as right sides and by using the same row operations as for the elimination of the unknown.

Eliminating From the Normal Equations

We want to make the above description more general and cogent by using the technique of ***block elimination***. Let the normal equations be split as follows (remember $B = C^{\mathrm{T}}$):

$$\begin{bmatrix} A & B \\ C & D \end{bmatrix} \begin{bmatrix} x_1 \\ x_2 \end{bmatrix} = \begin{bmatrix} b_1 \\ b_2 \end{bmatrix}.$$

Block elimination subtracts CA^{-1} times the first row $[\,A \quad B\,]$ and b_1. This is achieved by multiplying to the left with the elimination matrix:

$$\begin{bmatrix} I & 0 \\ -CA^{-1} & I \end{bmatrix} \begin{bmatrix} A & B \\ C & D \end{bmatrix} \begin{bmatrix} x_1 \\ x_2 \end{bmatrix} = \begin{bmatrix} I & 0 \\ -CA^{-1} & I \end{bmatrix} \begin{bmatrix} b_1 \\ b_2 \end{bmatrix}$$

or explicitly

$$\begin{bmatrix} A & B \\ 0 & D - CA^{-1}B \end{bmatrix} \begin{bmatrix} x_1 \\ x_2 \end{bmatrix} = \begin{bmatrix} b_1 \\ b_2 - CA^{-1}b_1 \end{bmatrix}.$$

The last row contains the wanted expression for the remaining unknown:

$$x_2 = (D - CA^{-1}B)^{-1}(b_2 - CA^{-1}b_1). \tag{11.51}$$

This formula is coded as the M-file elimnor.

Eliminating Parameters: A Reduced Estimation Problem

Suppose the vector $\boldsymbol{x}$ of modeling parameters is separated into an unimportant part $\boldsymbol{y}$ and an important part $\boldsymbol{z}$. Then we can eliminate $\hat{\boldsymbol{y}}$ from the normal equations and solve only for $\hat{\boldsymbol{z}}$. And we can return to find $\hat{\boldsymbol{y}}$ if we want. It is useful to describe those steps.

We will execute a standard elimination of $\hat{\boldsymbol{y}}$ from the normal equations for $\hat{\boldsymbol{x}} = [\,\hat{\boldsymbol{y}} \quad \hat{\boldsymbol{z}}\,]$. Start with the observation equations

$$\boldsymbol{b} = A\boldsymbol{x} + \boldsymbol{e} = B\boldsymbol{y} + G\boldsymbol{z} + \boldsymbol{e}. \tag{11.52}$$

Denote the weight matrix Σ_b^{-1} by C. The normal equations are

$$\begin{bmatrix} B^{\mathrm{T}} \\ G^{\mathrm{T}} \end{bmatrix} C \begin{bmatrix} B & G \end{bmatrix} \begin{bmatrix} \hat{\boldsymbol{y}} \\ \hat{\boldsymbol{z}} \end{bmatrix} = \begin{bmatrix} B^{\mathrm{T}} \\ G^{\mathrm{T}} \end{bmatrix} C\boldsymbol{b}. \tag{11.53}$$

This produces two block equations for $\hat{\boldsymbol{y}}$ and $\hat{\boldsymbol{z}}$:

$$\begin{bmatrix} B^{\mathrm{T}}CB & B^{\mathrm{T}}CG \\ G^{\mathrm{T}}CB & G^{\mathrm{T}}CG \end{bmatrix} \begin{bmatrix} \hat{\boldsymbol{y}} \\ \hat{\boldsymbol{z}} \end{bmatrix} = \begin{bmatrix} B^{\mathrm{T}}C\boldsymbol{b} \\ G^{\mathrm{T}}C\boldsymbol{b} \end{bmatrix}. \tag{11.54}$$

To eliminate $\hat{\boldsymbol{y}}$, multiply row 1 by $G^{\mathrm{T}}CB(B^{\mathrm{T}}CB)^{-1}$ and subtract from row 2. This produces a zero block in row 2, column 1. It leaves an equation for $\hat{\boldsymbol{z}}$ alone, with a ***modified weighting matrix*** C':

$$G^{\mathrm{T}}C'G\hat{\boldsymbol{z}} = G^{\mathrm{T}}C'\boldsymbol{b} \qquad \text{with} \qquad C' = C - CB(B^{\mathrm{T}}CB)^{-1}B^{\mathrm{T}}C. \tag{11.55}$$

Note that $C'B$ is the zero matrix. The algebra has confirmed what we could have expected: the reduced model for $\hat{\boldsymbol{z}}$ alone has a smaller weight matrix C' (and a larger covariance matrix) because the $B\boldsymbol{y}$ term is projected out. As always, back substitution in (11.54) yields $\hat{\boldsymbol{y}}$ when we know $\hat{\boldsymbol{z}}$:

$$\hat{\boldsymbol{y}} = (B^{\mathrm{T}}CB)^{-1}B^{\mathrm{T}}C(\boldsymbol{b} - G\hat{\boldsymbol{z}}). \tag{11.56}$$

Eliminating From the Observation Equations

If we use the above procedure directly on the observation equations it works formally, but we generally get a wrong result. The column space must be split into two subspaces conforming to the splitting of A into $[\,A_1 \quad A_2\,]$, and furthermore, the two subspaces must be orthogonal complements.

Let the observation equations be partitioned as follows

$$\begin{bmatrix} A_1 & A_2 \end{bmatrix} \begin{bmatrix} \boldsymbol{x}_1 \\ \boldsymbol{x}_2 \end{bmatrix} = \boldsymbol{b}. \tag{11.57}$$

The columns of $A = [A_1\ A_2]$ span $\boldsymbol{R}(A)$. We decompose $\boldsymbol{R}(A)$ into the space spanned by the columns of A_1 and into the orthogonal complement $\boldsymbol{R}^{\perp}$ of $\boldsymbol{R}(A)$. By definition we have: $\boldsymbol{R}(A)$ is spanned by $A = [A_1\ A_2]$ and $A_1^{\mathrm{T}}CA_2^* = 0$ where A_2^* span $\boldsymbol{R}^{\perp}$.

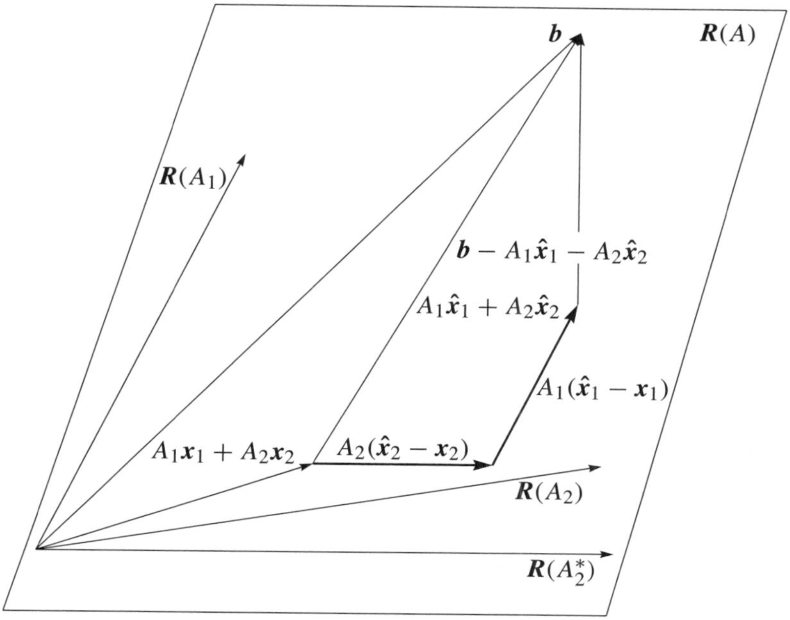

Figure 11.5 Geometry of the orthogonal decomposition in the column space of $A = [\,A_1 \;\; A_2\,]$.

The projector $P = I - A_1(A_1^{\mathrm{T}}CA_1)^{-1}A_1^{\mathrm{T}}C$ projects A_2 in $\boldsymbol{R}(A)$ to A_2^* in $\boldsymbol{R}^{\perp}$:

$$A_2^* = \bigl(I - A_1(A_1^{\mathrm{T}}CA_1)^{-1}A_1^{\mathrm{T}}C\bigr)A_2.$$

The reduced observation equations are

$$\bigl(I - A_1(A_1^{\mathrm{T}}CA_1)^{-1}A_1^{\mathrm{T}}C\bigr)A_2\boldsymbol{x}_2 = \bigl(I - A_1(A_1^{\mathrm{T}}CA_1)^{-1}A_1^{\mathrm{T}}C\bigr)\boldsymbol{b}$$

or abbreviated

$$A_2^*\boldsymbol{x}_2 = \boldsymbol{b}^*.$$

Next we solve for $\boldsymbol{x}_2$:

$$\boldsymbol{x}_2 = (A_2^{*\mathrm{T}}CA_2^*)^{-1}A_2^{*\mathrm{T}}C\boldsymbol{b}^*. \tag{11.58}$$

The above procedure has important applications. When processing GPS observations we may want to eliminate the ambiguity unknowns. We rewrite (11.57) as

$$A_1\boldsymbol{x}_1 + A_2\boldsymbol{x}_2 = \boldsymbol{b}.$$

The unknowns $\boldsymbol{x}_1$ are eliminated by multiplication by the projector P:

$$PA_1\boldsymbol{x}_1 + PA_2\boldsymbol{x}_2 = P\boldsymbol{b}.$$

As $PA_1 = 0$ the transformed observation equations become

$$A_2^* \boldsymbol{x}_2 = \boldsymbol{b}^*.$$

We depict the geometry of this orthogonal decomposition in Figure 11.5.

Finally we want to demonstrate the procedure on the observations from (11.46). The coefficient matrix A is split into the two columns $[\,A_1 \quad A_2\,]$. Hence the projector P becomes

$$P = \frac{1}{4}\begin{bmatrix} 3 & -1 & -1 & -1 \\ -1 & 3 & -1 & -1 \\ -1 & -1 & 3 & -1 \\ -1 & -1 & -1 & 3 \end{bmatrix}.$$

The transformed observation equations $A_2^* \boldsymbol{x}_2 = \boldsymbol{b}^*$ are

$$\begin{bmatrix} -2 \\ -1 \\ 1 \\ 2 \end{bmatrix} \boldsymbol{x}_2 = \begin{bmatrix} -9 \\ -1 \\ -1 \\ 11 \end{bmatrix}.$$

The normals are $10\boldsymbol{x}_2 = 40$ and the solution is again recovered as $\boldsymbol{x}_2 = 4$.

The procedure is coded in the M-file elimobs. The reader may find details on error calculation in Meissl (1982), Section A.9.

11.8 Decorrelation and Weight Normalization

Most computer programs for solving least-squares problems read the observation equations by rows and store the relevant contributions at the matching places of the normal equation matrix. The ith observation equation from the ith row of A can be immediately stored at the correct places in the normal equation matrix $A^{\mathrm{T}}CA$

$$(A^{\mathrm{T}}CA)_{\text{old}} + \frac{1}{\sigma_i^2} A_i^{\mathrm{T}} A_i \rightarrow (A^{\mathrm{T}}CA)_{\text{new}}. \tag{11.59}$$

This procedure works as long as the individual observations are uncorrelated. If they are dependent the procedure is inapplicable. We first have to decorrelate the observations. This happens by diagonalizing the C matrix:

$$A' = TA \tag{11.60}$$

and

$$(A')^{\mathrm{T}}A' = A^{\mathrm{T}}T^{\mathrm{T}}TA = A^{\mathrm{T}}CA. \tag{11.61}$$

Such a transformation T not only decorrelates the transformed observations but can also guarantee these equations have the weight 1.

As the covariance matrix C^{-1} is symmetric and positive definite we can use the method of Cholesky for factorization:

$$\Sigma_b = C^{-1} = W^{-1}W^{-\mathrm{T}} \tag{11.62}$$

or

$$C = W^{\mathrm{T}}W \tag{11.63}$$

By substituting (11.61) we get

$$T = W. \tag{11.64}$$

This transformation thus described is given as

$$A' = WA \qquad \text{and} \qquad \boldsymbol{b}' = W\boldsymbol{b} \qquad \text{and} \qquad \boldsymbol{r}' = W\boldsymbol{r}.$$

It is easy to show, see below, that $\boldsymbol{x}$ remains unchanged under this special transformation and that the a priori covariance matrix of the transformed observations is simplified to

$$\Sigma_{b'} = I. \tag{11.65}$$

This decomposition of the covariance matrix Σ_b and the transformation of A and $\boldsymbol{b}$ often is called ***decorrelation*** of the observations. Strictly speaking it is the transformed observations which are decorrelated. In practice you need only to calculate W^{-1} for each set of correlated observations and then solve the equations

$$W^{-1}A' = A \qquad \text{and} \qquad W^{-1}\boldsymbol{b}' = \boldsymbol{b} \tag{11.66}$$

for A' and $\boldsymbol{b}'$ by one forward reduction. Although A contains more correlated rows the calculation of A' continues column-wise. After the decomposition, A' and $\boldsymbol{b}'$ are added to the normal equations as decribed in (11.59). (The Cholesky factors W^{-1} in the covariance matrix of the observations are not to be confused with the usual L-factors for the normal equations.)

The transformation is a change of basis in the column space of T. This column space is determined uniquely for each set of correlated observations. Any such linear transformation leaves $A^{\mathrm{T}}CA$, $A^{\mathrm{T}}C\boldsymbol{b}$, and $\boldsymbol{x}$ unchanged. The transformations can be performed to secure simultaneously a unit weight matrix for the correlated observation equations as demonstrated in (11.65).

The weighted sum of squares of the residuals $\boldsymbol{r}^{\mathrm{T}}C\boldsymbol{r}$ equals the square sum of the transformed residuals

$$\boldsymbol{r}^{\mathrm{T}}C\boldsymbol{r} = \sum_{i=1}^{m} r_i'^2.$$

The actual residuals r_i are calculated as

$$\boldsymbol{r} = W^{-1}\boldsymbol{r}'.$$

Example 11.14 We give a numerical example demonstrating the theory in all details, and use a covariance matrix for double differenced phase observations. We generalize D_d as described in (14.29) to $r = 3$ and $s = 4$:

$$D_d = \left[\begin{array}{rrrr|rrrr|rrrr} -1 & 1 & 0 & 0 & 1 & -1 & 0 & 0 & 0 & 0 & 0 & 0 \\ -1 & 0 & 1 & 0 & 1 & 0 & -1 & 0 & 0 & 0 & 0 & 0 \\ -1 & 0 & 0 & 1 & 1 & 0 & 0 & -1 & 0 & 0 & 0 & 0 \\ \hline -1 & 1 & 0 & 0 & 0 & 0 & 0 & 0 & 1 & -1 & 0 & 0 \\ -1 & 0 & 1 & 0 & 0 & 0 & 0 & 0 & 1 & 0 & -1 & 0 \\ -1 & 0 & 0 & 1 & 0 & 0 & 0 & 0 & 1 & 0 & 0 & -1 \end{array}\right].$$

The covariance matrix $\Sigma_d = D_d(\sigma^2 I)D_d^{\mathrm{T}} = C^{-1}$ is

$$\sigma^2 \left[\begin{array}{rrr|rrr} 4 & 2 & 2 & 2 & 1 & 1 \\ 2 & 4 & 2 & 1 & 2 & 1 \\ 2 & 2 & 4 & 1 & 1 & 2 \\ \hline 2 & 1 & 1 & 4 & 2 & 2 \\ 1 & 2 & 1 & 2 & 4 & 2 \\ 1 & 1 & 2 & 2 & 2 & 4 \end{array}\right].$$

The first step is to factorize C^{-1} into $W^{-1}W^{-\mathrm{T}}$. The result is

$$W^{-1} = \begin{bmatrix} 2 & 0 & 0 & 0 & 0 & 0 \\ 1 & \sqrt{3} & 0 & 0 & 0 & 0 \\ 1 & \frac{1}{\sqrt{3}} & \frac{2\sqrt{2}}{\sqrt{3}} & 0 & 0 & 0 \\ 1 & 0 & 0 & \sqrt{3} & 0 & 0 \\ \frac{1}{2} & \frac{\sqrt{3}}{2} & 0 & \frac{\sqrt{3}}{2} & \frac{3}{2} & 0 \\ \frac{1}{2} & \frac{1}{2\sqrt{3}} & \frac{\sqrt{2}}{\sqrt{3}} & \frac{\sqrt{3}}{2} & \frac{1}{2} & \sqrt{2} \end{bmatrix}.$$

Notice that W^{-1} has only positive entries like Σ_d. The second step is to calculate the inverse:

$$W = \begin{bmatrix} \frac{1}{2} & 0 & 0 & 0 & 0 & 0 \\ -\frac{1}{2\sqrt{3}} & \frac{1}{\sqrt{3}} & 0 & 0 & 0 & 0 \\ -\frac{1}{2\sqrt{6}} & -\frac{1}{2\sqrt{6}} & \frac{\sqrt{3}}{2\sqrt{2}} & 0 & 0 & 0 \\ -\frac{1}{2\sqrt{3}} & 0 & 0 & \frac{1}{\sqrt{3}} & 0 & 0 \\ \frac{1}{6} & -\frac{1}{3} & 0 & -\frac{1}{3} & \frac{2}{3} & 0 \\ \frac{1}{6\sqrt{2}} & \frac{1}{6\sqrt{2}} & -\frac{1}{2\sqrt{2}} & -\frac{1}{3\sqrt{2}} & -\frac{1}{3\sqrt{2}} & \frac{1}{\sqrt{2}} \end{bmatrix}.$$

Note that the diagonal entries of W^{-1} and W are mutually inverse. The zero entries are placed at identical places in the two matrices and they share other properties which are common to all triangular matrices: The product of two lower (upper) triangular matrices is again lower (upper) triangular, and the inverse of a nonsingular, triangular matrix is also triangular.

After a left multiplication with W of both sides of the original correlated observation equations, the independent observations can be read in by the least-squares program like any other independent observations.

All least-squares problems treated so far only have involved one single type of observation: leveling, directions, or distances. Each observation has been assigned a weight according to the weight relation (9.44).

Next we shall demonstrate how to combine various types of observations into one least-squares problem. The key is to divide the single observation equation by its standard deviation. Or in other words to multiply the single observation equation by the square root of the pertinent weight. By this the observation equations become dimensionless; subsequently they enter a least-squares problem. We say that the observation equations are ***weight normalized***:

$$WA\boldsymbol{x} = W\boldsymbol{b} - W\boldsymbol{r}, \tag{11.67}$$

and by this the normal equations are

$$A^{\mathrm{T}}W^{\mathrm{T}}WA\boldsymbol{x} = A^{\mathrm{T}}W^{\mathrm{T}}W\boldsymbol{b} \tag{11.68}$$

or

$$A^{\mathrm{T}}CA\boldsymbol{x} = A^{\mathrm{T}}C\boldsymbol{b}. \tag{11.69}$$

They are the correct normal equations with weight matrix C.

A last remark on weights. If you compare the given description with Example 10.3 you notice that there is a slight difference. Our distance observations a priori are divided by the preliminary value for the distance. This leaves the opportunity to emphasize that any observation equation can be multiplied by any constant; but you must understand that this *changes the weight* and consequently the result of the least-squares problem.

12

CONSTRAINTS FOR SINGULAR NORMAL EQUATIONS

12.1 Rank Deficient Normal Equations

So far we have assumed that $A^{\mathrm{T}}A$ is invertible. The columns of A are independent. The normal equations have a unique solution. However, in geodetic practice we also encounter least-squares problems with dependent columns in A.

A basic example is a leveling network without any postulated heights. Then Ax involves only *differences* of heights—there is an arbitrary constant in all the heights, which cannot be determined. (It is removed when we postulate one height.) The constant height vector $\boldsymbol{x} = \boldsymbol{e} = (1, 1, \dots, 1)$ has differences $A\boldsymbol{x} = (0, 0, \dots, 0)$. Then $A^{\mathrm{T}}A\boldsymbol{x} = \boldsymbol{0}$ and $A^{\mathrm{T}}A$ is not invertible.

The same is true for two- and three-dimensional control networks, for the same reason. Still we can define a meaningful and unique solution, by fixing one or more heights. We shall study both geometrical and statistical properties for these types of solutions. They are defined via extensive use of the ***pseudoinverse*** matrix A^+.

For these rank deficient matrices A, there does not exist any unbiased linear estimator $\hat{\boldsymbol{x}} = P\boldsymbol{b}$. This would require $E\{\hat{\boldsymbol{x}}\} = PA\boldsymbol{x} = \boldsymbol{x}$ for all $\boldsymbol{x}$, or $PA = I$ which is impossible. ($A\boldsymbol{x} = \boldsymbol{0}$ would require $I\boldsymbol{x} = \boldsymbol{0}$.) But there do exist linear functions of $\boldsymbol{x}$ which allow unbiased estimates (expected value equal to true mean). Basically, we must have no component in the direction of the singular vector $(1, 1, \dots, 1)$. Our discussion focuses on suitable choices of such linear functions and how to interpret them geometrically.

In general, the purpose of using a least-squares procedure for geodetic networks is to determine coordinates of the unknown points. Points with postulated coordinates keep these values; the network is "pin-pointed" at those points.

Although the least-squares estimation of such a network furnishes us with a covariance matrix for the coordinates of the new points, this covariance matrix is greatly influenced by the presence and distribution of the postulated points. So the covariance matrix tells less about the general features of the network, because it also reflects possible internal errors between the known coordinates.

In 1962 Meissl proposed a least-squares procedure for geodetic networks that allowed for *singular normal equations* combined with certain linear constraints. The idea was to consider all network points of equal status. Then points with postulated coordinates can be assigned changes to their coordinates too.

There exists a close connection between the constrained least-squares problem and a similarity transformation of the same network.

12.2 Representations of the Nullspace

We start by deriving the singular vectors connected to the most common types of geodetic observations in case of a rank deficient matrix A. They solve $A\boldsymbol{x} = \boldsymbol{0}$.

Leveling Network Incidence matrices take differences. The nullspace contains

$$\boldsymbol{e} = (1, 1, \dots, 1). \tag{12.1}$$

Two-Dimensional Control Network A *translation* of the network in the x direction adds a common constant to all x coordinates. If the vector of unknowns is arranged as follows

$$\boldsymbol{X} = (X_1, Y_1, X_2, Y_2, \dots, X_n, Y_n) \tag{12.2}$$

then the nullspace contains x translations and y translations:

$$\boldsymbol{e}_x = (1, 0, 1, 0, \dots, 1, 0) \qquad \text{and} \qquad \boldsymbol{e}_y = (0, 1, 0, 1, \dots, 0, 1). \tag{12.3}$$

Furthermore the network can be *rotated* by an angle φ. We shall show in detail how to linearize the equations of a differential two-dimensional rotational transformation:

$$\begin{aligned} X_i' &= \cos\varphi\, X_i - \sin\varphi\, Y_i \\ Y_i' &= \sin\varphi\, X_i + \cos\varphi\, Y_i. \end{aligned} \tag{12.4}$$

Here (X_i, Y_i) is a set of given coordinates which the rotation transforms to (X_i', Y_i'). We linearize and keep only the first order terms ($\sin d\varphi \approx d\varphi$ and $\cos d\varphi \approx 1$):

$$\begin{aligned} X_i' + \xi_i &= 1\, X_i - d\varphi\, Y_i \qquad & \xi_i &= -Y_i\, d\varphi \\ Y_i' + \eta_i &= d\varphi\, X_i + 1\, Y_i \qquad & \eta_i &= X_i\, d\varphi. \end{aligned}$$

This small rotation gives rise to yet another vector in the nullspace of A:

$$\boldsymbol{e}_\varphi = (-Y_1, X_1, -Y_2, X_2, \dots, -Y_n, X_n). \tag{12.5}$$

Finally a *change in scale* dk is described by the vector

$$\boldsymbol{e}_k = (X_1, Y_1, X_2, Y_2, \dots, X_n, Y_n). \tag{12.6}$$

Our network is determined up to two translations, one rotation, and one change of scale. The nullspace $N(A)$ is spanned by the four rows of

$$G^{\mathrm{T}} = \left[\begin{array}{cc|cc|c|cc} 1 & 0 & 1 & 0 & & 1 & 0 \\ 0 & 1 & 0 & 1 & \cdots & 0 & 1 \\ -Y_1 & X_1 & -Y_2 & X_2 & & -Y_n & X_n \\ X_1 & Y_1 & X_2 & Y_2 & & Y_n & X_n \end{array}\right]. \tag{12.7}$$

The differential parameters are $df = (dt_x, dt_y, d\varphi, dk)$ and the transformation is

$$dx = (\xi_1, \eta_1, \ldots, \xi_n, \eta_n) = G\,df. \tag{12.8}$$

Three-Dimensional Control Network Three-dimensional networks can be subject to three infinitesimal translations dt_x, dt_y, dt_z, three infinitesimal rotations $d\varphi_x, d\varphi_y, d\varphi_z$, and one infinitesimal change of scale dk. The nullvectors are the rows of

$$G^{\mathrm{T}} = \left[\begin{array}{ccc|c|ccc} 1 & 0 & 0 & & 1 & 0 & 0 \\ 0 & 1 & 0 & & 0 & 1 & 0 \\ 0 & 0 & 1 & & 0 & 0 & 1 \\ 0 & -Z_1 & Y_1 & \cdots & 0 & -Z_n & Y_n \\ Z_1 & 0 & -X_1 & & Z_n & 0 & -X_n \\ -Y_1 & X_1 & 0 & & -Y_n & X_n & 0 \\ X_1 & Y_1 & Z_1 & & X_n & Y_n & Z_n \end{array}\right] \tag{12.9}$$

with

$$df = (dt_x, dt_y, dt_z, d\varphi_x, d\varphi_y, d\varphi_z, dk). \tag{12.10}$$

The columns of G (rows of G^{T}) span $N(A)$ and we have

$$AG = 0. \tag{12.11}$$

We close this section by suggesting a geometrical interpretation. The first three rows of $G^{\mathrm{T}}\boldsymbol{x} = \boldsymbol{0}$ means that the origin is fixed. For numerical reasons we translate the origin to the barycenter and provide coordinates relative to the barycenter with an *:

$$\sum_{i=1}^{n} X_i^* = 0, \qquad \sum_{i=1}^{n} Y_i^* = 0, \qquad \sum_{i=1}^{n} Z_i^* = 0.$$

The next three equations in $G^{\mathrm{T}}\boldsymbol{x} = \boldsymbol{g}$ lead to fixed rotations:

$$\sum_{i=1}^{n}(-Y_i^*\xi_i + X_i^*\eta_i) = g_4, \qquad \sum_{i=1}^{n}(Z_i^*\xi_i - X_i^*\zeta_i) = g_5, \qquad \sum_{i=1}^{n}(-Z_i^*\eta_i + Y_i^*\zeta_i) = g_6.$$

The final condition $\sum_{i=1}^{n}(X_i^*\xi_i + Y_i^*\eta_i + Z_i^*\zeta_i) = g_7$ secures that scale is kept fixed.

Table 12.1 Possible G-columns spanning the nullspace of A, see Teunissen (1985a). The number of unknowns at each node is denoted u, and d is the dimension of $\mathcal{N}(A)$

u	d	Actual observation type(s)	Components of G-column(s)		
			translation(s)	rotation(s)	scale(s)
1	1	height differences	1		
2	2	distances and azimuths	$\begin{matrix}1 & 0\\ 0 & 1\end{matrix}$		
	3	distances	$\begin{matrix}1 & 0\\ 0 & 1\end{matrix}$	$\begin{matrix}Y_i\\ -X_i\end{matrix}$	
	4	angles and/or distance ratios	$\begin{matrix}1 & 0\\ 0 & 1\end{matrix}$	$\begin{matrix}Y_i\\ -X_i\end{matrix}$	$\begin{matrix}X_i\\ Y_i\end{matrix}$
3	3	distances, azimuths, astronomical latitudes and longitudes	$\begin{matrix}1 & 0 & 0\\ 0 & 1 & 0\\ 0 & 0 & 1\end{matrix}$		
	6	distances	$\begin{matrix}1 & 0 & 0\\ 0 & 1 & 0\\ 0 & 0 & 1\end{matrix}$	$\begin{matrix}0 & -Z_i & Y_i\\ Z_i & 0 & -X_i\\ -Y_i & X_i & 0\end{matrix}$	
	7	angles and/or distance ratios	$\begin{matrix}1 & 0 & 0\\ 0 & 1 & 0\\ 0 & 0 & 1\end{matrix}$	$\begin{matrix}0 & -Z_i & Y_i\\ Z_i & 0 & -X_i\\ -Y_i & X_i & 0\end{matrix}$	$\begin{matrix}X_i\\ Y_i\\ Z_i\end{matrix}$

12.3 Constraining a Rank Deficient Problem

When A is rank deficient there are an infinite number of solutions (differing by the solutions of $A\boldsymbol{x} = \boldsymbol{0}$) to the singular least-squares problem

$$\min \|\boldsymbol{b} - A\boldsymbol{x}\|_2.$$

We look for a unique solution $\hat{\boldsymbol{x}}$ with the additional constraints

$$G^{\mathrm{T}}\boldsymbol{x} = \boldsymbol{g}. \tag{12.12}$$

The rows of the d by m matrix G span the d-dimensional $\mathcal{N}(A)$. Then A augmented by the d new rows from G^{T} has full rank n.

We formulate the problem as one consisting of singular normals $A^{\mathrm{T}}A$ with an *orthogonal bordering* matrix G:

$$\begin{bmatrix} A^{\mathrm{T}}A & G \\ G^{\mathrm{T}} & 0 \end{bmatrix} \begin{bmatrix} \boldsymbol{x} \\ \boldsymbol{0} \end{bmatrix} = \begin{bmatrix} A^{\mathrm{T}}\boldsymbol{b} \\ \boldsymbol{g} \end{bmatrix}. \tag{12.13}$$

The augmentation of $A^{\mathrm{T}}A$ by G^{T} makes the coefficient matrix invertible. The solution is unique, and it can be expressed in terms of the pseudoinverse A^+:

$$\boldsymbol{x}^+ = A^+\boldsymbol{b}. \tag{12.14}$$

Remark 12.1 There is another formulation of the problem. Let the normal equations be

$$A^{\mathrm{T}}A\boldsymbol{x} = A^{\mathrm{T}}\boldsymbol{b}.$$

$A^{\mathrm{T}}A$ is still singular and nonnegative definite. In order to make the problem uniquely solvable we add a suitable set of d fictitious *observation equations*

$$F\boldsymbol{x} = \boldsymbol{g} \tag{12.15}$$

and suppose these observation equations are weight normalized such that the normals become

$$(A^{\mathrm{T}}A + F^{\mathrm{T}}F)\boldsymbol{x} = A^{\mathrm{T}}\boldsymbol{b} + F^{\mathrm{T}}\boldsymbol{g}.$$

Such an addition to the problem is called a *soft postulation.* We shall see how the inverse of the coefficient matrix depends on this soft postulation.

If the fictitious observations are given infinite weight—*hard postulation*, which implies that (12.15) is strictly enforced—then those are regarded as *condition equations* and not as observation equations.

We demonstrate the technique on the transformation (12.37) below and modify it for a soft postulation as follows

$$\Sigma_{\text{transformed}} = S(\Sigma_{\hat{x}} + GG^{\mathrm{T}})S^{\mathrm{T}} \tag{12.16}$$

according to Krarup (1979).

Example 12.1 Let the oriented graph in Figure 12.1 and height differences along the edges be given. The incidence matrix corresponding to this graph is

$$A = \begin{bmatrix} -1 & 0 & 0 & 1 & 0 \\ -1 & 0 & 0 & 0 & 1 \\ 0 & 0 & -1 & 1 & 0 \\ 0 & -1 & 0 & 0 & 1 \\ 0 & 0 & 0 & 1 & -1 \end{bmatrix}$$

with the observations

$$\boldsymbol{b} = \begin{bmatrix} 1.978 \\ 0.732 \\ 0.988 \\ 0.420 \\ 1.258 \end{bmatrix}.$$

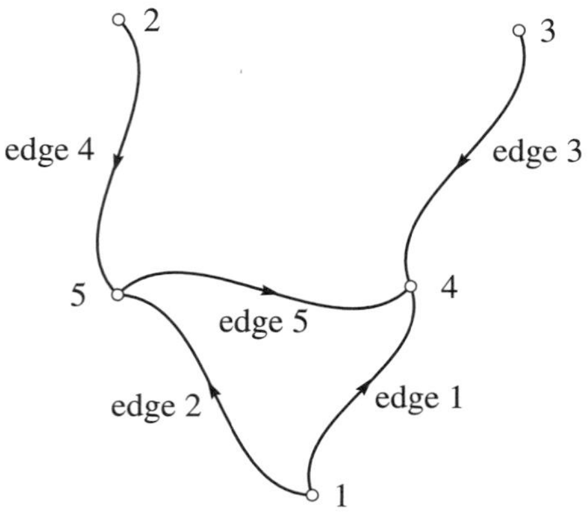

Figure 12.1 Graph for oriented free leveling network.

The matrix $A^{\mathrm{T}}A$ is singular. A unique solution can be calculated by means of the pseudo-inverse matrix A^+. This matrix is often determined by means of the SVD for A. This is the decomposition $A = U\Sigma V^{\mathrm{T}}$ and from here $A^+ = V\Sigma^+ U^{\mathrm{T}}$. The unique least squares solution of minimum norm is

$$\boldsymbol{x}^+ = A^+\boldsymbol{b} = \begin{bmatrix} -0.8024 \\ -0.4944 \\ 0.1916 \\ 1.1796 \\ -0.0744 \end{bmatrix}.$$

The same solution $\hat{\boldsymbol{x}} = \boldsymbol{x}^+$ can be achieved by augmenting $A^{\mathrm{T}}A$ by the row $\boldsymbol{e}^{\mathrm{T}}$ and the column $\boldsymbol{e}$ of all ones:

$$\begin{bmatrix} A^{\mathrm{T}}A & \boldsymbol{e} \\ \boldsymbol{e}^{\mathrm{T}} & 0 \end{bmatrix}\begin{bmatrix} \boldsymbol{x} \\ 0 \end{bmatrix} = \begin{bmatrix} A^{\mathrm{T}}\boldsymbol{b} \\ 0 \end{bmatrix}.$$

Note that the mean value of $\hat{\boldsymbol{x}} = \hat{\boldsymbol{x}}_0$ is zero. This gives the last equation $\boldsymbol{e}^{\mathrm{T}}\hat{\boldsymbol{x}} = 0$. To find the solution we did not form the normal equations but rather used the SVD.

If we want to change the solution $\hat{\boldsymbol{x}}_0$ from "0-level" to a level such as $l = 20$, we simply have to change g to $5 \times 20 = 100$ and we obtain the solution

$$\hat{\boldsymbol{x}}_{20} = \begin{bmatrix} 19.1976 \\ 19.5056 \\ 20.1916 \\ 21.1796 \\ 19.9256 \end{bmatrix}.$$

The mean value of this solution is 20 because $\boldsymbol{e}^{\mathrm{T}}\hat{\boldsymbol{x}}_{20} = 100$.

Example 12.2 We want to demonstrate how a projector can bring $\hat{x}_{20}$ from Example 12.1 back to $\hat{x}_0$. In this special case we project onto the plane perpendicular to e:

$$P = I - e(e^{\mathrm{T}}e)^{-1}e^{\mathrm{T}} = I - \tfrac{1}{5}ee^{\mathrm{T}} = \tfrac{1}{5}\begin{bmatrix} 4 & -1 & -1 & -1 & -1 \\ -1 & 4 & -1 & -1 & -1 \\ -1 & -1 & 4 & -1 & -1 \\ -1 & -1 & -1 & 4 & -1 \\ -1 & -1 & -1 & -1 & 4 \end{bmatrix}$$

or

$$\hat{x}_0 = P\hat{x}_{20} = P\begin{bmatrix} 19.1976 \\ 19.5056 \\ 20.1916 \\ 21.1796 \\ 19.9256 \end{bmatrix} = \begin{bmatrix} -0.8024 \\ -0.4944 \\ 0.1916 \\ 1.1796 \\ -0.0744 \end{bmatrix}.$$

Example 12.3 Continuing Example 12.1 we shall calculate the covariance matrix for the pseudoinverse solution. We shall determine the standard deviation (of unit weight)

$$\hat{r} = b - Ax^+ = b - AA^+b = (I - AA^+)b \quad \text{and} \quad \hat{\sigma}_0 = \|\hat{r}\|_2 = 0.0069 = 6.9\,\mathrm{mm}.$$

Thus the covariance matrix $\Sigma_+ = \hat{\sigma}_0^2 A^+(A^+)^{\mathrm{T}}$ is

$$\begin{bmatrix} 0.192 & -0.096 & -0.096 & -0.000 & -0.000 \\ -0.096 & 0.416 & -0.224 & -0.128 & 0.032 \\ -0.096 & -0.224 & 0.416 & 0.032 & -0.128 \\ -0.000 & -0.128 & 0.032 & 0.128 & -0.032 \\ -0.000 & 0.032 & -0.128 & -0.032 & 0.128 \end{bmatrix} \times 10^{-4}$$

with $\mathrm{tr}(\Sigma_+) = 1.280 \times 10^{-4}$. Next we have

$$Bx = \begin{bmatrix} A \\ e^{\mathrm{T}} \end{bmatrix} x = \begin{bmatrix} b \\ 0 \end{bmatrix}$$

and $\Sigma = \hat{\sigma}_0^2(B^{\mathrm{T}}B)^{-1}$ becomes

$$\begin{bmatrix} 0.211 & -0.077 & -0.077 & 0.019 & 0.019 \\ -0.077 & 0.435 & -0.205 & -0.109 & 0.051 \\ -0.077 & -0.205 & 0.435 & 0.051 & -0.109 \\ 0.019 & -0.109 & 0.051 & 0.147 & -0.013 \\ 0.019 & 0.051 & -0.109 & -0.013 & 0.147 \end{bmatrix} \times 10^{-4}.$$

Finally $\mathrm{tr}(\Sigma) = 1.376 \times 10^{-4}$. Obviously $\mathrm{tr}(\Sigma_+) < \mathrm{tr}(\Sigma)$.

Example 12.4 This is the example of a ***free network***. We augment the original $n = 2$ unknowns to include the unknowns of all other points in the network. Hence the number of unknowns increases to 8. As $\mathrm{rank}(A) = 8 - 3 = 5$ we have to include at least two

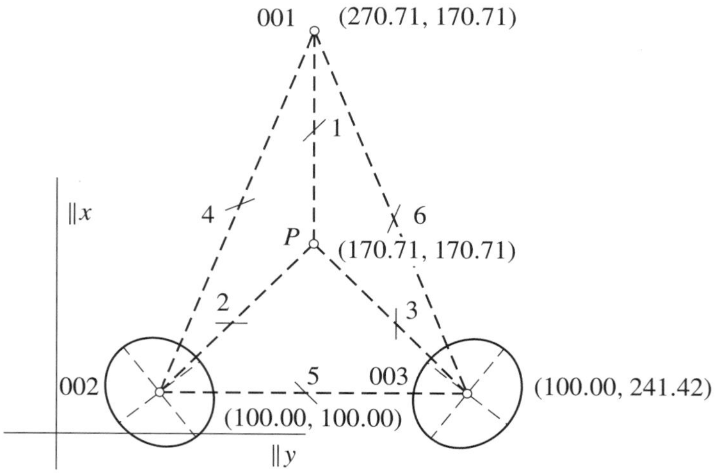

Figure 12.2 Free distance network with 6 observations. The confidence ellipses correspond to the postulated coordinates X_P, Y_P, X_1, and Y_1.

more observations compared to Example 10.1. For reasons of symmetry we include even three more observations, viz. distance observations between points 001-002, 002-003, and 003-001. The vector of unknowns is

$$\boldsymbol{x} = (x_P, y_P, x_{001}, y_{001}, x_{002}, y_{002}, x_{003}, y_{003});$$

the augmented coefficient matrix is

$$A = \begin{bmatrix} -1 & 0 & 1 & 0 & 0 & 0 & 0 & 0 \\ 0.707 & 0.707 & 0 & 0 & -0.707 & -0.707 & 0 & 0 \\ 0.707 & -0.707 & 0 & 0 & 0 & 0 & -0.707 & 0.707 \\ 0 & 0 & 0.924 & 0.383 & -0.924 & -0.383 & 0 & 0 \\ 0 & 0 & 0 & 0 & 0 & -1 & 0 & 1 \\ 0 & 0 & 0.924 & -0.383 & 0 & 0 & -0.924 & 0.383 \end{bmatrix}.$$

We choose a nullspace of dimension 4:

$$G^{\mathrm{T}} = \begin{bmatrix} 1 & 0 & 1 & 0 & 1 & 0 & 1 & 0 \\ 0 & 1 & 0 & 1 & 0 & 1 & 0 & 1 \\ -170.71 & 170.71 & -170.71 & 270.71 & -100 & 100 & -241.42 & 100 \\ 170.71 & 170.71 & 270.71 & 170.71 & 100 & 100 & 100 & 241.42 \end{bmatrix}$$

and the right side is

$$\boldsymbol{b} = \begin{bmatrix} 0.010 \\ 0.020 \\ 0.030 \\ 0.010 \\ 0.020 \\ 0.030 \end{bmatrix}.$$

The pseudoinverse solution $\boldsymbol{x}^+$ and the residuals $\boldsymbol{r}$ are

$$\boldsymbol{x}^+ = \begin{bmatrix} 0.0080 \\ 0.0023 \\ 0.0113 \\ -0.0068 \\ -0.0026 \\ -0.0087 \\ -0.0167 \\ 0.0132 \end{bmatrix} \quad \text{and} \quad \boldsymbol{r} = \begin{bmatrix} 0.0067 \\ 0.0047 \\ 0.0047 \\ -0.0036 \\ -0.0020 \\ -0.0036 \end{bmatrix}.$$

The standard deviation is $\hat{\sigma}_0 = 10.9\,\text{mm}$. The least-squares solution of the free network yields the following coordinates (X_i', Y_i'):

Point	X_i	ξ_i	X_i'	Y_i	η_i	Y_i'
P	170.71	0.008	170.718	170.71	0.002	170.712
1	270.71	0.011	270.721	170.71	−0.007	170.703
2	100.00	−0.003	99.997	100.00	−0.009	99.991
3	100.00	−0.017	99.983	241.42	0.013	241.433

Note that $\sum \xi_i = \sum \eta_i = 0$ within the computational accuracy.

12.4 Linear Transformation of Random Variables

In the case where A was singular we circumvented this problem by introducing the solution vector of shortest length: $\boldsymbol{x}^+$. That was the effect of augmenting $A^{\mathrm{T}}A$ and producing the pseudoinverse.

We shall try to analyze this situation from a random point of view. The nonunique solution was determined up to any additional solution from the nullspace. Any nonunique solution vector $\hat{\boldsymbol{x}}$ is connected to a covariance matrix $\Sigma_{\hat{x}}$. But as the vectors $\hat{\boldsymbol{x}}$ differ, so do the covariance matrices $\Sigma_{\hat{x}}$. Among all these possible covariance matrices we shall demonstrate that the covariance matrix corresponding to the pseudoinverse solution is the one with *smallest trace*. In statistical terms this means that the pseudoinverse solution $\boldsymbol{x}^+$ gives the smallest overall variances of the unknowns.

Let $\underset{m \text{ by } 1}{\boldsymbol{v}} = \underset{m \text{ by } n}{B} \underset{n \text{ by } 1}{\boldsymbol{u}}$ be given as a linear transformation between some random variables $\boldsymbol{u}$ and $\boldsymbol{v}$, and if $E\{\boldsymbol{u}\} = \boldsymbol{0}$ then $E\{\boldsymbol{u}\boldsymbol{u}^{\mathrm{T}}\} = \Sigma_u$.

Now we want to approximate $\boldsymbol{u}$ by $A\boldsymbol{v}$ and substitute

$$\underset{n \text{ by } 1}{\boldsymbol{u}} = \underset{n \text{ by } m}{A} \underset{m \text{ by } 1}{\boldsymbol{v}} + \underset{n \text{ by } 1}{\boldsymbol{w}}$$

where $\boldsymbol{w}$ is some residual vector. We solve for $\boldsymbol{w}$

$$\boldsymbol{w} = \boldsymbol{u} - A\boldsymbol{v} = (I - AB)\boldsymbol{u}. \tag{12.17}$$

The covariance matrix Σ_w for $\boldsymbol{w}$ is

$$\Sigma_w = (I - AB)\Sigma_u(I - AB)^{\mathrm{T}}. \tag{12.18}$$

We want to minimize the trace of Σ_w—B being the variable matrix—and shall prove that

$$\operatorname{tr}\Sigma_w = \min \qquad \text{for} \qquad B = (A^{\mathrm{T}}A)^{-1}A^{\mathrm{T}}. \tag{12.19}$$

The results are

$$\operatorname{tr}\Sigma_w = \operatorname{tr}\Sigma_u - \operatorname{tr}(A^{\mathrm{T}}A)^{-1}A^{\mathrm{T}}\Sigma_u A$$

and

$$\Sigma_w = \big(I - A(A^{\mathrm{T}}A)^{-1}A^{\mathrm{T}}\big)\Sigma_u\big(I - A(A^{\mathrm{T}}A)^{-1}A^{\mathrm{T}}\big)^{\mathrm{T}}$$

and

$$\Sigma_v = (A^{\mathrm{T}}A)^{-1}A^{\mathrm{T}}\Sigma_u A(A^{\mathrm{T}}A)^{-1}.$$

This theorem shows a result which is different from that following from the usual least-squares estimation of $\boldsymbol{v} = B\boldsymbol{u}$. Ordinary least-squares estimation, minimizing the square sum of weighted residuals, yields the result

$$B = (A^{\mathrm{T}}\Sigma_u^{-1}A)^{-1}A^{\mathrm{T}}\Sigma_u^{-1} \qquad \text{where} \qquad \operatorname{rank}\Sigma_u = n.$$

In the present theorem we minimize the sum of variances of residuals $\boldsymbol{w}$; Σ_u may have any rank. When $\Sigma_u = I$ there is no difference between the two cases.

Proof We start from (12.18) and get

$$\Sigma_w = (\Sigma_u - AB\Sigma_u)(I - B^{\mathrm{T}}A^{\mathrm{T}}) = \Sigma_u - \Sigma_u B^{\mathrm{T}}A^{\mathrm{T}} - AB\Sigma_u + AB\Sigma_u B^{\mathrm{T}}A^{\mathrm{T}}.$$

The first term is independent of B, so we get

$$\operatorname{tr}\Sigma_w = \text{const.} - \operatorname{tr}(\Sigma_u B^{\mathrm{T}}A^{\mathrm{T}}) - \operatorname{tr}(AB\Sigma_u) + \operatorname{tr}(AB\Sigma_u B^{\mathrm{T}}A^{\mathrm{T}})$$

and

$$\frac{\partial \operatorname{tr}\Sigma_w}{\partial B} = -2A^{\mathrm{T}}\Sigma_u + 2A^{\mathrm{T}}AB\Sigma_u = 0$$

or $B = (A^{\mathrm{T}}A)^{-1}A^{\mathrm{T}}$ which proves the theorem.

12.5 Similarity Transformations

In the previous section we suspended the rank defect of $A^{\mathrm{T}}A$ by a bordering matrix G. At the same time we added constraints on the least-squares problem. These constraints were also described by G! However, the constraints may be of a more general form. In this section we shall introduce constraints which transform the free network to one with at least d postulated coordinates and we shall derive the pertinent formulas. The related covariance matrix $\Sigma_{\text{transformed}}$ is correspondingly transformed, and consequently d columns and rows will be zeroed. We start by working through a simple example.

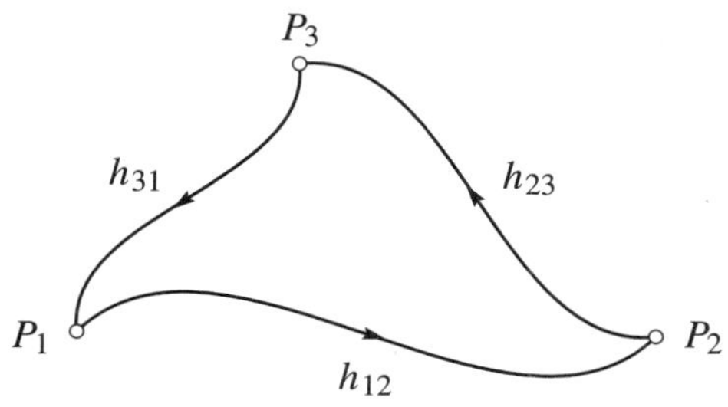

Figure 12.3 Simple free leveling network

Example 12.5 A simple leveling network has three nodes P_1, P_2, and P_3 and observed oriented differences of heights h_{12}, h_{23}, and h_{31}. According to the loop law we have

$$h_{12} + h_{23} + h_{31} = 0. \tag{12.20}$$

We cannot calculate the heights of the nodes from these differences. So we choose P_1 as reference point, i.e. we put $P_1 =$ given constant $= h_1$. The best estimate for h_2 and h_3 is now unique

$$h_i = h_{1i} + h_1 = h_i^1 + h_1, \qquad i = 1, 2, 3 \tag{12.21}$$

where the upper index refers to the reference point chosen. Yet there is a slight flaw because of the arbitrariness in which we are choosing P_1 as reference. In fact we can take any point of the network as reference point. We can even choose the barycenter P_b and assign it the height

$$h_b = \tfrac{1}{n} \sum_{i=1}^{n} h_i.$$

Consequently we get

$$h_i = \tfrac{1}{3}\big(h_{1i} + h_{2i} + h_{3i}\big) + h_b = h_i^b + h_b.$$

All sets of heights like h_i^1 or h_i^b can be looked upon as valid heights. So it is more a question of what set to choose. However, the statistical properties of the heights h_i^1 or h_i^b depend highly on the choice of reference point. From $E\{h_3^1\} = E\{h_{12} + h_{23}\}$ and $E\{h_3^b\} = E\big\{\frac{2}{3}h_{12} + h_{23} + \frac{1}{3}h_{31}\big\}$ follows for instance that $E\{h_3^1\} \neq E\{h_3^b\}$. Their covariances differ, too.

When comparing two sets of heights it is essential that they refer to the same level of reference. So in order to do this we shall learn how to transform heights from one system to another. We write (12.21) explicitly

$$\begin{bmatrix} h_1 \\ h_2 \\ h_3 \end{bmatrix} = \begin{bmatrix} h_1^1 \\ h_2^1 \\ h_3^1 \end{bmatrix} + \left[\begin{array}{c|cc} 1 & 0 & 0 \\ \hline 1 & 0 & 0 \\ 1 & 0 & 0 \end{array}\right] \begin{bmatrix} h_1 \\ h_1 \\ h_1 \end{bmatrix} \tag{12.22}$$

or

$$\begin{bmatrix} h_1^1 \\ h_2^1 \\ h_3^1 \end{bmatrix} = \left[\begin{array}{c|cc} 0 & 0 & 0 \\ \hline -1 & 1 & 0 \\ -1 & 0 & 1 \end{array}\right] \begin{bmatrix} h_1 \\ h_2 \\ h_3 \end{bmatrix}. \tag{12.23}$$

Equation (12.23) shows us how we can transform from one height system, say h_i^1, to the height system defined by P_1 as reference point. In a similar way we obtain

$$\begin{bmatrix} h_1^b \\ h_2^b \\ h_3^b \end{bmatrix} = \frac{1}{3} \begin{bmatrix} 2 & -1 & -1 \\ -1 & 2 & -1 \\ -1 & -1 & 2 \end{bmatrix} \begin{bmatrix} h_1^1 \\ h_2^1 \\ h_3^1 \end{bmatrix}$$

and in general

$$\begin{bmatrix} h_1^b \\ h_2^b \\ \vdots \\ h_n^b \end{bmatrix} = \frac{1}{n} \begin{bmatrix} n-1 & -1 & \dots & -1 \\ -1 & n-1 & \dots & -1 \\ \vdots & \vdots & \ddots & \vdots \\ -1 & -1 & \dots & n-1 \end{bmatrix} \begin{bmatrix} h_1^1 \\ h_2^1 \\ \vdots \\ h_n^1 \end{bmatrix}. \tag{12.24}$$

This square matrix is already seen for $n = 5$ in Example 12.2.

The 2-dimensional transformation model must allow for two translations t_x, t_y, a rotation φ and a change of scale k. The transformation from one system (x, y) (in geodesy often termed the "from" system) to another (ξ, η) (the "to" system) is given by

$$\begin{bmatrix} \xi_i \\ \eta_i \end{bmatrix} = k \begin{bmatrix} \cos\varphi & \sin\varphi \\ -\sin\varphi & \cos\varphi \end{bmatrix} \begin{bmatrix} x_i \\ y_i \end{bmatrix} + \begin{bmatrix} t_x \\ t_y \end{bmatrix}. \tag{12.25}$$

Equation (12.25) is not linear in φ, but by a trick we may make it linear. We introduce new slack variables $a = k\cos\varphi$, and $b = k\sin\varphi$ and get

$$\begin{bmatrix} \xi_i \\ \eta_i \end{bmatrix} = \begin{bmatrix} a & b \\ -b & a \end{bmatrix} \begin{bmatrix} x_i \\ y_i \end{bmatrix} + \begin{bmatrix} t_x \\ t_y \end{bmatrix}. \tag{12.26}$$

The observation equations resulting from p common points are

$$\begin{bmatrix} x_1 & y_1 & 1 & 0 \\ x_2 & y_2 & 1 & 0 \\ & \vdots & & \\ x_p & y_p & 1 & 0 \\ y_1 & -x_1 & 0 & 1 \\ y_2 & -x_2 & 0 & 1 \\ & \vdots & & \\ y_p & -x_p & 0 & 1 \end{bmatrix} \begin{bmatrix} a \\ b \\ t_x \\ t_y \end{bmatrix} = \begin{bmatrix} \xi_1 \\ \xi_2 \\ \vdots \\ \xi_p \\ \eta_1 \\ \eta_2 \\ \vdots \\ \eta_p \end{bmatrix} - \begin{bmatrix} r_{\xi 1} \\ r_{\xi 2} \\ \vdots \\ r_{\xi p} \\ r_{\eta 1} \\ r_{\eta 2} \\ \vdots \\ r_{\eta p} \end{bmatrix}$$

or symbolically

$$\underset{2p\text{ by }4}{A}\;\underset{4\text{ by }1}{\boldsymbol{f}} = \underset{2p\text{ by }1}{\boldsymbol{b}} - \underset{2p\text{ by }1}{\boldsymbol{r}}\,.$$

This linear least-squares problem does not show any difficulties.

Yet, we cannot resist from demonstrating the classical solution procedure. We set all weights to unity and the normal equations $A^{\mathrm{T}}A\boldsymbol{f} = A^{\mathrm{T}}\boldsymbol{b}$ can be written

$$\begin{bmatrix} \sum(x_i^2+y_i^2) & 0 & \sum x_i & \sum y_i \\ 0 & \sum(x_i^2+y_i^2) & \sum y_i & -\sum x_i \\ \sum x_i & \sum y_i & p & 0 \\ \sum y_i & -\sum x_i & 0 & p \end{bmatrix} \begin{bmatrix} a \\ b \\ t_x \\ t_y \end{bmatrix} = \begin{bmatrix} \sum(\xi_i x_i + \eta_i y_i) \\ \sum(\xi_i y_i - \eta_i x_i) \\ \sum \xi_i \\ \sum \eta_i \end{bmatrix}. \tag{12.27}$$

All summations run from 1 to p. For numerical reasons we reduce all coordinates in both systems to their respective barycenters (x_s, y_s) and (ξ_s, η_s). This leads to $\sum x_i = \sum y_i = \sum \xi_i = \sum \eta_i \equiv 0$ and $A^{\mathrm{T}}A$ becomes diagonal. We introduce coordinates relative to the barycenter $x_i^* = x_i - x_s$, $y_i^* = y_i - y_s$, and $x_s = \sum x_i/p$, $y_s = \sum y_i/p$:

Now the inversion reduces to solving individual equations:

$$\begin{bmatrix} \star & 0 & 0 & 0 \\ 0 & \star & 0 & 0 \\ 0 & 0 & p & 0 \\ 0 & 0 & 0 & p \end{bmatrix} \begin{bmatrix} a \\ b \\ t_x \\ t_y \end{bmatrix} = \begin{bmatrix} \star \\ \star \\ 0 \\ 0 \end{bmatrix}.$$

Nonzero entries are marked by $\star$. The explicit solution can be written

$$\hat{a} = \frac{\sum(\xi_i^* x_i^* + \eta_i^* y_i^*)}{\sum(x_i^{*2} + y_i^{*2})} \quad\text{and}\quad \hat{b} = \frac{\sum(\xi_i^* y_i^* - \eta_i^* x_i^*)}{\sum(x_i^{*2} + y_i^{*2})}. \tag{12.28}$$

Next, we can determine

$$\hat{k} = \sqrt{\hat{a}^2 + \hat{b}^2} \quad\text{and}\quad \hat{\varphi} = \arctan \hat{b}/\hat{a}. \tag{12.29}$$

Finally we get $\hat{t}_x$ and $\hat{t}_y$ from (12.27):

$$\hat{t}_x = \frac{\sum \xi_i - \hat{a}\sum x_i - \hat{b}\sum y_i}{p} = \xi_s - \hat{a}x_s - \hat{b}y_s, \tag{12.30}$$

$$\hat{t}_y = \frac{\sum \eta_i - \hat{a}\sum y_i + \hat{b}\sum x_i}{p} = \eta_s - \hat{a}y_s + \hat{b}x_s. \tag{12.31}$$

The covariance matrix for the estimated $\hat{\boldsymbol{f}}$ is

$$\Sigma_{\hat{f}} = \frac{1}{p\sum(x_i^2+y_i^2) - \left(\sum x_i\right)^2 - \left(\sum y_i\right)^2} \times \begin{bmatrix} p & 0 & -\sum x_i & -\sum y_i \\ 0 & p & -\sum y_i & \sum x_i \\ -\sum x_i & -\sum y_i & \sum(x_i^2+y_i^2) & 0 \\ -\sum y_i & \sum x_i & 0 & \sum(x_i^2+y_i^2) \end{bmatrix}.$$

The variance of unit weight which also is the variance of a transformed coordinate is according to (9.64)

$$\hat{\sigma}_0^2 = \frac{\sum(r_{\xi_i}^2 + r_{\eta_i}^2)}{2p-4}.$$

For the translations we have

$$\hat{\sigma}_{\hat{t}_x} = \hat{\sigma}_{\hat{t}_y} = \hat{\sigma}_0 \sqrt{\frac{\sum(x_i^2+y_i^2)}{p\sum(x_i^2+y_i^2) - \left(\sum x_i\right)^2 - \left(\sum y_i\right)^2}} = \hat{\sigma}_0 \sqrt{\frac{1}{p} + \frac{x_s^2+y_s^2}{\sum(x_i^{*2}+y_i^{*2})}}. \tag{12.32}$$

Observe that the standard deviations for $\hat{t}_x$ and $\hat{t}_y$ not only depend on p, as expected, but also on the position of the origin.

Thus we have described the most important, elementary circumstances about the similarity transformation. It includes p points whose coordinates are given in two systems, namely an original x, y system and a different ξ, η system. The transformation parameters $\hat{k}, \hat{\varphi}, \hat{t}_x$, and $\hat{t}_y$ are estimated through a least-squares procedure. Additionally, the solution makes it possible to transform any other point (x_j, y_j) into the corresponding (ξ_j, η_j) by means of the ***transformation equations***

$$\begin{aligned} \xi_j &= \hat{a}x_j + \hat{b}y_j + \hat{t}_x \\ \eta_j &= -\hat{b}x_j + \hat{a}y_j + \hat{t}_y. \end{aligned} \tag{12.33}$$

Example 12.6 We shall demonstrate the procedure just described by a numerical example. Suppose we know the exact coordinates $(N, E) = (\xi, \eta)$ and (x, y) of 7 common points. We want to determine the transformation parameters and to transform a single point with given x, y coordinates. The procedure is in fact a 2-dimensional interpolation.

For a start we list the postulated coordinates. Point s is the barycenter (its coordinates are in the last line). The second table gives the same data reduced to the barycenter. The latter are marked $*$.

Point	x_i [m]	y_i [m]	ξ_i [m]	η_i [m]
62-04-005	277 722.022	−230 855.152	6 310 000.527	562 940.820
62-04-801	275 956.869	−231 105.839	6 308 231.260	562 725.625
62-04-810	277 563.374	−235 447.400	6 309 749.964	558 354.121
62-04-811	278 608.525	−233 945.915	6 310 824.656	559 833.890
62-04-815	276 163.682	−236 471.626	6 308 330.475	557 358.463
63-01-002	273 578.801	−230 941.425	6 305 857.705	562 937.589
63-04-003	274 533.958	−235 063.723	6 306 729.799	558 798.283
s	276 303.890	−233 404.440	6 308 532.055	560 421.256

Point	x_i^* [m]	y_i^* [m]	ξ_i^* [m]	η_i^* [m]
62-04-005	1 418.132	2 549.288	1 468.472	2 519.564
62-04-801	−347.021	2 298.601	−300.795	2 304.369
62-04-810	1 259.484	−2 042.960	1 217.909	−2 067.135
62-04-811	2 304.635	−541.475	2 292.601	−587.366
62-04-815	−140.208	−3 067.186	−201.580	−3 062.793
63-01-002	−2 725.089	2 463.015	−2 674.350	2 516.333
63-04-003	−1 769.932	−1 659.283	−1 802.256	−1 622.973
Sum	.001	.000	.001	.001

The necessary sums of product are

$$\sum x_i^{*2} = 19\,607\,592.00 \qquad \sum y_i^{*2} = 34\,476\,609.52$$

$$\sum x_i^*\eta_i^* = -4\,739\,029.25 \qquad \sum y_i^*\xi_i^* = -3\,655\,603.21$$

$$\sum x_i^*\xi_i^* = 19\,510\,390.19 \qquad \sum y_i^*\eta_i^* = 34\,545\,930.52$$

$$\hat{a} = 0.999\,484\,49 \qquad \hat{b} = 0.020\,032\,21$$

$$\hat{k} = 0.999\,685\,22 \qquad \hat{\varphi} = 1.275\,777 \text{ gon}$$

$$\hat{t}_x = \xi_s - \hat{a}x_s - \hat{b}y_s = 6\,037\,046.208$$

$$\hat{t}_y = \eta_s + \hat{b}x_s - \hat{a}y_s = 799\,240.351.$$

The transformation equations (12.33) from (x_j, y_j) to (ξ_j, η_j) include a rotation and translation:

$$\xi_j = 0.999\,484\,49\,x_j + 0.020\,032\,21\,y_j + 6\,037\,046.208$$

$$\eta_j = -0.020\,032\,21\,x_j + 0.999\,484\,49\,y_j + 799\,240.351.$$

A point with coordinates $(x, y) = (276\,109.847, -233\,507.185)$ is transformed to $(\xi, \eta) = (6\,308\,336.054, 560\,322.451)$. The official values are $(6\,308\,336.054, 560\,322.449)$. Evidently, the accuracy is satisfactory. This is partly due to the fact that the transformed point is close to the barycenter. By means of the M-file simil we find $\hat{\sigma}_0 = 3$ mm.

For the sake of completeness we shall add that this procedure is called a *Helmert transformation*. A similar least-squares problem was mentioned by Helmert (1893).

Symmetric Similarity Transformation The thoughtful reader might ask why we leave (x_i, y_i) of the common points unchanged under the transformation. The formulation favors one point set above the other: Are the postulated coordinates (x_i, y_i) much better than (ξ_i, η_i) so that they can be assumed free of errors? Actually those (x_i, y_i) are transformed as a "stiff" point set while the points (ξ_i, η_i) adjust individually under the procedure. So why not allow both point sets to adjust under the transformation? In practice the (x_i, y_i)

coordinates are worse than the newly calculated (ξ_i, η_i). So the relevant formulation is to augment (12.25) by two extra types of observation equations:

$$\begin{bmatrix} \xi_i \\ \eta_i \\ X_i' \\ Y_i' \end{bmatrix} = \begin{bmatrix} a & b \\ -b & a \\ 1 & 0 \\ 0 & 1 \end{bmatrix} \begin{bmatrix} x_i \\ y_i \end{bmatrix} + \begin{bmatrix} t_x \\ t_y \\ 0 \\ 0 \end{bmatrix}. \tag{12.34}$$

where (ξ_i, η_i) and (X_i', Y_i') denote the postulated coordinates of the common points in the two systems. But a noteworthy change also has happened: the p sets of (x_i, y_i) have been introduced as additional unknowns which shall be estimated together with the earlier 4 unknowns: a, b, t_x, t_y.

We only quote the results given in Teunissen (1985b), 141–146. The scale is given by a rather complicated expression $\hat{k} = \lambda + \sqrt{1+\lambda^2}$ where

$$\begin{aligned}
\lambda &= \frac{\sum(\xi_i^{*2} + \eta_i^{*2}) - \sum(X_i'^{*2} + Y_i'^{*2})}{2\sqrt{\left(\sum \xi_i^* X_i'^* + \eta_i^* Y_i'^*\right)^2 + \left(\sum \xi_i^* Y_i'^* - \eta_i^* X_i'^*\right)^2}} \\
X_s &= \tfrac{1}{n}\sum X_i', \quad Y_s = \tfrac{1}{n}\sum Y_i', \quad X_i'^* = X_s - X_i', \quad Y_i'^* = Y_s - Y_i' \\
\hat{\varphi} &= \arctan \frac{\sum \xi_i^* Y_i'^* - \eta_i^* X_i'^*}{\sum \xi_i^* X_i'^* + \eta_i^* Y_i'^*} \\
\hat{t}_x &= x_s - X_s \hat{k} \cos\hat{\varphi} - Y_s \hat{k} \sin\hat{\varphi} \\
\hat{t}_y &= y_s + X_s \hat{k} \sin\hat{\varphi} - Y_s \hat{k} \cos\hat{\varphi} \\
\hat{x}_i &= X_s + \tfrac{1}{1+\hat{k}^2}\left(X_i'^* + \xi_i^* \hat{k}\cos\hat{\varphi} - \eta_i^* \hat{k} \sin\hat{\varphi}\right) \\
\hat{y}_i &= Y_s + \tfrac{1}{1+\hat{k}^2}\left(Y_i'^* + \xi_i^* \hat{k}\sin\hat{\varphi} + \eta_i^* \hat{k} \cos\hat{\varphi}\right).
\end{aligned} \tag{12.35}$$

A comparison of $\hat{k} = \lambda + \sqrt{1+\lambda^2}$ with the expression $\hat{k} = \sqrt{\hat{a}^2 + \hat{b}^2}$, where $\hat{a}$ and $\hat{b}$ are given by (12.28), shows that the latter expression systematically underestimates the scale compared to the value coming from (12.35).

When both sets of coordinates of the common points are adjusted the model is no longer linear and we get a biased estimate for $\hat{k}$. The estimates for the rotation $\hat{\varphi}$ and the translations $\hat{t}_x$ and $\hat{t}_y$ are still unbiased. *The bias* for $\hat{k}$ can be shown to be

$$b_{\hat{k}} = 3\frac{1+\hat{k}^2}{(n-1)n\hat{k}}\frac{\hat{\sigma}_0^2}{d^2} \tag{12.36}$$

where the common points are placed in a square lattice with side length d. The expression (12.36) shows that the bias is negligible for most practical problems. If $\hat{\sigma}_0/d = 10^{-5}$, $\hat{k} = 1$, and $n = 4$ we have $b_{\hat{k}} = 0.5 \times 10^{-10}$.

The example is interesting because even a simple nonlinear least-squares problem may introduce biased estimates for the unknowns.

12.6 Covariance Transformations

We repeat that any nonsingular least-squares problem results in estimated coordinates $\hat{x}$ and an a posteriori covariance matrix $\Sigma_{\hat{x}}$. This result implies the following two options:

- The *coordinate system* is defined by postulating four quantities. They can be: (1) All four coordinates of two points. (2) Two coordinates of one point, one orientation, and one distance. (3) Four linear functions of the coordinates. This coordinate system is the reference in which all other coordinates are calculated. ***The transition from one coordinate system to another*** is accomplished by a similarity transformation.

- The *variances* of the four fixed quantities *are set to zero* to serve as reference for confidence ellipses at all other points. ***The transition of one covariance system to another*** may be performed in two ways:

  1 Indirectly by repeating the above procedure with newly selected reference covariances. In many cases it is not of interest to transform the coordinate system itself, leaving the coordinate values unchanged.

  2 Directly by selecting a member of the family of covariance transformations

$$\Sigma_{\text{transformed}} = S\Sigma_{\hat{x}}S^{\mathrm{T}} \tag{12.37}$$

  having one covariance matrix available.

Some of those choices are of a subjective nature and follow from practical circumstances.

The matrix S can be derived from simple geometric considerations. As usual the least squares model is $Ax = \boldsymbol{b}$. The components of $\boldsymbol{x}$ do not necessarily represent coordinates. But what we like to find are unbiased estimable linear functions of $\boldsymbol{x}$ which can be *interpreted as coordinates*. We denote them by $\boldsymbol{x}^*$. Unbiased estimability implies that $\boldsymbol{x}^*$ is the expected value of a linear function of $\boldsymbol{b}$: $\boldsymbol{x}^* = E\{B\boldsymbol{b}\}$ or

$$\boldsymbol{x}^* = BA\boldsymbol{x}. \tag{12.38}$$

In Example 12.5 we already encountered a sample of $\boldsymbol{x}^* = E\{B\boldsymbol{b}\}$, namely the heights h_1^b, h_2^b, and h_n^b which are linear functions of the observed height differences.

Now we concentrate on two dimensional networks. The coordinates $\boldsymbol{x}^*$ should be transformable to another coordinate system $\boldsymbol{x}$ by adding the contribution from a differential similarity transformation (12.8):

$$\boldsymbol{x} = \boldsymbol{x}^* + G\boldsymbol{f}. \tag{12.39}$$

This is rewritten $\boldsymbol{f} = G^+(\boldsymbol{x}-\boldsymbol{x}^*)$ and substituted into (12.39): $\boldsymbol{x} = \boldsymbol{x}^* + GG^+(\boldsymbol{x}-\boldsymbol{x}^*)$ or

$$(I - GG^+)\boldsymbol{x} = (I - GG^+)\boldsymbol{x}^* \tag{12.40}$$

which is a consistency condition on the transformation. As G has full rank we have $G^+ = (G^{\mathrm{T}}G)^{-1}G^{\mathrm{T}}$. So the projector S is defined as

$$S = I - G(G^{\mathrm{T}}G)^{-1}G^{\mathrm{T}}. \tag{12.41}$$

Geometrically S projects onto a subspace $\boldsymbol{R}(S)$ complementary to $\boldsymbol{N}(A)$ and along the nullspace $\boldsymbol{N}(A) = \boldsymbol{R}(G)$. The matrix to invert is of small dimension equal to the dimension of $\boldsymbol{N}(A)$.

If we want to achieve minimum trace of $\Sigma_{\hat{x}}$ for p common points of the network, $2 \le p \le \frac{n}{2}$, we split the G matrix into two: G_1 containing the rows pertaining to the p selected common points and G_2 containing the rest:

$$G = \overset{d}{\begin{bmatrix} G_1 \\ G_2 \end{bmatrix}} \begin{matrix} 2p \\ n-2p \end{matrix}.$$

We assume that the coordinates for the common points are on the top of $\boldsymbol{x}$ (the rows of G may have to be interchanged). In this more general case we shall calculate $S = I - GG^+$, i.e. we must find an adequate expression for G^+.

We postulate that $G^+ = \begin{bmatrix} G_1^+ & 0 \end{bmatrix}$ and shall now prove this is correct. We start by the identity

$$GG^+G = \begin{bmatrix} G_1 \\ G_2 \end{bmatrix} \begin{bmatrix} G_1^+ & 0 \end{bmatrix} \begin{bmatrix} G_1 \\ G_2 \end{bmatrix} = \begin{bmatrix} G_1 G_1^+ & 0 \\ G_2 G_1^+ & 0 \end{bmatrix} \begin{bmatrix} G_1 \\ G_2 \end{bmatrix} = \begin{bmatrix} G_1 G_1^+ G_1 \\ G_2 G_1^+ G_1 \end{bmatrix}.$$

According to the definition of a pseudoinverse we have $G_1 G_1^+ G_1 = G_1$; and $G_2 G_1^+ G_1 = G_2 (G_1^{\mathrm{T}} G_1)^{-1} G_1^{\mathrm{T}} G_1 = G_2$. Therefore

$$GG^+G = \begin{bmatrix} G_1 \\ G_2 \end{bmatrix} = G.$$

Thus $G^+ = \begin{bmatrix} G_1^+ & 0 \end{bmatrix} = \begin{bmatrix} (G_1^{\mathrm{T}} G_1)^{-1} G_1^{\mathrm{T}} & 0 \end{bmatrix}$ is the correct pseudoinverse.

The transformation matrix for p common points with minimum variance is

$$S_p = I - GG^+ = I - \begin{bmatrix} G_1 \\ G_2 \end{bmatrix} \begin{bmatrix} (G_1^{\mathrm{T}} G_1)^{-1} G_1^{\mathrm{T}} & 0 \end{bmatrix} = I - \begin{bmatrix} G_1 (G_1^{\mathrm{T}} G_1)^{-1} G_1^{\mathrm{T}} & 0 \\ G_2 (G_1^{\mathrm{T}} G_1)^{-1} G_1^{\mathrm{T}} & 0 \end{bmatrix}.$$

Still we only have to invert a d by d matrix. For $2p = n$ we recover (12.41).

A special and important case is $p = 2$. We get d zero variances:

$$S_p = I - \begin{bmatrix} I & 0 \\ G_2 G_1^{-1} & 0 \end{bmatrix} = \begin{bmatrix} 0 & 0 \\ -G_2 G_1^{-1} & I \end{bmatrix}. \tag{12.42}$$

Example 12.7 We continue Example 12.4 and want to transform the adjusted network so that the coordinates X_P, Y_P, X_1, and Y_1 are kept fixed. This is achieved by

$$G_1 = \begin{bmatrix} 1 & 0 & -Y_P \\ 0 & 1 & X_P \\ 1 & 0 & -Y_1 \\ 0 & 1 & X_1 \end{bmatrix}.$$

Hence

$$S_p = \left[\begin{array}{cccc|cccc} 0 & 0 & 0 & 0 & 0 & 0 & 0 & 0 \\ 0 & 0 & 0 & 0 & 0 & 0 & 0 & 0 \\ 0 & 0 & 0 & 0 & 0 & 0 & 0 & 0 \\ 0 & 0 & 0 & 0 & 0 & 0 & 0 & 0 \\ \hline -1.707 & 0.707 & 0.707 & -0.707 & 1 & 0 & 0 & 0 \\ -0.707 & -1.707 & 0.707 & 0.707 & 0 & 1 & 0 & 0 \\ -1.707 & -0.707 & 0.707 & 0.707 & 0 & 0 & 1 & 0 \\ 0.707 & -1.707 & -0.707 & 0.707 & 0 & 0 & 0 & 1 \end{array}\right].$$

The transformed solution is

$$\hat{\boldsymbol{x}}_p = S_p \hat{\boldsymbol{x}} = \begin{bmatrix} 0 \\ 0 \\ 0 \\ 0 \\ -0.001\,9 \\ -0.015\,0 \\ -0.028\,8 \\ 0.002\,2 \end{bmatrix}.$$

The covariance matrix for the transformed solution is

$$\Sigma_{\hat{x}_p} = \begin{bmatrix} 0 & 0 & 0 & 0 & 0 & 0 & 0 & 0 \\ 0 & 0 & 0 & 0 & 0 & 0 & 0 & 0 \\ 0 & 0 & 0 & 0 & 0 & 0 & 0 & 0 \\ 0 & 0 & 0 & 0 & 0 & 0 & 0 & 0 \\ 0 & 0 & 0 & 0 & 0.004\,6 & -0.003\,0 & -0.006\,3 & -0.003\,9 \\ 0 & 0 & 0 & 0 & -0.003\,0 & 0.006\,1 & 0.004\,0 & 0.004\,9 \\ 0 & 0 & 0 & 0 & -0.006\,3 & 0.003\,9 & 0.004\,6 & 0.003\,0 \\ 0 & 0 & 0 & 0 & -0.003\,9 & 0.004\,9 & 0.003\,0 & 0.006\,1 \end{bmatrix}.$$

The confidence ellipses are shown in Figure 12.2. Remember the units of $\Sigma_{\hat{x}_p}$ are m^2. The confidence ellipses at points 002 and 003 in the figure are a tenth of true size.

12.7 Variances at Control Points

Repeatedly we have used the term "postulated coordinate." In daily terms this is a "given value" or a "fixed value." Even if these values are treated computationally as real numbers, most often they are created through a previous least-squares procedure and consequently are born with nonzero variances.

For practical reasons (or lack of knowledge) the covariances are often set to zero. Today this attitude to the problem is changing. We shall demonstrate a model that correctly treats nonzero covariances of postulated coordinates. An excellent reference is Schwarz (1994), and we shall benefit from it.

Frequently, the estimated observations are more accurate than the observed values. The coordinates estimated from a constrained network are more accurate than those estimated from a free network. We shall investigate the statistical implications of this.

We start by partitioning the usual observation equation $Ax = b - r$ into

$$\begin{bmatrix} A_1 \\ A_2 \\ A_3 \end{bmatrix} x = \begin{bmatrix} b_1 \\ b_2 \\ b_3 \end{bmatrix} - r. \tag{12.43}$$

The observations described by A_1 are usual geodetic observations as treated in Section 10.2. The matrix A_2 contains the necessary rows to span the nullspace of A, cf. Section 12.2. Finally A_3 contains coordinate observations as described in Section 10.2. The observation vector b is partitioned conformally, and the covariance matrices are denoted Σ_1, Σ_2, and Σ_3.

In case a least-squares problem involves "postulated values" with zero variance we have $\Sigma_2 = \Sigma_3 = 0$.

When the third group of observations A_3 is missing, we are left with a free problem indicated by a subscript f. The estimate for x is called $\hat{x}_f$ with covariance matrix Σ_f. If we subsequently add the third set of observations A_3 we have to constrain the solution x_f with the latter equation: $A_3 x = b_3 - r_3$. For reasons of curiosity we shall carry the derivation through in all detail.

Let the first two observation types be combined and the problem is now formulated as

$$\begin{bmatrix} A \\ A_3 \end{bmatrix} x = \begin{bmatrix} b \\ b_3 \end{bmatrix} - \begin{bmatrix} r \\ r_3 \end{bmatrix}. \tag{12.44}$$

We introduce Lagrange multipliers λ and λ_3 and define the extended function

$$L(r, r_3, x, \lambda, \lambda_3) = \tfrac{1}{2} r^{\mathrm{T}} \Sigma^{-1} r + \tfrac{1}{2} r_3^{\mathrm{T}} \Sigma_3^{-1} r_3 + \lambda^{\mathrm{T}}(Ax - b + r) + \lambda_3^{\mathrm{T}}(A_3 x - b_3 + r_3).$$

The partial derivatives of L with respect to the variables are put equal to zero:

$$\frac{\partial L}{\partial r} = \Sigma^{-1} r + \lambda = 0 \tag{12.45}$$

$$\frac{\partial L}{\partial r_3} = \Sigma_3^{-1} r_3 + \lambda_3 = 0 \tag{12.46}$$

$$\frac{\partial L}{\partial x} = A^{\mathrm{T}} \lambda + A_3^{\mathrm{T}} \lambda_3 = 0 \tag{12.47}$$

$$\frac{\partial L}{\partial \lambda} = Ax - b + r = 0 \tag{12.48}$$

$$\frac{\partial L}{\partial \lambda_3} = A_3 x - b_3 + r_3 = 0. \tag{12.49}$$

The residuals are obtained from (12.45) and (12.47)

$$r = -\Sigma\lambda \tag{12.50}$$

$$r_3 = -\Sigma_3 \lambda_3. \tag{12.51}$$

Combining (12.50) and (12.48) yields

$$Ax - b - \Sigma\lambda = 0$$

or

$$\boldsymbol{\lambda} = \Sigma^{-1}(A\boldsymbol{x} - \boldsymbol{b}). \tag{12.52}$$

From (12.47), and (12.49), and furthermore from (12.49) and (12.51) we get

$$A^{\mathrm{T}}\Sigma^{-1}(A\boldsymbol{x} - \boldsymbol{b}) + A_3^{\mathrm{T}}\boldsymbol{\lambda}_3 = 0 \tag{12.53}$$

$$A_3\boldsymbol{x} - \boldsymbol{b}_3 - \Sigma_3\boldsymbol{\lambda}_3 = 0. \tag{12.54}$$

In matrix form this is

$$\begin{bmatrix} A^{\mathrm{T}}\Sigma^{-1}A & A_3^{\mathrm{T}} \\ A_3 & -\Sigma_3 \end{bmatrix}\begin{bmatrix} \boldsymbol{x} \\ \boldsymbol{\lambda}_3 \end{bmatrix} = \begin{bmatrix} A^{\mathrm{T}}\Sigma^{-1}\boldsymbol{b} \\ \boldsymbol{b}_3 \end{bmatrix}. \tag{12.55}$$

This equation shows how the normal equation matrix of the first group must be augmented in order to find the solution of both groups together. Of course, it is possible to solve for $\boldsymbol{x}$ and $\boldsymbol{\lambda}_3$, but we prefer a recursive solution for the parameters. (In so doing we in fact derive the Kalman gain matrix and prediction of the covariance matrix in filtering, developed in Chapter 17.)

Next we need to know the inverse of a 2 by 2 block matrix. Let

$$S = \begin{bmatrix} A & B \\ B^{\mathrm{T}} & C \end{bmatrix}$$

where A and C are square but B can be rectangular. The inverse block matrix is

$$S^{-1} = \begin{bmatrix} R_{11} & R_{12} \\ R_{21} & R_{22} \end{bmatrix} = \begin{bmatrix} A^{-1} + A^{-1}B(C - B^{\mathrm{T}}A^{-1}B)^{-1}B^{\mathrm{T}}A^{-1} & -A^{-1}B(C - B^{\mathrm{T}}A^{-1}B)^{-1} \\ -(C - B^{\mathrm{T}}A^{-1}B)^{-1}B^{\mathrm{T}}A^{-1} & (C - B^{\mathrm{T}}A^{-1}B)^{-1} \end{bmatrix}.$$

This is easily verified by direct multiplication of A with A^{-1}. The matrix R is introduced for reasons of reference.

Hence the solution for $\boldsymbol{x}$ can be written

$$\boldsymbol{x} = R_{11}A^{\mathrm{T}}\Sigma^{-1}\boldsymbol{b} + R_{12}\boldsymbol{b}_3 \tag{12.56}$$

$$\boldsymbol{\lambda}_3 = R_{21}A^{\mathrm{T}}\Sigma^{-1}\boldsymbol{b} + R_{22}\boldsymbol{b}_3. \tag{12.57}$$

We seek a recursive solution so we introduce $\hat{\boldsymbol{x}}_f = (A^{\mathrm{T}}\Sigma^{-1}A)^{-1}A^{\mathrm{T}}\Sigma^{-1}\boldsymbol{b}$ and $\Sigma_f = (A^{\mathrm{T}}\Sigma^{-1}A)^{-1}$ for the solution of the first equation in (12.44). After lengthy calculations we get the (updated) constrained estimate

$$\hat{\boldsymbol{x}}_c = \hat{\boldsymbol{x}}_f + \Sigma_f A_3^{\mathrm{T}}\big(\Sigma_3 + A_3\Sigma_f A_3^{\mathrm{T}}\big)^{-1}(\boldsymbol{b}_3 - A_3\hat{\boldsymbol{x}}_f). \tag{12.58}$$

The covariance matrix of the constrained estimate $\hat{\boldsymbol{x}}_c$ is recognized as

$$\Sigma_c = \Sigma_f - \Sigma_f A_3^{\mathrm{T}}\big(\Sigma_3 + A_3\Sigma_f A_3^{\mathrm{T}}\big)^{-1}A_3\Sigma_f. \tag{12.59}$$

A nonnegative definite matrix is subtracted from Σ_f, so

$$\Sigma_c \le \Sigma_f. \tag{12.60}$$

This means that the variance of any scalar function of $\hat{\boldsymbol{x}}_c$ is less than or equal to the variance of the same function evaluated for $\hat{\boldsymbol{x}}_f$.

Intuitively, adding the new information $\boldsymbol{b}_3$ to an already existing set of observations cannot worsen things, and most often it improves things.

Until now a constrained network seems as good as or even better than a free network. This is only true if the least-squares procedure uses the correct covariance matrix. This assumption does not always hold when we fix the control points. These points are almost never known perfectly. So next we want to study the effect of fixing control points to values that are not optimal.

We start with the following model:

$$\begin{bmatrix} A_1 & A_2 \\ 0 & I \end{bmatrix}\begin{bmatrix} \boldsymbol{x}_n \\ \boldsymbol{x}^0 \end{bmatrix} = \begin{bmatrix} \boldsymbol{b} \\ \boldsymbol{b}^0 \end{bmatrix} - \begin{bmatrix} \boldsymbol{r} \\ \boldsymbol{r}^0 \end{bmatrix} \tag{12.61}$$

with covariance matrix

$$\begin{bmatrix} \Sigma & 0 \\ 0 & \Sigma^0 \end{bmatrix}. \tag{12.62}$$

The natural way to proceed would be a least-squares procedure for the whole problem. This is most often not done because it changes the coordinates of the control points. Besides the adjustment task often is so comprehensive that in daily life it never will be done.

So more often $\boldsymbol{r}^0$ is set to zero, and we substitute $\boldsymbol{x}^0 = \boldsymbol{b}^0$ into (12.61) to obtain

$$A_1\boldsymbol{x}_n = \boldsymbol{b} - A_2\boldsymbol{b}^0 - \boldsymbol{r}. \tag{12.63}$$

This equation is to be compared to (8.39): $A_2\boldsymbol{x} = \boldsymbol{b} - A_1\boldsymbol{x}^0 - \boldsymbol{r}$. The solution of (12.63) is

$$\hat{\boldsymbol{x}}_n = \left(A_1^{\mathrm{T}}\Sigma^{-1}A_1\right)^{-1}A_1^{\mathrm{T}}\Sigma^{-1}(\boldsymbol{b} - A_2\boldsymbol{b}^0). \tag{12.64}$$

This estimate is influenced by two error sources: the observational errors in $\boldsymbol{b}$ and the errors in the postulated coordinates $\boldsymbol{b}^0$. They are described by the covariance matrices Σ and Σ^0.

We want to find the covariance matrix for the estimate $\hat{\boldsymbol{x}}_n$. We know the covariance matrices of $\boldsymbol{b}$ and $\boldsymbol{b}^0$ so we calculate the partial derivatives

$$\begin{bmatrix} \dfrac{\partial\hat{\boldsymbol{x}}_n}{\partial\boldsymbol{b}} & \dfrac{\partial\hat{\boldsymbol{x}}_n}{\partial\boldsymbol{b}^0} \end{bmatrix} = \begin{bmatrix} \left(A_1^{\mathrm{T}}\Sigma^{-1}A_1\right)^{-1}A_1^{\mathrm{T}}\Sigma^{-1} & -\left(A_1^{\mathrm{T}}\Sigma^{-1}A_1\right)^{-1}A_1^{\mathrm{T}}\Sigma^{-1}A_2 \end{bmatrix}.$$

Now apply the law of covariance propagation and get

$$\begin{aligned} \Sigma_{\hat{x}_n} &= \begin{bmatrix} \left(A_1^{\mathrm{T}}\Sigma^{-1}A_1\right)^{-1}A_1^{\mathrm{T}}\Sigma^{-1} & -\left(A_1^{\mathrm{T}}\Sigma^{-1}A_1\right)^{-1}A_1^{\mathrm{T}}\Sigma^{-1}A_2 \end{bmatrix}\begin{bmatrix} \Sigma & 0 \\ 0 & \Sigma^0 \end{bmatrix} \\ &\qquad\times\begin{bmatrix} \Sigma^{-1}A_1\left(A_1^{\mathrm{T}}\Sigma^{-1}A_1\right)^{-1} \\ -A_2^{\mathrm{T}}\Sigma^{-1}A_1\left(A_1^{\mathrm{T}}\Sigma^{-1}A_1\right)^{-1} \end{bmatrix} \\ &= \left(A_1^{\mathrm{T}}\Sigma^{-1}A_1\right)^{-1} + \left(A_1^{\mathrm{T}}\Sigma^{-1}A_1\right)^{-1}A_1^{\mathrm{T}}\Sigma^{-1}A_2\Sigma^0A_2^{\mathrm{T}}\Sigma^{-1}A_1\left(A_1^{\mathrm{T}}\Sigma^{-1}A_1\right)^{-1}. \end{aligned} \tag{12.65}$$

The first term specifies the contribution from the observations $\boldsymbol{b}$. This might be called the *internal* error. The second term describes the *external* error coming from the control points. So we can write (12.65) as

$$\Sigma_{\hat{x}_n} = \Sigma_{\text{internal}} + \Sigma_{\text{external}}. \tag{12.66}$$

This equation explains statistically why control networks sometimes become inadequate. Usually the control network is supposed to be much more accurate than the new densification network. This means that Σ^0 should be so small compared to Σ that the second term in (12.65) is much smaller than the first term. As long as this is the case $\Sigma_{\hat{x}} = (A_1^{\mathrm{T}}\Sigma^{-1}A_1)^{-1}$ can be used as a reasonable approximation of (12.65).

So far this fits the traditional manner of evaluating control networks. It is supposed that the accuracy of the control network is better than that of the new observations. Then we can use (12.65) with only the first term on the right side. If the accuracy of the new observations approaches or exceeds that of the existing control points it is essential to including both terms from (12.65).

We can also determine the effect on the estimated observations when the control points are held fixed. We have

$$\hat{\boldsymbol{b}} = A_1\hat{\boldsymbol{x}}_n + A_2\boldsymbol{b}^0 = A_1\big(A_1^{\mathrm{T}}\Sigma^{-1}A_1\big)^{-1}A_1^{\mathrm{T}}\Sigma^{-1}\boldsymbol{b} + \big(I - A_1(A_1^{\mathrm{T}}\Sigma^{-1}A_1)^{-1}A_1^{\mathrm{T}}\Sigma^{-1}\big)A_2\boldsymbol{b}^0. \tag{12.67}$$

Again the covariance matrix consists of two terms:

$$\begin{aligned}\Sigma_{\hat{b}} = A_1\big(A_1^{\mathrm{T}}\Sigma^{-1}A_1\big)^{-1}A_1^{\mathrm{T}} &+ \big(I - A_1(A_1^{\mathrm{T}}\Sigma^{-1}A_1)^{-1}A_1^{\mathrm{T}}\Sigma^{-1}\big)\\ &\times A_2\Sigma^0A_2^{\mathrm{T}}\big(I - A_1(A_1^{\mathrm{T}}\Sigma^{-1}A_1)^{-1}A_1^{\mathrm{T}}\Sigma^{-1}\big)^{\mathrm{T}}.\end{aligned} \tag{12.68}$$

If the second term vanishes we are left with the usual expression

$$\Sigma_{\hat{b}} = A_1\big(A_1^{\mathrm{T}}\Sigma^{-1}A_1\big)^{-1}A_1^{\mathrm{T}}. \tag{12.69}$$

Now the difference between the covariance matrix of the actual observations and that of the estimated observations is

$$\begin{aligned}\Sigma_b - \Sigma_{\hat{b}} &= \Sigma_b - A_1\big(A_1^{\mathrm{T}}\Sigma^{-1}A_1\big)^{-1}A_1^{\mathrm{T}}\\ &= \big(I - A_1(A_1^{\mathrm{T}}\Sigma^{-1}A_1)^{-1}A_1^{\mathrm{T}}\Sigma^{-1}\big)A_2\Sigma^0A_2^{\mathrm{T}}\big(I - A_1(A_1^{\mathrm{T}}\Sigma^{-1}A_1)^{-1}A_1^{\mathrm{T}}\Sigma^{-1}\big)^{\mathrm{T}}.\end{aligned} \tag{12.70}$$

This is a nonnegative definite matrix, and we have

$$\Sigma_{\hat{b}} \leq \Sigma_b. \tag{12.71}$$

That is, the estimated observations $\hat{\boldsymbol{b}}$ have a smaller variance than the actual observations.

If the second term in (12.65) does not vanish, equation (12.71) does not necessarily hold. Nowadays it often happens that the estimated observations have larger variances than the actual observations. In other words, if we fix the control points we might cause the adjusted values of the observations to be worse than the actually observed values. ***This argument applies especially when we try to fit GPS vectors into existing control networks.***

Equation (12.65) also holds for a free network where we only add the necessary conditions to define the coordinate system. The usual least squares leads to the estimate $\Sigma_{\hat{x}} = (A_1^{\mathrm{T}}\Sigma^{-1}A_1)^{-1}$. This covariance matrix tells us how well the coordinates of the new points are estimated, but not how well they are known relative to the control points. The second term in (12.65) accounts for the uncertainty of the fixed control.

For a free network the columns of A_2 are linear combinations of the columns of A_1, say $A_2 = A_1 H$ for some matrix H. Then

$$\big(A_1^{\mathrm{T}}\Sigma^{-1}A_1\big)^{-1}A_1^{\mathrm{T}}\Sigma^{-1}A_2 = \big(A_1^{\mathrm{T}}\Sigma^{-1}A_1\big)^{-1}A_1^{\mathrm{T}}\Sigma^{-1}A_1H = H.$$

Equation (12.65) becomes

$$\Sigma_{\hat{x}_n} = \big(A_1^{\mathrm{T}}\Sigma^{-1}A_1\big)^{-1} + H\Sigma^0 H^{\mathrm{T}}. \tag{12.72}$$

Even more interesting, we then have

$$\big(I - A_1(A_1^{\mathrm{T}}\Sigma^{-1}A_1)^{-1}A_1^{\mathrm{T}}\Sigma^{-1}\big)A_2 = \big(A_1H - A_1(A_1^{\mathrm{T}}\Sigma^{-1}A_1)^{-1}A_1^{\mathrm{T}}\Sigma^{-1}A_1H\big) = 0 \tag{12.73}$$

so that the second term in (12.68) vanishes. This means that $\Sigma_{\hat{b}} \le \Sigma_b$ holds for all free networks, irrespective of the uncertainty of the fixed control. ***The coordinates obtained from a least-squares solution of a free network may be affected by the errors in the fixed control, but the adjusted observations are not.*** This is the sense in which these least-squares solutions are "free."

Example 12.8 This theory is now applied to a leveling network. We only need to know the covariance matrix for the postulated heights

$$\Sigma^0 = \begin{bmatrix} 0.010\,0 & 0.007\,5 & 0.007\,5 \\ 0.007\,5 & 0.010\,0 & 0.007\,5 \\ 0.007\,5 & 0.007\,5 & 0.010\,0 \end{bmatrix}.$$

We recall the data from Example 8.1:

$$\boldsymbol{b} = \begin{bmatrix} 1.978 \\ 0.732 \\ 0.988 \\ 0.420 \\ 1.258 \end{bmatrix} \qquad \text{and} \qquad \boldsymbol{b}^0 = \begin{bmatrix} 10.021 \\ 10.321 \\ 11.002 \end{bmatrix}.$$

The inverse covariance matrix ($\sigma_0 = 1$) for the observations $\boldsymbol{b}$ is

$$\Sigma^{-1} = \begin{bmatrix} 0.980\,4 & & & & \\ & 1.030\,9 & & & \\ & & 0.900\,9 & & \\ & & & 0.934\,6 & \\ & & & & 1.123\,6 \end{bmatrix}.$$

The observation equations (12.63) are $A_1 \boldsymbol{x}_n = \boldsymbol{b} - A_2 \boldsymbol{b}^0 - \boldsymbol{r}$:

$$\begin{bmatrix} 1 & 0 \\ 0 & 1 \\ 1 & 0 \\ 0 & 1 \\ 1 & -1 \end{bmatrix} \begin{bmatrix} H_D \\ H_E \end{bmatrix} = \begin{bmatrix} 1.978 \\ 0.732 \\ 0.988 \\ 0.420 \\ 1.258 \end{bmatrix} - \begin{bmatrix} -1 & 0 & 0 \\ -1 & 0 & 0 \\ 0 & 0 & -1 \\ 0 & -1 & 0 \\ 0 & 0 & 0 \end{bmatrix} \begin{bmatrix} 10.021 \\ 10.321 \\ 11.002 \end{bmatrix} - \boldsymbol{r}.$$

The solution is

$$\hat{\boldsymbol{x}}_n = \left(A_1^{\mathrm{T}} \Sigma^{-1} A_1\right)^{-1} A_1^{\mathrm{T}} \Sigma^{-1} (\boldsymbol{b} - A_2 \boldsymbol{b}^0) = \begin{bmatrix} H_D \\ H_E \end{bmatrix} = \begin{bmatrix} 11.997\,6 \\ 10.744\,5 \end{bmatrix}$$

and the covariance matrix is

$$\Sigma_{\hat{x}_n} = \begin{bmatrix} 0.007\,5 & 0.008\,0 \\ 0.008\,0 & 0.008\,4 \end{bmatrix} = \begin{bmatrix} 0.217 & 0.079 \\ 0.079 & 0.211 \end{bmatrix} \times 10^{-4} + \begin{bmatrix} 0.007\,4 & 0.008\,0 \\ 0.008\,0 & 0.008\,4 \end{bmatrix}.$$

We recognize that the last term—the variance of the postulated heights—by far dominates the result. The variances of the new points D and E are much larger than the variances of the observations themselves (computed in Example 8.1). The two heights H_D and H_E are strongly correlated, too, since they share the uncertainties of the control points A, B, and C.

According to (12.68) the covariance matrix of the estimated observations is

$$\Sigma_{\hat{b}} = \begin{bmatrix} -0.000\,1 & 0.000\,4 & 0.000\,9 & -0.001\,1 & -0.000\,6 \\ 0.000\,4 & 0.000\,8 & 0.000\,0 & -0.001\,3 & -0.000\,4 \\ 0.000\,9 & 0.000\,0 & -0.002\,0 & 0.001\,0 & 0.000\,9 \\ -0.001\,1 & -0.001\,3 & 0.001\,0 & 0.001\,7 & 0.000\,2 \\ -0.000\,6 & -0.000\,4 & 0.000\,9 & 0.000\,2 & -0.000\,2 \end{bmatrix}$$

$$= \begin{bmatrix} 0.217 & 0.079 & 0.217 & 0.079 & 0.138 \\ 0.079 & 0.211 & 0.079 & 0.211 & -0.132 \\ 0.217 & 0.079 & 0.217 & 0.079 & 0.138 \\ 0.079 & 0.211 & 0.079 & 0.211 & -0.132 \\ 0.138 & -0.132 & 0.138 & -0.132 & 0.269 \end{bmatrix} \times 10^{-4}$$

$$+ \begin{bmatrix} -0.000\,2 & 0.000\,4 & 0.000\,9 & -0.001\,1 & -0.000\,6 \\ 0.000\,4 & 0.000\,8 & 0.000\,0 & -0.001\,3 & -0.000\,4 \\ 0.000\,9 & 0.000\,0 & -0.002\,0 & 0.001\,0 & 0.000\,9 \\ -0.001\,1 & -0.001\,3 & 0.001\,0 & 0.001\,6 & 0.000\,2 \\ -0.000\,6 & -0.000\,4 & 0.000\,9 & 0.000\,2 & -0.000\,2 \end{bmatrix}.$$

The uncertainty of the control points A, B, and C—reflected in the second term—dominates this covariance matrix. Recall that the covariance matrix of the actual observations is

$$\Sigma_b = \Sigma = \begin{bmatrix} 0.57 & & & & \\ & 0.55 & & & \\ & & 0.62 & & \\ & & & 0.60 & \\ & & & & 0.50 \end{bmatrix} \times 10^{-4}.$$

We observe that the second term causes the covariance matrix of the estimated observations $\Sigma_{\hat{b}}$ to be larger than the covariance matrix of the actual observations Σ_b.

13

PROBLEMS WITH EXPLICIT SOLUTIONS

13.1 Free Stationing as a Similarity Transformation

Any *free stationing* can be perceived and calculated as a usual least-squares problem where the observations are distances and horizontal directions.

Normally, the distance and direction observations are taken in pairs. Yet, this is not required for carrying out the procedure.

Alternatively, the situation can be conceived in the following way: Introduce a local coordinate system x, y with origin at the free stationing point P and x-axis through the first object in the horizontal direction set. Any observation of distance and direction from P to a known point can be looked upon as a polar determination (θ, s) of this point in the x, y coordinate system. The task is now to *transform the locally observed x, y coordinates for the given points into ξ, η coordinates in a global system.* The translational parameters t_x, t_y of the transformation, and the rotation φ of the x, y system relative to the ξ, η system are the 3 unknowns.

Futhermore, we may introduce a change of scale k between the two coordinate systems. Then the similarity transformation works on unknowns t_x, t_y, φ, and k. The coordinates for P are $(\hat{t}_x, \hat{t}_y)$ and the orientation unknown is $\hat{\varphi}$.

We shall illustrate the theory by a calculational example, using the observations and coordinates presented in Table 13.1. We reduce the coordinates by $x_i^* = x_i - x_s, \dots,$ and get the values in Table 13.2.

In Section 12.5 the classical formulas for the least-squares solution of the problem were described. According to this, the first step is to calculate product sums:

$$\sum(x_i^{*2} + y_i^{*2}) = 18\,219.017\,32$$

$$\sum \xi_i^* x_i^* = 88.485\,5 \qquad \sum \eta_i^* x_i^* = -13\,575.232\,19$$

$$\sum \xi_i^* y_i^* = 3\,681.640\,980 \qquad \sum \eta_i^* y_i^* = -5\,926.219\,065$$

Table 13.1 Observations and given coordinates for free stationing

Observed station	reduced reading θ_i [gon]	horizontal distance s_i [m]	Cartesian coordinates x_i [m]	y_i [m]	given coordinates ξ_i [m]	η_i [m]
100	0.000	51.086	51.086	0.000	60.10	−926.73
101	107.548	50.771	−6.006	50.415	126.16	−888.81
102	191.521	65.249	−64.671	8.664	105.38	−819.88
103	231.709	110.362	−96.953	−52.725	57.60	−769.61
s		mean value	−29.136	1.589	87.310	−851.258

Next the unknowns are given by simpler formulas which emerge because of the translation to the barycenter:

$$\hat{a} = \frac{\sum \xi_i^* x_i^* + \sum \eta_i^* y_i^*}{\sum (x_i^{*2} + y_i^{*2})} = \frac{88.485\,5 - 5\,926.219\,066}{18\,219.017\,32} = -0.320\,420$$

$$\hat{b} = \frac{-\sum \eta_i^* x_i^* + \sum \xi_i^* y_i^*}{\sum (x_i^{*2} + y_i^{*2})} = \frac{13\,575.232\,19 + 3\,681.640\,980}{18\,219.017\,32} = 0.947\,190$$

$$\hat{k} = \sqrt{\hat{a}^2 + \hat{b}^2} = 0.999\,919 = 1 - 81 \text{ ppm}$$

$$\hat{\varphi} = 120.767 \text{ gon}.$$

We calculate $\hat{t}_x$ and $\hat{t}_y$ according to (12.30) and (12.31); in this special case we furthermore have $\hat{t}_x = \hat{\xi}_P$ and $\hat{t}_y = \hat{\eta}_P$

$$\hat{\xi}_P = \xi_s - \hat{a}x_s - \hat{b}y_s = 87.310 - 0.320\,420 \cdot 29.136 - 0.947\,190 \cdot 1.589 = 76.470 \text{ m}$$

$$\hat{\eta}_P = \eta_s - \hat{b}x_s - \hat{a}y_s = -851.258 + 0.947\,190 \cdot 29.136 - 0.320\,420 \cdot 1.589 = -878.346 \text{ m}.$$

Example 10.3 uses the same data and results in $(\hat{\xi}_P, \hat{\eta}_P) = (76.469, -878.344)$.

As in all least-squares problems we are furnished with a residual of each observation. In the present case we can interpret these quantities. The residual vector $\hat{\boldsymbol{r}} = \boldsymbol{b} - A\hat{\boldsymbol{x}}$ is

$$\begin{aligned} \hat{r}_{\xi_i} &= \xi_i - \hat{a}x_i - \hat{b}y_i - \hat{t}_x \\ \hat{r}_{\eta_i} &= \eta_i + \hat{b}x_i - \hat{a}y_i - \hat{t}_y. \end{aligned} \tag{13.1}$$

In our numerical example the residual vectors are shown in Table 13.2. The pair $(\hat{r}_{\xi_i}, \hat{r}_{\eta_i})$ can be looked upon as a residual vector between the transformed point (x_i, y_i) and the "observed" point (ξ_i, η_i). The variance factor is $\hat{\sigma}_0^2 = \sum(\hat{r}_{\xi_i}^2 + \hat{r}_{\eta_i}^2)/(2p-4) = (14 \text{ mm})^2$.

Remark 13.1 The present result deviates a little from the solution resulting from a usual least-squares procedure as described in Example 10.3. Looked upon as an ordinary least-squares problem we have 3 unknowns but the similarity transformation per se involves

Table 13.2 Coordinate sets reduced to respective barycenters and final residuals

Point	x_i^* [m]	y_i^* [m]	ξ_i^* [m]	η_i^* [m]	$\hat{r}_{\xi i}$ [mm]	$\hat{r}_{\eta i}$ [mm]
100	80.222	−1.589	−27.210	−75.472	−1	4
101	23.130	48.826	38.850	−37.552	13	1
102	−35.535	7.075	18.070	31.378	−18	−14
103	−67.817	−54.314	−29.710	81.648	5	9
Sum	0.000	−0.002	0.000	0.002	−1	0

4 unknowns. There is another difference between the two procedures. In the similarity transformation all observations get equal weights while in the ordinary least-squares procedure they get varying weights.

In case of the similarity transformation nothing can prevent us from introducing varying weights for the observations except that the formulas change so much that the procedure is no longer an attractive alternative.

Remark 13.2 Above we started by converting the polar observations (θ_{P_i}, s_{P_i}) into Cartesian coordinates. Is it possible to directly evaluate the polar observations? Actually, it is possible and the observation equations are

$$\begin{aligned}\xi_i &= \xi_P + k\, s_{P_i} \cos(\theta_{P_i} + z_P)\\ \eta_i &= \eta_P + k\, s_{P_i} \sin(\theta_{P_i} + z_P).\end{aligned} \tag{13.2}$$

They are nonlinear and hence the similarity transformation must be solved iteratively starting with preliminary values for ξ_P, η_P, $k \approx 1$ and $z_P \approx 0$. The orientation unknown is denoted z_P as in Section 13.3.

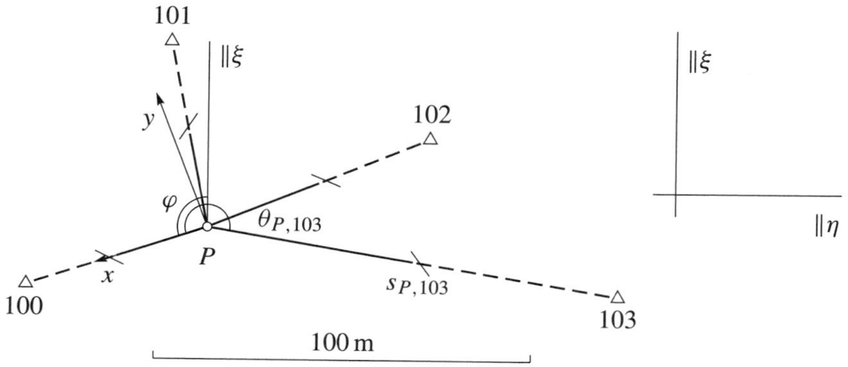

Figure 13.1 The free stationing

13.2 Optimum Choice of Observation Site

The coordinates (X, Y) of the observation site P in a free stationing are determined by observations of distances $s_1, s_2, \dots, s_n$ and directions $\varphi_1, \varphi_2, \dots, \varphi_n$. For every distance observation a direction observation is taken and vice versa. The set of n directions is connected to the coordinate axes through the orientation unknown Z, see Figure 13.2.

The unknowns of the least squares problem are the coordinates (X, Y) of point P and the orientation unknown Z. We introduce preliminary coordinates (X^0, Y^0) for P and put $X = X^0 + x$ and $Y = Y^0 + y$. The coordinates of the ith fixed point are called (X_i, Y_i). As observation we use the *observed distance* s_i divided by the calculated distance $s_i^0 = \sqrt{(X_i - X^0)^2 + (Y_i - Y^0)^2}$ and we have

$$\frac{s_i}{s_i^0} - r_i = \frac{\sqrt{(X_i - X)^2 + (Y_i - Y)^2}}{s_i^0} = f_i(X, Y).$$

In a first order linearization we have

$$\frac{s_i}{s_i^0} - r_i = f_i(X^0, Y^0) + \Big(\frac{\partial f_i}{\partial X}\Big)_0 x + \Big(\frac{\partial f_i}{\partial Y}\Big)_0 y.$$

Remember $f_i(X^0, Y^0) = s_i^0/s_i^0$. We differentiate and rearrange

$$r_i = -\frac{X_i - X^0}{(s_i^0)^2}x - \frac{Y_i - Y^0}{(s_i^0)^2}y - \frac{s_i - s_i^0}{s_i^0} = -u_i x - v_i y - b_i$$

where we have introduced the coefficients $u_i = (X_i - X^0)/(s_i^0)^2$ and $v_i = (Y_i - Y^0)/(s_i^0)^2$. The weight for the ith observation is $c_i = m/(\sigma_d^2 + \sigma_c^2)$ where m denotes the number of repeated observations, σ_d^2 is the variance of the distance and σ_c^2 is the variance of centricity.

The observation equation for the *horizontal direction* is

$$A_{Pi} = \arctan\frac{Y_i - Y}{X_i - X} = \varphi_i + Z.$$

We split the value for the orientation unknown Z into $Z^0 + z$ and linearize

$$r_i = -v_i x - u_i y - z - \big(-A_{Pi}^0 + \varphi_i + Z^0\big) = -v_i x - u_i y - 1 \cdot z - b_i.$$

The weight is given as $d_i = m/(\sigma_r^2 + (\sigma_c\omega/s_i)^2)$ where the variance of a direction observed with one set is called σ_r^2. The factor ω converts from units of arc to radians and m is the number of sets.

We rewrite the expressions for the coefficients u_i and v_i, and introduce the direction angle A_{Pi}:

$$u_i = \frac{X_i - X^0}{(s_i^0)^2} = \frac{\cos A_{Pi}}{s_i^0} \qquad \text{and} \qquad v_i = \frac{Y_i - Y^0}{(s_i^0)^2} = \frac{\sin A_{Pi}}{s_i^0}.$$

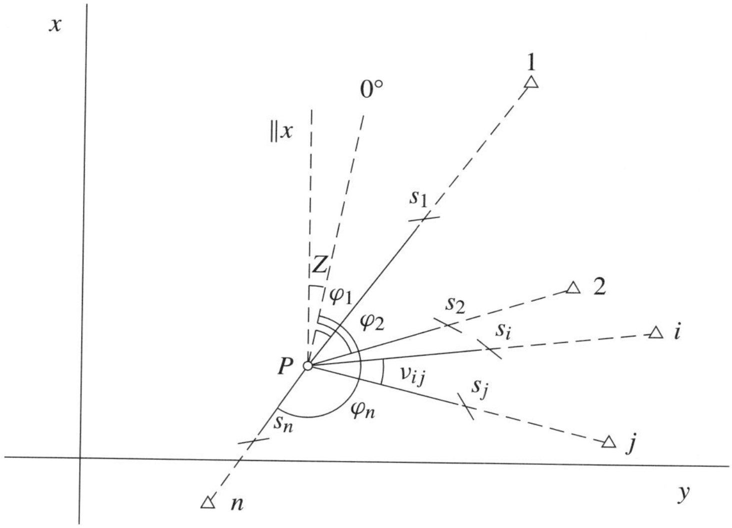

Figure 13.2 Free stationing at P with n fixed points.

In the sequel we omit the ^{0}. Note that $u_i^2 + v_i^2 = s_i^{-2}$. Next we gather all linearized observation equations in the following matrix equation:

$$Ax = \begin{bmatrix} A_1 \\ A_2 \end{bmatrix} x = \left[\begin{array}{ccc} -u_1 & -v_1 & 0 \\ -u_2 & -v_2 & 0 \\ & \vdots & \\ -u_n & -v_n & 0 \\ \hline v_1 & -u_1 & -1 \\ v_2 & -u_2 & -1 \\ & \vdots & \\ v_n & -u_n & -1 \end{array}\right] \begin{bmatrix} x \\ y \\ z \end{bmatrix} = b - r. \tag{13.3}$$

We assemble the weights into the diagonal matrix

$$C = \operatorname{diag}(c_1, \dots, c_n, d_1, \dots, d_n)$$

and the normal equations become

$$A^{\mathrm{T}}CA = \begin{bmatrix} \sum(c_i u_i^2 + d_i v_i^2) & \sum(c_i - d_i)u_i v_i & -\sum d_i v_i \\ \sum(c_i - d_i)u_i v_i & \sum(c_i v_i^2 + d_i u_i^2) & \sum d_i u_i \\ -\sum d_i v_i & \sum d_i u_i & \sum d_i \end{bmatrix}. \tag{13.4}$$

It is rather complicated to evaluate this matrix expression. However we may simplify by setting $c_i = d_i = 1$. For a modern total station this is in fact reasonable. We now obtain a much simpler expression

$$A^{\mathrm{T}}CA = \left[\begin{array}{cc|c} \sum s_i^{-2} & 0 & -\sum v_i \\ 0 & \sum s_i^{-2} & \sum u_i \\ \hline -\sum v_i & -\sum u_i & n \end{array}\right] = \begin{bmatrix} S & T \\ T^{\mathrm{T}} & U \end{bmatrix}.$$

We are aiming at the trace of the covariance matrix $\Sigma = (A^{\mathrm{T}}CA)^{-1}$. We shall use the following formula

$$\Sigma = \begin{bmatrix} S & T \\ T^{\mathrm{T}} & U \end{bmatrix}^{-1} = \begin{bmatrix} S^{-1}+S^{-1}T(U-T^{\mathrm{T}}S^{-1}T)^{-1}T^{\mathrm{T}}S^{-1} & -S^{-1}T(U-T^{\mathrm{T}}S^{-1}T)^{-1} \\ -(U-T^{\mathrm{T}}S^{-1}T)^{-1}T^{\mathrm{T}}S^{-1} & (U-T^{\mathrm{T}}S^{-1}T)^{-1} \end{bmatrix}. \tag{13.5}$$

This is easily verified by direct multiplication of Σ with Σ^{-1}. We want to evaluate the (1,1) entry of Σ and start with the expression in parenthesis:

$$\begin{aligned} U - T^{\mathrm{T}}S^{-1}T &= n - \begin{bmatrix} -\sum v_i & \sum u_i \end{bmatrix} \begin{bmatrix} (\sum s_i^{-2})^{-1} & 0 \\ 0 & (\sum s_i^{-2})^{-1} \end{bmatrix} \begin{bmatrix} -\sum v_i \\ \sum u_i \end{bmatrix} \\ &= n - \frac{(\sum u_i)^2 + (\sum v_i)^2}{\sum s_i^{-2}}. \end{aligned}$$

The inverse is

$$\left(U - T^{\mathrm{T}}S^{-1}T\right)^{-1} = \frac{\sum s_i^{-2}}{n\sum s_i^{-2} - (\sum u_i)^2 - (\sum v_i)^2}.$$

Furthermore

$$\begin{aligned} &T\left(U - T^{\mathrm{T}}S^{-1}T\right)^{-1}T^{\mathrm{T}}S^{-1} \\ &\qquad = \frac{1}{n\sum s_i^{-2} - (\sum u_i)^2 - (\sum v_i)^2} \begin{bmatrix} (\sum v_i)^2 & -\sum u_i \sum v_i \\ -\sum u_i \sum v_i & (\sum u_i)^2 \end{bmatrix}. \end{aligned}$$

Finally we get

$$\begin{aligned} &S^{-1}\Big(I + T\left(U - T^{\mathrm{T}}S^{-1}T\right)^{-1}T^{\mathrm{T}}S^{-1}\Big) \\ &\quad = \frac{(\sum s_i^{-2})^{-1}}{n\sum s_i^{-2} - (\sum u_i)^2 - (\sum v_i)^2} \begin{bmatrix} n\sum s_i^{-2} - (\sum u_i)^2 & -\sum u_i \sum v_i \\ -\sum u_i \sum v_i & n\sum s_i^{-2} - (\sum v_i)^2 \end{bmatrix}. \end{aligned}$$

The trace is

$$\operatorname{tr}\Sigma = \frac{2n\sum s_i^{-2} - \Big((\sum u_i)^2 + (\sum v_i)^2\Big)}{\sum s_i^{-2}\Big\{n\sum s_i^{-2} - \Big((\sum u_i)^2 + (\sum v_i)^2\Big)\Big\}}. \tag{13.6}$$

The square sum $\left(\sum u_i\right)^2 + \left(\sum v_i\right)^2$ can be evaluated as follows

$$\begin{aligned}\Big(\sum u_i\Big)^2 + \Big(\sum v_i\Big)^2 &= (u_1 + u_2 + \cdots + u_n)^2 + (v_1 + v_2 + \cdots + v_n)^2 \\ &= u_1^2 + \cdots + u_n^2 + 2\sum_{i \neq j} u_i u_j + v_1^2 + \cdots + v_n^2 + 2\sum_{i \neq j} v_i v_j \\ &= \sum s_i^{-2} + 2\sum_{i \neq j} \frac{\cos\alpha_{Pi}\cos\alpha_{Pj} + \sin\alpha_{Pi}\sin\alpha_{Pj}}{s_i s_j} \\ &= \sum s_i^{-2} + 2\sum_{i \neq j} \frac{\cos(\alpha_{Pi} - \alpha_{Pj})}{s_i s_j} = \sum s_i^{-2} + 2\sum_{i \neq j} \frac{\cos\nu_{ij}}{s_i s_j}.\end{aligned}$$

The angle between the directions of distances s_i and s_j is denoted ν_{ij}. The first sum contains n terms and the last sum $n(n-1)/2$ terms. The final formula is

$$\operatorname{tr}\Sigma = \Big(\sum s_i^{-2}\Big)^{-1}\left(1 + \frac{n\sum s_i^{-2}}{(n-1)\sum s_i^{-2} - 2\sum_{i \neq j}\frac{\cos\nu_{ij}}{s_i s_j}}\right). \tag{13.7}$$

In this type of plane geometric formulas one often can substitute some areas or fractions of areas. In the present case we introduce the area T_{ij} of triangle P, i, j and get

$$\operatorname{tr}\Sigma = \Big(\sum s_i^{-2}\Big)^{-1}\left(1 + \frac{n\sum s_i^{-2}}{(n-1)\sum s_i^{-2} - \sum_{i \neq j}\frac{\sin 2\nu_{ij}}{2T_{ij}}}\right).$$

If ν_{ij} equals 0 or π, then $\sin 2\nu_{ij}/2T_{ij}$ must be interpreted as 0.

It is not easy to interpret formula (13.7). We start by analysing the situation where P is located at the center of a unit circle and all n fixed points are equally distributed along the circumference. For this ideal case we get

$$\operatorname{tr}\Sigma = \tfrac{1}{n}\left(1 + \frac{n^2}{(n-1)n - 2\left(-\frac{n}{2}\right)}\right) = \tfrac{2}{n}.$$

We define the ***point error*** σ_P of P as $\sigma_P = \sqrt{\operatorname{tr}\Sigma/2} = n^{-1/2}$. Therefore the point error of P diminishes inversely proportional to the square root of the number n of fixed points.

Our example concentrated on the unit circle. Now we multiply the network scale with a constant k. Thus all distances are multiplied by k and we get a new covariance matrix Σ^* for which we have $\operatorname{tr}\Sigma^* = k^2 \operatorname{tr}\Sigma$ which also was to be expected.

Finally we return to the original model with all fixed points laying on the unit circle. But now we move the point P outside the circle at a distance k times the radius 1 from the center. We specialize to $n = 2$ and then it is possible to have $s_1 = s_2$:

$$\operatorname{tr}\Sigma = \frac{k}{2}\frac{3k^3 - k^2 + 2}{k^3 - k^2 + 2} \approx \tfrac{3}{2}k.$$

For large values of k we approximately have $s_i \approx k$ we get

$$\operatorname{tr}\Sigma \approx \frac{k}{n}\frac{(2n-1)k^3 - (n-1)c^2 + 2(n-1)}{(n-1)k^3 - (n-1)c^2 + 2(n-1)} \approx \left(\tfrac{1}{n} + \tfrac{1}{n-1}\right)k.$$

This expression tells that a multiplication by k can be compensated for by introducing more fixed points n. If we move further away from the area of the fixed points we must use more fixed points to maintain an unchanged trace of the covariance matrix.

Remark 13.3 The present problem can be viewed as a combination of distance and direction observations. Accordingly the coefficient matrix A in (13.3) can be split into $\begin{bmatrix} A_1 \\ A_2 \end{bmatrix}$. If we estimate the two observation types separately, the trace of the common covariance matrix is $\operatorname{tr}\big((A_1^{\mathrm{T}} A_1)^{-1} + (A_2^{\mathrm{T}} A_2)^{-1}\big)$. In general this value is different from $\operatorname{tr}\big((A_1^{\mathrm{T}} A_1 + A_2^{\mathrm{T}} A_2)^{-1}\big)$ based on a joint estimation of the problem. If we in (13.4) put $c_i = 1$ and $d_i = 0$ we get

$$\operatorname{tr}\big((A_1^{\mathrm{T}} A_1)^{-1}\big) = \frac{\sum u_i^2 + \sum v_i^2}{\sum u_i^2 \sum v_i^2 - \left(\sum u_i v_i\right)^2}$$

The similar expression for $\operatorname{tr}\big((A_1^{\mathrm{T}} A_1)^{-1}\big)$ is complicated.

The least squares problem of a free stationing is an elegant example of the application of the *Korn inequality*. This inequality yields bounds for how much the point variance changes if we change the weight of some distance or direction observations.

Free stationing is for the first time described by W. Snellius in 1617 in his book "Eratosthenes Batavus."

13.3 Station Adjustment

When observing directions between stations, one station is usually selected as a reference object. There are no conditions attached to the choice. Readings are subsequently taken to the other stations, swinging clockwise, and a second reading is taken beginning with the last station and swinging back counterclockwise. The observations are completed with a reading to the reference object. The observational result of a full round is the mean values b_i of the first and the second readings. The total number of rounds depends on the precision required.

We apply a least-squares procedure to the observations. It will be seen that an estimated direction is the mean of its corresponding observations reduced to the reference object. The directions have equal variance and are uncorrelated.

The number of directions is denoted by r, and the number of rounds by s. There are $m = rs$ observations in all. The r observations in the first round and the first observation of each subsequent round are the necessary ones. This yields a sum total of $n = r + s - 1$. The number of redundant observations then becomes $m - n = (r - 1)(s - 1)$.

Selecting the directions as unknowns x_i would give r unknowns. In order to compare rounds, the observations must be reduced to the same reference direction which may be chosen at will. Normally the value of the first direction is put equal to zero. For this purpose an *orientation unknown* z_i must be introduced as an unknown in each round. One of the s orientation unknowns is superfluous (any of them may be chosen as zero). It is easier to maintain symmetry and impose the condition

$$z_1 + z_2 + \cdots + z_s = 0$$

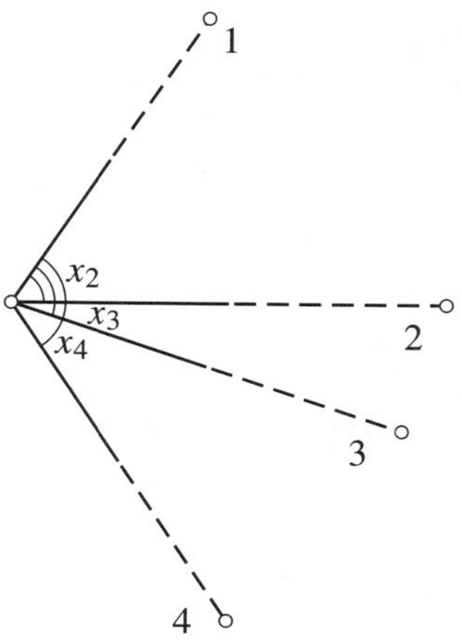

Figure 13.3 Station adjustment.

or

$$e^{\mathrm{T}}z = 0. \tag{13.8}$$

The observational equations express that the observation plus the orientation unknown plus the residual equals the unknown value for the direction:

$$Ay = \underbrace{\begin{bmatrix} 1 & 0 & 0 & 0 \\ 0 & 1 & 0 & 0 \\ 0 & 0 & 1 & 0 \\ 0 & 0 & 0 & 1 \\ 1 & 0 & 0 & 0 \\ 0 & 1 & 0 & 0 \\ 0 & 0 & 1 & 0 \\ 0 & 0 & 0 & 1 \\ 1 & 0 & 0 & 0 \\ 0 & 1 & 0 & 0 \\ 0 & 0 & 1 & 0 \\ 0 & 0 & 0 & 1 \end{bmatrix}}_{B} \underbrace{\begin{bmatrix} -1 & 0 & 0 \\ -1 & 0 & 0 \\ -1 & 0 & 0 \\ -1 & 0 & 0 \\ 0 & -1 & 0 \\ 0 & -1 & 0 \\ 0 & -1 & 0 \\ 0 & -1 & 0 \\ 0 & 0 & -1 \\ 0 & 0 & -1 \\ 0 & 0 & -1 \\ 0 & 0 & -1 \end{bmatrix}}_{C} \begin{bmatrix} x_1 \\ x_2 \\ x_3 \\ x_4 \\ \hline z_1 \\ z_2 \\ z_3 \end{bmatrix} = \boldsymbol{b} - \boldsymbol{r}. \tag{13.9}$$

Note that A is an incidence matrix. We know that $Ae = 0$. This rank deficiency of $A^{\mathrm{T}}A$ may be removed by orthogonal bordering to produce the augmented matrix

$$\begin{bmatrix} A^{\mathrm{T}}A & e \\ e^{\mathrm{T}} & 0 \end{bmatrix}.$$

This is nonsingular and then $(A^{\mathrm{T}}A)^{+}$ results from the inverse by deleting the last row and column. But we shall follow an alternative development.

The structure of the least-squares problem (13.9) suggests splitting it into two in the following manner—the notation C has nothing to do with a weight matrix:

$$Ay = [\,B \quad C\,]\begin{bmatrix} x \\ z \end{bmatrix} = b - r. \tag{13.10}$$

If we suppose equal weights for all observations, the normal equations are

$$B^{\mathrm{T}}Bx + B^{\mathrm{T}}Cz = B^{\mathrm{T}}b \tag{13.11}$$

$$C^{\mathrm{T}}Bx + C^{\mathrm{T}}Cz = C^{\mathrm{T}}b. \tag{13.12}$$

We solve (13.11) to find

$$\hat{x} = (B^{\mathrm{T}}B)^{-1}B^{\mathrm{T}}b - (B^{\mathrm{T}}B)^{-1}B^{\mathrm{T}}Cz.$$

An easy calculation reveals that $(B^{\mathrm{T}}B)^{-1} = \frac{1}{s}I$ and $B^{\mathrm{T}}Cz = -Ez$ where E denotes a matrix consisting of ones. Now, by (13.8) we have $Ez = \mathbf{0}$, and then $B^{\mathrm{T}}Cz = \mathbf{0}$. This yields the nice estimate

$$\hat{x} = \tfrac{1}{s}B^{\mathrm{T}}b. \tag{13.13}$$

Insertion into (13.12) yields

$$\tfrac{1}{s}C^{\mathrm{T}}BB^{\mathrm{T}}b + C^{\mathrm{T}}Cz = C^{\mathrm{T}}b.$$

Then

$$\hat{z} = (C^{\mathrm{T}}C)^{-1}C^{\mathrm{T}}\big(I - \tfrac{1}{s}BB^{\mathrm{T}}\big)b$$

and finally

$$\hat{z} = \tfrac{1}{r}C^{\mathrm{T}}\big(I - \tfrac{1}{s}BB^{\mathrm{T}}\big)b \tag{13.14}$$

where we have used $(C^{\mathrm{T}}C)^{-1} = \frac{1}{r}I$. However, the variance calculations remain. The covariance matrix for $\hat{x}$ becomes

$$\Sigma_{\hat{x}} = \hat{\sigma}_0^2(B^{\mathrm{T}}B)^{-1} = \tfrac{1}{s}\hat{\sigma}_0^2 I \tag{13.15}$$

which shows that all $\hat{x}_i$ have the same variance $\hat{\sigma}_0^2$ and furthermore they are independent! The only way to reduce the variances is by increasing s, by observing more rounds.

The sum of squared residuals is estimated by

$$\begin{aligned}
\hat{r}^{\mathrm{T}}\hat{r} &= b^{\mathrm{T}}b - b^{\mathrm{T}}A\hat{y} = b^{\mathrm{T}}\big(b - B\hat{x} - C\hat{z}\big) \\
&= b^{\mathrm{T}}\big(b - \tfrac{1}{s}BB^{\mathrm{T}}b - \tfrac{1}{r}CC^{\mathrm{T}}b + \tfrac{1}{rs}CC^{\mathrm{T}}BB^{\mathrm{T}}b\big) \\
&= b^{\mathrm{T}}\big(I - \tfrac{1}{s}BB^{\mathrm{T}} - \tfrac{1}{r}CC^{\mathrm{T}} + \tfrac{1}{rs}CC^{\mathrm{T}}BB^{\mathrm{T}}\big)b \\
&= b^{\mathrm{T}}\big(I - \tfrac{1}{s}BB^{\mathrm{T}} - \tfrac{1}{r}CC^{\mathrm{T}}\big(I - \tfrac{1}{s}BB^{\mathrm{T}}\big)\big)b \\
&= b^{\mathrm{T}}\big(I - \tfrac{1}{r}CC^{\mathrm{T}}\big)\big(I - \tfrac{1}{s}BB^{\mathrm{T}}\big)b.
\end{aligned}$$

Finally

$$\hat{\sigma}_0^2 = \frac{b^{\mathrm{T}}(I - \frac{1}{r}CC^{\mathrm{T}})(I - \frac{1}{s}BB^{\mathrm{T}})b}{(r-1)(s-1)}. \tag{13.16}$$

Given a set of rounds we want to estimate directions and orientation unknowns. The best estimate for the directions is—not surprisingly—the mean of the measured directions. All estimates of directions are independent and of equal variance $\hat{\sigma}_0^2$. This variance is given by the expression (13.16). Today station adjustment only is used for educational reasons to estimate $\hat{\sigma}_0^2$ for a given data set taken with a given theodolite.

In professional connections the original sets of rounds are entered into a least-squares procedure. The orientation unknowns z_i in most cases are of no further interest so they are eliminated from the normals according to the technique described in Section 11.7.

Example 13.1 Let us assume observations (in gons) for 3 rounds each containing 4 directions. The observations are given as *reduced rounds*, i.e. all observations in each round are diminished by the observational value of the first direction which consequently has value zero. This zero column is omitted. Hence there are only 3 significant columns. The numbers are arranged in a form suitable for input to MATLAB.

50.193	77.873	268.441
50.187	77.871	268.439
50.188	77.872	268.440

The estimated directions are

$$\begin{gathered} 0.000\,0 \text{ gon} \\ 50.189\,3 \text{ gon} \\ 77.872\,0 \text{ gon} \\ 268.440\,0 \text{ gon}. \end{gathered}$$

The estimated orientation unknowns are

$$\begin{aligned} \hat{z}_1 &= -1.4 \text{ mgon} \\ \hat{z}_2 &= 1.1 \text{ mgon} \\ \hat{z}_3 &= 0.3 \text{ mgon}. \end{aligned}$$

The standard deviation of unit weight is $\hat{\sigma}_0 = 1.4$ mgon.

The M-file sets computes the estimated directions $\hat{x}_i$ and orientation unknowns $\hat{z}_i$.

13.4 Fitting a Straight Line

Let p observations of coordinates be given as pairs (x_i, y_i). The pairs are not symmetric in the sense that only the x_i values are subject to errors, and the y_i values are considered observed exactly. (In the case where both coordinates are stochastic variables the situation

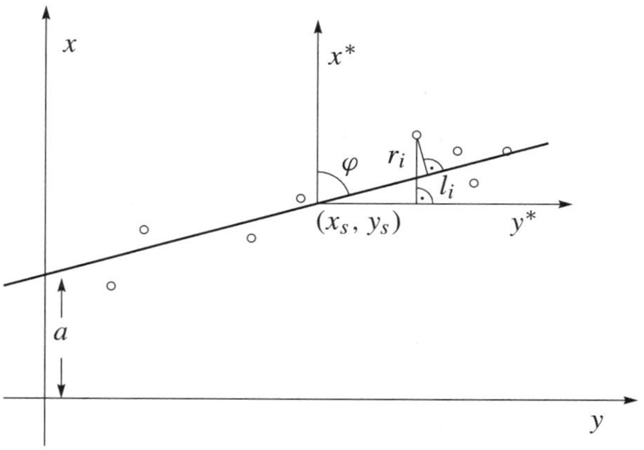

Figure 13.4 Fitting line.

is more complex.) Geometrically, we want to regard them as coordinates of p points in the plane. The least-squares problem may be described by a linear condition between x_i and y_i:

$$x_i = a + \cot\varphi\, y_i, \tag{13.17}$$

with a and φ as unknowns.

Some readers will probably argue that the problem is the usual linear regression. We minimize the sum of squares of l_i in the direction parallel to the x-axis. This is not the case; we want to minimize $l_i \sin\varphi$ which is a quantity orthogonal to the fitting line, see Figure 13.4. In statistics our problem is called ***linear orthogonal regression***. In numerical analysis it is an example of ***total least squares***. The observation equations are

$$(a + y_i \cot\varphi - x_i)\sin\varphi = a\sin\varphi + y_i\cos\varphi - x_i\sin\varphi = -r_i. \tag{13.18}$$

Thus the unknowns are a and φ and so we see that a simple least-squares problem may result in a not so elementary nonlinear problem.

Returning to the fundamental principle of least squares we find a surprisingly simple solution. By partial differentiation of the sum of squares $s = \sum r_i^2$ and equating to zero we get

$$\frac{\partial s}{\partial a} = 2\sin\varphi \sum_{i=1}^{p} (a\sin\varphi + y_i\cos\varphi - x_i\sin\varphi) = 0. \tag{13.19}$$

Similarly the second minimum condition may be expressed as

$$\frac{\partial s}{\partial \varphi} = 2\sum_{i=1}^{p} (a\sin\varphi + y_i\cos\varphi - x_i\sin\varphi)(a\cos\varphi - y_i\sin\varphi - x_i\cos\varphi) = 0. \tag{13.20}$$

We rewrite (13.19) and get

$$pa\sin\varphi + \cos\varphi \sum y_i - \sin\varphi \sum x_i = 0$$

or

$$\hat{a} = \tfrac{1}{p}\Big(\sum x_i - \cot\varphi \sum y_i\Big). \tag{13.21}$$

Once again we introduce the barycenter (x_s, y_s) defined by $x_s = \sum x_i/p$ and $y_s = \sum y_i/p$. We reduce the observations (x_i, y_i) to this center

$$x_i^* = x_i - x_s \qquad \text{and} \qquad y_i^* = y_i - y_s$$

and substitute in (13.21):

$$\hat{a} = \tfrac{1}{p}\big(px_s - py_s \cot\varphi\big) = x_s - \cot\varphi\, y_s \tag{13.22}$$

or $x_s = \hat{a} + \cot\varphi\, y_s$. In view of (13.17) we conclude that ***the fitting line contains the barycenter*** (x_s, y_s).

Finally we shall find an expression for $\hat{\varphi}$. Substituting the (x_i^*, y_i^*) pair into (13.20) we can put $a = 0$ and obtain

$$-\sum y_i^{*2} \sin\varphi\cos\varphi - \sum x_i^* y_i^* (\cos^2\varphi - \sin^2\varphi) + \sum x_i^{*2} \sin\varphi\cos\varphi = 0$$

or after introducing the double angle 2φ

$$\sin 2\varphi \sum (x_i^{*2} - y_i^{*2}) - 2\cos 2\varphi \sum x_i^* y_i^* = 0.$$

Finally the angle has

$$\tan 2\varphi = \frac{2\boldsymbol{x}^{*\mathrm{T}} \boldsymbol{y}^*}{\boldsymbol{x}^{*\mathrm{T}}\boldsymbol{x}^* - \boldsymbol{y}^{*\mathrm{T}}\boldsymbol{y}^*} \tag{13.23}$$

where $\boldsymbol{x}^* = (x_1^*, x_2^*, \dots, x_p^*)$ and $\boldsymbol{y}^* = (y_1^*, y_2^*, \dots, y_p^*)$. The expression yields two solutions for $\hat{\varphi}$; they differ by a right angle. For any given context it should be possible to select the correct value.

Example 13.2 Suppose $(x_i, y_i) = (0, 0), (3, 1)$ and $(12, 2)$. The barycenter is $(x_s, y_s) = (5, 1)$ and consequently $(x_i^*, y_i^*) = (-5, -1), (-2, 0)$ and $(7, 1)$. So

$$\tan 2\varphi = \frac{24}{78 - 2} = 0.315\,79$$

or $\hat{\varphi} = 9.736$ gon. Then $\cot\hat{\varphi} = 6.49$ and $\hat{a} = 5 - 6.49 \cdot 1 = -1.49$. The fitting line has the equation

$$x = -1.49 + 6.49y.$$

For comparison, the line of linear regression is $x = -1 + 6y$.

Example 13.3 The problem of fitting p pairs of coordinates (x_i, y_i) to a straight line

$$cx + sy = h \qquad \text{and} \qquad c^2 + s^2 = 1$$

can also be solved by means of a singular value decomposition of the matrix $\overline{Y}$ containing the reduced coordinates (x_i^*, y_i^*). The column of V pertaining to the largest singular value in the decomposition $\overline{Y} = U\Sigma V^{\mathrm{T}}$ plays a fundamental role. We finish with a MATLAB code for solving the problem in our particular coordinate system.

M-file: lor

```
function lor(Y)
% Linear orthogonal regression.
%    See Dahlquist & Björck, 2nd edition, (1995) Example 7.6.2.
%    The first column of Y contains x and the second column y
%    for p given points

[p,q] = size(Y);
e = ones(p,1);
m = mean(Y);
Y = Y-e*m;
[U,S,V] = svd(Y);
cs = V(:,1);
alpha = atan2(cs(2),cs(1));
t = tan(alpha)
x = tan(-alpha + pi/2);
     % necessary statement due to special orientation of coordinate system
a = m(2)-1/x
```

Part III

Global Positioning System (GPS)

14

GLOBAL POSITIONING SYSTEM

14.1 Positioning by GPS

GPS has revolutionized the science of positioning and Earth measurement. One part of that revolution is accuracy, another part is speed and simplicity. A third part is cost. All of these improvements are contributing to the growth of major applications. We frankly hope that our readers will develop new uses for GPS; the technology is ready, only imagination is needed. And the initiative to turn imagination into reality.

But this is a scientific book, not a brochure. We focus on one major advantage of GPS: *accuracy*. The inherent accuracy of a GPS receiver can be enhanced or degraded. It is enhanced by careful processing, it is degraded by accepting (instead of trying to eliminate) significant sources of error. We start by listing three critical techniques for achieving centimeter and even millimeter accuracy in positioning:

1 Work with two or more receivers. The key idea of *differential* GPS (DGPS) is to compute *differences of position* instead of absolute position. Errors that are shared by receivers will cancel when we form differences.

2 Repeat the measurements. A sequence of observations has a significantly smaller variance than a single observation. If the receiver is moving, the *Kalman filter* can account for changes of state as well as new observations.

3 Estimate each source of error in the observations. Section 14.2 describes the major errors and their approximate magnitudes.

The reader already knows about the dithering of the satellite clocks. President Clinton directed that the military should terminate this "Selective Availability" before the year 2006. All indications are favorable; we would not be surprised to see it end earlier. The 2-dimensional rms positioning error from this source can be close to 100 meters (and it is removed by DGPS, when two receivers are measuring their range from the same satellites).

We strongly emphasize the importance of time. *In GPS, time is the fourth dimension.* It is the reason we need at least four satellites, not three, to locate the receiver. The four coordinates to be computed are x, y, z, and $c\,dt$—the speed of light multiplies the clock discrepancy. This quantity $c\,dt$ has the units of distance. Since an ordinary receiver clock might only know the time within a few seconds, the elimination of error from $c\,dt$ is not an optional improvement—it is absolutely required!

In short: The key to the accuracy of GPS is a precise knowledge of *the satellite orbits and the time*. On the gound are computed Keplerian elements based on the actually observed orbits (each set of quasi-Keplerian elements is good for 2 hours) and these elements are uploaded to the satellites' memories. The satellites carry atomic clocks (cesium and rubidium). They broadcast their own quasi-Keplerian elements for position computation at receivers. They also broadcast with lower accuracy (in the almanac) the Keplerian elements of other satellites. But it is the precise ephemerides that locate one end of the line segment between satellite and receiver. The problem of GPS positioning is to locate the other end.

One basic fact about GPS deserves attention. Its measurements yield *distances* and not angles. We are dealing with *trilateration* and not triangulation. This has been desired for centuries, because angles are definitely awkward. Of course lengths are nonlinear too, in the position coordinates x, y, z, $c\,dt$. The receiver must solve nonlinear equations.

The purpose of this chapter is to explain in reasonable detail how GPS works, and where mathematics is involved. Then later chapters will describe the actual calculations in much more detail:

Chapter 15: Accurate processing of GPS observations.

Chapter 16: Random errors and their covariances.

Chapter 17: Kalman filtering of observations as the state changes.

We will also describe the MATLAB software that is freely available to the reader. For this introduction to GPS, the lecture by Ponsonby (1996) has been particularly helpful.

Clock Errors and Hyperbolas of Revolution

The goal is get a fix on the receiver's position. Suppose there were no clock errors (which is false). Then the distances from three satellites would provide a fix. Around each satellite, the known distance determines a sphere. The first two spheres intersect in a circle. Assuming that the satellites do not lie on a straight line, the third sphere will normally cut this circle at two points. One point is the correct receiver position, the other point is somewhere out in space. So three satellites are sufficient if all clocks are correct and all ranges are measured precisely.

In reality the receiver clock is typically inexpensive and inaccurate. When the clock error is dt, every range measured at that instant will be wrong by a distance $c\,dt$. We are measuring the arrival time of a signal that contains its own departure time. (The velocity of light is $c \approx 300\,\mathrm{m}/\mu\mathrm{sec}$. Of course we would use many more correct digits for c, which is slightly different in the ionosphere and the troposphere. These are among the errors to be modeled.) The incorrect range, which includes $c\,dt$ from the unknown clock error, is called a *pseudorange*.

From two satellites we have two pseudoranges ρ^1 and ρ^2. Their difference $d^{12} = \rho^1 - \rho^2$ has no error $c\,dt$ from the receiver's clock. The receiver must lie on a *hyperbola of revolution*, with the two satellites as the foci. This is the graph of all points in space whose distances from the satellites differ by d^{12}.

The third pseudorange locates the receiver on another hyperbola of revolution (a hyperboloid). It intersects the first in a curve. The fourth pseudorange contributes a third independent hyperboloid, which cuts the curve (normally twice). Provided the four satellites are not coplanar, we again get two possible locations for the receiver: the correct fix, and a second point in space that is far from correct and readily discarded. This is the geometry from the four pseudoranges, and the algebra is straightforward but nonlinear: for $k = 1, 2, 3, 4$ we know that

$$(x - X^k)^2 + (y - Y^k)^2 + (x - Z^k)^2 + (c\,dt)^2 = (\rho^k)^2.$$

The Bancroft algorithm to solve for x, y, z, and $c\,dt$ is described in Section 15.7.

Note the important step that remains. The receiver must convert this spatial fix into a position on a standard geodetic reference system. For GPS this reference is WGS 84. The Russian system GLONASS uses now the slightly different reference PZ-90. Then the receiver employs a model of the geoid to compute geographical coordinates and height above sea level. An ordinary receiver displays latitude and longitude, which allows the user to find the position on a map. Not taking into account the correction from WGS 84 to the map projection may be the error-prone of all! (Map projections would apply only for navigation.) Existing charts are unlikely to be accurate at the centimeter level. Still they are probably sufficient for the immediate purposes of typical users (to head toward their destination, to locate a landmark, to save their lives, . . .).

Radio Signals

The GPS satellites transmit radio signals that are phase modulated according to a known sequence of bits. These are DSSS signals (Direct Sequence Spread Spectrum). A very useful simplified model is given by Ponsonby (1996), showing a switch SW1 that reverses the phase of the sine wave according to a (known) pseudorandom bit stream. The time interval T is the *chipping time*, and over each interval the sine wave is multiplied by $+1$ or -1. Then the satellite transmits these positive and negative segments of sine waves.

The signal coming into the receiver is chopped up by a second switch SW2. This also follows a pseudorandom pattern. Everything depends on whether the two chip sequences are coherent. If not, the output signal is highly chopped in time and widely spread in frequency. Very little power passes through a filter. But if the switching sequences at SW1 and SW2 are exactly matched, the output signal is back to a perfect sine wave. The wave passes through the filter with maximum gain in the power level. This timing alignment is described by Ponsonby as the equivalent of fitting a key into a lock.

Once the sequences are aligned, a Delay Lock Loop maintains the synchronism. And since the receiver is moving with respect to the transmitter (the speed of the satellite is about 3900 m/sec), there will be a *Doppler shift* in the frequency of the pure sine wave. This frequency shift provides a good measurement of *velocity*. The receiver displays its velocity converted to Earth coordinates. Of course differences in position give a direct (non-Doppler and less accurate) estimate of velocity when divided by Δt.

Actual Doppler observations are very receiver-dependent, sometimes directly from the phase and sometimes by separate software. Phase tracking over several seconds will produce a much more accurate velocity than observations over milliseconds. So Doppler is application-dependent: ship navigation is very different from aircraft.

C/A Code and P Code

The actual modulated signal is divided into two independent components. They are modulated by different bit sequences. The slower Coarse/Acquisition (C/A) code has a chipping rate of 1.023 MHZ, and the Precision (P) code is 10 times faster. The C/A code is available to all users. The encrypted P code is referred to as the Y code which is reserved for the military (although it is partly useful to others also).

All GPS satellites use the same carrier frequencies. Each satellite has its own pseudorandom sequence of 1023 bits (its periodic C/A code which is a Gold Code). The repeat time is 1.5 s for the C/A code and one week for the P code. It is most important to know that there are (at present) two coherent transmission frequencies for the P code (and its secret encryption, the Y code):

$$L_1 \text{ is } 1575.42\,\text{MHz} \qquad \text{and} \qquad L_2 \text{ is } 1227.60\,\text{MHz}.$$

The two frequencies are differently delayed by the ionosphere. A receiver that accepts both frequencies can compute this delay (because it is known to be proportional to $1/f^2$). This correction to the speed of light through the ionosphere is essential. A single frequency receiver has to make do with estimates of the ionospheric correction whose parameters are broadcast by the satellites.

At this time there is wide discussion of the proposal for a new civilian frequency L_5 (then the military might remove L_2 from civilian use). The difficulties of agreeing on a new frequency *before* launching expensive satellites seem to have frustrated everyone.

Pseudorandom Noise Code Generation

The pseudorandom noise (PRN) codes transmitted by the GPS satellites are deterministic binary sequences with noise-like properties. Each C/A and P code is generated using a tapped linear feedback shift register (LFSR).

A shift register is a set of one bit storage or memory cells. When a clock pulse is applied to the register, the content of each cell shifts one bit to the right. The content of the last cell is "read out" as output. The special properties of such shift registers depend on how information is "read in" to cell 1.

For a tapped linear feedback shift register, the input to cell 1 is determined by the state of the other cells. For example, the binary sum from cells 3 and 10 in a 10-cell register could be the input. If cells 3 and 10 have different states (one is 1 and the other 0), a 1 will be read into cell 1 on the next clock pulse. If cells 3 and 10 have the same state, 0 will be read into cell 1. If we start with 1 in every cell, 12 clock pulses later the contents will be 0010001110. The next clock pulse will take the 1 in cell 3 and the 0 in cell 10, and place their sum (1) in cell 1. Meanwhile, all other bits have shifted cell to the right, and the 0 in cell 10 becomes the next bit in the output. A shorthand way of denoting this particular design is by the modulo 2 polynomial $f(x) = 1 + x^3 + x^{10}$. Such a polynomial representation is particularly useful because if $1/f(x) = h_0 + h_1x + h_2x^2 + h_3x^3 + \cdots$, then the coefficients $h_0, h_1, h_2, \ldots$, form the binary output sequence.

The C/A code is generated by two 10-bit LFSRs of maximal length $(2^{10} - 1)$. One is the $1 + x^3 + x^{10}$ register already described, and is referred to as $G1$. The other has $f(x) = 1 + x^2 + x^3 + x^6 + x^8 + x^9 + x^{10}$. Cells 2, 3, 6, 8, 9, and 10 are tapped and binary added to get the new input to cell 1. In this case, the output comes not from cell 10 but from a second set of taps. Various pairs of these second taps are binary added. The different pairs yield the same sequence with different delays or shifts (as given by the 'shift and add' or 'cycle and add' property: a chip-by-chip sum of a maximal length register sequence and any shift of itself is the same sequence except for a shift). The delayed version of the $G2$ sequence is binary added to the output of $G1$. That becomes the C/A code. The $G1$ and $G2$ shift registers are set to the all ones state in synchronism with the epoch of the $X1$ code used in the generation of the P code (see below). The various alternative pairs of $G2$ taps (delays) are used to generate the complete set of 36 unique PRN C/A codes. These are Gold codes, Gold (1967), Dixon (1984), and any two have a very low cross correlation (are nearly orthogonal).

There are actually 37 PRN C/A codes, but two of them (34 and 37) are identical. The first 32 codes are assigned to satellites. Codes 33 through 37 are reserved for other uses including ground transmitters.

The P code generation follows the same principles as the C/A code, except that four shift registers with 12 cells are used. Two registers are combined to produce the $X1$ code, which is 15 345 000 chips long and repeats every 1.5 seconds; and two registers are combined to produce the $X2$ code, which is 15 345 037 chips long. The $X1$ and $X2$ codes can be combined with 37 different delays on the $X2$ code to produce 37 different one-week segments of the P code. Each of the first 32 segments is associated with a different satellite.

Special and General Relativity

GPS positioning is one of the very few everyday events in which relativistic effects *must* be accounted for. The whole system is based on clocks, and those clocks are moving. The satellite clock is moving with respect to the receiver clock, so time is dilated and Special Relativity enters. All the clocks are in a gravitational field (of the Earth), so General Relativity is significant. And the Earth is rotating; light follows a spiral path. We cannot perfectly synchronize the clocks!

The Sagnac Effect from the rotation is also fascinating. It destroys Einstein Synchronization, which depends on a constant speed of light. That constancy is restricted to inertial frames (no relative acceleration). The rotation of the Earth means that clock A can be synchronized with B, and B with C, but clock C is *not* synchronized with A. So we need a universal time that goes at a different rate from local time. This coordinated Universal Time is maintained at the GPS control center in Colorado Springs.

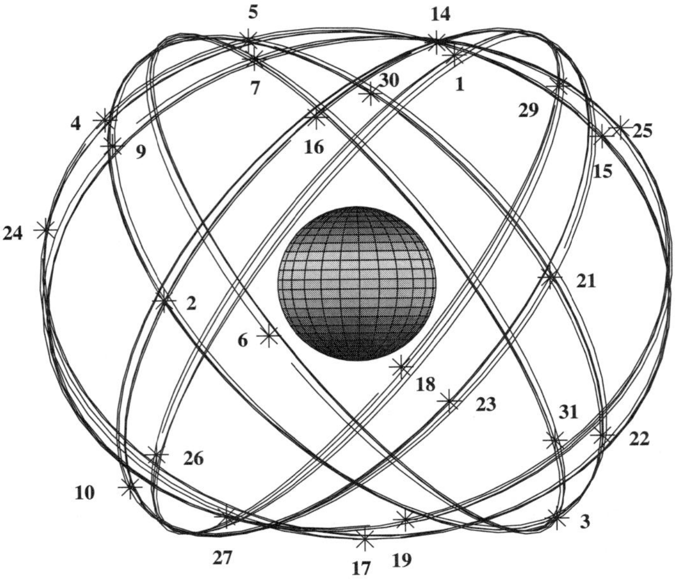

Figure 14.1 The GPS constellation: 6 planes with 4 satellites. The 2 orbits/day fall 15 minutes short of completion, producing the small gaps at the end of each orbit. The configuration corresponds to GPS week number 907 starting at 324 000 seconds of week.

Orbits of the Satellites

The satellites stay at an altitude of approximately 3 Earth radii. They travel in nearly circular orbits, two complete orbits in each sidereal day. The inclinations to the Equatorial plane are 55°. In practice a pseudorange is most reliable when the satellite is at least 10° or better 15° above the horizon. Figure 14.1 shows in two dimensions the GPS constellation of 24 satellites, with four satellites spaced around each of six orbital planes. Actually we count 25 satellites in Figure 14.1. Tomorrow it might be 24 or 26!

Differential GPS

Suppose two receivers are reasonably close together (say less than 100 kilometers). Then the signals from a satellite that is 20 000+ kilometers away will reach the two receivers along very close paths. The delays in the ionosphere will be nearly identical. The errors due to an incorrect satellite clock (which has been dithered to achieve Selective Availability) are the same. So are the errors in the satellite orbits. Those errors will cancel in the *difference* of travel times to the two receivers.

We emphasize that only frequency standard variations cancel exactly in differencing. Time synchronization, atmospheric effects, and orbit errors are all *proportional to station separation*. Ionospheric gradients are typically 1 part in 10^6, for example. For precise positioning at a single frequency, a 100 km separation is much too large to make the effects negligible (in fact they may reach 10 mm even over 1–2 km). With two frequencies, they nearly cancel for any distance. Tropospheric effects are also important in precise positioning for separations over 1–2 km. Without SA the orbital errors are 20 m, which is 1 ppm, or 5 cm over 100 km. This is important for tectonics but not for routine surveying or navigation.

DGPS is based on this simple idea. If the position of one "home" receiver is exactly known, then its distance from each satellite is easily computed. The travel time at the speed of light is then available (divide the distance by c). This theoretical travel time can be compared with the measured travel time. The error is found for each satellite in view, and these errors are transmitted by the home receiver. The second receiver (or any number of roving receivers) will pick up this information and correct their own travel times from the satellite, by that same amount.

Thus the small investment in an extra receiver-transmitter enables us to cancel part of the natural effects of the ionosphere and the unnatural effects of SA.

The principle of working with differences is even more used in post processing of GPS data. *Double differences* (2 satellites as well as 2 receivers) are explained and constantly applied in Chapter 15. The accuracy achieved by DGPS is remarkable.

New Uses

This extremely brief section is for the reader to complete! If our book can encourage new ideas and applications for GPS, its authors will be very pleased. Note especially that the microprocessor in a GPS receiver has *unused capacity*. The power is there. It can be coupled to other tools. Give it some thought.

14.2 Errors in the GPS Observables

There are two fundamental observables, the pseudorange P_i^k and the carrier phase Φ_i^k. As usual k specifies the satellite and i identifies the receiver. The phase is much more accurate than the pseudorange (even using the P code). Both observables will include errors from many sources! Some errors can be removed, others can be reduced, and others are just neglected. This section estimates the magnitudes of the errors and our options for dealing with them.

We deal first with Selective Availability. When SA is imposed for military reasons, all receivers and users encounter identical errors in the satellite clock. This affects the C/A and P code and carrier phase measurements equally. The technique of ***differential*** GPS (using one receiver in a known position as a reference) can essentially eliminate this artificial degrading of the system. If it is not eliminated, the error from SA will dominate all others. Parkinson & Spilker Jr. (1996) estimate the rms error (the clock offset multiplied by the speed of light) as 20 meters.

A solitary individual will see this error in the range. When multiplied by the Dilution of Precision factors in Section 14.4 (say VDOP $= 2.5$ in the vertical and HDOP $= 2$ in the horizontal), this implies rms positioning errors of 50 meters vertically and 40 meters horizontally. Those are so large compared to the intrinsic GPS errors that we hope SA will be turned off before this book is a year old. SA is guaranteed to disappear eventually.

Delay of the Signal

Now suppose that SA is eliminated. It is either removed by the military or cancelled by forming differences in DGPS. We follow Parkinson & Spilker Jr. (1996) and Kleusberg & Teunissen (1996) in estimating the sources of error that remain.

The largest errors (until we compensate for them) come from the delay when a signal travels through the atmosphere. So we first recall the connections between velocity and refraction and travel time. Then we estimate the range errors due to incorrect travel times and other sources.

In a vacuum the speed of an electromagnetic wave is c. In the atmosphere, this speed (the *phase velocity*) is reduced to v. The dimensionless ratio $n = \frac{c}{v}$ is the refractive index. The number v is related to the angular frequency ω and wave number k by $v = \frac{\omega}{k}$.

In a dispersive medium, these numbers are functions of ω. A packet of waves with frequencies near ω will travel not with the phase velocity v but with the *group velocity*:

$$v_{\text{group}} = \frac{d\omega}{dk} = v + k\frac{dv}{dk}. \tag{14.1}$$

That is the travel speed of a modulation superimposed onto the carrier wave. The refractive index becomes $n_{\text{group}} = c/v_{\text{group}}$.

The wave travels from satellite to receiver along the quickest path S (by Fermat's principle of least time). At a planar interface in the medium, this principle yields Snell's law $n_1 \sin z_1 = n_2 \sin z_2$ for the change in the angle z (between the path and the normal to the discontinuity). The delay along S of the carrier, in comparison to the straight-line path L in a vacuum, is dt_Φ:

$$dt_\Phi = \int_S \frac{dS}{v} - \int_L \frac{dL}{c}.$$

Most of this delay comes from change of speed, a smaller part comes from change of path. Multiplying by c, the two parts are

$$c\,dt_\Phi = \int_L (n-1)\,dL + \left(\int_S n\,dS - \int_L n\,dL\right). \tag{14.2}$$

For the modulation, which carries the important signal for GPS, v becomes v_{group} and the refractive index becomes n_{group}.

The scientific problem is to determine n and n_{group} from properties of the atmosphere, the electron density in the ionosphere and the air/water densities in the troposphere.

Table 14.1 Relative accuracy of clocks

Clock type	relative accuracy
Crystal wrist watch	10^{-6}
Geodetic GPS receiver	10^{-5}–10^{-7}
TI, crystal in oven	10^{-8}–10^{-9}
GPS satellite clock, SA on	10^{-9}
Rubidium	10^{-11}–10^{-12}
Cesium	10^{-12}–10^{-13}
Hydrogen maser	10^{-15}–10^{-16}

Error Budget for GPS

Here are the principal errors in GPS positioning—their sources and also their approximate magnitudes. We estimate errors in the range and multiply by the DOP factors.

Ephemeris Errors The satellite transmits its Keplerian elements, almost exactly but with a small error. This grows from the time of upload by a control station until the next upload. The error growth is slow and smooth, and only the projection of the ephemeris error along the line of sight produces an error in the range. Parkinson & Spilker Jr. (1996) estimate the rms ranging error as 2.1 meters (and the estimate now might be smaller).

Satellite Clock Errors An atomic clock, with a rubidium or cesium oscillator, is correct to about 1 part in 10^{12}. In a day the offset could reach 10^{-7} seconds; multiplied by c this represents 26 meters. With clock corrections every 12 hours, an average error of 1 meter is reasonably conservative.

Ionosphere Errors GPS signals are delayed as they pass through the ionosphere, which starts 50 km above the Earth and extends to 1000 km or more. The delay is proportional to the number of electrons (integrated density along the signal path) and inversely proportional to f^2. Thus the effect is dispersive; it depends on the frequency f. The density of free electrons varies strongly with the time of day and the latitude. The variations from solar cycles and seasons and especially short-term effects are less strong but less predictable. If the delay were not accounted for at all, the ranging errors on the L_1 frequency in the zenith direction could reach 30 meters. *The effects on the pseudorange P and phase Φ are opposite in sign*; the carrier phase is advanced.

So we *must* estimate the ionospheric delay. A dual-frequency receiver can measure the pseudoranges P_1 and P_2 on both frequencies L_1 and L_2, and solve for the delay:

$$dP_{\text{ion}} = \frac{f_2^2}{f_2^2 - f_1^2}(P_1 - P_2) + \text{random/unmodelled errors}. \tag{14.3}$$

This should be removed from P_1. Similarly the phase correction for ionospheric delay is

$$d\Phi_{\text{ion}} = \frac{f_2^2}{f_2^2 - f_1^2}\big((\lambda_1 N_1 - \lambda_2 N_2) - (\Phi_1 - \Phi_2)\big) + \text{random/unmodelled errors.} \quad (14.4)$$

Equations (14.3) and (14.4) have equal value (in delay units) but opposite signs, so the ionosphere can be unambiguously calibrated by a combination of pseudorange and phase at both frequencies.

The ambiguities N_1 and N_2 remain constant (but possibly unknown) if there are no cycle slips. So at least a differential delay is known. This estimate is good but there is often a better way. If you have a dual frequency phase receiver, the P code observations allow you to estimate the ionospheric correction. Then the improved pseudoranges can help resolve the ambiguities N_1 and N_2, completing the circle.

For measurements at only one frequency, these formulas for dP_{ion} and $d\Phi_{\text{ion}}$ are useless. In DGPS the ionospheric delay at two receivers is cancelled when we compute the (sufficiently short!) baseline between them. The difference in signal paths produces a slight baseline shortening, proportional to electron content and baseline length. One receiver at one frequency can use the prediction model for dP_{ion} and $d\Phi_{\text{ion}}$ contained in the GPS broadcast message. Tests show better results than promised (but not great).

Troposphere Errors The troposphere is the lower part of the atmosphere, thickest over the Equator (about 16 km). The temperature and pressure and humidity alter the speed of radio waves. Their effects are nearly independent of the radio frequency, but they depend on the time of passage. For a flat Earth we would divide the zenith delay (the delay at elevation angle $El = \frac{\pi}{2}$) by $\sin El$. There are a number of good *mapping functions* to improve this to a spherical-surface model. The M-file tropo uses a mapping function proposed by Goad & Goodman (1974) to compute the reduction.

In the zenith direction, the total tropospheric delay is estimated as about 2.3 meters. The hydrostatic component (responsible for 90%) is the path integral of the density of moist air. The *wet* component is a function of water vapor density which is highly variable. It is questionable how descriptive actual measurements can be. The classical example is sitting in a fog bank only 50 m high. The other extreme is sitting in relatively dry air below a dark thundercloud. Both of these conditions are met in GPS surveys.

In fact Duan et al. (1996) proposed and successfully demonstrated that in the reverse direction, the water vapor density could be measured by GPS! This is a beautiful example of an unexpected contribution coming from accurate measurements of time and distance. The electron content of the ionosphere can also be studied by GPS.

The delay from liquid water in clouds and rain is well below 1 cm. But models of the wet delay (water vapor) using surface meteorology are often wrong by more than 1 cm. Again we recommend the discussion by Langley in the Kleusberg-Teunissen book.

Multipath Errors A GPS signal might follow several paths to a receiver's antenna. The same signal arrives at different times and interferes with itself. This produces ghost images on TV (before cable) and corresponds to echos for our voice. In GPS, the signal can be reflected from buildings or the ground and create a range error of several meters or more. Some authors would allow 10 meters for multipath error in C/A code measurements.

Table 14.2 Standard Errors without SA

	Single frequency	Double frequency
Ephemeris data	2 m	2 m
Satellite clock	2 m	2 m
Ionosphere	4 m	0.5–1 m
Troposphere	0.5–1 m	0.5–1 m
Multipath	0–2 m	0–2 m
UERE: rms	5 m	2–4 m

Multipath is a serious problem because it is so difficult to model. Sometimes we can improve the site for the receiver. The design of the antenna is also critical. Large groundplanes, with various antenna elements (dipoles, microstrip), are the most common antidote for multipath. The receiver can be built with a narrow correlator to block the reflection, or with multiple correlators to allow estimation on several paths. And for a given satellite/static receiver pair, at a given time of day, we could try to estimate repeatable paths.

The multipath errors in the *phase* observations Φ are much smaller, at the centimeter level. We receive the sum of two signals. The reflected signal has a phase shift $\Delta\Phi$ and its magnitude is attenuated by a factor α:

$$\text{received signal} = A\cos\Phi + \alpha A\cos(\Phi + \Delta\Phi).$$

The multipath error, comparing the phase of this sum to the correct value Φ, is

$$d\Phi = \arctan\left(\frac{\sin\Phi}{\alpha^{-1} + \cos\Phi}\right).$$

The worst case has no attenuation ($\alpha = 1$) and $d\Phi = 90°$. This means only a quarter-wavelength error (5 cm) from multipath.

Other sources of error exist! Receivers are not perfect but they are continually improving. Of course the Earth is moving up and down (there are tides on dry ground just as in the sea, but smaller). These tides can be accounted for almost completely. The actual position of the Earth's surface and of sea level affects vertical positioning by GPS, which is less accurate than horizontal positioning. As a limitation on vertical accuracy for precise positioning, tides are less difficult than the troposphere and antenna problems (multipath and phase-center variations). But still we can land an airplane.

Summary of the Error Budget

Table 14.2 gives approximate rms errors ignoring selective availability. (Otherwise SA would dominate everything.) The error sources are reasonably independent, so the square root of sum of squares is the UERE—*the user equivalent range error*. This is multiplied by the *Dilution of Precision* (say VDOP $= 2.5$ and HDOP $= 2$) to give the standard deviation in position—in other words, the one sigma error.

This overall range error multiplied by the DOP factor in Section 14.4 is roughly 10 meters for a single frequency civilian C/A receiver. A dual-frequency P/Y code receiver would experience roughly half that error, when ionospheric delay is cancelled and the ephemeris and clock errors become the largest. Elsewhere we discuss the double differences and long-time averages and Kalman filtering that reduce the position error to centimeters and even millimeters.

14.3 Description of the System

The traditional observable in positioning and navigation has been the angle. This quantity is fundamental to geodesy. But since the 1960's geodesists additionally have used the electronically measured *distance* in their science. For GPS, distance is the basic observable. Thus triangulation is being largely replaced by *trilateration*.

From a geodetical point of view a key feature of GPS is that the user can calculate coordinates to all visible satellites at any time. The calculation is done according to a complex algorithm, described in Chapter 15. The input is derived from the satellite signals; the receiver immediately finds the satellite coordinates. These coordinates relate to an Earth centered and Earth fixed (ECEF) system. The geocentric coordinates of the receiver are also related to this system. Any good receiver provides the position within 2 min. from being switched on!

Conceptually the satellites may be looked upon as points with known coordinates in space, continuously emitting information on their position. The receiver measures its distance to each satellite. In the field, the receiver coordinates (X_i, Y_i, Z_i) can be determined with an accuracy varying from 20 m to 100 m. The accuracy can be greatly improved by using offset values from a fixed receiver, whose position is known. This method is known as *differential* GPS or DGPS. Geodesy deals exclusively with DGPS and typically it achieves accuracies at the level of centimeters.

The frequency $f_0 = 10.23$ Mhz is fundamental for GPS. Each satellite transmits carrier waves on two frequencies (this is in 1997—a new civilian frequency L_5 is under discussion). The L_1 signal uses the frequency $f_1 = 154 f_0 = 1575.42$ Mhz with wavelength $\lambda_1 = 0.1905$ m and the L_2 signal uses frequency $f_2 = 120 f_0 = 1227.60$ Mhz with wavelength $\lambda_2 = 0.2445$ m. The two frequencies are coherent because 154 and 120 are integers. L_1 carries a precise code and a coarse/acquisition code. L_2 carries the precise P code only. A navigation data message is superimposed on all these codes.

The most accurate distances are computed from ***phase observation***. This consists of the difference in phase of the incoming satellite signal and a receiver generated signal with the same frequency. The phase difference is measured by a phase meter and often with a resolution of 10^{-3} cycle or better. The initial observation only consists of the *fractional part* of the phase difference. When tracking is continued without loss of lock we still record a fractional part plus the integer number of cycles since the initial epoch. But the observation *does not provide us with the initial integer*—the ***ambiguity*** N.

The resolution of this ambiguity—counting the cycles without slips—is a crucial problem for GPS. This is particularly true for short observations, a few hours or less, of phases that are *doubly differenced.* The ambiguity problem is irrelevant for one-ways and single differenced observations because of the non-cancelling term $\varphi_i(t_0)$ in (14.16).

Precise GPS solutions depend highly on how well clock errors in satellites and receiver are eliminated. Realizing that light travels 3 mm in 0.01 nanoseconds it is evident that *usable GPS observations* on millimeter level require that we can keep time to within nanosecond level.

A standard observational method consists of phase observations by two or more receivers of four or more satellites at receiver ***epochs***. An epoch is an instant of time. For double differenced observations we only need to *know the sampling epoch to within a half microsecond.* This can almost always be obtained from P code pseudoranges, even with SA. Double differences require only good short-term stability of the receiver oscillator; the long-term stability need be no better than your wrist watch, see Table 14.1.

The original observation is a (one-way) difference between a receiver and a satellite. Next we make differences between two receivers and one satellite to eliminate errors common to the satellite. The most important error is from the satellite clock because of the dithering by SA. Next we make differences between two receivers and two satellites—***double differences***. This double difference eliminates errors that are common to both receivers. The largest errors are receiver clock offsets. Receiver clock drifts are not eliminated.

The differencing technique is effective for repairing a not too perfect synchronization between the two receivers. The serious multipath errors do not cancel as they depend on the specific reflecting surface at the receiver. A better antenna design anda better signal processing is the goal of current research.

The differencing may be continued: We can form the difference of two double differences between two epochs. The resulting ***triple differences*** (over receivers, satellites, and time) eliminate the ambiguities N. These are constant over time. But triple differences lose geometric strength because of the differencing over time. They are highly correlated and numerically less stable.

The GPS signals may be interrupted by buildings or other constructions causing *cycle slips.* When lock of signal is established again, the integer ambiguity most likely is wrong and has to be determined anew. Today detection of cycle slips is based on efficient algorithms which assume that the epoch-to-epoch ionospheric changes are small. The combination $\lambda_1 N_1 - \lambda_2 N_2$ therefore is sensitive to cycle slips. For time intervals smaller than a few seconds, the ionospheric variations are at the subcentimeter level. Cycle slips result in distance errors that are multiples of the wavelength λ_i—too large for serious science.

Signals at the high frequencies L_1 and L_2 propagate relatively easily through the ionosphere. The ionospheric delay is inversely proportional to the square of the frequency. For dual frequency receivers this is utilized to eliminate most of the ionospheric delay. If we succeed in estimating the correct integer ambiguity value N we talk about a ***fixed solution*** (of the ambiguity). A ***float solution*** will have an incorrect and non-integer N.

The final step of the data processing is a least squares estimation of point coordinates based on the estimated vectors. The estimation yields an assessment of quality of the observations and a possible detection of gross errors. The result of a GPS survey is a network of points whose position is controlled through a least squares estimation.

On these few pages we tried to give a brief introduction to how GPS is used for positional purposes. The following pages focus on computational aspects of GPS.

14.4 Receiver Position From Code Observations

The basic equation for a ***code observation*** (distance but not phase) looks like

$$P_i^k(t) = \rho_i^k + I_i^k + T_i^k + c\big(dt^k(t-\tau_i^k) - dt_i(t)\big) - e_i^k. \tag{14.5}$$

The ionospheric delay is I_i^k, the tropospheric delay is T_i^k, and c denotes the vacuum speed of light. The clock offsets are dt^k for the satellite and dt_i for the receiver; and e denotes an error. The distance between satellite k and receiver i—corrected for Earth rotation—is defined by

$$\rho_i^k = \big\| R_3(\omega_e\tau_i^k)\boldsymbol{r}^k(t-\tau_i^k)_{\text{geo}} - \boldsymbol{r}_i(t)_{\text{ECEF}} \big\|. \tag{14.6}$$

The matrix R_3 accounts for rotation by the angle $\omega_e\tau_i^k$ while the signal is traveling:

$$R_3(\omega_e\tau_i^k) = \begin{bmatrix} \cos(\omega_e\tau_i^k) & \sin(\omega_e\tau_i^k) & 0 \\ -\sin(\omega_e\tau_i^k) & \cos(\omega_e\tau_i^k) & 0 \\ 0 & 0 & 1 \end{bmatrix}. \tag{14.7}$$

The rotation matrix is necessary when using vectors referenced to an Earth centered and Earth fixed system (ECEF). The travel time from (the signal generator in) the satellite k to (the signal correlator in) the receiver i is denoted τ_i^k. The rotation rate of the Earth is ω_e. Position vectors in the ECEF system are denoted $\boldsymbol{r}(t)_{\text{ECEF}}$. The argument t emphasizes dependence on time.

Later we only deal with differenced observations from which most of the systematic errors are eliminated. With the eventual application in mind, we simply omit those errors.

Example 14.1 We want to demonstrate the practical use of the code observation equation (14.5). We linearize it. The Jacobian acts several times as observation matrix in a least squares estimation of receiver coordinates and receiver clock offset. Let

$$\begin{bmatrix} X^k \\ Y^k \\ Z^k \end{bmatrix} = R_3(\omega_e\tau_i^k)\boldsymbol{r}^k(t-\tau_i^k)_{\text{geo}} \qquad \text{and} \qquad \begin{bmatrix} X_i \\ Y_i \\ Z_i \end{bmatrix} = \boldsymbol{r}_i(t)_{\text{ECEF}}.$$

Omitting the refraction terms I_i^k and T_i^k, equation (14.5) now linearizes as

$$-\frac{X^k - X_i^0}{(P_i^k)^0}x_i - \frac{Y^k - Y_i^0}{(P_i^k)^0}y_i - \frac{Z^k - Z_i^0}{(P_i^k)^0}z_i + 1(c\,dt_i) = P_{i\,\text{obs}}^k - (P_i^k)^0 - \epsilon_i^k = b_i - \epsilon_i^k. \tag{14.8}$$

In a first approximation we put $\rho_i^k \approx (\rho_i^k)^0 = (P_i^k)^0$, which is the geometric distance as calculated from the preliminary coordinates of satellite and receiver. The number b_i denotes the correction to this preliminary value.

We assume that the code observation P_i^k is corrected for the clock offset dt^k of the satellite according to the broadcast ephemerides. The preliminary value of P_i^k is calculated from the preliminary coordinates (X_i^0, Y_i^0, Z_i^0) of the receiver. It is corrected for receiver clock offset dt_i, with first guess $dt_i = 0$. If the preliminary values are good the right side b_i is small. Note that the direction cosines use the components described by (14.6). The satellite coordinates must be rotated by the angle $\omega_e \tau_i^k$ around the 3-axis before they can be used as ECEF coordinates.

When more than four observations are available we estimate the position through a least squares procedure. It is reasonable to assume that the observations are uncorrelated. Then the weight matrix $C = \sigma_0^{-2} I$ is diagonal.

The linearized observation equations are of the type described by (14.8) and we arrange the unknowns as $\boldsymbol{x} = (x_i, y_i, z_i, c\,dt_i)$:

$$
A\boldsymbol{x} = \begin{bmatrix} -\dfrac{X^1 - X_i}{\rho_i^1} & -\dfrac{Y^1 - Y_i}{\rho_i^1} & -\dfrac{Z^1 - Z_i}{\rho_i^1} & 1 \\ -\dfrac{X^2 - X_i}{\rho_i^2} & -\dfrac{Y^2 - Y_i}{\rho_i^2} & -\dfrac{Z^2 - Z_i}{\rho_i^2} & 1 \\ \vdots & & & \\ -\dfrac{X^m - X_i}{\rho_i^m} & -\dfrac{Y^m - Y_i}{\rho_i^m} & -\dfrac{Z^m - Z_i}{\rho_i^m} & 1 \end{bmatrix} \begin{bmatrix} x_i \\ y_i \\ z_i \\ c\,dt_i \end{bmatrix} = \boldsymbol{b} - \boldsymbol{\epsilon}. \tag{14.9}
$$

Note that the factors c and dt_i stay together in one product. This is done for numerical reasons. The unknown $c\,dt_i$ has the same dimension as the other unknowns, namely length. However some people like to estimate the term as time. This can be done by using the unknown dt_i together with a reasonably small coefficient like $c \times 10^{-9}$ yielding dt_i in nanoseconds.

Row-wise the first three columns of A contain the direction cosines for the vector between satellite and receiver. Note that all entries of A have magnitude less than or equal to one. The least squares solution is

$$
\begin{bmatrix} \hat{x}_i \\ \hat{y}_i \\ \hat{z}_i \\ \widehat{c\,dt_i} \end{bmatrix} = (A^{\mathrm{T}} \Sigma^{-1} A)^{-1} A^{\mathrm{T}} \Sigma^{-1} \boldsymbol{b}. \tag{14.10}
$$

The code observations are considered independent with equal variance so $\boldsymbol{\epsilon} \sim N(0, \sigma^2 I)$. In other words the vector $\boldsymbol{\epsilon}$ has zero mean and covariance matrix $\sigma^2 I$. If this assumption is correct, (14.10) is simplified to

$$
\begin{bmatrix} \hat{x}_i \\ \hat{y}_i \\ \hat{z}_i \\ \widehat{c\,dt_i} \end{bmatrix} = (A^{\mathrm{T}} A)^{-1} A^{\mathrm{T}} \boldsymbol{b}. \tag{14.11}
$$

The final receiver coordinates are $\widehat{X}_i = X_i^0 + \hat{x}_i$, $\widehat{Y}_i = Y_i^0 + \hat{y}_i$, and $\widehat{Z}_i = Z_i^0 + \hat{z}_i$.

Example 14.2 (Dilution of Precision) The covariance matrix Σ_{ECEF} for $(x_i, y_i, z_i, c\,dt_i)$ contains information about the geometric quality of the position determination. It is smaller (and $\hat{x}$ is more accurate) when the satellites are well spaced. The trace of Σ_{ECEF} is a very compressed form of information—only a single number—and we cannot recover the confidence ellipsoid from it. According to its definition the trace is the sum of the four variances $\sigma_X^2+\sigma_Y^2+\sigma_Z^2+\sigma_{c\,dt}^2$, and it is independent of the coordinate system. In addition to the information on the geometry these variances involve the accuracy of the observations. This can be eliminated by division by σ_0^2.

We start from the covariance matrix of the least squares problem (14.9):

$$\Sigma_{\mathrm{ECEF}} = \left[\begin{array}{ccc|c} \sigma_X^2 & \sigma_{XY} & \sigma_{XZ} & \sigma_{X,c\,dt} \\ \sigma_{YX} & \sigma_Y^2 & \sigma_{YZ} & \sigma_{Y,c\,dt} \\ \sigma_{ZX} & \sigma_{ZY} & \sigma_Z^2 & \sigma_{Z,c\,dt} \\ \hline \sigma_{c\,dt,X} & \sigma_{c\,dt,Y} & \sigma_{c\,dt,Z} & \sigma_{c\,dt}^2 \end{array}\right]. \tag{14.12}$$

The propagation law transforms Σ_{ECEF} into the covariance matrix expressed in a local system with coordinates (E, N, U). The interesting 3 by 3 submatrix S of Σ_{ECEF} is shown in (14.12). After the transformation F, the submatrix becomes

$$\Sigma_{\mathrm{ENU}} = \begin{bmatrix} \sigma_E^2 & \sigma_{EN} & \sigma_{EU} \\ \sigma_{NE} & \sigma_N^2 & \sigma_{NU} \\ \sigma_{UE} & \sigma_{UN} & \sigma_U^2 \end{bmatrix} = F^{\mathrm{T}} S F. \tag{14.13}$$

The matrix F^{T} connects Cartesian coordinate differences in the local system (at latitude φ and longitude λ) and the ECEF system. The sequence (E, N, U) assures that both the local and the ECEF systems shall be right handed:

$$\begin{aligned} F^{\mathrm{T}} &= R_3(\pi) R_2(\varphi - \tfrac{\pi}{2}) R_3(\lambda - \pi) \\ &= \begin{bmatrix} 0 & -1 & 0 \\ 1 & 0 & 0 \\ 0 & 0 & 1 \end{bmatrix} \begin{bmatrix} \sin\varphi & 0 & \cos\varphi \\ 0 & 1 & 0 \\ -\cos\varphi & 0 & \sin\varphi \end{bmatrix} \begin{bmatrix} -\cos\lambda & -\sin\lambda & 0 \\ \sin\lambda & -\cos\lambda & 0 \\ 0 & 0 & 1 \end{bmatrix} \\ &= \begin{bmatrix} -\sin\lambda & \cos\lambda & 0 \\ -\sin\varphi\cos\lambda & -\sin\varphi\sin\lambda & \cos\varphi \\ \cos\varphi\cos\lambda & \cos\varphi\sin\lambda & \sin\varphi \end{bmatrix}. \end{aligned} \tag{14.14}$$

In practice we meet several forms of the ***dilution of precision*** (abbreviated DOP):

Geometric: $\quad \mathrm{GDOP} = \sqrt{\dfrac{\sigma_E^2+\sigma_N^2+\sigma_U^2+\sigma_{c\,dt}^2}{\sigma_0^2}} = \sqrt{\dfrac{\mathrm{tr}(\Sigma_{\mathrm{ECEF}})}{\sigma_0^2}}$

Horizontal: $\quad \mathrm{HDOP} = \sqrt{\dfrac{\sigma_E^2+\sigma_N^2}{\sigma_0^2}}$

Position: $\quad \mathrm{PDOP} = \sqrt{\dfrac{\sigma_E^2 + \sigma_N^2 + \sigma_U^2}{\sigma_0^2}} = \sqrt{\dfrac{\sigma_X^2 + \sigma_Y^2 + \sigma_Z^2}{\sigma_0^2}} = \sqrt{\dfrac{\mathrm{tr}(\Sigma_{\mathrm{ENU}})}{\sigma_0^2}}$

Time: $\quad \mathrm{TDOP} = \sigma_{c\,dt}/\sigma_0$

Vertical: $\quad \mathrm{VDOP} = \sigma_U/\sigma_0.$

Note that all DOP values are dimensionless. They multiply the range errors to give the position errors (approximately). Furthermore we have

$$\mathrm{GDOP}^2 = \mathrm{PDOP}^2 + \mathrm{TDOP}^2 = \mathrm{HDOP}^2 + \mathrm{VDOP}^2 + \mathrm{TDOP}^2.$$

The DOP values are especially useful when planning the observational periods. For this purpose it is better to use almanacs without high accuracy, rather than transmitted ephemerides. Almanac data allow for precomputation of satellite positions over several months and with sufficient accuracy (the ephemerides repesentation of the orbits is valid over a short period of time). Some satellite constellations are better than others and the knowledge of the time of best satellite coverage is a useful tool for anybody using GPS.

Experience shows that *good observations are achieved when PDOP < 5 and measurements come from at least five satellites.*

14.5 Combined Code and Phase Observations

GPS observations are characterized by a multitude of data collected at short time intervals varying from 1 s to say 30 s. The data are processed either in real time or in post processing mode by means of least squares or the filtering techniques in Chapter 15. Some computational methods focus on achieving high accuracies; they require longer times. Other methods concentrate on short processing time or real time availability.

Today the best geodetic receivers deliver dual frequency P code and phase observations. We shall restrict ourselves to the latest methods for processing data observed with static antennas. The accuracy is greatest with post processing.

A popular procedure for processing GPS observations uses ***double differences***. Such double differences are quite insensitive to shared changes of position of the two receivers, but they are very sensitive to changes of one receiver relative to the other. Therefore, ***double differences resemble classical distance and direction observations***.

In order to separate ***geometry***, i.e. the vector, from the ***ambiguities*** in the cycle count, we want a lot of data. With no a priori knowledge about the vector, it is difficult within a short period of time to distinguish the vector from the ambiguities. But as time goes by, hopefully only one vector fits the double differenced observations. The more satellites that are observed, the faster the vector can be determined. As soon as the vector is uniquely determined within a fraction of a cycle, an ambiguity can be fixed to its integer value. This is the key to optimal use of double differenced observations.

Usually the ambiguities and the vector are estimated according to the least-squares principle. That is, the best estimates of the ambiguities and the vector are those values which minimize the squared sum of residuals. Ambiguities are often treated as reals! As

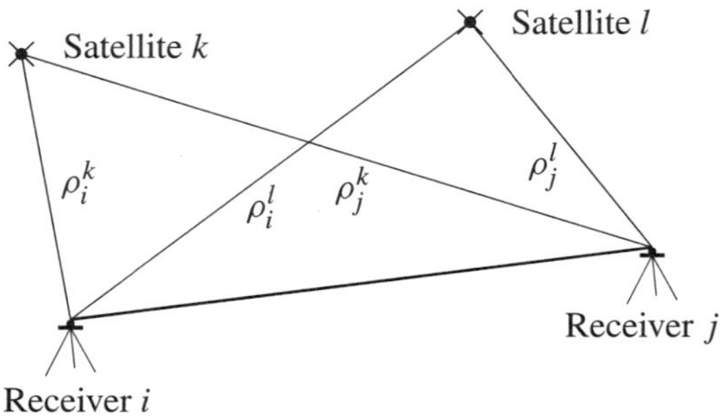

Figure 14.2 Double difference $\rho_i^k - \rho_i^l - \rho_j^k + \rho_j^l$. Two receivers observe two satellites at the same time.

the systematic errors are eliminated, the ambiguities tend to integer values. The classical case involves short baselines for which ambiguities are safely determined.

For reference, we shall use the terminology already introduced by Yang & Goad & Schaffrin (1994). The ***ideal pseudorange*** ρ^* is a combination of all nondispersive clock-based terms:

$$\rho_i^{*k} = \rho_i^k(t, t - \tau_i^k) + T_i^k + c\big(dt_i(t) - dt^k(t - \tau_i^k)\big). \tag{14.15}$$

If the ionospheric effect I_i^k were zero, ρ^* would be identical to the pseudorange ρ. The term T_i^k denotes the tropospheric delay. The terms in parentheses cancel when we use double differenced observations, because these are clock errors.

The basic equation for a ***phase observation*** at the same frequency as pseudorange is

$$\Phi_i^k(t) = \rho_i^k - I_i^k + T_i^k + c\big(dt_i(t) - dt^k(t - \tau_i^k)\big) + \lambda\big(\varphi_i(t_0) - \varphi^k(t_0)\big) + \lambda N_i^k - \epsilon_i^k. \tag{14.16}$$

The new terms are the ambiguities N_i^k between satellite k and receiver i, and the non-zero initial phases $\varphi^k(t_0)$ and $\varphi_i(t_0)$. Again we introduce the ideal pseudorange given above:

$$\Phi_i^k(t) = \rho_i^{*k} + \lambda N_i^{*k} \tag{14.17}$$

where $N_i^{*k} = N_i^k + \varphi_i(t_0) - \varphi^k(t_0)$. For double differences the two φ terms and the two dt terms cancel. This means that in case of double differences $N_{ij}^{*kl} = N_{ij}^{kl}$. This applies below in (14.20) and (14.21).

We shall demonstrate in detail how to make double differences of the observations. The original observation is a (one-way) difference between a receiver and a satellite. To eliminate the satellite clock error we make differences between two receivers and one satellite. Finally the double difference between two receivers and two satellites also eliminates receiver clock errors:

$$\text{on } L_1: \quad P_{1,ij}^{kl} = \rho_i^k - \rho_i^l - \rho_j^k + \rho_j^l + I_{ij}^{kl} + T_{ij}^{kl} - e_{1,ij}^{kl} \tag{14.18}$$

$$\text{on } L_2: \quad P_{2,ij}^{kl} = \rho_i^k - \rho_i^l - \rho_j^k + \rho_j^l + (f_1/f_2)^2 I_{ij}^{kl} + T_{ij}^{kl} - e_{2,ij}^{kl}. \tag{14.19}$$

Explicitly $T_{ij}^{kl} = (T_i^k - T_i^l) - (T_j^k - T_j^l)$, and similarly for I_{ij}^{kl}, N_{ij}^{kl}, and ϵ_{ij}^{kl}. Subscripts 1 and 2 refer to L_1 and L_2 with frequencies f_1 and f_2. In order to emphasize the influence of geometry we have left the ρ terms uncombined; they are also double differences. So are the phase observations:

$$\Phi_{1,ij}^{kl} = \rho_i^k - \rho_i^l - \rho_j^k + \rho_j^l - I_{ij}^{kl} + T_{ij}^{kl} + \lambda_1 N_{1,ij}^{kl} - \epsilon_{1,ij}^{kl} \tag{14.20}$$

$$\Phi_{2,ij}^{kl} = \rho_i^k - \rho_i^l - \rho_j^k + \rho_j^l - (f_1/f_2)^2 I_{ij}^{kl} + T_{ij}^{kl} + \lambda_2 N_{2,ij}^{kl} - \epsilon_{2,ij}^{kl}. \tag{14.21}$$

The ionospheric delay is frequency-dependent (dispersive); the factor $(f_1/f_2)^2$ multiplies I for the L_2 observations. The *group delay* is connected to the distances P while the *phase advance* is connected to the phase observations Φ. Thus, we see a reversed sign for I in (14.20) and (14.21). All observational errors are included in the e_{ij}^{kl} and ϵ_{ij}^{kl} terms.

The rest of this section deals exclusively with ***double differenced*** observations. We also omit the subscript and superscripts related to the receivers and satellites, since there are exactly two of each:

$$\begin{aligned} P_1 &= \rho^* + I - e_1 \\ \Phi_1 &= \rho^* - I + \lambda_1 N_1 - \epsilon_1 \\ P_2 &= \rho^* + (f_1/f_2)^2 I - e_2 \\ \Phi_2 &= \rho^* - (f_1/f_2)^2 I + \lambda_2 N_2 - \epsilon_2. \end{aligned} \tag{14.22}$$

Actually, we have $f_1/f_2 = 154/120 = 1.283\,333\ldots$. Equation (14.22) is transformed by Yang & Goad & Schaffrin (1994) into the elegant matrix equation

$$\begin{bmatrix} P_1 \\ \Phi_1 \\ P_2 \\ \Phi_2 \end{bmatrix} = \begin{bmatrix} 1 & 1 & 0 & 0 \\ 1 & -1 & \lambda_1 & 0 \\ 1 & (f_1/f_2)^2 & 0 & 0 \\ 1 & -(f_1/f_2)^2 & 0 & \lambda_2 \end{bmatrix} \begin{bmatrix} \rho^* \\ I \\ N_1 \\ N_2 \end{bmatrix} - \begin{bmatrix} e_1 \\ \epsilon_1 \\ e_2 \\ \epsilon_2 \end{bmatrix}. \tag{14.23}$$

When all e and ϵ values are set to zero, we can solve the four equations to find the four unknowns. This determines the ideal pseudorange ρ^*, the instantaneous ionospheric delay I, and the ambiguities N_1 and N_2.

The standard deviation of phase noise ϵ is a few millimeters, while standard deviation of the pseudorange error e depends on the quality of the receiver. L_1 C/A code pseudoranges can have noise values up to 2–3 m. This is due to the slow chipping rate (which is another term for frequency). The P code has a frequency of 10.23 Mhz, i.e. a sequence of 10.23 million binary digits or *chips* per second. The chipping rate of the P code is ten times more frequent and this implies a noise level of 10–30 cm. A small pseudorange standard deviation is critical to quickly determining the ambiguities on L_1 or L_2.

14.6 Weight Matrix for Differenced Observations

Suppose we are observing three (or m) satellites at an epoch. We can set up two (or $m-1$) linearly independent double differenced observations. Since the observations are correlated, *the weight matrix is no longer diagonal.* Therefore this weight matrix must be handled with special care when the normal equations are formed. We now study this problem in detail.

Let the covariance matrix for the original phase observations be $\Sigma_b = \sigma_0^2 I$. The observations are given with equal weight for phases (and they are independent). The unit for the standard deviation σ is length; as an example $\sigma = 0.01$ m.

First of all we have the one-way phase difference observations

$$\boldsymbol{b}_s = \begin{bmatrix} \Phi_i^k \\ \Phi_j^k \\ \Phi_i^l \\ \Phi_j^l \end{bmatrix}. \tag{14.24}$$

We shall determine the covariance matrix of two distinct single differences:

$$\boldsymbol{s} = \begin{bmatrix} \Phi_{ij}^k \\ \Phi_{ij}^l \end{bmatrix} = \begin{bmatrix} -1 & 1 & 0 & 0 \\ 0 & 0 & -1 & 1 \end{bmatrix} \begin{bmatrix} \Phi_i^k \\ \Phi_j^k \\ \Phi_i^l \\ \Phi_j^l \end{bmatrix} = D_s \boldsymbol{b}_s. \tag{14.25}$$

The law of variance propagation (using s to indicate single differences) yields

$$\Sigma_s = D_s \Sigma_b D_s^{\mathrm{T}} = \sigma_0^2 D_s D_s^{\mathrm{T}} = \sigma_0^2 \begin{bmatrix} 2 & 0 \\ 0 & 2 \end{bmatrix} = 2\sigma_0^2 I. \tag{14.26}$$

Single differences over a single baseline are uncorrelated. Single differences between different epochs are uncorrelated, too. However single differences can be correlated when more than one baseline is considered and they contain the same one-way phase observations (using subscript c to indicate the common receiver j):

$$\boldsymbol{c} = \begin{bmatrix} \Phi_{ij}^k \\ \Phi_{jg}^k \end{bmatrix} = \begin{bmatrix} -1 & 1 & 0 \\ 0 & -1 & 1 \end{bmatrix} \begin{bmatrix} \Phi_i^k \\ \Phi_j^k \\ \Phi_g^k \end{bmatrix} = D_c \boldsymbol{b}_s. \tag{14.27}$$

Now the covariance matrix is not diagonal:

$$\Sigma_c = D_c \Sigma_b D_c^{\mathrm{T}} = \sigma_0^2 D_c D_c^{\mathrm{T}} = \sigma_0^2 \begin{bmatrix} 2 & -1 \\ -1 & 2 \end{bmatrix}. \tag{14.28}$$

Then we compute the covariance for double differences between satellites k, l, and m. Satellite k is chosen as ***reference satellite***:

$$\boldsymbol{d} = \begin{bmatrix} \Phi_{ij}^{kl} \\ \Phi_{ij}^{km} \end{bmatrix} = \begin{bmatrix} 1 & -1 & -1 & 1 & 0 & 0 \\ 1 & -1 & 0 & 0 & -1 & 1 \end{bmatrix} \begin{bmatrix} \Phi_i^k \\ \Phi_j^k \\ \Phi_i^l \\ \Phi_j^l \\ \Phi_i^m \\ \Phi_j^m \end{bmatrix} = D_d \boldsymbol{b}_s. \tag{14.29}$$

The covariance matrix (using subscript d to indicate double difference) is

$$\Sigma_d = D_d \Sigma_b D_d^{\mathrm{T}} = \sigma_0^2 D_d D_d^{\mathrm{T}} = \sigma_0^2 \begin{bmatrix} 4 & 2 \\ 2 & 4 \end{bmatrix}. \tag{14.30}$$

Again we see a nondiagonal covariance matrix: ***double difference observations even over a single baseline are correlated***.

Finally comes the correlation of ***triple differences*** between epochs t and $t+1$:

$$\boldsymbol{t} = D_t \boldsymbol{b}_s = \left[\begin{array}{cccc|cccc|cccc} 1 & -1 & -1 & 1 & -1 & 1 & 1 & -1 & 0 & 0 & 0 & 0 \\ 0 & 0 & 0 & 0 & 1 & -1 & -1 & 1 & -1 & 1 & 1 & -1 \end{array}\right] \boldsymbol{b}_s$$

where now $\boldsymbol{b}_s$ involves three epochs (time differences at two epochs):

$$\boldsymbol{b}_s = \Big[\ \Phi_i^k(t) \quad \Phi_j^k(t) \quad \Phi_i^l(t) \quad \Phi_j^l(t) \quad \Phi_i^k(t+1) \quad \Phi_j^k(t+1) \quad \Phi_i^l(t+1)$$
$$\Phi_j^l(t+1) \quad \Phi_i^k(t+2) \quad \Phi_j^k(t+2) \quad \Phi_i^l(t+2) \quad \Phi_j^l(t+2)\ \Big]^{\mathrm{T}}.$$

The covariance matrix for the triple difference is

$$\Sigma_t = D_t \Sigma_b D_t^{\mathrm{T}} = \sigma_0^2 D_t D_t^{\mathrm{T}} = \sigma_0^2 \begin{bmatrix} 8 & -4 \\ -4 & 8 \end{bmatrix}. \tag{14.31}$$

So we conclude that ***triple differences are also correlated***. Unlike single and double differences, this correlation extends over epochs as well as within a single epoch.

14.7 Geometry of the Ellipsoid

Ellipsoidal System

In geodesy the reference surface is an *ellipsoid of revolution*. This is a reasonable approximation to the Earth (but of course not perfect). It is described by the rotation of an ellipse around its minor axis (the Z-axis). The ellipse is given by

$$\frac{p^2}{a^2} + \frac{Z^2}{b^2} = 1 \qquad \text{with} \qquad a > b. \tag{14.32}$$

This *meridian ellipse* in Figure 14.3 is a slice taken through the North Pole and the South Pole. Often the dimensionless quantities of *flattening* f or *second eccentricity* e' are used to describe the shape of the ellipse:

$$f = \frac{a-b}{a} \qquad \text{and} \qquad e'^2 = \frac{a^2-b^2}{b^2} = \frac{(2-f)f}{(1-f)^2}. \tag{14.33}$$

The ratio of the semi axes is

$$\frac{b}{a} = 1 - f = \frac{1}{\sqrt{1+e'^2}}. \tag{14.34}$$

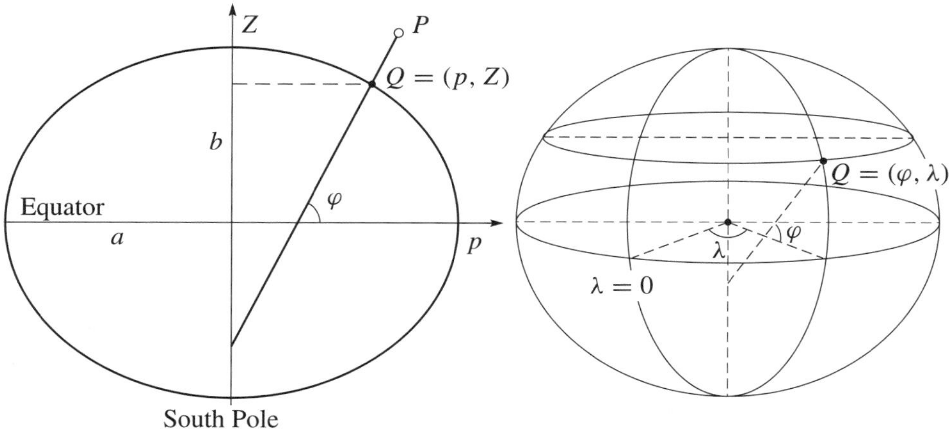

Figure 14.3 The meridian ellipse and the ellipsoid of revolution.

Finally we often use the radius of curvature $c = a^2/b$ at the poles:

$$c = \frac{a^2}{b} = a\sqrt{1 + e'^2} = b\left(1 + e'^2\right). \tag{14.35}$$

The flattening has a very small value $f = 0.003\,35$. A globe of radius $a = 1$ m would only be flattened by 7 mm. In this case $e' \approx 0.082\,09$.

A point Q on the ellipsoid is determined by (φ, λ) = (latitude, longitude). The *geographical latitude* φ is the angle between the normal at Q and the plane of the Equator. For an ellipse, the normal at Q does *not* go through the center point. (This is the price we must pay for flattening; φ is not the angle from the center of the ellipse.) The *geographical longitude* λ is the angle between the plane of the meridian of Q and the plane of a reference meridian (through Greenwich).

If we connect all points on the ellipsoid with equal latitude φ, this closed curve is φ = constant; it is called a *parallel*. In a similar way all points with equal longitude lie on the parameter curve λ = constant, which is a *meridian*. Meridians and parallels constitute the geographical net (or grid). The meridians are ellipses while the parallels are circles. The geographical coordinates can be understood both as angles and as surface coordinates.

Table 14.3 Parameters for reference ellipsoids used in North America

Datum	Reference Ellipsoid	a (in meters)	$f = (a - b)/a$
NAD 27	Clarke 1866	6 378 206.4	1/294.9786982
NAD 83	Geodetic Reference System 1980	6 378 137	1/298.257222101
WGS 84	World Geodetic System 1984	6 378 137	1/298.257223563

Geographical longitude is reckoned from the meridian $\lambda_0 = 0$ of Greenwich. A surface curve through the point Q makes an angle with the meridian; this angle is called the *azimuth* A of the curve. The azimuth of a parallel is $A = \pi/2$ or $3\pi/2$. Normally A is computed *clockwise* from the northern branch of the meridian.

The reader will remember that it was Ptolemy, looking at the shape of the Mediterranean Sea, who gave us the words *longitude* (in the long direction) and *latitude* (across).

The Ellipsoidal System Extended to Outer Space

To determine a point P above the surface of the Earth, we use geographical coordinates φ, λ and the *height* h. The height above the ellipsoid is measured along a perpendicular line. Let a point Q on the ellipsoid have coordinates φ, λ. In a geocentric X, Y, Z-system, with X-axis at longitude $\lambda = 0$, the point Q with height $h = 0$ has coordinates

$$X = N\cos\varphi\cos\lambda$$

$$Y = N\cos\varphi\sin\lambda$$

$$Z = (1-f)^2 N\sin\varphi.$$

The distance to the Z-axis along the normal at Q is $N = a\big/\sqrt{1 - f(2-f)\sin^2\varphi}$. This is the radius of curvature in the direction perpendicular to the meridian (the prime vertical).

The formula for N results from substitution of $p = N\cos\varphi$ and (14.34) into (14.32). When P is above the ellipsoid, we must add to Q a vector of length h along the normal. From spatial geographical (φ, λ, h) we get spatial Cartesian (X, Y, Z):

$$X = (N+h)\cos\varphi\cos\lambda \tag{14.36}$$

$$Y = (N+h)\cos\varphi\sin\lambda \tag{14.37}$$

$$Z = \big((1-f)^2 N + h\big)\sin\varphi. \tag{14.38}$$

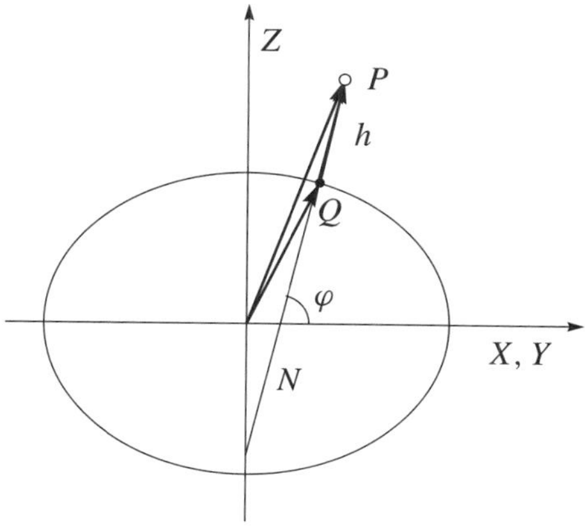

Figure 14.4 Conversion between (φ, λ, h) and Cartesian (X, Y, Z).

Example 14.3 Let P be given by the coordinates (φ, λ, h) in the WGS 84 system:

$$\varphi = 40^\circ\, 07'\, 04.595\,51''$$
$$\lambda = 277^\circ\, 01'\, 10.221\,76''$$
$$h = 231.562\,\text{m}$$

We seek the (X, Y, Z) coordinates of P. The result is achieved by the M-file g2c:

$$X = 596\,915.961\,\text{m}$$
$$Y = -4\,847\,845.536\,\text{m}$$
$$Z = 4\,088\,158.163\,\text{m}.$$

The reverse problem—compute (φ, λ, h) from (X, Y, Z)—requires an iteration for φ and h. Directly $\lambda = \arctan(Y/X)$. There is quick convergence for $h \ll N$, starting at $h = 0$:

φ from h (14.38):
$$\varphi = \arctan\Bigl(\frac{Z}{\sqrt{X^2+Y^2}}\Bigl(1 - \frac{(2-f)fN}{N+h}\Bigr)^{-1}\Bigr) \tag{14.39}$$

h from φ (14.36)–(14.37):
$$h = \frac{\sqrt{X^2+Y^2}}{\cos\varphi} - N. \tag{14.40}$$

For large h (or φ close to $\pi/2$) we recommend the procedure given in the M-file c2gm.

Example 14.4 Given the same point as in Example 14.3, the M-file c2gm solves the reverse problem. The result agrees with the original values for (φ, λ, h) in Example 14.3.

Using GPS for positioning we have to convert coordinates from Cartesian (X, Y, Z) to spatial geographical (φ, λ, h) and vice versa as described in (14.39)–(14.40) and (14.36)–(14.38). The immediate result of a satellite positioning is a set of X, Y, Z-values which we most often want to convert to φ, λ, h-values.

14.8 The Direct and Reverse Problems

A basic GPS output is the coordinates of a single point. Between two points we compute distance and azimuth. Or from a known point we calculate "polar coordinates" to another given point.

Traditionally the problem of determining (φ_2, λ_2) for a new point, from a known (φ_1, λ_1) and distance S and azimuth A, has been called the *direct problem*. The *reverse problem* is to compute distance S and the two azimuths A_1 and A_2 between known points with given coordinates (φ_1, λ_1) and (φ_2, λ_2).

Through ages those two problems have attracted computationally oriented persons. We shall only mention a solution given by Gauss. Its limitation is that the distance S should be smaller than 50 km. For longer distances we recommend the formulas of Bessel.

We quote the basic equations called the ***mid latitude formulas***:

$$S \sin A = N l \cos\varphi\Big(1 - \frac{1}{24}(l \sin\varphi)^2 + \frac{1+\eta^2-9\eta^2t^2}{24V^4}b^2\Big) \tag{14.41}$$

$$S \cos A = N b' \cos(l/2)\Big(1 + \frac{1-2\eta^2}{24}(l \cos\varphi)^2 + \frac{\eta^2(1-t^2)}{8V^4}b'^2\Big) \tag{14.42}$$

$$\Delta A = l \sin\varphi\Big(1 + \frac{1+\eta^2}{12}(l \cos\varphi)^2 + \frac{3+8\eta^2}{24V^4}b^2\Big). \tag{14.43}$$

Here $\lambda_2 - \lambda_1 = l$ and $\varphi_2 - \varphi_1 = b$ and $b/V^2 = b'$. Furthermore according to geodetic tradition we have set $t = \tan\varphi$, $\eta = e'\cos\varphi$, $V^2 = 1+\eta^2$, $V = c/N$ and $\Delta A = A_2 - A_1 \pm \pi$.

In the direct problem we seek φ_2 and λ_2 and A_2. The mid latitude formulas (14.41)–(14.43) can only be solved iteratively for l and b. The correction terms $l\cos\varphi$ and $l\sin\varphi$ remain unchanged during the iteration. The changing values for l and b can be determined with sufficient accuracy from the main terms outside the parentheses:

$$l = \frac{S}{N}\frac{\sin A}{\cos\varphi}\Big(1 + \cdots (l\sin\varphi)^2 - \cdots b^2\Big)$$

$$b' = \frac{S}{N}\frac{\cos A}{\cos(l/2)}\Big(1 - \cdots (l\cos\varphi)^2 - \cdots b'^2\Big)$$

$$\Delta A = l\sin\varphi\big(1 + \cdots (l\cos\varphi)^2 + \cdots b^2\big).$$

In the second iteration φ_2 and λ_2 have accuracy better than $0.000\,1''$ and A_2 better than $0.001''$.

The M-file gauss1 solves the direct problem. Equations (14.41) and (14.42) allow for an easy solution of the reverse problem, because they directly yield $S\sin A$ and $S\cos A$. The M-file gauss2 does the job.

14.9 Geodetic Reference System 1980

Table 14.3 lists values for a and f that have been used throughout time to describe the reference ellipsoid. For global use the latest values (adopted at the General Assembly of the International Union of Geodesy and Geophysics, Canberra 1980) are

$$\begin{aligned} a &= 6\,378\,137\,\text{m} \\ kM &= 3.986\,005 \times 10^{14}\,\text{m}^3/\text{s}^2 \\ J_2 &= 108\,263 \times 10^{-8} \\ \omega &= 7.292\,115 \times 10^{-5}\text{rad/s}. \end{aligned}$$

Here kM is the product of the universal constant of attraction and the mass of the Earth. The coefficient J_2 is given from the spherical harmonic series expansion of the normal gravity field of the Earth. The number J_2 is closely related to f and is also named the dynamic shape factor of the Earth. Finally ω denotes the mean rotation rate of the Earth. For the flattening we may derive the value $1/f = 298.257\,222\,101$ from these values.

Hence we have described the surface to which the ellipsoidal height h is related.

Example 14.5 In Example 14.3 we considered a point with coordinates

$$(X, Y, Z) = (596\,915.961,\ -4\,847\,845.536,\ 4\,088\,158.163).$$

For this terrain point, c2g.m computes (φ, λ) on Clarke's ellipsoid of 1866:

$$\varphi = 40^\circ\,07'\,12.192\,79''$$
$$\lambda = 277^\circ\,01'\,10.221\,77''$$
$$h = 260.754\,\text{m}.$$

We compare with the WGS 84 values given earlier:

$$\varphi = 40^\circ\,07'\,04.595\,52''$$
$$\lambda = 277^\circ\,01'\,10.221\,77''$$
$$h = 231.562\,\text{m}.$$

Due to rotational symmetry there is no difference in λ. The height h and the latitude φ are changed substantially and of the same order of magnitude as the change in a, namely 69.4 m. To the accuracy we are using here, the computations in relation to GRS 80 would agree with the WGS 84 result.

14.10 Geoid, Ellipsoid, and Datum

The geoid is a surface that is defined physically, not geometrically. Contour maps show the geoid as an irregular surface with troughs and ripples caused by local mass distribution.

The center of the geoid coincides with the center of the Earth. It is an equipotential surface. One can imagine the geoid if the Earth was totally covered by water. Theoretically this sea surface would be at constant potential, since the water would stream if there were any difference of height. The actual sea surface deviates a little from this true equipotential surface because of variation in sea temperature, salinity, and sea currents.

The geoid is the reference surface for leveling. In most countries the mean of the geoid is fixed to coincide with mean sea level (registrated by mareographs). This does not imply that height zero will coincide with mean sea level at the sea shore. In reality there can be up to 1 m in difference between zero level of the leveling network and the mean sea level at a given point. The geoid is very irregular and it is impossible to model it exactly. Geodesists have spent much time on developing approximations by spherical harmonics.

At a terrain point P there exists a basic connection between the *ellipsoidal height* h which is a GPS determined quantity, the *geoidal undulation* N which is the height of the geoid above the ellipsoid, and the *orthometric height* H:

$$h = H + N. \tag{14.44}$$

In daily life we hardly think about the assumptions made for measuring distances and directions. It seems obvious to refer all our measurements to a "horizontal" plane. This

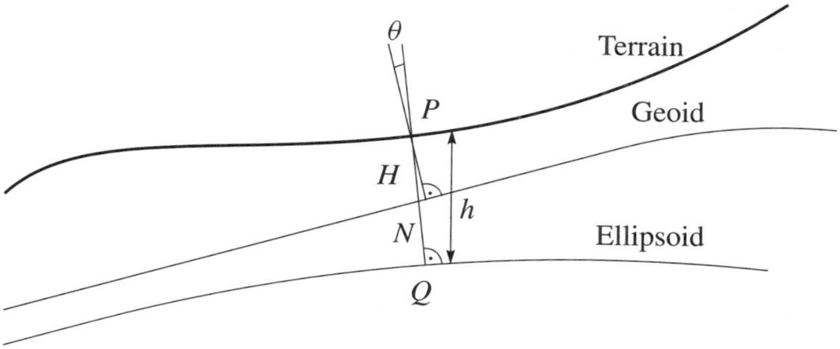

Figure 14.5 Ellipsoidal height h, geoid undulation N and orthometric height H.

plane is perpendicular to the (local) plumb line. However this is no unique description; a point just 100 m away has a different tangent plane. This is because the plumb lines at two neighboring points converge. They are *not parallel*.

Unfortunately, this convergence is not easy to describe by a simple formula. Its variation is irregular. But a major part can be separated. This regular part corresponds to introducing a reference ellipsoid given by adequate parameters and with a reasonable position in relation to the Earth. Note that our original problem was of *local* nature and we now give a solution of *global* nature.

Most geodetic observations refer directly to the plumb line, because our instruments are set up by means of levels. One exception is electronically measured distances. To start we most often reduce the actual observations to the reference ellipsoid. This requires a knowledge of the orientation of the ellipsoid in relation to the plumb line. This description may be given by the angle between the plumb line at P and the ellipsoidal normal, and the distance of P from the ellipsoid. The angle is called the ***deflection of the vertical***. Usually it is split into north-south and east-west components ξ and η. Finally the geoid undulation N between ellipsoid and geoid is related to the distance from the ellipsoid h and orthometric height H of the point as given in (14.44). Often N is an unknown quantity that has a global variation up to one hundred meters. GPS measurements furnish us with h. So if N is known we can determine the height H. That is leveling!

The relative position between the geoid and the ellipsoid at P is described by three quantities: ξ, η, and N. If we do not know better they often are put equal to zero.

Additionally we have to describe the size and shape of the ellipsoid. This requires two parameters, usually the semi major axis a and the flattening f. All in all five parameters describe the relation between the rotational ellipsoid and the plumb line at a given terrain point P. Such a set of parameters is called a ***datum***.

Today the three parameters ξ, η, N are often replaced by equivalent quantities t_X, t_Y, t_Z. These are translations of the center of the ellipsoid relative to the center of the Earth. So modern datum descriptions include the five quantities: a, f, t_X, t_Y, and t_Z.

Such a datum can be extended to connected areas of land. But if the geodesist has to cross larger areas covered by water, it is impossible by classical methods to transfer knowledge about the deflection of the vertical. So a datum can at most cover a continent.

However, GPS opens the possibility for merging all continental data. A realization of this is the WGS 84 geodetic system, described below.

To be concrete, we introduce the unit vector $\boldsymbol{u}$ along the plumb line:

$$\boldsymbol{u} = \begin{bmatrix} \cos\Phi\cos\Lambda \\ \cos\Phi\sin\Lambda \\ \sin\Phi \end{bmatrix}$$

where (Φ, Λ) denote the astronomical coordinates of the surface point P. Those coordinates can be determined by classical astronomical observations.

Any vector $\delta\boldsymbol{x}$ from P can be described by its slope distance s, zenith distance z and azimuth α (from the northern branch of the meridian). The representation in Cartesian coordinates is

$$\delta\boldsymbol{x} = \begin{bmatrix} \delta x_1 \\ \delta x_2 \\ \delta x_3 \end{bmatrix} = s \begin{bmatrix} \sin\alpha\sin z \\ \cos\alpha\sin z \\ \cos z \end{bmatrix}$$

where δx_1 is east, δx_2 is north, and δx_3 is in direction of $\boldsymbol{u}$. This local ***topocentric system*** is based on the *surface* of the Earth (the topography). It can be related to a ***geocentric system*** with origin at the *center* of the Earth. This is an Earth centered, Earth fixed system—***ECEF*** for short. Now the axes point to the Greenwich meridian at the Equator, to the east, and to the conventional terrestrial pole (CTP).

The transformation from topocentric $\delta\boldsymbol{x} = F\delta\boldsymbol{x}$ to geocentric $\delta\boldsymbol{X}$ is by an orthogonal matrix F:

$$F = P_2 R_3(\pi - \Lambda) R_2\left(\tfrac{\pi}{2} - \Phi\right). \tag{14.45}$$

Explicitly this product of plane rotations gives the unit vectors $\boldsymbol{e}, \boldsymbol{n}, \boldsymbol{u}$:

$$F = \begin{bmatrix} -\sin\Lambda & -\sin\Phi\cos\Lambda & \cos\Phi\cos\Lambda \\ \cos\Lambda & -\sin\Phi\sin\Lambda & \cos\Phi\sin\Lambda \\ 0 & \cos\Phi & \sin\Phi \end{bmatrix} = \begin{bmatrix} \boldsymbol{e} & \boldsymbol{n} & \boldsymbol{u} \end{bmatrix}. \tag{14.46}$$

The inverse transformation is

$$\delta\boldsymbol{x} = F^{-1}(\delta\boldsymbol{X}) = F^{\mathrm{T}}(\delta\boldsymbol{X}). \tag{14.47}$$

Over and over again equations (14.45) and (14.47) are used to combine GPS and terrestrial observations. Note that we use small letters $\boldsymbol{x}$ in the topocentric system (small coordinate values around the surface point) and capital letters $\boldsymbol{X}$ in the geocentric system (large coordinate values around the Earth's center). In (14.14), we use F with geographical rather than astronomical coordinates. We shall now describe this difference.

To perform computations relative to the ellipsoid, the situation is a little different. We repeat the connection between astronomical and geographical coordinates and the definition of *components of the deflection of the vertical*:

$$\xi = \Phi - \varphi \tag{14.48}$$

$$\eta = (\Lambda - \lambda)\cos\varphi. \tag{14.49}$$

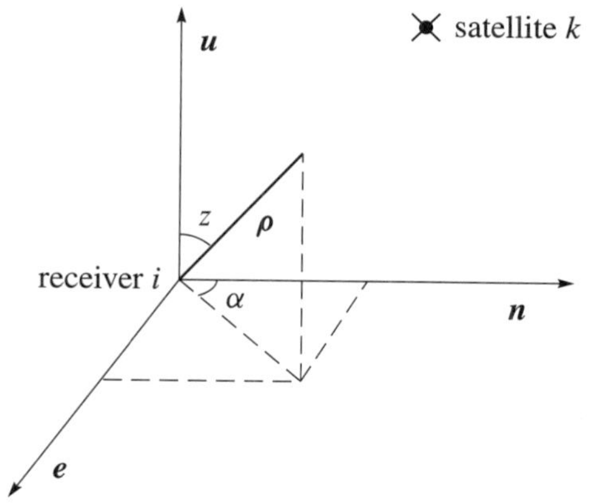

Figure 14.6 Zenith distance z and azimuth α in the topocentric system $(\boldsymbol{e}, \boldsymbol{n}, \boldsymbol{u})$.

The geographical coordinates (φ, λ) and the height h above the ellipsoid can be transformed to Cartesian coordinates. The new (X, Y, Z) system has origin at the center of the ellipsoid, 3-axis coinciding with the polar axis of revolution, 1-axis in the plane of the zero meridian, and 2-axis pointing eastwards. According to (14.37) the transformation is

$$\begin{aligned} X &= (N + h)\cos\varphi\cos\lambda \\ Y &= (N + h)\cos\varphi\sin\lambda \\ Z &= \bigl((1 - f)^2 N + h\bigr)\sin\varphi. \end{aligned} \tag{14.50}$$

From now on we will use geographical coordinates (φ, λ).

Example 14.6 To emphasize the fundamental impact of the matrix F we shall determine the elevation angle for a satellite. The local topocentric system uses three unit vectors $(\boldsymbol{e}, \boldsymbol{n}, \boldsymbol{u})$ = (east, north, up). Those are the columns of F. The vector $\boldsymbol{r}$ between satellite k and receiver i is

$$\boldsymbol{r} = \bigl(X^k - X_i, Y^k - Y_i, Z^k - Z_i\bigr).$$

The unit vector in this satellite direction is $\boldsymbol{\rho} = \boldsymbol{r}/\|\boldsymbol{r}\|$. Then Figure 14.6 gives

$$\begin{aligned} \boldsymbol{\rho}^{\mathrm{T}}\boldsymbol{e} &= \sin\alpha\sin z \\ \boldsymbol{\rho}^{\mathrm{T}}\boldsymbol{n} &= \cos\alpha\sin z \\ \boldsymbol{\rho}^{\mathrm{T}}\boldsymbol{u} &= \cos z. \end{aligned}$$

From this we determine α and z. Especially we have $\sin h = \cos z = \boldsymbol{\rho}^T\boldsymbol{u}$ for the elevation angle h. The angle h or rather $\sin h$ is an important parameter for any procedure calculating the tropospheric delay, cf. the M-file tropo.

Furthermore the quantity $\sin h$ has a decisive role in planning observations: When is h larger than 15°, say? Those are the satellites we prefer to use in GPS. Many other computations and investigations involve this elevation angle h.

14.11 World Geodetic System 1984

The ellipsoid in WGS 84 is defined through four parameters, with specified variances:

1. the semi major axis $a = 6\,378\,137\,\text{m}$ ($\sigma_a = 2\,\text{m}$)
2. the Earth's gravitational constant (including the mass of the Earth's atmosphere) $kM = 3\,986\,005 \times 10^8\,\text{m}^3/\text{s}^3$ ($\sigma_{kM} = 0.6 \times 10^8\,\text{m}^3/\text{s}^3$)
3. the normalized second degree zonal coefficient of the gravity potential $\overline{C}_{20}$
4. the Earth's rotational rate $\omega = 7\,292\,115 \times 10^{-11}\text{rad/s}$ ($\sigma_\omega = 15 \times 10^{-11}\text{rad/s}$).

The International Astronomical Union uses $\omega_e = 7\,292\,115.146\,7 \times 10^{-11}\text{rad/s}$, with four extra digits, together with a new definition of time. In order to maintain consistency with GPS it is necessary to use ω_e instead of ω. The speed of light in vacuum is taken as

$$c = 299\,792\,458\,\text{m/s} \qquad \text{with} \qquad \sigma_c = 1.2\,\text{m/s}.$$

Conceptually WGS 84 is a very special datum as it includes a model for the gravity field. The description is given by spherical harmonics up to degree and order 180. This adds 32 755 more coefficients to WGS 84 allowing for determination of the global features of the geoid. A truncated model ($n = m = 18$) of the geoid is shown in Figure 14.7. For a more detailed description see Department of Defense (1991).

In North America the transformation from NAD 27 to WGS 84 is given as

$$\begin{bmatrix} X_{\text{WGS}\,84} \\ Y_{\text{WGS}\,84} \\ Z_{\text{WGS}\,84} \end{bmatrix} = \begin{bmatrix} X_{\text{NAD}\,27} + 9\,\text{m} \\ Y_{\text{NAD}\,27} - 161\,\text{m} \\ Z_{\text{NAD}\,27} - 179\,\text{m} \end{bmatrix}.$$

A typical datum transformation into WGS 84 only includes changes in the semi major axis of the ellipsoid and its flattening and three translations of the origin of the ellipsoid.

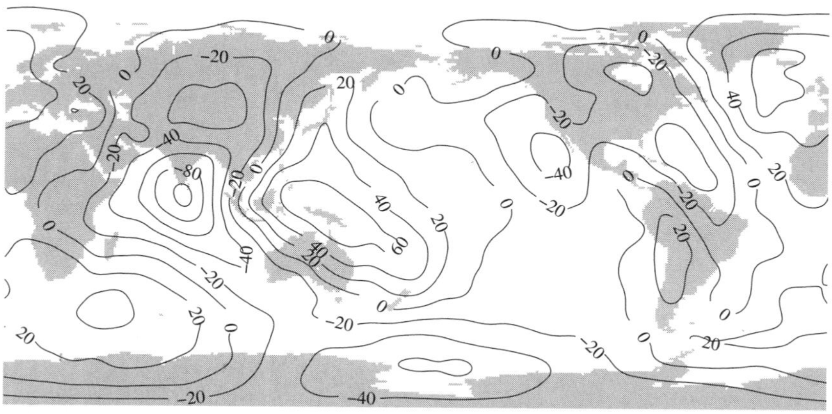

Figure 14.7 The WGS 84 geoid for $n = m = 18$. The contour interval is 20 m.

WGS 84 is a global datum, allowing us to transform between regions by means of GPS. Yet, we finish by a warning about such transformation formulas. They shall only be used with caution. They are defined to an accuracy of a few meters, though the mathematical relationship is exact. The importance of WGS 84 is undoubtedly to provide a *unified global datum.*

14.12 Coordinate Changes From Datum Changes

In geodetic practice it is today common to have coordinates derived from GPS and also from traditional terrestrial methods. For computations we convert the local topocentric coordinates to geocentric coordinates, which can be compared with the GPS derived coordinates. Thus physical control points may have two sets of coordinates, one derived from classical methods and one derived from GPS.

The connection between topocentric and geocentric coordinates is established by transformation formulas. The most general transformation includes rotations, translations, and a change of scale. *This type of transformation is only established between Cartesian systems.* It is described by seven parameters: three translations t_X, t_Y, t_Z, three rotations $\epsilon_X, \epsilon_Y, \epsilon_Z$, and a change of scale k. We assume that the transformation is infinitesimal. Our starting point is (14.50) which for small changes becomes

$$\boldsymbol{T} = \begin{bmatrix} 1 & \epsilon_Z & -\epsilon_Y \\ -\epsilon_Z & 1 & \epsilon_X \\ \epsilon_Y & -\epsilon_X & 1 \end{bmatrix} \begin{bmatrix} (N+h)\cos\varphi\cos\lambda \\ (N+h)\cos\varphi\sin\lambda \\ \big((1-f)^2N+h\big)\sin\varphi \end{bmatrix} + k \begin{bmatrix} X \\ Y \\ Z \end{bmatrix} + \begin{bmatrix} t_X \\ t_Y \\ t_Z \end{bmatrix}. \tag{14.51}$$

A first order Taylor expansion of the radius of curvature in the prime vertical (that is the direction orthogonal to the meridian) yields

$$(1-f)^2N \approx (1-2f+f^2)a(1+f\sin^2\varphi) \approx a(1+f\sin^2\varphi-2f).$$

Equation (14.51) can be linearized using $\boldsymbol{t} = (t_X, t_Y, t_Z)$ to

$$\begin{aligned} \boldsymbol{T} &= \begin{bmatrix} 1 & \epsilon_Z & -\epsilon_Y \\ -\epsilon_Z & 1 & \epsilon_X \\ \epsilon_Y & -\epsilon_X & 1 \end{bmatrix} \begin{bmatrix} \big(a(1+f\sin^2\varphi)+h\big)\cos\varphi\cos\lambda \\ \big(a(1+f\sin^2\varphi)+h\big)\cos\varphi\sin\lambda \\ \big(a(1+f\sin^2\varphi-2f)+h\big)\sin\varphi \end{bmatrix} + k \begin{bmatrix} X \\ Y \\ Z \end{bmatrix} + \boldsymbol{t} \\ &= R\,\boldsymbol{X} + k\,\boldsymbol{X} + \boldsymbol{t}. \end{aligned} \tag{14.52}$$

To find the linear dependency between the variables we differentiate $\boldsymbol{T}$:

$$d\boldsymbol{T} = R\,d\boldsymbol{X} + dR\,\boldsymbol{X} + dk\,\boldsymbol{X} + k\,d\boldsymbol{X} + d\boldsymbol{t}.$$

The product $k\,d\boldsymbol{X}$ is of second order. To first order we have $R\,d\boldsymbol{X} \approx d\boldsymbol{X}$ and thus

$$d\boldsymbol{T} = d\boldsymbol{X} + dR\,\boldsymbol{X} + dk\,\boldsymbol{X} + d\boldsymbol{t}.$$

This yields the corrections $d\lambda$, $d\varphi$, dh from changes in the size a and shape f of the ellipsoid, rotation by the small angles $d\epsilon_X, d\epsilon_Y, d\epsilon_Z$, and translations by dt_X, dt_Y, dt_Z.

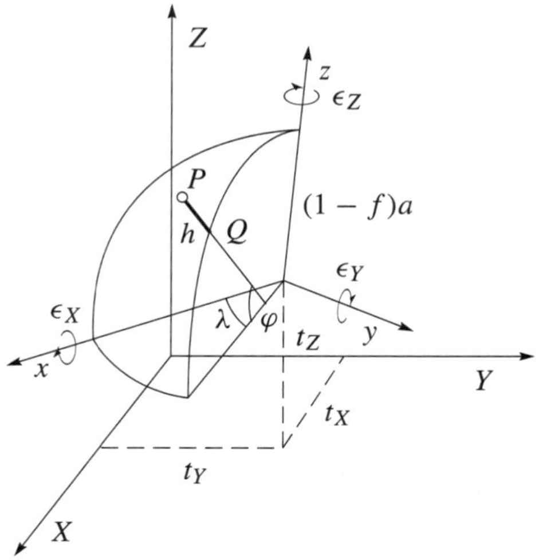

Figure 14.8 Geocentric datum parameters and ellipsoidal coordinates φ, λ and h.

This can be done by setting $d\boldsymbol{T} = \boldsymbol{0}$. Then a point before and after the transformation remains physically the same:

$$d\boldsymbol{X} = -dR\,\boldsymbol{X} - dk\,\boldsymbol{X} - d\boldsymbol{t}. \tag{14.53}$$

The left side $d\boldsymbol{X}$ depends on λ, φ, h, a and f. The *total differential* of $\boldsymbol{X}$ in (14.52) is

$$d\boldsymbol{X} = \begin{bmatrix} -\sin\lambda & -\sin\varphi\cos\lambda & \cos\varphi\cos\lambda \\ \cos\lambda & -\sin\varphi\sin\lambda & \cos\varphi\sin\lambda \\ 0 & \cos\varphi & \sin\varphi \end{bmatrix} \begin{bmatrix} a\cos\varphi\,d\lambda \\ a\,d\varphi \\ dh \end{bmatrix}$$

$$+ \begin{bmatrix} \cos\varphi\cos\lambda & \sin^2\varphi\cos\varphi\cos\lambda \\ \cos\varphi\sin\lambda & \sin^2\varphi\cos\varphi\sin\lambda \\ \sin\varphi & (\sin^2\varphi - 2)\sin\varphi \end{bmatrix} \begin{bmatrix} da \\ a\,df \end{bmatrix} = F\boldsymbol{\alpha} + G\boldsymbol{\beta}. \tag{14.54}$$

Substitution into (14.53) gives

$$F\boldsymbol{\alpha} + G\boldsymbol{\beta} = -dR\,\boldsymbol{X} - dk\,\boldsymbol{X} - d\boldsymbol{t}.$$

We look for $\boldsymbol{\alpha}$ and isolate it—remembering that orthogonality implies $F^{-1} = F^{\mathrm{T}}$:

$$\boldsymbol{\alpha} = -F^{\mathrm{T}}dR\,\boldsymbol{X} - dk\,F^{\mathrm{T}}\boldsymbol{X} - F^{\mathrm{T}}\,d\boldsymbol{t} - F^{\mathrm{T}}G\boldsymbol{\beta}. \tag{14.55}$$

Now follows a calculation of each component in $\boldsymbol{\alpha}$:

$$-F^{\mathrm{T}}dR\,\boldsymbol{X} = -F^{\mathrm{T}} \begin{bmatrix} d\epsilon_Z\,Y - d\epsilon_Y\,Z \\ -d\epsilon_Z\,X + d\epsilon_X\,Z \\ d\epsilon_Y\,X - d\epsilon_X\,Y \end{bmatrix} = -F^{\mathrm{T}} \begin{bmatrix} 0 & -Z & Y \\ Z & 0 & -X \\ -Y & X & 0 \end{bmatrix} \begin{bmatrix} d\epsilon_X \\ d\epsilon_Y \\ d\epsilon_Z \end{bmatrix}$$

$$= \begin{bmatrix} -a\sin\varphi\cos\lambda & -a\sin\varphi\sin\lambda & a\cos\varphi \\ a\sin\lambda & -a\cos\lambda & 0 \\ 0 & 0 & 0 \end{bmatrix} \begin{bmatrix} d\epsilon_X \\ d\epsilon_Y \\ d\epsilon_Z \end{bmatrix},$$

$$-F^{\mathrm{T}}G = \begin{bmatrix} 0 & 0 \\ 0 & -\sin\varphi\cos^3\varphi \\ -1 & \sin^2\varphi \end{bmatrix}, \qquad \text{and} \qquad -F^{\mathrm{T}}\boldsymbol{X} = -\begin{bmatrix} 0 \\ 0 \\ a \end{bmatrix}.$$

This allows us to give an explicit expression for $\boldsymbol{\alpha}$:

$$\begin{aligned}\begin{bmatrix} a\cos\varphi\, d\lambda \\ a\, d\varphi \\ dh \end{bmatrix} &= \begin{bmatrix} \sin\lambda & -\cos\lambda & 0 \\ \sin\varphi\cos\lambda & \sin\varphi\sin\lambda & -\cos\varphi \\ -\cos\varphi\cos\lambda & -\cos\varphi\sin\lambda & -\sin\varphi \end{bmatrix} \begin{bmatrix} dt_X \\ dt_Y \\ dt_Z \end{bmatrix} \\ &\quad + \begin{bmatrix} -a\sin\varphi\cos\lambda & -a\sin\varphi\sin\lambda & a\cos\varphi \\ a\sin\lambda & -a\cos\lambda & 0 \\ 0 & 0 & 0 \end{bmatrix} \begin{bmatrix} d\epsilon_X \\ d\epsilon_Y \\ d\epsilon_Z \end{bmatrix} \\ &\quad - \begin{bmatrix} 0 \\ 0 \\ a \end{bmatrix} dk - \begin{bmatrix} 0 & 0 \\ 0 & a\sin\varphi\cos^3\varphi \\ 1 & -a\sin^2\varphi \end{bmatrix} \begin{bmatrix} da \\ df \end{bmatrix}. \end{aligned} \tag{14.56}$$

The number of datum parameters has been augmented by two, namely the changes in the ellipsoidal parameters a and f. The total number of parameters is thus nine. How does that agree with the fact that a datum is defined by five parameters? Well, we have increased this number by three rotations and one change of scale. By that we again come to the number nine and our bookkeeping is in balance again.

Example 14.7 An example of a transformation of the type decribed by (14.51) is the transformation between European Datum 1950 ED 50 and WGS 84:

$$\begin{bmatrix} X_{\mathrm{WGS\,84}} \\ Y_{\mathrm{WGS\,84}} \\ Z_{\mathrm{WGS\,84}} \end{bmatrix} = \begin{bmatrix} 0.0\,\mathrm{m} \\ 0.0\,\mathrm{m} \\ 4.5\,\mathrm{m} \end{bmatrix} + k \begin{bmatrix} 1 & \alpha & 0 \\ -\alpha & 1 & 0 \\ 0 & 0 & 1 \end{bmatrix} \begin{bmatrix} X_{\mathrm{ED\,50}} - 89.5\,\mathrm{m} \\ Y_{\mathrm{ED\,50}} - 93.8\,\mathrm{m} \\ Z_{\mathrm{ED\,50}} - 127.6\,\mathrm{m} \end{bmatrix}.$$

The inverse transformation back to the European Datum is

$$\begin{bmatrix} X_{\mathrm{ED\,50}} \\ Y_{\mathrm{ED\,50}} \\ Z_{\mathrm{ED\,50}} \end{bmatrix} = \frac{1}{k} \begin{bmatrix} 1 & -\alpha & 0 \\ \alpha & 1 & 0 \\ 0 & 0 & 1 \end{bmatrix} \begin{bmatrix} X_{\mathrm{WGS\,84}} \\ Y_{\mathrm{WGS\,84}} \\ Z_{\mathrm{WGS\,84}} - 4.5\,\mathrm{m} \end{bmatrix} + \begin{bmatrix} 89.5\,\mathrm{m} \\ 93.8\,\mathrm{m} \\ 127.6\,\mathrm{m} \end{bmatrix}.$$

The variables have the following values: $\alpha = 0.156'' = 0.756 \times 10^{-6}$ rad and $k = 1 + 1.2 \times 10^{-6}$ or $1/k = 0.999\,998\,8$. The transformation looks a little more involved than earlier expressions. This is due to inclusion of some minor differences between WGS 84 and the original Doppler datum NSWC 9Z-2.

15

PROCESSING OF GPS DATA

15.1 Baseline Computation and M-Files

The previous chapter described the principles of global positioning. This chapter will present and discuss a series of MATLAB M-files that are freely available to the reader. *Their objective is to compute the baseline vector between two stationary receivers.* This is the essence of GPS. Large systems like GAMIT and GYPSY and Bernese extend to a network of baselines. We believe that the essential problems are best understood by actual experiments with relatively simple subroutines.

The outline of our presentation is given next. It could also be the outline of a lecture series or a hands-on GPS workshop. This outline imposes a structure on the 70 or more GPS M-files that are associated with this book. The baseline computation is achieved in five steps:

Step 1: The positions of the satellites The satellite positions are to be computed in Earth-fixed x, y, z coordinates, from the ephemerides that are transmitted in Keplerian coordinates. We are using the raw navigation data files converted to `*.*n` files in RINEX format. New ephemerides will come every 60 to 90 minutes, so there may be multiple versions (covering different spans of time) for each satellite.

This coordinate change (from space to Earth) will be discussed in Section 15.2. The Keplerian elements are stable and slowly varying. The x, y, z coordinates change rapidly. The applications of GPS need and use Earth coordinates.

Step 2: Preliminary position of the receiver This is based on pseudorange data from each satellite. Recall that the pseudorange is the measurement of travel time from each satellite to each receiver, not accounting for clock errors and not highly accurate. The C/A code pseudorange has an accuracy of 30 meters. Our recpos code for the receiver positions is a simple and straightforward search. This is for pedagogical reasons; an alternative code is bancroft. Later they are both replaced by upgraded software which will use further information from RINEX observation files (P code distances and phases at the L_1 and the L_2 frequencies).

Step 3: Separate estimation of ambiguities and the baseline vector This is our first serious estimate of the baseline (dx, dy, dz) between the two receivers. It is reasonably efficient for short baselines (less than 20 km). We first solve the ambiguity problem to find the integers $N^{kl}_{1,ij}$ and $N^{kl}_{2,ij}$ for double differenced observations.

In this one instance, we have based the calculations directly on a binary data format used by Ashtech receivers. The code produces a number of figures to illustrate the residuals—it is good for demonstrational purposes.

Step 4: Joint estimation of ambiguities and baseline vector Section 15.5 describes this main step and it yields high precision. We find the preliminary receiver position more accurately with bancroft. Then we compute the least squares solution to the complete system of observational equations, taking as unknowns the differences dx, dy, dz and all the ambiguities. Naturally this solution does not give integer ambiguities. The LAMBDA method presented in Section 15.6 iteratively determines the optimal solution (in integers for the ambiguities and in reals for the distances). It was devised by Peter Teunissen, and we thank Christian Tiberius for generous help in adding it to our M-files. Our complete MATLAB subroutine is not quite state-of-the-art but relatively close.

Step 5: Updates of ambiguities and baseline with new observations This step is often referred to as "Kalman filtering." Strictly speaking we are doing "sequential least squares" on a static problem. New rows are added to the observation matrix, while the set of unknowns remains the same. A series of M-files will demonstrate how the preliminary receiver position and the ambiguities and the final baseline vector are corrected to account for each new set of observations. We indicate several of those M-files by name, using the letter k or b to indicate a Kalman or Bayes (covariance before gain matrix) update: k_point, b_point, k_ud, k_dd3, k_dd4.

15.2 Coordinate Changes and Satellite Position

This section connects Earth centered and Earth fixed (ECEF) coordinates X, Y, Z to a satellite position described in space by Keplerian orbit elements. First we recall those orbit elements a, e, ω, Ω, i, and μ, shown in Figure 15.1. This is unavoidably somewhat technical; many readers will proceed, assuming ECEF coordinates are found.

The X-axis points towards the *vernal equinox*. This is the direction of the intersection between equator and *ecliptic*. For our purpose this direction can be considered fixed. The Z-axis coincides with the spin axis of the Earth. The Y-axis is orthogonal to these two directions and forms a right-handed coordinate system.

The orbit plane intersects the Earth equator plane in the *nodal line*. The direction in which the satellite moves from south to north is called the *ascending node* K. The angle between the equator plane and the orbit plane is the *inclination* i. The angle at the Earth's center C between the X-axis and the ascending node K is called Ω; it is a right ascension. The angle at C between K and the perigee P is called *argument of perigee* ω; it increases counter-clockwise viewed from the positive Z-axis.

Figure 15.2 shows a coordinate system in the orbital plane with origin at the Earth's center C. The ξ-axis points to the perigee and the η-axis towards the descending node. The ζ-axis is perpendicular to the orbit plane. From Figure 15.2 we read the eccentric

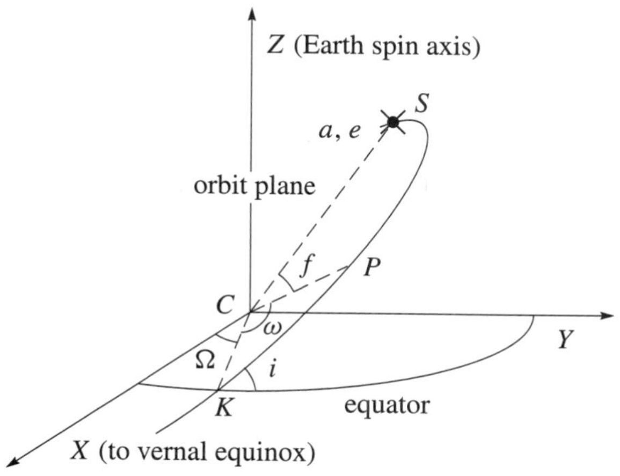

Figure 15.1 The Keplerian orbit elements: semi major axis a, eccentricity e, inclination of orbit i, right ascension Ω of ascending node K, argument of perigee ω, and true anomaly f. Perigee is denoted P. The center of the Earth is denoted C.

anomaly E and the true anomaly f. Also immediately we have

$$\xi = r\cos f = a\cos E - ae = a(\cos E - e)$$
$$\eta = r\sin f = \tfrac{b}{a}a\sin E = b\sin E = a\sqrt{1-e^2}\sin E.$$

Hence the position vector $\boldsymbol{r}$ of the satellite with respect to the center of the Earth C is

$$\boldsymbol{r} = \begin{bmatrix} \xi \\ \eta \\ \zeta \end{bmatrix} = \begin{bmatrix} a(\cos E - e) \\ a\sqrt{1-e^2}\sin E \\ 0 \end{bmatrix}. \tag{15.1}$$

Simple trigonometry leads to the following expression for the norm

$$\|\boldsymbol{r}\| = a(1 - e\cos E). \tag{15.2}$$

In general E varies with time t while a and e are nearly constant. (There are long and short periodic perturbations to e, only short for a.) Recall that $\|\boldsymbol{r}\|$ is the geometric distance between satellite S and the Earth center $C = (0, 0)$.

For later reference we introduce the mean motion n which is the mean angular satellite velocity. If the period of one revolution of the satellite is T we have

$$n = \frac{T}{2\pi} = \sqrt{\frac{GM}{a^3}}. \tag{15.3}$$

Let t_0 be the time the satellite passes perigee, so that $\mu(t) = n(t - t_0)$. Kepler's famous equation relates the mean anomaly μ and the eccentric anomaly E:

$$E = \mu + e\sin E. \tag{15.4}$$

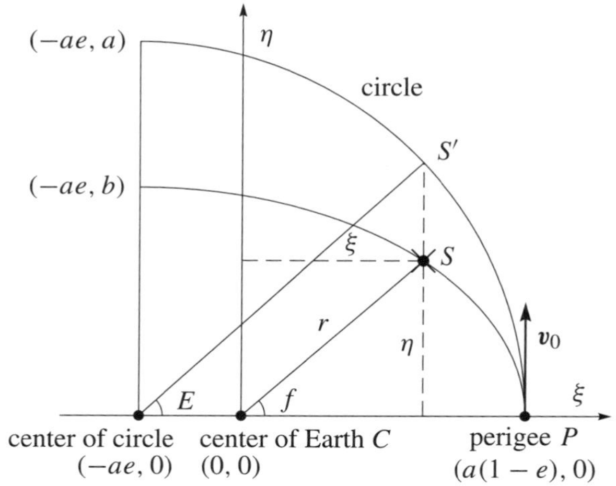

Figure 15.2 The elliptic orbit with (ξ, η) coordinates. The eccentric anomaly E at C.

From equation (15.1) we finally get

$$f = \arctan\frac{\eta}{\xi} = \arctan\frac{\sqrt{1-e^2}\sin E}{\cos E - e}. \tag{15.5}$$

By this we have connected the true anomaly f, the eccentric anomaly E, and the mean anomaly μ. These relations are basic for every calculation of a satellite position.

The six Keplerian orbit elements constitute an important description of the orbit, so they are repeated in schematic form in Table 15.1.

It is important to realize that the orbital plane remains fairly stable in relation to the geocentric X, Y, Z-system. In other words: seen from space the orbital plane remains fairly fixed in relation to the equator. The Greenwich meridian plane rotates around the Earth spin axis in accordance with Greenwich apparent sidereal time (GAST), that is with a speed of approximately 24 h/day. A GPS satellite performs two revolutions a day in its orbit having a speed of 3.87 km/s.

The geocentric coordinates of satellite k at time t_j are given as

$$\begin{bmatrix} X^k(t_j) \\ Y^k(t_j) \\ Z^k(t_j) \end{bmatrix} = R_3(-\Omega_j^k)R_1(-i_j^k)R_3(-\omega_j^k)\begin{bmatrix} r_j^k \cos f_j^k \\ r_j^k \sin f_j^k \\ 0 \end{bmatrix} \tag{15.6}$$

where $r_j^k = \|\boldsymbol{r}(t_j)\|$ comes from (15.2) with a, e, and E evaluated for $t = t_j$. However, GPS satellites do not follow the presented normal orbit theory. We have to use time dependent, more accurate orbit values. They come to us as the socalled broadcast ephemerides (practically during the downloading of receiver data). We insert those values in a procedure given below and finally we get a set of variables to be inserted into (15.6).

Obviously the vector is time-dependent and one speaks about the *ephemeris* (plural: *ephemerides*, accent on "phem") of the satellite. These are the parameter values at a specific time. Each satellite transmits its unique ephemeris data.

Table 15.1 Keplerian orbit elements: Satellite position

a	semi major axis	size and shape of orbit
e	eccentricity	
ω	argument of perigee	the orbital plane in the apparent system
Ω	right ascension of ascending node	
i	inclination	
μ	mean anomaly	position in the plane

The parameters chosen for description of the actual orbit of a GPS satellite and its perturbations are similar to the Keplerian orbital elements. The broadcast ephemerides are calculated using the immediate previous part of the orbit and they predict the following part of the orbit. The broadcast ephemerides are accurate to $\approx$ 10 m only. For some geodetic applications better accuracy is needed. One possibility is to obtain post processed *precise ephemerides* which are accurate at dm-level.

An ephemeris is intended for use from the epoch t_{oe} of reference counted in seconds of GPS week. It is nominally at the center of the interval over which the ephemeris is useful. The broadcast ephemerides are intended for use during this period. However they describe the orbit to within the specified accuracy for 1.5–5 hours afterward. They are predicted by curve fit to 4–6 hours data. The broadcast ephemerides include

$$\mu_0,\ \Delta n,\ e,\ \sqrt{a},\ \Omega_0,\ i_0,\ \omega,\ \dot{\Omega},\ \dot{i},\ C_{\omega c},\ C_{\omega s},\ C_{rc},\ C_{rs},\ C_{ic},\ C_{is},\ t_{oe}$$

where $\dot{\Omega} = \partial\Omega/\partial t$ and $\dot{i} = \partial i/\partial t$. The coefficients C_ω, C_r, and C_i correct argument of perigee, orbit radius, and orbit inclination due to inevitable perturbations of the theoretical orbit caused by variations in the Earth's gravity field, albedo and sun pressure, and attraction from sun and moon.

Given the transmit time t (in GPS system time) the following procedure gives the necessary variables to use in (15.6):

Time elapsed since t_{oe} $\qquad t_j = t - t_{oe}$

Mean anomaly at time t_j $\qquad \mu_j = \mu_0 + \left(\sqrt{GM/a^3} + \Delta n\right)t_j$

$$GM = 3.986\,005 \cdot 10^{14}\ \mathrm{m^3/s^2}$$

Iterative solution for E_j $\qquad E_j = \mu_j + e \sin E_j$

True anomaly $\qquad f_j = \arctan \dfrac{\sqrt{1-e^2}\sin E_j}{\cos E_j - e}$

Longitude for ascending node	$\Omega_j = \Omega_0 + (\dot{\Omega}_0 - \omega_e)t_j - \omega_e t_{oe}$
	$\omega_e = 7.292\,115\,147 \cdot 10^{-5}\,\text{rad/s}$
Argument of perigee	$\omega_j = \omega + f_j + C_{\omega c}\cos 2(\omega + f_j) + C_{\omega s}\sin 2(\omega + f_j)$
Radial distance	$r_j = a(1 - e\cos E_j) + C_{rc}\cos 2(\omega + f_j) + C_{rs}\sin 2(\omega + f_j)$
Inclination	$i_j = i_0 + \dot{i}t_j + C_{ic}\cos 2(\omega + f_j) + C_{is}\sin 2(\omega + f_j).$

As usual the mean Earth rotation is denoted ω_e. This algorithm is coded as the M-file satpos. The function calculates the position of any GPS satellite at any time. It is fundamental to every position calculation.

In an attempt to realign WGS 84 with the more accurate terrestrial reference frame of the International Earth Rotation Service the ten tracking stations have been recoordinated at the epoch 1994.0. This redefined frame has been designated WGS 84 (G730); G stands for GPS derived and 730 for the GPS week number when these modifications were implemented. In WGS 84 (G730) the value for GM has been changed to $GM = 3.986\,004\,418 \cdot 10^{14}\,\text{m}^3/\text{s}^2$.

The M-file satposin computes satellite positions in an inertial frame where $\omega_e = 0$.

All information about the Kepler elements is contained in the ephemerides. Files containing broadcast ephemerides are created when downloading the GPS observations to a PC. These files most often are in a receiver specific binary format. Fortunately these formats can be converted into the RINEX format (described in the back of this book). The third character in the extension of such files is always n (for navigation).

Often we only need part of the information in the navigation file for the ephemeris file. The selection and reformatting is done via the following commands:

```
rinexe(ephemerisfile, outputfile);
eph = get_eph(outputfile);
satp = satpos(t,eph);
```

Some comments should be added on timekeeping in the GPS environment. GPS time counts in weeks and seconds of week starting on January 6, 1980. Each week has its own number. Time within a week is counted in seconds from the beginning at midnight between Saturday and Sunday (day 1 of the week).

The count of seconds goes up to $7 \times 24 \times 60 \times 60 = 604\,800$. GPS calculations need to keep track of nanoseconds, so professional software often splits the second into an integer and a decimal.

In the sample code above, satpos needs a time parameter t in seconds of week. All epoch times in the RINEX observation files use time in the format: year, month, day, hour, minutes, and seconds. So basically we need a conversion from this date-format to GPS time (week number and seconds of week). Traditionally geodesists and astronomers solve this problem by introducing the (modified) Julian Day Number. This is an elegant way to circumvent the problem created by months of various lengths. JD simply counts the days consecutively since January 1, 4713 BC. The modified Julian Date MJD, starting at midnight, is defined as

$$\mathrm{MJD} = \mathrm{JD} - 2\,400\,000.5.$$

The M-file gps_time finds GPS time (week w and seconds of week sow):

```
t = julday(1997,2,10,20); % year, month, day, hour
[w,sow] = gps_time(t)
w =
      892
sow =
        158400
```

To avoid under- or overflow at the beginning or end of a week we use the M-file check_t.

15.3 Receiver Position from Pseudoranges

We start by a nice and very useful application which combines the least-squares method and a searching technique. The procedure solves a least-squares problem without forming normal equations. Here we find the minimum of the sum of squared residuals through searching.

Assume four or more pseudoranges $P_1, P_2, \dots, P_m$ are given in the same epoch. We want to determine the position of our receiver, with no a priori knowledge of its location.

First we compute the ECEF coordinates of all m satellites from the ephemeris file. Then we transform the Cartesian coordinates of each satellite at transmission time (which can be obtained from the pseudorange) to (φ_i, λ_i) = (latitude, longitude) and average those coordinates. This average is our first guess of the receiver location.

Taking this point as a center we introduce a grid covering a hemisphere. There may be ten equal radial subdivisions out to $\pi/2$ and possibly 16 equal subdivisions of the 2π angle. At each grid point we calculate a residual $R_i = P_i - P_1$, $i = 2, \dots, m$, as the difference between the observed pseudorange and the first value to eliminate the receiver clock offset. Next we difference these differences:

$$\begin{aligned}
r_1 &= R_2 - R_2^0 = (P_2 - P_1) - (P_2^0 - P_1^0)\\
r_2 &= R_3 - R_3^0 = (P_3 - P_1) - (P_3^0 - P_1^0)\\
&\vdots\\
r_{m-1} &= R_m - R_m^0 = (P_m - P_1) - (P_m^0 - P_1^0).
\end{aligned}$$

This gives a first approximation to the sum $S = \sum r_i^2$. Among all the possible values for S, we seek the smallest one; the gridpoint connected to this value is the new guess for the location of the receiver. The radial grid is subdivided for each iteration. The procedure is repeated until no further improvement in position is achieved.

Finally we show how to calculate the new position by spherical trigonometry, as in Figure 15.3. Let the coordinates of the original point be (φ_1, λ_1). We want to move ψ

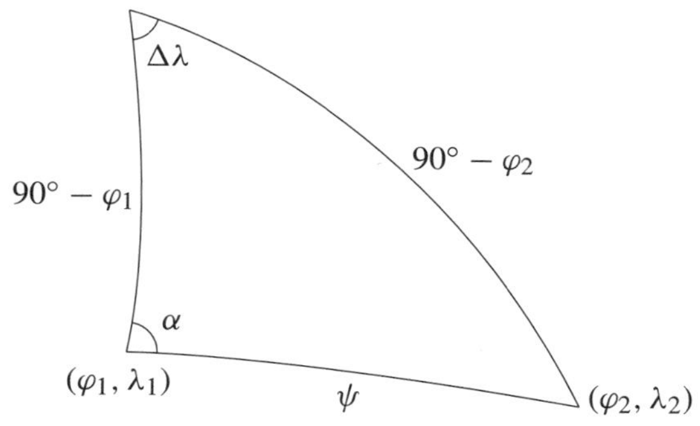

Figure 15.3 Spherical triangle for calculation of receiver latitude and longitude (φ, λ).

degrees along the azimuth α to the new point (φ_2, λ_2). Computing φ_2 and λ_2 is a classical geodetic problem and the following basic equations yield a solution on the sphere:

$$\sin\varphi_2 = \sin\varphi_1 \cos\psi + \cos\varphi_1 \sin\psi \cos\alpha \qquad \text{and} \qquad \sin\Delta\lambda = \frac{\sin\alpha \sin\psi}{\cos\varphi_2}. \tag{15.7}$$

These equations give the latitude φ_2 of the new point and the *increment* $\Delta\lambda$ in longitude. The M-file recpos shows a fast implementation of the code. Eventually one finds a receiver position consistent with precision of the orbit, the refraction model, and the pseudoranges.

This is an example of another look at least squares. Here, since P_1 is subtracted from every other pseudorange, the differences are obviously correlated. Any error in P_1 is present in each of the differences. In the search we ignore these correlations since the procedure is designed only to give a first guess of the receiver location.

A least-squares procedure using the proper covariance matrix should follow our search technique. The resulting guess is almost always within the (linear) convergence region. Obtain the navigation data from rinexe('ohiostat.96n','rinex_n.dat'). You may zoom in on the figure by pressing your mouse button. After many zooms you may read off the preliminary position on the x- and y-labels.

The M-file get_eph opens, reads and reshapes an ephemerides file. Such a file usually contains several data sets for a particular satellite. It is common practice to select the ephemeris that is immediately before the epoch of use. The file find_eph does this. Keeping to this practice, however, eventually leads to a change of ephemeris data. This most probably will introduce a jump in the calculated orbit. More advanced programs smooth orbits, if they have to be exploited over longer periods of time. An even better solution is to change to *precise ephemerides*.

15.4 Separate Ambiguity and Baseline Estimation

After downloading the recorded data stored in a GPS-receiver we most often have a file containing observations as well as a file containing the ephemerides. Sometimes there is also an auxiliary file with meteorological data and information about the site.

One procedure is to use these receiver dependent files in the further data processing. Another procedure is to convert to the RINEX format. Here we start out from a set of receiver dependent files (from Ashtech receivers).

The M-file bdata reads two observation files, extracts the information necessary, and stores it in the binary file bdata.dat. Only data contemporary for master and rover are stored; they are receiver time, satellite number, code on L_1, phase on L_1, code on L_2, phase on L_2, and elevation angle. Typical calls for binary and ephemeris data are

bdata('b0810a94.076','b0005a94.076') and edata('e0810a94.076').

To reformat edata.dat into a matrix with 21 rows and as many columns as there are ephemerides make the call eph = get_eph('rinex_n.dat'). Probably a given satellite has more than one ephemeris. The proper column in the matrix picks the ephemeris just before or equal to time. For this column use icol = find_eph(eph,sv,time).

At the end of the main M-file ash_base we need information about the antenna heights. This information is contained in the original s-files. They may be converted to MATLAB-format by sdata('s0810a94.076','s0005a94.076').

The M-file ash_dd starts by computing the means of the elevation angles of the individual satellites. It counts the number of epochs in which the satellite appears. Then a cut-off angle is selected and all satellites with mean elevation smaller than this angle are deleted. Next a reference satellite is chosen as the one which appears in most epochs. Finally all data are ***double differenced***. The M-file ash_dd is common to several M-files using the supplied data set.

The first step in finding receiver position is to estimate the ambiguities N_1 and N_2. The observation equations are

$$\begin{aligned} P_1 &= \rho^* + I - e_1 \\ \Phi_1 &= \rho^* - I + \lambda_1 N_1 - \epsilon_1 \\ P_2 &= \rho^* + (f_1/f_2)^2 I - e_2 \\ \Phi_2 &= \rho^* - (f_1/f_2)^2 I + \lambda_2 N_2 - \epsilon_2. \end{aligned} \tag{15.8}$$

Actually, we have $f_1/f_2 = 77/60 = 1.283\,333\ldots$. We assume we are processing a short baseline and consequently put $I = 0$. So equation (15.8) can be transformed into the elegant matrix equation

$$\begin{bmatrix} 1 & 0 & 0 \\ 1 & \lambda_1 & 0 \\ 1 & 0 & 0 \\ 1 & 0 & \lambda_2 \end{bmatrix} \begin{bmatrix} \rho^* \\ N_1 \\ N_2 \end{bmatrix} = \begin{bmatrix} P_1 \\ \Phi_1 \\ P_2 \\ \Phi_2 \end{bmatrix} - \text{noise}. \tag{15.9}$$

This is four equations in three unknowns: the ideal pseudorange and the two ambiguities on L_1 and L_2. The weight matrix is

$$C = \begin{bmatrix} 1/0.3^2 & & & \\ & 1/0.005^2 & & \\ & & 1/0.3^2 & \\ & & & 1/0.005^2 \end{bmatrix}.$$

The standard deviation of phase error is a few millimeters; for the pseudorange this is receiver dependent. C/A code pseudoranges on L_1 can have noise values up to 2–3 m. This is due to the slow chipping rate which just is another term for frequency. The P code has a frequency of 10.23 Mhz, i.e. a sequence of 10.23 million binary digits or *chips* per second. The chipping rate of the P code is ten times more frequent and this implies an error of 10–30 cm.

For each epoch we add the relevant contribution to the normals and only solve the system when all observations are read. We use a little more sophisticated code as we eliminate the ideal pseudorange according to the method described in Section 11.7.

The least squares solution consists of two reals n_1 and n_2 from which we have to recover two integers N_1 and N_2. Here we follow a method indicated by Clyde Goad. The estimated difference $n_1 - n_2$ is rounded to the nearest integer and named K_1. The rounded value of $60n_1 - 77n_2$ we call K_2. The best integer estimates for N_1 and N_2 are then found as the solution

$$\widehat{N}_2 = (60K_1 - K_2)/17 \tag{15.10}$$

$$\widehat{N}_1 = \widehat{N}_2 + K_1. \tag{15.11}$$

The values for K_1 and K_2 are not free of error, but only particular combinations yield integer solutions for N_1 and N_2. Gradually these estimates improve as more epochs are processed. The numbers K_1 and K_2, in theory, become more reliable.

Above we put $I = 0$. Now we are more familiar with the observables and we can investigate this assumption. Let us emphasize once more that we are dealing with double differenced observations. We repeat the second and fourth observation equations in (15.8):

$$\Phi_1 = \rho^* - I + \lambda_1 N_1 - \epsilon_1$$

$$\Phi_2 = \rho^* - (f_1/f_2)^2 I + \lambda_2 N_2 - \epsilon_2.$$

Ignoring the error terms and eliminating ρ^* gives an expression for the ionospheric delay

$$I = \frac{(\Phi_2 - \lambda_2 N_2) - (\Phi_1 - \lambda_1 N_1)}{1 - (f_1/f_2)^2}. \tag{15.12}$$

This variable is plotted in Figure 15.4. It scatters within a few cm. So it is a matter of testing if the condition $I = 0$ is accepted or not.

We can eliminate the ionospheric term I from the two equations. The result is

$$60\Phi_1/\lambda_1 - 77\Phi_2/\lambda_2 = 60N_1 - 77N_2. \tag{15.13}$$

The coefficients 60 and 77 appear because $\frac{f_1}{f_2} = \frac{154}{120} = \frac{77}{60}$. However, a drawback of the large coefficients 60 and 77 is that they amplify the noise.

The final step is the ***baseline estimation***. In order to operate with correct geometry we need good approximate coordinates for the master station. We also need to set preliminary values for the baseline vector. Amazingly enough it works to set it to the zero vector.

We start from equation (14.20):

$$\rho_i^k - \rho_i^l - \rho_j^k + \rho_j^l = \Phi_{q,ij}^{kl} - T_{ij}^{kl} - \lambda_q N_{q,ij}^{kl} - \text{noise}. \tag{15.14}$$

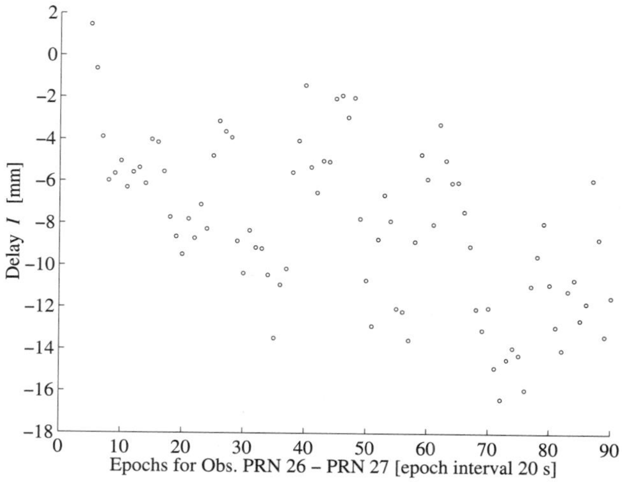

Figure 15.4 Ionospheric delay for double differenced phases from (15.12).

The wavelength λ_q and the ambiguities N_q are assumed known; $q = 1$ *refers to frequency* L_1 *and* $q = 2$ *refers to* L_2. The M-file tropo models the tropospheric delay T. We denote the double differenced phase observations multiplied by the wavelength λ_q by Φ_q.

Our model assumes that the observational data do not suffer from *cycle slips* or *reset of receiver clocks*. Some receiver manufacturers typically reset clocks by 1 ms to avoid clock drifts larger than this amount. Both sources cause code and phase observations to violate the condition

$$\Delta\Phi(j) = \Delta P(j) \qquad \text{or} \qquad \Phi(j) - \Phi(1) = P(j) - P(1)$$

where Δ means change over time. Apart from the random errors of the pseudorange observation this condition should be fulfilled for all values of j with

$$P(j) = \alpha_1 P_1(j) + \alpha_2 P_2(j)$$

$$\Phi(j) = \alpha_1 \Phi_1(j) + \alpha_2 \Phi_2(j) \qquad \alpha_1 = \frac{f_1^2}{f_1^2 - f_2^2} = 1 - \alpha_2.$$

For concreteness we take point i (= 810) as a known reference. The coordinates of the other point j (= 005) are unknown. The reference satellite is $k = 26$ and the index l runs over five more satellites 2, 9, 16, 23, and 27. Linearizing equation (15.14) brings the Jacobian matrix J from the derivatives of the double difference on the left side:

$$Jx = \begin{bmatrix} \boldsymbol{u}_i^2 - \boldsymbol{u}_j^{26} \\ \boldsymbol{u}_i^9 - \boldsymbol{u}_j^{26} \\ \boldsymbol{u}_i^{16} - \boldsymbol{u}_j^{26} \\ \boldsymbol{u}_i^{23} - \boldsymbol{u}_j^{26} \\ \boldsymbol{u}_i^{27} - \boldsymbol{u}_j^{26} \end{bmatrix} \begin{bmatrix} x_j \\ y_j \\ z_j \end{bmatrix} \tag{15.15}$$

where

$$u_i^k = \left(\frac{X_{\text{ECEF}}^k - X_i}{\rho_i^k}, \frac{Y_{\text{ECEF}}^k - Y_i}{\rho_i^k}, \frac{Z_{\text{ECEF}}^k - Z_i}{\rho_i^k} \right).$$

The original satellite coordinates are rotated by $\omega\tau_i^k$. Here ω_e denotes the rotation rate of the Earth ($7.292\,115\,147 \cdot 10^{-5}$ rad/s) and $\tau_i^k = \rho_i^k/c$. According to (14.7) we get:

$$\begin{bmatrix} X_{\text{ECEF}}^k \\ Y_{\text{ECEF}}^k \\ Z_{\text{ECEF}}^k \end{bmatrix} = \begin{bmatrix} \cos(\omega_e\tau_i^k) & \sin(\omega_e\tau_i^k) & 0 \\ -\sin(\omega_e\tau_i^k) & \cos(\omega_e\tau_i^k) & 0 \\ 0 & 0 & 1 \end{bmatrix} \begin{bmatrix} X^k \\ Y^k \\ Z^k \end{bmatrix} \tag{15.16}$$

On the right side of the linearized equations are the zero-order terms

$$c_{q,ij}^{kl} = \Phi_{q,ij}^{kl} - \lambda_q N_{q,ij}^{kl} - (\rho_i^k)^0 + (\rho_i^l)^0 + (\rho_j^k)^0 - (\rho_j^l)^0. \tag{15.17}$$

The index $q = 1, 2$ corresponds to observations on frequencies L_1 and L_2. There is an equation for each receiver j and satellite l, except the reference $i = 810$ and $k = 26$.

The least-squares solution of the complete system $Jx = c$ yields the position vector. The covariance for both λ_1 and λ_2 is $\Sigma_d = D_d D_d^{\mathrm{T}}$ in accordance with (14.30).

The algorithm yields exciting (and good) results with good P code and phase data. Nevertheless we have included a simple test for outliers. In each epoch we ask if the weighted square sum of residuals of the observations exceeds a given multiple of the previous iteration mean square. If that is the case we ignore data from that epoch.

The initial estimate x for the baseline from site 810 to site 005 has components

$$\begin{aligned} \delta\widehat{X} &= -941.313\,\text{m} & \sigma_X &= 0.002\,\text{m} \\ \delta\widehat{Y} &= 4\,496.358\,\text{m} & \sigma_Y &= 0.001\,\text{m} \\ \delta\widehat{Z} &= 112.053\,\text{m} & \sigma_Z &= 0.002\,\text{m}. \end{aligned}$$

The observed slope distance of the antenna h_s must be reduced through a vertical distance h_a. A radius r of the antenna yields $h_a^2 = h_s^2 - r^2$. In the present situation $h_s = 1.260$ m at site 810, and $h_s = 1.352$ m at site 005. The corresponding values for h_a are 1.254 m and 1.346 m with $r = 0.135$ m. The vertical distances refer to the topocentric coordinate system and it must be transformed into differences in X, Y, and Z. That happens by introducing the vertical vector u as defined in equation (10.30):

$$h_a u = h_a \begin{bmatrix} \cos\varphi\cos\lambda \\ \cos\varphi\sin\lambda \\ \sin\varphi \end{bmatrix}. \tag{15.18}$$

Remember that u is the third column of R. The difference in antenna heights contributes

$$\begin{bmatrix} \delta X_a \\ \delta Y_a \\ \delta Z_a \end{bmatrix} = (h_m - h_r)u = \begin{bmatrix} -0.050 \\ -0.008 \\ -0.077 \end{bmatrix}.$$

This has to be added to the estimated difference vector, so the final baseline estimate is

$$X_{\text{ECEF}} = \begin{bmatrix} \delta\widehat{X} + \delta X_a \\ \delta\widehat{Y} + \delta Y_a \\ \delta\widehat{Z} + \delta Z_a \end{bmatrix} = \begin{bmatrix} -941.263 \\ 4\,496.350 \\ 111.976 \end{bmatrix}$$

This compares closely to the correct vector from site 810 to site 005. That is a fairly elementary treatment of positioning, to be improved.

Tropospheric delay

A simple empirical model for the tropospheric delay dT of GPS signals received at a position with the latitude φ, at a zenith distance of $z = 90° - h$ and with air pressure P_0 in millibars at height H in km, at temperature T_0 °K and partial pressure of water vapor e_0 in millibars is

$$dT = 0.002\,277 \frac{1 + 0.002\,6\cos 2\varphi + 0.000\,28H}{\cos z}\left(P_0 + \left(\frac{1255}{T_0} + 0.05\right)e_0\right). \tag{15.19}$$

This simple model can been extended in various ways. As we only intend to determine short baselines, and to an accuracy of cm-level we have implemented the M-file tropo which easily fulfills this demand. We compute the tropospheric zenith delay to be 2.4 m. And we read from the formula that the delay grows inversely proportional to $\cos z$. In processing GPS observations you often select a cut-off angle of 15°.

The M-file tropo needs the following parameters: sinel sine of elevation angle of satellite, hsta height of station in km, p atmospheric pressure in mb at height hp, tkel surface temperature in degrees Kelvin at height htkel, hum humidity in % at height hhum, hp height of pressure measurement in km, htkel height of temperature measurement in km, and hhum height of humidity measurement in km. Our code may compete even with the most precise and recent ones published.

The zenith delay is known to within about 2% uncertainty or better. For zenith distances smaller than 75° this uncertainty does not increase very much. If all pseudodistances have a constant bias this would not affect the positon calculation but just add to the reciever clock offset. If the tropospheric delay is not constant there is a tendency that it affects the height of the point. Yet, in double differencing for short baselines this is not a serious error.

Adjustment of Vectors

When the individual vectors have been processed they have to be tied together through a least-squares procedure to establish a network. Each vector estimates a difference of Cartesian coordinates between two points. The observation equations are decribed by an incidence matrix and therefore are very much like a three-dimensional leveling problem. We only need to know coordinates for the fixed points. The M-file v_loops described in Example 8.5 identifies all possible loops of three-dimensional vectors. This is useful when calculating the closure errors of loops and for a later least-squares estimation of station coordinates.

The processing of the individual vectors results in a covariance matrix for the three components of the vector. All 3 by 3 covariance matrices make up a block diagonal weight matrix for the least-squares adjustment.

Often there are many vectors and sometimes also multiple observations of the same vector. The software has to detect and handle gross errors automatically. The most frequent errors are wrong antenna heights and wrong point identifications. As a minumium the output must consist of WGS 84 coordinates, a plot of the network possibly containing confidence ellipses, and a labelling of non-checked vectors.

If geoidal heights are available, the output also can be in form of projection coordinates and heights above sea level. Scale of projection and meridian convergence at newly estimated points can be useful.

In practice it is also very useful if the software incorporates terrestrial observations. A few commercial softwares do this.

15.5 Joint Ambiguity and Baseline Estimation

Our most comprehensive and flexible code for baseline estimation performs the estimation of baseline as well as ambiguities at the same time. The code is activated by the call proc_dd('pta.96o','ptb.96o') where the files pta.96o and ptb.96o are RINEX observation files. ***We start from files in RINEX format.***

However, as always we have to bring the ephemerides file in order. This is done by the call rinexe('pta.96n','pta.nav').

We start by analyzing the header of the observation file from the master receiver: anheader('pta.96o'). The result is a list composed of some of the following abbreviations:

L1, L2 Phase measurements on L_1 and L_2

C1 Pseudorange using C/A Code on L_1

P1, P2 Pseudorange using P Code on L_1, L_2

D1, D2 Doppler frequency on L_1 and L_2

Next we open the master observation file and read until we find the first epoch flag equal to 0. This indicates *static observations* (flag 2 means start of moving antenna, and flag 3 is new site occupation). The file fepoch_0('pta.96o') reads continuously epoch times in the master file and compares with epoch times in the rover file until they match. When this happens we read observations of NoSv satellites by

```
grabdata(fid1,NoSv1,NoObs_types1)
grabdata(fid2,NoSv2,NoObs_types2)
```

The observations are transformed into equations which in turn are contributing to the normals. Double differenced observations are highly correlated, but the technique described in Section 11.8 decorrelates them before adding to the normals. That is the only valid method for adding the individual contributions from a single observation. Finally the normals are solved to produce estimates for the baseline components and the ambiguities.

In this code we deal with the *antenna offsets* in an untraditional way. A common procedure is to compute the baseline between the two antennas and correct for the antenna offset at each station. However we introduce a nominal phase center related to the actual marker by the offset vector $\boldsymbol{dx} = (dH, dE, dN)$. The vector $\boldsymbol{dx}$ is expressed in topographic coordinates but has to be converted to the ECEF system. Next we calculate the distance D from the nominal phase center rather than the marker point:

$$D = \left\| \boldsymbol{X}_{\text{sat}} - (\boldsymbol{X}_{\text{marker}} + \boldsymbol{dx}) \right\|.$$

The distance D has to be subtracted from the calculated pseudorange ρ_{j1} when forming the right side. Let the observed pseudorange be P_1, and correct the calculated geometric distance from satellite j to phase center 1 by D:

$$\text{corrected\_obsj1} = P_1 + (\rho_{j1} - D).$$

The last step of setting up the observation equations and normal equations is quite complex. This is partly due to the fact that we do not want to omit any valid observation. In the M-file ash_dd we shaped the observations with brute force as to include only those satellites which were observed from the first to the last epoch. Here we want to be more flexible and include all observations recorded in the RINEX file in any epochs.

This flexibility implies that we must be very careful in building the normals. The first three unknowns correspond to coordinate increments dx, dy, and dz of the baseline; the subsequent unknowns are ambiguities. Each time we encounter a new ambiguity we allocate a new position and increment the number of unknowns by one. All the necessary book keeping is done by the M-file locate. The file accum0 adds the individual contributions from the observation equations to the normal equations.

15.6 The LAMBDA Method for Ambiguities

The letters in LAMBDA stand for Least-squares AMBiguity Decorrelation Adjustment. The problem arises in highly accurate positioning by GPS. The carrier phase measurements give a very precise value for the *fraction* (in the number of wavelengths from satellite to receiver), but these measurements do not directly yield the *integer*. For each satellite-receiver pair, the problem is to determine this "ambiguity." Once we find the integer, we can generally track it; a cycle slip is large enough to detect. (Recall that the wavelengths for the L_1 and L_2 frequencies are 0.190 5 m and 0.244 5 m.) The problem is to find the integer in the first place.

Example 15.3 describes the "one-way" method of Euler and Goad, using the M-file one_way. In many cases this calculates the ambiguities, quickly and easily. The more powerful LAMBDA method was developed by Peter Teunissen (1993–1996) especially for networks with long baselines. Its MATLAB implementation has been clearly described by de Jonge & Tiberius (1996). We are grateful to Christian Tiberius for his help in adding the M-files to the library for this book.

The information at our disposal is from phase observations. To remove contamination by clock errors, we generally form *double differences*. Each component of the observation vector $\boldsymbol{b}$ involves two receivers i, j and two satellites k, l. The double difference of carrier phase measurements at a specific time (epoch t) is

$$b_{ij}^{kl}(t) = \left(\Phi_i^k(t) - \Phi_i^l(t)\right) - \left(\Phi_j^k(t) - \Phi_j^l(t)\right). \tag{15.20}$$

The vector $\boldsymbol{b}$ contains m measurements, and the vector $\boldsymbol{I}$ contains the n integer unknowns:

$$I_{ij}^{kl} = \left(I_i^k - I_i^l\right) - \left(I_j^k - I_j^l\right). \tag{15.21}$$

There are other unknowns of great importance! Those are the baseline coordinates $X_i - X_j$ and $Y_i - Y_j$ and $Z_i - Z_j$ that we set out to compute. They are real numbers, not integers, and they go into a vector $\boldsymbol{x}$ with p components. For one baseline $p = 3$. Then the linearized double difference equations are

$$\boldsymbol{b} = A\boldsymbol{x} + G\boldsymbol{I} + \text{noise}. \tag{15.22}$$

A is an m by p matrix and G is m by n. The matrix A relates baselines (the coordinate differences in $\boldsymbol{x}$) to the phase measurements in $\boldsymbol{b}$. Thus A involves the geometry of positions and lengths, as it does for code observations. The matrix G picks out from each double difference $b_{ij}^{kl}(t)$ the contribution I_{ij}^{kl} from the integers. These ambiguities I_{ij}^{kl} are *fixed in time*, nominally at their starting values. The other term $A\boldsymbol{x}$ accounts for all fractions at the start and all phase changes as the observations proceed.

We suppose that enough observations have been made to determine (usually they *over*determine) $\boldsymbol{x}$ and $\boldsymbol{I}$. Algebraically, the combined matrix $[\,A \quad G\,]$ has full column rank $p + n$. The ordinary normal equations could be solved, but $\hat{\boldsymbol{I}}$ won't contain integers. We are hoping to achieve good precision from a short series of observations. (But not too short. The n observations at *one* instant are not enough to determine the n integers in $\boldsymbol{I}$ and also the baseline coordinates.)

The covariance matrix Σ_b of the observations is assumed known. Since differences are correlated, Σ_b is not at all a diagonal matrix. This is what makes our problem more difficult. This also explains the letter D in LAMBDA, for Decorrelation. The method consists in decorrelating errors (diagonalizing Σ_b by a change of variables) as far as possible. The limitation is that the change of variables and its inverse must take integers to integers.

The usual problem in weighted least squares is to minimize $\|\boldsymbol{b} - A\boldsymbol{x} - G\boldsymbol{I}\|^2$. The weighting matrix Σ_b^{-1} determines the norm, as in $\|\boldsymbol{e}\|^2 = \boldsymbol{e}^{\mathrm{T}}\Sigma_b^{-1}\boldsymbol{e}$. The minimization gives real numbers $\hat{\boldsymbol{x}}$ and $\hat{\boldsymbol{I}}$ (not integers!). This estimate for $\boldsymbol{x}$ and $\boldsymbol{I}$ is called the ***float solution***, and it comes from the ordinary normal equations:

$$\begin{bmatrix} A & G \end{bmatrix}^{\mathrm{T}} \Sigma_b^{-1} \begin{bmatrix} A & G \end{bmatrix} \begin{bmatrix} \hat{\boldsymbol{x}} \\ \hat{\boldsymbol{I}} \end{bmatrix} = \begin{bmatrix} A & G \end{bmatrix}^{\mathrm{T}} \Sigma_b^{-1} \boldsymbol{b}. \tag{15.23}$$

We have two block equations, and the left side can be processed first (set $\Sigma_b^{-1} = C$):

$$\begin{bmatrix} A^{\mathrm{T}} \\ G^{\mathrm{T}} \end{bmatrix} C \begin{bmatrix} A & G \end{bmatrix} = \begin{bmatrix} A^{\mathrm{T}}CA & A^{\mathrm{T}}CG \\ G^{\mathrm{T}}CA & G^{\mathrm{T}}CG \end{bmatrix}. \tag{15.24}$$

A triangular factorization comes directly from elimination. The unknowns $\hat{\boldsymbol{x}}$ are first to be eliminated, leaving the reduced normal equations for $\hat{\boldsymbol{I}}$. Exactly as in Section 11.7, we are multiplying the upper block row by $G^{\mathrm{T}}CA(A^{\mathrm{T}}CA)^{-1}$ and subtracting from the lower block row. The new coefficient matrix in the (2, 2) block is $G^{\mathrm{T}}C'G$, with the reduced weight matrix C' in Section 11.7 and (15.28). Elimination of $\hat{\boldsymbol{x}}$ reaches an equation for $\hat{\boldsymbol{I}}$:

$$G^{\mathrm{T}}C'G\hat{\boldsymbol{I}} = G^{\mathrm{T}}C'\boldsymbol{b}. \tag{15.25}$$

When elimination continues, the matrix on the left is factored into $G^{\mathrm{T}}C'G = LDL^{\mathrm{T}}$. Then ordinary forward elimination and back substitution yield $\hat{\boldsymbol{I}}$. The triangular L (with ones on the diagonal) and the diagonal matrix D are actually the (2, 2) blocks in a factorization of the complete matrix in (15.24).

Our problem is that $\hat{\boldsymbol{I}}$ is not a vector of integers. This float solution minimizes a quadratic over all real vectors, while our problem is really one of ***integer least squares***:

$$\text{Minimize} \quad (G\boldsymbol{I} - \boldsymbol{b})^{\mathrm{T}}C'(G\boldsymbol{I} - \boldsymbol{b}) \quad \text{over integer vectors } \boldsymbol{I}. \tag{15.26}$$

The integer solution will be denoted by $\bar{\boldsymbol{I}}$ and called the ***final solution***. After $\bar{\boldsymbol{I}}$ is found (this is our real problem), the corresponding $\bar{\boldsymbol{x}}$ will come from back-substitution in (15.23):

$$A^{\mathrm{T}}CA\bar{\boldsymbol{x}} = A^{\mathrm{T}}C\boldsymbol{b} - A^{\mathrm{T}}CG\bar{\boldsymbol{I}}. \tag{15.27}$$

The right side is known, and $A^{\mathrm{T}}CA$ on the left side was already factored at the start of elimination. So $\bar{\boldsymbol{x}}$ is quickly found.

Integer Least Squares

We are minimizing a quadratic expression over integer variables. The absolute minimum occurs at $\hat{\boldsymbol{I}}$; the best integer vector is $\bar{\boldsymbol{I}}$. The coefficient matrix of the second-degree term in (15.26) is $G^{\mathrm{T}}C'G$, which is just the (2, 2) block in (15.24) after elimination. For blocks A, B, C, D that (2, 2) entry will be $Q = D - CA^{-1}B$. (In mathematics this is called the Schur complement.) With the blocks that appear in (15.24), this matrix Q is

$$Q = G^{\mathrm{T}}CG - G^{\mathrm{T}}CA(A^{\mathrm{T}}CA)^{-1}A^{\mathrm{T}}CG \qquad (= G^{\mathrm{T}}C'G). \tag{15.28}$$

Now consider the minimization for $\bar{\boldsymbol{I}}$. The quadratic expression in (15.26) has its absolute minimum at $\hat{\boldsymbol{I}}$. By adding a constant, the minimum value can be moved to zero. Knowing that the coefficient matrix is Q, the problem (15.26) can be stated in a very clear and equivalent form:

$$\text{Minimize} \quad (\boldsymbol{I} - \hat{\boldsymbol{I}})^{\mathrm{T}}Q(\boldsymbol{I} - \hat{\boldsymbol{I}}) \quad \text{over integer vectors } \boldsymbol{I}. \tag{15.29}$$

This is the problem we study, and one case is especially simple. ***If Q is a diagonal matrix, the best vector $\bar{\boldsymbol{I}}$ comes from rounding each component of $\hat{\boldsymbol{I}}$ to the nearest integer***. The components are uncoupled when Q is diagonal. The quadratic in (15.29) is purely a sum of squares, $\sum Q_{jj}(\boldsymbol{I}_j - \hat{\boldsymbol{I}}_j)^2$. The minimum comes by making each term as small as possible. So the best $\bar{\boldsymbol{I}}_j$ is the integer nearest to $\hat{\boldsymbol{I}}_j$.

Unfortunately, the actual Q may be far from diagonal. If we could change variables at will, we could diagonalize Q. But we are not completely free, since $\boldsymbol{I}$ is restricted to integers. ***A change of variables to $\boldsymbol{J} = Z^{-1}\boldsymbol{I}$ is only allowed if Z and Z^{-1} are matrices***

of integers. Then $\boldsymbol{J}$ is integer exactly when $\boldsymbol{I}$ is integer. The transformed quadratic has absolute minimum at $\hat{\boldsymbol{J}} = Z^{-1}\hat{\boldsymbol{I}}$, and we search for its integer minimum $\bar{\boldsymbol{J}}$:

$$\text{Minimize} \quad (\boldsymbol{J} - \hat{\boldsymbol{J}})^{\mathrm{T}}(Z^{\mathrm{T}}QZ)(\boldsymbol{J} - \hat{\boldsymbol{J}}) \quad \text{over integer vectors } \boldsymbol{J}. \tag{15.30}$$

The search is easier if $Z^{\mathrm{T}}QZ$ is nearly diagonal; its off-diagonal entries should be small.

We can describe in one paragraph the idea behind the choice of Z in the LAMBDA method. "Integer elimination" will be done in the natural order, starting with the first row of Q. There will certainly be row exchanges, and you will see why. The essential idea was given by Lenstra & Lenstra & Lovász (1982), and the algorithm is sometimes called L^3. The actual LAMBDA implementation might operate on columns instead of rows, and might go right to left. But to make the following paragraph clear, we just ask ourselves how to create near-zeros off the diagonal with integer elimination.

The first pivot is Q_{11} and the entry below it is Q_{21}. Normally we multiply the pivot row by the ratio $l_{21} = Q_{21}/Q_{11}$ and subtract from the second row. This produces a zero in the (2, 1) position. ***Our algorithm chooses instead the integer n_{21} that is nearest to l_{21}.*** That choice produces a *near-zero* in the (2, 1) position, not larger than $\frac{1}{2}Q_{11}$:

$$|Q_{21} - n_{21}Q_{11}| = |l_{21}Q_{11} - n_{21}Q_{11}| \le \tfrac{1}{2}Q_{11}. \tag{15.31}$$

If elimination continues in this usual order, the entry in each off-diagonal (i, j) position becomes ***not larger than half of the jth pivot d_j***. Together with the row operations to reduce the subdiagonal, we are including the corresponding column operations to reduce the superdiagonal. This produces a symmetric $Q' = Z^{\mathrm{T}}QZ$. The integers n_{ij} yield Z^{T} and Z as products of "integer Gauss steps" like

$$\begin{bmatrix} 1 & 0 \\ -n_{21} & 1 \end{bmatrix} \quad \text{with inverse} \quad \begin{bmatrix} 1 & 0 \\ n_{21} & 1 \end{bmatrix}.$$

So Z and Z^{-1} are integral, and they are assembled in the usual way. But a crucial point is still to be considered: ***smaller pivots d_j lead to smaller off-diagonal entries*** ($\le \frac{1}{2}d_j$). We prefer a row ordering in which the small pivots come first. The pivots are not known in advance, so the LAMBDA algorithm exchanges rows when a small pivot appears later. Then it recalculates the elimination to achieve (after iteration) the desired ordering.

A row exchange comes from a permutation matrix. The "decorrelating matrices" Z and Z^{-1} are no longer triangular, but they still contain integers. The new form (15.30) of the minimization has a more nearly diagonal matrix $Q' = Z^{\mathrm{T}}QZ$. This greatly reduces the number of candidate vectors $\boldsymbol{J}$. We display here a typical matrix Z from a specific calculation (and Z^{-1} also contains integers!):

$$Z = \begin{bmatrix} -2 & 3 & 1 \\ 3 & -3 & -1 \\ -1 & 1 & 0 \end{bmatrix} \quad \text{and} \quad Z^{-1} = \begin{bmatrix} 1 & 1 & 0 \\ 1 & 1 & 1 \\ 0 & -1 & -3 \end{bmatrix}.$$

You can see how Q moves significantly toward a diagonal matrix:

$$Q = \begin{bmatrix} 6.290 & 5.978 & 0.544 \\ 5.978 & 6.292 & 2.340 \\ 0.544 & 2.340 & 6.288 \end{bmatrix} \quad \text{decorrelates to} \quad Q' = \begin{bmatrix} 4.476 & 0.334 & 0.230 \\ 0.334 & 1.146 & 0.082 \\ 0.230 & 0.082 & 0.626 \end{bmatrix}.$$

The original float ambiguities are

$$\hat{\boldsymbol{I}} = \begin{bmatrix} 5.450 \\ 3.100 \\ 2.970 \end{bmatrix}.$$

The pivots = (conditional variances)$^{-1}$ for Q and Q' are

$$\begin{bmatrix} 0.090 \\ 5.421 \\ 6.288 \end{bmatrix} \quad \text{and} \quad \begin{bmatrix} 4.310 \\ 1.135 \\ 0.626 \end{bmatrix}.$$

Notice that LAMBDA reverses the elimination order, up instead of down! The final ambiguities are not rounded values of $\hat{\boldsymbol{I}}$, but close:

$$\bar{\boldsymbol{I}} = \begin{bmatrix} 5 \\ 3 \\ 4 \end{bmatrix}.$$

The matrix Q' is more nearly diagonal than Q. The integer least squares problem (= shortest lattice vector problem) is easier because the ellipsoids $\boldsymbol{J}^{\mathrm{T}} Q' \boldsymbol{J}$ = constant are not so elongated. But we still have to search for the $\bar{\boldsymbol{J}}$ that minimizes $(\boldsymbol{J} - \hat{\boldsymbol{J}})^{\mathrm{T}} Q' (\boldsymbol{J} - \hat{\boldsymbol{J}})$. We could search in a ball around the float solution $\hat{\boldsymbol{J}}$, but the actual LAMBDA implementation is more subtle.

Somewhere the algorithm will factor Q' into LDL^{T}. These factors come from elimination, and they indicate search ranges for the different components of $\boldsymbol{J}$. The off-diagonal entries of L reflect correlation that has not been removed. The search ellipsoid around $\hat{\boldsymbol{J}}$ has its volume controlled by the constant c:

$$(\boldsymbol{J} - \hat{\boldsymbol{J}})^{\mathrm{T}} Q' (\boldsymbol{J} - \hat{\boldsymbol{J}}) = \sum d_i \Big(\boldsymbol{J}_i - \hat{\boldsymbol{J}}_i + \sum l_{ki} (\boldsymbol{J}_k - \hat{\boldsymbol{J}}) \Big)^2 \le c^2. \tag{15.32}$$

de Jonge & Tiberius (1996) search for the components $\boldsymbol{J}_i$ in the order $n, n-1, \ldots, 1$. When index i is reached, a list of possibilities has been created for all $\boldsymbol{J}_k$ with $k > i$. For each of those possibilities, the bound (15.32) allows a finite search interval (probably small and possibly empty) of integer candidates $\boldsymbol{J}_i$. When we successfully reach $i = 1$, a complete candidate vector $\boldsymbol{J}$ satisfying (15.32) has been found. The search terminates when all candidates are known.

We chose c large enough to be certain that there is at least one candidate (for example, $\hat{\boldsymbol{J}}$ rounded to nearest integers). Then there is an efficient recursion to compute $(\boldsymbol{J} - \hat{\boldsymbol{J}})^{\mathrm{T}} Q' (\boldsymbol{J} - \hat{\boldsymbol{J}})$ for all candidates.

15.7 Sequential Filter for Absolute Position

To find the absolute position of a point is a *very* fundamental problem in positional GPS. We already have mentioned several methods to achieve the goal. We shall deal with one more method which is described by Bancroft (1985).

The implementation is done via filters to demonstrate the effect of each additional pseudorange. After three pseudoranges the covariance is expected to be very large. The fourth one is crucial because in the absence of other error sources, four pseudoranges determine the exact position and time.

We start by exposing the method in detail. It is implemented in the M-file bancroft. As always the raw pseudoranges have to be corrected for the tropospheric delay. This is a function of the satellite's elevation angle, so the correction for tropospheric delay needs at least two iterations. (Recently Jin (1996) has introduced a Taylor series expansion of the observation equations to avoid iteration.)

The observation equation of a single pseudorange P^k is

$$P^k = \sqrt{(X^k - X)^2 + (Y^k - Y)^2 + (Z^k - Z)^2} + c\,dt. \qquad (15.33)$$

We substitute the receiver clock offset $c\,dt$ by b. Move this to the left side, and square both sides:

$$\begin{aligned} P^k P^k - 2P^k b + b^2 &= (X^k - X)^2 + (Y^k - Y)^2 + (Z^k - Z)^2 \\ &= X^k X^k - 2X^k X + X^2 + \cdots + Z^k Z^k - 2Z^k Z + Z^2. \end{aligned}$$

We rearrange and get

$$\begin{aligned} (X^k X^k + Y^k Y^k + Z^k Z^k - P^k P^k) - 2(X^k X + Y^k Y + Z^k Z - P^k b) \\ = -(X^2 + Y^2 + Z^2 - b^2). \end{aligned}$$

This expression asks for the Lorentz inner product (which is computed by lorentz):

$$\langle \boldsymbol{g}, \boldsymbol{h} \rangle = \boldsymbol{g}^{\mathrm{T}} M \boldsymbol{h} \qquad \text{with} \qquad M = \begin{bmatrix} 1 & & & \\ & 1 & & \\ & & 1 & \\ & & & -1 \end{bmatrix}.$$

Using this inner product the equation above becomes

$$\tfrac{1}{2}\left\langle \begin{bmatrix} \boldsymbol{r}^k \\ P^k \end{bmatrix}, \begin{bmatrix} \boldsymbol{r}^k \\ P^k \end{bmatrix} \right\rangle - \left\langle \begin{bmatrix} \boldsymbol{r}^k \\ P^k \end{bmatrix}, \begin{bmatrix} \boldsymbol{r} \\ b \end{bmatrix} \right\rangle + \tfrac{1}{2}\left\langle \begin{bmatrix} \boldsymbol{r} \\ b \end{bmatrix}, \begin{bmatrix} \boldsymbol{r} \\ b \end{bmatrix} \right\rangle = 0. \qquad (15.34)$$

Every pseudorange gives rise to an equation of the type (15.34). Four equations are sufficient to solve for the receiver coordinates (X, Y, Z) and the receiver clock offset $b = c\,dt$. All our known quantities are in the matrix

$$B = \begin{bmatrix} X^1 & Y^1 & Z^1 & P^1 \\ X^2 & Y^2 & Z^2 & P^2 \\ X^3 & Y^3 & Z^3 & P^3 \\ X^4 & Y^4 & Z^4 & P^4 \end{bmatrix}.$$

Here X^k, Y^k, and Z^k denote the *geocentric coordinates* for the kth satellite at time of transmission and P^k is the observed pseudorange. The four observation equations are

$$\boldsymbol{\alpha} - BM \begin{bmatrix} \boldsymbol{r} \\ b \end{bmatrix} + \Lambda \boldsymbol{e} = \boldsymbol{0} \qquad (15.35)$$

where $\boldsymbol{e} = (1, 1, 1, 1)$, with

$$\alpha_k = \tfrac{1}{2}\left\langle \begin{bmatrix} \boldsymbol{r}^k \\ P^k \end{bmatrix}, \begin{bmatrix} \boldsymbol{r}^k \\ P^k \end{bmatrix} \right\rangle \qquad \text{and} \qquad \Lambda = \tfrac{1}{2}\left\langle \begin{bmatrix} \boldsymbol{r} \\ b \end{bmatrix}, \begin{bmatrix} \boldsymbol{r} \\ b \end{bmatrix} \right\rangle.$$

We solve (15.35) and get

$$\begin{bmatrix} \boldsymbol{r} \\ b \end{bmatrix} = MB^{-1}(\Lambda \boldsymbol{e} + \boldsymbol{\alpha}). \tag{15.36}$$

Since $\boldsymbol{r}$ and b also enter Λ, we insert (15.36) into (15.35) and use $\langle M\boldsymbol{g}, M\boldsymbol{h}\rangle = \langle \boldsymbol{g}, \boldsymbol{h}\rangle$:

$$\langle B^{-1}\boldsymbol{e}, B^{-1}\boldsymbol{e}\rangle \Lambda^2 + 2\big(\langle B^{-1}\boldsymbol{e}, B^{-1}\boldsymbol{\alpha}\rangle - 1\big)\Lambda + \langle B^{-1}\boldsymbol{\alpha}, B^{-1}\boldsymbol{\alpha}\rangle = 0. \tag{15.37}$$

This equation is quadratic in Λ. There are two possible Λ's, which give two solutions by (15.36). One of these solutions is the correct one.

Often five or more pseudoranges are observed. We advocate to use all available observations for calculating the receiver position. This changes equation (15.35) to a set of normal equations, i.e. we multiply to the left with B^{T}:

$$B^{\mathrm{T}}\boldsymbol{\alpha} - B^{\mathrm{T}}BM\begin{bmatrix} \boldsymbol{r} \\ b \end{bmatrix} + B^{\mathrm{T}}\Lambda \boldsymbol{e} = \boldsymbol{0}. \tag{15.38}$$

The same development as above gives $(B^{\mathrm{T}}B)^{-1}B^{\mathrm{T}} = B^{+}$ in the equation for Λ:

$$\big\langle B^{+}\boldsymbol{e}, B^{+}\boldsymbol{e}\big\rangle \Lambda^2 + 2\big(\big\langle B^{+}\boldsymbol{e}, B^{+}\boldsymbol{\alpha}\big\rangle - 1\big)\Lambda + \big\langle B^{+}\boldsymbol{\alpha}, B^{+}\boldsymbol{\alpha}\big\rangle = 0. \tag{15.39}$$

This expression comprises all observations in the sense of least squares. That completes the theory, now we turn to the MATLAB implementation b_point. Any comprehensive GPS code must be able to calculate a satellite position at a given time. This happens by calls of the M-files get_eph, find_eph, and satpos.

For more precise GPS calculations one needs to correct phase observations for the delay through the troposphere. A lot of procedures have been proposed; we implemented a method due to Goad & Goodman (1974) as the M-file tropo. The tropospheric delay mainly depends on the elevation angle to the satellite. (The M-file tropp contains hints on handling graphics of axes, contour labels and lines.)

The M-file topocent yields a value for the elevation angle El of a satellite. This is the most important parameter in the function tropo. The function also needs to know the receiver's elevation above sea level, by a call of togeod.

We transform a topocentric vector $\boldsymbol{x}$ into a local $\boldsymbol{e}$, $\boldsymbol{n}$, $\boldsymbol{u}$ coordinate system, with $\boldsymbol{u}$ in the direction of the plumb line, $\boldsymbol{n}$ pointing north, and $\boldsymbol{e}$ pointing east. The topocenter is given by the geocentric vector $\boldsymbol{X}$, and the three unit vectors go into the orthogonal matrix

$$F = \begin{bmatrix} \boldsymbol{e} & \boldsymbol{n} & \boldsymbol{u} \end{bmatrix} = \begin{bmatrix} -\sin\lambda & -\sin\varphi\cos\lambda & \cos\varphi\cos\lambda \\ \cos\lambda & -\sin\varphi\sin\lambda & \cos\varphi\sin\lambda \\ 0 & \cos\varphi & \sin\varphi \end{bmatrix}. \tag{15.40}$$

Let $(E, N, U) = F^{\mathrm{T}}\boldsymbol{x}$. Immediately we have azimuth, elevation angle, and length:

$$\begin{aligned} &\text{Azimuth:} && Az = \arctan(E/N) \\ &\text{Elevation angle:} && El = \arctan\big(U/\sqrt{N^2+E^2}\big) \\ &\text{Length:} && s = \|\boldsymbol{x}\|. \end{aligned}$$

The central part of the b_point code uses also togeod and of course the important M-file bancroft. The received time t_R^{GPS} and transmit time t_x^{GPS} are

$$t_R^{\mathrm{GPS}} = t_R - dt_R \qquad \text{and} \qquad t_x^{\mathrm{GPS}} = t_x + dt_x.$$

The pseudorange P is $c(t_R - t_x)$ with corrections trop and ion:

$$P = \underbrace{c(t_R^{\mathrm{GPS}} - t_x^{\mathrm{GPS}})}_{\rho} + c\,dt_R + c\,dt_x + \text{trop} + \text{ion} = \underbrace{\rho + c\,dt_R}_{\text{Bancroft}} + c\,dt_x + \text{trop} + \text{ion}.$$

The Bancroft model handles the two first terms. Additionally the pseudorange must be corrected by the term $c\,dt_x$. Remember to subtract trop. We set ion $= 0$. The auxiliary code lorentz, called by bancroft, calculates $\boldsymbol{x}^{\mathrm{T}}M\boldsymbol{x} = x_1^2 + x_2^2 + x_3^2 - x_4^2$.

The final result from b_point is the very best position one can deduce from a given set of pseudoranges. All possible corrections have been taken into account.

The Bayes filter yields one solution through the final update, but the Kalman filter is more illustrative. We can follow the contribution of each individual observation. Of course the calculations are similar for both methods. Yet the M-file k_point contains a few special calls. One of those is the M-file e_r_corr. It corrects the satellite position for Earth rotation during signal travel time: Let ω_e denote the Earth's rotation rate, x^S the satellite position before rotation and x_sat_rot the position after rotation. The rotation matrix is R_3 and the angle is $\omega\tau = \omega_e\rho/c$. Hence x_sat_rot $= R_3(\omega\tau)x^S$.

The file normals adds a given contribution to the normals and k_ud performs the Kalman update after reading a new observation.

We conclude with a numerical example. The epoch contains PRN's 23, 9, 5, 1, 21, and 17. The M-file satpos computes the following positions at corrected time of epoch:

PRN	23	9	5
X [m]	14 177 553.47	15 097 199.81	23 460 342.33
Y [m]	−18 814 768.09	−4 636 088.67	−9 433 518.58
Z [m]	12 243 866.38	21 326 706.55	8 174 941.25
P [m]	21 119 278.32	22 527 064.18	23 674 159.88

PRN	1	21	17
X [m]	−8 206 488.95	1 399 988.07	6 995 655.48
Y [m]	−18 217 989.14	−17 563 734.90	−23 537 808.26
Z [m]	17 605 231.99	19 705 591.18	−9 927 906.48
P [m]	20 951 647.38	20 155 401.42	24 222 110.91

Knowing the satellite positions and the measured pseudoranges we estimate the receiver coordinates by the Bancroft procedure. The M-file abs_pos yields the following result

$$\begin{aligned}\widehat{X} &= 596\,889.19\,\text{m}\\ \widehat{Y} &= -4\,847\,827.33\,\text{m}\\ \widehat{Z} &= 4\,088\,207.25\,\text{m}\\ \widehat{dt} &= -5.28\,\text{ns}.\end{aligned}$$

The value $\widehat{dt}$ is the estimated off-set of the receiver clock compared to GPS time. We find it useful to state explicitly the expression for the corrected pseudorange:

$$\text{corrected\_pseudorange} = P + c\,dt^k - \text{trop}. \tag{15.41}$$

The standard deviation of the receiver position is less than 27 m. This calculation is continued in a filter setting as Examples 17.7 and 17.8.

An Alternative Algorithm for Receiver Position

We mention another method for finding preliminary receiver coordinates (X, Y, Z) from four pseudoranges P^k, see Kleusberg (1994). The geometrically oriented reader may find this method easier to understand. Our point of departure is again the basic equation (15.33):

$$P^k = \sqrt{(X^k - X)^2 + (Y^k - Y)^2 + (Z^k - Z)^2} + c\,dt, \qquad k = 1, 2, 3, 4. \tag{15.42}$$

We subtract P^1 from P^2, P^3, and P^4. This eliminates the receiver clock offset dt:

$$\begin{aligned} d_l = P^l - P^1 &= \sqrt{(X^l - X)^2 + (Y^l - Y)^2 + (Z^l - Z)^2}\\ &\quad - \sqrt{(X^1 - X)^2 + (Y^1 - Y)^2 + (Z^1 - Z)^2} = \rho_l - \rho_1,\\ &\qquad l = 2, 3, 4. \end{aligned} \tag{15.43}$$

Once the coordinates (X, Y, Z) have been computed, dt can be determined from (15.42).

The three quantities d_2, d_3, d_4 are differences between distances to known positions of satellites. Points with correct d_l lie on one sheet of a hyperboloid (points on the other sheet have $-d_l$). Its axis of symmetry is the line between satellites l and 1. The hyperboloids intersect at the receiver position, and normally there exist two solutions. So at the end we have to identify which one is the correct solution.

Let b_2, b_3, b_4 be the known distances from satellite 1 to satellites 2, 3, 4, along unit vectors $\boldsymbol{e}_2$, $\boldsymbol{e}_3$, $\boldsymbol{e}_4$. From the cosine law for triangle 1-l-R follows

$$\rho_l^2 = b_l^2 + \rho_1^2 - 2b_l\rho_1\boldsymbol{e}_1 \cdot \boldsymbol{e}_l. \tag{15.44}$$

We rewrite (15.43) as $\rho_l = d_l + \rho_1$ and square the expression

$$\rho_l^2 = d_l^2 + \rho_1^2 + 2d_l\rho_1. \tag{15.45}$$

Equating (15.44) and (15.45) yields

$$2\rho_1 = \frac{b_l^2 - d_l^2}{d_l + b_l\boldsymbol{e}_1 \cdot \boldsymbol{e}_l}. \tag{15.46}$$

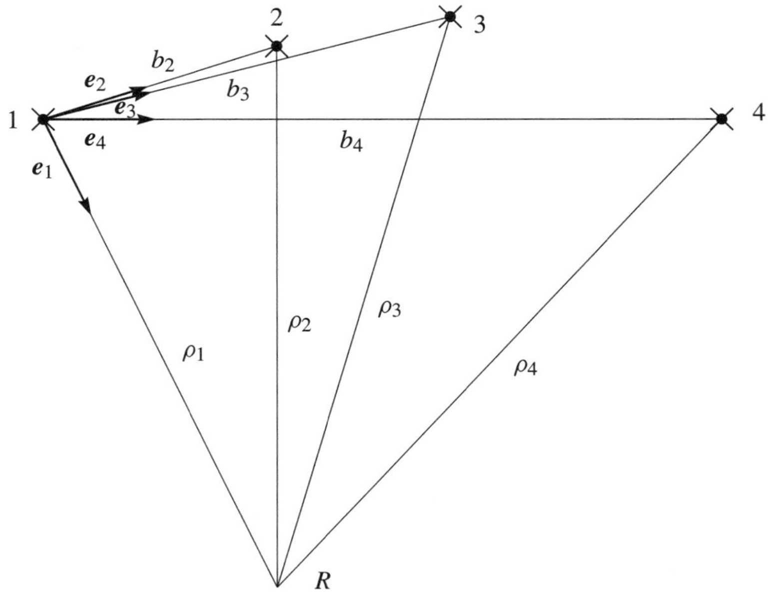

Figure 15.5 Four satellites and receiver R. Geometric distance between R and satellite k is ρ_k. Inter-satellite distance b_l along the unit vector e_l, $l = 2, 3, 4$.

This is three equations for ρ_1, or equivalently the following equations:

$$\frac{b_2^2 - d_2^2}{d_2 + b_2 \boldsymbol{e}_1 \cdot \boldsymbol{e}_2} = \frac{b_3^2 - d_3^2}{d_3 + b_3 \boldsymbol{e}_1 \cdot \boldsymbol{e}_3} = \frac{b_4^2 - d_4^2}{d_4 + b_4 \boldsymbol{e}_1 \cdot \boldsymbol{e}_4}. \tag{15.47}$$

Now ρ_1 is eliminated and the only unknown is the 3 by 1 unit vector $\boldsymbol{e}_1$.

Some rewritings result in the two scalar equations

$$\boldsymbol{e}_1 \cdot \boldsymbol{f}_m = u_m, \qquad m = 2, 3. \tag{15.48}$$

Here we have used the following abbreviations:

$$\boldsymbol{F}_m = \frac{b_m}{b_m^2 - d_m^2}\boldsymbol{e}_m - \frac{b_{m+1}}{b_{m+1}^2 - d_{m+1}^2}\boldsymbol{e}_{m+1}$$

$$\boldsymbol{f}_m = \frac{\boldsymbol{F}_m}{\|\boldsymbol{F}_m\|}$$

$$u_m = \frac{1}{\|\boldsymbol{F}_m\|}\left(\frac{d_{m+1}}{b_{m+1}^2 - d_{m+1}^2} - \frac{d_{m+1}}{b_{m+1}^2 - d_{m+1}^2}\right).$$

The unit vector $\boldsymbol{f}_2$ lies in the plane through satellites 1, 2, 3. This plane is spanned by $\boldsymbol{e}_2$ and $\boldsymbol{e}_3$. Similarly $\boldsymbol{f}_3$ is in the plane determined by satellites 1, 3, and 4. The two vectors $\boldsymbol{f}_2$ and $\boldsymbol{f}_3$ and the right sides u_2 and u_3 may be computed from the known coordinates of the satellites and the measured differences of distances.

However we want to present a geometric procedure based on vector algebra. Equation (15.48) determines the cosine of the two given unit vectors $\boldsymbol{f}_2$ and $\boldsymbol{f}_3$ and the sought unit vector $\boldsymbol{e}_1$. In general the problem has two solutions, one above and one below the plane spanned by $\boldsymbol{f}_2$ and $\boldsymbol{f}_3$. In case $\boldsymbol{f}_2$ and $\boldsymbol{f}_3$ are parallel their inner product is zero and there are infinitly many solutions.

In principle the solution of (15.48) may be found after a parametrization into spherical coordinates of the unit vectors $\boldsymbol{e}_1$, $\boldsymbol{f}_2$, and $\boldsymbol{f}_3$. The two solutions for $\boldsymbol{e}_1$ can then follow as plane solutions on the unit sphere. However this procedure leads to problems with determination of signs. In order to circumvent this problem we proceed from the basic formula for double vector products:

$$\boldsymbol{e}_1 \times (\boldsymbol{f}_1 \times \boldsymbol{f}_2) = \boldsymbol{f}_1(\boldsymbol{e}_1 \cdot \boldsymbol{f}_2) - \boldsymbol{f}_2(\boldsymbol{e}_1 \cdot \boldsymbol{f}_1). \tag{15.49}$$

Comparing with (15.48) the scalar products on the right side can immediately be identified with the unknowns u_2 and u_3. We substitute $\boldsymbol{h}$ for the right side and $\boldsymbol{g}$ for $\boldsymbol{f}_1 \times \boldsymbol{f}_2$. Then (15.49) becomes

$$\boldsymbol{e}_1 \times \boldsymbol{g} = \boldsymbol{h}. \tag{15.50}$$

The two solutions of (15.48) are

$$\boldsymbol{e}^{\pm} = \frac{1}{\boldsymbol{g} \cdot \boldsymbol{g}}\Big(\boldsymbol{g} \times \boldsymbol{h} \pm \boldsymbol{g}\sqrt{\boldsymbol{g} \cdot \boldsymbol{g} - \boldsymbol{h} \cdot \boldsymbol{h}}\Big). \tag{15.51}$$

As long as $\boldsymbol{g} \cdot \boldsymbol{g} \neq 0$ this gives two solutions. When $\boldsymbol{f}_1$ and $\boldsymbol{f}_2$ are parallel this product equals zero!

Knowing the solutions $\boldsymbol{e}^+$ and $\boldsymbol{e}^-$ we insert into one of the three equations (15.46) and get

$$\rho_1^{\pm} = \frac{b_l^2 - d_l^2}{2(d_l + b_l \boldsymbol{e}_1^{\pm} \cdot \boldsymbol{e}_l)}. \tag{15.52}$$

Now different solution situations may appear. The two distances ρ_1^+ and ρ_1^- may or may not both be positive. In case they both are positive we have two points of intersection of the hyperboloids: one above and one below the plane spanned by the vectors $\boldsymbol{f}_1$ and $\boldsymbol{f}_2$. The solution wanted must have a distance of about 6 700 km from the origin. In case one distance $\rho_1^{\pm}$ is negative this one can be omitted because we have only one intersection point and the solution corresponding to the non existing intersection point has a negative denominator which will not be compatibel with equation (15.46). With the correctly identified solution ρ_1^{correct} from (15.52) and the corresponding unit vector $\boldsymbol{e}_1^{\text{correct}}$ we finally get the receiver coordinates

$$\boldsymbol{X} = \boldsymbol{X}_1 + \rho_1^{\text{correct}} \boldsymbol{e}_1^{\text{correct}}. \tag{15.53}$$

The solution is implemented as the M-file kleus.

15.8 Additional Useful Filters

To demonstrate what filters can really do, we start with simple examples relevant to GPS. Suppose P_1, Φ_1, P_2, Φ_2 are double differenced observations between two receivers and two satellites. The observation equation for each epoch is

$$\begin{bmatrix} 1 & 0 & 0 \\ 1 & \lambda_1 & 0 \\ 1 & 0 & 0 \\ 1 & 0 & \lambda_2 \end{bmatrix} \begin{bmatrix} \rho^* \\ N_1 \\ N_2 \end{bmatrix} = \begin{bmatrix} P_1 \\ \Phi_1 \\ P_2 \\ \Phi_2 \end{bmatrix} - \boldsymbol{e}.$$

This is four equations in three unknowns: the ideal pseudorange ρ^* and the two ambiguities on frequencies L_1 and L_2. The covariance matrix for the observations is

$$\Sigma_e = \begin{bmatrix} 0.3^2 & & & \\ & 0.005^2 & & \\ & & 0.3^2 & \\ & & & 0.005^2 \end{bmatrix}. \tag{15.54}$$

The covariance matrix for the errors ϵ in the state equation is

$$\Sigma_\epsilon = \begin{bmatrix} 10 & & \\ & 0 & \\ & & 0 \end{bmatrix}.$$

The variance of ρ^* equals $10\,\text{m}^2$ while the ambiguities have zero variance. The initial value for $\boldsymbol{x}_0$ is found as a least squares solution of the four equations in the three unknowns at epoch 5. The early data are quite noisy due to a cold start of the receiver, with

$$P_{0|0} = \begin{bmatrix} 10 & & \\ & 10 & \\ & & 10 \end{bmatrix}.$$

The output of the M-file k_dd3 is a plot as well as filtered values for N_1 and $N_1 - N_2$. Of course, the filtered values are not integers. We have added the Goad algorithm for rounding to integers as described in equations (15.10) and (15.11). The output indicates the effectiveness of this algorithm. For the given data samples the Goad algorithm finds the correct ambiguities every time.

Ionospheric delay

We mentioned earlier that the tropospheric delay has its minumum in direction of zenith. In zenith the ionospheric delay can vary form a few meters to many tens of meters. Fortunately the ionosphere is dispersive: the refraction index depends on the frequency.

From equation (15.8) we derive a dual frequency ionospheric correction. We repeat the two pseudorange observations

$$P_1 = \rho^* + I - e_1$$
$$P_2 = \rho^* + (f_1/f_2)^2 I - e_2.$$

Ignoring the error terms e_i, and eliminating the ionospheric delay I, we get an expression for the ideal pseudorange ρ^* freed from I:

$$\rho^* = P_1 - \frac{P_1 - P_2}{1-\beta} = P_1 + 1.545\,727\,802(P_1 - P_2). \tag{15.55}$$

Here $\beta = (f_1/f_2)^2 = (77/60)^2$.

Next we describe a modified version k_dd4 that works with four unknowns: the ideal pseudorange ρ^*, the ionospheric delay I (*now included*), and the integer ambiguities N_1 and N_2. We still observe pseudorange and code on both frequencies:

$$\begin{bmatrix} 1 & 1 & 0 & 0 \\ 1 & -1 & \lambda_1 & 0 \\ 1 & \beta & 0 & 0 \\ 1 & \beta & 0 & \lambda_2 \end{bmatrix} \begin{bmatrix} \rho^* \\ I \\ N_1 \\ N_2 \end{bmatrix} = \begin{bmatrix} P_1 \\ \Phi_1 \\ P_2 \\ \Phi_2 \end{bmatrix} + \boldsymbol{e}.$$

The covariance matrix for P_1, Φ_1, P_2, Φ_2 is again the Σ_e described in equation (15.54). The initial value for $\boldsymbol{x}_0$ is again found as a solution to the four equations in the four unknowns at epoch 5. The early data are quite noisy due to a cold start of the receiver. The covariance matrix for the errors $\boldsymbol{\epsilon}$ in the state equation is set to

$$\Sigma_\epsilon = \begin{bmatrix} 100 & & & \\ & 10 & & \\ & & 0 & \\ & & & 0 \end{bmatrix}.$$

The standard deviation is 10 m for the range ρ^* and $\sqrt{10}$ m for ionospheric delay. The ambiguities N_1 and N_2 again have zero variances.

The output contains the filtered values for ρ^*, I, N_1 and the difference dN_w between the widelane value $N_w = N_1 - N_2$ found by k_dd3 and the actual value. The ambiguities are evidently more unreliable using this filter than the k_dd3 values. This is due to the inclusion of the ionospheric delay as unknown.

The ionospheric delay can change rapidly in absolute value. Variations depend on season, latitude, time of day, and other parameters. Extensive studies of the ionospheric delay have been made by Klobuchar, see Klobuchar (1996).

Example 15.1 (Estimation of receiver clock offset) In GPS surveying, the code observations b_i (pseudoranges) are often used to estimate the coordinates X, Y, Z of a single point and the sequence of receiver clock offsets (changing with time). Suppose that epoch i contributes the following linearized observation

$$A_i \begin{bmatrix} x \\ y \\ z \end{bmatrix} + \boldsymbol{e}_i^{\mathrm{T}} c\, dt_i = b_i - \epsilon_i, \qquad i = 1, 2, \ldots, n. \tag{15.56}$$

The matrix A_i contains the partial derivatives of the observation with respect to the coordinates of the point at the epoch i, and $\boldsymbol{e}_i = (1, 1, \ldots, 1)$ is a vector of as many ones as there are satellites in the epoch.

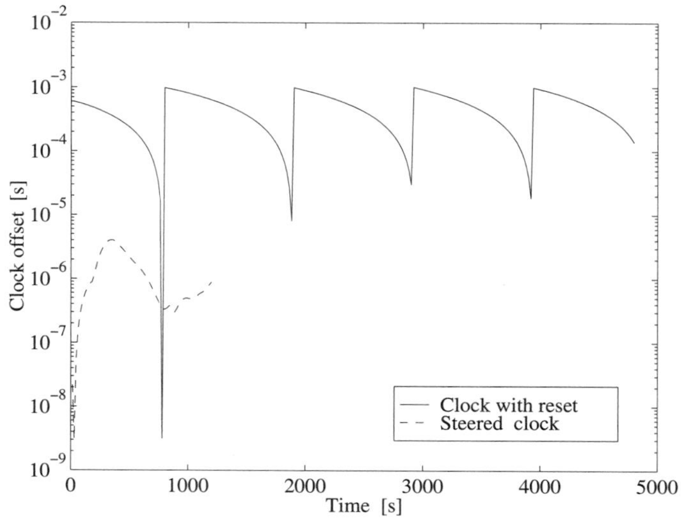

Figure 15.6 Offsets for different receiver types. The clock reset is 1 millisecond.

We gather the observations from all n epochs into a unified least squares problem

$$\begin{bmatrix} A_1 \\ A_2 \\ \vdots \\ A_n \end{bmatrix} \begin{bmatrix} x \\ y \\ z \end{bmatrix} + \begin{bmatrix} e_1^{\mathrm{T}} & & & \\ & e_2^{\mathrm{T}} & & \\ & & \ddots & \\ & & & e_n^{\mathrm{T}} \end{bmatrix} \begin{bmatrix} c\,dt_1 \\ c\,dt_2 \\ \vdots \\ c\,dt_n \end{bmatrix} = \begin{bmatrix} b_1 \\ b_2 \\ \vdots \\ b_n \end{bmatrix} - \boldsymbol{\epsilon}. \tag{15.57}$$

The normal equations are

$$\begin{bmatrix} e_1^{\mathrm{T}} e_1 & & & & e_1^{\mathrm{T}} A_1 \\ & e_2^{\mathrm{T}} e_2 & & & e_2^{\mathrm{T}} A_2 \\ & & \ddots & & \vdots \\ & & & e_n^{\mathrm{T}} e_n & e_n^{\mathrm{T}} A_n \\ A_1^{\mathrm{T}} e_1 & A_2^{\mathrm{T}} e_2 & \dots & A_n^{\mathrm{T}} e_n & \sum_{j=1}^{n} A_j^{\mathrm{T}} A_j \end{bmatrix} \begin{bmatrix} c\,dt_1 \\ c\,dt_2 \\ \vdots \\ c\,dt_n \\ x \\ y \\ z \end{bmatrix} = \begin{bmatrix} e_1^{\mathrm{T}} b_1 \\ e_2^{\mathrm{T}} b_2 \\ \vdots \\ e_n^{\mathrm{T}} b_n \\ \sum_{j=1}^{n} A_j^{\mathrm{T}} b_j \end{bmatrix}. \tag{15.58}$$

By ordinary Gauss elimination we subtract multiples of the first n equations from the last block row. We write E_j for the matrix $e_j(e_j^{\mathrm{T}} e_j)^{-1} e_j^{\mathrm{T}}$. The correction (x, y, z) of the preliminary position (X^0, Y^0, Z^0) is determined by the matrix that appears in the last corner:

$$\begin{bmatrix} x \\ y \\ z \end{bmatrix} = \left(\sum_{j=1}^{n} \left(A_j^{\mathrm{T}} A_j - A_j^{\mathrm{T}} E_j A_j \right) \right)^{-1} \sum_{j=1}^{n} \left(A_j^{\mathrm{T}} b_j - A_j^{\mathrm{T}} E_j b_j \right).$$

The estimate of the receiver clock offsets $c\,dt_i$ is found by back substitution.

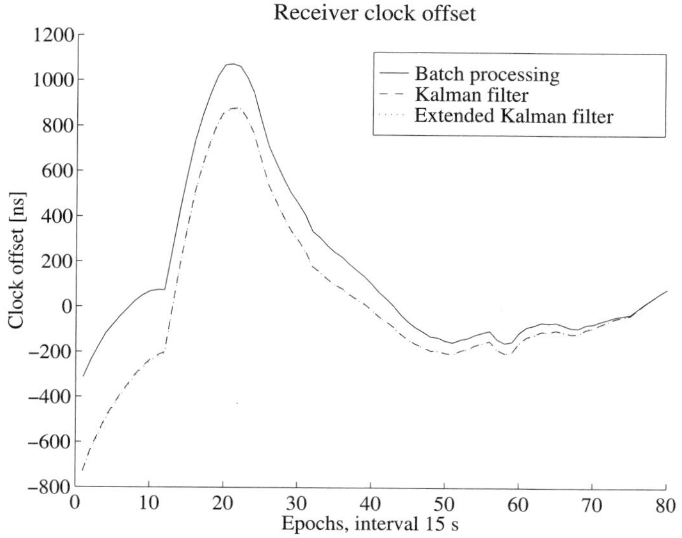

Figure 15.7 Receiver clock offsets computed by batch processing and Kalman filtering.

The estimation model clearly demonstrates why it is not necessary to collect four observations at all epochs for static observations. But a sufficient number of observations is needed to keep (15.58) invertible. If only one observation is available at a particular epoch, we can estimate the receiver clock offset, but not the position.

We recommend to use the described procedure, as some manufacturers introduce discontinuous changes in the clock time to keep the offsets within prescribed tolerances. Certain receivers have their clocks reset when the offset approaches one millisecond. Figure 15.6 demonstrates the jumps in the offset for this receiver type as well as another receiver type which has a steered clock.

Figure 15.7 shows the offset for a steered receiver clock with continual corrections to reduce the offset. The plot is made by the M-file recclock. The code iterates three times to get the correct receiver position—the clock estimation is linear!

Extended Kalman Filter

All state vectors considered so far have been differential corrections $\delta \boldsymbol{x}_k$ to some starting value $\boldsymbol{X}^0$. Especially for a position filter the vector $\boldsymbol{X}^0$ denotes the preliminary coordinates (X^0, Y^0, Z^0) of a (receiver) position. The matrix A collects all the partial derivatives of the observations $\boldsymbol{b}_k$ with respect to the coordinates. This is all well known.

Sometimes it is also necessary to keep track of the total state vector $\boldsymbol{x}_k = \boldsymbol{X}^0 + \delta \boldsymbol{x}_k$. So we proceed to demonstrate how this is done. Linearize the observation equation:

$$\boldsymbol{b} = g(\boldsymbol{X}^0 + \delta \boldsymbol{x}) + \boldsymbol{e} = g(\boldsymbol{X}^0) + \frac{\partial g}{\partial \boldsymbol{X}}\bigg|_{\boldsymbol{X}=\boldsymbol{X}^0} \delta \boldsymbol{x} + \boldsymbol{e} = g(\boldsymbol{X}^0) + G\delta \boldsymbol{x} + \boldsymbol{e}.$$

For the original state $\boldsymbol{X}$, the observation is $\boldsymbol{b}$ rather than $\boldsymbol{b} - g(\boldsymbol{X}^0)$, and we get

$$\delta\hat{\boldsymbol{x}}_k = \delta\hat{\boldsymbol{x}}_{k|k-1} + K_k\big(\boldsymbol{b}_k - g(\boldsymbol{X}^0) - G_k\,\delta\boldsymbol{x}_{k|k-1}\big).$$

The update equation, when we add $\boldsymbol{X}_k^0$ to both sides, is now

$$\begin{aligned}\underbrace{\boldsymbol{X}_k^0 + \delta\hat{\boldsymbol{x}}_k}_{\hat{\boldsymbol{x}}_k} &= \underbrace{\boldsymbol{X}_k^0 + \delta\hat{\boldsymbol{x}}_{k|k-1}}_{\hat{\boldsymbol{x}}_{k|k-1}} + K_k\big(\boldsymbol{b}_k - \underbrace{g(\boldsymbol{X}^0) - G_k\,\delta\boldsymbol{x}_{k|k-1}}_{\hat{\boldsymbol{b}}_{k|k-1}}\big).\end{aligned}$$

This is the usual linear update equation, for *total* rather than incremental quantities. It simply says that we correct the a priori estimate by adding the observation *residual*, weighted by K_k. Note that after the correction is made in the ***extended Kalman filter***, the increment $\delta\hat{\boldsymbol{x}}_k$ is reduced to zero. The prediction is then trivial. The only non-trivial prediction is $\hat{\boldsymbol{x}}_{k|k-1}$ (which has become the nominal $\boldsymbol{X}$ at t_k). This must be done through the nonlinear dynamics of $\hat{\boldsymbol{x}}_{k|k-1} = g(\boldsymbol{X}^0 + \hat{\boldsymbol{x}}_k)$. Then we can form the predicted $\hat{\boldsymbol{b}}_{k|k-1} = g(\hat{\boldsymbol{x}}_{k|k-1})$. The residual is $\boldsymbol{b}_k - \hat{\boldsymbol{b}}_{k|k-1}$ and we are prepared for the next loop.

Example 15.2 (Estimation of receiver clock offset by extended Kalman filter) The M-file kalclock uses an extended filter. After filtering of all observations in each epoch we do the following

```
if extended_filter == 1
  pos(1:3,1) = pos(1:3,1) + x(1:3,1);
  x(1:3,1)= [0; 0; 0];
end
rec_clk_offset = [rec_clk_offset x(4,1)];
```

This code implies that at the end of the first iteration we have an updated position which deviates only a small amount from the position computed in the batch run. For the file pta.96o the discrepancy in position is $(0.12, 0.54, -0.19)$ m. So this small deviation is what we pay for a much faster computation using only one iteration instead of three. The result is shown in Figure 15.7.

Example 15.3 We want to study a single one-way range between a satellite and a receiver. We assume that P code pseudoranges and phase observations are taken on both frequencies L_1 and L_2 for 50 epochs. As usual we denote the observations by P_1, Φ_1, P_2, and Φ_2. Unlike all earlier instances, the observations are *un*differenced! Our goal is to study how well P code pseudoranges can help to estimate ambiguities $N_1 - N_2$ and N_1 for undifferenced observations. We describe ideas published in Euler & Goad (1991).

An appropriate filter is a sequential formulation of the *Bayes version*:

State prediction:	$\hat{\boldsymbol{x}}_{k|k-1} = F_k\hat{\boldsymbol{x}}_{k-1|k-1} + \boldsymbol{\epsilon}_k$	(15.59)
Covariance prediction:	$P_{k|k-1} = F_k P_{k-1|k-1} F_k^{\mathrm{T}} + \Sigma_{\epsilon,k}$	(15.60)
Covariance update:	$P_{k|k} = \big(P_{k|k-1}^{-1} + A_k^{\mathrm{T}}\Sigma_{e,k}^{-1}A_k\big)^{-1}$	(15.61)
Gain matrix (after $P_{k|k}$!):	$K_k = P_{k|k}A_k^{\mathrm{T}}\Sigma_{e,k}^{-1}$	(15.62)
State update:	$\hat{\boldsymbol{x}}_{k|k} = \hat{\boldsymbol{x}}_{k|k-1} + K_k(\boldsymbol{b}_k - A_k\hat{\boldsymbol{x}}_{k|k-1})$.	(15.63)

A remark is needed on the matrix $P_{k|k-1}^{-1}$ in (15.61). In case the state transition noise is large, it becomes difficult to predict the corresponding entry of the state vector. This situation is handled well by the predicted covariance matrix $P_{k|k-1} = \left[\begin{smallmatrix} \infty & \sigma_{12} \\ \sigma_{21} & \sigma_{22} \end{smallmatrix}\right]$. Here ∞ denotes one very large variance, or a submatrix with almost infinite diagonal elements. The inverse of this (block) matrix has the following form by the identity (17.47):

$$P_{k|k-1}^{-1} = \begin{bmatrix} 0 & 0 \\ 0 & \sigma_{22}^{-1} \end{bmatrix}.$$

This implies that the first entries of the previous state vector $\boldsymbol{x}$ will have no influence on the new state. The filter process that we now describe takes advantage of this behavior.

Our observation equations $\boldsymbol{b}_k = A_k \boldsymbol{x}_k - \boldsymbol{e}_k$ are identical to (15.30):

$$\begin{bmatrix} P_1 \\ \Phi_1 \\ P_2 \\ \Phi_2 \end{bmatrix} = \begin{bmatrix} 1 & 1 & 0 & 0 \\ 1 & -1 & \lambda_1 & 0 \\ 1 & (f_1/f_2)^2 & 0 & 0 \\ 1 & -(f_1/f_2)^2 & 0 & \lambda_2 \end{bmatrix} \begin{bmatrix} \rho^* \\ I \\ N_1 \\ N_2 \end{bmatrix} + \text{noise}.$$

The system equation is the steady model $\hat{\boldsymbol{x}}_{k|k-1} = \hat{\boldsymbol{x}}_{k-1|k-1}$, so we use the filter with $F_k = I$. Hence (15.60) becomes

$$P_{k|k-1} = P_{k-1|k-1} + \Sigma_{\epsilon,k}. \tag{15.64}$$

The transition covariance matrix $\Sigma_{\epsilon,k}$ is diagonal. The $(3,3)$ and $(4,4)$ entries must be zero to prevent N_1 and N_2 from changing. We have to allow for large changes of ρ^* and I:

$$\Sigma_{\epsilon,k} = \begin{bmatrix} \infty & & & \\ & \infty & & \\ & & 0 & \\ & & & 0 \end{bmatrix}.$$

Again ∞ symbolizes a very large but finite number. The update from (15.60) is

$$P_{k|k-1} = P_{k-1|k-1} + \Sigma_{\epsilon,k} = \left[\begin{array}{cc|cc} \infty & \sigma_{12} & \sigma_{13} & \sigma_{14} \\ \sigma_{21} & \infty & \sigma_{23} & \sigma_{24} \\ \hline \sigma_{31} & \sigma_{32} & \sigma_{33} & \sigma_{34} \\ \sigma_{41} & \sigma_{42} & \sigma_{43} & \sigma_{44} \end{array}\right].$$

According to the Schur identity (17.47) we get

$$P_{k|k-1}^{-1} = \begin{bmatrix} 0 & 0 & 0 & 0 \\ 0 & 0 & 0 & 0 \\ 0 & 0 & \multicolumn{2}{c}{\multirow{2}{*}{$\begin{bmatrix} \sigma_{33} & \sigma_{34} \\ \sigma_{43} & \sigma_{44} \end{bmatrix}^{-1}$}} \\ 0 & 0 & & \end{bmatrix}.$$

This means that previous information of ρ^* and I is effectively neglected in the update of the new state vector $\hat{\boldsymbol{x}}_{k|k}$. The covariance matrix of the observations is

$$\Sigma_{e,k} = \begin{bmatrix} \sigma_{P_1}^2 & & & \\ & \sigma_{\Phi_1}^2 & & \\ & & \sigma_{P_2}^2 & \\ & & & \sigma_{\Phi_2}^2 \end{bmatrix} = \begin{bmatrix} \sigma_{P_1}^2 & & & \\ & 0.005^2 & & \\ & & \sigma_{P_2}^2 & \\ & & & 0.005^2 \end{bmatrix}.$$

Table 15.2 Standard deviation σ_P for one-way ranges as function of elevation angle h

h (in °)	0	10	20	30	40	50	60	70	80	90
σ_P (in m)	4.58	1.73	0.69	0.30	0.16	0.11	0.09	0.08	0.08	0.08

Uncertainties depending on elevation can be modeled as an exponential expression for the standard error $\sigma_P = a_0 + a_1 e^{-h/h_0}$, where h_0 is a scaled value of the elevation error. In Section 16.3 we demonstrate how to estimate a_0, a_1, and h_0. Reasonable values are $a_0 = 0.08$ m, $a_1 = 4.5$ m and $h_0 = 10°$. This gives $\sigma_P = 0.08 + 4.5e^{-h/10}$ in meters and h in degrees, tabulated in Table 15.2. To repeat, σ_P^2 yields the entries (1,1) and (3,3) of $\Sigma_{e,k}$. Phase measurements are considered to be independent of elevation angle.

The estimated values for the wide lane ambiguity $N_w = N_1 - N_2$ were used to form double difference ambiguities $N_{w,ij}^{kl} = (N_{w,i}^k - N_{w,i}^l) - (N_{w,j}^k - N_{w,j}^l)$. In all cases the computed values were in agreement with the "exact" values. So a good way of estimating double difference ambiguities is *to start from the one-way ambiguities*. The estimates of one-way ambiguities are independent of the length of the baseline. These cases can avoid a computational need for a LAMBDA method.

A similar procedure for N_1 ambiguity shows that the computed double difference ambiguities never deviate more than two cycles from the true values. This is a really promising procedure.

From the filter covariance matrix $P_{N|N}$ we can compute the covariance matrix for the combinations $N_w = N_1 - N_2$ and $N_n = N_1 + N_2$. The smallest eigenvalue $\lambda_{\min}$ in Table 15.3 shows the standard deviation of $N_1 - N_2$. The smaller it is, the more reliable is the computation. Recall that the wide lane has wave length $\lambda_w = 0.863$ m (from $1/\lambda_w = 1/\lambda_1 - 1/\lambda_2$) and the narrow lane wave has $\lambda_n = 0.107$ m (from $1/\lambda_n = 1/\lambda_1 + 1/\lambda_2$). To estimate the standard deviation σ_w of the wide lane ambiguity we use $\sigma_w = \sqrt{\lambda_{\min}}\,\lambda_w = 0.05$ m. The largest eigenvalue $\lambda_{\max}$ measures the difficulty in estimating $N_1 + N_2$. A similar calculation yields $\sigma_n = \sqrt{\lambda_{\max}}\,\lambda_n = 0.49$ m which is more than four times λ_n! This explains why it is always more difficult to calculate the narrow lane ambiguity $N_1 + N_2$.

The M-file one_way allows the reader to experiment with the enclosed data sets.

Table 15.3 Eigenvalues of covariance matrix for narrow and wide lane ambiguities

PRN	mean of h	$\lambda_{\max}$	$\lambda_{\min}$
2	60	0.276	0.000 07
9	18	15.594	0.003 48
16	20	7.424	0.001 66
23	20	12.010	0.002 68
26	70	0.235	0.000 06
27	30	3.241	0.000 73

Real-time Positioning Using Differential Carrier Phase

This book does not include the topic of real-time positioning. Yet we cannot resist to mention a promising recent development.

Most published models for computing real-time differential corrections for the rover are based on simultaneity. The corrections calculated at the master i are directly transmitted to the rover j. But this procedure inherits especially a latency problem.

Lapucha & Barker & Liu (1996) give up the time matching at master and rover. To describe their idea we repeat the basic equation (14.16) for a phase observation using satellite k:

$$\Phi_i^k(t) = \rho_i^k - I_i^k + T_i^k + c\big(dt^k(t-\tau_i^k) - dt_i(t)\big) + \lambda\big(\varphi_i(t_0) - \varphi^k(t_0)\big) + \lambda N_i^k + \epsilon_i^k.$$

Double frequency receivers will eliminate the ionospheric delay I_i^k. The tropospheric delay T_i^k can be modeled and largely removed. Anyway the distance between master and rover can never extend the distance over which the correction signal can be transmitted. This is typically less than 25 km. The tropospheric correction over such distances nearly cancels out in double differences. We also assume the ambiguity N_i^k to be solved on-the-fly (typical duration 15–30 s).

All information about a receiver position and dynamics is contained in the range ρ_i^k. The remaining terms on the right generally are unknown and need to be accounted for in the processing. However at the master we can determine a phase correction:

$$\Phi_{c,i}^k(t) = \Phi_i^k - \big(\rho_i^k + c\,dt^{\mathrm{Broadcast},k}(t-\tau_i^k) - c\,dt_i(t)\big). \tag{15.65}$$

The approximate master clock offset $c\,dt_i(t)$ is computed using an appropriate algorithm. The phase correction $\Phi_{c,i}^k(t)$ is thus equivalent to

$$\Phi_{c,i}^k(t) = c\,dt^{\mathrm{SA},k}(t) - I_i^k + T_i^k + \lambda\big(\varphi_i(t_0) - \varphi^k(t_0)\big) + \lambda N_i^k + \epsilon_i^k. \tag{15.66}$$

The term $c\,dt^{\mathrm{SA},k}(t) = c\,dt^k(t-\tau_i^k) - c\,dt^{\mathrm{Broadcast},k}(t-\tau_i^k)$ represents the unknown satellite clock dithering due to SA (selective availability).

The phase corrections given in (15.65) and (15.66) refer to the past time $t = t_0$. But these corrections have to be extrapolated to the current user time t_e. Any second order extrapolation involves errors that depend on the correction rates and their accelerations. The changes in the observed carrier phase are mainly due to the satellite clock dithering $c\,dt^{\mathrm{SA},k}(t)$. The orbit and atmosphere errors vary slowly by comparison.

Experiments show that the correction accelerations can be as large as 0.01 m/s^2. Neglecting these accelerations would cause an error of several centimeters in the extrapolation. So the second order extrapolation model must be as follows:

$$\Phi_{c,i}^k(t_e) = \Phi_{c,i}^k(t_0) + \dot\Phi_{c,i}^k(t_0)(t_e - t_0) + \tfrac{1}{2}\ddot\Phi_{c,i}^k(t_0)(t_e - t_0)^2. \tag{15.67}$$

The phase rate $\dot\Phi_{c,i}^k(t_0)$ and the acceleration $\ddot\Phi_{c,i}^k(t_0)$ are estimated from the past observations at the master i. At the rover j one applies the extrapolated corrections (15.67):

$$\widetilde\Phi_j^k(t_e) = \Phi_j^k(t_e) - \Phi_{c,i}^k(t_e). \tag{15.68}$$

The actual observable used in the rover differential phase positioning filter is the difference of the corrected phase observations with respect to a reference satellite k. The observation model for the rover is thus derived by combining equations (15.65) and (15.68):

$$\widetilde{\Phi}_j^{kl}(t_e) = \rho_j^{kl}(t_e) + N_{ij}^{kl}(t_e). \tag{15.69}$$

If $t_0 = t_e$ the formulation in (15.69) is equivalent to the double difference kinematic model:

$$\Phi_{ij}^{kl}(t_0) = \rho_{ij}^{kl}(t_0) + N_{ij}^{kl}(t_0).$$

Experiments show that *the phase prediction error is on average below 5 cm at a correction update interval of 5 seconds*. At this rate the positioning accuracy should be maintained at the several-centimeter level. This opens the possibility of using slower data links than those required for real time kinematic positioning to maintain the same position output rate. Applications of differential phase positioning include construction machine guidance and high resolution hydrographic surveying, where continuous output with minimum latency is a must.

16

RANDOM PROCESSES

16.1 Random Processes in Continuous Time

So far the order of observations has been of no concern. We collected all the observations and by least squares we estimated the parameters in $\boldsymbol{x}$. However, there are important situations where the ***time*** for the observation does play a role. The observation made at time t is denoted $\boldsymbol{x}(t)$. The sequence of observations taken at times $t = t_1, t_2, \ldots, t_n$ is denoted $\boldsymbol{x}(t_1), \boldsymbol{x}(t_2), \ldots, \boldsymbol{x}(t_n)$. We are observing (and then estimating) a function of time.

Unavoidably there are errors in the observations. Each observation is a random variable and the whole sequence $\boldsymbol{x}(t),\ t = t_1, t_2, \ldots, t_n$ is a ***random process***. A process is the evolution over time of a dynamic system. We must and shall develop a statistical theory for these functions. Classical statistical theory aims to infer the probability law of a random variable X from a finite number of independent observations $X_1, X_2, \ldots, X_n$. In this chapter we are observing a function that changes with time. We have a probability distribution for ***functions***.

The system consisting of the Earth and a GPS satellite is an example of a dynamic system. Their motions are governed by laws that depend only on current relative positions and velocities. Such dynamic systems are often modeled by differential equations. We solve differential equations to obtain formulas for predicting the future behavior of dynamic systems.

The vector $\boldsymbol{x}(t)$ is the ***state*** of the process. The original process $\boldsymbol{b}(t)$ is required to be a linear combination of the system variables via $\boldsymbol{b}(t) = A(t)\boldsymbol{x}(t) + \boldsymbol{e}(t)$. Except for this error $\boldsymbol{e}(t)$, the process $\boldsymbol{b}(t)$ can be recovered from the model $\boldsymbol{x}(t)$ by a linear combination of state variables.

A linear random process in ***continuous time*** with state $\boldsymbol{x}(t)$ and state covariance $\Sigma(t)$ has the model equations

$$\dot{\boldsymbol{x}}(t) = F(t)\boldsymbol{x}(t) + G(t)\boldsymbol{\epsilon}(t) \tag{16.1}$$

$$\boldsymbol{b}(t) = A(t)\boldsymbol{x}(t) + \boldsymbol{e}(t) \tag{16.2}$$

$$\dot{\Sigma}(t) = F(t)\Sigma(t) + \Sigma(t)F^{\mathrm{T}}(t) + G(t)\Sigma_\epsilon(t)G^{\mathrm{T}}(t). \tag{16.3}$$

The observation noise is measured by $\boldsymbol{e}(t)$ and the system noise by $\boldsymbol{\epsilon}(t)$ with covariance $\Sigma_\epsilon(t)$. Often initial values for the state are given.

Example 16.1 (Random ramp) A process with random initial value a_0 and random slope a_1 may be written as

$$b(t) = a_0 + a_1 t. \tag{16.4}$$

The differential equation corresponding to (16.4) is

$$\ddot{b}(t) = 0 \quad \text{with initial conditions} \quad b(0) = a_0 \quad \text{and} \quad \dot{b}(0) = a_1.$$

This is a second order differential equation so the state vector $\boldsymbol{x}$ for the process b must have two components. The dimension of the state vector equals the number of degrees of freedom of the system. Using phase variables $\boldsymbol{x}(t) = (x_1, x_2) = \big(b(t), \dot{b}(t)\big)$ in the vector model leads to

$$\begin{bmatrix} \dot{x}_1 \\ \dot{x}_2 \end{bmatrix} = \begin{bmatrix} 0 & 1 \\ 0 & 0 \end{bmatrix} \begin{bmatrix} x_1 \\ x_2 \end{bmatrix} + \begin{bmatrix} 0 \\ 0 \end{bmatrix} \epsilon, \qquad \begin{bmatrix} x_1(0) \\ x_2(0) \end{bmatrix} = \begin{bmatrix} a_0 \\ a_1 \end{bmatrix}$$

$$b = \begin{bmatrix} 1 & 0 \end{bmatrix} \begin{bmatrix} x_1 \\ x_2 \end{bmatrix} + e(t).$$

Frequently random errors exhibit a definite time-growing behavior. A function which grows ***linearly*** with time can be used to describe them. The growth rate a_1 is a random quantity with a given probability density. Two state elements are necessary to describe the model which is called a ***random ramp***:

$$\dot{x}_1 = x_2 \qquad \text{and} \qquad \dot{x}_2 = 0. \tag{16.5}$$

The state x_1 is the random ramp process; x_2 is an auxiliary variable whose initial condition provides the slope of the ramp. The solution of (16.5) is $x_1(t) = t\,x_2(0)$. The variance of x_1 is seen to grow quadratically with time. So the covariance matrix is

$$\Sigma(t) = \begin{bmatrix} t^2\sigma^2 & t\sigma^2 \\ t\sigma^2 & \sigma^2 \end{bmatrix}, \qquad \text{hence} \qquad \dot{\Sigma}(t) = \begin{bmatrix} 2t\sigma^2 & \sigma^2 \\ \sigma^2 & 0 \end{bmatrix}.$$

We want to check this result by means of equation (16.3)

$$\begin{aligned} \dot{\Sigma}(t) &= F(t)\Sigma(t) + \Sigma(t)F^{\mathrm{T}}(t) + G(t)\Sigma_\epsilon(t)G^{\mathrm{T}}(t) \\ &= \begin{bmatrix} 0 & 1 \\ 0 & 0 \end{bmatrix} \begin{bmatrix} t^2\sigma^2 & t\sigma^2 \\ t\sigma^2 & \sigma^2 \end{bmatrix} + \begin{bmatrix} t^2\sigma^2 & t\sigma^2 \\ t\sigma^2 & \sigma^2 \end{bmatrix} \begin{bmatrix} 0 & 0 \\ 1 & 0 \end{bmatrix} + \begin{bmatrix} 0 \\ 0 \end{bmatrix} \sigma_\epsilon^2 \begin{bmatrix} 0 & 0 \end{bmatrix} \\ &= \begin{bmatrix} t\sigma^2 & \sigma^2 \\ 0 & 0 \end{bmatrix} + \begin{bmatrix} t\sigma^2 & 0 \\ \sigma^2 & 0 \end{bmatrix} = \begin{bmatrix} 2t\sigma^2 & \sigma^2 \\ \sigma^2 & 0 \end{bmatrix}. \end{aligned}$$

This small computation verifies the validity of (16.3).

Mean and Correlation

In analogy with a single random variable we define the mean of an n-dimensional random process. The mean is a vector $\boldsymbol{\mu}$:

$$E\{\boldsymbol{x}(t)\} = \boldsymbol{\mu} = \int_{-\infty}^{\infty} \boldsymbol{x}(t) p\big(\boldsymbol{x}(t)\big)\, dt \tag{16.6}$$

or component-wise

$$E\{x_i(t)\} = \mu_i = \int_{-\infty}^{\infty} x_i(t)p\big(\boldsymbol{x}(t)\big)\,dt, \qquad i = 1, 2, \dots, n. \tag{16.7}$$

A random process is called ***Gaussian*** or ***normal*** if its probability density function is normal. The ***autocorrelation function*** for a random process $\boldsymbol{x}(t)$ is defined as the expected value of the product $\boldsymbol{x}(t_1)\boldsymbol{x}(t_2)^{\mathrm{T}}$:

$$\textit{\textbf{Autocorrelation}} \qquad R_x(t_1, t_2) = E\big\{\boldsymbol{x}(t_1)\boldsymbol{x}(t_2)^{\mathrm{T}}\big\} \tag{16.8}$$

where t_1 and t_2 are arbitrary observation times. Sometimes you find the correlation properties of a random process described by means of the *autocovariance function* which is defined as $E\big\{\big(\boldsymbol{x}(t_1) - \boldsymbol{\mu}(t_1)\big)\big(\boldsymbol{x}(t_2) - \boldsymbol{\mu}(t_2)\big)^{\mathrm{T}}\big\}$. The two functions are obviously related. The mean is included in the autocorrelation and the mean is subtracted in the autocovariance. The two functions are identical for processes with zero mean.

The autocorrelation tells how well the process is correlated with itself at two different times. A rapidly decreasing autocorrelation function has a "short memory" and allows the process to jump. A function with "long memory" entails a more smooth process.

Stationarity

A random process is ***stationary*** if the density functions $p\big(\boldsymbol{x}(t)\big)$ describing the process are invariant under *translation* of time. This means that

$$p\big(\boldsymbol{x}(t_1)\big) = p\big(\boldsymbol{x}(t_1 + t)\big).$$

In this stationary case the autocorrelation function depends only on the time difference $\tau = t_2 - t_1$. Thus R_x reduces to a function of just one variable τ:

$$\textit{\textbf{Stationary autocorrelation}} \qquad R_x(\tau) = E\big\{\boldsymbol{x}(t)\boldsymbol{x}(t+\tau)^{\mathrm{T}}\big\}. \tag{16.9}$$

Stationarity assures us that the expectation does not depend separately on $t_1 = t$ and $t_2 = t + \tau$, but only on the difference τ.

Example 16.2 (Random walk) The process is the result of integrating uncorrelated signals. The name indicates fixed-length steps in arbitrary directions. When the number of steps n is large and the individual steps are short, the distance travelled in a particular direction resembles the random walk process.

The random walk then is described as $x(t_n) = l_1 + l_2 + \cdots + l_n$. By linearity the mean value is $E\{x(t)\} = 0$. Let each l_i have variance $\sigma_i^2 = 1$. Independent steps yield

$$\operatorname{Var}\big(x(t_n)\big) = 1 + 1 + \cdots + 1 = n.$$

For $n \to \infty$ the variance of x tends to ∞; so the process is *not* stationary! For the autocorrelation we have

$$R_x(t_1, t_2) = E\{x(t_1)x(t_2)\} = \int_0^{t_2}\int_0^{t_1} \delta(\tau - \sigma)\,d\tau d\sigma = \begin{cases} t_2 & \text{for } t_1 \ge t_2 \\ t_1 & \text{for } t_1 < t_2. \end{cases} \tag{16.10}$$

The continous random walk process is also known as the Wiener process. The Wiener process defined by (16.10) is often taken as the definition of the white noise process. So Gaussian white noise is that hypothetical process, when integrated, which yields a Wiener process. See also Example 16.6.

Cross-correlation

Suppose we have two random processes $\boldsymbol{x}(t)$ and $\boldsymbol{y}(t)$. Their cross-correlation function gives the expected values of all the products $x_i(t_1)y_j(t_2)$. In matrix form:

$$R_{xy}(t_1, t_2) = E\{\boldsymbol{x}(t_1)\boldsymbol{y}(t_2)^{\mathrm{T}}\}. \tag{16.11}$$

If both processes are stationary, only the time difference $\tau = t_2 - t_1$ between sample points is relevant. We again have a (matrix) function of τ alone:

$$R_{xy}(\tau) = E\{\boldsymbol{x}(t)\boldsymbol{y}(t+\tau)^{\mathrm{T}}\}. \tag{16.12}$$

The cross-correlation function gives information on the mutual correlation between the two processes. Note that $R_{xy}(\tau) = R_{yx}(-\tau)$. The sum of two stationary random processes $\boldsymbol{z}(t) = \boldsymbol{x}(t) + \boldsymbol{y}(t)$ has an autocorrelation function defined as

$$\begin{aligned} R_z(\tau) &= E\{(\boldsymbol{x}(t)+\boldsymbol{y}(t))(\boldsymbol{x}(t+\tau)+\boldsymbol{y}(t+\tau))^{\mathrm{T}}\} \\ &= E\{\boldsymbol{x}(t)\boldsymbol{x}(t+\tau)^{\mathrm{T}}\}+E\{\boldsymbol{x}(t)\boldsymbol{y}(t+\tau)^{\mathrm{T}}\}+E\{\boldsymbol{y}(t)\boldsymbol{x}(t+\tau)^{\mathrm{T}}\}+E\{\boldsymbol{y}(t)\boldsymbol{y}(t+\tau)^{\mathrm{T}}\} \\ &= R_x(\tau)+R_{xy}(\tau)+R_{yx}(\tau)+R_y(\tau). \end{aligned} \tag{16.13}$$

If $\boldsymbol{x}$ and $\boldsymbol{y}$ are uncorrelated processes with zero mean, then $R_{xy} = 0$ and $R_{yx} = 0$ and

$$R_z(\tau) = R_x(\tau) + R_y(\tau). \tag{16.14}$$

The sum can be extended to more stationary and uncorrelated processes.

We summarize some properties for ***stationary processes with zero mean***:

1 $R_x(0)$ is the mean-square value of the process $\boldsymbol{x}(t)$.

2 $R_x(\tau)$ is an even function: $R_x(\tau) = R_x(-\tau)$. This follows from stationarity.

3 $|R_x(\tau)| \le R_x(0)$ for all τ. The mean-square value of $\boldsymbol{x}(t)$ equals that of $\boldsymbol{x}(t+\tau)$. The correlation coefficient between these two random variables is never greater than unity, by the Schwarz inequality.

4 If $\boldsymbol{x}(t)$ contains no periodic component, then $R_x(\tau)$ tends to zero as $\tau \to \infty$. Therefore $\boldsymbol{x}(t+\tau)$ becomes uncorrelated with $\boldsymbol{x}(t)$ for large τ, provided there are no hidden periodicities in the process. Note that a constant is a special case of a periodic function. Thus $R_x(\infty) = 0$ implies zero mean for the process.

5 The Fourier transform of $R_x(\tau)$ is real, symmetric, and nonnegative. The symmetry follows directly from $R_x(\tau) = R_x(-\tau)$. The nonnegative property is not obvious at this point. It will be justified in the section on the spectral density function for the process.

The autocorrelation function is an important descriptor of a random process and is relatively easy to obtain. *Often the autocorrelation is all we know!* Since there is always a Gaussian random process that has this given autocorrelation, we generally assume that our process is Gaussian. This uses the information we have (which is $R(\tau)$), and it makes the optimal estimators *linear*.

Ergodicity

In order to understand ergodicity we have to focus on *averaging*. In equation (9.24) we introduced a ***sample average*** as

$$\widehat{X} = \frac{X_1 + X_2 + \cdots + X_N}{N}.$$

This is the arithmetic mean. Note that $X_1, X_2, \ldots, X_N$ are numbers. The ***ensemble average*** is the expectation of $\widehat{X}$. It is not an average $\widehat{X}$ of an observed set of numbers. Finally a ***time average*** is defined as

$$\mathcal{R}(\tau) = \lim_{T\to\infty} \frac{1}{2T} \int_{-T}^{T} \boldsymbol{x}(t)\boldsymbol{x}(t+\tau)^{\mathrm{T}}\, dt. \tag{16.15}$$

We want to stress that $\widehat{X}$ and $E\{\widehat{X}\}$ are estimators of the mean μ while $\mathcal{R}(\tau)$ is a time average. The two types of average estimate different objects.

A random process is ***ergodic*** if time averaging is equivalent to ensemble averaging. This implies that a single sample time signal of the process contains all possible statistical variations of the process. Hence observations from several sample times do not give more information than is obtained from a single sample time signal.

Note that the autocorrelation function is the expectation of the product of $\boldsymbol{x}(t_1)$ and $\boldsymbol{x}(t_2)$. It can formally be written as

$$\mathcal{R}_x(t_1, t_2) = E\{\boldsymbol{x}(t_1)\boldsymbol{x}(t_2)^{\mathrm{T}}\} = \int_{-\infty}^{\infty}\int_{-\infty}^{\infty} \boldsymbol{x}(t_1)\boldsymbol{x}(t_2)^{\mathrm{T}} p_{x_1x_2}\big((\boldsymbol{x}(t_1), \boldsymbol{x}(t_2)\big)\, dt_1 dt_2. \tag{16.16}$$

However (16.16) is often not the simplest way of determining $\mathcal{R}_x$ because the joint density function $p_{x_1x_2}(x_1, x_2)$ must be known explicitly in order to evaluate the integral. If ergodicity applies it is often easier to compute $\mathcal{R}_x$ as a time average rather than as an ensemble average.

Example 16.3 We want to study the concept of ergodicity for a stationary process which is the ensemble of sinusoids of given amplitude A and frequency f and with a uniform distribution of phase φ. All ensemble members are of the form

$$x(t) = A\sin(ft + \varphi).$$

The phase φ is uniformly distributed over the interval $(0, 2\pi)$. Any average taken over the members of this ensemble at any fixed time would find all phase angles represented with equal probability density. The same is true of an average over all time on any one member.

The density function is $p_{x_1x_2} = \frac{1}{2\pi}$ for $0 \leq \varphi \leq 2\pi$. According to (16.16) the *ensemble average* autocorrelation function is

$$R_x(\tau) = \int_0^{2\pi} A\sin(ft+\varphi)A\sin(ft+f\tau+\varphi)\tfrac{1}{2\pi}\,d\varphi$$
$$= \frac{A^2}{4\pi}\int_0^{2\pi}\big(\cos(f\tau)-\cos(2ft+f\tau+2\varphi)\big)\,d\varphi = \tfrac{1}{2}A^2\cos(f\tau).$$

The *time average* autocorrelation function according to (16.15) is

$$\mathcal{R}(\tau) = \lim_{T\to\infty}\frac{1}{2T}\int_{-T}^{T} A\sin(ft+\varphi)A\sin(ft+f\tau+\varphi)\,dt$$
$$= \tfrac{1}{2}A^2\lim_{T\to\infty}\frac{1}{2T}\int_{-T}^{T}\big(\cos(f\tau)-\cos(2ft+f\tau+2\varphi)\big)\,dt = \tfrac{1}{2}A^2\cos(f\tau).$$

The two results are equal, thus $x(t)$ is an ergodic process.

Note that any distribution of the phase angle φ other than the uniform distribution over an integral number of cycles would define a nonergodic process.

Power Spectral Density

The Fourier transform of the autocorrelation function

$$S_x(\omega) = \int_{-\infty}^{\infty} R_x(\tau)e^{-j\omega\tau}\,d\tau \tag{16.17}$$

is called the ***power spectral density function*** or power density spectrum of the random process $x(t)$. The variable ω denotes the frequency in Hertz (cycles per second).

Example 16.4 We shall study a ***Gauss-Markov process*** described by the autocorrelation function

$$R_x(\tau) = \sigma^2 e^{-\alpha|\tau|}. \tag{16.18}$$

We calculate its power spectral density in two parts $\tau \leq 0$ and $\tau \geq 0$:

$$S_x(\omega) = \int_{-\infty}^{0} \sigma^2 e^{\alpha\tau}e^{-j\omega\tau}\,d\tau + \int_0^{\infty}\sigma^2 e^{-\alpha\tau}e^{-j\omega\tau}\,d\tau$$
$$= \sigma^2\Big(\frac{1}{\alpha-j\omega}+\frac{1}{\alpha+j\omega}\Big) = \sigma^2\frac{\alpha-j\omega+\alpha+j\omega}{\alpha^2+\omega^2} = \frac{2\sigma^2\alpha}{\alpha^2+\omega^2}. \tag{16.19}$$

The M-file gmproc allows the reader to experiment with the shape of $R_x(\tau)$ and $S_x(\omega)$ for various values of σ and α. The functions $R_x(\tau)$ and $S_x(\omega)$ are shown in Figure 16.1

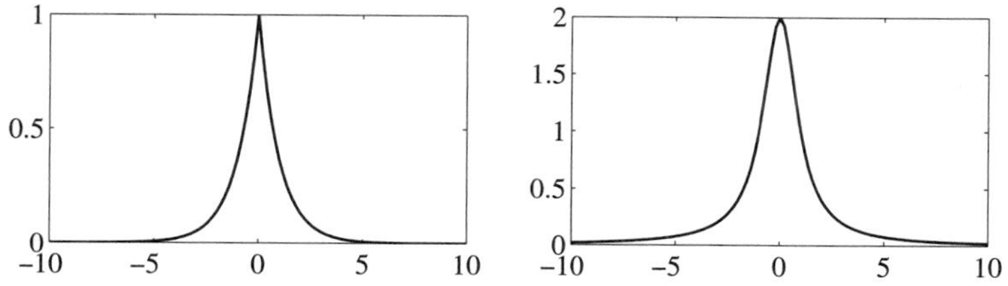

Figure 16.1 Autocorrelation $R_x(\tau)$ and spectral density function $S_x(\omega)$ for a Gauss-Markov process with $\sigma = 1$ and $\alpha = 1$.

for $\sigma = 1$ and $\alpha = 1$. The *correlation time* is $1/\alpha$. Many physical phenomena are well described by a Gauss-Markov process.

The process may also be characterized as

$$x(t+\tau) = e^{-\alpha|\tau|}x(t) + \epsilon_\tau.$$

We assume the noise ϵ_τ is normally distributed with zero mean and calculate the variance:

$$\begin{aligned} E\{\epsilon_\tau^2\} &= E\big\{\big(x(t+\tau) - e^{-\alpha|\tau|}x(t)\big)^2\big\} = E\big\{\big(x(t+\tau)^2 - 2e^{-\alpha|\tau|}x(t+\tau)x(t) + e^{-2\alpha|\tau|}x(t)^2\big)\big\} \\ &= \sigma^2 - 2\sigma^2 e^{-\alpha|\tau|}e^{-\alpha|\tau|} + \sigma^2 e^{-2\alpha|\tau|} = \sigma^2\big(1 - e^{-2\alpha|\tau|}\big). \end{aligned} \quad (16.20)$$

We assumed that ϵ_τ is independent of $x(t)$. This process is called an ***autoregression***.

If $\{\epsilon_k\}$ are i.i.d. and we sample at $k = 1, 2, 3, \ldots$ the process is described as

$$x_k = px_{k-1} + \epsilon_k \quad (16.21)$$

where $p = e^{-\alpha}$ when the sampling interval is $\tau = 1$.

The ***inverse transform*** of the spectral density $S_x(\omega)$ reconstructs the autocorrelation:

$$R_x(\tau) = \frac{1}{2\pi}\int_{-\infty}^{\infty} S_x(\omega)e^{j\omega\tau}\,d\omega. \quad (16.22)$$

For $\tau = 0$ this is

$$R_x(0) = E\big\{x^2(t)\big\} = \frac{1}{2\pi}\int_{-\infty}^{\infty} S_x(\omega)\,d\omega. \quad (16.23)$$

As $R_x(\tau) = R_x(-\tau)$ we also get $S_x(\omega) = S_x(-\omega)$, so the power spectral density function is a symmetric function in ω.

Example 16.5 (Gauss-Markov process) We consider the spectral function (16.19)

$$S_x(\omega) = \frac{2\sigma^2\alpha}{\alpha^2 + \omega^2}.$$

Using (16.23) we should recover σ^2. So we perform the integration:

$$E\{x^2\} = \frac{1}{2\pi}\int_{-\infty}^{\infty} \frac{2\sigma^2\alpha}{\alpha^2+\omega^2}\,d\omega = \frac{\sigma^2\alpha}{\pi}\left[\frac{1}{\alpha}\arctan\left(\frac{\omega}{\alpha}\right)\right]_{-\infty}^{\infty} = \sigma^2.$$

Taking the inverse Fourier transform of the Fourier transform we have recovered the original function.

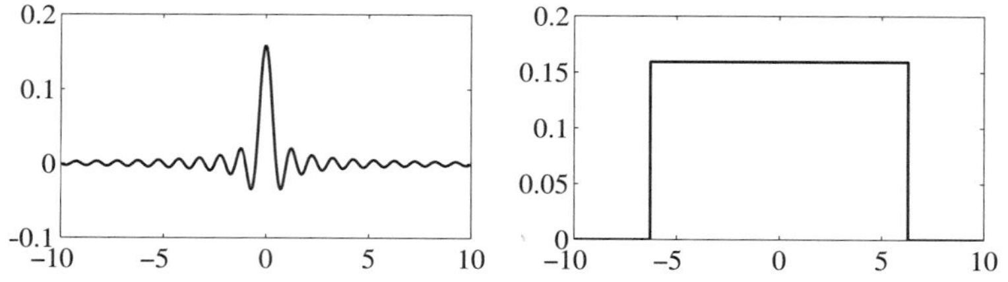

Figure 16.2 Autocorrelation $R_x(\tau)$ and spectral density function $S_x(\omega)$ for bandlimited white noise.

We should like to comment on the interpretation of the power spectral density. ***White noise*** is similar to the buzz in the radio and it contains all tones (frequencies). Its power spectral density function is $S_x(\omega) = \text{constant}$. On the contrary the power spectral density of a pure cosine waveform has the delta function (actually two delta functions, at $+\omega$ and $-\omega$) as power spectral density. The spectral density reflects the composition of frequencies in the signal $\boldsymbol{x}(t)$. The following example discusses the subject in a little more detail.

Example 16.6 Which autocorrelation function corresponds to a constant power spectral density S_0 from $-\infty$ to ∞? The power is distributed uniformly over all frequencies. In analogy to the case of white light such a random process is called ***white noise***:

$$R_x(\tau) = \frac{1}{2\pi}\int_{-\infty}^{\infty} S_0 e^{-j\omega\tau}\, d\omega = S_0\,\delta(\tau) \qquad \text{(informally).}$$

At $\tau = 0$, the Dirac delta $\delta(\tau)$ yields $R_x(0) = S_0\,\delta(0) = \infty$. This is $E\{(x(t))^2\}$. Thus white noise is an idealized process.

Another characteristic of sound is the ***bandwidth***. Often the bandwidth of the noise is made wide compared to the bandwidth of the system. We define a *bandlimited white noise* as constant in a finite range of frequencies and zero elsewhere:

$$S(\omega) = \begin{cases} A & \text{for } |\omega| \le 2\pi W \\ 0 & \text{for } |\omega| > 2\pi W. \end{cases}$$

W is the physical bandwidth in Hertz. The autocorrelation of $S(\omega)$ is the inverse transform of this box function, which produces the ***sinc function*** $\frac{\sin x}{x}$:

$$R(\tau) = 2WA\frac{\sin(2\pi W\tau)}{2\pi W\tau}.$$

This is *not* bandlimited. It is impossible for both $S(\omega)$ and $R(\omega)$ to be supported on finite intervals. Heisenberg's Uncertainty Principle, see page 225, gives a precise limitation $\sigma_S\sigma_R \ge \frac{1}{2}$ on their variances.

Both the autocorrelation and spectral density functions for bandlimited white noise are sketched in Figure 16.2. The function $R_x(\tau)$ is zero for $\tau = \frac{1}{2W}, \frac{2}{2W}, \frac{3}{2W}, \ldots$. Thus if the process is sampled at the Nyquist rate of $2W$ samples/sec, the resulting discrete random variables are uncorrelated. Since this usually simplifies the analysis, the white bandlimited assumption is frequently made.

Table 16.1 System models of continuous random processes

Random process	Autocorrelation $R_x(\tau)$	Power spectral density $S_x(\omega)$
White noise	$\sigma^2\,\delta(\tau)$	σ^2 (constant)
Random walk[a]	undefined	$\propto \sigma^2/\omega^2$
Random constant	σ^2	$2\pi\sigma^2\,\delta(\omega)$
Exponentially correlated or Gauss-Markov	$\sigma^2 e^{-\alpha\|\tau\|}$, where $1/\alpha =$ correlation time	$\frac{2\sigma^2\alpha}{\omega^2+\alpha^2}$

[a]Not a stationary process, hence undefined $R_x(\tau)$.

16.2 Random Processes in Discrete Time

Chapter 9 started with *continuous* random variables. In the present chapter we proceeded similarly. However in geodesy most applications involve discrete time and we must now discretize the time parameter. A ***discrete-time*** linear random process replaces the differential equation $\dot{\boldsymbol{x}}(t) = F(t)\boldsymbol{x}(t) - G(t)\boldsymbol{\epsilon}(t)$ for the state $\boldsymbol{x}(t)$ by a difference equation for the state $\boldsymbol{x}_k = \boldsymbol{x}(k)$:

$$\begin{aligned} \boldsymbol{x}_k &= F_{k-1}\boldsymbol{x}_{k-1} + G_k\boldsymbol{\epsilon}_k \\ \boldsymbol{b}_k &= A_k\boldsymbol{x}_k + \boldsymbol{e}_k. \end{aligned} \tag{16.24}$$

Suppose the uncorrelated process noise $\boldsymbol{\epsilon}_k$ in the state equation has covariance matrix $\Sigma_{\epsilon,k}$. If the observation noise $\boldsymbol{e}_k$ is uncorrelated and with zero mean, we have by covariance propagation the following recursion formula for the covariance of $\boldsymbol{x}_k$:

$$\Sigma_k = F_{k-1}\Sigma_{k-1}F_{k-1}^{\mathrm{T}} + G_k\Sigma_{\epsilon,k}G_k^{\mathrm{T}}. \tag{16.25}$$

However, the model (16.24) allows for time correlation in the random process noise $\boldsymbol{\epsilon}_k$. Such correlation often occurs in models from practice. It can be handled correctly by an ***augmentation of the state vector*** $\boldsymbol{x}_k$. Suppose $\boldsymbol{\epsilon}_k$ can be split into correlated quantities $\boldsymbol{\epsilon}_{1,k}$ and uncorrelated quantities $\boldsymbol{\epsilon}_{2,k}$: $\boldsymbol{\epsilon}_k = \boldsymbol{\epsilon}_{1,k} + \boldsymbol{\epsilon}_{2,k}$. We suppose that $\boldsymbol{\epsilon}_{1,k}$ can be modelled as a difference equation

$$\boldsymbol{\epsilon}_{1,k} = G_\epsilon\boldsymbol{\epsilon}_{1,k-1} + \boldsymbol{\epsilon}_{3,k-1}$$

where $\boldsymbol{\epsilon}_3$ is a vector of uncorrelated noises. Then the augmented state vector $\boldsymbol{x}_k'$ is given by

$$\boldsymbol{x}_k' = \begin{bmatrix} \boldsymbol{x}_k \\ \boldsymbol{\epsilon}_{1,k} \end{bmatrix}$$

and the augmented state equation, driven only by uncorrelated disturbances, is

$$\boldsymbol{x}_k' = \begin{bmatrix} \boldsymbol{x}_k \\ \boldsymbol{\epsilon}_{1,k} \end{bmatrix} = \begin{bmatrix} F & G \\ 0 & G_\epsilon \end{bmatrix}\begin{bmatrix} \boldsymbol{x}_{k-1} \\ \boldsymbol{\epsilon}_{1,k-1} \end{bmatrix} + \begin{bmatrix} G & 0 \\ 0 & I \end{bmatrix}\begin{bmatrix} \boldsymbol{\epsilon}_{2,k-1} \\ \boldsymbol{\epsilon}_{3,k-1} \end{bmatrix}. \tag{16.26}$$

Next we consider four specific correlation models for system disturbances. In each case scalar descriptions are presented.

Example 16.7 (Random constant) The random constant is a non-dynamic quantity with a fixed random amplitude. The process is described by

$$x_k = x_{k-1}.$$

The random constant may have a random initial condition x_0.

Example 16.8 (Random walk) The process is described by

$$x_k = x_{k-1} + \epsilon_k.$$

The variance of the noise is

$$E\{\epsilon_k^2\} = E\{(x_k - x_{k-1})^2\} = E\{x_k^2\} + E\{x_{k-1}^2\} - 2E\{x_k x_{k-1}\} = 2\sigma_x^2.$$

Example 16.9 (Random ramp) The random ramp is a process growing linearly with time. The growth rate of the random ramp is a random quantity with given variance. We need two state elements to decribe the random ramp:

$$\begin{aligned} x_{1,k} &= x_{1,k-1} + (t_k - t_{k-1})x_{2,k-1} + \epsilon_{1,k} \\ x_{2,k} &= x_{2,k-1} + \epsilon_{2,k}. \end{aligned}$$

Example 16.10 (Exponentially correlated random variable)

$$x_k = e^{-\alpha(t_k - t_{k-1})} x_{k-1} + \epsilon_k.$$

We have $\epsilon_k = x_k - e^{-\alpha(t_k - t_{k-1})}$. The time difference is $\Delta t = t_k - t_{k-1}$. According to (16.20) we have $E\{\epsilon_k^2\} = \sigma^2(1 - e^{-2\alpha\Delta t})$.

The random processes in Examples 16.7–16.10 make the basis for a lot of linear filters. Next we bring three examples related to GPS applications, see Axelrad & Brown (1996).

Example 16.11 (Discrete random ramp) Often random errors exhibit a definite time-growing behavior. The ***discrete random ramp***, a function that grows *linearly* with time, can often be used to describe them. The growth rate of the random ramp is a random quantity with a given variance. A good example of this model is the behavior of the offset b and the drift d of a GPS ***receiver clock***.

Two state components are necessary to describe the random ramp. So we use a vector $\boldsymbol{x}_k$ and a matrix equation:

$$\textbf{\textit{State equation}} \quad \boldsymbol{x}_k = F\boldsymbol{x}_{k-1} + \boldsymbol{\epsilon}_k \quad \text{with} \quad \boldsymbol{x}_k = \begin{bmatrix} b_k \\ d_k \end{bmatrix} \quad \text{and} \quad F = \begin{bmatrix} 1 & \Delta t \\ 0 & 1 \end{bmatrix}. \tag{16.27}$$

The offset b is the random ramp process. The drift d describes the slope of the ramp. The second row of F gives random changes of slope from d_{k-1} to d_k. The first row gives the random ramp:

$$b_k = b_{k-1} + \Delta t\, d_{k-1} + \text{random error}.$$

Next we shall estimate the covariance matrix of observation errors $\Sigma_\epsilon = E\{\epsilon\epsilon^{\mathrm{T}}\}$. We start with a continuous system formulation, and integrate over a time step:

$$\epsilon_k = \int_{t_{k-1}}^{t_k} F(t_k, \tau)\, \epsilon(\tau)\, d\tau$$

which yields

$$E\{\epsilon_k\epsilon_k^{\mathrm{T}}\} = E\left\{\int_{t_{k-1}}^{t_k}\int_{t_{k-1}}^{t_k} F(t_k, \tau)\, \epsilon(\tau)\, \epsilon(\sigma)^{\mathrm{T}} F(t_k, \sigma)\, d\tau d\sigma\right\}$$

$$= \int_{t_{k-1}}^{t_k} F(t_k, \tau)\, \Sigma(\tau)\, F(t_k, \tau)^{\mathrm{T}}\, d\tau.$$

The matrix $\Sigma(\tau)$ is the spectral density matrix. Let the spectral amplitudes for the offset and drift be s_b and s_d:

$$\Sigma = \begin{bmatrix} s_b & 0 \\ 0 & s_d \end{bmatrix}.$$

Then the integrand is a 2 by 2 matrix:

$$F\Sigma F^{\mathrm{T}} = \begin{bmatrix} s_b + s_d\tau^2 & s_d\tau \\ s_d\tau & s_d \end{bmatrix}.$$

Therefore we get a formula for the covariance matrix:

$$E\{\epsilon\epsilon^{\mathrm{T}}\} = \int_{t_{k-1}}^{t_k} \begin{bmatrix} s_b + s_d\tau^2 & s_d\tau \\ s_d\tau & s_d \end{bmatrix} d\tau = \begin{bmatrix} s_b\Delta t + s_d(\Delta t)^3/3 & s_d(\Delta t)^2/2 \\ s_d(\Delta t)^2/2 & s_d\Delta t \end{bmatrix}. \tag{16.28}$$

Typical values for the white noise spectral amplitudes s_b and s_d in GPS receiver clocks are 4×10^{-19} and 15×10^{-19}. A typical time step is $\Delta t = 20\,\mathrm{s}$. In this case the covariance matrix is Σ_{clock}:

$$\Sigma_{\text{clock}} = E\{\epsilon\epsilon^{\mathrm{T}}\} = \begin{bmatrix} 40\,004 & 300 \\ 300 & 3\,000 \end{bmatrix} \times 10^{-19}.$$

Example 16.12 A process model for a ***GPS receiver*** includes the three coordinates of the receiver position combined with the clock offset and the clock drift. The dynamic model is still given by (16.27) and the state vector $\boldsymbol{x}$ has five components:

$$\boldsymbol{x}_k = \begin{bmatrix} x \\ y \\ z \\ \hline b \\ d \end{bmatrix} \quad \text{and} \quad F = \left[\begin{array}{ccc|cc} 1 & 0 & 0 & 0 & 0 \\ 0 & 1 & 0 & 0 & 0 \\ 0 & 0 & 1 & 0 & 0 \\ \hline 0 & 0 & 0 & 1 & \Delta t \\ 0 & 0 & 0 & 0 & 1 \end{array}\right].$$

The covariance matrix for this static receiver is

$$\Sigma_{\text{static}} = E\{\epsilon\epsilon^{\mathrm{T}}\} = \left[\begin{array}{c|c} \Sigma_{\text{position}} & 0 \\ \hline 0 & \Sigma_{\text{clock}} \end{array}\right]. \tag{16.29}$$

The matrix Σ_{clock} reflects the random contribution from the receiver clock. The covariance matrix Σ_{position} describes the model noise related to the position. When the receiver is kept at a fixed position (***static receiver***) it would be natural to set $\Sigma_{\text{position}} = 0$. This however would imply that all new position information is ignored and that is not meaningful. So "artificially" we let the position have small variances in order that the filter does not get stuck.

Example 16.13 A ***kinematic receiver*** is a GPS receiver that is moved around. Often it goes on board a vehicle with low velocity and without sudden shifts. Now the state vector includes eight components: three coordinates, three velocities, and two clock terms:

$$\boldsymbol{x}_k = \left[\begin{array}{c} x \\ y \\ z \\ \hline \dot{x} \\ \dot{y} \\ \dot{z} \\ \hline b \\ d \end{array}\right] \quad \text{and} \quad F = \left[\begin{array}{ccc|ccc|cc} 1 & 0 & 0 & \Delta t & 0 & 0 & 0 & 0 \\ 0 & 1 & 0 & 0 & \Delta t & 0 & 0 & 0 \\ 0 & 0 & 1 & 0 & 0 & \Delta t & 0 & 0 \\ \hline 0 & 0 & 0 & 1 & 0 & 0 & 0 & 0 \\ 0 & 0 & 0 & 0 & 1 & 0 & 0 & 0 \\ 0 & 0 & 0 & 0 & 0 & 1 & 0 & 0 \\ \hline 0 & 0 & 0 & 0 & 0 & 0 & 1 & \Delta t \\ 0 & 0 & 0 & 0 & 0 & 0 & 0 & 1 \end{array}\right].$$

The covariance matrix is

$$\Sigma_{\text{kinematic}} = E\{\boldsymbol{\epsilon}\boldsymbol{\epsilon}^{\mathrm{T}}\} = \left[\begin{array}{cc|c} \Sigma_{\text{position}} & \Sigma_{\text{position, velocity}} & 0 \\ \Sigma_{\text{position, velocity}} & \Sigma_{\text{velocity}} & 0 \\ \hline 0 & 0 & \Sigma_{\text{clock}} \end{array}\right]. \tag{16.30}$$

The matrix Σ_{velocity} often uses different values for the horizontal and vertical components. A car does not substantially change its vertical velocity. But it can accelerate or decelerate rapidly. Of course if the variances in the diagonal terms of $\Sigma_{\text{kinematic}}$ are large, a filtering process—as described in the next chapter—will not improve the accuracy of the position very much.

Example 16.14 (Gauss-Markov process) Let x_k be a stationary random process with zero mean and exponentially decreasing autocorrelation:

$$R_x(t_2 - t_1) = \sigma^2 e^{-\alpha|t_2 - t_1|}.$$

This type of random process can be modeled as the output of a linear system, when the input ϵ_k is zero-mean white noise with power spectral density equal to unity. (In standard time series literature this is called an AR(1) model. AR(1) means autoregressive of order 1.) A difference equation model for this type of process is

$$\begin{aligned} x_k &= F x_{k-1} + G\epsilon_k \\ b_k &= x_k. \end{aligned} \tag{16.31}$$

In order to use this model we need to solve for the unknown scalar parameters F and G as functions of the parameter α. To do so we multiply (16.31) by x_{k-1} on both sides and take

Table 16.2 System models of discrete-time random processes

Process type	Autocorrelation R_x	State-space model
Random constant	$R_x(\Delta t) = \sigma^2$	$x_k = x_{k-1}, \quad \sigma^2\{x_0\} = \sigma^2$
Random walk	$\to +\infty$	$x_k = x_{k-1} + \epsilon_k, \quad \sigma^2\{x_0\} = 0$
Random ramp		$x_{1,k} = x_{1,k-1} + \Delta t x_{2,k-1}$
		$x_{2,k} = x_{2,k-1}$
Exponentially correlated	$R_x(\Delta t) = \sigma^2 e^{-\alpha\lvert\Delta t_k\rvert}$	$x_k = e^{-\alpha\lvert\Delta t\rvert} x_{k-1} + \epsilon_k$
		$\sigma^2\{x_0\} = \sigma^2, \quad \Delta t_k = t_k - t_{k-1}$

expected values to obtain the equations

$$\begin{aligned} E\{x_k x_{k-1}\} &= F E\{x_{k-1}x_{k-1}\} + G E\{\epsilon_k x_{k-1}\} \\ \sigma^2 e^{-\alpha} &= F\sigma^2 \end{aligned} \tag{16.32}$$

assuming that the ϵ_k are uncorrelated and $E\{\epsilon_k\} = 0$ so that $E\{\epsilon_k x_k\} = 0$. The transition matrix (factor) is $F = e^{-\alpha}$. Next square the state variable defined by (16.31) and take its expected value:

$$\begin{aligned} E\{x_k^2\} &= F^2 E\{x_{k-1}x_{k-1}\} + G^2 E\{\epsilon_k \epsilon_k\} \\ \sigma^2 &= \sigma^2 F^2 + G^2 \end{aligned} \tag{16.33}$$

because the system variance is $E\{\epsilon_k^2\} = 1$. We insert $F = e^{-\alpha}$ into (16.33) and get $G = \sigma\sqrt{1 - e^{-2\alpha}}$. The complete model is then

$$\begin{aligned} x_k &= e^{-\alpha} x_{k-1} + \sigma\sqrt{1 - e^{-2\alpha}}\,\epsilon_k \\ b_k &= x_k \end{aligned}$$

with $E\{\epsilon_k\} = 0$ and $E\{\epsilon_k \epsilon_j\} = \delta_{jk}$.

Ideally a random process should be based on the physical laws that govern the noise of the system errors. An exact representation is often impossible, either because the underlying physics is not well understood or because implementing the ideal random process would yield a cumbersome solution. The Gauss-Markov model (exponential decay in correlation) is extremely useful, requiring only one parameter α.

16.3 Modeling

In applied works it only rarely happens that the given physical problem is in the exact form

$$\begin{aligned} \boldsymbol{x}_k &= F_{k-1}\boldsymbol{x}_{k-1} + \boldsymbol{\epsilon}_k \\ \boldsymbol{b}_k &= A_k \boldsymbol{x}_k + \boldsymbol{e}_k. \end{aligned}$$

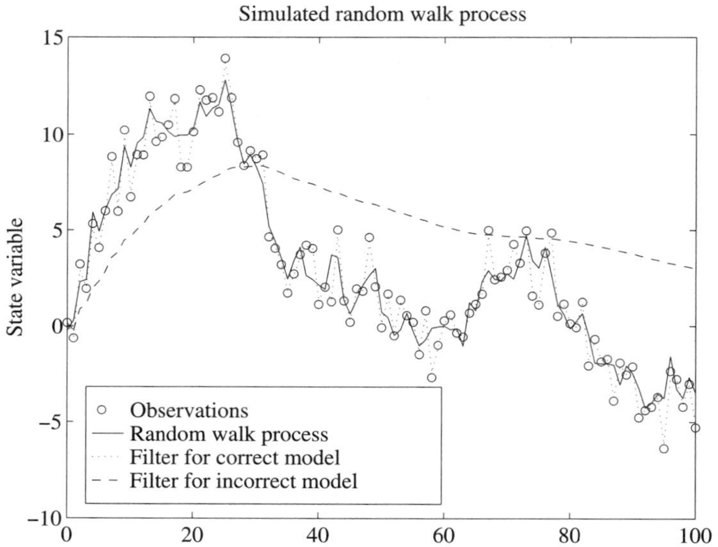

Figure 16.3 Correctly and incorrectly filtered random walk process.

Most often the original problem has to be modified to fit into the appropriate form. This twisting around is often referred to as modeling and it is all-important in Kalman filtering applications. Good modeling leads to good results; bad modeling leads to bad results. It is as simple as that. There are no set rules for the modeling procedure, and it often requires some imagination. Perhaps the best way to become adept at modeling is to look at a wide variety of examples.

We start by an example from Brown & Hwang (1997) demonstrating the effects of mis-modeling.

Example 16.15 Consider a process that is actually random walk but is *incorrectly* modeled as a random constant c. We have then for the true model

$$\begin{aligned} x_k &= x_{k-1} + \epsilon_k, & \epsilon_k &= \text{unity Gaussian white noise, } \sigma^2\{x_0\} = 1 \\ b_k &= x_k + e_k, & k &= 0, 1, 2, \dots \qquad \sigma^2\{e_k\} = 0.1 \end{aligned}$$

and for the incorrect model

$$\begin{aligned} x_k &= c, & c &\sim N(0, 1) \\ b_k &= x_k + e_k, & k &= 0, 1, 2, \dots \qquad \sigma^2\{e_k\} = 0.1. \end{aligned}$$

The wrong model has $F_k = 1$, $\Sigma_{\epsilon,k} = 0$, $A_k = 1$, $\Sigma_{e,k} = 0.1$, $\hat{x}_0 = 0$, and $P_0^{-1} = 1$. For the true model the parameters are the same except that $\Sigma_{\epsilon,k} = 1$, rather than zero.

The random walk $x_0, x_1, \dots$ was simulated using Gaussian random numbers with zero mean and unity variance. The resulting process for 100 seconds is shown in Figure 16.3. A measurement sequence b_k of this sample process was also generated using another set of $N(0, 1)$ random numbers for e_k. This measurement sequence was first processed using the incorrect model ($\Sigma_{\epsilon,k} = 0$), and again with the correct model ($\Sigma_{\epsilon,k} = 1$). The results are shown along with the sample process in Figure 16.3. In this case the measurement noise is relatively small ($\sigma \approx 0.3$), and we note that the estimate of the modeled filter does very poorly after the first few steps. This is due to the filter's gain decreasing with each succeeding step. At the 100th step the gain is more than two orders of magnitude less than at the beginning. Thus the filter becomes very sluggish and will not follow the random walk. Had the simulation been allowed to go further on, it would have become even more sluggish.

The moral to Example 16.15 is simply this. Any model that assumes the process, or any facet of the process, to be absolutely constant forever and ever is a risky model. In the physical world, very few things remain absolutely constant. The obvious remedy for this type of divergence problem is always to insert some process noise into each of the state variables. Do this even at the risk of some degree of suboptimality; it makes for a much safer filter than otherwise. It also helps with potential roundoff problems. Often a random walk model is a safer model where the time span is large, and it is usually preferred over the truly constant model.

Choosing an appropriate process model is always an important consideration. A certain amount of common sense judgement is called for in deciding on a model that will fit the situation at hand reasonably well, but at the same time will not be too complicated.

For instance no process is a random walk to infinity. At some point it is (band) limited somehow: The troposphere for a short time looks like random walk. But suppose there were no measurements for an entire day? Would our lack of knowledge about the troposphere approach infinity? No, of course not. We know that the troposphere zenith delay can be predicted to be 2.4 meters with about 2% uncertainty or better. So infinity is not the limit.

Computing the Autocorrelation

It is simple to compute the autocorrelation for an ordered set of data $a_0, a_1, \ldots, a_{n-1}$, taken at a constant time interval. First we compute the mean m (the average). Second we may imagine the data arranged in two rows:

$$\text{Shift} = 0: \quad \begin{matrix} a_0 & a_1 & a_2 & a_3 & a_4 & \cdots \\ a_0 & a_1 & a_2 & a_3 & a_4 & \cdots \end{matrix} \qquad \text{auto}(0) = \sum_0^{n-1} a_i a_i / n.$$

We multiply elements $a_i a_i$ above each other, and add. Now shift the lower row:

$$\text{Shift} = 1: \quad \begin{matrix} a_0 & a_1 & a_2 & a_3 & a_4 & \cdots & \\ & a_0 & a_1 & a_2 & a_3 & a_4 & \cdots \end{matrix} \qquad \text{auto}(1) = \sum_1^{n-1} a_i a_{i-1} / n.$$

Again we multiply elements above each other. This time the number of terms is diminished by one. We continue shifting, and each time we divide the sum by the number of data n. The M-file has the following core code:

```
auto = autocorr(a)
m = mean(a)
```

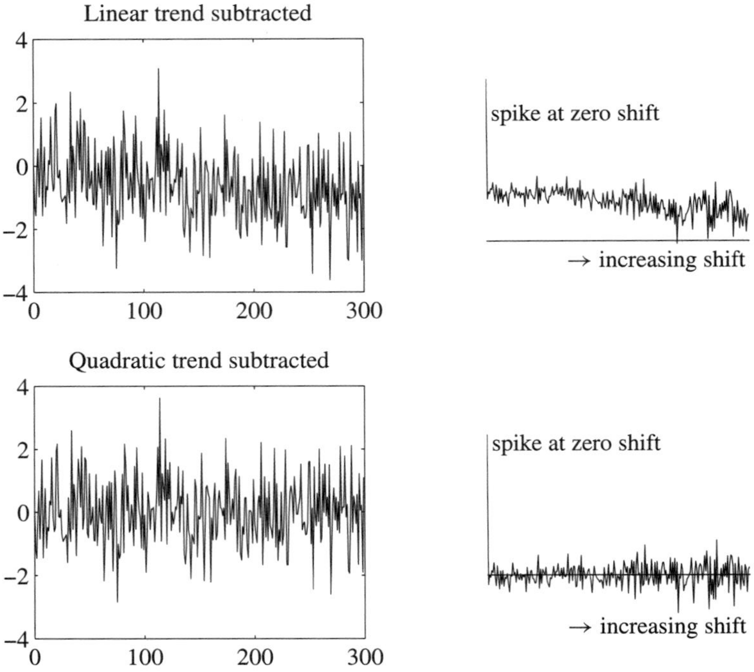

Figure 16.4 Autocorrelation for random data. The peak at zero measures the variance.

```
for shift = 0:n−2
  q = 0;
  for t = 1:n−shift
    q = q + (a(t)−m) ∗ (a(t + shift)−m);
  end
  auto(shift + 1) = q/n;
end
```

The sums for large shifts contain only a small number of terms; the overlap count n − shift − 1 is small. It is statistical practice to omit the last 20% (or a similar fraction) of the shifted product sums; they are not so reliable. Luckily enough we are most interested in the autocorrelation for small shifts as those reveal the nature of the data. So the result of autocorr is most important for small shifts.

We divided the sums by n, and not n − shift − 1. This happens to secure that

$$R = \begin{bmatrix} \text{auto}(0) & \text{auto}(1) & \dots & \text{auto}(n-1) \\ \text{auto}(1) & \text{auto}(0) & \dots & \text{auto}(n-2) \\ \vdots & \vdots & \ddots & \vdots \\ \text{auto}(n-1) & \text{auto}(n-2) & \dots & \text{auto}(0) \end{bmatrix} \tag{16.34}$$

is positive semi-definite.

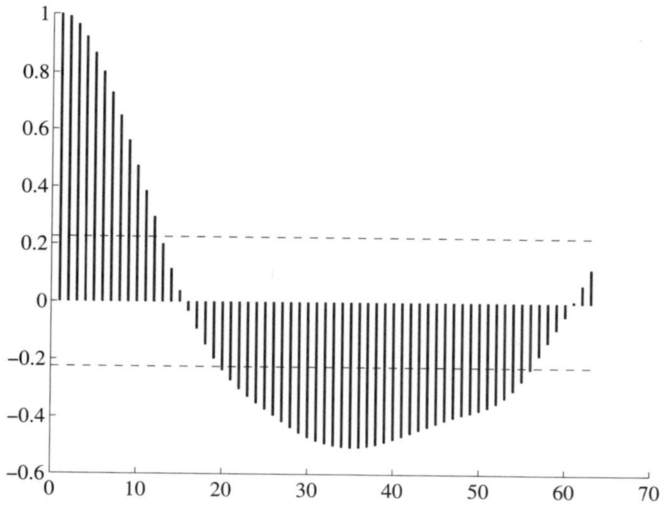

Figure 16.5 Correlogram for an autocovariance function. The dashed horizontal lines represent the limits $\pm 2/\sqrt{n}$, n being 80.

Variance Component Model

Once again we assume that our observations (signal) are stationary. Furthermore that observations are made at equidistant time invervals. In the following we shall introduce some useful tools for analyzing autocorrelation functions.

Often we are furnished with a normalized autocorrelation function. The ***autocorrelation coefficient*** r_k is defined as

$$r_k = \frac{\mathcal{R}_x(k)}{\mathcal{R}_x(0)}, \qquad k = 0, 1, \ldots, n-1. \tag{16.35}$$

The plot of r_k for varying k is called the ***correlogram*** for the random process $\boldsymbol{x}_k$. One simple use of the correlogram is to check whether there is evidence of any serial dependence in an observed time series. To do this, we use a result due to Bartlett who showed that for a white noise sequence a_k and for large n, r_k is approximately normally distributed with mean zero and variance $1/n$. Thus values of r_k greater than $2/\sqrt{n}$ in absolute value can be regarded as significant at about the 5% level. If a large number of r_k are computed it is likely that some will exceed this treshold even if a_k is a white noise sequence.

Figure 16.5 shows a correlogram for a receiver clock offsets. The limits $\pm 2/\sqrt{n}$ are exceeded for shift 0 to 12 and a longer sequence from $k = 20$ to $k = 56$. So the correlogram indicates correlation between observations even with shifts up to 56 units of time.

In practice random processes often show a non-random trend μ_t. With the noise term ϵ_t we have

$$Z_t = \mu_t + \epsilon_t.$$

We shall demonstrate how to handle this circumstance by splitting the observation $Z_i(t)$ of the ith series into a starting level $L_i(t)$, a random stationary part $M_i(t)$, and noise $N_i(t)$:

$$Z_i(t_k) = L_i(t_0) + M_i(t_k) + N_i(t_k). \tag{16.36}$$

All observations for all experimental subjects $i = 1, 2, \dots, m$ are taken at equidistant times $t_0, t_1, \dots, t_k$. L_i is defined at time t_0 while the random variables depend on time t_k. L_i is independent of L_j for $i \neq j$ and is distributed as $N(0, \lambda^2)$. The noise $N_i(t_k)$ is distributed as $N(0, \nu^2)$ and independent in time as well as between subjects. Finally $M_i(t_k)$ is distributed as $N(0, \mu^2)$ and independent between subjects and with $R(k) = \mu^2 r(k)$. We remember that $r(0) = 1$ and $r(k) \to 0$ for $k \to \infty$. The three components L_i, M_i, and N_i are assumed mutually independent. Hence the observational variance is

$$R_z(0) = \mathrm{Var}(Z_i) = \lambda^2 + \mu^2 + \nu^2.$$

The covariance between $Z_i(t_l)$ and $Z_j(t_m)$ is

$$\begin{aligned}
\sigma\big(Z_i(t_l), Z_j(t_m)\big) &= E\{Z_i(t_l) Z_j(t_m)\} \\
&= E\big\{\big(L_i(t_0) + M_i(t_l) + N_i(t_l)\big)\big(L_j(t_0) + M_j(t_m) + N_j(t_m)\big)\big\} \\
&= E\{L_i L_j\} + E\{M_i(t_l) M_j(t_m)\} = \big(\mathrm{Var}(L_i) + \mu^2 r_z(t_m - t_l)\big)\delta_{ij} \\
&= \big(\lambda^2 + \mu^2 r_z(k)\big)\delta_{ij}.
\end{aligned}$$

Note that especially for $i \neq j$ the covariance is zero due to the independence of subjects.

For a moment we dwell on the autocorrelation $R_z(k)$. The observational error $N_i(t_k)$ is not equal to zero; this implies that $R_z(k)$ does not approach $R_z(0)$ for $k \to 0$. Also $R_z(k)$ does not approach 0 for $k \to \infty$. This is caused by the subject specific random variable L_i.

From (16.34) we remember R and let E denote a matrix with all ones, then the covariance matrix can be written as

$$\Sigma = \lambda^2 E + \mu^2 R + \nu^2 I.$$

Now we are ready to introduce the ***variogram function***

$$V(k) = \tfrac{1}{2} E\big\{\big(Z_i(t_l) - Z_i(t_m)\big)^2\big\}. \tag{16.37}$$

We have

$$Z_i(t_l) - Z_i(t_m) = M_i(t_l) - M_i(t_m) + N_i(t_l) - N_i(t_m)$$

and remembering that M_i and N_i are independent we get

$$V(k) = \tfrac{1}{2} E\big\{\big(M_i(t_l) - M_i(t_m)\big)^2\big\} + \tfrac{1}{2} E\big\{\big(N_i(t_l) - N_i(t_m)\big)^2\big\} = \mu^2\big(1 - r_z(k)\big) + \nu^2.$$

Figure 16.6 shows a variogram with the three variance components λ^2, μ^2, and ν^2. Remember $r_z(0) = 1$ and hence $\lim_{k \to 0} V(k) = \nu^2$. We can read ν^2 from the figure as the intercept with the ordinate axis.

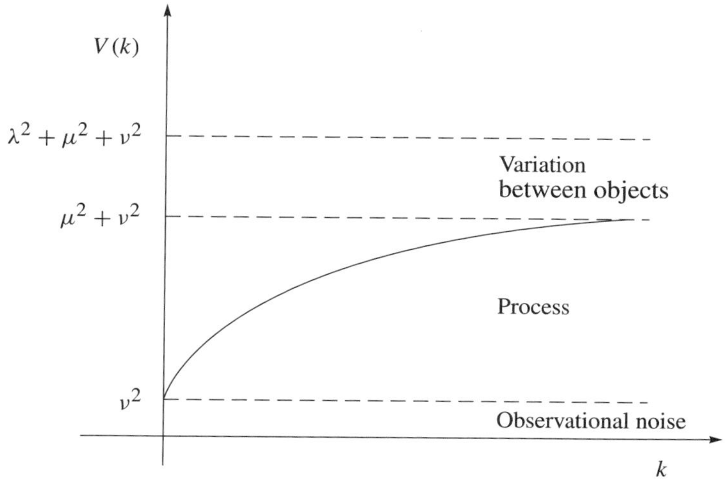

Figure 16.6 Variogram illustrating the three variance components.

For any stationary, random process we have $r_z(k) \to 0$ for $k \to \infty$. Thus $V(\infty) = \mu^2 + \nu^2$. So the variance μ^2 within the individual subjects is found as $V(\infty)$ minus ν^2.

The total variance for all observations Z_i is estimated by taking data across experimental subjects. Again under the assumption of mutually independent processes the total variance is computed as the mean value of $\bigl(Z_i(t_l) - Z_j(t_m)\bigr)^2/2$ for all l and m and all i and j with $i \neq j$. This makes $M = \sum_i \binom{n_i}{2}$ terms, where n_i is the number of observations in the ith series:

$$\lambda^2 + \mu^2 + \nu^2 = \frac{1}{2M} \sum_{i \neq j} \bigl(Z_i(t_l) - Z_j(t_m)\bigr)^2.$$

The total variance also is illustrated in Figure 16.6. The variance λ^2 for the population can be read at the ordinate axis, too.

Two common examples of autocorrelation functions are the *exponential correlation function*

$$r(k) = e^{-\alpha k} \tag{16.38}$$

and the *Gaussian correlation function*

$$r(k) = e^{-\alpha k^2}. \tag{16.39}$$

In Figure 16.7 we have sketched the autocorrelation function for (16.38) and (16.39) and the corresponding variograms. The exponential correlation function decreases strongly for small time differences k while the Gaussian correlation function has a strong correlation over a larger time. Later it decreases rapidly.

Example 16.16 For a demonstration of the theory we use one-way differences between a receiver and a satellite. We use satellites with pseudo random noise (PRN) codes 2, 9,

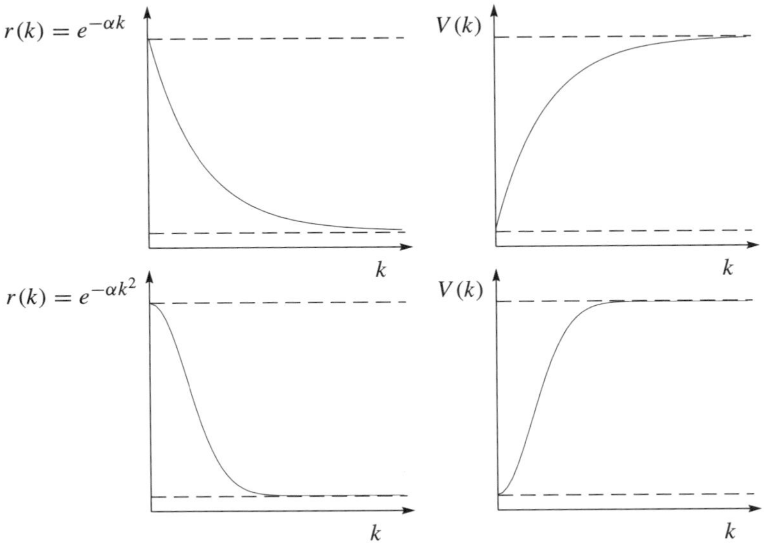

Figure 16.7 Exponential and Gaussian autocorrelation functions and variograms.

16, 23, 26, and 27. We concentrate on ionospheric delays for epoch k. We observe on two frequencies to cancel the main error in the delay I_k:

$$I_k = \frac{(\Phi_{2,k} - \lambda_2 N_2) - (\Phi_{1,k} - \lambda_1 N_1)}{1 - (f_1/f_2)^2}.$$

Figure 16.8 depicts I_k as calculated for one-ways, their differences and their differences again which are the so-called double differences. The actual baseline length is 4.6 km. For one-ways I_k typically varies from 5–15 m. Single differences have I_k values of 2.5–3 m and double differences between -0.2–0.2 m. Note that the mean values of elevation angle for the individual PRN's are listed in Table 16.3. From this you observe that I_k depends strongly on the elevation angle for the single PRN.

Now we turn to *autocorrelation functions* for I_k. To eliminate a possible trend in an observation series one often starts the investigations from differences in time: $I_k - I_{k-1}$. The autocorrelation for the differenced ionospheric delay for the one-way to PRN 2 is shown in Figure 16.9 (upper left). The spike at zero equals the variance of a difference $I_k - I_{k-1}$. Next we compute the autocorrelation for the undifferenced I_k. The result is shown at the upper right. This evidently reflects a remaining systematic term in the delay. Our ultimate goal is to model this part and subsequently subtract it from the actual delay, hopefully leaving only white or nearly white noise. Knowing a good model we can with a high degree of accuracy *predict* the ionospheric delay in time Δt and also with distance d.

We continue computing the autocorrelation for I_k for a single difference. When computing autocorrelation for differences we must remember the rules given in (16.13). This implies that we have to compute cross-correlations. The lower left of Figure 16.8 shows the result for PRN 2.

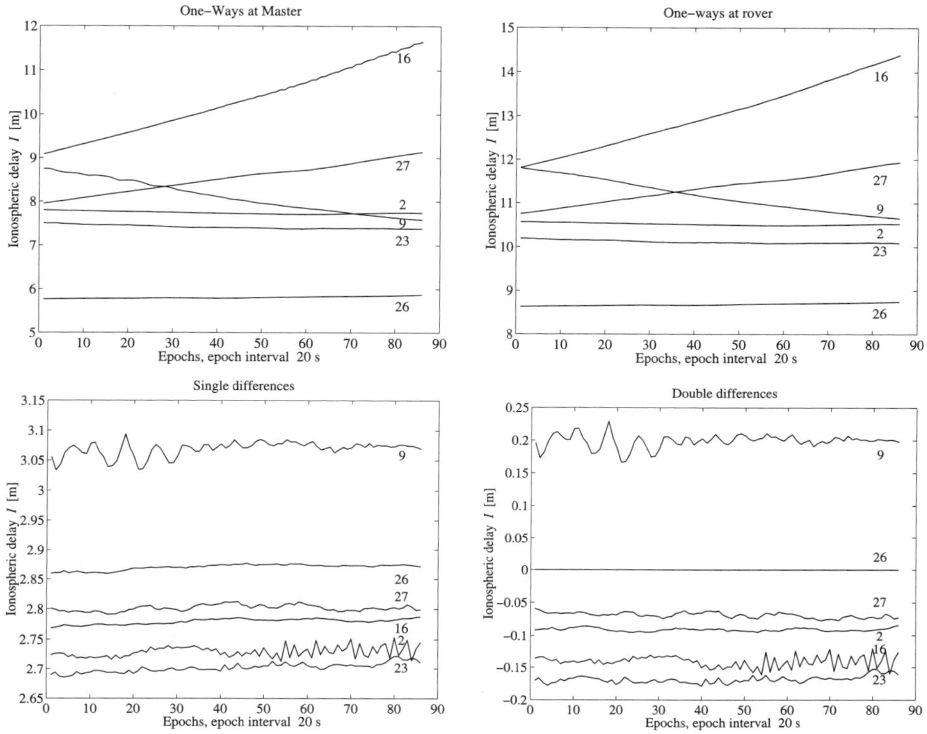

Figure 16.8 Ionospheric delays for one-ways, single and double differences.

In Table 16.3 we have $\sigma_I = 2\,\text{cm}$ for PRN 2 and $\sigma_I = 77\,\text{cm}$ for PRN 16. The numbers in Table 16.3 are produced by calls like

one_way (m) (27) and autocorr(x(2,:)′)

However these numbers diminish to 4 mm and 9 mm for single differences and 2 mm and 9 mm for double differences. The general M-file for this example is called oneway_i.

Table 16.3 Autocorrelation of ionosphere delay for one-ways

PRN	Elevation (in °)	σ_I (in m) master	rover	Shift for first zero master	rover
26	68.9	0.08	0.11	35	30
2	59.0	0.08	0.04	15	35
27	28.0	0.39	0.35	30	32
16	22.8	0.77	0.71	30	30
23	20.4	0.17	0.19	12	20
9	18.5	0.48	0.19	15	30

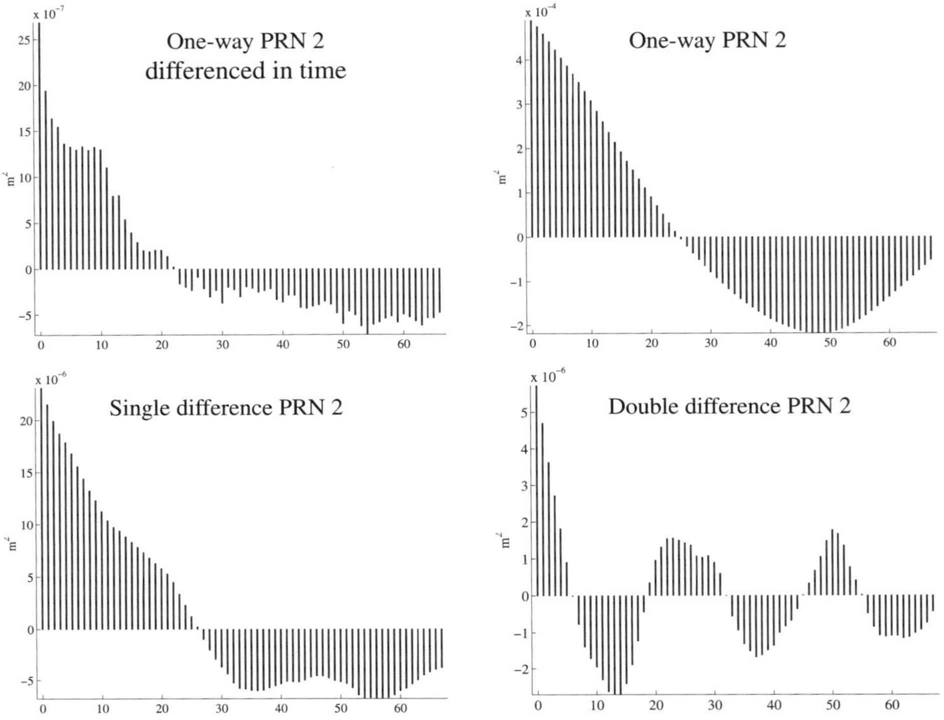

Figure 16.9 Autocorrelation for ionospheric delay to PRN 2. Upper left shows difference in time: $I_k - I_{k-1}$, upper right the one-way, lower left the single difference, and lower left the double difference. Note the various orders of magnitude.

Because of varying geometry, I_k will vary with different baseline lengths $d = 5$, 25, 50, 100, 500, 1000 km, say. So we suggest the interested reader to explore this sort of investigation with the final goal of determining the variance of differential ionosphere σ_0^2, correlation time T, and correlation length D in an expression for the autocorrelation with sample interval Δt and baseline length d:

$$R_x(k,d) = \sigma_0^2 e^{-|\Delta t|/T} e^{-d/D}. \tag{16.40}$$

This function R_x has been determined from double differenced phase observations in Goad & Yang (1994). The authors assume stationarity and an exponentially correlated process: they estimated T to 64 min., $\sigma_0^2 = 2\,\text{m}^2$, and $D \approx 1\,500\,\text{km}$. In general you may adjust a Gauss-Markov process to the parameters σ^2 and α. At $k = 0$ we fit σ^2, and at the $1/e$-point we fit α.

It is known that the ionospheric delay has a distinct *daily variation*. In Klobuchar (1996) we find an approximate expression for I in meters as function of local time t in hours:

$$I = 2.1 + 0.75\cos\bigl((t-14)2\pi/28\bigr).$$

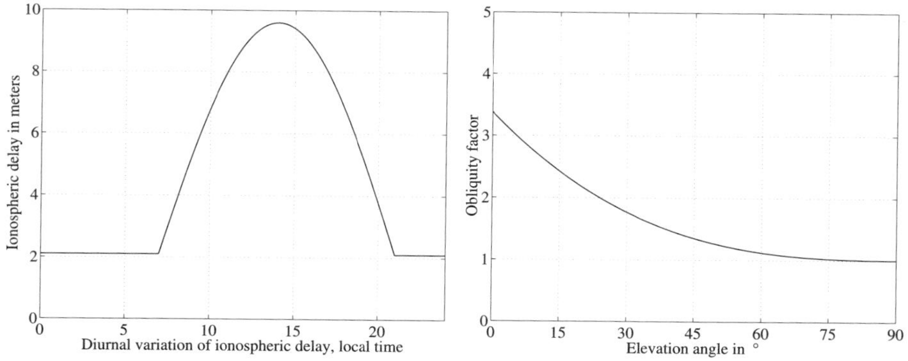

Figure 16.10 Ionospheric delay model as function of local time and elevation angle.

This value is valid in the direction of zenith. To get the increased value of I in a direction with elevation angle El in half circles (the range of El is 0–0.5, half circle times π equals radians) we have to multiply by the *obliquity factor*

$$F(El) = 1 + 16(0.53 - El)^3.$$

This expression is based on a formula given in ICD-GPS-200 (1991). Figure 16.10 shows the graphs of I and $F(El)$. The small constant level of delay equal to 2.1 m represents the delay at night. As the sun rises and sets, the ionospheric model gives rise to the cosine-shaped pulse for daytime.

When interpreting an autocorrelation function it is useful to remember the following important properties:

Maximum value The autocorrelation function has a maximum at zero shift:

$$E\{x^2(t)\} = R_x(0) \geq |R_x(k)|.$$

Symmetry conditions The autocorrelation function is an even function

$$R_x(k) = R_x(-k)$$

while the cross-correlation satisfies

$$R_{xy}(k) = R_{yx}(-k).$$

Mean-square value The cross-correlation is bounded by

$$|R_{xy}(k)|^2 \leq R_x(0)R_y(0) \leq \tfrac{1}{2}\Big(\big(R_x(0)\big)^2 + \big(R_y(0)\big)^2\Big).$$

Periodic component The autocorrelation has a periodic component if x has one.

Remark 16.1 (Filter theoretical interpretation of the technique of differencing) We may look at double differenced data D as made up of two single difference measurements S_i: $D = S_1 - S_2$. Each of these is made up simplistically as $S_i = \rho_i + t$ where ρ_i is pseudorange and t is receiver clock offset converted to length by multiplication by c. Note that t is same in both. So double differencing does remove this clock error.

We illustrate a few basic facts about differencing. Start with two single differences:

$$S_1 = \rho_1 + t$$
$$S_2 = \rho_2 + t$$

and their double difference:

$$S_2 - S_1 = \rho_2 - \rho_1.$$

For the double difference the situation has changed. We have one less equation and cannot estimate t. You could estimate from single differences but the a priori clock variance σ^2_{clock} must be infinite; otherwise you introduce a third measurement, which you do not have in the double difference case. So one case has S_1, S_2, and σ^2_{clock}. This is equivalent to S_1, D, and σ^2_{clock}. If $\sigma^2_{\text{clock}} = \infty$ it really means it is of no value. So this is really equivalent to S_1 and D. Now since the double differences do not involve the clock explicitly, the only use to be made of S_1 is to estimate the clock. Since a clock estimate is not needed by double differencing, D stands by itself in estimating position.

However if $\sigma^2_{\text{clock}} \neq \infty$ then this couples all three measurements in which case all three should be processed together. It does not matter whether S_1, S_2, t or S_1, D, t or S_2, D, t is used. If you treat clocks as having infinite variance at each epoch and then plot their estimates, you will see that clocks do not drift with infinite variance. So theoretically it is correct that we do know something about clock drift, and using double differences alone does not allow us to take advantage of this knowledge (because t cancels).

Then the difficult question is "how to model (receiver) clock drift randomly?" Clearly different receivers have vastly differing (usually quartz) clocks. The best opportunity would be for the International GPS Service for Geodynamics (IGS) to drive their receivers used for orbit calculations with rubidium or, even better, cesium oscillators. Then a more descriptive random model with smaller variances could prove useful. Quartz clocks drift so wildly relative to centimeter positioning requirements, that trying to model them would add little to improving positions. So we may conclude:

- Double differences can be conceived as a filter where the clock behavior is modeled as white noise with large variances.
- Using single differences with analytical prediction of receiver clock offset is much better than using a filter.

Aspects of Random Processes

We have presented three basic models: random walk, random ramp, and an exponentially correlated process. Each one introduces a special pattern for how random errors interact with the model. If the model is chosen correctly the residuals should be white noise. ***A good measure for the randomness of errors is to look at their autocorrelation***. Or in other words: A given autocorrelation function is based on a given model. ***If the model fits the data, the error autocorrelation function ideally is white noise.***

Pure observational errors will show a very clear spike at zero. Systematic errors will give an autocorrelation that remains large, even far away from zero. To demonstrate the autocorrelation for random data we call the M-file model_g(randn(1:300)). The result was shown in Figure 16.4. Here we have an autocorrelation function which has a spike at zero. This spike is a measure for the variance of observation. The global features of the autocorrelation function (away from shift $= 0$) originate from the model. If the autocorrelation drops off at a certain distance from zero we speak of a band limited process.

Sometimes we are lucky to know the physics behind a problem. Then it is often given in terms of a differential equation.

Example 16.17 We describe a procedure starting from a differential equation, leading to the discrete state equation. The z-transform exhibits poles of the solution and these poles determine whether the filter is stable or not.

The actual *differential equation* describes a forced motion for a body with mass m, damping constant d, and spring modulus c, all under the influence of an external force f:

$$m\ddot{x} + d\dot{x} + cx = f. \tag{16.41}$$

The actual displacement of the body is $x(t) = \int \dot{x}(t)\,dt$ and we differentiate to get $x_k \approx x_{k-1} + \dot{x}_{k-1}$ with $\Delta t = 1$. Furthermore

$$\dot{x}(t) = \int \ddot{x}(t)\,dt \qquad \text{and} \qquad \dot{x}_k \approx \dot{x}_{k-1} + \ddot{x}_{k-1}. \tag{16.42}$$

The discrete form of equation (16.41) is

$$\ddot{x}_{k-1} = -\frac{d}{m}\dot{x}_{k-1} - \frac{c}{m}x_{k-1} + \frac{f}{m}. \tag{16.43}$$

Insertion of (16.43) into (16.42) yields

$$\dot{x}_k = \dot{x}_{k-1} + \left(-\frac{d}{m}\dot{x}_{k-1} - \frac{c}{m}x_{k-1} + \frac{f}{m}\right).$$

Hence in matrix form the *state equation* becomes

$$\begin{bmatrix} x_k \\ \dot{x}_k \end{bmatrix} = \begin{bmatrix} 1 & 1 \\ -c/m & 1-d/m \end{bmatrix} \begin{bmatrix} x_{k-1} \\ \dot{x}_{k-1} \end{bmatrix} + \begin{bmatrix} 0 \\ f/m \end{bmatrix}.$$

Random walk is the special case with $c = 0$ and $d = 0$. There is no spring force and no damping:

$$\begin{bmatrix} x_k \\ \dot{x}_k \end{bmatrix} = \begin{bmatrix} 1 & 1 \\ 0 & 1 \end{bmatrix} \begin{bmatrix} x_{k-1} \\ \dot{x}_{k-1} \end{bmatrix} + \begin{bmatrix} 0 \\ f/m \end{bmatrix}. \tag{16.44}$$

We substitute the *state variables* by $X_{1,k} = x_k,\ X_{2,k} = \dot{x}_k,\ F(z) = f/m$:

$$\begin{bmatrix} X_{1,k} \\ X_{2,k} \end{bmatrix} = \begin{bmatrix} 1 & 1 \\ -c/m & 1 - d/m \end{bmatrix} \begin{bmatrix} X_{1,k-1} \\ X_{2,k-1} \end{bmatrix} + \begin{bmatrix} 0 \\ F(z) \end{bmatrix}. \tag{16.45}$$

The displacement of the body from static equilibrium is $X_{1,k}$, and $X_{2,k}$ is the instantaneous velocity of the body.

The *solution of this discrete system* is found via z-transforms $X_1(z) = \sum X_{1,k} z^{-k}$ and $X_2(z) = \sum X_{2,k} z^{-k}$. The transform of (16.45) is

$$zX_1(z) = X_1(z) + X_2(z)$$
$$zX_2(z) = -\frac{c}{m} X_1(z) + \left(1 - \frac{d}{m}\right) X_2(z) + F(z).$$

We rewrite the last equation as

$$X_2(z)\left(z - \left(1 - \frac{d}{m}\right)\right) = -\frac{c}{m} X_1(z) + F(z).$$

Then insertion of $X_2(z)$ into the first equation gives

$$zX_1(z) = X_1(z) + \frac{-(c/m)X_1(z) + F(z)}{z - (1 - d/m)}$$

or

$$\left(z^2 - \left(2 - \frac{d}{m}\right) z + \left(1 - \frac{d}{m} + \frac{c}{m}\right)\right) X_1(z) = F(z).$$

The position $X_1(z)$ of the body is given by

$$\frac{X_1(z)}{F(z)} = \frac{1}{z^2 - (2 - d/m)z + (1 - d/m + c/m)}.$$

The random walk case with $c = 0$ and $d = 0$ gives

$$\frac{X_1(z)}{F(z)} = \frac{1}{z^2 - 2z + 1} = \frac{1}{z - 1} \frac{1}{z - 1}.$$

The function $X_1(z)$ has a multiple pole at $z = 1$. In general if all poles are within the unit circle we deal with a *stable filter*. If one or more poles are outside the unit circle we have an unstable filter. The present case with $z = 1$ is most delicate. The neighborhood of $z = 1$ is a true "minefield".

The z-transform of the impulse response is the transfer function $H(z)$. Input and output are connected by $H(z)$ and it is the key to an optimal control. This leads further to explicit expressions for autocorrelation and power spectral density. However, this relatively simple example is at the limit of complexity for finding closed form solutions to algebraic equations by purely algebraic means. So we do not want to present more theory, because in most real world examples numbers rather than formulas have to describe reality.

Example 16.18 We want to *combine a random walk and white noise* in the dynamic linear model

System equation:	$x_k = x_{k-1} + \epsilon_k,$	$\epsilon_k \sim N_1(0, \sigma_\epsilon^2)$
Observation equation:	$b_k = x_k + e_k,$	$e_k \sim N_1(0, \sigma_e^2).$

This is the simplest possible model which has a surprisingly wide range of applications.

Example 16.19 Extended filter for the phase observation $x = \varphi$:

System equation:	$x_k = x_{k-1} + \epsilon_k,$	$\epsilon_k \sim N_1(0, \sigma_\epsilon^2)$
Observation equation:	$b_k = \text{constant} + \sin(f_1 \Delta t + x_k) + e_k,$	$e_k \sim N_1(0, \sigma_e^2).$

Our first development of this example is in the M-file mangouli.

We summarize this chapter by repeating the essential points for a Gauss-Markov process

Autocorrelation $\mathcal{R}_x(k) = \sigma^2 e^{-\alpha|k|}$

Process $x_k = e^{-\alpha|\Delta t|} x_{k-1} + \epsilon_k$

Variance $\sigma_{\epsilon_k}^2 = E\{\epsilon_k^2\} = \sigma^2(1 - e^{-2\alpha|\Delta t|})$ to be used on the diagonal of the filter covariance matrix Σ_{ϵ_k}, see next chapter!

17

KALMAN FILTERS

17.1 Updating Least Squares

The key to the Kalman filter is the idea of *updating*. New observations come in, and they change our best least squares estimate of the parameters $\boldsymbol{x}$. We want to compute that change—to update the estimate $\hat{\boldsymbol{x}}$. The process will be efficient if we can express the new estimate as *a linear combination of the old estimate* $\hat{\boldsymbol{x}}_{\text{old}}$ *and the new observation* $\boldsymbol{b}_{\text{new}}$:

$$\hat{\boldsymbol{x}}_{\text{new}} = L\hat{\boldsymbol{x}}_{\text{old}} + K\boldsymbol{b}_{\text{new}}. \tag{17.1}$$

This section will make four key points as it introduces the Kalman filter. Then the following sections explain and illustrate these points in detail.

The first point is that the process is *recursive*. We do not store the old observations $\boldsymbol{b}_{\text{old}}$! Those measurements were already used in the estimate $\hat{\boldsymbol{x}}_{\text{old}}$. Built into equation (17.1) is the expectation (or hope) that all the information from $\boldsymbol{b}_{\text{old}}$ that we need for $\hat{\boldsymbol{x}}_{\text{new}}$ is available in $\hat{\boldsymbol{x}}_{\text{old}}$. Since $\boldsymbol{x}$ is a much shorter vector than $\boldsymbol{b}$ (which is growing in length with each new measurement) the filter is efficient.

The second point is that the update formula (17.1) can be written (and derived) in many equivalent ways. This makes a lot of expositions of the Kalman filter difficult to follow. The frustrated reader finally just asks for the damn formula. But some of the variations help the intuition, for example by separating the "prediction" from the "correction":

$$\hat{\boldsymbol{x}}_{\text{new}} = \hat{\boldsymbol{x}}_{\text{old}} + K(\boldsymbol{b}_{\text{new}} - A_{\text{new}}\hat{\boldsymbol{x}}_{\text{old}}). \tag{17.2}$$

Direct comparison with (17.1) gives the relation $L = I - KA_{\text{new}}$. The *Kalman gain matrix* K becomes the crucial matrix to identify. It multiplies the mismatch $\boldsymbol{b}_{\text{new}} - A_{\text{new}}\hat{\boldsymbol{x}}_{\text{old}}$ (the *innovation*) between old estimates and new measurements. From the estimate $\hat{\boldsymbol{x}}_{\text{old}}$ we would have predicted measurements $A_{\text{new}}\hat{\boldsymbol{x}}_{\text{old}}$. The difference between this prediction and the actual $\boldsymbol{b}_{\text{new}}$ is multiplied by K to give the correction in $\hat{\boldsymbol{x}}$. That is (17.2).

This gain matrix K involves the statistics of $\boldsymbol{b}_{\text{new}}$ and $\hat{\boldsymbol{x}}_{\text{old}}$, because they tell how much weight to give to this mismatch. (Thus the crucial covariance matrix P must also be updated! This will be our third point.) Variations of the updating formula are introduced for the sake of numerical stability, when this formula is to be applied many times. All the equivalent forms must be related by matrix identities and we will try to make those clear.

The third key point is that the reliability of $\hat{\boldsymbol{x}}$ is a crucial part of the output. This is the *error covariance matrix* $P = \Sigma_{\hat{x}}$ for the estimate. P tells us the statistical properties of $\hat{\boldsymbol{x}}$ based on the statistical properties of $\boldsymbol{b}$. This matrix does not depend on the particular measurements $\boldsymbol{b}$ or measurement errors $\boldsymbol{e}$. Those are only random samples from a whole population of possible measurements and errors. We assume a Gaussian (normal) distribution of this error population, with a known covariance matrix Σ_e. (Well, barely known. We may struggle to find a realistic Σ_e.) The least-squares solution $\hat{\boldsymbol{x}}$ to a static equation $A\boldsymbol{x} = \boldsymbol{b}$ is weighted by Σ_e^{-1}, and $P = (A^{\mathrm{T}}\Sigma_e^{-1}A)^{-1}$ is the error covariance matrix for $\hat{\boldsymbol{x}}$. It is this covariance that we update when measurements $\boldsymbol{b}_{\text{new}}$ arrive with $\Sigma_{e,\text{new}}$:

$$P_{\text{new}} = (I - KA)P_{\text{old}}(I - KA)^{\mathrm{T}} + K\Sigma_{e,\text{new}}K^{\mathrm{T}}. \tag{17.3}$$

To obtain this formula from (17.1) with $L = I - KA$ *we assume that the errors* $\boldsymbol{e}_{\text{new}}$ *in* $\boldsymbol{b}_{\text{new}}$ *are statistically independent from the errors* $\boldsymbol{e}_{\text{old}}$:

$$\text{The covariance matrix for} \quad \boldsymbol{b} = \begin{bmatrix} \boldsymbol{b}_{\text{old}} \\ \boldsymbol{b}_{\text{new}} \end{bmatrix} \quad \text{is} \quad \Sigma_e = \begin{bmatrix} \Sigma_{e,\text{old}} & 0 \\ 0 & \Sigma_{e,\text{new}} \end{bmatrix}.$$

Note We will go through these steps again in detail. The goal of this first discussion is to see how the pieces of the Kalman filter come together. The reader will not forget that the inverse covariance matrix Σ_e^{-1} enters as a weight in the least-squares problem. Thus $\hat{\boldsymbol{x}}_{\text{old}}$ involved the covariance $\Sigma_{e,\text{old}}$ for $\boldsymbol{b}_{\text{old}}$. Similarly $\hat{\boldsymbol{x}}_{\text{new}}$ will also involve the covariance $\Sigma_{e,\text{new}}$ for $\boldsymbol{b}_{\text{new}}$. Again we hope and expect that it will not be necessary to store $\Sigma_{e,\text{old}}$, since it has already been used to compute $\hat{\boldsymbol{x}}_{\text{old}}$ and P_{old}.

We come to the fourth point, which is fundamentally important. The least-squares problems described up to now have been *static*. We have been updating our estimate $\hat{\boldsymbol{x}}$ of the *same* parameter vector $\boldsymbol{x}$. It is recursive least squares and we could have done it earlier in the book, without mentioning Kalman. With exact measurements, $\boldsymbol{x} = \hat{\boldsymbol{x}}_{\text{old}} = \hat{\boldsymbol{x}}_{\text{new}}$ would solve both the old and new overdetermined systems in this static situation:

$$A_{\text{old}}\boldsymbol{x} = \boldsymbol{b}_{\text{old}} \qquad \text{and} \qquad \begin{bmatrix} A_{\text{old}} \\ A_{\text{new}} \end{bmatrix} \begin{bmatrix} \boldsymbol{x} \end{bmatrix} = \begin{bmatrix} \boldsymbol{b}_{\text{old}} \\ \boldsymbol{b}_{\text{new}} \end{bmatrix}. \tag{17.4}$$

We have been adding *new rows* to the system *but not new columns*. If the best estimate $\hat{\boldsymbol{x}}_{\text{old}}$ happened to be exactly consistent with the new measurements, so that $A_{\text{new}}\hat{\boldsymbol{x}}_{\text{old}} = \boldsymbol{b}_{\text{new}}$, then the measurements $\boldsymbol{b}_{\text{new}}$ would not change that old estimate. This is clear from (17.2): $\hat{\boldsymbol{x}}_{\text{new}}$ would equal $\hat{\boldsymbol{x}}_{\text{old}}$ because the gain matrix K will be multiplying the zero vector to give a zero correction.

The Kalman filter operates on *dynamic* problems. The new state vector $\boldsymbol{x}_k$ (for example, the position of a GPS receiver) is not generally the same as the state vector $\boldsymbol{x}_{k-1}$. The state is changing, the receiver is moving, and we assume a linear equation for that change in state—with its own error $\boldsymbol{\epsilon}_k$ (some texts call it $\boldsymbol{\epsilon}_{k-1}$):

$$\boldsymbol{x}_k = F_{k-1}\boldsymbol{x}_{k-1} + \boldsymbol{\epsilon}_k. \tag{17.5}$$

This is a discrete-time linear random process. The index k indicates the new state $\boldsymbol{x}_k$ and the newly measured quantities $\boldsymbol{b}_k$ and the errors $\boldsymbol{e}_k$ in those new measurements:

$$\boldsymbol{b}_k = A_k \boldsymbol{x}_k + \boldsymbol{e}_k. \tag{17.6}$$

Thus $\boldsymbol{e}_k$ still shows the mismatch in the prediction, but the prediction is using (of course) the state equation (17.5) for $\boldsymbol{x}_k$. And that state equation includes an error (or noise) $\boldsymbol{\epsilon}_k$.

Since (17.5) is a one-step equation (a *first-order process*), involving only $\boldsymbol{x}_{k-1}$ and not the earlier states $\boldsymbol{x}_0, \ldots, \boldsymbol{x}_{k-2}$, we look for an estimate of $\boldsymbol{x}_k$ that does not involve the very old estimates $\hat{\boldsymbol{x}}_0, \ldots, \hat{\boldsymbol{x}}_{k-2}$. The update comes in two stages, a ***prediction*** using the state equation (17.5), and a ***correction*** using the observation (17.6):

$$\begin{aligned} &\textit{Prediction:} \qquad & \hat{\boldsymbol{x}}_{k|k-1} &= F_{k-1}\hat{\boldsymbol{x}}_{k-1|k-1} \\ &\textit{Correction:} \qquad & \hat{\boldsymbol{x}}_{k|k} &= \hat{\boldsymbol{x}}_{k|k-1} + K_k(\boldsymbol{b}_k - A_k \hat{\boldsymbol{x}}_{k|k-1}). \end{aligned} \tag{17.7}$$

We also want a two-stage update of the error covariance matrix P (or its inverse). That will be an extremely important equation. We present the prediction $P_{k|k-1}$, from which (17.3) gives the correction $P_{\text{new}} = P_{k|k}$. The prediction comes (of course) from the state equation:

$$P_{k|k-1} = F_{k-1} P_{k-1} F_{k-1}^{\mathrm{T}} + \Sigma_{\epsilon,k}. \tag{17.8}$$

The combined step to $P_{k|k}$ uses the new covariances $\Sigma_{e,k}$ and $\Sigma_{\epsilon,k}$ that describe the error populations from which $\boldsymbol{e}_k$ and $\boldsymbol{\epsilon}_k$ are drawn.

Important We assume that $\boldsymbol{\epsilon}_k$ is statistically independent of all other $\boldsymbol{\epsilon}_j$ and all $\boldsymbol{e}_j$. The whole covariance matrix Σ is block-diagonal, with blocks for every $\boldsymbol{\epsilon}_k$ and $\boldsymbol{e}_k$. One block describes errors in the state equation, the other describes errors in measurement.

The Complete System $\mathcal{A}x = \mathscr{b} - \mathscr{e}$

It may be useful to the reader if we write down the complete system of equations through time t_k. The initial value $\boldsymbol{x}_0 = \hat{\boldsymbol{x}}_0$ is given, with its own error covariance matrix $P_{0|0}$. Then each time step introduces *two* new block rows in the big matrix $\mathcal{A}$ and *one* new block column—corresponding to the new $\boldsymbol{x}_k$ in the list x_k of unknown state vectors:

$$\begin{bmatrix} A_0 & & & & \\ -F_0 & I & & & \\ & A_1 & & & \\ & -F_1 & I & & \\ & & \ddots & & \\ & & -F_{k-1} & I \\ & & & A_k \end{bmatrix} \begin{bmatrix} \boldsymbol{x}_0 \\ \boldsymbol{x}_1 \\ \vdots \\ \boldsymbol{x}_{k-1} \\ \boldsymbol{x}_k \end{bmatrix} = \begin{bmatrix} \boldsymbol{b}_0 \\ \boldsymbol{0} \\ \boldsymbol{b}_1 \\ \boldsymbol{0} \\ \vdots \\ \boldsymbol{0} \\ \boldsymbol{b}_k \end{bmatrix} - \begin{bmatrix} \boldsymbol{e}_0 \\ \boldsymbol{\epsilon}_1 \\ \boldsymbol{e}_1 \\ \boldsymbol{\epsilon}_2 \\ \vdots \\ \boldsymbol{\epsilon}_k \\ \boldsymbol{e}_k \end{bmatrix}. \tag{17.9}$$

Looking at those last two rows, you see the state equation (17.5) and the observation equation (17.6). It is this complete system that the Kalman filter solves recursively by weighted least squares. We will write equation (17.9) as $\mathcal{A}x = \mathcal{b} - \mathcal{e}$ or more precisely $\mathcal{A}_k x_k = \mathcal{b}_k - \mathcal{e}_k$. The script letters $\mathcal{A}$, x, $\mathcal{b}$, $\mathcal{e}$ indicate the complete system up to t_k.

The weights are the covariance matrices of the errors $\boldsymbol{e}_k$ and $\boldsymbol{\epsilon}_k$. The output is the set of state estimates $\hat{\boldsymbol{x}}_0, \ldots, \hat{\boldsymbol{x}}_k$ and their covariance matrices $P_0, \ldots, P_k$. Those could come from the usual formulas like $\mathcal{P} = (\mathcal{A}^{\mathrm{T}}\Sigma^{-1}\mathcal{A})^{-1}$, but they don't. *The recursive formulas determine $\hat{\boldsymbol{x}}_k$ and P_k from $\boldsymbol{b}_k$ and $\hat{\boldsymbol{x}}_{k-1}$ and P_{k-1} and $\Sigma_{e,k}$ and $\Sigma_{\epsilon,k}$.*

The possibility of a simple update is built into the structure of that complete matrix $\mathcal{A}$ (which is block bidiagonal). The block diagonal weight matrix $\mathcal{C} = \Sigma^{-1}$ contains the inverses of all the blocks Σ_e and Σ_ϵ. The large matrix $\mathcal{A}^{\mathrm{T}}\Sigma^{-1}\mathcal{A}$ is block tridiagonal—and tridiagonal matrices form the classical situation for recursive formulas. Without a tridiagonal matrix (if the new state equation for $\boldsymbol{x}_k$ involved states earlier than $\boldsymbol{x}_{k-1}$), a one-step update would be impossible.

The variety of Kalman filter formulas comes from the variety of ways that we can represent the matrix $\mathcal{A}^{\mathrm{T}}\mathcal{C}\mathcal{A}$. This block tridiagonal matrix is symmetric positive definite. It has a Cholesky factorization LL^{T} into block bidiagonal matrices. We can update L for a more stable "square root algorithm." Or we can update the Gram-Schmidt factorization QR that orthogonalizes the columns of $\mathcal{A}$.

All those updates can be derived by manipulating the complete matrix $\mathcal{A}$. It is important to recognize that (17.9) is the system we are solving (by weighted least squares). But the derivation is much easier if we assume an expression involving only the most recent times $k-1$ and k, and determine the matrices (especially Kalman's gain matrix K_k) in that update formula. Unlike the static case, the gain matrix K_k will now depend on the discrete time index k.

Example 17.1 (Recursive update for a *static* one-parameter problem) A doctor takes m independent and equally reliable readings $b_1, \ldots, b_k$ of your pulse rate x. Each measurement error $(x - b_j)^2$ has expected value σ^2. We want to compute the best estimate $\hat{x}_k$ and its variance $P_k = E\{(x - \hat{x}_k)^2\}$. First we compute $\hat{x}_k$ and P_k directly, then we compute *recursively* from $\hat{x}_{k-1}$ and P_{k-1}.

The k measurement equations are simply

$$x = b_1, \quad x = b_2, \quad \ldots, \quad x = b_k \qquad \text{or} \qquad \begin{bmatrix} 1 \\ \vdots \\ 1 \end{bmatrix} \begin{bmatrix} x \end{bmatrix} = \begin{bmatrix} b_1 \\ \vdots \\ b_k \end{bmatrix}.$$

Thus A (we reserve $\mathcal{A}$ for the dynamic case including a state equation) is a column of ones. The covariance matrix is $\Sigma_b = \sigma^2 I$, diagonal because the readings were independent. The 1 by 1 matrix $P^{-1} = A^{\mathrm{T}}\Sigma_b^{-1}A$ and the right side $A^{\mathrm{T}}Cb$ are

$$P^{-1} = \begin{bmatrix} 1 & \ldots & 1 \end{bmatrix} \frac{I}{\sigma^2} \begin{bmatrix} 1 \\ \vdots \\ 1 \end{bmatrix} = \frac{k}{\sigma^2} \quad \text{and} \quad \begin{bmatrix} 1 & \ldots & 1 \end{bmatrix} \frac{I}{\sigma^2} \begin{bmatrix} b_1 \\ \vdots \\ b_k \end{bmatrix} = \frac{b_1 + \cdots + b_k}{\sigma^2}.$$

The best unbiased estimate, as everyone expects, is the average of the readings:

$$\hat{x}_k = P(A^{\mathrm{T}}Cb) = \frac{\sigma^2}{k}\left(\frac{b_1 + \cdots + b_k}{\sigma^2}\right). \tag{17.10}$$

The 1 by 1 matrix $E\{(x-\hat{x}_k)^2\}$ is $P = (A^{\mathrm{T}}CA)^{-1} = \sigma^2/k$. By taking k independent measurements with variance σ^2, the variance of the average value $\hat{x}_k$ is reduced to σ^2/k. We can show that without matrices and without Kalman!

But now do those computations recursively. The step begins with the correct values at time $k-1$:

$$\hat{x}_{\text{old}} = \tfrac{1}{k-1}(b_1 + \cdots + b_{k-1}) \qquad \text{and} \qquad P_{\text{old}} = \tfrac{\sigma^2}{k-1}. \tag{17.11}$$

The new observation b_k brings a correction = prediction error = innovation, and remember that A_{new} is just $[\,1\,]$:

$$\hat{x}_k = \hat{x}_{k-1} + K_k(b_k - \hat{x}_{k-1}).$$

The extra measurement reduces P and increases P^{-1}:

$$P_k^{-1} = P_{k-1}^{-1} + \tfrac{1}{\sigma^2} = \tfrac{k-1}{\sigma^2} + \tfrac{1}{\sigma^2} = \tfrac{k}{\sigma^2}. \tag{17.12}$$

We will explain that increase $1/\sigma^2$. It is computed as $A_{\text{new}}^{\mathrm{T}}\Sigma_k^{-1}A_{\text{new}} = [\,1\,][\,1/\sigma^2\,][\,1\,]$. By working with P^{-1} instead of P, the update (17.12) is very simple. From P_k we obtain the gain matrix K_k and the correction to $\hat{x}$:

$$K_k = P_k A_k^{\mathrm{T}}\Sigma_k^{-1} = [\,\tfrac{\sigma^2}{k}\,][\,1\,][\,\tfrac{1}{\sigma^2}\,] = [\,\tfrac{1}{k}\,] \tag{17.13}$$

$$\hat{x}_k = \hat{x}_{k-1} + \tfrac{1}{k}(b_k - \hat{x}_{k-1}) = \tfrac{k-1}{k}\left(\tfrac{b_1+\cdots+b_{k-1}}{k-1}\right) + \tfrac{1}{k}b_k. \tag{17.14}$$

The key point is that final combination of the old average and the new measurement. You see how they combine to give the new average $(b_1+\cdots+b_k)/k$. We did not have to add all the old and new b_j to compute their average $\hat{x}_k$! The first $k-1$ measurements were already averaged in $\hat{x}_{k-1}$, and the correction gave the new average with only one more addition.

Example 17.2 (Same static problem but change to $\Sigma_{\text{new}} = \sigma_k^2$ for the kth measurement) We can show quickly that the update is still correct when the variance for the kth observation changes from σ^2 to σ_k^2. The matrix A is the same column of ones. The last entry of Σ is now σ_k^2 and a direct calculation gives $P_k^{-1} = A^{\mathrm{T}}\Sigma^{-1}A$:

$$P_k^{-1} = \begin{bmatrix} 1 & \dots & 1 \end{bmatrix} \begin{bmatrix} \sigma^2 & & & \\ & \ddots & & \\ & & \sigma^2 & \\ & & & \sigma_k^2 \end{bmatrix}^{-1} \begin{bmatrix} 1 \\ \vdots \\ 1 \end{bmatrix} = \frac{k-1}{\sigma^2} + \frac{1}{\sigma_k^2}. \tag{17.15}$$

This shows the update from P_{k-1}^{-1}, no problem. Then $\hat{x}_k$ can come from $(A^{\mathrm{T}}CA)^{-1}A^{\mathrm{T}}Cb$, with $C = \Sigma^{-1}$:

$$\text{Directly:} \quad \hat{x}_k = P_k(A^{\mathrm{T}}\Sigma^{-1}b) = P_k\left(\frac{b_1}{\sigma^2} + \cdots + \frac{b_{k-1}}{\sigma^2} + \frac{b_k}{\sigma_k^2}\right). \tag{17.16}$$

Compare with the recursive form. The kth gain K_k is $P_k A^{\mathrm{T}}\Sigma_k^{-1} = P_k/\sigma_k^2$. Then the updated $\hat{x}_k$ is just

$$\hat{x}_k = \hat{x}_{k-1} + P_k(b_k - \hat{x}_{k-1})/\sigma_k^2. \tag{17.17}$$

To verify the equivalence to (17.15), introduce $P_k P_k^{-1} = 1$ into (17.15) and rewrite using (17.17):

$$\hat{x}_k = P_k\left(P_k^{-1}\hat{x}_{k-1} + \frac{b_k - \hat{x}_{k-1}}{\sigma_k^2}\right) = P_k\left(\frac{k-1}{\sigma^2}\frac{b_1 + \cdots + b_{k-1}}{k-1} + \frac{\hat{x}_{k-1}}{\sigma_1^2} + \frac{b_k}{\sigma_k^2} - \frac{\hat{x}_{k-1}}{\sigma_k^2}\right).$$

The key point is that $\hat{x}_{k-1}$ already captured the needed information from $b_1, \ldots, b_{k-1}$. Those measurements are no longer wanted! And we still owe the reader an explanation of the gain matrix K_k.

17.2 Static and Dynamic Updates

The first measurement is $\boldsymbol{b}_0 = A_0\boldsymbol{x}_0 + \boldsymbol{e}_0$. We estimate $\boldsymbol{x}_0$ using this set of observations $\boldsymbol{b}_0$. The estimate is $\hat{\boldsymbol{x}}_0$. The equation $A_0\boldsymbol{x}_0 = \boldsymbol{b}_0$ is probably overdetermined. The least-squares solution—weighted according to the inverse $C = \Sigma_{e,0}^{-1}$ of the covariance matrix for the residuals of $\boldsymbol{b}_0$—is the usual one:

$$\hat{\boldsymbol{x}}_0 = \left(A_0^{\mathrm{T}}\Sigma_{e,0}^{-1}A_0\right)^{-1}A_0^{\mathrm{T}}\Sigma_{e,0}^{-1}\boldsymbol{b}_0. \tag{17.18}$$

The error $\boldsymbol{x}_0 - \hat{\boldsymbol{x}}_0$ has expectation zero and its covariance matrix (which is a minimum for this choice of weight) is

$$\Sigma_0 = E\left\{(\boldsymbol{x}_0 - \hat{\boldsymbol{x}}_0)(\boldsymbol{x}_0 - \hat{\boldsymbol{x}}_0)^{\mathrm{T}}\right\} = \left(A_0^{\mathrm{T}}\Sigma_{e,0}^{-1}A_0\right)^{-1}. \tag{17.19}$$

Now we ask: *If more data are available, is it possible to estimate* $\boldsymbol{x}$ *for the total system* $A_0\boldsymbol{x} = \boldsymbol{b}_0$, $A_1\boldsymbol{x} = \boldsymbol{b}_1$ *without starting the computation from the beginning with* $\boldsymbol{b}_0$*?*

It is easy to imagine this situation. We want to estimate a static position—latitude, longitude and height—by means of a GPS receiver. At an arbitrary moment we estimate the best position. The next observation is likely to deviate a little from the earlier estimate and we should generate a new estimate based on the increased data set. If the error of the new estimate is independent of earlier errors there must be a way to work recursively. We want to find $\hat{\boldsymbol{x}}_1$ based on the earlier estimate $\hat{\boldsymbol{x}}_0$ and the new observation $\boldsymbol{b}_1$. (Point of notation: $\hat{\boldsymbol{x}}_k$ depends on all measurements $\boldsymbol{b}_0, \ldots, \boldsymbol{b}_k$. It does not come only from $\boldsymbol{b}_k$.)

The final result should be identical to the one we get by calculating $\hat{\boldsymbol{x}}_1$ from the beginning. The right choice for the weights is $C = \Sigma_e^{-1}$ where

$$\Sigma_e = \begin{bmatrix}\Sigma_{e,0} & 0\\ 0 & \Sigma_{e,1}\end{bmatrix} \quad \text{is the covariance matrix of the residuals} \quad \begin{bmatrix}\boldsymbol{e}_0\\ \boldsymbol{e}_1\end{bmatrix}.$$

The matrix Σ_e is block diagonal because $\boldsymbol{e}_1$ is independent of $\boldsymbol{e}_0$. So the coefficient matrix $A^{\mathrm{T}}\Sigma_e^{-1}A$ in the equation for $\hat{\boldsymbol{x}}_1$ is

$$\Sigma_1^{-1} = \begin{bmatrix}A_0^{\mathrm{T}} & A_1^{\mathrm{T}}\end{bmatrix}\begin{bmatrix}\Sigma_{e,0} & 0\\ 0 & \Sigma_{e,1}\end{bmatrix}^{-1}\begin{bmatrix}A_0\\ A_1\end{bmatrix} = A_0^{\mathrm{T}}\Sigma_{e,0}^{-1}A_0 + A_1^{\mathrm{T}}\Sigma_{e,1}^{-1}A_1. \tag{17.20}$$

Remember that $\hat{x}_1$ is best for the combined system $A_0\hat{x}_1 = b_0$, $A_1\hat{x}_1 = b_1$. Now the normal equations are $A^{\mathrm{T}}\Sigma_e^{-1}A\hat{x} = A^{\mathrm{T}}\Sigma_e^{-1}b$. We write P or P_1 or P_{new} for the inverse matrix $(A^{\mathrm{T}}\Sigma_e^{-1}A)^{-1}$. Then the optimal solution is

$$\hat{x}_1 = P_1\begin{bmatrix} A_0^{\mathrm{T}} & A_1^{\mathrm{T}} \end{bmatrix}\Sigma_e^{-1}\begin{bmatrix} b_0 \\ b_1 \end{bmatrix} = P_1\big(A_0^{\mathrm{T}}\Sigma_{e,0}^{-1}b_0 + A_1^{\mathrm{T}}\Sigma_{e,1}^{-1}b_1\big). \tag{17.21}$$

It is this solution that we hope to find recursively by using the already computed $\hat{x}_0$ instead of b_0. The difficulty is that the b_0 term is now multiplied by P_1 instead of P_0. So *we update the coefficient matrix in the normal equations* by means of (17.20):

$$P_1^{-1} = P_0^{-1} + A_1^{\mathrm{T}}\Sigma_{e,1}^{-1}A_1. \tag{17.22}$$

We realize that the covariance matrix $P_1 = P_{\text{new}}$ for the updated estimate is "smaller" than $P_0 = P_{\text{old}}$. The addition of new information should reduce the covariance matrix. The last term describes the increase in information originating from the latest observations. Note again that (17.22) does not depend on the individual observations b_0 and b_1. It only exploits their statistical characteristics to determine the characteristics of $\hat{x}_1$.

Of course, the estimate $\hat{x}_1$ is based upon the actual observations b_0 and b_1. It is calculated according to (17.21) and the entire recursive least-squares procedure is about rewriting this expression:

$$\begin{aligned}\hat{x}_1 = P_1\big(P_0^{-1}\hat{x}_0 + A_1^{\mathrm{T}}\Sigma_{e,1}^{-1}b_1\big) &= P_1\big(P_1^{-1}\hat{x}_0 - A_1^{\mathrm{T}}\Sigma_{e,1}^{-1}A_1\hat{x}_0 + A_1^{\mathrm{T}}\Sigma_{e,1}^{-1}b_1\big) \\ &= \hat{x}_0 + K_1(b_1 - A_1\hat{x}_0).\end{aligned} \tag{17.23}$$

This formula identifies the gain matrix that multiplies the innovation:

$$\textit{Gain Matrix:} \qquad K_1 = P_1A_1^{\mathrm{T}}\Sigma_{e,1}^{-1}. \tag{17.24}$$

By this manipulation the formula for $\hat{x}_1$ has become recursive; it uses $\hat{x}_0$ instead of b_0. This is the result we want and it is easy to explain.

If the new observations are consistent with the original $\hat{x}_0$ then $b_1 = A_1\hat{x}_0$. In this case the mismatch is zero. Consequently there is no reason for changing our estimate of $\hat{x}$. The best choice is $\hat{x}_1 = \hat{x}_0$ when the new observations b_1 still lead to the old positions $\hat{x}_0$. In general this is not the case. There will be an error of prediction $b_1 - A_1\hat{x}_0$ which is called the ***innovation***. That is the unexpected part of b_1 and (17.23) multiplies by the gain matrix K_1 in order to give the correction. Now the new P_1 and $\hat{x}_1$ contain all we know. We are ready to receive another new observation b_2.

When b_2 arrives we look again at (17.22) and (17.23). The algebra that provided those formulas can be used in the new situation. Indices 0 and 1 are changed to 1 and 2 (and later to $k-1$ and k). We obtain the fundamental theorem for "static" *recursive least-squares* (expressed for P^{-1} where the "dynamic" Kalman filter updates P):

The least-squares estimate $\hat{x}_k$ and its variance P_k are determined recursively by

$$\begin{aligned} P_k^{-1} &= P_{k-1}^{-1} + A_k^{\mathrm{T}}\Sigma_{e,k}^{-1}A_k \\ \hat{x}_k &= \hat{x}_{k-1} + K_k(b_k - A_k\hat{x}_{k-1}) \quad \text{with} \quad K_k = P_kA_k^{\mathrm{T}}\Sigma_{e,k}^{-1}. \end{aligned} \tag{17.25}$$

Dynamic Problems and State Equation

Now comes the big step. The state of our system is modeled by the ***state vector*** $\boldsymbol{x}(t)$. This vector ***changes with time***. Part of that change is described by a known differential equation or difference equation (taken to be linear). This state equation—we will work in discrete time with a one-step difference equation—also allows for an error $\boldsymbol{\epsilon}_k$ at each time step:

$$\boldsymbol{x}_k = F_{k-1}\boldsymbol{x}_{k-1} + \boldsymbol{\epsilon}_k. \tag{17.26}$$

The known matrix F_{k-1} is the state transition matrix. The unknown vector $\boldsymbol{x}_k$ is the state at time k, and its estimate will be $\hat{\boldsymbol{x}}_k$. In addition to information about $\boldsymbol{x}_k$ from the state equation, we make direct measurements $\boldsymbol{b}_k$ at time k:

$$\boldsymbol{b}_k = A_k\boldsymbol{x}_k + \boldsymbol{e}_k. \tag{17.27}$$

It is this system of equations (17.26) and (17.27) at $t = 0, 1, \ldots, k$ that the Kalman filter solves by weighted least squares.

The weights are as always the inverses of the covariance matrices. We assume that the errors $\boldsymbol{\epsilon}_k$ and $\boldsymbol{e}_k$ are independent and Gaussian, with mean zero and known covariance matrices $\Sigma_{\epsilon,k}$ and $\Sigma_{e,k}$:

$$E\{\boldsymbol{\epsilon}_k\boldsymbol{\epsilon}_k^{\mathrm{T}}\} = \Sigma_{\epsilon,k} \qquad \text{and} \qquad E\{\boldsymbol{e}_k\boldsymbol{e}_k^{\mathrm{T}}\} = \Sigma_{e,k}. \tag{17.28}$$

Now Kalman's problem is completely stated. One way to solve it would be to construct long column vectors $\mathcal{b}$ and $\mathcal{x}$ and $\mathcal{e}$ (different lengths!) and a big rectangular matrix $\mathcal{A}$:

$$\mathcal{b} = \begin{bmatrix} \boldsymbol{b}_0 \\ \boldsymbol{0} \\ \boldsymbol{b}_1 \\ \boldsymbol{0} \\ \vdots \\ \boldsymbol{b}_k \end{bmatrix}, \quad \mathcal{x} = \begin{bmatrix} \boldsymbol{x}_0 \\ \boldsymbol{x}_1 \\ \vdots \\ \boldsymbol{x}_{k-1} \\ \boldsymbol{x}_k \end{bmatrix}, \quad \mathcal{e} = \begin{bmatrix} \boldsymbol{e}_0 \\ \boldsymbol{\epsilon}_1 \\ \vdots \\ \boldsymbol{e}_{k-1} \\ \boldsymbol{\epsilon}_k \\ \boldsymbol{e}_k \end{bmatrix}, \quad \mathcal{A} = \begin{bmatrix} A_0 & & & \\ -F_0 & I & & \\ & A_1 & & \\ & -F_1 & I & \\ & & \ddots & \\ & & -F_{k-1} & I \\ & & & A_k \end{bmatrix}.$$

Our model is exactly $\mathcal{A}\mathcal{x} = \mathcal{b} - \mathcal{e}$. The optimal estimate $\hat{\mathcal{x}}$ solves the normal equations $\mathcal{A}^{\mathrm{T}}\Sigma^{-1}\mathcal{A}\hat{\mathcal{x}} = \mathcal{A}^{\mathrm{T}}\Sigma^{-1}\mathcal{b}$. The big covariance matrix Σ is block diagonal, because we assume that the error vectors $\boldsymbol{e}_0, \boldsymbol{\epsilon}_1, \ldots, \boldsymbol{e}_{k-1}, \boldsymbol{\epsilon}_k, \boldsymbol{e}_k$ are independent of each other (but might be correlated within those vectors). The blocks of Σ are $\Sigma_{e,0}, \Sigma_{\epsilon,1}, \ldots, \Sigma_{e,k-1}, \Sigma_{\epsilon,k}, \Sigma_{e,k}$. Then the solution is a long vector $\hat{\mathcal{x}}$ containing our best estimate of the whole history:

$$\hat{\mathcal{x}}_k = \begin{bmatrix} \hat{\boldsymbol{x}}_{0|k} \\ \vdots \\ \hat{\boldsymbol{x}}_{k-1|k} \\ \hat{\boldsymbol{x}}_{k|k} \end{bmatrix}.$$

The second subscript indicates that we have used all information up to and including time k. This is a point to emphasize: early state estimates are eventually affected by the later measurements. It is useful to think separately of those corrected (often called *smoothed*) values for the earlier states $\boldsymbol{x}_0, \ldots, \boldsymbol{x}_{k-1}$. In using the Kalman filter we *may or may not* go back in time to compute them. The one thing we are sure to compute is the estimate $\hat{\boldsymbol{x}}_{k|k}$ for the current state:

Smoothed values (of earlier states $j < k$):	$\hat{\boldsymbol{x}}_{j\|k}$
Filtered value (of the current state):	$\hat{\boldsymbol{x}}_{k\|k}$
Predicted value (of the next state):	$\hat{\boldsymbol{x}}_{k+1\|k} = F_k\hat{\boldsymbol{x}}_{k\|k}$.

The prediction comes directly from the state equation (17.26). It does not account for the next measurement $\boldsymbol{b}_{k+1}$. The Kalman filter will operate recursively, starting with the prediction and correcting it (based on the mismatch betwen prediction and observation). It will do this for the next state $k+1$, and it did this for the current state k:

$$\textit{Prediction:} \qquad \hat{\boldsymbol{x}}_{k|k-1} = F_{k-1}\hat{\boldsymbol{x}}_{k-1|k-1} \tag{17.29}$$

$$\textit{Correction:} \qquad \hat{\boldsymbol{x}}_{k|k} = \hat{\boldsymbol{x}}_{k|k-1} + K_k(\boldsymbol{b}_k - A_k\hat{\boldsymbol{x}}_{k|k-1}). \tag{17.30}$$

The correction is needed because $\boldsymbol{b}_k$ contains new information. The gain K_k involves the covariance matrices, which tell the reliability of the inputs and how much weight to give to them. The covariance matrices $\Sigma_{e,k}$ and $\Sigma_{\epsilon,k}$ for the observation and state equations contribute to the covariance matrix $P_{k|k}$ for the error in $\hat{\boldsymbol{x}}_{k|k}$. And this $P_{k|k}$ is also found by prediction and correction, rather than computing $(A\Sigma^{-1}A)^{-1}$ from big matrices:

$$\textit{Predicted Covariance:} \qquad P_{k|k-1} = F_{k-1}P_{k-1|k-1}F_{k-1}^{\mathrm{T}} + \Sigma_{\epsilon,k}$$

$$\textit{Gain Matrix:} \qquad K_k = P_{k|k-1}A_k^{\mathrm{T}}\left(A_kP_{k|k-1}A_k^{\mathrm{T}} + \Sigma_{e,k}\right)^{-1}$$

$$\textit{Corrected Covariance:} \qquad P_{k|k} = (I - K_kA_k)P_{k|k-1}.$$

Notice again that these covariance calculations do not depend on the actual measurements $\boldsymbol{b}_k$. We often compute the reliability $P_{k|k}$ of $\hat{\boldsymbol{x}}_{k|k}$ in advance. (And if we are not satisfied, we might never compute $\hat{\boldsymbol{x}}_{k|k}$ from the observations.) In *on-the-fly filtering* the recursions for $P_{k|k}$ and $\hat{\boldsymbol{x}}_{k|k}$ go forward in parallel. We store earlier values only if we plan to return and smooth them.

It remains to verify the correctness of these update equations, and to express them in different forms. The symmetry of $P_{k|k}$ is not so clear from the expression above, and the gain matrix K_k can take a simpler form. First we comment on the notation. Then the reader can quickly match our (standard) symbols with other (also standard) symbols in the literature.

Notation

Everybody agrees that $\boldsymbol{x}$ is the state and K is the gain matrix. But there are several alternatives for the covariance matrices. Up to now we have consistently used Σ, indicating by Σ_b or $\Sigma_{\hat{x}}$ which covariance matrix we mean. With more subscripts needed in this chapter, we introduce the letter P to replace $\Sigma_{\hat{x}}$. Thus the covariance matrices are $\Sigma_{e,k}$ (observation),

Table 17.1 Commonly used notations for a discrete Kalman filter

Definition	This book	Gelb et al.	Grewal et al.
Matrix in observation equations	A_k	H_k	H_k
State transition matrix	F_k	Φ_k	Φ_k
Vector of observations at time k	$\boldsymbol{b}_k$	z_k	z_k
State vector at time k	$\boldsymbol{x}_k$	x_k	x_k
Error in observation $\boldsymbol{b}_k$	$\boldsymbol{e}_k$	$\underline{v}_k$	v_k
System error in state equation	$\boldsymbol{\epsilon}_k$	$\underline{w}_{k-1}$	w_{k-1}
Kalman gain matrix	K_k	K_k	$\bar{K}_k$
Covariance matrix of error in $\boldsymbol{b}_k$	$\Sigma_{e,k}$	R_k	R_k
Covariance matrix of system	$\Sigma_{\epsilon,k}$	Q_k	Q_k
Predicted covariance for $\hat{\boldsymbol{x}}_{k\|k-1}$	$P_{k\|k-1}$	$P_k(-)$	$P_k(-)$
Corrected covariance for $\hat{\boldsymbol{x}}_{k\|k}$	$P_{k\|k}$	$P_k(+)$	$P_k(+)$
Predicted estimate of $\boldsymbol{x}_k$	$\hat{\boldsymbol{x}}_{k\|k-1}$	$\hat{\underline{x}}_k(-)$	$\hat{x}_k(-)$
Corrected estimate of $\boldsymbol{x}_k$	$\hat{\boldsymbol{x}}_{k\|k}$	$\hat{\underline{x}}_k(+)$	$\hat{x}_k(+)$

$\Sigma_{\epsilon,k}$ (state equation), $P_{k|k}$ (estimate $\hat{\boldsymbol{x}}_{k|k}$ of $\boldsymbol{x}_k$). Other authors use Q_k and R_k for those first two covariance matrices. We found it simplest to use $\boldsymbol{\epsilon}_k$ (not $\boldsymbol{\epsilon}_{k-1}$) for the error in the state equation for $\boldsymbol{x}_k$.

Now for the subscripts. They must indicate the time k to which the estimate applies. They must also indicate the time when the estimate is made. This is often $k-1$ (for the prediction) and k (for the correction). We could use the words *old* and *new*, or the symbols $(-)$ and $(+)$, or the subscripts $k-1$ and k. We chose the subscripts because they are the most explicit:

$$\text{Predictions:} \quad P_k(-) = P_{k,\text{old}} = P_{k|k-1} \quad \text{and} \quad \hat{\boldsymbol{x}}_k(-) = \hat{\boldsymbol{x}}_{k,\text{old}} = \hat{\boldsymbol{x}}_{k|k-1}$$

$$\text{Corrections:} \quad P_k(+) = P_{k,\text{new}} = P_{k|k} \quad \text{and} \quad \hat{\boldsymbol{x}}_k(+) = \hat{\boldsymbol{x}}_{k,\text{new}} = \hat{\boldsymbol{x}}_{k|k}.$$

Table 17.1 shows our notation along with the notations in two of our favorite books. There are more notations in use but you would not want to know them.

17.3 The Steady Model

We can compare the Kalman filter with ordinary least squares on a problem where they might be expected to give the same answer—but they don't. It is known as a *steady model*, since all observations $b_1, \ldots, b_m$ measure the same scalar quantity x. Furthermore all errors are independent with variance $\sigma^2 = 1$. The two problems look identical:

1 The least-squares solution of the n equations $x = b_k$ is the average

$$\hat{x}_n = \frac{b_1 + b_2 + \cdots + b_n}{n} \qquad \text{with variance} \qquad P = \tfrac{1}{n}.$$

This was checked earlier by the formulas of recursive least squares.

2 The Kalman filter solution comes from the observation equations $x_k = b_k$ together with the steady model $x_k = x_{k-1}$. Thus every $F_k = 1$ in this state equation.

Nevertheless there is a difference. It comes from the presence of errors ϵ_k in the state equation. We have $x_k = x_{k-1} + \epsilon_k$ as well as $x_k = b_k - e_k$. Thus we are at the same time assuming that x does not change, and allowing for the possibility that it does. When it drifts away from x_0, the latest $\hat{x}_{k-1}$ is the predicted value $\hat{x}_{k|k-1}$ (since $F = 1$). Then after we measure b_k, the corrected estimate $\hat{x}_{k|k}$ is a combination of prediction and measurement. As in least squares, this must use all the measurements—but the recent b_k are weighted more heavily. In ordinary least squares, which does not allow for drift and just computes the average, the b_k are weighted equally.

The difference can be seen after two measurements and again after three. The equations that combine $x_k = b_k$ with $x_k = x_{k-1}$ are

$$\begin{bmatrix} 1 & \\ -1 & 1 \\ & 1 \end{bmatrix} \begin{bmatrix} x_0 \\ x_1 \end{bmatrix} \begin{bmatrix} b_0 \\ 0 \\ b_1 \end{bmatrix} \quad \text{and} \quad \begin{bmatrix} 1 & & \\ -1 & 1 & \\ & 1 & \\ & -1 & 1 \\ & & 1 \end{bmatrix} \begin{bmatrix} x_0 \\ x_1 \\ x_2 \end{bmatrix} = \begin{bmatrix} b_0 \\ 0 \\ b_1 \\ 0 \\ b_2 \end{bmatrix}. \qquad (17.31)$$

If we force $x_k = x_{k-1}$ to hold exactly, we are back to ordinary least squares with one unknown. But if we allow drift errors ϵ_k, also of variance one, then it is the equations in (17.31) that are solved by least squares. The unknowns are the states $x_0, \ldots, x_k$. The matrix will be called A instead of $\mathcal{A}$. The normal equations $A^{\mathrm{T}}A\hat{\boldsymbol{x}} = A^{\mathrm{T}}\boldsymbol{b}$ are

$$\begin{bmatrix} 2 & -1 \\ -1 & 2 \end{bmatrix} \begin{bmatrix} \hat{x}_{0|1} \\ \hat{x}_{1|1} \end{bmatrix} = \begin{bmatrix} b_0 \\ b_1 \end{bmatrix} \quad \text{and} \quad \begin{bmatrix} 2 & -1 & 0 \\ -1 & 3 & -1 \\ 0 & -1 & 2 \end{bmatrix} \begin{bmatrix} \hat{x}_{0|2} \\ \hat{x}_{1|2} \\ \hat{x}_{2|2} \end{bmatrix} = \begin{bmatrix} b_0 \\ b_1 \\ b_2 \end{bmatrix}. \qquad (17.32)$$

You see how the lower corner is changed (a diagonal entry changes from 2 to 3) as the new row and column are added. This pattern would continue: $A^{\mathrm{T}}A$ has 3's along its diagonal, except for 2's at the top and bottom. The factors in LDL^{T} or in QR change only at the bottom—which is the reason the Kalman filter works!

We can see what happens without complicated formulas. The inverses in (17.32) are

$$(A^{\mathrm{T}}A)^{-1} = \tfrac{1}{3}\begin{bmatrix} 2 & 1 \\ 1 & 2 \end{bmatrix} \quad \text{and} \quad (A^{\mathrm{T}}A)^{-1} = \tfrac{1}{8}\begin{bmatrix} 5 & 2 & 1 \\ 2 & 4 & 2 \\ 1 & 2 & 5 \end{bmatrix}. \qquad (17.33)$$

After the measurements b_0 and b_1, the best estimates come from the first equation in (17.32). Multiplying by $(A^{\mathrm{T}}A)^{-1}$ gives, not the ordinary average, but

$$\hat{x}_{0|1} = \frac{2b_0 + b_1}{3} \qquad \text{and} \qquad \hat{x}_{1|1} = \frac{b_0 + 2b_1}{3}.$$

The new data b_1 has a heavier weight $2/3$ in the filtered estimate $\hat{x}_{1|1}$ of x_1. And b_1 also appears in the smoothed estimate $\hat{x}_{0|1}$ of x_0. There it is weighted less heavily, but still we know more about x_0 after measuring b_1.

With b_2 included, the smoothed and filtered estimates change to

$$\hat{x}_{0|2} = \frac{5b_0 + 2b_1 + b_2}{8}, \qquad \hat{x}_{1|2} = \frac{2b_0 + 4b_1 + 2b_2}{8}, \qquad \hat{x}_{2|2} = \frac{b_0 + 2b_1 + 5b_2}{8}.$$

The last is the most important. It is the best estimate of x_2, using b_0 and b_1 but emphasizing b_2. The possibility of drift produces an exponential decay of the weight attached to old measurements.

The other half of the Kalman filter computes P_k (here it is a scalar), which appears as the last entry of $(A^{\mathrm{T}}A)^{-1}$: $P_1 = 2/3$ and $P_2 = 5/8$. From b_0 alone we would have had $P_0 = 1$. Each P_k gives the reliability of the estimate $\hat{x}_{k|k}$. Thus the estimation errors are steadily decreasing. The reciprocals 1, 3/2, 8/5 are the information matrices and they increase with every new measurement of the model.

Asymptotics of the Steady Model

The 2 by 2 and 3 by 3 examples above bring out the difference between the static problem with fixed x and the dynamic problem with evolving x. Out of curiosity, we find the limiting behavior of the n by n dynamic problem as n becomes large. We will keep $\Sigma_e = 1$ and also $\Sigma_\epsilon = 1$. It would be interesting to allow different variances σ_e^2 and σ_ϵ^2 in the observation and state equations, and recompute the asymptotic limits. This model will give numerical examples of the Kalman filter in Section 17.6, where x is the offset in the receiver clock. The steady model is quite relevant to GPS (and the figures will show rapid convergence to the limiting value).

The crucial matrix is $V = A^{\mathrm{T}}A$. It is tridiagonal with entries -1, 3, -1 in every row (*except* $V_{11} = V_{nn} = 2$). This is because every column of A has 3 entries except the first and last columns; all entries are 1 or -1. Since $A^{\mathrm{T}}A$ is positive definite, it has a Cholesky factorization into LL^{T}. Since $A^{\mathrm{T}}A$ is tridiagonal, the triangular factors L and L^{T} are bidiagonal.

Consider the first matrix T that has the steady -1, 3, -1 pattern (including the ends $T_{11} = 3$ and $T_{nn} = 3$). We may expect the entries of *its* triangular factors to approach limits a on the main diagonal and b on the off-diagonal. It is those limits a and b that we now compute, by equating to the entries of T:

$$a^2 + b^2 = 3 \quad \text{on the diagonal} \qquad \text{and} \qquad ab = -1 \quad \text{off the diagonal.}$$

Substituting $b^2 = 1/a^2$ into the first equation leads to $a^4 - 3a^2 + 1 = 0$. This quadratic equation in a^2 yields the values of the limits a and b:

$$a^2 = \frac{3 + \sqrt{5}}{2} \qquad \text{and} \qquad b^2 = \frac{3 - \sqrt{5}}{2}.$$

The diagonal a dominates the off-diagonal b since T is positive definite. We take the positive square root for a and the negative square root for b. Now we need the *inverse* matrix T^{-1}. In particular we want its last entry T^{-1}_{nn}, in the lower right corner. For upper triangular times lower triangular, T^{-1}_{nn} comes from the last entries (asymptotically $1/a$):

$$(T^{-1})_{nn} \approx \frac{1}{a^2} = b^2 = \frac{3-\sqrt{5}}{2}. \tag{17.34}$$

Now we must account for the changes from 3 to 2 in the corners, when T becomes V. The change in the $(1,1)$ entry has exponentially small effect on the (n,n) entry of $V^{-1} = (A^{\mathrm{T}}A)^{-1}$. But the change from $T_{nn} = 3$ to $V_{nn} = 2$ is significant. If $\boldsymbol{e}^{\mathrm{T}} = [\,0 \ \ldots \ 0 \ 1\,]$ then this is a *rank-one change* by $\boldsymbol{e}\boldsymbol{e}^{\mathrm{T}}$. The Sherman-Morrison-Woodbury-Schur formula in Section 17.4 gives the rank-one change in the inverse:

$$T^{-1} = \left(V + \boldsymbol{e}\boldsymbol{e}^{\mathrm{T}}\right)^{-1} = V^{-1} - \frac{V^{-1}\boldsymbol{e}\boldsymbol{e}^{\mathrm{T}}V^{-1}}{1 + \boldsymbol{e}^{\mathrm{T}}V^{-1}\boldsymbol{e}}. \tag{17.35}$$

Take the (n,n) entries and denote $(V^{-1})_{nn} = \boldsymbol{e}^{\mathrm{T}}V^{-1}\boldsymbol{e}$ by P:

$$\frac{3-\sqrt{5}}{2} = P - \frac{P^2}{1+P} = \frac{P}{1+P}.$$

This yields $P = \frac{\sqrt{5}-1}{2} \approx 0.618$ which is confirmed by MATLAB experiment. It is the weight assigned to the latest measurement b_n in the formula for $\hat{x}_{n|n}$, in the limit as $n \to \infty$. This is the "golden mean" for the steady model.

Fibonacci Numbers

We want to offer more about this steady model, which is the simplest of all Kalman filters. Each matrix F_k in the state equation and each A_k in the observation equation and each variance $\Sigma_{e,k}$ and $\Sigma_{\epsilon,k}$ is the single number 1. Part of our motivation is the innocent pleasure of meeting Fibonacci numbers. They enter the explicit formulas for $\hat{x}$ and its covariance P at every step. You will appreciate the contrast with the matrix manipulations in the next section, where the Kalman formulas are derived.

Of course those general update formulas, when applied to this steady model, will yield the Fibonacci numbers. (F_4 and F_5 and F_6 appeared above, in the fractions 3/5 and 5/8.) But here we can provide complete detail. These discoveries were made jointly with Steven L. Lee. We will compute the $A = QR$ factorization (Gram-Schmidt) and also the $A^{\mathrm{T}}A = LDL^{\mathrm{T}}$ factorization (Cholesky or symmetric Gauss).

Recall that Fibonacci's numbers $0, 1, 1, 2, 3, 5, 8, \ldots$ arise from $F_k = F_{k-1} + F_{k-2}$. They start with $F_0 = 0$ and $F_1 = 1$. Our first step is now to identify the determinants of the $-1, 3, -1$ tridiagonal matrices T_n:

$$T_1 = [\,3\,], \qquad T_2 = \begin{bmatrix} 3 & -1 \\ -1 & 3 \end{bmatrix}, \qquad T_3 = \begin{bmatrix} 3 & -1 & \\ -1 & 3 & -1 \\ & -1 & 3 \end{bmatrix}, \qquad \cdots$$

Those have determinants 3, 8, and 21. *The n by n matrix T_n has determinant F_{2n+2}*. The natural induction proof is to use the cofactors of the first row to find the recursion formula

$$(\det T_n) = 3(\det T_{n-1}) - (\det T_{n-2}). \tag{17.36}$$

The Fibonacci numbers F_{2n+2} satisfy the same recursion, because $F_{2n+2} = F_{2n+1} + F_{2n} = (F_{2n} + F_{2n-1}) + F_{2n} = 2F_{2n} + (F_{2n} - F_{2n-2}) = 3F_{2n} - F_{2n-2}$. Thus F_{2n+2} is the determinant.

The matrices $V = A^{\mathrm{T}}A$ are slightly different from T, because A has only *two* nonzeros in its first and last columns. (All nonzero entries are 1 or -1 from the state and observation equations. The 1 by 1 matrix has only a single 1 from the first observation.) Thus V differs from T by subtracting 1 from its first entry and also from its last entry. Here are the matrices $V = A^{\mathrm{T}}A$:

$$V_1 = \begin{bmatrix} 1 \end{bmatrix}, \qquad V_2 = \begin{bmatrix} 2 & -1 \\ -1 & 2 \end{bmatrix}, \qquad V_3 = \begin{bmatrix} 2 & -1 & \\ -1 & 3 & -1 \\ & -1 & 2 \end{bmatrix}, \qquad \cdots$$

The determinants are now 1, 3, and 8. ***The determinant of V_n is F_{2n}.*** For proof we first subtract 1 from the $(1, 1)$ entry of T_n to reach U_n. This reduces the determinant by $\det T_{n-1} = F_{2n}$. Therefore $\det U_n = F_{2n+2} - F_{2n} = F_{2n+1}$. Now subtract 1 from the (n, n) entry of U_n to reach V_n. This reduces $\det U_n$ by $\det U_{n-1} = F_{2n-1}$. So $\det V_n = F_{2n+1} - F_{2n-1} = F_{2n}$.

For the reader's convenience we display four matrices T, U, V, W of order $n = 3$:

$$\begin{bmatrix} 3 & -1 & \\ -1 & 3 & -1 \\ & -1 & 3 \end{bmatrix} \quad \begin{bmatrix} 2 & -1 & \\ -1 & 3 & -1 \\ & -1 & 3 \end{bmatrix} \quad \begin{bmatrix} 2 & -1 & \\ -1 & 3 & -1 \\ & -1 & 2 \end{bmatrix} \quad \begin{bmatrix} 2 & -1 & \\ -1 & 3 & -1 \\ & -1 & 1 \end{bmatrix}$$

$$\det T_n = F_{2n+2} \qquad \det U_n = F_{2n+1} \qquad \det V_n = F_{2n} \qquad \det W_n = F_{2n-2}$$

The matrix W_n appears at the prediction step of the Kalman filter, before the observation row is included. At that point A has only a single "1" in its last column, so the last entry of W_n is 1. Note that the order n is $k + 1$ in Kalman's numbering, which starts at $k = 0$.

Now we factor V_n and W_n into LDL^{T}. Since V_n differs from W_n only in the last entry, their lower triangular factors L will be the same. L contains the number that multiplies row i when we subtract it from row $i + 1$. The pivots agree for V_n and W_n until the nth pivot d_{nn}. The determinants immediately give these pivots and multipliers:

$$d_{ii} = \frac{F_{2i+1}}{F_{2i-1}} \qquad \text{and} \qquad l_{i+1,i} = -\frac{F_{2i-1}}{F_{2i+1}} \qquad \text{for } i < n. \tag{17.37}$$

The last pivot d_{nn} is F_{2n}/F_{2n-1} for the matrix V_n and F_{2n-2}/F_{2n-1} for W_n. The entries in D^{-1} are the reciprocals of the pivots d_{ii}. The really attractive formula appears in L^{-1}. This inverse matrix is lower triangular with Fibonacci ratios:

$$(L^{-1})_{ij} = \frac{F_{2j-1}}{F_{2i-1}} \qquad \text{for } j \le i. \tag{17.38}$$

Thus for $n = 3$, the factorization $V^{-1} = L^{-\mathrm{T}}D^{-1}L^{-1}$ is

$$V_3^{-1} = \begin{bmatrix} 1 & \frac{1}{2} & \frac{1}{5} \\ & 1 & \frac{2}{5} \\ & & 1 \end{bmatrix} \begin{bmatrix} \frac{1}{2} & & \\ & \frac{2}{5} & \\ & & \frac{5}{8} \end{bmatrix} \begin{bmatrix} 1 & & \\ \frac{1}{2} & 1 & \\ \frac{1}{5} & \frac{2}{5} & 1 \end{bmatrix}.$$

Every entry of V^{-1} is positive. All the row and column sums are 1. This is a Markov matrix.

Now go through the steps of the Kalman filter. The nth state equation takes us from V_{n-1} to W_n. Then the observation equation takes us to V_n. Here are the updates:

$$\text{Last entry in } W_n^{-1}: \qquad P_{n|n-1} = P_{n-1|n-1} + 1 = \frac{F_{2n-3}}{F_{2n-2}} + 1 = \frac{F_{2n-1}}{F_{2n-2}}$$

$$\text{Kalman gain}: \qquad K_n = P_{n|n-1}(1 + P_{n|n-1})^{-1} = \frac{F_{2n-1}}{F_{2n-2} + F_{2n-1}} = \frac{F_{2n-1}}{F_{2n}}$$

$$\text{Last entry in } V_n^{-1}: \qquad P_{n|n} = (1 - K_n)P_{n|n-1} = \left(1 - \frac{F_{2n-1}}{F_{2n}}\right)\frac{F_{2n-1}}{F_{2n-2}} = \frac{F_{2n-1}}{F_{2n}}.$$

The last entry agrees with $1/d_{nn}$ as it should. The prediction $\hat{x}_{n|n-1}$ is just $\hat{x}_{n-1|n-1}$ from our simple state equation. Then the correction is

$$\hat{x}_{n|n} = (1 - K_n)\hat{x}_{n|n-1} + K_n b_n = \frac{F_{2n-2}}{F_{2n}}\hat{x}_{n|n-1} + \frac{F_{2n-1}}{F_{2n}} b_n. \tag{17.39}$$

In the totally steady case of equal measurements $b_n = b$, this correctly gives all $\hat{x}_{n|n} = b$. The Kalman recursion (17.39) can be unrolled to see how b_{n-1} is multiplied by F_{2n-3}/F_{2n}:

$$\hat{x}_{n|n} = \frac{F_{2n-2}}{F_{2n}}\left(\frac{F_{2n-4}}{F_{2n-2}}\hat{x}_{n-1|n-1} + \frac{F_{2n-3}}{F_{2n-2}} b_{n-1}\right) + \frac{F_{2n-1}}{F_{2n}} b_n.$$

The explicit matrix inverses multiply b_{n-1} by F_{2n-3}/F_{2n-1} and then F_{2n-1}/F_{2n}. So F_{2n-3}/F_{2n} is correct:

$$\begin{bmatrix}\hat{x}_{1|n}\\ \vdots \\ \hat{x}_{n|n}\end{bmatrix} = \begin{bmatrix} \cdot & \cdots & \cdot \\ & \ddots & \vdots \\ & & 1\end{bmatrix}\begin{bmatrix}\ddots & \\ & \dfrac{F_{2n-1}}{F_{2n}}\end{bmatrix}\begin{bmatrix}\cdot & & \\ \vdots & \ddots & \\ \cdot & \dfrac{F_{2n-3}}{F_{2n-1}} & 1\end{bmatrix}\begin{bmatrix}\vdots\\ b_{n-1}\\ b_n\end{bmatrix}.$$

The Gram-Schmidt Factorization

We mentioned earlier an alternative to the LDL^{T} factorization of $A^{\mathrm{T}}A$. We can orthogonalize the columns of A itself. This Gram-Schmidt procedure leads to $A = QR$, where R is upper triangular because of the order of the steps (columns are subtracted from later columns). In the application to Kalman filtering, only neighboring columns of A have nonzero inner products. Therefore R has only *two nonzero diagonals*, the main diagonal and the one above. All further entries of R are zero, because column j is already orthogonal to column k for $|j - k| > 1$.

Our convention will have diagonal entries $r_{ii} = 1$. The entries of $R = L^{\mathrm{T}}$ and its inverse were computed in (17.37) and (17.38). So it only remains to find Q (with orthogonal but not orthonormal columns). Once again these columns contain ratios of

Fibonacci numbers! We display $A = QR$ with $n = 3$ columns:

$$A = \begin{bmatrix} 1 & & \\ -1 & 1 & \\ & 1 & \\ & -1 & 1 \\ & & 1 \end{bmatrix} = \begin{bmatrix} \frac{1}{1} & \frac{1}{2} & \frac{1}{5} \\ -1 & \frac{1}{2} & \frac{1}{5} \\ & \frac{2}{2} & \frac{2}{5} \\ & -1 & \frac{3}{5} \\ & & \frac{5}{5} \end{bmatrix} \begin{bmatrix} 1 & -\frac{1}{2} & 0 \\ & 1 & -\frac{2}{5} \\ & & 1 \end{bmatrix} = QR.$$

The fractions in Q are F_i/F_{2j-1} for $i \le 2j-1$. The orthogonality of columns 2 and 3 depends on $1^2 + 1^2 + 2^2 = 2 \cdot 3$. The orthogonality of columns j and $j+1$ requires

$$F_1^2 + F_2^2 + \cdots + F_{2j-1}^2 = F_{2j-1}F_{2j}. \tag{17.40}$$

For proof by induction, add F_{2j}^2 to both sides. The right side becomes $(F_{2j-1}+F_{2j})F_{2j} = F_{2j}F_{2j+1}$ which completes the induction. It is pleasant to see the Fibonacci ratios in Q.

The Kalman equations can be derived from $A = QR$ instead of $A = LDL^{\mathrm{T}}$. This brings to light new orthogonalities, involving the innovations. Without developing the recursive part (the essence of Kalman), $A^{\mathrm{T}}A\hat{\boldsymbol{x}} = A^{\mathrm{T}}\boldsymbol{b}$ gives

$$A^{\mathrm{T}}(\boldsymbol{b} - A\hat{\boldsymbol{x}}) = \boldsymbol{0} \qquad \text{and} \qquad Q^{\mathrm{T}}(\boldsymbol{b} - A\hat{\boldsymbol{x}}) = \boldsymbol{0}.$$

The columns of A are orthogonal to the innovations $\boldsymbol{b} - A\hat{\boldsymbol{x}}$. So are the columns of Q. (The equations above differ only by an invertible matrix R^{T}.) In stochastic terms we have *zero correlation*. The next section approaches the Kalman equations through LDL^{T}.

17.4 Derivation of the Kalman Filter

All authors try to find a clear way to derive the formulas for the Kalman filter. Those formulas are certainly correct. When we look at the reasoning already supplied by earlier authors, it looks much too complicated. Often we don't completely understand it. There is a definite feeling that there must be a better way. This produces almost as many new explanations as new books.

The present authors are no exception. We will base all steps on two matrix identities. Those identities come from the inverse of a 2 by 2 *block matrix*. The problem is to update the last entries of $(\mathcal{A}^{\mathrm{T}}\Sigma^{-1}\mathcal{A})^{-1}$, when new rows are added to the big matrix $\mathcal{A}$. Those new rows will be of two kinds, coming from the *state equation* and the *observation equation*. A typical filtering step has two updates, from $\mathcal{A}_{k-1}$ to $\mathcal{S}_k$ (by including the state equation) and then to $\mathcal{A}_k$ (by including the observation equation):

$$\mathcal{A}_1 = \begin{bmatrix} A_0 & \\ -F_0 & I \\ & A_1 \end{bmatrix} \to \mathcal{S}_2 = \begin{bmatrix} A_0 & & \\ -F_0 & I & \\ & A_1 & \\ & -F_1 & I \end{bmatrix} \to \mathcal{A}_2 = \begin{bmatrix} A_0 & & \\ -F_0 & I & \\ & A_1 & \\ & -F_1 & I \\ & & A_2 \end{bmatrix}.$$

The step to $\mathcal{S}_k$ adds a new row and column (this is *bordering*). The second step only adds a new row (this is *updating*). The least-squares solution $\hat{\boldsymbol{x}}_{k-1|k-1}$ for the old state leads to $\hat{\boldsymbol{x}}_{k|k-1}$ (using the state equation) and then to $\hat{\boldsymbol{x}}_{k|k}$ (using the new observation $\boldsymbol{b}_k$).

Remember that least squares works with the symmetric block tridiagonal matrices $\mathcal{T}_k = \mathcal{A}_k^{\mathrm{T}} \Sigma_k^{-1} \mathcal{A}_k$. So we need the change in $\mathcal{T}_{k-1}$ when $\mathcal{A}_{k-1}$ is bordered and then updated:

$$\text{Bordering by the row} \quad R = \begin{bmatrix} 0 & \dots & -F_{k-1} & I \end{bmatrix} \quad \text{adds} \quad R^{\mathrm{T}} \Sigma_{\epsilon,k}^{-1} R \tag{17.41}$$

$$\text{Updating by the row} \quad W = \begin{bmatrix} 0 & \dots & 0 & A_k \end{bmatrix} \quad \text{adds} \quad W^{\mathrm{T}} \Sigma_{e,k}^{-1} W. \tag{17.42}$$

We are multiplying matrices using "columns times rows." Every matrix product BA is the sum of (column j of B) times (row j of A). In our case B is the transpose of A, and these are *block* rows and columns. The matrices $\Sigma_{\epsilon,k}^{-1}$ and $\Sigma_{e,k}^{-1}$ appear in the middle, because this is *weighted* least squares.

The complete system, including all errors up to ϵ_k and e_k in the error vector $\boldsymbol{e}_k$, is $\mathcal{A}_k x_k = \boldsymbol{b}_k - \boldsymbol{e}_k$:

$$\mathcal{A}_2 x_2 = \begin{bmatrix} A_0 & & \\ -F_0 & I & \\ & A_1 & \\ & -F_1 & I \\ & & A_2 \end{bmatrix} \begin{bmatrix} \boldsymbol{x}_0 \\ \boldsymbol{x}_1 \\ \boldsymbol{x}_2 \end{bmatrix} = \begin{bmatrix} \boldsymbol{b}_0 \\ \boldsymbol{0} \\ \boldsymbol{b}_1 \\ \boldsymbol{0} \\ \boldsymbol{b}_2 \end{bmatrix} - \begin{bmatrix} \boldsymbol{e}_0 \\ \epsilon_1 \\ \boldsymbol{e}_1 \\ \epsilon_2 \\ \boldsymbol{e}_2 \end{bmatrix} = \boldsymbol{b}_2 - \boldsymbol{e}_2. \tag{17.43}$$

The weight matrix Σ_k^{-1} for the least-squares solution is block diagonal:

$$\Sigma_2^{-1} = \begin{bmatrix} \Sigma_{e,0}^{-1} & & & \\ & \ddots & & \\ & & \Sigma_{\epsilon,2}^{-1} & \\ & & & \Sigma_{e,2}^{-1} \end{bmatrix}.$$

Our task is to compute the ***last corner block*** $P_{k|k}$ *of the matrix* $\left(\mathcal{A}_k^{\mathrm{T}} \Sigma_k^{-1} \mathcal{A}_k\right)^{-1}$. We do that in two steps. The "prediction" finds $P_{k|k-1}$ from the bordering step, when the state row is added to $\mathcal{A}_{k-1}$ (producing $\mathcal{S}_k$). Then the "correction" finds $P_{k|k}$ when the observation row is included too.

The second task is to compute the prediction $\hat{\boldsymbol{x}}_{k|k-1}$ and correction $\hat{\boldsymbol{x}}_{k|k}$ for the new state vector $\boldsymbol{x}_k$. This is the last component of $\hat{x}_k$. If we compute *all* components of $\hat{x}_k$, then we are smoothing old estimates as well as filtering to find the new $\hat{\boldsymbol{x}}_{k|k}$.

Naturally Σ affects all the update formulas. The derivation of these formulas will be simpler if we begin with the case $\Sigma = I$, in which all noise is "white." Then we adjust the formulas to account for the weight matrices.

Remark 17.1 Our derivation will be systematic. Probably you will not study every step, but you will know the underlying reasoning. We compute the last entries in $(\mathcal{A}^{\mathrm{T}} \Sigma^{-1} \mathcal{A})^{-1}$ and then in $(\mathcal{A}^{\mathrm{T}} \Sigma^{-1} \mathcal{A})^{-1} \mathcal{A}^{\mathrm{T}} \Sigma^{-1} \boldsymbol{b}$. These entries are $P_{k|k}$ and $\hat{\boldsymbol{x}}_{k|k}$. Block matrices will be everywhere.

There is another approach to the same result. Instead of working with $\mathcal{A}^{\mathrm{T}}\mathcal{A}$ (which here is block tridiagonal) we could ***orthogonalize*** the columns of $\mathcal{A}$. This Gram-Schmidt process factors $\mathcal{A}$ into a block orthogonal $\mathcal{Q}$ times a block bidiagonal $\mathcal{R}$. Those factors are updated at each step of the ***square root information filter***. This $\mathcal{Q}\mathcal{R}$ approach adds new insight about "orthogonalizing the innovations."

Block Matrix Identities

The key formulas give the inverse of a 2 by 2 block matrix, assuming T is invertible:

$$\begin{bmatrix} T & U \\ V & W \end{bmatrix}^{-1} = \begin{bmatrix} L & M \\ N & P \end{bmatrix}. \tag{17.44}$$

Our applications have symmetric matrices. All blocks on the diagonal are symmetric; the blocks off the diagonal are $U = V^{\mathrm{T}}$ and $M = N^{\mathrm{T}}$. The key to Kalman's success is that the matrix to be inverted is *block tridiagonal*. Thus U and V are nonzero only in their last block entries, and T itself is block tridiagonal. But the general formula gives the inverse without using any special properties of T, U, V, W:

$$\begin{aligned} L &= T^{-1} + T^{-1}UPVT^{-1} \\ M &= -T^{-1}UP \\ N &= -PVT^{-1} \\ P &= (W - VT^{-1}U)^{-1}. \end{aligned} \tag{17.45}$$

The simplest proof is to multiply matrices and obtain I. The actual derivation of (17.45) is by block elimination. Multiply the row $[\,T \quad U\,]$ by VT^{-1} and subtract from $[\,V \quad W\,]$:

$$\begin{bmatrix} I & 0 \\ -VT^{-1} & I \end{bmatrix} \begin{bmatrix} T & U \\ V & W \end{bmatrix} = \begin{bmatrix} T & U \\ 0 & W - VT^{-1}U \end{bmatrix}. \tag{17.46}$$

The two triangular matrices are easily inverted, and then block multiplication produces (17.45). This is only Gaussian elimination with blocks. In the scalar case the last corner is $W - VU/T$. In the matrix case we keep the blocks in the right order! The inverse of that last entry is P.

Now make a trivial but valuable observation. We could eliminate in the opposite order. This means that we subtract UW^{-1} times the *second* row $[\,V \quad W\,]$ from the *first* row $[\,T \quad U\,]$:

$$\begin{bmatrix} I & -UW^{-1} \\ 0 & I \end{bmatrix} \begin{bmatrix} T & U \\ V & W \end{bmatrix} = \begin{bmatrix} T - UW^{-1}V & 0 \\ V & W \end{bmatrix}.$$

Inverting this new right side yields different (but still correct) formulas for the blocks L, M, N, P in the inverse matrix. We pay particular attention to the (1, 1) block. It becomes $L = (T - UW^{-1}V)^{-1}$. That is completely parallel to $P = (W - VT^{-1}U)^{-1}$, just changing letters.

Now compare the new form of L with the form in (17.45). Their equality is the most important formula in matrix update theory. We only mention four of its originators.

Sherman–Morrison–Woodbury–Schur formula

$$(T - UW^{-1}V)^{-1} = T^{-1} + T^{-1}U(W - VT^{-1}U)^{-1}VT^{-1}. \qquad (17.47)$$

We are looking at this as an update formula, when the matrix T is *perturbed* by $UW^{-1}V$. Often this is a perturbation of low rank. Previously we looked at (17.45) as a bordering formula, when the matrix T was *bordered* by U and V and W. Well, matrix theory is beautiful.

Updates of the Covariance Matrices

We now compute the blocks $P_{k|k-1}$ and $P_{k|k}$ in the lower right corners of $(\mathcal{S}_k^{\mathrm{T}}\mathcal{S}_k)^{-1}$ and $(\mathcal{A}_k^{\mathrm{T}}\mathcal{A}_k)^{-1}$. Then we operate on the right side, to find the new state $\hat{\boldsymbol{x}}_{k|k}$.

Remember that $\mathcal{S}_k$ comes by adding the new row $[\,V \quad I\,] = [\,0 \ \ldots \ -F_{k-1} \quad I\,]$. Then $\mathcal{A}_{k-1}^{\mathrm{T}}\mathcal{A}_{k-1}$ grows to $\mathcal{S}_k^{\mathrm{T}}\mathcal{S}_k$ by adding $[\,V \quad I\,]^{\mathrm{T}}[\,V \quad I\,]$:

$$\mathcal{S}_k^{\mathrm{T}}\mathcal{S}_k = \begin{bmatrix} \mathcal{A}_{k-1}^{\mathrm{T}}\mathcal{A}_{k-1} + V^{\mathrm{T}}V & V^{\mathrm{T}} \\ V & I \end{bmatrix}. \qquad (17.48)$$

Because V is zero until its last block, equation (17.48) is really the addition of a 2 by 2 block in the lower right corner. This has two effects at once. It *perturbs* the existing matrix $T = \mathcal{A}_{k-1}^{\mathrm{T}}\mathcal{A}_{k-1}$ and it *borders* the result T_{new}. The perturbation is by $V^{\mathrm{T}}V$. The bordering is by the row V and the column V^{T} and the corner block I. Therefore the update formula for T_{new}^{-1} goes inside the bordering formula:

Update to $T + V^{\mathrm{T}}V$: $\quad T_{\text{new}}^{-1} = T^{-1} - T^{-1}V^{\mathrm{T}}(I + VT^{-1}V^{\mathrm{T}})^{-1}VT^{-1}$

Border by V, V^{T}, I: $\quad P = (I - VT_{\text{new}}^{-1}V^{\mathrm{T}})^{-1}$.

Substitute the update into the bordering formula and write Z for the block $VT^{-1}V^{\mathrm{T}}$. This gives the great simplification $P = I + Z$:

$$P = \left(I - V\left(T^{-1} - T^{-1}V^{\mathrm{T}}(I+Z)^{-1}VT^{-1}\right)V^{\mathrm{T}}\right)^{-1}$$

$$= \left(I - Z + Z(I+Z)^{-1}Z\right)^{-1} = I + Z. \qquad (17.49)$$

This block P is $P_{k|k-1}$. It is the corner block in $(\mathcal{S}_k^{\mathrm{T}}\mathcal{S}_k)^{-1}$. The row $[\,V \quad I\,]$ has been included, and the matrix Z is

$$VT^{-1}V^{\mathrm{T}} = \begin{bmatrix} 0 & \ldots & -F_{k-1} \end{bmatrix} \begin{bmatrix} \cdot & \cdots & \cdot \\ \vdots & \ddots & \vdots \\ \cdot & \cdots & P_{k-1|k-1} \end{bmatrix} \begin{bmatrix} 0 \\ \vdots \\ -F_{k-1}^{\mathrm{T}} \end{bmatrix} = F_{k-1}P_{k-1|k-1}F_{k-1}^{\mathrm{T}}.$$

Therefore $P = I + Z$ in equation (17.49) is exactly the Kalman update formula

$$P_{k|k-1} = I + F_{k-1}P_{k-1|k-1}F_{k-1}^{\mathrm{T}}. \qquad (17.50)$$

The identity matrix I will change to $\Sigma_{\epsilon,k}$ when the covariance of the state equation error $\boldsymbol{\epsilon}_k$ is accounted for.

Now comes the second half of the update. The new row $W = [\,0 \ \ \ldots \ \ 0 \ \ A_k\,]$ enters from the observation equations. This row is placed below $\mathcal{S}_k$ to give $\mathcal{A}_k$. Therefore $W^{\mathrm{T}}W$ is added to $\mathcal{S}_k^{\mathrm{T}}\mathcal{S}_k$ to give $\mathcal{A}_k^{\mathrm{T}}\mathcal{A}_k$. We write $\mathcal{Y}$ for the big tridiagonal matrix $\mathcal{S}_k^{\mathrm{T}}\mathcal{S}_k$ before this update, and we use the update formula (17.47) for the new inverse:

$$(\mathcal{A}_k^{\mathrm{T}}\mathcal{A}_k)^{-1} = (\mathcal{Y} + W^{\mathrm{T}}IW)^{-1} = \mathcal{Y}^{-1} - \mathcal{Y}^{-1}W^{\mathrm{T}}(I + W\mathcal{Y}^{-1}W^{\mathrm{T}})^{-1}W\mathcal{Y}^{-1}. \quad (17.51)$$

Look at $W\mathcal{Y}^{-1}W^{\mathrm{T}}$. The row $W = [\,0 \ \ \ldots \ \ 0 \ \ A_k\,]$ is zero until the last block. The last corner of $\mathcal{Y}^{-1}$ is $P_{k|k-1}$, found above. Therefore $W\mathcal{Y}^{-1}W^{\mathrm{T}}$ reduces immediately to $A_kP_{k|k-1}A_k^{\mathrm{T}}$. Similarly the last block of $\mathcal{Y}^{-1}W^{\mathrm{T}}$ is $P_{k|k-1}A_k^{\mathrm{T}}$.

Now concentrate on the *last block row* in equation (17.51). Factoring out $\mathcal{Y}^{-1}$ on the right, this last row of $(\mathcal{A}_k^{\mathrm{T}}\mathcal{A}_k)^{-1}$ is

$$\begin{aligned}&\big(I - P_{k|k-1}A_k^{\mathrm{T}}(I + A_kP_{k|k-1}A_k^{\mathrm{T}})^{-1}A_k\big)(\text{last row of } \mathcal{Y}^{-1})\\ &\qquad = (I - K_kA_k)\big(\text{last row of } (\mathcal{S}_k^{\mathrm{T}}\mathcal{S}_k)^{-1}\big).\end{aligned} \quad (17.52)$$

In particular the final entry of this last row is the corner entry $P_{k|k}$ in $(\mathcal{A}_k^{\mathrm{T}}\mathcal{A}_k)^{-1}$:

$$P_{k|k} = (I - K_kA_k)P_{k|k-1}. \quad (17.53)$$

This is the second half of the Kalman filter. It gives the error covariance $P_{k|k}$ from $P_{k|k-1}$. The new P uses the observation equation $\boldsymbol{b}_k = A_k\boldsymbol{x}_k + \boldsymbol{e}_k$, whereas the old P doesn't use it. The matrix K that was introduced to simplify the algebra in (17.52) is Kalman's *gain matrix*:

$$K_k = P_{k|k-1}A_k^{\mathrm{T}}\big(I + A_kP_{k|k-1}A_k^{\mathrm{T}}\big)^{-1}. \quad (17.54)$$

This identity matrix I will change to $\Sigma_{e,k}$ when the covariance of the observation error is accounted for. You will see in equation (17.63) why we concentrated on the whole last row of $(\mathcal{A}_k^{\mathrm{T}}\mathcal{A}_k)^{-1}$, as well as its final entry $P_{k|k}$ in (17.53).

The Gain Matrix and the Weights

In a moment we will complete the filter by estimating the new state vector $\boldsymbol{x}_k$:

$$\begin{aligned}&\text{Prediction:} \qquad \hat{\boldsymbol{x}}_{k|k-1} = F_{k-1}\hat{\boldsymbol{x}}_{k-1|k-1}\\ &\text{Correction:} \qquad \hat{\boldsymbol{x}}_{k|k} = \hat{\boldsymbol{x}}_{k|k-1} + K_k(\boldsymbol{b}_k - A_k\hat{\boldsymbol{x}}_{k|k-1}).\end{aligned}$$

We do not put boxes around those formulas until they are derived. But we want to show that these state estimates are consistent with the prediction $P_{k|k-1}$ and correction $P_{k|k}$ for their error covariance matrices.

The prediction $\hat{\boldsymbol{x}}_{k|k-1}$ is easy. Actually it could come before or after $P_{k|k-1}$; you will see that the reasoning is very straightforward. The point to note is that by the ordinary propagation law, the state $\boldsymbol{x}_k = F_{k-1}\boldsymbol{x}_{k-1} + \boldsymbol{\epsilon}_k$ has the error covariance matrix

$$P_{k|k-1} = F_{k-1}P_{k-1|k-1}F_{k-1}^{\mathrm{T}} + \Sigma_{\epsilon,k}. \quad (17.55)$$

This is our formula (17.50), corrected by including $\Sigma_{\epsilon,k}$ instead of the identity matrix I. We could also have included $\Sigma_{\epsilon,k}$ in our first derivation of (17.50). The update matrix added in equation (17.48) would become $[\,V \quad I\,]^{\mathrm{T}}\Sigma^{-1}[\,V \quad I\,]$. The long matrix (17.49) changes to

$$P = \left(\Sigma_{\epsilon,k}^{-1} - \Sigma_{\epsilon,k}^{-1} Z \Sigma_{\epsilon,k}^{-1} + \Sigma_{\epsilon,k}^{-1} Z (\Sigma_{\epsilon,k} + Z)^{-1} Z \Sigma_{\epsilon,k}^{-1}\right)^{-1}.$$

And this happily reduces to $Z + \Sigma_{\epsilon,k}$ which is the covariance (17.55) with weight matrix included.

Now turn to the corrected state estimate $\hat{\boldsymbol{x}}_{k|k} = (I - K_k A_k)\hat{\boldsymbol{x}}_{k|k-1} + K_k \boldsymbol{b}_k$. Again we use the ordinary propagation law to find the variance of $\hat{\boldsymbol{x}}_{k|k}$:

$$P_{k|k} = (I - K_k A_k) P_{k|k-1} (I - K_k A_k)^{\mathrm{T}} + K_k \Sigma_{e,k} K_k^{\mathrm{T}}. \tag{17.56}$$

This is a new form of the covariance update. Unlike equation (17.53), this "Joseph form" is clearly symmetric. It is a sum of two positive definite matrices (so that property is never lost by numerical error). The computation is a little slower, but the form (17.56) is preferred in many calculations. Kalman and Bucy noted that a perturbation ΔK in the gain gives only a *second-order* change in the Joseph form, while it gives a first-order change $-\Delta K A_k P_{k|k-1}$ in the simpler form of equation (17.53):

$$P_{k|k} = (I - K_k A_k) P_{k|k-1}. \tag{17.57}$$

The equality of those two forms of $P_{k|k}$ must come from manipulation of the expression for Kalman's matrix K_k. So we turn to that gain matrix. We also insert the correct weight matrix $\Sigma_{e,k}$ into its formula.

We introduced the matrix K_k in (17.54) to simplify equation (17.52). If the covariance matrix $\Sigma_{e,k}$ had been included in (17.52), then it would have been included in the gain matrix: The correct form is

$$K_k = P_{k|k-1} A_k^{\mathrm{T}} \left(\Sigma_{e,k} + A_k P_{k|k-1} A_k^{\mathrm{T}}\right)^{-1}. \tag{17.58}$$

The reason is that the first identity I in (17.51) should have been $\Sigma_{e,k}^{-1}$. Then the second I would have been $\Sigma_{e,k}$. Now we manipulate that expression (17.58). It is equivalent to

$$P_{k|k-1} A_k^{\mathrm{T}} = K_k \left(\Sigma_{e,k} + A_k P_{k|k-1} A_k^{\mathrm{T}}\right). \tag{17.59}$$

Shifting terms to the left side this becomes

$$(I - K_k A_k) P_{k|k-1} A_k^{\mathrm{T}} = K_k \Sigma_{e,k}. \tag{17.60}$$

Now substitute (17.60) into the Joseph form (17.56) for $P_{k|k}$:

$$(17.56) = (I - K_k A_k) P_{k|k-1} - [(17.60)] K_k^{\mathrm{T}} + K_k \Sigma_{e,k} K_k^{\mathrm{T}} = (I - K_k A_k) P_{k|k-1}.$$

This is the "unsymmetric" form (17.53) of $P_{k|k}$. But that product $P_{k|k}$ must be symmetric!

Note The gain matrix K also has an important optimality property. The state update formula $(I - K_k A_k)\hat{\boldsymbol{x}}_{k|k-1} + K_k \boldsymbol{b}_k$ led directly to the Joseph form (17.56) for $P_{k|k}$. *We could choose the gain matrix in that update so as to minimize* $P_{k|k}$. Since Gauss chose weights for the same purpose, the gain matrix K brings Kalman into agreement with Gauss.

Recall from Section 9.4 the proof by Gauss that the covariance of $\boldsymbol{x}$ is smallest when the weight is Σ^{-1}. A similar proof shows that K_k minimizes (17.56) because it satisfies (17.60). The first-order change in $P_{k|k}$ due to ΔK in (17.56) is

$$\big((I - K_k A_k) P_{k|k-1}(-A_k^{\mathrm{T}} \Delta K)\big) + \big(\ldots\big)^{\mathrm{T}} = 0.$$

This is exactly the comment made earlier, that a perturbation ΔK produces only a second-order change in $P_{k|k}$. When we examine that change $\Delta P = \Delta K\big(A_k^{\mathrm{T}} P_{k|k-1} A_k + \Sigma_{e,k}\big)\Delta K^{\mathrm{T}}$ we see that it is upwards (positive definite). So Kalman's gain matrix K_k does minimize the covariance $P_{k|k}$. This is perhaps the neatest proof of the Kalman update formulas.

We note that the quantity most often minimized in the Kalman filter literature is the *trace* of P. This is the expected value of the scalar $(\boldsymbol{x} - \hat{\boldsymbol{x}})^{\mathrm{T}}(\boldsymbol{x} - \hat{\boldsymbol{x}})$; its minimization leads to the same gain matrix K. But it seems more informative (and just as easy) to minimize the matrix $P = E\{(\boldsymbol{x} - \hat{\boldsymbol{x}})(\boldsymbol{x} - \hat{\boldsymbol{x}})^{\mathrm{T}}\}$. We just proved that any other choice of K (that is, any movement ΔK) would increase P by a positive definite matrix ΔP.

Now we systematically derive the state updates $\hat{\boldsymbol{x}}_{k|k-1}$ and $\hat{\boldsymbol{x}}_{k|k}$ as solutions of the overall least-squares problem. And we mention that all our derivations could be expressed in stochastic terms, with expectations instead of matrix equations.

State Updates $\hat{\boldsymbol{x}}_{k|k-1}$ and $\hat{\boldsymbol{x}}_{k|k}$

The left side of the big least-squares problem is now dealt with. We know the last row of $(\mathcal{A}_k^{\mathrm{T}} \Sigma_k^{-1} \mathcal{A}_k)^{-1}$. Now we multiply by the right side $\mathcal{A}_k^{\mathrm{T}} \Sigma^{-1} \boldsymbol{b}_k$, when $\boldsymbol{b}_k$ includes the newest observation $\boldsymbol{b}_k$.

The predicted value of $\boldsymbol{x}_k$ (before that observation) is simple to understand. Only the state equation has been added to the system. We can solve it exactly by

$$\hat{\boldsymbol{x}}_{k|k-1} = F_{k-1} \hat{\boldsymbol{x}}_{k-1|k-1}. \tag{17.61}$$

This is our best estimate of $\boldsymbol{x}_k$, based on the state equation and the old observations. It solves the new state equation exactly, and it keeps the best solution to the earlier equations. So it maintains the correct least-squares solution, when the new row and column are added to the system.

Now we include the new observation. This changes everything. The earlier estimates $\hat{\boldsymbol{x}}_{i|k-1}$ are "smoothed" in the new $\hat{\boldsymbol{x}}_{i|k}$. We leave those smoothing formulas (for $i < k$) until later. The predicted $\hat{\boldsymbol{x}}_{k|k-1}$ in (17.61) changes to a corrected value $\hat{\boldsymbol{x}}_{k|k}$. This is what we compute now. It is the last component of the weighted least-squares solution to the complete system $\mathcal{A}_k x_k \approx \boldsymbol{b}_k$:

$$\begin{bmatrix} A_0 & & & \\ -F_0 & I & & \\ & A_1 & & \\ & & \ddots & \\ & & A_{k-1} & \\ & & -F_{k-1} & I \\ & & & A_k \end{bmatrix} \begin{bmatrix} \boldsymbol{x}_0 \\ \vdots \\ \boldsymbol{x}_k \end{bmatrix} \approx \begin{bmatrix} \boldsymbol{b}_0 \\ \boldsymbol{0} \\ \boldsymbol{b}_1 \\ \vdots \\ \boldsymbol{b}_{k-1} \\ \boldsymbol{0} \\ \boldsymbol{b}_k \end{bmatrix} = \boldsymbol{b}_k. \tag{17.62}$$

Table 17.2 The equations for the Kalman filter

System equation Observation equation	$x_k = F_{k-1}x_{k-1} + \epsilon_k, \quad \epsilon_k \sim N(\mathbf{0}, \Sigma_{\epsilon,k})$ $b_k = A_k x_k + e_k, \quad e_k \sim N(\mathbf{0}, \Sigma_{e,k})$
Initial conditions Other conditions	$E\{x_0\} = \hat{x}_0$ $E\{(x_0 - \hat{x}_{0\|0})(x_0 - \hat{x}_{0\|0})^{\mathrm{T}}\} = P_{0\|0}$ $E\{\epsilon_k e_j^{\mathrm{T}}\} = 0, \quad$ for all k, j
Prediction of the state vector Prediction of the covariance matrix	$\hat{x}_{k\|k-1} = F_{k-1}\hat{x}_{k-1\|k-1}$ $P_{k\|k-1} = F_{k-1}P_{k-1\|k-1}F_{k-1}^{\mathrm{T}} + \Sigma_{\epsilon,k}$
The Kalman gain matrix Filtering of state vector Covariance matrix for filtering	$K_k = P_{k\|k-1}A_k^{\mathrm{T}}\big(A_k P_{k\|k-1}A_k^{\mathrm{T}} + \Sigma_{e,k}\big)^{-1}$ $\hat{x}_{k\|k} = \hat{x}_{k\|k-1} + K_k(b_k - A_k\hat{x}_{k\|k-1})$ $P_{k\|k} = (I - K_k A_k)P_{k\|k-1}$

The least-squares solution is always $\hat{x} = (\mathcal{A}_k^{\mathrm{T}}\Sigma^{-1}\mathcal{A}_k)^{-1}\mathcal{A}_k^{\mathrm{T}}\Sigma^{-1}\mathscr{b}_k$. We want the last block $\hat{x}_{k|k}$ in this least-squares solution. So we use the last block row of $(\mathcal{A}_k^{\mathrm{T}}\Sigma^{-1}\mathcal{A}_k)^{-1}$ from equation (17.52):

$$\begin{aligned}\hat{x}_{k|k} &= \big(\text{last row of } (\mathcal{A}_k^{\mathrm{T}}\Sigma^{-1}\mathcal{A}_k)^{-1}\big)\mathcal{A}_k^{\mathrm{T}}\Sigma^{-1}\mathscr{b}_k \\ &= (I - K_k A_k)\big(\text{last row of } (\mathcal{S}_k^{\mathrm{T}}\Sigma^{-1}\mathcal{S}_k)^{-1}\big)\mathcal{A}_k^{\mathrm{T}}\Sigma^{-1}\mathscr{b}_k. \end{aligned} \tag{17.63}$$

We start with $\mathscr{b}_k$ on the right side, and carry out each multiplication in this equation. Separate the old observations in $\mathscr{b}_{k-1}$ from the new b_k, and multiply by Σ^{-1}:

$$\Sigma^{-1}\mathscr{b}_k = \begin{bmatrix} \Sigma_{k-1}^{-1}\mathscr{b}_{k-1} \\ \Sigma_{e,k}^{-1}b_k \end{bmatrix}.$$

Multiply next by $\mathcal{A}_k^{\mathrm{T}} = [\, \mathcal{S}_k^{\mathrm{T}} \quad W^{\mathrm{T}} \,]$ and recall that $W = [\, 0 \ \ \ldots \ \ 0 \ \ A_k \,]$:

$$\mathcal{A}_k^{\mathrm{T}}\Sigma^{-1}\mathscr{b}_k = \mathcal{S}_k^{\mathrm{T}}\Sigma_{k-1}^{-1}\mathscr{b}_{k-1} + A_k^{\mathrm{T}}\Sigma_{e,k}^{-1}b_k. \tag{17.64}$$

Now multiply by the last row of $(\mathcal{S}_{k-1}^{\mathrm{T}}\Sigma^{-1}\mathcal{S}_k)^{-1}$. This produces the least-squares solution $\hat{x}_{k|k-1}$ in the old $k-1$ part. Watch what it produces in the new part:

$$\big(\text{last row of } (\mathcal{S}_k^{\mathrm{T}}\Sigma^{-1}\mathcal{S}_k)^{-1}\big)\big(\mathcal{S}_k^{\mathrm{T}}\Sigma_{k-1}^{-1}\mathscr{b}_{k-1} + A_k^{\mathrm{T}}\Sigma_{e,k}^{-1}b_k\big) = \hat{x}_{k|k-1} + P_{k|k-1}A_k^{\mathrm{T}}\Sigma_{e,k}^{-1}b_k. \tag{17.65}$$

Finally equation (17.63) multiplies this by $(I - K_k A_k)$ to yield $\hat{x}_{k|k}$:

$$\hat{x}_{k|k} = (I - K_k A_k)\hat{x}_{k|k-1} + K_k b_k. \tag{17.66}$$

That final term used the identity (17.60) to replace $(I - K_k A_k)P_{k|k-1}A_k^{\mathrm{T}}$ by $K_k\Sigma_{e,k}$.

This completes the sequence of Kalman filter update equations. As we hoped, formula (17.66) for $\hat{x}_{k|k}$ can be expressed as the prediction $\hat{x}_{k|k-1}$ plus a correction:

$$\hat{x}_{k|k} = \hat{x}_{k|k-1} + K_k(b_k - A_k\hat{x}_{k|k-1}). \tag{17.67}$$

The correction is Kalman's gain matrix times the ***innovation***.

17.5 Bayes Filter for Batch Processing

The steps of the Kalman filter are *prediction-gain matrix-correction*. For the state vector this is the natural order. From the prediction $\hat{x}_{k|k-1}$ and the gain matrix K_k and the new observation b_k, we compute the filtered estimate $\hat{x}_{k|k}$. This uses b_k from the right side of the normal equations. But the hard part (the computationally intensive part) is always on the left side, where we are factoring the big matrix $\mathcal{A}^{\mathrm{T}}\Sigma^{-1}\mathcal{A}$ and updating the covariance to $P_{k|k}$. The Kalman filter creates the gain matrix K before the covariance P, but this is not the only possible order. We want to consider if it is the best order.

Suppose we compute $P_{k|k}$ before the gain matrix. Then we use the gain matrix for $\hat{x}_{k|k}$. This is the one point where K is actually needed, to multiply the innovation $b - Ax$ and update the state vector in (17.67). The new observation is that we can go directly from the prediction $P_{k|k-1}$ to the correction $P_{k|k}$:

$$P_{k|k} = \left(P_{k|k-1}^{-1} + A_k^{\mathrm{T}}\Sigma_{e,k}^{-1}A_k\right)^{-1}. \tag{17.68}$$

This is straightforward when we remember that the matrices P^{-1} are the (block) pivots of $\mathcal{A}^{\mathrm{T}}\Sigma^{-1}\mathcal{A}$. The new row $[\,0 \ \ \dots \ \ 0 \ \ A_k\,]$ simply adds $A_k^{\mathrm{T}}\Sigma_{e,k}^{-1}A_k$ to the (k,k) block and therefore to that pivot. This is equation (17.68).

P^{-1} is called the ***information matrix***. It increases as the covariance P decreases (then the information gets better). It is sometimes more economical to work directly with the pivot, which is P^{-1}, than with P. The inverse of (17.68) is

$$P_{k|k}^{-1} = P_{k|k-1}^{-1} + A_k^{\mathrm{T}}\Sigma_{e,k}^{-1}A_k. \tag{17.69}$$

In either case the gain matrix (it now comes *after* P or P^{-1}) has the new formula

$$K_k = P_{k|k}A_k^{\mathrm{T}}\Sigma_{e,k}^{-1}. \tag{17.70}$$

To verify that this is the correct gain matrix, remember that the least-squares solution is $(\mathcal{A}^{\mathrm{T}}\Sigma^{-1}\mathcal{A})^{-1}\mathcal{A}^{\mathrm{T}}\Sigma^{-1}\mathcal{b}$. The matrix multiplying b_k in the last block is $P_{k|k}$ (from the big inverse matrix) times A_k^{T} (from $\mathcal{A}^{\mathrm{T}}$) times $\Sigma_{e,k}^{-1}$. This is exactly formula (17.70) for the gain matrix.

To repeat: ***We can compute the covariance $P_{k|k}$ before the gain.*** Morrison (1969) calls this the "Bayes filter," although this name is perhaps not widely used. The formulas can be derived from the Bayes theorem about conditional expectation (just as the Kalman formulas can be derived from maximum likelihood). We could see the two forms more directly, as coming from the two sides of the Sherman-Morrison-Woodbury-Schur matrix identity. And we could see them in a deeper way as coming from *dual* optimization problems.

We now focus on the practical questions. Which form is more efficient? Which form is more stable? Since the Kalman form is the most famous and frequently used, we expect that it usually wins. But not always.

The computational cost is greatest when we invert matrices. The Bayes form inverts P, whose order equals the number of state variables. The Kalman form inverts $\Sigma_{e,k} + A_k P_{k|k-1} A_k^{\mathrm{T}}$, whose order equals the number of new observations. If frequent views of the state vector are required, then Bayes is expensive and Kalman is better.

The Kalman form is best for immediate updates. The Bayes form has advantages for batch processing. If we want to update the state as soon as possible—to give a better prediction in a nonlinear state equation, or to reach a quicker control decision—then we need Kalman. If we can collect a larger batch of observations before using them to update the state, we can choose Bayes. The gain matrix is only needed for the state update.

Let us mention some other advantages of the Bayes form (covariance before gain). The significance depends on the particular application:

1 The Bayes filter can start with *no a priori information* on $\boldsymbol{x}_0$, by setting $P_{0|0}^{-1} = 0$.

In the Kalman form this would require $P_{0|0} = \infty$. Substituting a large initial variance is certainly possible, and very common. Morrison (1969), page 472, discusses the difficulties with this approach; it affects the later estimates, which does not happen with $P_{0|0}^{-1} = 0$ in Bayes. And there is a further difficulty to tune the large $P_{0|0}$ to the word-length of the computer.

2 If the covariance matrix P becomes too small, there are again numerical difficulties. (It may become indefinite, for one.) And Kalman updates run into trouble when the data is *highly precise* and its covariance $\Sigma_{e,k}$ is nearly zero. This matrix could be lost to roundoff error when $\Sigma_{e,k} + A_k P_{k|k-1} A_k^{\mathrm{T}}$ is inverted. The Bayes form remains successful provided A_k has full column rank. Its new estimate uses heavily the new data, which is reliable because $\Sigma_{e,k}$ is small.

Recall that the ***square-root filter*** was created to avoid the worst of these possible difficulties. The positive-definiteness of a matrix P is guaranteed if we express it in the form $R^{\mathrm{T}}R$. The reason is that $\boldsymbol{x}^{\mathrm{T}}R^{\mathrm{T}}R\boldsymbol{x}$ could never be negative, because it equals $\|R\boldsymbol{x}\|^2$. So the numerically stable approach to filtering is to use a "square root" of P or P^{-1}.

Note that $A = QR$ gives $A^{\mathrm{T}}A = R^{\mathrm{T}}R$. Thus the factorization into QR by Gram-Schmidt orthogonalization is the entrance to the square root filter. But the overhead of working with this Q matrix is serious! So the square-root approach is more expensive (and may not be needed), compared to plain Kalman.

And we must also note disadvantages of Bayes:

1 The number of unknown states (the dimension of $\boldsymbol{x}_k$) is *fixed*. We cannot conveniently account for a satellite that rises or sets. If a parameter becomes unimportant during the processing, we have to stay with it.

2 For quick updates, Bayes is more expensive.

Both filters must run into difficulty (and they do) if observation errors are ***perfectly correlated***. In this case $\Sigma_{e,k}$ is singular. Bayes requires its inverse, so at least it identifies the problem immediately. Kalman proceeds forward, without trying to invert, but it finds singular covariance matrices $P_{k|k}$. It is this danger (and also near singularity) that the square root filter or the Deyst filter, Deyst (1969), can handle better.

We don't mean to be such pessimists! The Kalman filter gives state estimates (and position estimates for GPS) of high accuracy in a wide variety of important problems. It reduces the error while it computes the covariance. Those are both extremely valuable in practice.

Remark 17.2 After writing these sections we discovered an interesting paper by Duncan & Horn (1972). That paper was written to overcome a "communications block" between engineering and statistics. We paraphrase their introduction: Despite a *basic closeness to regression* and a *high utility* for providing simple solutions, little of Kalman's work has reached the average statistician—because it was developed from conditional probability theory and not regression theory.

Duncan and Horn verify the Kalman filter equations and establish the key property that $E\{(\boldsymbol{x} - \hat{\boldsymbol{x}}_{k|k})\boldsymbol{b}_i^{\mathrm{T}}\} = 0$. Their argument extends the Gauss-Markov Theorem from regression theory, which identifies optimal estimators. Their reasoning will be familiar to statisticians. To a non-expert in both fields (wide sense conditional probability and regression), the latter approach is simpler but still far from transparent. We hope that our straightforward and more ponderous derivation (slogging through the normal equations) will be helpful to a third group. We do not attempt to define this third group, who prefer block tridiagonal matrices to subtle and useful insights about uncorrelated variables.

The common step in all derivations is our favorite matrix identity:

$$P_{k|k} = (T + A^{\mathrm{T}}\Sigma_e^{-1}A)^{-1} = T^{-1} - T^{-1}A^{\mathrm{T}}(\Sigma_e + AT^{-1}A^{\mathrm{T}})^{-1}AT^{-1}. \tag{17.71}$$

Duncan and Horn choose the Bayes form as more intuitive. That form uses the left side of (17.71), with $T = P_{k|k-1}^{-1}$ as the predicted inverse covariance. The Kalman form uses the right side of (17.71). You see again how Kalman inverts a matrix of size given by the number of observations, where the left side involves the number of state components (just count rows in A and A^{T}).

Remark 17.3 Paige & Saunders (1977) have proposed a Kalman algorithm based on ***orthogonalization*** of the columns of the big matrix $M = \mathcal{A}\Sigma^{-1/2}$. By including $\Sigma^{-1/2}$, which is found one block at a time since Σ is block diagonal, they can simplify to unit covariance. The main point is to create $\mathcal{Q}$ from a sequence of rotations so that

$$Q^{\mathrm{T}}M = \begin{bmatrix} Q_1^{\mathrm{T}} \\ Q_2^{\mathrm{T}} \end{bmatrix} M = \begin{bmatrix} R \\ \mathbf{0} \end{bmatrix} \qquad \text{and then} \qquad \hat{x} = R^{-1}Q_1^{\mathrm{T}}\mathcal{b}.$$

The factor R is block bidiagonal. Its last block R_{kk} reveals everything about $P_{k|k}$ and $\hat{\boldsymbol{x}}_{k|k}$. Because Q is orthogonal and R is triangular, the overall covariance matrix P and its last block $P_{k|k}$ are given by

$$P = (R^{\mathrm{T}}R)^{-1} \qquad \text{and} \qquad P_{k|k}^{-1} = R_{kk}^{\mathrm{T}}R_{kk}. \tag{17.72}$$

Thus R is the Cholesky factor ("square root") of the "information matrix" P^{-1} and R_{kk} is the Cholesky factor of $P_{k|k}^{-1}$. This is a ***square root information filter***. By working with Q and R it is numerically stable; see also Bierman (1977). The Paige-Saunders constructions are a little slower than necessary, so today's implementation is different. But for the third group described above, who understand matrix algebra better than statistics and probability, this paper will be helpful.

17.6 Smoothing

The forward process of filtering produces the values $\hat{x}_{k|k}$. These are the best state estimates from the observations up to time k. Smoothing produces a better estimate $\hat{x}_{k|N}$ for the state at time k by using the observations up to the later time N.

Actually, ***filtering is forward elimination and smoothing is back substitution***. Both are to be executed recursively. The normal equations have the block tridiagonal matrix $\mathcal{T}$ as coefficient matrix. As always, the steps of elimination factor that matrix into lower triangular times upper triangular. There will be a block diagonal matrix in the middle (containing the pivots), if L and L^{T} have I as their diagonal:

$$\mathcal{T} = LDL^{\mathrm{T}} = \begin{bmatrix} I & & & \\ L_{1|0} & I & & \\ & \ddots & \ddots & \\ & & L_{N|N-1} & I \end{bmatrix} \begin{bmatrix} P_0^{-1} & & \\ & \ddots & \\ & & P_N^{-1} \end{bmatrix} \begin{bmatrix} I & L_{1|0} & & \\ & \ddots & \ddots & \\ & & I & L_{N|N-1} \\ & & & I \end{bmatrix}. \tag{17.73}$$

This factorization combines in each row the "double step" of prediction followed by correction (bordering by the state equation and update by the observation equation). We look at the net result in these simple terms:

1 The off-diagonal block $L_{k|k-1}$ gives the off-diagonal block $T_{k|k-1}$ in $\mathcal{T}$:
Directly from (17.73): $L_{k|k-1}P_{k-1}^{-1}I = T_{k|k-1}$.

2 The diagonal block P_k^{-1} completes the diagonal block $T_{k|k}$ in $\mathcal{T}$:
Directly from (17.73): $L_{k|k-1}P_{k-1}^{-1}L_{k|k-1}^{\mathrm{T}} + P_k^{-1} = T_{k|k}$.

This direct computation of the pivot P_k^{-1} is what we call the Bayes filter. The Kalman filter computes the inverse pivot P_k. The steps become completely explicit when we write the blocks $T_{k|k-1}$ and $T_{k|k}$ in terms of F_{k-1} and A_k and $\Sigma_{\epsilon,k}^{-1}$ and $\Sigma_{e,k}^{-1}$. What we are interested in now is the solution $\hat{x}$ to the normal equations. Let us focus on that.

The right side of the normal equations is a long vector $\boldsymbol{v} = \mathcal{A}^{\mathrm{T}}\Sigma^{-1}\boldsymbol{b}$. The equations themselves are just $\mathcal{T}\hat{\boldsymbol{x}} = \boldsymbol{v}$. The point of elimination is to break this into two triangular systems, and the forward filtering algorithm puts the pivots P_k^{-1} into the lower triangular factor:

Forward filtering:	Solve	$LD\hat{\boldsymbol{x}}_{\text{filtered}} = \boldsymbol{v}$
Backward smoothing:	Solve	$L^{\mathrm{T}}\hat{\boldsymbol{x}}_{\text{smoothed}} = \hat{\boldsymbol{x}}_{\text{filtered}}$.

Again, this is forward elimination and back substitution. The intermediate vector $\hat{\boldsymbol{x}}_{\text{filtered}}$ contains the estimates $\hat{\boldsymbol{x}}_{k|k}$. Then $\hat{\boldsymbol{x}} = \hat{\boldsymbol{x}}_{\text{smoothed}}$ is the actual solution to $\mathcal{T}\hat{\boldsymbol{x}} = \boldsymbol{v}$.

Now we look at smoothing algorithms. Row k of the equation $L^{\mathrm{T}}\hat{\boldsymbol{x}} = \hat{\boldsymbol{x}}_{\text{filtered}}$ is

$$I\hat{\boldsymbol{x}}_{k|N} + L_{k+1,k}^{\mathrm{T}}\hat{\boldsymbol{x}}_{k+1|N} = \hat{\boldsymbol{x}}_{k|k}. \tag{17.74}$$

Thus back substitution is a backward recursion that starts from $\hat{\boldsymbol{x}}_{N|N}$, which is the last output from forward filtering:

$$\hat{\boldsymbol{x}}_{k|N} = \hat{\boldsymbol{x}}_{k|k} - L_{k+1|k}^{\mathrm{T}}\hat{\boldsymbol{x}}_{k+1|N} = \hat{\boldsymbol{x}}_{k|k} - P_k^{-1}T_{k+1|k}^{\mathrm{T}}\hat{\boldsymbol{x}}_{k+1|N}. \tag{17.75}$$

This is the RTS recursion of Rauch & Tung & Striebel (1965), equation (3.30). Those authors noted that the pivots could be stored or else computed recursively by a backward filter. We emphasize that P_k is the covariance matrix for the filtered (forward) estimate $\hat{\boldsymbol{x}}_{k|k}$. The covariance matrix for $\hat{\boldsymbol{x}}_{k|N}$ will be smaller, since more information has been used. The covariance matrix for the complete system is always $(\mathcal{A}^{\mathrm{T}}\Sigma^{-1}\mathcal{A})^{-1}$.

Example 17.3 We bring a numerical example from the original paper by Rauch & Tung & Striebel (1965). Consider a dynamical system with four state variables:

$$\boldsymbol{x}_k = F_{k-1}\boldsymbol{x}_{k-1} + \boldsymbol{\epsilon}_k = \begin{bmatrix} 1 & 1 & 0.5 & 0.5 \\ 0 & 1 & 1 & 1 \\ 0 & 0 & 1 & 0 \\ 0 & 0 & 0 & 0.606 \end{bmatrix} \boldsymbol{x}_{k-1} + \boldsymbol{\epsilon}_k.$$

Suppose $\boldsymbol{b}_k = [\,1 \;\; 0 \;\; 0 \;\; 0\,]\,\boldsymbol{x}_k + \boldsymbol{e}_k$ is the 1 by 1 output vector that measures x_1 (with noise). The errors $\boldsymbol{\epsilon}_k$ and $\boldsymbol{e}_k$ are independent Gaussian with covariances

$$\Sigma_{\epsilon,k} = \begin{bmatrix} 0 & 0 & 0 & 0 \\ 0 & 0 & 0 & 0 \\ 0 & 0 & 0 & 0 \\ 0 & 0 & 0 & \epsilon \end{bmatrix} \quad \text{and} \quad \Sigma_{e,k} = 1.$$

The initial condition $\hat{\boldsymbol{x}}_0$ is a Gaussian vector with covariance given by $P_{0|0}$.

The entire system is a linearized version of the in-track motion of a satellite traveling in a circular orbit. The satellite motion is affected by both constant and random drag. The state variables x_1, x_2, and x_3 can be considered as angular position, velocity, and (constant) acceleration. The state variable x_4 is a random component of acceleration generated by a first-order Gauss-Markov process.

Three cases will be considered, with different choices for ϵ and $P_{0|0}$:

1 $\quad \epsilon = 0.0063 \quad$ and $\quad P_{0|0} = \begin{bmatrix} 1 & 0 & 0 & 0 \\ 0 & 1 & 0 & 0 \\ 0 & 0 & 1 & 0 \\ 0 & 0 & 0 & 0.01 \end{bmatrix}$

2 $\quad \epsilon = 0.000063 \quad$ and $\quad P_{0|0} = \begin{bmatrix} 1 & 0 & 0 & 0 \\ 0 & 1 & 0 & 0 \\ 0 & 0 & 1 & 0 \\ 0 & 0 & 0 & 0.0001 \end{bmatrix}$

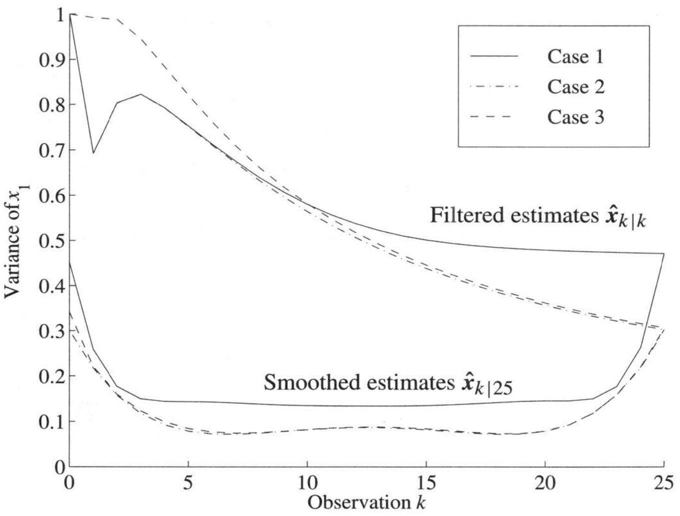

Figure 17.1 Variance history of x_1 for different values of ϵ and $P_{0|0}$.

3 $\epsilon = 0.0063$ and $P_{0|0} = \begin{bmatrix} 100 & 0 & 0 & 0 \\ 0 & 100 & 0 & 0 \\ 0 & 0 & 100 & 0 \\ 0 & 0 & 0 & 0.01 \end{bmatrix}.$

In each case $N = 25$ measurements are taken starting with $\boldsymbol{b}_1$. In Figure 17.1 the variances of the filtered and smoothed estimates of the state variable x_1 are plotted for all three cases. Smoothing the estimate decreases the errors. Reducing the variance of the random disturbance reduces the variance of the estimates. The effect of initial conditions (the a priori information about the state) rapidly dies out.

The code generating Figure 17.1 is named rts.

Backward Filter

Smoothing at epoch k involves data from before and *after* epoch k. The accuracy generally is superior to that of an unsmoothed filter because the estimate involves more observations. It is easy to imagine a filter working on the data from the last epoch N "backwards" to 0. So one could hope that a weighted mean of the forward $\hat{\boldsymbol{x}}_k$ and backward $\hat{\boldsymbol{x}}_k^*$ filters yields an optimal smoother:

$$\hat{\boldsymbol{x}}_{k|N} = A_k\hat{\boldsymbol{x}}_k + (I - A_k)\hat{\boldsymbol{x}}_k^*. \tag{17.76}$$

The weighting matrix A_k is still to be determined, by *minimizing the covariance* $P_{k|N}$ of the estimation error. This gives a new approach to smoothing.

Let P_k be the covariance matrix for the forward optimal filter error and P_k^* the covariance matrix for the backward optimal filter. The law of covariance propagation yields

$$P_{k|N} = E\{\boldsymbol{v}_{k|N}\boldsymbol{v}_{k|N}^{\mathrm{T}}\} = A_k P_k A_k^{\mathrm{T}} + (I - A_k)P_k^*(I - A_k)^{\mathrm{T}}. \tag{17.77}$$

We determine $A = A_k$ so that the trace of $P_{k|N}$ becomes a minimum:

$$\frac{\partial}{\partial A}\big(\mathrm{tr}(P_{k|N})\big) = 2AP + 2(I - A)P^*(-I) = 0.$$

This leads to

$$A = P^*(P + P^*)^{-1} \qquad \text{and} \qquad I - A = P(P + P^*)^{-1}. \tag{17.78}$$

Substituting into equation (17.77), matrix algebra gives the neat result

$$P_{k|N}^{-1} = P_k^{-1} + (P_k^*)^{-1}. \tag{17.79}$$

From this equation follows $P_{k|N} \le P_k$. ***The smoothed estimate of*** $\boldsymbol{x}_k$ ***is always better than the filtered estimate***, when we consider their variances.

Next we insert the expressions A and $I - A$ into equation (17.76) and simplify:

$$\hat{\boldsymbol{x}}_{k|N} = \big(P_k^{-1} + (P_k^*)^{-1}\big)^{-1}\big(P_k^{-1}\hat{\boldsymbol{x}}_k + (P_k^*)^{-1}\hat{\boldsymbol{x}}_k^*\big). \tag{17.80}$$

Equation (17.80) is a matrix generalization of the scalar equation (9.53) for weighted means. The same equation is also the basis for all expressions for smoothing estimates. But back-substitution is the faster way to evaluate $\hat{\boldsymbol{x}}_{k|N}$.

Usually one distinguishes between three types of smoothers:

- **fixed-interval:** Smoothing of data from epoch 0 to N; we seek $\hat{\boldsymbol{x}}_k$ from 0 to N.
- **fixed-point:** Smoothing from 0 to an increasing N; we seek $\hat{\boldsymbol{x}}_k$ for a fixed k.
- **fixed-lag:** Smoothing from $k - n$ to k as k increases; n is fixed and we seek $\hat{\boldsymbol{x}}_{k-n}$.

The first type is only for post processing. The last two types may also be used in real time.

Fixed-interval smoothing Given observations from epochs 0 to N, we filter forward and keep the results $\hat{\boldsymbol{x}}_{k|k-1}$, $\hat{\boldsymbol{x}}_{k|k}$, $\Sigma_{k|k-1}$, and $\Sigma_{k|k}$. Next we filter backward from epoch N to 0 by these recursive formulas (starting from $N - 1$):

$$\hat{\boldsymbol{x}}_{k|N} = \hat{\boldsymbol{x}}_{k|k} + A_k(\hat{\boldsymbol{x}}_{k+1|N} - \hat{\boldsymbol{x}}_{k+1|k}) \tag{17.81}$$

$$A_k = \Sigma_{k|k}F_k^{\mathrm{T}}\Sigma_{k+1|k}^{-1} \tag{17.82}$$

Example 17.4 Equations (17.81) and (17.82) are coded as the M-file smoother. Figure 17.2 shows the result of a forward filtering and a smoothing of the filtered values. (The call was smoother(1,2), this means that $\Sigma_e = 1$ and $\Sigma_\epsilon = 2$.) The data describe the drift of a steered clock in a GPS receiver. The second graph shows the variances of the filter and the smoother. The variance curve for smoothing is minimum in the middle and then gets

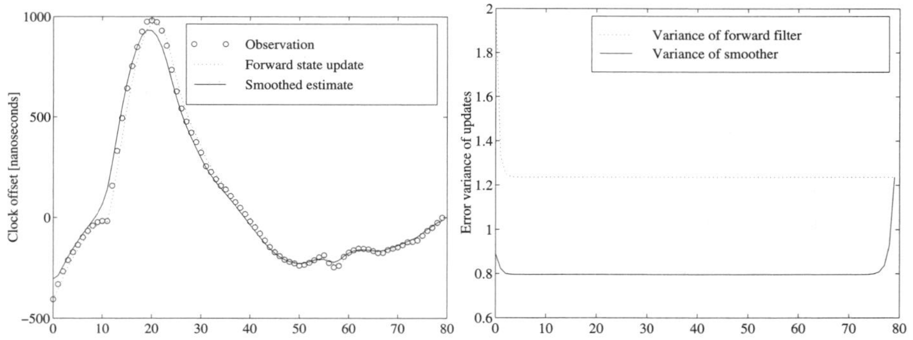

Figure 17.2 Forward filtering and smoothing of receiver clock offset. The epoch interval is 15 seconds. Notice the reduction of variance by smoothing.

larger as either end point is approached. This indicates that the best estimation occurs in the interior region where there is an abundance of observation data in either direction from the point of estimation. This is exactly what we would expect intuitively.

We note the steady state behavior in the middle of the plot. This happens often when the data span is large and the process is stationary. In the steady state region, the gains of the filter and smoother and error variance are constant. The calculations are fast and the storage is greatly reduced. Sometimes a problem that appears to be quite formidable at first glance works out to be feasible because most of the data can be processed in the steady state region.

Often the variance of the innovation due to the predicted state is tested. If it is less than a tolerance, the measurement is thrown away since it will have no effect anyway. There is no need to activate the filter update.

Fixed-point smoothing

$$\hat{\boldsymbol{x}}_{k|N} = \hat{\boldsymbol{x}}_{k|N-1} + B_N\big(\hat{\boldsymbol{x}}_{N|N} - \hat{\boldsymbol{x}}_{N|N-1}\big) \tag{17.83}$$

$$B_N = \prod_{i=k}^{N-1} S_i, \qquad S_i = \Sigma_{i|i} F_i^{\mathrm{T}} \Sigma_{i+1|i}^{-1} \tag{17.84}$$

with initial value $\hat{\boldsymbol{x}}_{k|k} = \hat{\boldsymbol{x}}_k$ for $N = k+1, k+2, \ldots$.

Fixed-lag smoothing

$$\begin{aligned}\hat{\boldsymbol{x}}_{k+1|k+1+N} = {} & F_k\hat{\boldsymbol{x}}_{k|k+N} + \Sigma_{e,k}\big(F_k^{\mathrm{T}}\big)^{-1}\Sigma_{k|k}^{-1}\big(\hat{\boldsymbol{x}}_{k|k+N} - \hat{\boldsymbol{x}}_{k|k}\big) \\ & + B_{k+1+N}K_{k+1+N}\big(\boldsymbol{b}_{k+1+N} - A_{k+1+N}F_{k+N}\hat{\boldsymbol{x}}_{k+N|k+N}\big)\end{aligned} \tag{17.85}$$

$$B_{k+1+N} = \prod_{i=k+1}^{k+N} S_i, \qquad S_i = \Sigma_{i|i} F_i^{\mathrm{T}} \Sigma_{i+1|i}^{-1} \tag{17.86}$$

for $k = 0, 1, 2, \ldots$ and $\hat{\boldsymbol{x}}_{0|N}$ is the initial condition.

17.7 An Example from Practice

The filters described so far have been of a simple nature. We now describe an example from daily life. It is a model of the 2-dimensional motion of a vessel. Our model is partly based on Tiberius (1991).

Circular motion is characterized by a constant speed v and a constant acceleration a in the direction perpendicular to the motion. Let x_k and y_k designate the x- and y-coordinates at time t_k. If α_k is the heading, the discretized equations of motion are

$$\begin{aligned}
x_k &= x_{k-1} + \sin\alpha_{k-1}\Delta t_k\, v_{k-1} + \tfrac{1}{2}\cos\alpha_{k-1}(\Delta t_k)^2 a_{k-1} + \tfrac{1}{6}\cos\alpha_{k-1}(\Delta t_k)^3 \dot{a}_{k-1}\\
y_k &= y_{k-1} + \cos\alpha_{k-1}\Delta t_k\, v_{k-1} - \tfrac{1}{2}\sin\alpha_{k-1}(\Delta t_k)^2 a_{k-1} - \tfrac{1}{6}\sin\alpha_{k-1}(\Delta t_k)^3 \dot{a}_{k-1}\\
\alpha_k &= \alpha_{k-1} + \tfrac{a_{k-1}}{v_{k-1}}\Delta t_k + \tfrac{1}{2v_{k-1}}(\Delta t_k)^2,
\end{aligned} \tag{17.87}$$

where $\Delta t_k = t_k - t_{k-1}$. We linearize this system and augment the state vector with the speed v_k and acceleration a_k. Thus $\boldsymbol{x}_k = (x_k, y_k, \alpha_k, v_k, a_k)$. The three equations (17.87), with v_k and a_k constant, are $\boldsymbol{x}_k = F_{k-1}\boldsymbol{x}_{k-1} + \boldsymbol{\epsilon}_k$. Here ϵ_k includes the error of linearization and

$$F_{k-1}\boldsymbol{x}_{k-1} = \begin{bmatrix} 1 & 0 & f_{13} & \sin\alpha^0\Delta t_k & \frac{1}{2}\cos\alpha^0(\Delta t_k)^2\\ 0 & 1 & f_{23} & \cos\alpha^0\Delta t_k & -\frac{1}{2}\sin\alpha^0(\Delta t_k)^2\\ 0 & 0 & 1 & -a^0\Delta t_k/(v^0)^2 & \Delta t_k/v^0\\ 0 & 0 & 0 & 1 & 0\\ 0 & 0 & 0 & 0 & 1 \end{bmatrix} \begin{bmatrix} x_{k-1}\\ y_{k-1}\\ \alpha_{k-1}\\ v_{k-1}\\ a_{k-1} \end{bmatrix}.$$

We have introduced the following abbreviations

$$\begin{aligned}
f_{13} &= -\big(-\cos\alpha^0 v^0 + \tfrac{1}{2}\sin\alpha^0\Delta t_k\, a^0\big)\Delta t_k\\
f_{23} &= -\big(\sin\alpha^0 v^0 + \tfrac{1}{2}\cos\alpha^0\Delta t_k\, a^0\big)\Delta t_k.
\end{aligned}$$

For α^0, v^0, and a^0 we use the estimates $\hat{\alpha}_{k-1|k-1}$, $\hat{v}_{k-1|k-1}$, and $\hat{a}_{k-1|k-1}$.

Now we make the model random by introducing small random fluctuations in the motion. Obviously, sudden large shifts in position and velocity are not likely. So we model small random fluctuations into the acceleration. The influence of these fluctuations on the state vector is described through $\Sigma_{\epsilon,k}$.

Covariance matrix for the system errors For system noise, the first step is to introduce

$$G_k = \begin{bmatrix} \frac{1}{2}\sin\alpha^0(\Delta t_k)^2 & \frac{1}{6}\cos\alpha^0(\Delta t_k)^3\\ \frac{1}{2}\cos\alpha^0(\Delta t_k)^2 & -\frac{1}{6}\sin\alpha^0(\Delta t_k)^3\\ 0 & (\Delta t_k)^2/2v^0\\ \Delta t_k & 0\\ 0 & \Delta t_k \end{bmatrix}. \tag{17.88}$$

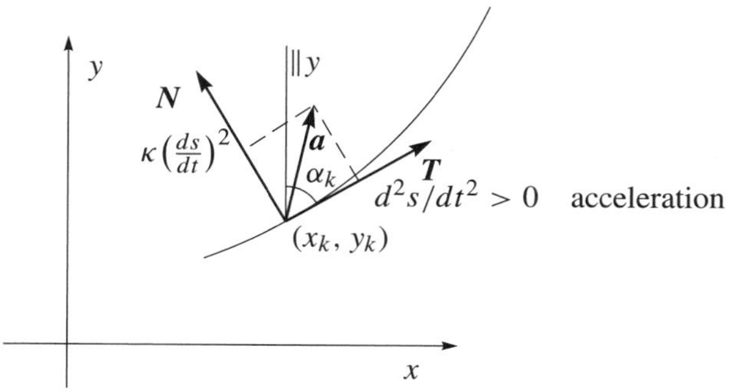

Figure 17.3 Dynamic model of the motion of a vessel. If $d^2s/dt^2 < 0$ we are dealing with a deceleration, and the tangent vector $\boldsymbol{T}$ changes direction.

G_k determines the influence of acceleration on the state vector $\boldsymbol{x}_k = (x_k, y_k, \alpha_k, v_k, a_k)$. The first column is for tangential acceleration (in the direction of motion). The second column comes from the normal acceleration (across the track). We shall describe this in detail. Any acceleration $\boldsymbol{a}$ can be split up into components along the track and across the track, cf. Strang (1991), page 462. Let $\boldsymbol{N}$ denote the normal vector and $\boldsymbol{T}$ the tangent vector and κ the curvature:

$$\boldsymbol{a} = \frac{d^2s}{dt^2}\boldsymbol{T} + \kappa\left(\frac{ds}{dt}\right)^2\boldsymbol{N}.$$

For a straight course the only acceleration is d^2s/dt^2. The term $\kappa(ds/dt)^2$ handles the acceleration in turning. Both have dimension length/(time)^2. The change in course is

$$d\alpha = \kappa\, ds = \frac{|\boldsymbol{a}|}{|\boldsymbol{v}|}dt. \tag{17.89}$$

The acceleration across the track may have a linear change in time.

Our procedure corresponds to the along track acceleration a_l and the across track acceleration a being uncorrelated random variables with zero mean value:

$$\Sigma_w = \begin{bmatrix} \sigma_{a_l}^2 & 0 \\ 0 & \sigma_a^2 \end{bmatrix} \qquad \text{and} \qquad \Sigma_{\epsilon,k} = G_k \Sigma_w G_k^{\mathrm{T}}. \tag{17.90}$$

In pure determination of position only a part of the state vector is used, e.g. only x_k and y_k. The observation equation becomes

$$\boldsymbol{b}_k = A_k \boldsymbol{x}_k + \boldsymbol{e}_k \qquad \text{with} \qquad A_k = \begin{bmatrix} 1 & 0 & 0 & 0 & 0 \\ 0 & 1 & 0 & 0 & 0 \end{bmatrix}.$$

Realistic values for the (uncorrelated) covariances in $\boldsymbol{x}_k = (x_k, y_k, \alpha_k, v_k, a_k)$ are $(0.5\,\mathrm{m})^2$, $(0.5\,\mathrm{m})^2$, $(0.0001\,\mathrm{rad})^2$, $(0.6\,\mathrm{m/s})^2$, and $(1.5\,\mathrm{m/s^2})^2$.

Example 17.5 We return to the steady model of Section 17.3 (*very* steady with no error ϵ in the scalar state equation). The observation errors are normally distributed with zero mean and variance σ_e^2. In other words we have

$$\begin{array}{rlll} \text{System equation:} & x_k = x_{k-1} & F_{k-1} = 1, \ \epsilon_k = 0 \\ \text{Observation equation:} & b_k = x_k + e_k & A_k = 1, \ e_k \sim N(0, \sigma_e^2). \end{array}$$

The covariance prediction is $\sigma_{k|k-1}^2 = \sigma_{k-1|k-1}^2$. (We write σ^2 for P.) The Bayes update is

$$\sigma_{k|k}^2 = \sigma_{k|k-1}^2 - \sigma_{k|k-1}^2\big(\sigma_{k|k-1}^2 + \sigma_e^2\big)^{-1}\sigma_{k|k-1}^2 = \frac{\sigma_{k|k-1}^2}{1 + \sigma_{k|k-1}^2/\sigma_e^2} = \frac{\sigma_{k-1|k-1}^2}{1 + \sigma_{k-1|k-1}^2/\sigma_e^2}.$$

Starting with $\sigma_{0|0}^2 = \sigma_0^2$ we solve the difference equation recursively to get

$$\sigma_{k|k}^2 = \frac{\sigma_0^2}{1 + (\sigma_0^2/\sigma_e^2)k}. \tag{17.91}$$

Thus the updated state vector is

$$x_{k|k} = x_{k|k-1} + \frac{\sigma_0^2}{1 + (\sigma_0^2/\sigma_e^2)k}\big(b_k - x_{k|k-1}\big). \tag{17.92}$$

For sufficiently large k we have $x_{k|k} \approx x_{k|k-1}$; this agrees well with the fact that new observations do not contribute much further information.

Example 17.6 Now we add an error ϵ_k to this steady model:

$$\begin{array}{rl} \text{System equation:} & x_k = x_{k-1} + \epsilon_k \\ \text{Observation equation:} & b_k = x_k + e_k. \end{array}$$

The presence of the error term ϵ_k changes drastically the behavior of the filter.

We use simulated data with normally distributed errors (uncorrelated). For demonstrational purpose the data are subdivided into epochs 1–25, 26–50, 51–75, 76–100, and 101–125. The observation variance is $\sigma_{e,k}^2 = 1$ for epochs 1–75 and $\sigma_{e,k}^2 = 100$ for epochs 76–125. The system variances are $\sigma_{\epsilon,k}^2 = 100$ for epochs 1–50 and $\sigma_{\epsilon,k}^2 = 0.5$ for the rest.

According to Table 17.2 the gain factor K (here a scalar) is

$$K_k = \frac{\sigma_{k|k-1}^2}{\sigma_{k|k-1}^2 + \sigma_{e,k}^2} = \frac{\sigma_{k-1|k-1}^2 + \sigma_{\epsilon,k}^2}{\sigma_{k-1|k-1}^2 + \sigma_{\epsilon,k}^2 + \sigma_{e,k}^2}$$

$$\sigma_{k|k}^2 = (1 - K_k)\big(\sigma_{k-1|k-1}^2 + \sigma_{\epsilon,k}^2\big).$$

The gain factor K_k is only large when the observation variance $\sigma_{e,k}^2$ is small. Then the observation b_k is weighted heavily; this makes the predicted estimate depend primarily on the latest observation. Figure 17.4 shows the results of the filtering.

Prediction curves are identical (because $F_{k-1} = 1$) to the graphs of $x_{k-1|k-1}$, except they are shifted one epoch. The calculation starts with $x_0 = 25$ and $\sigma_{0|0}^2 = 0.001$. (If no information about x_0 were available we would set $\sigma_{0|0}^2 = \infty$ and consequently $K_0 = 1$.)

In epochs 1–25 the system variance $\sigma_{\epsilon,k}^2 = 100$ is large compared to $\sigma_{e,k}^2 = 1$. This causes the gain factor K_k to be close to unity which again implies that the current obser-

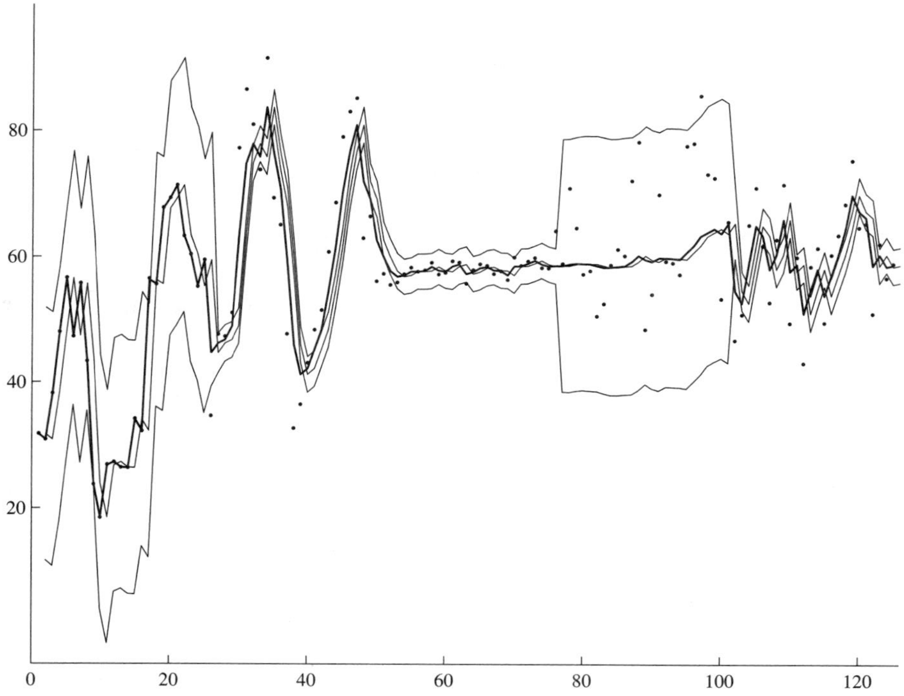

Figure 17.4 Filtering of a simulated steady model. The observations are marked with dots. The thick line indicates the filtered values $x_{k-1|k-1}$, the thin line the predicted values $x_{k|k-1} = x_{k-1|k-1}$. The two outer lines indicate a confidence band calculated from the predicted filter by adding and subtracting $2\sigma_{e,k}$. Initial values $x_0 = 25$ and $\sigma^2_{0|0} = 0.001$.

vation b_k determines the state x_k. The filtering variance $\sigma^2_{k|k}$ equals unity and expresses the uncertainty of $x_{k|k-1}$. The observational variance is small and the state values fluctuate quite a bit. All observations lie inside the confidence band.

Epochs 26–50 have $\sigma^2_{\epsilon,k} = 0.5$ and $\sigma^2_{e,k} = 1$. *The Kalman filter has a wrongly specified value* 100 *for* $\sigma^2_{\epsilon,k}$. The confidence band becomes very narrow. The gain factor drops to 0.5 as $\sigma^2_{\epsilon,k} = 0.5$ is slightly smaller than $\sigma^2_{e,k} = 1$. Now b_k and $x_{k|k-1}$ are equally weighted and the predictions are lagging one step behind the observed values.

The predicted values are as if the state hardly changes. The wrong specification largely influences this prediction. An earlier state is weighted too heavily. A very large weight K_k for b_k would be preferable due to the large uncertainty in x_k.

The filtering variance converges quickly to 0.5. The state estimates are more accurate than for epochs 1–25. The state variance σ^2_ϵ is smaller so the filtering variance σ^2_k is calculated as if the state is fairly constant, which obviously is not the case.

Epochs 51–75 have $\sigma^2_{\epsilon,k} = 0.5$ and $\sigma^2_{e,k} = 1$ and the gain factor converges to 0.5. The predicted values are close to the observations. Due to low observational variance they vary only a little. Small system variances cause fairly constant state values.

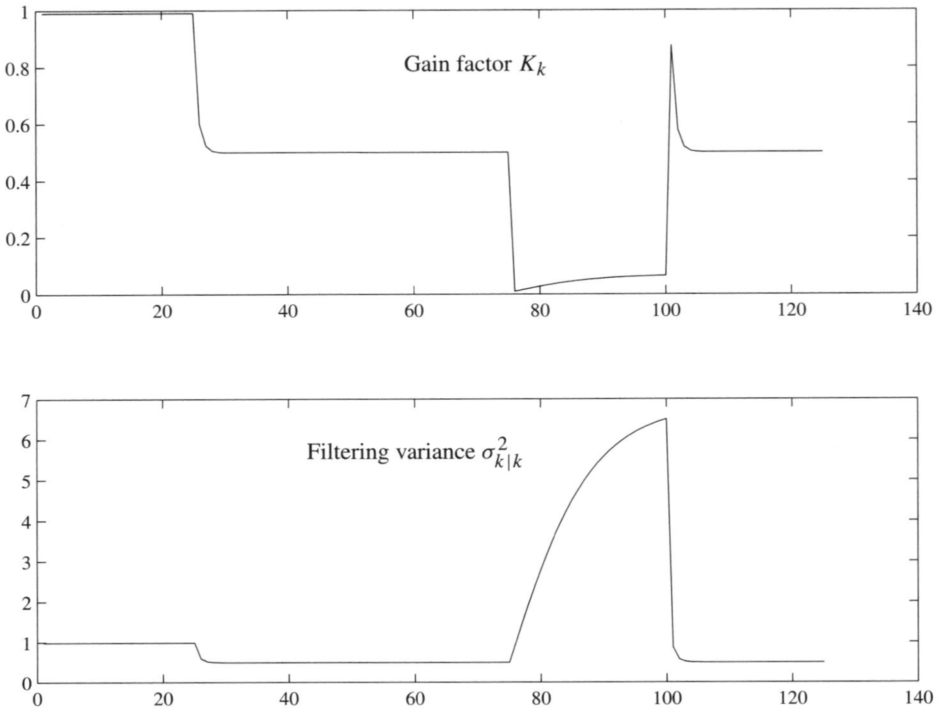

Figure 17.5 Gain factor K_k and filtering variance $\sigma^2_{k|k}$ for the model in Figure 17.4.

Epochs 76–100 have $\sigma^2_{\epsilon,k} = 0.5$ and $\sigma^2_{e,k} = 100$. This yields a small gain factor. The predictions are fairly constant with wide band limits because $x_{k|k-1}$ is weighted heavily compared to b_k. The predictions are not affected by the observations. The filtering variance converges to a large value in accordance with the values $\sigma^2_{\epsilon,k} = 0.5$ and $\sigma^2_{e,k} = 100$.

The state level becomes more and more uncertain as time goes by. The observations fluctuate largely and thus become bad for predicting the state level.

Epochs 101–125 have $\sigma^2_{\epsilon,k} = 0.5$ and $\sigma^2_{e,k} = 1$. *In the Kalman filter $\sigma^2_{e,k}$ is wrongly specified to 100.* The gain factor converges to 0.5. Due to the large value for $\sigma^2_{e,k}$ a small gain would have been preferable. This would have weighted the observation very little.

Due to the wrongly specified variances for epochs 26–50 and 101-125 the bandwidth is too narrow in these periods leaving many observations outside the band limits.

Figure 17.4 shows that when the Kalman filter is applied for a time series all observations are situated within the 95% band of confidence. The band is determined by $x_{k|k} \pm 2\sqrt{q_k}$ where $q_k = \sigma^2_{k-1|k-1} + \sigma^2_{\epsilon,k} + \sigma^2_{e,k}$. The large values for $\sigma^2_{\epsilon,k}$ and $\sigma^2_{e,k}$ give rise to a large q_k making the confidence band wider around the observations (and more uncertainty).

The M-file k_simul produces Figures 17.4 and 17.5. The example demonstrates the importance of correct variances when a filter is used for prediction and monitoring.

Remark 17.4 In most applications the observation vector $\boldsymbol{b}_k$ has a *diagonal* covariance matrix Σ_e. The individual components of b_k are uncorrelated. *Diagonal matrices give m scalar equations.* So it may be advantageous to filter the components of b_k as individual observations. Treating the problem this way leads to

1 improved numerical properties. Usually the filter must invert $A_k\Sigma_{k|k-1}A_k^{\mathrm{T}} + \Sigma_{e,k}$. This matrix inversion is replaced by m scalar divisions.

2 improved computational time. The vector implementation time grows proportionally to m^3 whereas the row implementation has a growth proportional to m.

The implementation involves an extra loop over the rows of the A matrix. We quote the essential part of the Kalman filter version:

```
% Coefficient matrix for observations
beta = (f1/f2)^2;  Big = 10^10;
A(1,:) = [1  1  0  0];
A(2,:) = [1  -1  lambda1  0];
A(3,:) = [1  beta  0  0];
A(4,:) = [1 -beta  0 lambda2];

for k = 1:number_of_satellites-1;
  ...
  b = [P1  Phi1  P2  Phi2];   % Double differenced observations
  x = [];
  x_plus = ... ;          % Initialization of state vector
  Sigma_plus = eye(4) * Big;  Q = Big;
  var = [.3^2 .005^2 .3^2 .005^2]; % Variance of observations
  for i = first_epoch:last_epoch       % Kalman filter
    Sigma_plus(1,1) = Sigma_plus(1,1) + Q;
    Sigma_plus(2,2) = Sigma_plus(2,2) + Q;
    for j = 1:4
      x_minus = x_plus;
      Sigma_minus = Sigma_plus;
      A_row = A(j,:);
      SigmaA_rowt = Sigma_minus * A_row';
      Innovation_variance = A_row * SigmaA_rowt + var(j);
      K = SigmaA_rowt/Innovation_variance;
      omc = b(i,j) - A_row * x_minus;
      x_plus = x_minus + K * omc;
      Sigma_plus = Sigma_minus - K * SigmaA_rowt';
    end;
    x = [x x_plus];
  end
end
```

Table 17.3 Kalman filter for point position

PRN	23	9	5	1	21	17
innovation [m]	−4.72	−11.82	0.33	3.15	−88.63	−1.11
X [m]	2.79	5.66	−4.15	14.77	51.68	14.72
Y [m]	−2.87	5.15	0.34	−6.89	−85.45	−59.97
Z [m]	1.67	9.59	18.12	30.89	139.55	46.34
$c\,dt$ [m]	−1.30	−2.63	−2.06	70.92	372.72	112.34
PDOP []	1443.1	1046.6	311.7	4.8	4.1	2.9

The row-versions of the Kalman and the Bayes filters are implemented in the M-files k_row and b_row.

Example 17.7 Example 15.7 used six pseudoranges to locate the receiver. The extra two pseudoranges were used in a least-squares manner to improve the calculation.

With the filter technique at hand it is natural to find a filter based solution to the problem. We want to state the right side b_i of the observation equation—or the innovation:

$$\text{innovation} = (P - \text{tropo}) - \left(\rho_i^k + c(dt_i - dt^k)\right). \tag{17.93}$$

For independent pseudorange observations, we update after each observation:

M-file: kud

```
function [x,P] = kud(x,P,H,b,var)
% KUD  Kalman update, one measurement per call
innovation = b-H'*x;
HP = H'*P;
innovation_variance = HP*H + var;
K = HP'/innovation_variance;
x = x + K*innovation;
P = P-K*HP;
```

The data from Example 15.7 yield the results in Table 17.3. Note that the fourth satellite drastically decreases the value of PDOP. This is because the fourth satellite is necessary for a solution of X, Y, Z, and $c\,dt$. Afterwards PDOP decreases much more slowly!

Example 17.8 A Bayes filter adds on to the normal equations one observation per call. *The normals are only solved at the end of input.* The following simple function is the core:

M-file: normals

```
function [AtA,Atb] = normals(AtA,Atb,H,innovation,var)
% NORMALS One observation equation is added to the coefficient
%          matrix AtA and the right side Atb.
Atb = Atb + H*innovation/var;
AtA = AtA + H*H'/var;
```

The final change in estimates is $(x, y, z, c\,dt) = (14.7, -60.0, 46.3, 112.3)$ with a standard deviation of 26 m. The implementation can also be studied in the M-file k_point.

Example 17.9 The least-squares problem in Example 11.11 was solved by the normal equations. It may as well be solved by means of a filter. The final $\hat{x}_{3|3} = \hat{x}$ must agree with $\hat{x}$ from the normal equations. The code is contained in the M-file wc.

The three observations have weights 1, 2, and $1 + \delta C_2$. Below we quote results from the filter process ($*$ denotes a large number). For $\delta C_2 = 0$ the final $\hat{x}_{3|3} = \hat{x}$ agrees with $\hat{x} = (2/3, 1/3)$:

$$\hat{x}_{1|1} = \begin{bmatrix} 1 \\ 1 \end{bmatrix} \qquad \hat{x}_{2|2} = \begin{bmatrix} 3 \\ -1 \end{bmatrix} \qquad \hat{x}_{3|3} = \begin{bmatrix} \frac{2}{3} \\ \frac{1}{3} \end{bmatrix}$$

$$P_{1|1} = \begin{bmatrix} * & * \\ * & * \end{bmatrix} \qquad P_{2|2} = \begin{bmatrix} \frac{9}{2} & -\frac{5}{2} \\ -\frac{5}{2} & \frac{3}{2} \end{bmatrix} \qquad P_{3|3} = \begin{bmatrix} \frac{5}{12} & -\frac{1}{6} \\ -\frac{1}{6} & \frac{1}{6} \end{bmatrix}.$$

For $\delta C_2 = -1$ (which is the same as omitting the last observation) we find

$$\hat{x}_{1|1} = \begin{bmatrix} 1 \\ 1 \end{bmatrix} \qquad \hat{x}_{2|2} = \begin{bmatrix} 3 \\ -1 \end{bmatrix} \qquad \hat{x}_{3|3} = \begin{bmatrix} 3 \\ -1 \end{bmatrix}$$

$$P_{1|1} = \begin{bmatrix} * & * \\ * & * \end{bmatrix} \qquad P_{2|2} = \begin{bmatrix} \frac{9}{2} & -\frac{5}{2} \\ -\frac{5}{2} & \frac{3}{2} \end{bmatrix} \qquad P_{3|3} = \begin{bmatrix} \frac{9}{2} & -\frac{5}{2} \\ -\frac{5}{2} & \frac{3}{2} \end{bmatrix}.$$

Finally for $\delta C_2 = \infty$ we still have $\hat{x}_{3|3} = \hat{x}$ (and notice the new $P_{3|3}$):

$$\hat{x}_{1|1} = \begin{bmatrix} 1 \\ 1 \end{bmatrix} \qquad \hat{x}_{2|2} = \begin{bmatrix} 3 \\ -1 \end{bmatrix} \qquad \hat{x}_{3|3} = \begin{bmatrix} \frac{5}{11} \\ \frac{5}{11} \end{bmatrix}$$

$$P_{1|1} = \begin{bmatrix} * & * \\ * & * \end{bmatrix} \qquad P_{2|2} = \begin{bmatrix} \frac{9}{2} & -\frac{5}{2} \\ -\frac{5}{2} & \frac{3}{2} \end{bmatrix} \qquad P_{3|3} = \begin{bmatrix} \frac{1}{22} & \frac{1}{22} \\ \frac{1}{22} & \frac{1}{22} \end{bmatrix}.$$

We finish this chapter by mentioning some M-files that solve earlier least-squares problems but now by means of Kalman filters.

An important fact to note is that ***setting the measurement error equal to zero in a Kalman filter is the same as a constraint in a standard least-squares formulation.***

Example 17.10 The M-file fixing1 demonstrates all the computations in Examples 12.1, 12.2, and 12.3 in a filter version.

```
%FIXING1  Filter version of Examples 12.1, 12.2, and 12.3
%         Shows the impact on introducing a constraint with
%         zero variance for the observation

big = 1.e6;
A = [-1 0 0 1 0; -1 0 0 0 1; 0 0 -1 1 0; 0 -1 0 0 1; 0 0 0 1 -1];
b = [1.978; 0.732; 0.988; 0.420; 1.258];
Cov = eye(5);
x = zeros(5,1);
P = big * eye(5);
```

```
% Regular update
for i = 1:5
  [x,P] = k_update(x,P,A(i,:),b(i),Cov(i,i))
  pause
end

% Update with constraint with variance one
A_aug = [1 1 1 1 1];
b_aug = 100;
Cov_aug = 1;
[x,P] = k_update(x,P,A_aug,b_aug,Cov_aug)
Sigma = (norm(b-A * x))^2 * P
pause

% Update with constraint with variance zero
Cov_aug = 0;
[x,P] = k_update(x,P,A_aug,b_aug,Cov_aug)
Sigma_plus = (norm(b-A * x))^2 * P
```

Example 17.11 The *M*-file fixing2 demonstrates all computations in Examples 12.4, and 12.7 in a filter version.

```
% FIXING2  Filter version of Examples 12.4 and 12.7.
%          Shows the impact on introducing constraints
%          as observations with zero variance

format short
A = [-1 0 1 0 0 0 0 0;
      0.707 0.707 0 0 -0.707 -0.707 0 0
         0.707 -0.707 0 0 0 0 -0.707 0.707;
      0 0 0.924 0.383 -0.924 -0.383 0 0;
      0 0 0 0 0 -1 0 1;
      0 0 0.924 -0.383 0 0 -0.924 0.383];
b = [0.01; 0.02; 0.03; 0.01; 0.02; 0.03];
Cov = eye(6);
x = zeros(8,1);
P = 1.e6 * eye(8);

% Regular update
for i = 1:6
  [x,P] = k_update(x,P,A(i,:),b(i),Cov(i,i));
  for j = 1:8
    fprintf('%8.4f',x(j))
  end
  fprintf('\n')
end
fprintf('\n')
```

```
% Update with constraints, that is observations with variance zero
G = [1  0  1  0  1  0  1  0;
     0  1  0  1  0  1  0  1;
     -170.71  170.71  -170.71  270.71  -100  100  -241.42  100;
     170.71 170.71  270.71 170.71  100 100  100   241.42]';
% Baarda theory
Sp = [zeros(4,8); -G(5:8,:) * inv(G(1:4,:)) eye(4)];
Cov = (norm(b-A * x))^2 * pinv(A' * A);
Sigma_xp = Sp * Cov * Sp';

% Fixing the first four coordinates to zero values
H = zeros(1,8);
b_fix = 0;
Cov_fix = 0;
for i = 1:4
  H(i) = 1;
  [x,P] = k_update(x,P,H,b_fix,Cov_fix);
  fprintf('\n')
  for j = 1:8
    fprintf('%8.4f',x(j))
  end
end
fprintf('\n')

Sigma = (norm(b-A * x))^2 * P;
```

THE RECEIVER INDEPENDENT EXCHANGE FORMAT

Werner Gurtner
Astronomical Institute
University of Berne

0 Introduction

This paper* is a revised version of the one published by W. Gurtner and G. Mader in the CSTG GPS Bulletin of September/October 1990. The main reason for a revision is the new treatment of antispoofing data by the RINEX format (see chapter 7). Chapter 4 gives a recommendation for data compression procedures, especially useful when large amounts of data are exchanged through computer networks. In Table A3 in the original paper the definition of the `PGM / RUN BY / DATE` navigation header record was missing, although the example showed it. The redefinition of `AODE/AODC` to `IODE/IODC` also asks for an update of the format description. For consistency reasons we also defined a Version 2 format for the Meteorological Data files (inclusion of a `END OF HEADER` record and an optional `MARKER NUMBER` record).†

In order to have all the available information about RINEX in one place we also included parts of earlier papers and a complete set of format definition tables and examples.

URA Clarification (10-Dec-93) The user range accuracy in the Navigation Message file did not contain a definition of the units: There existed two ways of interpretation: Either the 4 bit value from the original message or the converted value in meters according to GPS ICD-200. In order to simplify the interpretation for the user of the RINEX files I propose the bits to be converted into meters prior to RINEX file creation.

1 The Philosophy of RINEX

The first proposal for the "Receiver Independent Exchange Format" RINEX has been developed by the Astronomical Institute of the University of Berne for the easy exchange of

*Version 2 has the following development: Revision, April 1993; Clarification December 1993; Doppler Definition: Jan. 1994; PR Clarification: Oct. 1994; Wlfact Clarification: Feb. 1995; Event Time Frame Clarification: May 1996; Minor errors in the examples A7/A8: May 1996.

†The slight modification (or rather the definition of a bit in the Loss of Lock Indicator unused so far) to flag AS data is so small a change that we decided to NOT increase the version number!

the GPS data to be collected during the large European GPS campaign EUREF 89, which involved more than 60 GPS receivers of 4 different manufacturers. The governing aspect during the development was the following fact:

Most geodetic processing software for GPS data use a well-defined set of observables:

- the carrier-phase measurement at one or both carriers (actually being a measurement on the beat frequency between the received carrier of the satellite signal and a receiver-generated reference frequency).
- the pseudorange (code) measurement, equivalent to the difference of the time of reception (expressed in the time frame of the receiver) and the time of transmission (expressed in the time frame of the satellite) of a distinct satellite signal.
- the observation time being the reading of the receiver clock at the instant of validity of the carrier-phase and/or the code measurements.

Usually the software assumes that the observation time is valid for both the phase *and* the code measurements, *and* for all satellites observed.

Consequently all these programs do not need most of the information that is usually stored by the receivers: They need phase, code, and time in the above mentioned definitions, and some station-related information like station name, antenna height, etc.

2 General Format Description

Currently the format consists of three ASCII file types:

1. Observation Data file
2. Navigation Message file
3. Meteorological Data file

Each file type consists of a header section and a data section. The header section contains global information for the entire file and is placed at the beginning of the file. The header section contains header labels in columns 61–80 for each line contained in the header section. These labels are mandatory and must appear exactly as given in these descriptions and examples.

The format has been optimized for mimimum space requirements independent from the number of different observation types of a specific receiver by indicating in the header the types of observations to be stored. In computer systems allowing variable record lengths the observation records may then be kept as short as possible. The maximum record length is 80 bytes per record.

Each Observation file and each Meteorological Data file basically contain the data from one site and one session. RINEX Version 2 also allows to include observation data from more than one site subsequently occupied by a roving receiver in rapid static or kinematic applications.

If data from more than one receiver has to be exchanged it would not be economical to include the identical satellite messages collected by the different receivers several times. Therefore the Navigation Message file from one receiver may be exchanged or a composite Navigation Message file created containing non-redundant information from several receivers in order to make the most complete file.

The format of the data records of the RINEX Version 1 Navigation Message file is identical to the former NGS exchange format.

The actual format descriptions as well as examples are given in the Tables at the end of the paper.

3 Definition of the Observables

GPS observables include three fundamental quantities that need to be defined: Time, Phase, and Range.

Time The time of the measurement is the receiver time of the received signals. It is identical for the phase and range measurements and is identical for all satellites observed at that epoch. It is expressed in GPS time (not Universal Time).

Pseudo-Range The pseudo-range (PR) is the distance from the receiver antenna to the satellite antenna including receiver and satellite clock offsets (and other biases, such as atmospheric delays):

$$\text{PR} = \text{distance} + c * (\text{receiver clock offset} - \text{satellite clock offset} + \text{other biases})$$

so that the pseudo-range reflects the actual behavior of the receiver and satellite clocks. The pseudo-range is stored in units of meters.

Phase The phase is the carrier-phase measured in whole cycles at both L_1 and L_2. The half-cycles measured by squaring-type receivers must be converted to whole cycles and flagged by the wavelength factor in the header section.

The phase changes in the same sense as the range (negative doppler). The phase observations between epochs must be connected by including the integer number of cycles. The phase observations will not contain any systematic drifts from intentional offsets of the reference oscillators.

The observables are not corrected for external effects like atmospheric refraction, satellite clock offsets, etc.

If the receiver or the converter software adjusts the measurements using the real-time-derived receiver clock offsets dT(r), the consistency of the 3 quantities phase/pseudo-range/epoch must be maintained, i.e. the receiver clock correction should be applied to all

3 observables:

$$\text{Time(corr)} = \text{Time(r)} - \text{dT(r)}$$
$$\text{PR(corr)} = \text{PR(r)} - \text{dT(r)} * c$$
$$\text{phase(corr)} = \text{phase(r)} - \text{dT(r)} * \text{freq}$$

Doppler The sign of the doppler shift as additional observable is defined as usual: Positive for approaching satellites.

4 The Exchange of RINEX Files

We recommend using the following naming convention for RINEX files:

ssssdddf.yyt	ssss:	4-character station name designator
	ddd:	day of the year of first record
	f:	file sequence number within day 0 : file contains all the existing data of the current day
	yy:	year
	t:	file type: O : Observation file N : Navigation file M : Meteorological data file

To exchange RINEX files on magnetic tapes we recommend using the following tape format:

– Non-label; ASCII; fixed record length: 80 characters; block size: 8000

– First file on tape contains list of files using above-mentioned naming conventions

When data transmission times or storage volumes are critical we recommend compressing the files prior to storage or transmission using the UNIX "compress" and "uncompress" programs. Compatible routines are available on VAX/VMS and PC/DOS systems, as well.

Proposed naming conventions for the compressed files:

System	Observation files	Navigation files
UNIX	ssssdddf.yyO.Z	ssssdddf.yyN.Z
VMS	ssssdddf.yyO_Z	ssssdddf.yyN_Z
DOS	ssssdddf.yyY	ssssdddf.yyX

5 RINEX Version 2 Features

The following section contains features that have been introduced for RINEX Version 2.

5.1 Satellite Numbers

Version 2 has been prepared to contain GLONASS or other satellite systems' observations. Therefore we have to be able to distinguish the satellites of the different systems: We precede the 2-digit satellite number with a system denominator.

snn	s:	satellite system
		blank : system as defined in header record
		G : GPS
		R : GLONASS
		T : Transit
	nn:	PRN (GPS), almanac number (GLONASS) or two-digit satellite number

Note: G, R, and T are mandatory in mixed files.

5.2 Order of the Header Records

As the record descriptors in columns 61–80 are mandatory, the programs reading a RINEX Version 2 header are able to decode the header records with formats according to the record descriptor, provided the records have been first read into an internal buffer.

We therefore propose to allow free ordering of the header records, with the following exceptions:

- The `RINEX VERSION / TYPE` record must be the first record in a file
- The default `WAVELENGTH FACT L1/2` record (if present) should precede all records defining wavelength factors for individual satellites
- The `# OF SATELLITES` record (if present) should be immediately followed by the corresponding number of `PRN / # OF OBS` records. (These records may be handy for documentary purposes.) However, since they may only be created after having read the whole raw data file we define them to be optional.

5.3 Missing Items, Duration of the Validity of Values

Items that are not known at the file creation time can be set to zero or blank or the respective record may be completely omitted. Consequently items of missing header records will be set to zero or blank by the program reading RINEX files. Each value remains valid until changed by an additional header record.

5.4 Event Flag Records

The "number of satellites" also corresponds to the number of records of the same epoch followed. Therefore it may be used to skip the appropriate number of records if certain event flags are not to be evaluated in detail.

5.5 Receiver Clock Offset

A large number of users asked to optionally include a receiver-derived clock offset into the RINEX format. In order to prevent confusion and redundancy, the receiver clock offset (if present) should report the value that has been used to correct the observables according to the formulae under item 1. It would then be possible to reconstruct the original observations if necessary. As the output format for the receiver-derived clock offset is limited to nanoseconds the offset should be rounded to the nearest nanosecond before it is used to correct the observables in order to guarantee correct reconstruction.

6 Additional Hints and Tips

Programs developed to read RINEX Version 1 files have to verify the version number. Version 2 files may look different (version number, `END OF HEADER` record, receiver and antenna serial number alphanumeric) even if they do not use any of the new features.

We propose that routines to read RINEX Version 2 files automatically delete leading blanks in any CHARACTER input field. Routines creating RINEX Version 2 files should also left-justify all variables in the CHARACTER fields.

DOS, and other, files may have variable record lengths, so we recommend to first read each observation record into a 80-character blank string and decode the data afterwards. In variable length records, empty data fields at the end of a record may be missing, especially in the case of the optional receiver clock offset.

7 RINEX Under Antispoofing (AS)

Some receivers generate code delay differences between the first and second frequency using cross-correlation techniques when AS is on and may recover the phase observations on L_2 in full cycles. Using the C/A code delay on L_1 and the observed difference it is possible to generate a code delay observation for the second frequency.

Other receivers recover P code observations by breaking down the Y code into P and W code.

Most of these observations may suffer from an increased noise level. In order to enable the postprocessing programs to take special actions, such AS-infected observations are flagged using bit number 2 of the Loss of Lock Indicators (i.e. their current values are increased by 4).

8 References

Evans, A. (1989): *Summary of the Workshop on GPS Exchange Formats*. Proceedings of the Fifth International Geodetic Symposium on Satellite Systems, pp. 917ff, Las Cruces.

Gurtner, W., G. Mader, D. Arthur (1989): *A Common Exchange Format for GPS Data.* CSTG GPS Bulletin Vol. 2 No. 3, May/June 1989, National Geodetic Survey, Rockville.

Gurtner, W., G. Mader (1990): *The RINEX Format: Current Status, Future Developments.* Proceedings of the Second International Symposium of Precise Positioning with the Global Positioning System, pp. 977ff, Ottawa.

Gurtner, W., G. Mader (1990): *Receiver Independent Exchange Format Version 2.* CSTG GPS Bulletin Vol. 3 No. 3, Sept/Oct 1990, National Geodetic Survey, Rockville.

9 RINEX Version 2 Format Definitions and Examples

TABLE A1

OBSERVATION DATA FILE - HEADER SECTION DESCRIPTION

	HEADER LABEL (Columns 61-80)	DESCRIPTION	FORMAT	
	RINEX VERSION / TYPE	- Format version (2) - File type ('O' for Observation Data) - Satellite System: blank or 'G': GPS 'R': GLONASS 'T': NNSS Transit 'M': Mixed	I6,14X, A1,19X, A1,19X	
	PGM / RUN BY / DATE	- Name of program creating current file - Name of agency creating current file - Date of file creation	A20, A20, A20	
*	COMMENT	Comment line(s)	A60	*
	MARKER NAME	Name of antenna marker	A60	
*	MARKER NUMBER	Number of antenna marker	A20	*
	OBSERVER / AGENCY	Name of observer / agency	A20,A40	
	REC # / TYPE / VERS	Receiver number, type, and version (Version: e.g. Internal Software Version)	3A20	
	ANT # / TYPE	Antenna number and type	2A20	
	APPROX POSITION XYZ	Approximate marker position (WGS84)	3F14.4	
	ANTENNA: DELTA H/E/N	- Antenna height: Height of bottom surface of antenna above marker - Eccentricities of antenna center relative to marker to the east and north (all units in meters)	3F14.4	
	WAVELENGTH FACT L1/2	- Wavelength factors for L1 and L2 1: Full cycle ambiguities	2I6,	

```
 |                    |   2:  Half cycle ambiguities (squaring)  |            |
 |                    |   0 (in L2): Single frequency instrument |            |
 |                    | - Number of satellites to follow in list |     I6,    |
 |                    |   0 or blank: Default wavelength factors |            |
 |                    |   Maximum 7. If more than 7 satellites:  |            |
 |                    |   Repeat record.                         |            |
 |                    | - List of PRNs (satellite numbers)       | 7(3X,A1,I2)|
 +--------------------+------------------------------------------+------------+
 |# / TYPES OF OBSERV | - Number of different observation types  |     I6,    |
 |                    |   stored in the file                     |            |
 |                    | - Observation types                      |   9(4X,A2) |
 |                    |                                          |            |
 |                    | The following observation types are      |            |
 |                    | defined in RINEX Version 2:              |            |
 |                    |                                          |            |
 |                    | L1, L2: Phase measurements on L1 and L2  |            |
 |                    | C1    : Pseudorange using C/A-Code on L1 |            |
 |                    | P1, P2: Pseudorange using P-Code on L1,L2|            |
 |                    | D1, D2: Doppler frequency on L1 and L2   |            |
 |                    | T1, T2: Transit Integrated Doppler on    |            |
 |                    |         150 (T1) and 400 MHz (T2)        |            |
 |                    |                                          |            |
 |                    | Observations collected under Antispoofing|            |
 |                    | are converted to "L2" or "P2" and flagged|            |
 |                    | with bit 2 of loss of lock indicator     |            |
 |                    | (see Table A2).                          |            |
 |                    |                                          |            |
 |                    | Units : Phase       : full cycles        |            |
 |                    |         Pseudorange : meters             |            |
 |                    |         Doppler     : Hz                 |            |
 |                    |         Transit     : cycles             |            |
 |                    |                                          |            |
 |                    | The sequence of the types in this record |            |
 |                    | has to correspond to the sequence of the |            |
 |                    | observations in the observation records  |            |
 +--------------------+------------------------------------------+------------+
*|INTERVAL            | Observation interval in seconds          |     I6     |*
 +--------------------+------------------------------------------+------------+
 |TIME OF FIRST OBS   | Time of first observation record         | 5I6,F12.6  |
 |                    |   year (4 digits), month,day,hour,min,sec|            |
 +--------------------+------------------------------------------+------------+
*|TIME OF LAST OBS    | Time of last  observation record         | 5I6,F12.6  |*
 |                    |   year (4 digits), month,day,hour,min,sec|            |
 +--------------------+------------------------------------------+------------+
*|# OF SATELLITES     | Number of satellites, for which          |     I6     |*
 |                    | observations are stored in the file      |            |
 +--------------------+------------------------------------------+------------+
*|PRN / # OF OBS      | PRN (sat.number), number of observations |3X,A1,I2,9I6|*
 |                    | for each observation type indicated      |            |
 |                    | in the "# / TYPES OF OBSERV" - record.   |            |
 |                    | This record is repeated for each         |            |
 |                    | satellite present in the data file       |            |
 +--------------------+------------------------------------------+------------+
 |END OF HEADER       | Last record in the header section.       |    60X     |
 +--------------------+------------------------------------------+------------+

                       Records marked with * are optional
```

```
+----------------------------------------------------------------------------+
|                                 TABLE A2                                   |
|                OBSERVATION DATA FILE - DATA RECORD DESCRIPTION             |
+-------------+--------------------------------------------------+------------+
| OBS. RECORD | DESCRIPTION                                      |   FORMAT   |
+-------------+--------------------------------------------------+------------+
| EPOCH/SAT   | - Epoch :                                        | 5I3,F11.7, |
|     or      |    year (2 digits), month,day,hour,min,sec       |            |
| EVENT FLAG  | - Epoch flag 0: OK                               |    I3,     |
|             |              1: power failure between            |            |
|             |                 previous and current epoch       |            |
|             |             >1: Event flag                       |            |
|             | - Number of satellites in current epoch          |    I3,     |
|             | - List of PRNs (sat.numbers) in current epoch    | 12(A1,I2), |
|             |   If more than 12 satellites: Continued in       |            |
|             |   next line with n(A1,I2)                        |            |
|             | - receiver clock offset (seconds, optional)      |   F12.9    |
|             |                                                  |            |
|             | If EVENT FLAG record (epoch flag > 1):           |            |
|             |   - Event flag:                                  |            |
|             |     2: start moving antenna                      |            |
|             |     3: new site occupation (end of kinem. data)  |            |
|             |        (at least MARKER NAME record follows)     |            |
|             |     4: header information follows                |            |
|             |     5: external event (epoch is significant,     |            |
|             |        same time frame as observation time tags) |            |
|             |     6: cycle slip records follow to optionally   |            |
|             |        report detected and repaired cycle slips  |            |
|             |        (same format as OBSERVATIONS records;     |            |
|             |         slip instead of observation; LLI and     |            |
|             |         signal strength blank)                   |            |
|             |   - "Number of satellites" contains number of    |            |
|             |     records to follow (0 for event flags 2,5)    |            |
+-------------+--------------------------------------------------+------------+
|OBSERVATIONS | - Observation      | rep. within record for      |  m(F14.3,  |
|             | - LLI              | each obs.type (same seq     |     I1,    |
|             | - Signal strength  | as given in header)         |     I1)    |
|             | This record is repeated for each satellite       |            |
|             | given in EPOCH/SAT - record.                     |            |
|             | If more than 5 observation types (=80 char):     |            |
|             | Continue observations in next record.            |            |
|             |                                                  |            |
|             | Observations:                                    |            |
|             |   Phase  : Units in whole cycles of carrier      |            |
|             |   Code   : Units in meters                       |            |
|             | Missing observations are written as 0.0          |            |
|             | or blanks.                                       |            |
|             |                                                  |            |
|             | Loss of lock indicator (LLI). Range: 0-7         |            |
|             |  0 or blank: OK or not known                     |            |
|             |  Bit 0 set : lost lock between previous and      |            |
|             |     current observation: cycle slip possible     |            |
|             |  Bit 1 set : Inverse wavelength factor to        |            |
|             |     default (does NOT change default)            |            |
|             |  Bit 2 set : observation under Antispoofing      |            |
|             |              (may suffer from increased noise)   |            |
|             |                                                  |            |
```

	Bits 0 and 1 for phase only. Signal strength projected into interval 1-9: 1: minimum possible signal strength 5: threshold for good S/N ratio 9: maximum possible signal strength 0 or blank: not known, don't care	

TABLE A3
NAVIGATION MESSAGE FILE - HEADER SECTION DESCRIPTION

	HEADER LABEL (Columns 61-80)	DESCRIPTION	FORMAT	
	RINEX VERSION / TYPE	- Format version (2) - File type ('N' for Navigation data)	I6,14X, A1,19X	
	PGM / RUN BY / DATE	- Name of program creating current file - Name of agency creating current file - Date of file creation	A20, A20, A20	
*	COMMENT	Comment line(s)	A60	*
*	ION ALPHA	Ionosphere parameters A0-A3 of almanac (page 18 of subframe 4)	2X,4D12.4	*
*	ION BETA	Ionosphere parameters B0-B3 of almanac	2X,4D12.4	*
*	DELTA-UTC: A0,A1,T,W	Almanac parameters to compute time in UTC (page 18 of subframe 4) A0,A1: terms of polynomial T : reference time for UTC data W : UTC reference week number	3X,2D19.12, 2I9	*
*	LEAP SECONDS	Delta time due to leap seconds	I6	*
	END OF HEADER	Last record in the header section.	60X	

Records marked with * are optional

TABLE A4
NAVIGATION MESSAGE FILE - DATA RECORD DESCRIPTION

OBS. RECORD	DESCRIPTION	FORMAT
PRN / EPOCH / SV CLK	- Satellite PRN number	I2,
	- Epoch: Toc - Time of Clock	
	year (2 digits)	5I3,
	month	
	day	
	hour	
	minute	
	second	F5.1,
	- SV clock bias (seconds)	3D19.12
	- SV clock drift (sec/sec)	
	- SV clock drift rate (sec/sec2)	
BROADCAST ORBIT - 1	- IODE Issue of Data, Ephemeris	3X,4D19.12
	- Crs (meters)	
	- Delta n (radians/sec)	
	- M0 (radians)	
BROADCAST ORBIT - 2	- Cuc (radians)	3X,4D19.12
	- e Eccentricity	
	- Cus (radians)	
	- sqrt(A) (sqrt(m))	
BROADCAST ORBIT - 3	- Toe Time of Ephemeris	3X,4D19.12
	(sec of GPS week)	
	- Cic (radians)	
	- OMEGA (radians)	
	- Cis (radians)	
BROADCAST ORBIT - 4	- i0 (radians)	3X,4D19.12
	- Crc (meters)	
	- omega (radians)	
	- OMEGA DOT (radians/sec)	
BROADCAST ORBIT - 5	- IDOT (radians/sec)	3X,4D19.12
	- Codes on L2 channel	
	- GPS Week # (to go with TOE)	
	- L2 P data flag	
BROADCAST ORBIT - 6	- SV accuracy (meters)	3X,4D19.12
	- SV health (MSB only)	
	- TGD (seconds)	
	- IODC Issue of Data, Clock	
BROADCAST ORBIT - 7	- Transmission time of message	3X,4D19.12
	(sec of GPS week, derived e.g.	
	from Z-count in Hand Over Word (HOW)	
	- spare	
	- spare	
	- spare	

TABLE A5
METEOROLOCICAL DATA FILE - HEADER SECTION DESCRIPTION

	HEADER LABEL (Columns 61-80)	DESCRIPTION	FORMAT	
	RINEX VERSION / TYPE	- Format version (2) - File type ('M' for Meteorological Data)	I6,14X, A1,39X	
	PGM / RUN BY / DATE	- Name of program creating current file - Name of agency creating current file - Date of file creation	A20, A20, A20	
*	COMMENT	Comment line(s)	A60	*
	MARKER NAME	Station Name (preferably identical to MARKER NAME in the associated Observation File)	A60	
*	MARKER NUMBER	Station Number (preferably identical to MARKER NUMBER in the associated Observation File)	A20	*
	# / TYPES OF OBSERV	- Number of different observation types stored in the file - Observation types The following meteorological observation types are defined in RINEX Version 2: PR : Pressure (mbar) TD : Dry temperature (deg Celsius) HR : Relative Humidity (percent) ZW : Wet zenith path delay (millimeters) (for WVR data) The sequence of the types in this record must correspond to the sequence of the measurements in the data records	I6, 9(4X,A2)	
	END OF HEADER	Last record in the header section.	60X	

TABLE A6
METEOROLOGICAL DATA FILE - DATA RECORD DESCRIPTION

OBS. RECORD	DESCRIPTION	FORMAT
EPOCH / MET	- Epoch in GPS time (not local time!) year (2 digits), month,day,hour,min,sec - Met data in the same sequence as given in the header	6I3, mF7.1

```
+--------------------------------------------------------------------------+
|                                TABLE A7                                  |
|                     OBSERVATION DATA FILE - EXAMPLE                      |
+--------------------------------------------------------------------------+

----|---1|0---|---2|0---|---3|0---|---4|0---|---5|0---|---6|0---|---7|0---|---8|

     2              OBSERVATION DATA    M (MIXED)           RINEX VERSION / TYPE
BLANK OR G = GPS,  R = GLONASS,  T = TRANSIT,  M = MIXED    COMMENT
XXRINEXO V9.9       AIUB                22-APR-93 12:43     PGM / RUN BY / DATE
EXAMPLE OF A MIXED RINEX FILE                               COMMENT
A 9080                                                      MARKER NAME
9080.1.34                                                   MARKER NUMBER
BILL SMITH          ABC INSTITUTE                           OBSERVER / AGENCY
X1234A123           XX                  ZZZ                 REC # / TYPE / VERS
234                 YY                                      ANT # / TYPE
  4375274.       587466.      4589095.                      APPROX POSITION XYZ
         .9030         .0000         .0000                  ANTENNA: DELTA H/E/N
     1     1                                                WAVELENGTH FACT L1/2
     1     2     6   G14   G15   G16   G17   G18   G19      WAVELENGTH FACT L1/2
     4    P1    L1    L2    P2                              # / TYPES OF OBSERV
    18                                                      INTERVAL
  1990     3    24    13    10   36.000000                  TIME OF FIRST OBS
                                                            END OF HEADER
 90  3 24 13 10 36.0000000  0  3G12G 9G 6                            -.123456789
  23629347.915            .300 8         -.353  23629364.158
  20891534.648           -.120 9         -.358  20891541.292
  20607600.189           -.430 9          .394  20607605.848
 90  3 24 13 10 50.0000000  4  3
     1     2     2   G 9   G12                                WAVELENGTH FACT L1/2
  *** WAVELENGTH FACTOR CHANGED FOR 2 SATELLITES ***          COMMENT
                                                              COMMENT
 90  3 24 13 10 54.0000000  0  5G12G 9G 6R21R22                      -.123456789
  23619095.450      -53875.632 8    -41981.375  23619112.008
  20886075.667      -28688.027 9    -22354.535  20886082.101
  20611072.689       18247.789 9     14219.770  20611078.410
  21345678.576       12345.567 5
  22123456.789       23456.789 5
 90  3 24 13 11  0.0000000  2
                            4  1
            *** FROM NOW ON KINEMATIC DATA! ***               COMMENT
 90  3 24 13 11 48.0000000  0  4G16G12G 9G 6                         -.123456789
  21110991.756       16119.980 7     12560.510  21110998.441
  23588424.398     -215050.557 6   -167571.734  23588439.570
  20869878.790     -113803.187 8    -88677.926  20869884.938
  20621643.727       73797.462 7     57505.177  20621649.276
                            3  4
A 9080                                                        MARKER NAME
9080.1.34                                                     MARKER NUMBER
         .9030         .0000         .0000                    ANTENNA: DELTA H/E/N
          --> THIS IS THE START OF A NEW SITE <--             COMMENT
 90  3 24 13 12  6.0000000  0  4G16G12G 6G 9                         -.123456987
  21112589.384       24515.877 6     19102.763 3  21112596.187
  23578228.338     -268624.234 7   -209317.284 4  23578244.398
  20625218.088       92581.207 7     72141.846 4  20625223.795
  20864539.693     -141858.836 8   -110539.435 5  20864545.943
 90  3 24 13 13  1.2345678  5  0
```

```
                            4  1
        (AN EVENT FLAG WITH SIGNIFICANT EPOCH)              COMMENT
 90  3 24 13 14 12.0000000  0  4G16G12G 9G 6                         -.123456012
  21124965.133       89551.30216     69779.62654  21124972.2754
  23507272.372     -212616.150 7   -165674.789 5  23507288.421
  20828010.354     -333820.093 6   -260119.395 5  20828017.129
  20650944.902      227775.130 7    177487.651 4  20650950.363
                            4  1
           *** ANTISPOOFING ON G 16 AND LOST LOCK           COMMENT
 90  3 24 13 14 12.0000000  6  2G16G 9
                 123456789.0      -9876543.5
                         0.0            -0.5
                            4  2
           ---> CYCLE SLIPS THAT HAVE BEEN APPLIED TO       COMMENT
                THE OBSERVATIONS                            COMMENT
 90  3 24 13 14 48.0000000  0  4G16G12G 9G 6                         -.123456234
  21128884.159      110143.144 7     85825.18545  21128890.7764
  23487131.045     -318463.297 7   -248152.728 4  23487146.149
  20817844.743     -387242.571 6   -301747.22925  20817851.322
  20658519.895      267583.67817    208507.26234  20658525.869
                            4  3
            *** SATELLITE G 9 THIS EPOCH ON WLFACT 1 (L2)   COMMENT
            *** G 6 LOST LOCK AND ON WLFACT 2 (L2)          COMMENT
                (INVERSE TO PREVIOUS SETTINGS)              COMMENT

----|---1|0---|---2|0---|---3|0---|---4|0---|---5|0---|---6|0---|---7|0---|---8|
```

```
+------------------------------------------------------------------------------+
|                                  TABLE A8                                    |
|                      NAVIGATION MESSAGE FILE - EXAMPLE                       |
+------------------------------------------------------------------------------+

----|---1|0---|---2|0---|---3|0---|---4|0---|---5|0---|---6|0---|---7|0---|---8|

     2              N: GPS NAV DATA                         RINEX VERSION / TYPE
XXRINEXN V2.0       AIUB                12-SEP-90 15:22     PGM / RUN BY / DATE
EXAMPLE OF VERSION 2 FORMAT                                 COMMENT
     .1676D-07   .2235D-07  -.1192D-06  -.1192D-06          ION ALPHA
     .1208D+06   .1310D+06  -.1310D+06  -.1966D+06          ION BETA
     .133179128170D-06  .107469588780D-12   552960       39 DELTA-UTC: A0,A1,T,W
     6                                                      LEAP SECONDS
                                                            END OF HEADER
 6 90  8  2 17 51 44.0 -.839701388031D-03 -.165982783074D-10  .000000000000D+00
     .910000000000D+02  .934062500000D+02  .116040547840D-08  .162092304801D+00
     .484101474285D-05  .626740418375D-02  .652112066746D-05  .515365489006D+04
     .409904000000D+06 -.242143869400D-07  .329237003460D+00 -.596046447754D-07
     .111541663136D+01  .326593750000D+03  .206958726335D+01 -.638312302555D-08
     .307155651409D-09  .000000000000D+00  .551000000000D+03  .000000000000D+00
     .000000000000D+00  .000000000000D+00  .000000000000D+00  .910000000000D+02
     .406800000000D+06
13 90  8  2 19  0  0.0  .490025617182D-03  .204636307899D-11  .000000000000D+00
     .133000000000D+03 -.963125000000D+02  .146970407622D-08  .292961152146D+01
    -.498816370964D-05  .200239347760D-02  .928156077862D-05  .515328476143D+04
     .414000000000D+06 -.279396772385D-07  .243031939942D+01 -.558793544769D-07
     .110192796930D+01  .271187500000D+03 -.232757915425D+01 -.619632953057D-08
    -.785747015231D-11  .000000000000D+00  .551000000000D+03  .000000000000D+00
     .000000000000D+00  .000000000000D+00  .000000000000D+00  .389000000000D+03
     .410400000000D+06

----|---1|0---|---2|0---|---3|0---|---4|0---|---5|0---|---6|0---|---7|0---|---8|
```

```
+------------------------------------------------------------------------------+
|                                  TABLE A9                                    |
|                    METEOROLOGICAL DATA FILE - EXAMPLE                        |
+------------------------------------------------------------------------------+

----|---1|0---|---2|0---|---3|0---|---4|0---|---5|0---|---6|0---|---7|0---|---8|

     2              METEOROLOGICAL DATA                     RINEX VERSION / TYPE
XXRINEXM V9.9       AIUB                22-APR-93 12:43     PGM / RUN BY / DATE
EXAMPLE OF A MET DATA FILE                                  COMMENT
A 9080                                                      MARKER NAME
     3    PR    TD    HR                                    # / TYPES OF OBSERV
                                                            END OF HEADER
 90  3 24 13 10 15  987.1   10.6   89.5
 90  3 24 13 10 30  987.2   10.9   90.0
 90  3 24 13 10 45  987.1   11.6   89.0

----|---1|0---|---2|0---|---3|0---|---4|0---|---5|0---|---6|0---|---7|0---|---8|
```

GLOSSARY

Almanac Approximate location information on all satellites.

Ambiguity The phase measurement when a receiver first locks onto a GPS signal is ambiguous by an integer number of cycles (because the receiver has no way of counting the cycles between satellite and receiver). This ambiguity remains constant until loss of lock.

Analog A clock with moving hands is an analog device; a clock with displayed numbers is digital. Over telephone lines, digital signals must be converted to analog using a modem (a modulator/demodulator).

Antispoofing (AS) Encrypting the P code by addition (modulo 2) of a secret W code. The resulting Y code prevents a receiver from being "spoofed" by a false P code signal.

Atomic clock A highly precise clock based on the behavior of elements such as cesium, hydrogen, and rubidium.

Azimuth Angle at your position between the meridian and the direction to a target.

Bandwidth The range of frequencies in a signal.

Bearing The angle at your position between grid north and the direction to a target.

Binary Biphase Modulation The phase of a GPS carrier signal is shifted by 180° when there is a transition from 0 to 1 or 1 to 0.

Block I, II, IIR, IIF satellites The generations of GPS satellites: Block I were prototypes first launched in 1978; 24 Block II satellites made up the fully operational GPS constellation declared in 1995; Block IIR are replacement satellites. Block IIF refers to the follow-on generation.

C/A code The coarse/acquisition or clear/acquisition code modulated onto the GPS L_1 signal. This Gold code is a sequence of 1023 pseudorandom binary biphase modulations on the GPS carrier at a chipping rate of 1.023 MHz, thus repeating in one millisecond. This "civilian code" was selected for good acquisition properties.

Carrier A radio wave whose frequency, amplitude, or phase may be varied by modulation.

Carrier-aided tracking A strategy that uses the GPS carrier signal to achieve an exact lock on the pseudorandom code.

Carrier frequency The frequency of the unmodulated output of a radio transmitter. The L_1 carrier frequency is 1575.42 MHz.

Carrier phase The accumulated phase of the L_1 or L_2 carrier, measured since locking onto the signal. Also called *integrated Doppler*.

Channel The circuitry in a receiver to receive the signal from a single satellite.

Chip The length of time to transmit 0 or 1 in a binary pulse code. The chip rate is the number of chips per second.

Circular error probable (CEP) In a circular normal distribution, the radius of the circle containing 50% of the measurements.

Clock bias The difference between the clock's indicated time and GPS time.

Clock offset Constant difference in the time reading between two clocks.

Code Division Multiple Access (CDMA) A method whereby many senders use the same frequency but each has a unique code. GPS uses CDMA with Gold's codes for their low cross-correlation properties.

Code phase GPS Measurements based on the coarse C/A code.

Code-tracking loop A receiver module that aligns a PRN code sequence in a signal with an identical PRN sequence generated in the receiver.

Control segment A world-wide network of GPS stations that ensure the accuracy of satellite positions and their clocks.

Cycle slip A discontinuity in the measured carrier beat phase resulting from a loss-of-lock in the tracking loop of a GPS receiver.

Data message A message in the GPS signal which reports the satellite's location, clock corrections and health (and rough information about the other satellites).

Differential positioning (DGPS) A technique to improve accuracy by determining the positioning error at a known location and correcting the calculations of another receiver in the same area.

Dilution of Precision (DOP) The purely geometrical contribution to the uncertainty in a position fix. PDOP multiplies rms range error to give rms position error. Standard terms for GPS are GDOP: Geometric (3 coordinates plus clock offset), PDOP: Position (3 coordinates), HDOP: Horizontal (2 coordinates), VDOP: Vertical (height only), TDOP: Time (clock offset only). PDOP is inversely proportional to the volume of the pyramid from the receiver to four observed satellites. A value near PDOP $= 3$ is associated with widely separated satellites and good positioning.

Dithering The introduction of digital noise. This is the process that adds inaccuracy to GPS signals to induce Selective Availability.

Doppler-aiding Using a measured Doppler shift to help the receiver track the GPS signal. Allows more precise velocity and position measurement.

Doppler shift The apparent change in frequency caused by the motion of the transmitter relative to the receiver.

Double difference A GPS observable formed by differencing carrier phases (or pseudoranges) measured by a pair of receivers i, j tracking the same pair of satellites k, l. The double difference $(\varphi_i^k - \varphi_j^k) - (\varphi_i^l - \varphi_j^l)$ removes essentially all of the clock errors.

Earth-Centered Earth-Fixed (ECEF) Cartesian coordinates that rotate with the Earth. Normally the X direction is the intersection of the prime meridian (Greenwich) with the equator. The Z axis is parallel to the spin axis of the Earth.

Elevation Height above mean sea level. Vertical distance above the geoid.

Elevation mask angle Satellites below this angle (often 15°) are not tracked, to avoid interference by buildings, trees and multipath errors, and large atmospheric delays.

Ellipsoid A surface whose plane sections are ellipses. Geodesy generally works with an ellipsoid of revolution (two of the three principal axes are equal): it is formed by revolving an ellipse around one of its axes.

Ellipsoidal height The vertical distance above the ellipsoid (not the same as elevation above sea level).

Ephemeris Accurate position (the orbit) as a function of time. Each GPS satellite transmits a predicted ephemeris for its own orbit valid for the current hour. The ephemeris (repeated every 30 seconds) is a set of 16 Keplerian-like parameters with corrections for the Earth's gravitational field and other forces. Available as "broadcast ephemeris" or as post-processed "precise ephemeris."

Epoch Measurement time or measurement interval or data frequency.

Fast-switching channel A single channel that rapidly samples a number of satellite signals. The switching time is sufficiently fast (2 to 5 milliseconds) to recover the data message.

Frequency spectrum The distribution of signal amplitudes as a function of frequency.

Geodesy The disciplin of point position determination, the Earth's gravity field and temporal variations of both.

Geodetic datum An ellipsoid of revolution designed to approximate part or all of the geoid. A point on the topographic surface is established as the origin of datum. The datum has five parameters: two for the axis lengths of the ellipse and three to give position. If the ellipsoid is not aligned with the coordinate axes we must add three rotations.

Geoid The undulating, but smooth, equipotential surface of the Earth's gravity field. The geoid is the primary reference surface, everywhere perpendicular to the force of gravity (plumb line).

Gigahertz (GHz) One billion cycles per second = 1000 Mhz.

GLONASS Russia's Global Navigation Satellite System (Globalnaya Navigatsionnaya Sputnikovaya Sistema).

Global Navigation Satellite System (GNSS) A European system that would incorporate GPS, GLONASS, and other segments to support navigation.

Global Positioning System (GPS) A constellation of 24 satellites orbiting the Earth at high altitude. GPS satellites transmit signals that allow one to determine, with great accuracy, the locations of GPS receivers.

GPS ICD-200 The GPS Interface Control Document contains the full technical description of the interface between satellites and user.

GPS Time The time scale to which GPS signals are referenced, steered to keep within about 1 microsecond of UTC, ignoring the UTC leap seconds. GPS Time equalled UTC in 1980, but 10 leap seconds have been inserted into UTC.

GPS Week The number of elapsed weeks (modulo 1024) since the week beginning January 6, 1980. The week number increments at Saturday/Sunday midnight in GPS Time.

Handover word (HOW) The second word in each subframe of the navigation message. It contains the Z-count at the leading edge of the next subframe and is used by a GPS receiver to determine where in its generated P code to start the correlation search.

Hertz (Hz) One cycle per second.

Ionosphere The band of charged particles 80 to 120 miles (or often wider) above the Earth.

Ionospheric delay A wave propagating through the ionosphere experiences delay by refraction. Phase delay depends on electron content and affects carrier signals. Group delay depends on dispersion and affects signal modulation (codes). The phase and group delay are of the same magnitude but opposite sign.

Kalman filter A recursive numerical method to track a time-varying signal and the associated covariance matrix in the presence of noise. The filter combines observation equations and state equations. The "Bayes filter" does the calculation of state estimate and covariance estimate in a different order.

Keplerian elements Six parameters that describe position and velocity in a purely elliptical (Keplerian) orbit: the semimajor axis and eccentricity, the inclination of the orbit plane to the celestial equator, the right ascension of the ascending node, the argument of perigee, and the time the satellite passes through perigee.

L-band The radio frequencies from 390 to 1550 MHz, or sometimes 1 to 2 GHz, including the GPS carrier frequencies.

L_1 signal The primary GPS signal at 1572.42 MHz. The L_1 broadcast is modulated with the C/A and P codes and the navigation message.

L_2 signal The second L-band signal is centered at 1227.60 MHz and carries the P code and navigation message.

Microstrip antenna (patch antenna) A type of antenna in many GPS receivers, constructed of (typically rectangular) elements that are photoetched on one side of double-coated, printed-circuit board.

Modem A modulator/demodulator that converts digital to analog (for telephone transmission) and back to digital.

Monitor stations Data collection points linked to a master control station, where corrections are calculated and uploaded to the satellites.

Multichannel receiver A receiver with multiple channels, each tracking one satellite continuously.

Multipath Interference caused by reflected GPS signals arriving at the receiver. Signals reflected from nearby structures travel longer paths than line-of-sight and produce higher positioning errors.

Multiplexing The technique of rapidly sequencing the signals from two or more satellites through one tracking channel. This ensures that navigation messages are acquired simultaneously.

NAD 83 North American Datum 1983.

Nanosecond One billionth of a second.

Narrow lane The sum of carrier-phase observations on the L_1 and L_2 frequencies. The effective wavelength is 10.7 centimeters.

Nav message The 1500-bit navigation message broadcast by each GPS satellite at 50 bps. This message contains satellite position, system time, clock correction, and ionospheric delay parameters.

Orthometric height The height of a point above the geoid.

P code The precise code of the GPS signal, typically used by military receivers. A very long sequence of pseudo-random binary biphase modulations on the GPS carrier at a chip rate of 10.23 MHz which repeats about every 267 days. Each segment is unique to one GPS satellite and is reset each Saturday/Sunday midnight.

Position Dilution Of Precision (PDOP) (see DOP) The factor that multiplies errors in ranges to give approximate errors in position.

Phase lock loop (Carrier tracking loop) The receiver compares the phases of an oscillator signal and reference signal. The reference frequency is adjusted to eliminate phase difference and achieve locking.

Point positioning A geographical position produced from one receiver.

Precise Positioning Service (PPS) The highest level of dynamic positioning accuracy by single-receiver GPS, with access to the dual-frequency P code and removal of SA.

Pseudolite (shortened form of pseudo-satellite) A ground-based receiver that simulates the GPS signal. The data may also contain differential corrections so other receivers can correct for GPS errors.

Pseudorandom Noise (PRN) A sequence of 1's and 0's that appears to be random but can be reproduced exactly. Their most important property is a low correlation with their own delays. Each GPS satellite has unique C/A and P pseudorandom noise codes.

Pseudorange A distance measurement from satellite to receiver (by the delay lock loop) that has not been corrected for clock differences.

Range rate The rate of change of distance between satellite and receiver. Range rate is determined from the Doppler shift of the satellite carrier.

Relative positioning DGPS measures relative position of two receivers by observing the same set of satellites at the same times.

Reliability The probability of performing a specified function without failure under given conditions for a specified period of time.

Receiver INdependent EXchange format (RINEX) A set of standard definitions and formats for time, phase, and range that permits interchangeable use of GPS data: pseudorange, carrier phase, and Doppler. The format also includes meteorological data and site information.

Root mean square (rms) The square root of the sum of squares. This quantity $\|A\boldsymbol{x} - \boldsymbol{b}\|$ is minimized by the least-squares solution $\hat{\boldsymbol{x}}$.

Satellite constellation The GPS constellation has six orbital planes, each containing four satellites. GLONASS has three orbital planes containing eight satellites.

Selective Availability (SA) A program that limits the accuracy of GPS pseudorange measurements, degrading the signal by dithering the time and position in the navigation message. The error is guaranteed to be below 100 meters, 95% of the time. It will soon be discontinued.

Spherical Error Probable (SEP) The radius of a sphere within which there is a 50% probability of locating the true coordinates of a point.

Spread spectrum A signal normally requiring a narrow transmission bandwidth but spread over a much larger bandwidth. The 50 bits-per-second GPS navigation message could use a bandwidth of 50 Hz, but it is spread by modulating with the pseudorandom C/A code. Then all satellites can be received unambiguously.

Squaring channel A GPS receiver channel that multiplies the received signal by itself to remove the code modulation (since $(-1)^2 = 1$).

Standard deviation (σ) A measure of the dispersion of random errors about their mean. Experimentally, the standard deviation is the square root of the sum of the squares of deviations from the mean divided by the number of observations less one.

Standard Positioning Service (SPS) The normal civilian positioning accuracy obtained by using the single frequency C/A code.

Triple difference The difference in time of doubly differenced carrier-phase observations. The triple difference is free of integer ambiguities. It is useful for detecting cycle slips.

Universal Time Coordinated (UTC) A highly accurate and stable atomic time system kept very close, by inserting leap seconds, to the universal time corrected for seasonal variations in the Earth's rotation rate. Maintained by the U. S. Naval Observatory. (The changing constant from GPS time to UTC is 10 seconds today.)

User Equivalent Range Error (UERE) Any positioning error expressed as an equivalent error in the range between receiver and satellite. The total UERE is the square root of the sum of squares of the individual errors, assumed to be independent. A prediction of maximum total UERE (minus ionospheric error) is in each satellite's navigation message as the *user range accuracy* (URA).

Wide Area Augmentation System (WAAS) A form of "Wide Area DGPS" with corrections from a network of reference stations.

WGS 84 A geocentric reference ellipsoid, a coordinate system, and a gravity field model (in terms of harmonic coefficients). The GPS satellite orbits have been referenced to WGS 84 since January 1987. Redefined at the epoch 1994.0 and designated WGS 84 (G730).

Y code The encrypted version of the P code.

Z-Count The GPS time unit (29 bits). GPS week number and time-of-week in units of 1.5 seconds. A truncated TOW with 6-second epochs is included in the *handover word*.

Zenith distance Angle at your position between the direction to zenith and the direction to a target.

REFERENCES

Abramowitz, Milton & Stegun, Irene A., editors (1972). *Handbook of Mathematical Functions*. Dover Publications, New York.

Axelrad, P. & Brown, R. G. (1996). GPS navigation algorithms. In Parkinson, Bradford W. & James J. Spilker, Jr., editors, *Global Positioning System: Theory and Applications*, number 163 in Progress In Astronautics and Aeronautics, chapter 9. American Institute of Aeronautics and Astronautics, Inc., Washington, DC.

Babai, L. (1986). On Lovasz lattice reduction and the nearest lattice point problem. *Combinatorica*, 6:1–13.

Bancroft, S. (1985). An algebraic solution of the GPS equations. *IEEE Transactions on Aerospace and Electronic Systems*, 21:56–59.

Bartelme, N. & Meissl, P. (1974). *Strength Analysis of Distance Networks*, volume 15. Mitteilungen der geodätischen Institute der Technischen Hochschule in Graz.

Bierman, G. J. (1977). *Factorization Methods for Discrete Sequential Estimation*. Academic Press, San Diego.

Björck, Åke (1996). *Numerical Methods for Least Squares Problems*. Society for Industrial and Applied Mathematics, Philadelphia.

Borre, Kai (1989). Investigation of geodetic networks by means of partial differential operators. *manuscripta geodaetica*, 14:247–275.

Brown, Robert Grover & Hwang, Patrick Y. C. (1997). *Introduction to Random Signals and Applied Kalman Filtering*. John Wiley & Sons, Inc., New York, 3rd edition.

de Jonge, P. J. & Tiberius, C. C. J. M. (1996). *The LAMBDA Method for Integer Ambiguity Estimation: Implementation Aspects*, volume 12 of *LGR-Series*. Publication of the Delft Geodetic Computing Centre.

Department of Defense (1991). *World Geodetic System 1984, Its Definition and Relationships with Local Geodetic Systems*. Defense Mapping Agency, Fairfax, Virginia, second edition.

Deyst, J. J. (1969). A derivation of the optimum continuous linear estimator for systems with correlated measurement noise. *American Institute of Aeronautics and Astronautics Journal*, 7:2116–2119.

Diggle, Peter J. (1990). *Time Series. A Biostatistical Introduction.* Clarendon Press, Oxford.

Dixon, R. C. (1984). *Spread Spectrum Systems.* John Wiley & Sons, New York, 2nd edition.

Duan, Jingping, Bevis, Michael, Fang, Peng, Bock, Yehuda, Chiswell, Steven, Businger, Steven, Rocken, Christian, Solheim, Frederick, van Hove, Teresa, Ware, Randolph, McClusky, Simon, Herring, Thomas A., & King, Robert W. (1996). GPS meteorology: Direct estimation of the absolute value of precipitable water. *Journal of Applied Meteorology*, 35:830–838.

Duncan, D. B. & Horn, S. D. (1972). Linear dynamic recursive estimation from the viewpoint of regression analysis. *Journal of American Statistical Association*, 67:815–821.

Euler, Hans-Jürgen & Goad, Clyde C. (1991). On optimal filtering of GPS dual frequency observations without using orbit information. *Bulletin Géodésique*, 65:130–143.

Forsythe, George E. & Moler, Cleve B. (1967). *Computer Solution of Linear Algebraic Systems.* Prentice-Hall, Englewood Cliffs.

Gantmacher, F. R. (1959). *Applications of the Theory of Matrices.* Chelsea, New York.

Gantmacher, F. R. & Krein, M. G. (1960). *Oszillationsmatrizen, Oszillationskerne und kleine Schwingungen mechanischer Systeme.* Akademie-Verlag, Berlin.

Gauss, Carl Friedrich (1995). *Theoria Combinationis Observationum Erroribus Minimis Obnoxiae, Pars Prior, Pars Posterior, Supplementum*, volume 11 of *Classics in Applied Mathematics.* Society for Industrial and Applied Mathematics, Philadelphia. Translated into English by G. W. Stewart.

Goad, Clyde & Yang, Ming (1994). On automatic precision airborne GPS positioning. In *International Symposium on Kinematic Systems in Geodesy, Geomatics and Navigation KIS'94*, pages 131–138, Banff.

Goad, C. C. & Goodman, L. (1974). A modified tropospheric refraction correction model. Presented at the American Geophysical Union Annual Fall Meeting, San Francisco.

Gold, R. (1967). Optimal binary sequences for spread spectrum multiplexing. *IEEE Transactions on Information Theory*, IT-13:619–621.

Golomb, S. (1982). *Shift Register Sequences.* Aegean Park Press, Laguna Hills, CA.

Golub, Gene H. & van Loan, Charles F. (1996). *Matrix Computations.* The Johns Hopkins University Press, Baltimore and London, third edition.

Helmert, F. R. (1893). *Die Europäische Längengradmessung in 52 Grad Breite von Greenwich bis Warschau*, volume I. Veröffentlichung des Königl. Preuß. Geodätischen Institutes, Berlin.

Herring, Thomas A., Davis, James L., & Shapiro, Irwin I. (1990). Geodesy by radio interferometry: The application of Kalman filtering to the analysis of very long baseline interferometry data. *Journal of Geophysical Research*, 95, No. B8:12561–12581.

Hofmann-Wellenhof, B., Lichtenegger, H., & Collins, J. (1997). *GPS Theory and Practice*. Springer-Verlag, Wien New York, fourth, revised edition.

ICD-GPS-200 (1991). Interface control document. ICD-GPS-200, Arinc Research Corporation, 11 770 Warner Ave., Suite 210, Fountain Valley, CA 92 708.

Jin, Xin-Xiang (1996). *Theory of Carrier Adjusted DGPS Positioning Approach and Some Experimental Results*. Delft University Press, Stevinweg 1, 2628 CN Delft, The Netherlands.

Kaplan, Elliott D., editor (1996). *Understanding GPS, Principles and Applications*. Artech House, Boston London.

Kleusberg, A. (1994). Direkte Lösung des räumlichen Hyperbelschnitts. *Zeitschrift für Vermessungswesen*, 119:188–192.

Kleusberg, Alfred & Teunissen, Peter, editors (1996). *GPS for Geodesy*. Number 60 in Lecture Notes in Earth Sciences. Springer Verlag, Heidelberg.

Klobuchar, J. A. (1996). Ionospheric effects on GPS. In Parkinson, Bradford W. & James J. Spilker, Jr., editors, *Global Positioning System: Theory and Applications*, number 163 in Progress In Astronautics and Aeronautics, chapter 12. American Institute of Aeronautics and Astronautics, Inc., Washington, DC.

Krarup, Torben (1972). On the geometry of adjustment. *Zeitschrift für Vermessungswesen*, 97:440–445.

Krarup, Torben (1979). S-transformation or How to live without generalized inverse—almost. Geodætisk Institut, Charlottenlund.

Krarup, Torben (1982). Non-linear adjustment and curvature. Daar heb ik veertig jaar over nagedacht .... Feestbundel ter gelegenheid van de 65ste verjaardag van Professor Baarda, 1:146–159, Delft.

Lapucha, Dariusz, Barker, Richard, & Liu, Ziwen (1996). High-rate precise real-time positioning using differential carrier phase. *Navigation*, 43:295–305.

Leick, Alfred (1995). *GPS Satellite Surveying*. John Wiley and Sons, Inc., New York Chichester Brisbane Toronto Singapore, 2nd edition.

Lenstra, A. K., Lenstra, H. W., & Lovász, L. (1982). Factoring polynomials with rational coefficients. *Mathematische Annalen*, 261:515–534.

Meissl, P. (1981). *Strength Analysis of Angular Anblock Networks With Distance Measurements Along the Perimeter*, volume 258/V of *Deutsche Geodätische Kommission, Reihe B*. C. H. Beck, München. 6th International Symposium on Geodetic Networks and Computations, Munich.

Meissl, Peter (1982). *Least Squares Adjustment. A Modern Approach*, volume 43. Mitteilungen der geodätischen Institute der Technischen Universität Graz.

Morrison, N. (1969). *Introduction to Sequential Smoothing and Prediction*. McGraw-Hill, New York.

Paige, C. C. & Saunders, M. A. (1977). Least squares estimation of discrete linear dynamic systems using orthogonal transformations. *SIAM Journal of Numerical Analysis*, 14:180–193.

Parkinson, Bradford W. & Spilker Jr., James J., editors (1996). *Global Positioning System: Theory and Applications*, volume 163 of *Progress In Astronautics and Aeronautics*. American Institute of Aeronautics and Astronautics, Inc., Washington, DC.

Ponsonby, John E. B. (1996). Global satellite navigation systems. *Radio Science Bulletin*, 277.

Rauch, H. E., Tung, F., & Striebel, C. T. (1965). Maximum likelihood estimates of linear dynamic systems. *American Institute of Aeronautics and Astronautics Journal*, 3:1445–1450.

Scheffé, Henry (1959). *The Analysis of Variance*. John Wiley and Sons, New York.

Schwarz, Charles R. (1994). The trouble with constrained adjustments. *Surveying and Land Information Systems*, 54:202–209.

Seal, Hilary L. (1967). Studies in the history of probability and statistics. xv The historical development of the Gauss linear model. *Biometrika*, 54:1–24.

Seeber, Günter (1993). *Satellite Geodesy. Foundations, Methods, and Applications*. Walter de Gruyter, Berlin New York.

Sobel, Dava (1995). *Longitude*. Walker and Co, New York.

Sorenson, Harold W., editor (1985). *Kalman Filtering: Theory and Application*. Institute of Electrical and Electronics Engineers, New York.

Strang, Gilbert (1986). *Introduction to Applied Mathematics*. Wellesley-Cambridge Press, Wellesley MA.

Strang, Gilbert (1991). *Calculus*. Wellesley-Cambridge Press, Wellesley MA.

Strang, Gilbert (1993). *Introduction to Linear Algebra.* Wellesley-Cambridge Press, Wellesley MA.

Teunissen, P. J. G. (1985a). Generalized inverses, adjustment, the datum problem, and S-transformations. In Grafarend, E. & Sanso, F., editors, *Optimization and Design of Geodetic Networks*, Berlin. Springer.

Teunissen, P. J. G. (1985b). *The Geometry of Geodetic Inverse Linear Mapping and Non-Linear Adjustment*, volume 8, Number 1 of *Publications on Geodesy. New Series.* Netherlands Geodetic Commission, Delft.

Teunissen, P. J. G. (1995a). The invertible GPS ambiguity transformations. *manuscripta geodaetica*, 20:489–497.

Teunissen, P. J. G. (1995b). The least-squares ambiguity decorrelation adjustment: a method for fast GPS integer ambiguity estimation. *Journal of Geodesy*, 70:65–82.

Tiberius, C. C. J. M. (1991). Quality control and integration aspects of vehicle location systems. Publications of the Delft Geodetic Computing Centre.

Trefethen, Lloyd N. & David Bau, III (1997). *Numerical Linear Algebra.* Society for Industrial and Applied Mathematics, Philadelphia.

Wolf, Paul R. & Ghilani, Charles D. (1997). *Adjustment Computations, Statistics and Least Squares in Surveying and GIS.* John Wiley & Sons, Inc., New York Chichester Brisbane Toronto Singapore Weinheim.

Yang, Ming, Goad, Clyde, & Schaffrin, Burkhard (1994). Real-time on-the-fly ambiguity resolution over short baselines in the presence of anti-spoofing. In *Proceedings of ION GPS-94*, pages 519–525, Salt Lake City.

INDEX OF M-FILES

On this and the following page, we list the names of all the M-files mentioned in the book. All these files (and others not referred to in this book) can be obtained from the Web page http://www.i4.auc.dk/borre/matlab

abs_pos 503
accum0 495
anheader 494
ash_dd 489, 495
autocorr 529
b_point 501, 502
b_row 580
bancroft 481, 482, 500, 502
bdata 489
c2g 470, 472
c2gm 470
check_t 487
corrdemo 326
dw 393
e_r_corr 502
edata 489
elimnor 397
elimobs 400
ellaxes 342
errell 337
fepoch_0 494
find_eph 488, 489, 501
findloop 302
findnode 302
fixing1 581
fixing2 582
frgeod 368
g2c 470
gauss1 471
gauss2 471
get_eph 486, 488, 489, 501
gmproc 520
gps_time 487
grabdata 494
julday 487
k_dd3 506, 507
k_dd4 507
k_point 502, 580
k_row 580
k_simul 578
k_ud 502
kalclock 510
kleus 505
kud 580
lev 286
locate 495
looplist 302
lor 444
lorentz 500, 502
mangouli 541
model_g 539
normals 502, 580
null 302
one_way 495, 512
oned 330
oneway_i 535
plu 302
proc_dd 494
qr 371
recclock 509
recpos 481, 488
ref 302
relellip 342
repair 322

rinexe 486, 494
rts 571
satpos 486, 501, 502
satposin 486
sdata 489
sets 441
simil 419
smoother 572
support 339
togeod 368, 501, 502
topocent 501
tropo 456, 475, 491, 493, 501
tropp 501
twod 311
v_loops 303, 493
vectors 303
wc 581

INDEX

$-1, 2, -1$ matrix 69, 77, 98, 204, 242, 378, 394

$A = LDL^{\mathrm{T}}$ 89, 241
$A = LDU$ 78, 83, 85
$A = LU$ 76, 83
$A = PLU$ 116
$A = Q\Lambda Q^{\mathrm{T}}$ 243, 244
$A = QR$ 190, 191, 194
$A = S\Lambda S^{-1}$ 222, 231
$A = U\Sigma V^{\mathrm{T}}$ 258, 263
AA^{T} 264
$A^{\mathrm{T}}A$ 171, 245
acceleration 574
adjustment 300
almanac 601
ambiguity 458, 463, 489, 495, 601
analog 601
angle 14
antenna 492, 495
antispoofing 601
AR(1) 526
arrow 4, 7
ascending node 482, 485, 486
Ashtech 482, 489
associative law 49, 57, 62
asymptotic 554
atomic clock 448, 455, 601
augmented matrix 50, 54, 69, 123
autocorrelation 517, 519, 523, 530, 531
autoregression 521, 526
axes 244
azimuth 354, 363, 469, 470, 475, 488, 502, 601

back substitution 37, 79, 82
backward filter 571
bandwidth 522, 601
barycenter 407, 415, 418, 419, 432, 443
baseline 490
basepoint 347
basis 137, 139
Bayes filter 510, 566, 567
bearing 343, 601
Bernese 481
binary biphase modulation 601
block elimination 397
block inverse matrix 560
block matrix 58
block multiplication 58
block tridiagonal matrix 560
BLUE 323
bordering 559
breakdown 31, 38, 43

C/A code 450, 451, 601
canonical vector 376
carrier frequency 601
carrier phase 602
carrier-aided tracking 601
Cayley-Hamilton 232
Central Limit Theorem 319
change of basis 259, 260
channel 602
characteristic equation 215
checkerboard matrix 156
χ^2 distribution 312
chip 449, 602
chipping rate 490
Cholesky 86, 247, 372, 395, 401
Cholesky factorization 336, 372–374, 554
circular error probable 602
clock 9, 455
clock bias 602

clock offset 507–509, 524, 525, 573, 602
clock reset 509
closest line 175, 176, 178, 182
code division multiple access 602
code observation 460
code phase 602
code-tracking loop 602
coefficient matrix 29
cofactor matrix 209
column picture 29, 31, 39
column space 104, 148
column vector 7
columns times rows 59, 64
combination of columns 31, 47, 56
commutative law 49, 57, 62
complete solution 112, 124, 129
complete system 545
complex conjugate 235, 237
complex matrix 64
components 4
condition equation 300, 409
condition number 378, 382, 386, 388
confidence ellipse 317, 337, 338, 341, 359, 390
confidence interval 311, 316
consistent 126
constellation 452
control network 423–428
control points 426
control segment 602
coordinate measurement 354
correction 543, 545, 551, 562
correlate equation 301
correlation 322, 331
correlation time 521
correlogram 531
cosine 16, 20
covariance 320
covariance function 389
covariance matrix 321, 324, 544
covariance propagation 329
Cramer's rule 206, 207
cross-correlation 518
cube 9
cumulative density 310, 315
Current Law 285, 293
cycle slip 459, 602

data message 602
datum 473, 477, 479
decorrelation 332, 401, 496, 498
degrees of freedom 315, 334
Delay Lock Loop 449
dependent 134, 136
derivation 558
determinant 66, 197, 215, 217, 556
determinant of inverse 202
determinant of product 202
determinant of transpose 202
diagonalizable 224
diagonalization 222
differential GPS 330, 458
differential positioning 602
Dilution of Precision 458, 462, 463, 602
dimension 114, 139, 140, 148, 150, 290
direction measurement 434
dispersive 454, 455
distance 22, 27, 171, 345, 350, 366
distance measurement 434
distributive law 57
dithering 447, 602
Doppler 449, 602
Doppler-aiding 602
dot product 11, 15, 16, 88
double difference 459, 463, 467, 535, 538, 603
dual 297, 300

eccentric anomaly 483
ECEF 462, 474, 482, 487, 603
echelon matrix 113
ecliptic 482
eigenvalue 211, 215
eigenvalues
complex 237
positive 237, 238, 242
product of 217, 219
real 233, 235

repeated 223
sum of 217, 219
eigenvalues of A^2 212
eigenvector 211, 216
Einstein 47, 451
elevation 603
elevation angle 502, 536
elevation mask angle 603
elimination 37, 80, 111, 394, 496
elimination matrix 48, 51, 76
ellipse 243, 244, 338, 467
ellipsoid 467, 473, 603
ellipsoidal height 603
ensemble average 519, 520
entry 47
enu system 363, 474, 475, 501
ephemerides 448, 481, 484, 488, 603
epoch 459, 603
equinox 482
ergodic 519
error ellipse 332, 337
error vector 168
Euler's formula 284, 294, 305
even permutation 198
exponential correlation 527, 533
extended Kalman filter 509

F distribution 318
factorization 76
fast-switching channel 603
Fibonacci 225, 229, 555
finite element method 299
fitting line 442, 443
fixed solution 459
fixed value 423
flattening 467
float solution 459, 496, 499
formula for A^{-1} 208
FORTRAN 8, 17
forward elimination 79, 82
four fundamental subspaces 147, 150, 265, 268, 290
Fourier series 187
fractile 315
Fredholm 163
free 39, 110
free network 411, 415, 428
free stationing 356, 361, 431, 433, 435
free variable 112, 114
frequency spectrum 603
full rank 128, 131
function space 102, 142, 146
Fundamental Theorem 151, 160, 163, 259

gain matrix 543, 549, 551, 562–564
GAMIT 481
GAST 484
Gauss 323, 392, 470, 563
Gauss-Jordan 68, 69, 74, 115, 133
Gauss-Markov process 520, 521, 526, 541
Gauss-Markov Theorem 568
Gaussian elimination 41, 115
Gaussian process 517, 519
Gelb 552
geodesy 603
geodetic datum 603
geographical coordinates 362, 468, 475
geoid 472, 476, 603
geometric mean 17, 19
Global Navigation Satellite System 604
Global Positioning System 604
GLONASS 449, 603
Goad 367, 490, 506
Gold code 450, 451
GPS ICD-200 604
GPS Time 604
GPS Week 604
Gram-Schmidt 188, 189, 371, 372, 557
graph 282
group delay 465
GYPSY 481

handover word 604
hard postulation 409
heading 574
height 469, 472
height differences 283
height measurement 354

heights 275
  postulated 276
Heisenberg 225, 232
Helmert 419
Hilbert matrix 75
homogeneity 390
homogeneous solution 124, 129
horizontal direction 354, 364
house 253, 256, 257
hyperbola 449
hyperboloid 503

identity matrix 30, 48
incidence matrix 282, 283, 287, 305
inclination 482, 485, 486
income 12
increments 343, 344
independent 134, 137
independent columns 135, 140, 150, 161
independent variables 321
infinite entries 511
information filter 560, 569
information matrix 324, 554, 566
inner product 11
innovation 543, 549
integer least squares 497
inverse block matrix 425
inverse matrix 65, 69
inverse of AB 67, 73
invertible 66, 71
ionosphere 455, 457, 604
ionospheric delay 456, 490, 506, 604
isotropy 390
iteration 248, 347

Jacobi 249
Jacobian matrix 330, 491
Jordan form 250
Joseph form 563

Kalman 563, 567
Kalman filter 482, 543, 552, 555, 604
Kepler's equation 483
Keplerian elements 448, 481, 482, 484, 485, 604
kernel 252, 255
kinematic receiver 526
Krarup 383, 394, 409
Kronecker product 384

L-band 604
L_1 signal 604
L_2 signal 604
Lagrange multiplier 297, 424
Lagrangian function 298, 424
LAMBDA 482, 495, 498
latency 513
Law of Cosines 20
Law of Propagation 329, 333
least-squares 174, 176
left nullspace 146, 149, 151, 154
left-inverse 66
length 13, 186
leveling 275, 367, 378, 405
linear combination 6, 10, 30, 104
linear independence 134
linear programming 277
linear regression 387, 442, 443
linear transformation 251, 253
linearization 344, 349, 357, 381, 382, 460
LINPACK 79
loop 291, 302
loop law 285
LU factorization 77

magic matrix 36
mapping function 456
marginal distribution 312
Markov matrix 35, 213, 220, 224
MATLAB 8, 17, 81
matrix 29, 54
  $-1, 2, -1$ 69, 77, 98, 204, 242, 378, 394
  augmented 50, 54, 69, 123
  block 58
  block inverse 560
  block tridiagonal 560

checkerboard 156
coefficient 29
cofactor 209
complex 64
covariance 321, 324, 544
echelon 113
elimination 48, 51, 76
gain 543, 549, 551, 562–564
Hilbert 75
identity 48
incidence 282, 283, 287, 305
information 324, 554, 566
inverse 65, 69
inverse block 425
Jacobian 330, 491
magic 36
Markov 35, 213, 220, 224
nullspace 117, 118
orthogonal 185, 214, 233, 263, 375
permutation 50, 90, 97
positive definite 239, 240, 242, 246, 281, 322
projection 167, 169, 173, 213, 236
reflection 185, 196, 213, 261
reverse identity 203
rotation 185, 214
semidefinite 240, 241
similar 259, 260
singular 201
skew-symmetric 98, 204, 214
square root 247, 569
symmetric 89, 214, 233
triangular 201
tridiagonal 70, 80, 85, 546
weight 323, 466
zero 55
matrix multiplication 49, 55
matrix notation 30, 32
matrix space 103, 107, 141, 145, 256
matrix times vector 32, 33, 47
mean 182, 309, 319, 516
mean anomaly 483, 485
mean motion 483
mean sea level 472
meridian 468
microstrip antenna 605
minimum 239
modem 605
moment 11
monitor station 605
multichannel receiver 605
multipath 456, 457, 605
multiplexing 605
multiplicity 228
multiplier 38, 76, 556

NAD 83 605
narrow lane 512, 605
nav message 605
navigation data 481
network 284, 379, 384, 386, 389, 406
node law 285, 303
nondiagonalizable 223, 228
nonlinearity 343
nonsingular 40
norm 13, 17, 277
normal distribution 309, 311, 312, 337
normal equation 169, 177, 269, 280, 435
notation 551
nullspace 109, 113, 117, 148
nullspace matrix 117, 118
nullspace of $A^T A$ 171
numerical precision 348, 349

observation equation 349, 355, 545
odd permutation 198
Ohm's law 296
operation count 56, 68, 80, 81
orbit 484
orientation unknown 355, 431, 438, 441
orthogonal 20
orthogonal complement 159, 160, 164
orthogonal matrix 185, 214, 233, 263, 375
orthogonal subspaces 157–159
orthogonalization 568
orthometric height 605
orthonormal 184

orthonormal basis 258, 262, 266
orthonormal columns 184, 186, 192
outer product 56, 59

P code 450, 451, 605
$PA = LU$ 91–93
parabola 179, 181
parallel machine 81
parallelogram 5, 7
particular solution 25, 113, 124, 129
Pascal 86
PDOP 463
perigee 482, 485, 486
permutation matrix 50, 90, 97
perpendicular 15, 18
perpendicular eigenvectors 214, 233, 235
phase 458
phase lock loop 605
phase observation 464
phase velocity 454
pivot 38, 39, 556
pivot column 117, 138
pivot variable 112
plane 20, 21, 24
plumb line 364, 473
point error 437
point positioning 605
polar decomposition 267
Position Dilution of Precision 605
position error 447, 454
positive definite matrix 239, 240, 242, 246, 281, 322
postulated coordinates 276, 423, 424
power spectral density 520, 523
powers of a matrix 211, 224, 227, 231, 250
Precise Positioning Service 605
preconditioner 248
prediction 543, 545, 551, 562, 564
President Clinton 447
principal axis 233, 244
probability density 309
product of determinants 202
product of pivots 201
projection 165, 166, 169, 281, 373, 393
projection matrix 167, 169, 173, 213, 236
projection onto a subspace 168
propagation law 329, 333, 462, 466, 523, 562
pseudodistance 352
pseudoinverse 161, 267, 270, 405, 411, 413
pseudolite 605
pseudorandom 450, 606
pseudorange 330, 448, 449, 500, 606
Ptolemy 469
pulse rate 546
Pythagoras 9

QR factorization 335
quasi-distance 351

random process 320, 515, 523
random ramp 516, 524, 527
random variable 309, 413
random walk 517, 524, 527–529
range 105, 252, 255
range error 457
range rate 606
rank 122, 127, 150
rank deficient 408
rank one 132, 152, 167, 270
ratios of distances 353
real-time 513
recursive least squares 183, 543, 545, 549
reduced echelon form 69, 70, 114, 117, 125
reference ellipsoid 362, 471
reflection matrix 185, 196, 213, 261
refraction 454
relative confidence ellipse 341, 342
relative positioning 606
relativity 451
reliability 606
repeated eigenvalues 223
residual 248, 282, 348
reverse identity matrix 203
reverse order 67, 87